Handbook of laser welding technologies

Related titles:

Hybrid laser-arc welding
(ISBN 978-1-84569-370-1)

Semiconductor lasers: Fundamentals and applications
(ISBN 978-0-85709-121-5)

Handbook of solid-state lasers: materials, systems and applications
(ISBN 978-0-85709-272-4)

Details of these books and a complete list of titles from Woodhead Publishing can be obtained by:

- visiting our web site at www.woodheadpublishing.com
- contacting Customer Services (e-mail: sales@woodheadpublishing.com; fax: +44 (0) 1223 832819; tel.: +44 (0) 1223 499140 ext. 130; address: Woodhead Publishing Limited, 80 High Street, Sawston, Cambridge CB22 3HJ, UK)
- in North America, contacting our US office (e-mail: usmarketing@woodheadpublishing.com; tel.: (215) 928 9112; address: Woodhead Publishing, 1518 Walnut Street, Suite 1100, Philadelphia, PA 19102-3406, USA

If you would like e-versions of our content, please visit our online platform: www.woodheadpublishingonline.com. Please recommend it to your librarian so that everyone in your institution can benefit from the wealth of content on the site.

We are always happy to receive suggestions for new books from potential editors. To enquire about contributing to our Electronic and Optical Materials series, please send your name, contact address and details of the topic/s you are interested in to laura.pugh@woodheadpublishing.com. We look forward to hearing from you.

The team responsible for publishing this book:

Commissioning Editor: Laura Pugh
Publications Coordinator: Anneka Hess
Project Editor: Sarah Lynch
Editorial and Production Manager: Mary Campbell
Production Editor: Richard Fairclough
Cover Designer: Terry Callanan

Woodhead Publishing Series in Electronic and Optical Materials: Number 41

Handbook of laser welding technologies

Edited by

Seiji Katayama

WOODHEAD PUBLISHING

Oxford Cambridge Philadelphia New Delhi

Published by Woodhead Publishing Limited,
80 High Street, Sawston, Cambridge CB22 3HJ, UK
www.woodheadpublishing.com
www.woodheadpublishingonline.com

Woodhead Publishing, 1518 Walnut Street, Suite 1100, Philadelphia,
PA 19102-3406, USA

Woodhead Publishing India Private Limited, G-2, Vardaan House, 7/28 Ansari Road,
Daryaganj, New Delhi – 110002, India
www.woodheadpublishingindia.com

First published 2013, Woodhead Publishing Limited

British Library Cataloguing in Publication Data
A catalogue record for this book is available from the British Library.

Library of Congress Control Number: 2013937868

ISBN 978-0-85709-264-9 (print)
ISBN 978-0-85709-877-1 (online)
ISSN 2050-1501 Woodhead Publishing Series in Electronic and Optical Materials (print)
ISSN 2050-151X Woodhead Publishing Series in Electronic and Optical Materials (online)

The publisher's policy is to use permanent paper from mills that operate a sustainable forestry policy, and which has been manufactured from pulp which is processed using acid-free and elemental chlorine-free practices. Furthermore, the publisher ensures that the text paper and cover board used have met acceptable environmental accreditation standards.

Typeset by Replika Press Pvt Ltd, India
Printed by Lightning Source

Contents

Contributor contact details

(* = main contact)

Editor and Chapters 1 and 12

Professor Seiji Katayama
Joining and Welding Research Institute
Osaka University
11-1 Mihohaoka, Ibaraki
Osaka 567-0047, Japan

E-mail: katayama@jwri.osaka-u.ac.jp

Chapter 2

Dr S. Tsukamoto
Structural Materials Unit
National Institute for Materials Science
1-2-1 Sengen
Tsukuba 305-0047, Japan

E-mail: tsukamoto.susumu@nims.go.jp

Chapter 3

Dr R. Fabbro
Laboratoire PIMM (Arts et Métiers ParisTech – CNRS)
151, Bd. de l'Hôpital
75013 Paris, France

E-mail: remy.fabbro@gmail.com

Chapter 4

Klaus Löffler
International Sales Laser Systems
TRUMPF Laser and Systems GmbH
Johann-Maus-Straße 2
71254 Ditzingen, Germany

E-mail: klaus.loeffler@de.trumpf.com

Chapter 5

Dr J. Zhou*
Pennsylvania State University
The Behrend College
Erie, PA 16509, USA

E-mail: juz17@psu.edu

Dr H. L. Tsai
Department of Mechanical and Aerospace Engineering
Missouri University of Science and Technology
335 Toomey Hall
Rolla, MO 65409, USA

E-mail: tsai@mst.edu

Chapter 6

Professor L. Quintino*
Instituto Superior Técnico
Departamento de Engenharia Mecânica
Área Científica de Tecnologia Mecânica e Gestão Industrial
Av. Rovisco Pais
1049-001 Lisboa, Portugal

E-mail: lquintino@ist.utl.pt

Dr Eurico Assunção
European Federation for Welding, Joining and Cutting
Av. Prof. Cavaco Silva, 33 – TagusPark
2740-120 Porto Salvo, Portugal

E-mail: egassuncao@isq.pt

Chapter 7

Dr Mohammed Naeem
JK Lasers
Cosford Lane
Swift Valley
Rugby CV21 1QN, UK

E-mail: mnaeem@gsig.com

Chapter 8

Dr J. M. Sánchez-Amaya*
Titania
Ensayos y Proyectos Industriales S.L.
Parque Tecnológico TecnoBahía-Edif.
RETSE Nave 4
Ctra Sanlúcar A-2001 Km 7,5.
11500- El Puerto de Santa María
Cádiz, Spain

E-mail: josemaria.sanchez@uca.es

Mrs M. R. Amaya-Vázquez and Professor F. J. Botana
Laboratory of Corrosion and Protection (LABCYP)
Department of Materials Science and Metallurgical Engineering and Inorganic Chemistry (CASEM)
University of Cadiz
Avda. República Saharaui s/n
11510-Puerto Real
Cádiz, Spain

E-mail: margarita.amaya@uca.es; javier.botana@uca.es

Chapter 9

Mr Felix Möller* and Dr Claus Thomy
BIAS – Bremer Institut für angewandte Strahltechnik GmbH
Klagenfurter Strasse 2
D-28359 Bremen, Germany

E-mail: moeller@bias.de

Chapter 10

Mr Ian Jones
TWI Ltd
Granta Park
Great Abington
Cambridge CB21 6AL, UK

E-mail: ian.jones@twi.co.uk

Chapter 11

Professor Isamu Miyamoto
Osaka University
2-1 Yamadaoka, Suita
Osaka 565-0871, Japan

E-mail: miyamoto-i@ares.eonet.ne.jp

Chapter 13

Professor Hidekazu Murakawa
Osaka University
11-1 Mihogaoka
Ibaraki
Osaka 567-0047, Japan

E-mail: murakawa@jwri.osaka-u.ac.jp

Chapter 14

Dr Menno de Graaf*
Cargotec ACT
and
University of Twente
Laboratory of Mechanical Automation
P.O. Box 217
7500 AE Enschede
The Netherlands

E-mail: m.w.degraaf@alumnus.utwente.nl

Dr Ronald Aarts
University of Twente
Laboratory of Mechanical Automation
P.O. Box 217
7500 AE Enschede
The Netherlands

Chapter 15

Dr Andreas Wetzig
Ablation and Cutting Department
Fraunhofer IWS Dresden
Germany

E-mail: andreas.wetzig@iws.fraunhofer.de

Chapter 16

Dr Yukio Miyashita
Department of Mechanical Engineering
Nagaoka University of Technology
1603-1 Kamitomioka
Nagaoka 940-2188
Niigata, Japan

E-mail: miyayuki@mech.nagaokaut.ac.jp

Chapter 17

Dr Xudong Zhang
Materials Research Center
Hitachi Research Laboratory
Hitachi Ltd
1-1, Omika-cho 7-chome
Hitachi
Ibaraki 319-1292, Japan

E-mail: Xudong.zhang.en@hitachi.com

Chapter 18

Dr Dirk Petring
Fraunhofer-Institute for Laser Technology (ILT)
Steinbachstrasse 15
52074 Aachen, Germany

E-mail: dirk.petring@ilt.fraunhofer.de

Chapter 19

Dr A. Gumenyuk* and Professor M. Rethmeier
Division of Welding Technology
BAM Federal Institute for Materials Research and Testing
Unter den Eichen 87
12205 Berlin, Germany

E-mail: Andrey.gumenyuk@bam.de

Chapter 20

Professor Suck-Joo Na* and Dr Won-Ik Cho
Department of Mechanical Engineering
KAIST
291 Daehak-ro, Yusoonggu
Daejeon, 305-701, South Korea

E-mail: sjna@kaist.ac.kr

Chapter 21

Dr Matthias Graudenz*
Technology Development – Laser Joining
Audi AG
85045 Ingolstadt, Germany

E-mail: matthias.graudenz@audi.de

Dr Markus Baur
Body Shop Audi A4/A5/Q5
Audi AG
85045 Ingolstadt, Germany

E-mail: markus.baur@audi.de

Chapter 22

Hongxiao Wang
Changchun Railway Vehicles Co. Ltd
Qingyin Road 435
Changchun City
Jilin Province, China

E-mail: magie527@hotmail.com

Chapter 23

Professor Jens Klæstrup Kristensen
Maglehoej 62
3520 Farum, Denmark

E-mail: jens@klaestrup.dk

Woodhead Publishing Series in Electronic and Optical Materials

1 **Circuit analysis**
J. E. Whitehouse

2 **Signal processing in electronic communications: For engineers and mathematicians**
M. J. Chapman, D. P. Goodall and N. C. Steele

3 **Pattern recognition and image processing**
D. Luo

4 **Digital filters and signal processing in electronic engineering: Theory, applications, architecture, code**
S. M. Bozic and R. J. Chance

5 **Cable engineering for local area networks**
B. J. Elliott

6 **Designing a structured cabling system to ISO 11801: Cross-referenced to European CENELEC and American Standards**
Second edition
B. J. Elliott

7 **Microscopy techniques for materials science**
A. Clarke and C. Eberhardt

8 **Materials for energy conversion devices**
Edited by C. C. Sorrell, J. Nowotny and S. Sugihara

9 **Digital image processing: Mathematical and computational methods**
Second edition
J. M. Blackledge

10 **Nanolithography and patterning techniques in microelectronics**
Edited by D. Bucknall

11 **Digital signal processing: Mathematical and computational methods, software development and applications**
Second edition
J. M. Blackledge

12 **Handbook of advanced dielectric, piezoelectric and ferroelectric materials: Synthesis, properties and applications**
Edited by Z.-G. Ye

13 **Materials for fuel cells**
Edited by M. Gasik

14 **Solid-state hydrogen storage: Materials and chemistry**
Edited by G. Walker

15 **Laser cooling of solids**
S. V. Petrushkin and V. V. Samartsev

16 **Polymer electrolytes: Fundamentals and applications**
Edited by C. A. C. Sequeira and D. A. F. Santos

17 **Advanced piezoelectric materials: Science and technology**
Edited by K. Uchino

18 **Optical switches: Materials and design**
Edited by S. J. Chua and B. Li

19 **Advanced adhesives in electronics: Materials, properties and applications**
Edited by M. O. Alam and C. Bailey

20 **Thin film growth: Physics, materials science and applications**
Edited by Z. Cao

21 **Electromigration in thin films and electronic devices: Materials and reliability**
Edited by C.-U. Kim

22 ***In situ* characterization of thin film growth**
Edited by G. Koster and G. Rijnders

23 **Silicon-germanium (SiGe) nanostructures: Production, properties and applications in electronics**
Edited by Y. Shiraki and N. Usami

24 **High-temperature superconductors**
Edited by X. G. Qiu

25 **Introduction to the physics of nanoelectronics**
S. G. Tan and M. B. A. Jalil

26 **Printed films: Materials science and applications in sensors, electronics and photonics**
Edited by M. Prudenziati and J. Hormadaly

27 **Laser growth and processing of photonic devices**
Edited by N. A. Vainos

28 **Quantum optics with semiconductor nanostructures**
Edited by F. Jahnke

29 **Ultrasonic transducers: Materials and design for sensors, actuators and medical applications**
Edited by K. Nakamura

30 **Waste electrical and electronic equipment (WEEE) handbook**
Edited by V. Goodship and A. Stevels

31 **Applications of ATILA FEM software to smart materials: Case studies in designing devices**
Edited by K. Uchino and J.-C. Debus

32 **MEMS for automotive and aerospace applications**
Edited by M. Kraft and N. M. White

33 **Semiconductor lasers: Fundamentals and applications**
Edited by A. Baranov and E. Tournie

34 **Handbook of terahertz technology for imaging, sensing and communications**
Edited by D. Saeedkia

35 **Handbook of solid-state lasers: Materials, systems and applications**
Edited by B. Denker and E. Shklovsky

36 **Organic light-emitting diodes: Materials, devices and applications**
Edited by A. Buckley

37 **Lasers for medical applications: Diagnostics, therapy and surgery**
Edited by H. Jelínková

38 **Semiconductor gas sensors**
Edited by R. Jaaniso and O. K. Tan

39 **Handbook of organic materials for optical and optoelectronic devices: Properties and applications**
Edited by O. Ostroverkhova

40 **Metallic films for electronic, optical and magnetic applications: Structure, processing and properties**
Edited by K. Barmak and K. Coffey

41 **Handbook of laser welding technologies**
Edited by S. Katayama

42 **Nanolithography: The art of fabricating nanoelectronics, nanophotonics and nanobiology devices and systems**
Edited by M. Feldman

43 **Laser spectroscopy for sensing: Fundamentals, techniques and applications**
Edited by M. Baudelet

44 **Chalcogenide glasses: Preparation, properties and applications**
Edited by J.-L. Adam and X. Zhang

45 **Handbook of MEMS for wireless and mobile applications**
Edited by D. Uttamchandani

46 **Subsea optics and imaging**
Edited by J. Watson and O. Zielinski

47 **Carbon nanotubes and graphene for photonic applications**
Edited by S. Yamashita, Y. Saito and J. H. Choi

48 **Optical biomimetics: Materials and applications**
Edited by M. Large

49 **Optical thin films and coatings**
Edited by Angela Piegari and François Flory

50 **Computer design of diffractive optics**
Edited by V. A. Soifer

51 **Smart sensors and MEMS: Intelligent devices and microsystems for industrial applications**
Edited by S. Nihtianov and A. L. Estepa

52 **Fundamentals of femtosecond optics**
V. Kozlov and V. Samartsev

Preface

More than 50 years have passed since a laser beam was first recognized by the irradiation of a pulsed ruby laser. Since then many characteristic lasers such as high-power CO_2 lasers, fiber-delivered YAG lasers and short-wavelength Excimer lasers have been utilized in various industries. Recently, high-efficiency diode lasers, high-quality disk lasers and high-power, high-brightness and compact-sized fiber lasers have been developed and are receiving considerable attention. Moreover, the second or third harmonic solid-state laser, called the green or ultra-violet laser, and the ultra-short pulse laser, often referred to as a femtosecond (fs) or nanosecond (ns) laser, have been developed to search for new applications. Accordingly, a variety of laser materials processing techniques including removal processes such as cutting and drilling, joining processes such as welding, brazing and soldering, surface modification treatments such as transformation hardening, cladding and annealing, have now been developed. Especially recently, peripheral apparatuses such as jigs and fixtures, monitoring systems and focusing optics, have been incorporated into the automation lines for welding, cutting and drilling to improve these processes. Thus laser processing is expected to continue to be an indispensable technology for the twenty-first century.

Among laser materials processing technologies, laser welding is especially noted in conjunction with recent advances in laser apparatuses. An understanding of laser welding mechanisms and phenomena is required for its proper application. Consequently, this book, which includes various laser or hybrid welding processes, welding of various kinds of materials, metallurgical, chemical and mechanical aspects, is planned to appeal to students, engineers, scientists and professors concerned with, or interested in, laser welding. The book is divided into four parts: established laser welding technologies, laser welding of various materials, emerging laser welding technologies and applications of laser welding, covering almost all topics on laser welding.

The editor would like to express his appreciation for the cooperation and contribution of all the authors. Also, the editor would like to thank Ms Rachel Cox, Ms Anneka Hess, Ms Laura Pugh and Ms Sarah Lynch at Woodhead

Publishing Limited for their encouragement, help and hard work. The editor really hopes that this book is essential and useful to everyone who is concerned with laser welding, from beginners to experts or professionals, and that laser welding technologies will be more and more utilized and advanced all over the world.

Seiji Katayama

Part I

Developments in established laser welding technologies

1 Introduction: fundamentals of laser welding

S. KATAYAMA, JWRI, Osaka University, Japan

DOI: 10.1533/9780857098771.1.3

Abstract: This chapter describes characteristics of laser welding, features of main lasers used for welding, factors affecting weld penetration, laser welding phenomena including behavior of laser-induced plume, keyhole behavior and melt flows in a molten pool during laser welding. It also refers to welding defects and the recent trend for laser welding.

Key words: laser welding, CO_2 laser, YAG laser, fiber laser, diode laser, fiber delivery, welding phenomena, welding defects, laser-induced plume, deep penetration, welding conditions.

1.1 Characteristics of laser welding

Welding is the most versatile and realistic joining method applicable to the construction of products in every industrial field. A laser is a high-power-density heat source. Therefore, 'laser welding' is recognized as an advanced process to join materials with a laser beam of high-power, high-energy density. Representative power density profiles of laser, electron beam, plasma and arc, and the consequent weld bead geometries are schematically illustrated in Fig. 1.1.

The power density of a laser beam, which is equivalent to that of an electron beam, is much higher than that of arc or plasma. Consequently a deep, narrow keyhole is formed during welding with a high-power-density beam of laser or electron, and a deep, narrow penetration weld can be effectively produced. In electron beam welding, the chamber for a vacuum environment and X-ray protection should be used, and for steel plates, demagnetization treatment is required before welding. Arc and plasma welding cannot be employed in a vacuum; however, laser welding can be performed and a sound, deep weld bead can be produced in a similar way to electron beam welding, which is described in Section 1.5. Generally the travelling speed of laser welding is higher than those of arc and plasma welding.

Among all the welding processes, laser welding can produce a variety of

The content of this chapter has been reprinted with permission from the article 'laser welding' in the *McGraw Hill Encyclopedia of Science & Technology*, Vol. 9 (2012), pp. 707–714, written by Seiji Katayama. The author would like to express his gratitude to this company.

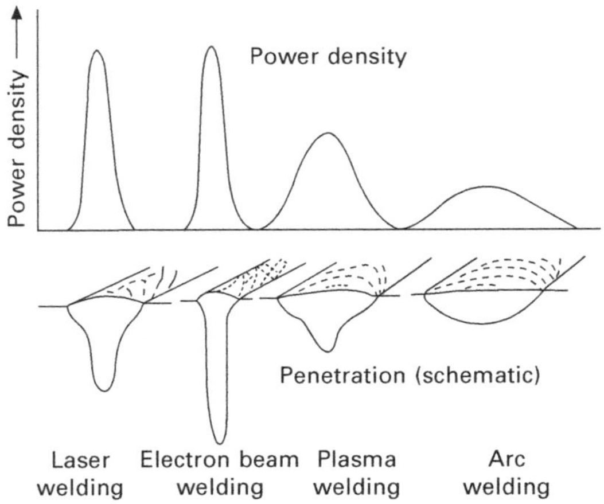

1.1 Power densities for typical welding heat sources, and geometry of weld beads obtained at respective densities.

joints of metals or plastics ranging from very thin sheets of about 0.01 mm thickness to thick plates of about 50 mm in the shielding gas such as helium (He), argon (Ar) or sometimes nitrogen (N_2) under the air environment, and has gained great popularity as a promising joining technology with high quality, high precision, high performance, high speed, good flexibility and low distortion.[1-4] It can also achieve robotization, reduced manpower, full automation, and systematization in production lines. Consequently, applications of laser welding are increasing together with the development of novel laser apparatuses and joining processes. In order to properly apply pulsed or continuous wave (PW or CW) lasers to welding or joining, it is important to know the specifications and capability of laser apparatuses, the factors affecting weld penetration and welding defects, and the mechanisms and behaviour of welding as well as to evaluate the weldability of materials and the mechanical properties of welded joints. For further details on the contents of this chapter, see the references in Section 1.6.[1–4]

1.2 Lasers for welding

The kind and characteristics of the main lasers used for welding are summarized in Table 1.1. Typical CO_2 and Nd:YAG laser systems are schematically represented in Fig. 1.2. CO_2 lasers, 10.6 μm in wavelength, are developed as high beam quality and easy achievement of higher power, the maximum laser power reaches up to 50 kW, and 1–15 kW power levels are mainly utilized in welding of steels, automotive components, airplanes, ships, etc.

The following inconvenient problems should be considered in CO_2 laser welding:

Table 1.1 Types and characteristics of laser for welding

Types of laser	Laser characteristics
CO_2 laser	Wavelength: 10.6 μm; far-infrared ray Laser media: CO_2-N_2-He mixed gas (gas) Average power [CW]: 50 kW (Maximum) (Normal): 1–15 kW
Lamp-pumped YAG laser	Wavelength: 1.06 μm; near-infrared ray Laser media: Nd^{3+}:$Y_3Al_5O_{12}$ garnet (solid) Average power [CW]: 10 kW (cascade-type max & fiber-coupling max) (Normal): 50 W–7 kW (Efficiency: 1–4%)
Laser diode (LD)	Wavelength: 0.8–1.1 μm; near-infrared ray Laser media: InGaAsP, etc. (solid) Average power [CW]: 10 kW (stack-type max.), 15 kW (fiber-delivery max.) Merits: Compact, and high efficiency (20–60%)
LD-pumped solid-state laser	Wavelength: about 1 μm; near-infrared ray Laser media: Nd^{3+}:$Y_3Al_5O_{12}$ garnet (solid), etc. Average power [CW]: 13.5 kW (fiber-coupling max.) [PW]: 6 kW (slab-type max.)
Disk laser	Wavelength: 1.03 μm; near-infrared ray Laser media: Yb^{3+}:YAG or YVO_4 (solid), etc. Average power [CW]: 16 kW (cascade-type Max.) Merits: Fiber delivery, high brightness, high efficiency (15–25%)
Fiber laser	Wavelength: 1.07 μm; near-infrared ray Laser media: Yb^{3+}:SiO_2 (solid), etc. Average power [CW]: 100 kW (fiber-coupling max.) Merits: Fiber delivery, high brightness, high efficiency (20–30%)

- High power lasers should be delivered by mirrors but not by an optical fiber, and
- an Ar plasma, which reduces weld penetration depth, can be easily formed in Ar shielding gas.

Nd:YAG lasers, 1.06 μm in wavelength, can be delivered through an optical fiber, and are operated in CW or PW mode. PW lasers are used in welding of small parts such as battery cases, electric components, glass frames, etc. CW lasers of 2–7 kW power are employed for laser welding of tailored blanks, aluminum cars, Zn-coated steel sheets, stainless steel pipes and tanks, etc. Such lamp-pumped Nd:YAG lasers have a drawback of low electrical efficiency (the ratio of laser output to electrical input) of less than 4%. Therefore, the development of high power CO_2 and YAG lasers has already been stopped. Instead, laser-diode (LD)-pumped solid-state (YAG) lasers have been developed up to 6–10 kW. However, the development of

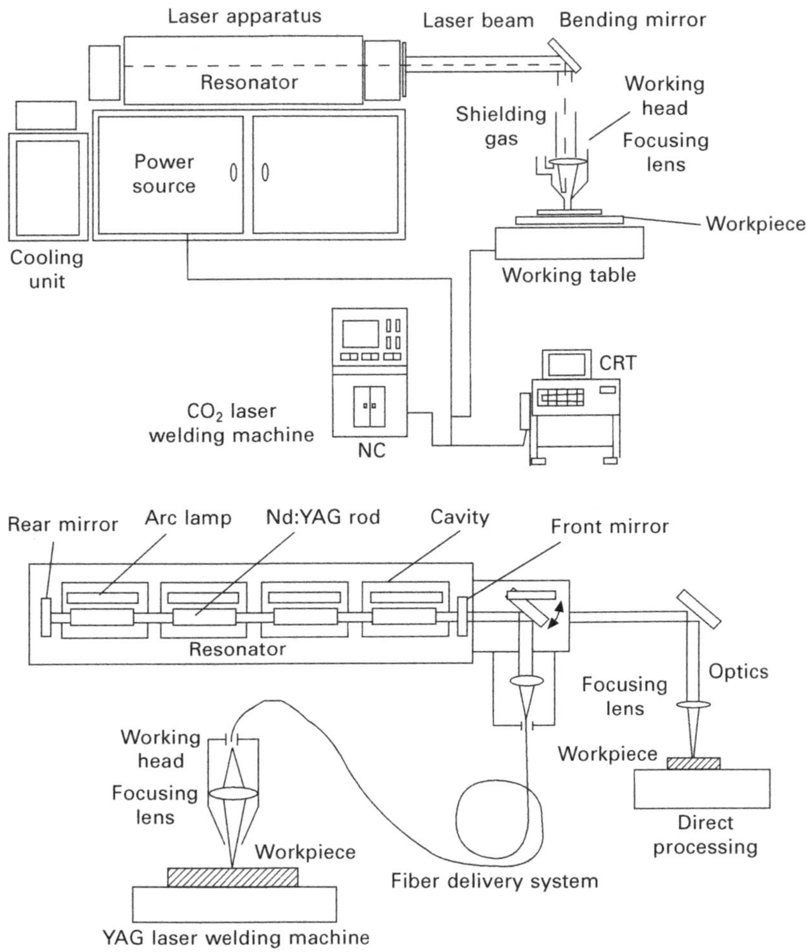

1.2 Schematic representation of CO_2 and YAG laser welding systems.

these lasers has also been stopped. Recently, more is expected of disk and fiber lasers in terms of high power, high efficiency and high beam quality than of lamp- or LD-pumped YAG lasers. Fiber-delivery, which is used for lamp- or LD-pumped Nd:YAG, diode, Yb:disk and Yb:fiber lasers, is representative of good flexibility.

During the present century, thereafter, developments of lasers with high power and high beam quality have been concentrated upon laser diodes (LD) themselves, and LD-pumped disk and fiber lasers. LD has high electrical efficiency of 30–60%. Direct LD and fiber-delivered LD mounted to robots are used for welding of thin sheets of aluminum alloys, steels, stainless steels, plastics, etc., and for brazing of Zn-coated steels, etc. The drawback of LD is poor beam quality. High power, high electrical efficiency and high beam quality are achieved by disk and fiber lasers. Disk lasers and fiber lasers with a high electrical efficiency of 25% and maximum power of 16 kW and 100 kW, respectively, are commercially available. Their beam qualities are extremely high as the beam parameter products (BPP) are smaller than 10 mm·mrad. Disk and fiber lasers are both utilized as heat sources for remote welding, as shown in Fig. 1.3. Remote laser welding with a robot and a scanner is the most promising joining technology of high speed and high production rate. Further applications of these lasers are evolved to welding for electrical components, cars, trains, bridges, pipelines, ships, airplanes, etc.

1.3 Laser welding phenomena

There are various kinds of laser-welded joints, as shown in Fig. 1.4. Butt- and lap-joints are commonly welded with PW or CW lasers. Typical phenomena during PW and CW laser welding are schematically illustrated in Fig. 1.5. Depending on the laser irradiation time and the power density, a spot or bead weld is formed in the morphology of a heat-conduction type or a keyhole type. When a laser beam is shot on the metallic plate, the absorption of laser energy is caused by the interaction against free electrons in the metal, the electrons transfer inside the band, and the interaction of moving electrons against metal lattice, defects, imperfection and potential perturbation. The temperature of the plate surface rises by the transfer from laser energy to thermal energy. The laser absorption increases slightly with an increase in the temperature in the solid and considerably above the melting temperature, and the temperature of the laser-irradiated area rises higher to the boiling temperature. A cavity or a keyhole is formed by recoil pressure due to evaporation. A coupling (absorption) coefficient of laser energy into

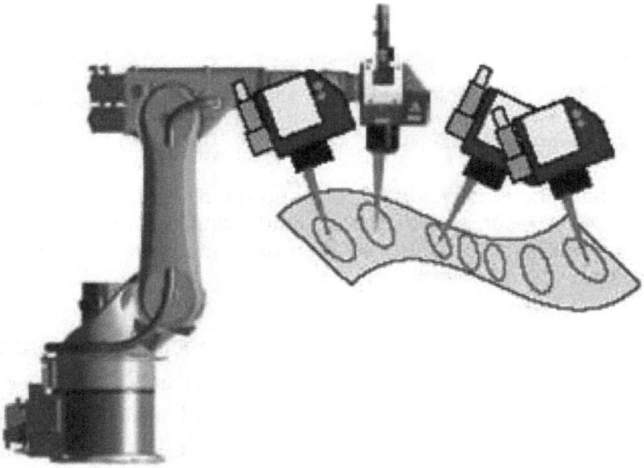

1.3 Schematic of remote welding system using solid-state laser.

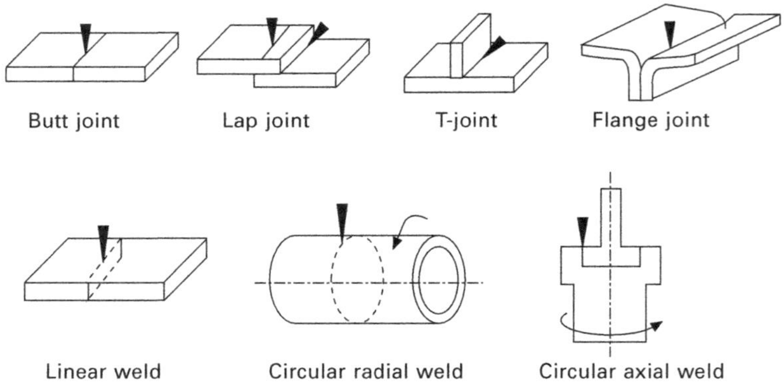

1.4 Schematic of typical examples of laser-welded joints.

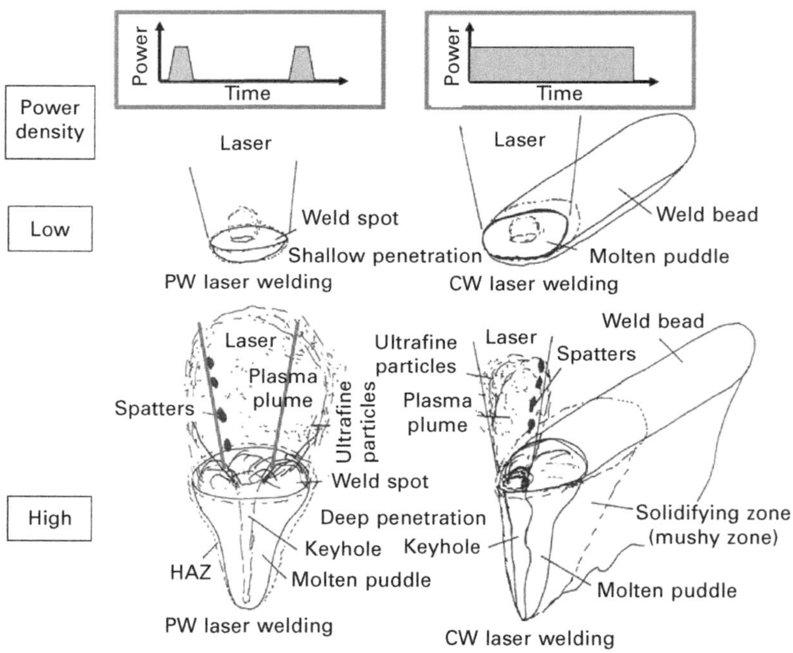

1.5 Schematic illustration of spot and bead welding phenomena with PW and CW laser, showing heat-conduction type shallow penetration and keyhole type deep penetration depending on laser power density.

the metal is schematically shown in Fig. 1.6. It increases according to the temperature and surface conditions. Since the laser absorption is extremely high due to multi-reflection inwards in the case of a keyhole formation, a keyhole type of deep penetration welding is regarded as an efficient joining process. A bright plume of evaporated metallic atoms and vapors is ejected

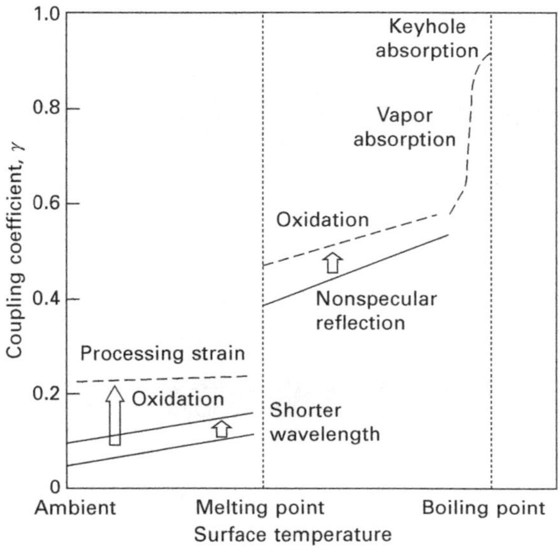

1.6 Coupling coefficient of laser energy into metal such as steel or aluminum alloy as a function of temperature .

from laser-irradiated part, especially a keyhole. Spattering of melt droplets, caused by a strong stream of the ejected plume, sometimes occurs from the inlet of a keyhole.

The cross sections of laser weld beads in an aluminum alloy produced by CO_2 laser welding at 5 kW in the shielding gas of He, N_2 or Ar are shown in Fig. 1.7. The penetration depths at both welding speeds are shallower in the order of He, N_2 and Ar. Examples of high speed video pictures during CO_2 laser welding under the above-mentioned same conditions are shown in Fig. 1.8. Ar and N plasma are seen in the respective gases, although He plasma is not observed during welding. It is well understood that the decrease in CO_2 laser weld penetration, especially in Ar or N_2 gas, is attributed to the stable formation tendency and size of Ar and N plasma. In welding with high power CO_2 lasers, He ratio of more than 50% is required to produce a deep penetration weld.

The formation situations of a laser-induced plume and/or plasma and their effects on the weld penetration are summarized in Fig. 1.9. The formation of Ar or N plasma is easy in the case of CO_2 laser welding, while a laser-induced plume only is formed from a keyhole during YAG, disk and fiber laser welding. The interaction between the CO_2 laser beam and the gas plasma is interpreted in terms of the absorption due to Inverse Bremsstrahlung, whose effect is about 100 times greater than that of YAG, disk or fiber laser. The temperature of a plume during welding with the laser of about 1 μm wavelength is estimated to be about 3,000–6,000 K depending on the high

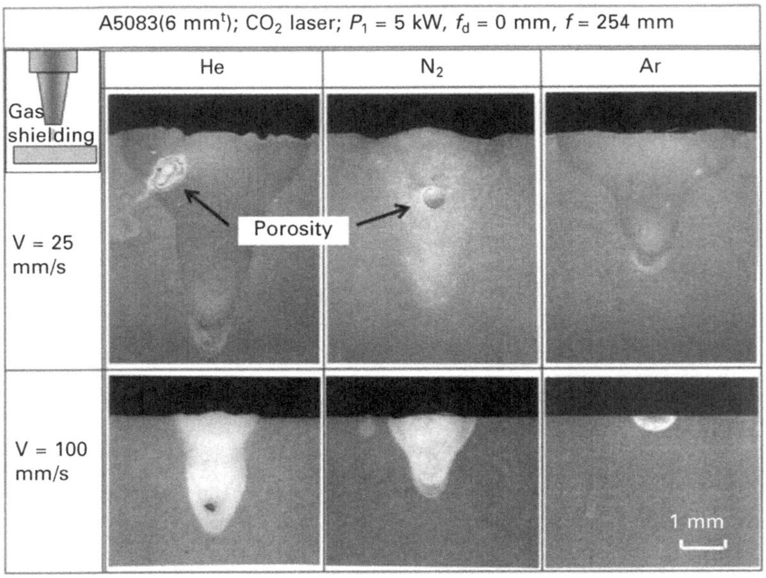

1.7 Cross-sectional photographs of CO_2 laser weld beads, made at 25 and 100 mm/s in He, N_2 and Ar coaxial shielding gas.

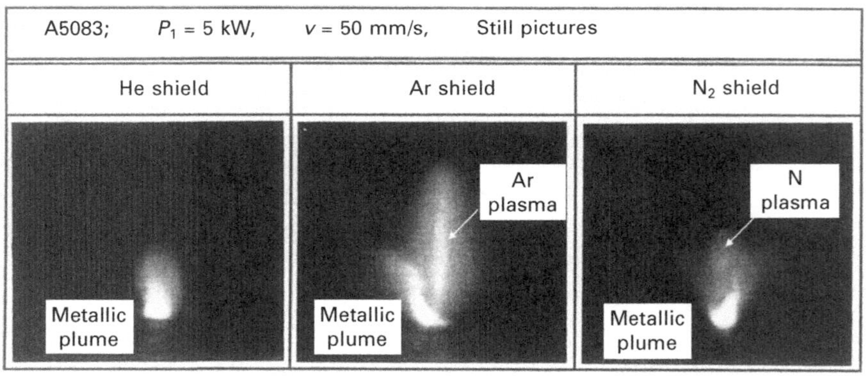

1.8 High speed video pictures of induced plume and gas plasma during CO_2 laser welding in He, N_2 and Ar coaxial shielding gas.

power density from about 1–1,000 kW/mm^2. The effects of the plume are attributed to refraction due to the difference in density between the plume and the environment as well as Rayleigh scattering due to the formation of ultrafine particles. In the case of a tall plume and the consequent wide formation range of low refraction index, the weld penetration is drastically changed to be shallower from the keyhole type to the heat-conduction mode (at the focal point) due to the beam refraction and the downward shift of the focal point. As long as the height of the plume or low refractive index area

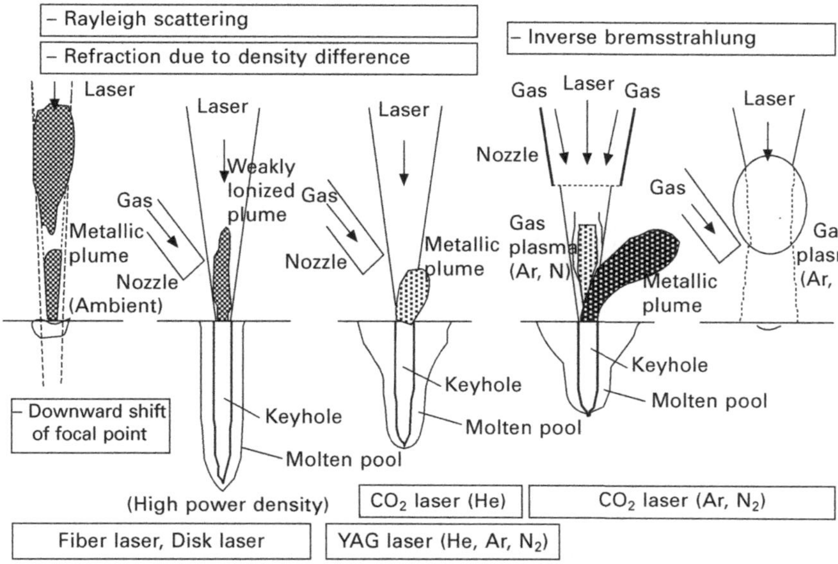

1.9 Schematic illustration of plume and plasma formation during laser welding and their respective physical effects on weld penetration in no gas or He, N_2 and Ar shielding gas.

(namely, the interaction length of the plume to a laser beam) is suppressed, a deep penetration weld can be produced.

Melt flows in the molten pool and near the surface are interpreted by observing using an X-ray transmission *in-situ* imaging system and high speed video cameras. It is consequently understood that the weld penetration depth is determined chiefly by the keyhole depth and partly by the melt downward flow around the keyhole tip, as shown in Fig. 1.10. Melt flows of the surface in the molten pool are affected by the surface tension of the melt and the shear stream due to a plume ejected strongly from a keyhole inlet, resulting in the typical shapes of wide or narrow weld beads near the surface.

1.4 Laser weld penetration and welding defects

Small parts are made by spot welding with a pulsed laser. Deep penetration welds are easily formed with the increase in the pulse width near the focal point of a focusing lens. However, the maximum depths of sound laser spot welds should normally be less than 1.5 mm or 3 mm under controlled conditions, because porosity is easily formed especially even in the shallow weld with extremely short time irradiation and deep welds made with a laser of rectangular pulse shape. Pulse shaping of slowly rising and falling power is carried out to reduce spattering and porosity, respectively. These results and data will be demonstrated in Chapter 12.

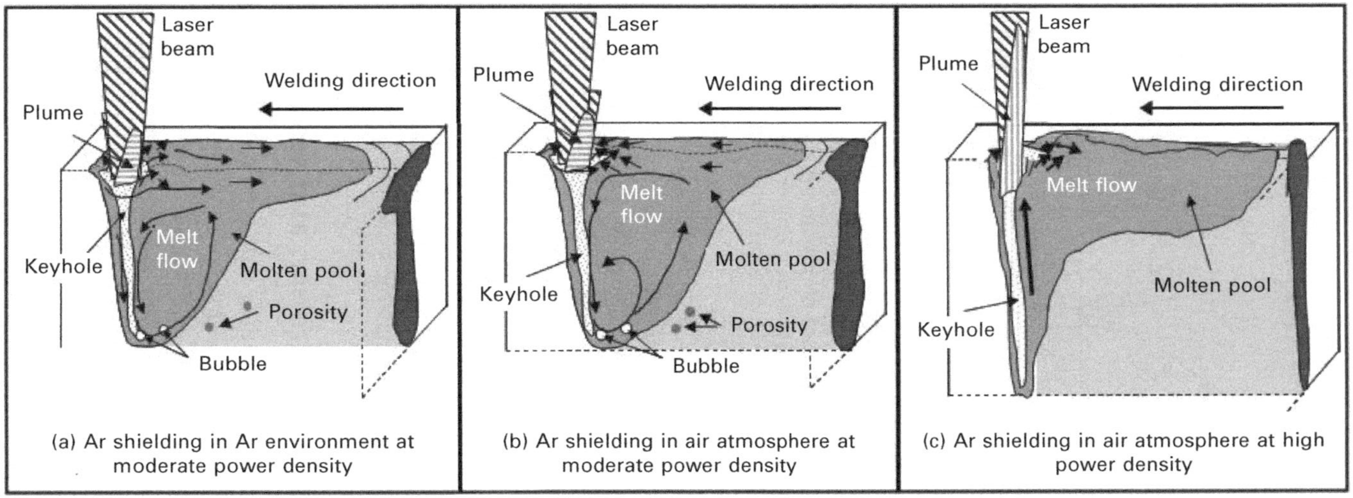

(a) Ar shielding in Ar environment at moderate power density

(b) Ar shielding in air atmosphere at moderate power density

(c) Ar shielding in air atmosphere at high power density

1.10 Schematic representation of melt flows in molten pool and weld bead geometry, showing effects of (a) outer or (b) inner surface flows due to surface tension and (c) high power density on weld penetration.

Deeply penetrated welds are effectively produced in high power CW laser welding. The penetration depths of stainless steel welds made with fiber lasers of different beam diameters at 6 and 10 kW in Ar shielding gas are shown in Fig. 1.11. In fiber laser welding at high power, deep penetration can be achieved even with Ar shielding gas, although the penetration is reduced due to Inverse Bremsstrahlung absorption by the formation of Ar gas plasma in the case of Ar shielding in CO_2 laser welding.[3] The penetration is shallower with the increase in the welding speed, and at high speeds, is deeper at higher power density of a smaller beam diameter. It is understood that the effect of laser power is more dominant at low speeds. The penetration is deeper at 10 kW than at 6 kW at a speed of less than 3 m/min (50 mm/s). At low welding speeds, porosity is easily formed, while at high speeds, porosity is prevented, but humping or spattering leading to underfilling occurs depending on the smaller or wider beam diameter, respectively. Under some conditions, welding defects such as porosity, cracks, humping, underfilling, and so on are easily formed. These mechanisms and preventive procedures will be described in Chapter 12.

1.5 Evolution of laser welding

In low and high vacuum, sound extremely-deep penetration welds can be produced, as shown in Fig. 1.12. The penetration depths of laser welds produced at low speeds in vacuum (under even low vacuum conditions)

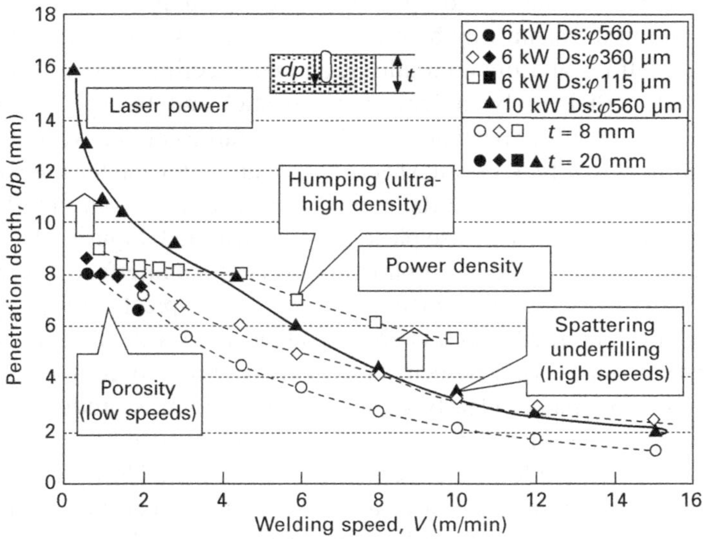

1.11 Effect of beam diameter (power density) and welding speed on penetration depth of Type 304 stainless steel weld made with 6 and 10 kW fiber laser.

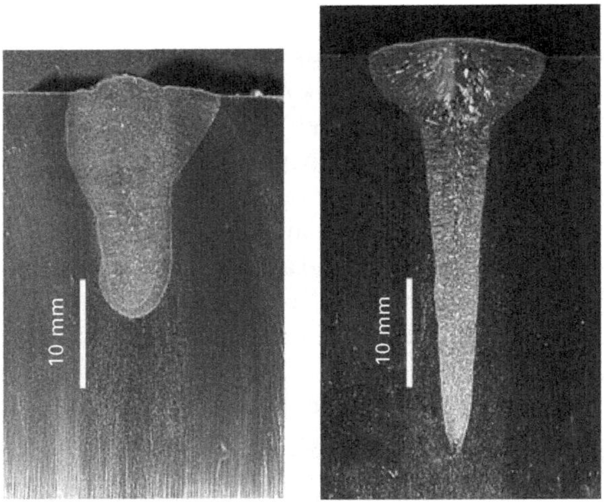

(a) Ar shielding (b) 10 kPa

1.12 Cross-sectional photos of laser welds in Type 304 steel produced with 10 kW fiber laser at 0.1 m/min in Ar shielding gas at 1 atm (a) and under low vacuum of 10 kPa (b).

are comparable to those of electron beam welds.[5] Laser welding is also applicable to the joining of similar or dissimilar plastics, dissimilar metals such as cast irons and steels, steels and aluminum alloys, or metals and plastics.[6] It has recently been demonstrated that strong lap joints could be produced between steels and light metals such as aluminum or magnesium alloys, and in the same way between metals such as steels, stainless steels, aluminum alloys and titanium alloys, and engineering plastics such as PA (polyamide), PET (polyethylene terephthalate) and PC (polycarbonate).[6] Figure 1.13[7] shows appearances of the laser lap joints between Type 304 stainless steel and PET plastic sheets before and after the tensile shear test. Small bubbles are formed in the melted plastic near the joint interface, and the elongation of the PET base plastic is demonstrated. Figure 1.14[7] exhibits cross-sectional SEM and TEM photographs near the interface of the joint at higher magnification and diffraction patterns from typical points. Cr_2O_3 type oxide film is identified, and it is observed that such oxide films always exist as an intermediate layer between plastic and metal for any combinations. The SEM and TEM images clearly show that a tight joint is produced, and that laser joining is performed partly by chemical or physical bonding of melted plastic on a thin oxide film covering the base metal. These results signify a feasibility of the production of a strong joint between metal and plastic sheet. Laser direct joining mechanisms of metal and plastic lap sheets are considered as follows: Laser is shot through the transparent plastic sheet or directly on the metal plate, and then the plate is heated to melt the plastic

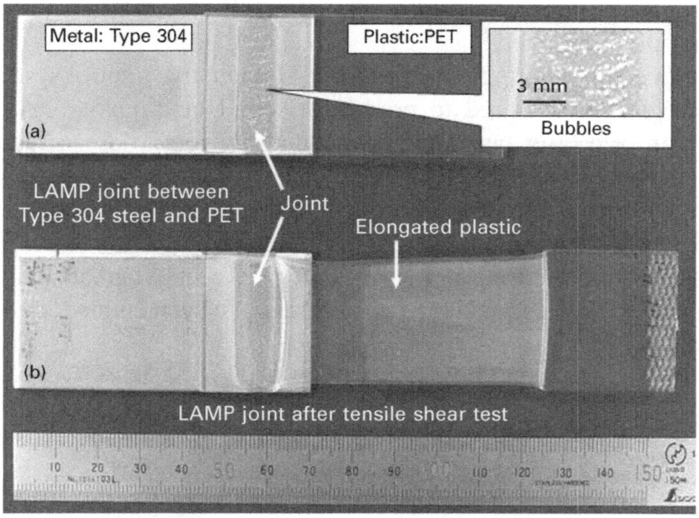

1.13 Laser lap joints of 3 mm thick Type 304 steel plate and 2 mm thick PET plastic sheet before and after tensile shear test, showing that bubbles are formed in melted zone in PET during joining and that elongation of PET base plastic occurs during testing.

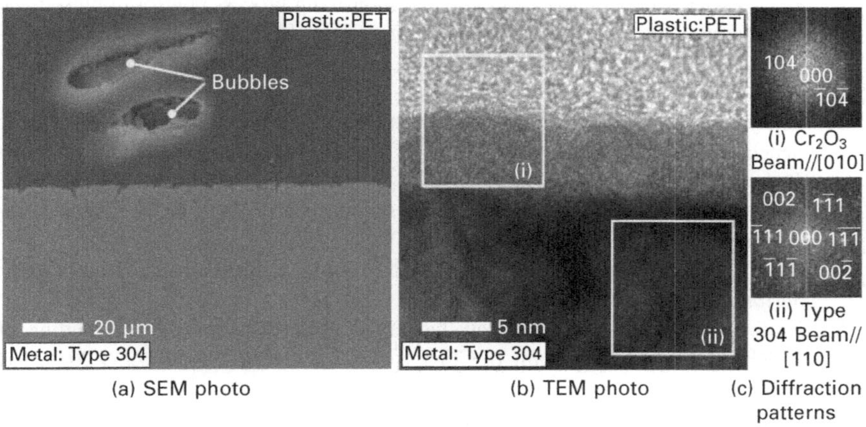

1.14 SEM and TEM photos of cross sections near joint interface between Type 304 steel plate and PET plastic sheet, and diffraction patterns and analyzed results of gray and dark parts, showing Cr_2O_3 oxide and Type 304 steel, respectively.

near the joint interface. Some bubbles are generated in order to force the activated melted plastic to flow over the metal surface. Tight joining of metal to plastic is performed through mechanisms of chemical, physical (van de Waals force) and mechanical (anchor effect) bonding. Laser joining of metal plate to GFRP (glass fiber reinforced plastics; PA matrix) or CFRP (carbon fiber reinforced plastics; PA matrix) sheet was also confirmed to be successfully performed.[8]

In-process monitoring, on-line sensing or adaptive control system during laser welding is required to produce a sound high-quality laser weld.[3,4] Reflection of a laser beam and heat radiation from a laser irradiated part and a molten pool are important candidates for monitoring signals.[9,10] The latter is also used for feedback or adaptive control.[10] Direct observation of phenomena during laser welding is also important in terms of understanding of formation mechanisms of welding defects. Recently, great advances have been made in this research and the systems have been employed in practical applications.

1.6 References

1 Katayama S (2012), 'Laser welding', *Ferrum (Bulletin of the Iron and Steel Institute of Japan)*, **17**(1) 18–29 (in Japanese).

2 Katayama S (2012), 'Laser welding of aluminum alloys', *Keikinzoku (Journal of Japan Institute of Light Metals)*, **62**(2) 75–83 (in Japanese).

3 Katayama S (2005), 'New development in laser welding', in *New Developments in Advanced Welding*, ed. by N. Ahmed. Cambridge: Woodhead Publishing pp. 158–197.

4 Katayama S (2010), 'Understanding and improving process control in pulsed and continuous wave laser welding', in *Advances in Laser Materials Processing – Technology, Research and Applications*, eds. by J. Lawrence, J. Pou, D. K. Y. Low. Cambridge: Woodhead Publishing, pp. 181–210.

5 Katayama S, Abe Y, Mizutani M and Kawahito Y (2011), 'Development of deep penetration welding technology with high brightness laser under vacuum', *Physics Procedia*, **12**(1) 75–80.

6 Katayama S (2010), 'Laser welding of dissimilar materials', *The Review of Laser Engineering*, **38**(8) 594–602 (in Japanese).

7 Katayama S and Kawahito Y (2008), 'Laser direct joining of metal and plastic', *Scripta Materialia*, **59**(12) 1247–1250.

8 Katayama S, Jung K-W and Kawahito Y (2010), 'High power laser cutting of CFRP, and laser direct joining of CFRP to metal, *Proc. of 29th ICALEO 2010*, 103 333–338.

9 Katayama S, Kawaguchi S, Mizutani M, Kawahito Y and Tarui T (2008), 'Welding phenomena and in-process monitoring in high power YAG laser welding of aluminum alloy', *J. of Light Metal Welding & Construction*, **46**(10) 480–490 (in Japanese).

10 Kawahito Y and Katayama S (2006), 'In-process monitoring and adaptive control for prevention of through-holes in full-penetration lap welds of aluminum alloy sheets', *Journal of Laser Applications*, **18**(2) 93–100.

2
Developments in CO_2 laser welding

S. TSUKAMOTO, National Institute for Materials Science, Japan

DOI: 10.1533/9780857098771.1.17

Abstract: This chapter describes carbon dioxide laser (CO_2 laser) welding. Some types of CO_2 lasers and special features of CO_2 lasers are introduced in the first sections. Laser welding phenomena, which include the interaction of the laser beam with materials and gases, keyhole behaviour, formation mechanism and prevention of weld defects, are also described. Finally, some practical applications of CO_2 laser welding are introduced.

Key words: CO_2 laser welding, resonator, interaction, welding phenomenon, application.

2.1 Introduction

C.K.N. Patel, working for Bell Laboratories, invented the CO_2 laser in 1964. He succeeded in producing 1 mW pure CO_2 lasers. The power and efficiency was much improved by addition of nitrogen and helium in the medium gas. The first commercial 100 W CO_2 laser was put on the market in 1966 by Coherent. Since then, CO_2 lasers have undergone continuous development and at the end of twentieth century, high power lasers operating at up to 50 kW were used for materials processing.

The wavelength of the CO_2 laser beam is normally 10.6 μm and it is about 10 times longer than that of solid-state laser beams. The long wavelength significantly affects the interaction of the laser beam with materials, shielding gases and plasmas and determines the characteristics of laser welding behaviour and laser systems. In this chapter, the principle and types of CO_2 lasers, characteristics of CO_2 laser beam, CO_2 laser–materials interaction, CO_2 laser welding behaviour and practical applications of CO_2 laser welding are described.

2.2 Laser principles and types of lasers

A mixture of carbon dioxide, nitrogen, and helium is used as a laser medium in CO_2 lasers. It is excited by DC or high frequency AC (generally RF) discharge to generate the laser beam with normally 10.6 μm wavelength.

There are some types of CO_2 lasers that are classified mainly by a heat removal method and flow direction of the laser medium. The first type is slow flow lasers. The laser gas slowly flows during discharge and the gas is cooled mainly through the cavity wall. A high-quality stable laser beam can be produced. However, the laser output power is normally small, because a very long cavity is required to achieve high power. Fast axis flow lasers increase the gas flow rate by circulating the gas with pumps or blower systems. The gas is cooled by heat exchangers set in the gas circulation path as shown in Fig. 2.1. The gas flow and laser beam have a common axis. High power lasers up to 20 kW with high beam quality are available in this type of laser. In transverse flow lasers, the laser beam axis is orthogonal to the discharge and gas flow as shown in Fig. 2.2. The laser gas is also cooled by circulation of the gas flow. This type of laser can achieve the highest laser power, although the beam quality is lower than the fast axis flow laser. Industrial lasers up to 50 kW are available. Slab lasers are classified as slow flow lasers. The laser medium is excited in a narrow space between two wide water-cooled electrodes as shown in Fig. 2.3. The wide discharge area and efficient heat removal through the wide electrodes enable high power and high beam quality laser beam to be generated. High power lasers up to 8kW with $M^2 = 1.05$ are available.

Relatively low power lasers up to 6 kW have stable resonators, where the output mirror is partially transparent to 10.6 μm infrared radiation. On the other hand, high power lasers have unstable resonators as shown in Fig. 2.4 to avoid distortion and damage to the mirror. The unstable resonator consists of concave and convex mirrors with a scraper mirror. The scraper mirror is located in front of the convex mirror and reflects the outer annulus of the beam from the cavity to produce the annular output beam. Zinc selenide (ZnSe) is normally used as a material of output windows. For high

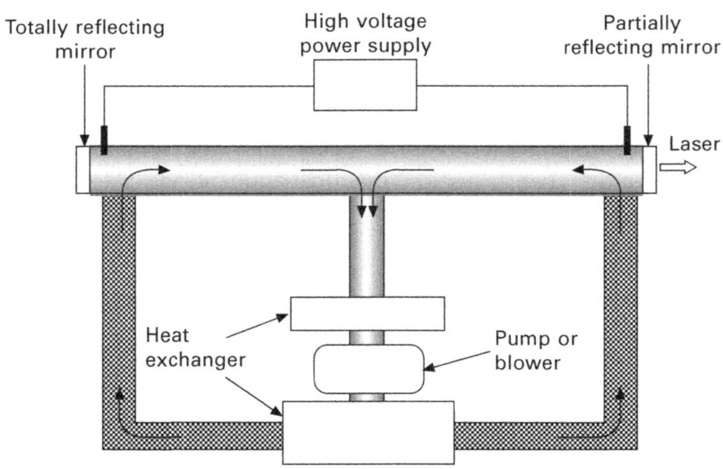

2.1 Fast axis flow laser.

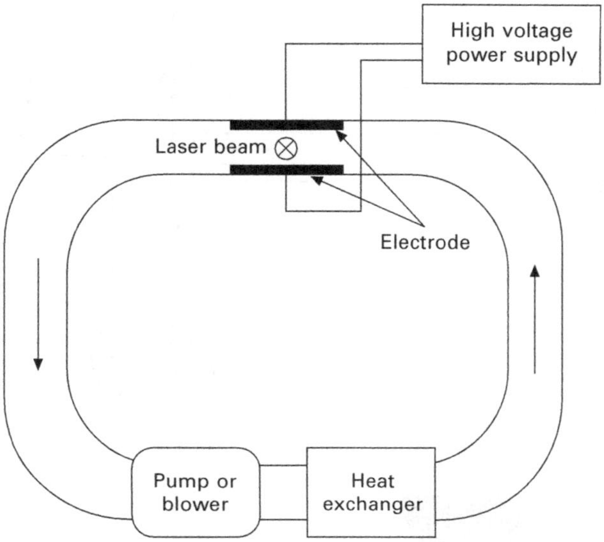

2.2 Transverse flow laser.

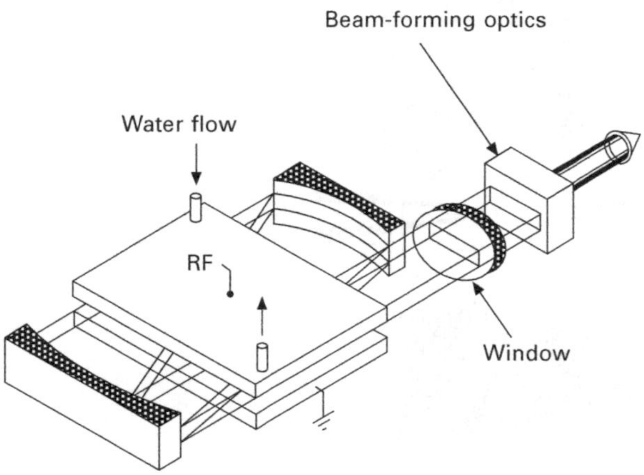

2.3 RF excited slab laser (from Davis, 2001).

power CO_2 lasers of more than 10 kW, an aerodynamic window is used to avoid thermal distortion and damage to the ZnSe window. As shown in Fig. 2.5, a rapid flow of dry air isolates the laser gas in the resonator from the atmosphere.

2.3 Characteristics of CO_2 laser beams

The wavelength of CO_2 laser beams is 10.6 μm, which is ten times longer than that in solid-state lasers. Electrical to optical beam conversion efficiency

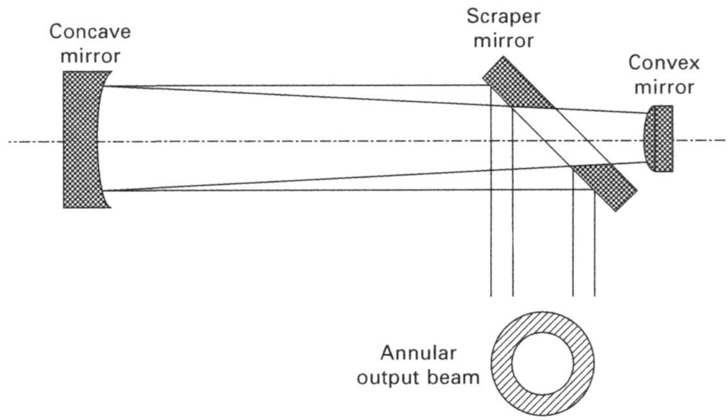

2.4 Unstable resonator.

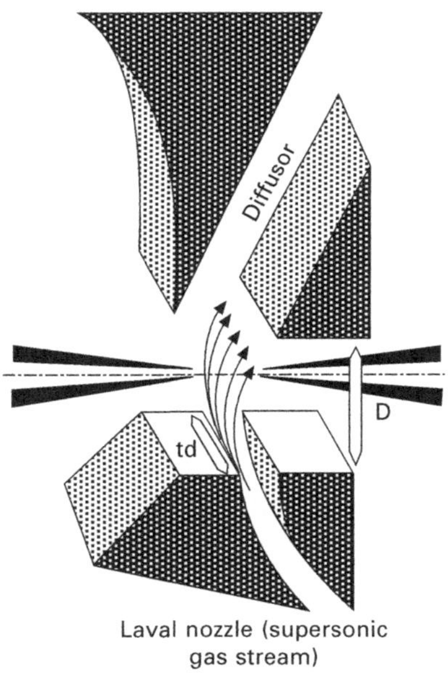

2.5 Aerodynamic window (source: TRUMPF Group).

is 8–15%, which is higher than rod type Nd:YAG lasers, but lower than diode lasers and recent brilliant lasers such as fibre and disc lasers. The main advantages of CO_2 laser systems compared with the other laser systems are:

- easily scaled to high powers
- high beam quality
- relatively low machine cost
- relatively stable welding.

The main disadvantages are:

- opaque to common transparent materials (expensive special materials are required for optics)
- no use of fibre-optic delivery
- high reflectivity of metals
- high absorption of laser energy in laser-induced plasma.

Development of high output machines was first progressed in CO_2 lasers, because it is not so difficult to increase the output power. 50 kW industrial machines were available by the end of twentieth century. In addition, high-quality laser beams can be attained in high power ranges. Figure 2.6 shows beam parameter product (BPP), which is the product of beam divergence angle and beam radius and quantifies beam quality. Lower values of BPP mean higher beam quality. In the low output power range of less than 10 W, BPP in the rod-type lamp and LD pumped Nd:YAG laser is one tenth of that in CO_2 lasers, because the minimum spot size d [mm] of the Gaussian beam at the focal point is proportional to the wavelength λ [mm] as expressed by the following equation:

$$d = 4\lambda f/\pi D \quad [2.1]$$

where f is the focal length (mm) and D is the input beam diameter at the lens (mm). Laser welding is normally performed using multi kW class lasers. In this high power range, BPP in the rod type Nd:YAG laser significantly increases with the output power due to thermal distortion of the cylindrical laser medium. On the other hand, CO_2 lasers can keep high beam quality

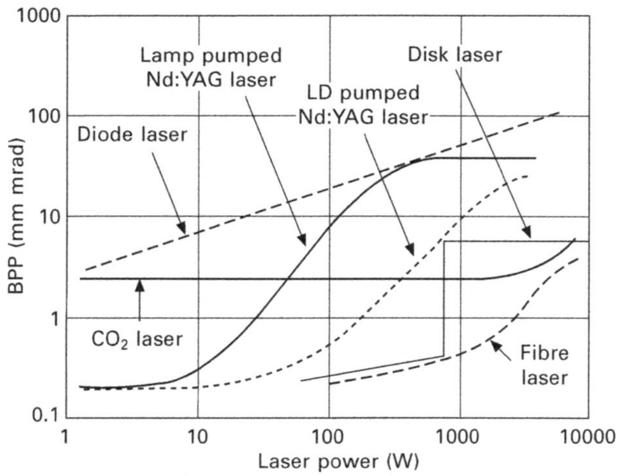

2.6 BPP of some lasers as a function of laser output power (from Katayama, 2009).

even in the high power range. The BPP is almost similar to that in disk and fibre lasers.

Relatively lower machine cost per unit power is also an advantage of the CO_2 laser. In heavy section welding, the welding phenomenon is relatively more stable in CO_2 laser welding compared with disk and fibre laser welding.

Most of the disadvantages in CO_2 laser welding are caused by the long wavelength (10.6 μm) of the laser beam. Common transparent materials such as glass and quartz are opaque to 10.6 μm wavelength laser radiation. Therefore, optical fibre delivery systems cannot be used. The laser beam must be transferred to the workstation by reflection using mirrors. Accurate beam alignment is required and flexibility of the laser welding is limited for practical purposes. This is the biggest disadvantage of CO_2 laser welding. In addition, special and expensive materials such as zinc selenide (ZnSe) must be used for optics such as windows and lenses.

2.4 Laser–materials interactions

2.4.1 Absorption of laser energy on material surfaces

When the laser beam irradiates materials, some of the energy is reflected on the material surface. The absorptivity may depend on the kind of materials, surface condition, surface temperature and laser wavelength. Table 2.1 shows

Table 2.1 Absorptivity of some materials for three wavelengths at room temperature (adapted from Modest, 2001)

Material	300–600 nm	1.06 μm	10.6 μm
Aluminium, smooth		0.06–0.2	0.03–0.06
rough		0.2–0.4	0.1–0.4
Copper, polished	0.05	0.04	0.01–0.03
rough	0.05	0.1–0.3	0.05–0.10
oxidized	0.85	0.5	
Gold		0.02–0.04	0.01–0.02
Iron, polished	0.37–0.40	0.25–0.32	0.12
Molybdenum, polished	0.4–0.5	0.25–0.35	0.05–0.15
Nickel	0.5	0.15–0.35	0.05–0.15
Platinum		0.25–0.30	0.03–0.08
Silver		0.03	0.02–0.10
Tungsten	0.5	0.35	0.03–0.3
Carbon (graphite)	0.75	0.8–0.9	0.7–0.9
Alumina (Al_2O_3)		0.05–0.1	0.90–0.99
Magnesia (MgO)		0.2	0.93–0.98
Silica (SiO_2)	Transparent	Transparent	0.9
Zirconia (ZrO_2)		0.1–0.2	0.85–0.98
Silicon carbide (SiC)	0.8–0.9	0.85–0.95	0.8–0.9
Silicon nitride (Si_3N_4)	0.6–0.7	0.6–0.8	0.9

the absorptivity of some materials at room temperature for three wavelengths corresponding to CO_2 laser (10.6 μm), Nd:YAG laser (1.06 μm) and short wavelength laser (300–600 nm) irradiation (Modest, 2001). The absorptivity generally tends to decrease with the wavelength. Only a small amount of CO_2 laser energy is absorbed on the metal surface. Conversely, CO_2 laser beams are significantly absorbed on most ceramics surfaces.

The absorptivity normally increases with temperature. During laser welding, formation of molten pools enhances the absorption of laser energy. In keyhole mode laser welding, multiple reflection of the laser beam on a narrow and deep keyhole wall significantly increases energy transfer efficiency due to multiple absorption. Thus, the absorptivity does not affect so much the weld geometry once the keyhole is formed.

2.4.2 Absorption of laser energy in laser-induced plasma

Attenuation of incident laser energy by inverse Bremsstrahlung

During CO_2 laser welding, a high power density laser beam irradiates the materials and causes intense evaporation. The laser beam interacts with the vapour jet and a laser plume or laser-induced plasma which is formed just above and inside the keyhole. The laser energy is absorbed in the plasma by inverse Bremsstrahlung and it attenuates the incident laser energy. The laser energy passed through the plasma I(W) with length x(m) and absorption coefficient α (m^{-1}) is

$$I = I_0 \exp(-\alpha x) \qquad [2.2]$$

where I_0 is the incident laser energy (W). For hot plasma ($h\omega/\pi << kT_e$), the absorption coefficient α(m^{-1}) is given (Hughes, 1975; Allmen and Blatter, 1995) by:

$$\alpha = \frac{n_i n_e Z^2 e^6 \lambda^2 \ln(2.25kT_e\lambda/hc)}{24\pi^3 \varepsilon_0^3 c^3 m_e kT_e \sqrt{2\pi m_e kT_e}} \qquad [2.3]$$

where ω is angular frequency, n_i and n_e are density of electron and ion (m^{-3}), respectively, Z is the ion charge number, e is electron charge (C), k is Boltzmann's constant (JK^{-1}), T_e is the electron temperature (K), h is Planck's constant (Js), c is the speed of light (ms^{-1}), ε_0 is the dielectric constant (C/Vm), and m_e is the mass of electron [kg]. From eq. [2.3], we can know that the absorption coefficient α is roughly proportional to square of the wavelength λ. Therefore, attenuation of laser energy is more significant in CO_2 laser welding than solid-state laser welding and this affects the weld geometry.

Assuming that the plasma is in local thermodynamic equilibrium (LTE) and $Z = 1$, the absorption coefficient α for CO_2 laser beam is given by:

$$a = 1.22 \times 10^{39} \frac{n_e^2}{T_e^{2/3}} \ln(1.655 \times 10^{-3} T_e) \quad [2.4]$$

Equation [2.4] indicates that a higher absorption coefficient is obtained for a higher degree of ionization (higher electron density n_e) at lower electron temperature T_e. The tendency of ionization is determined by the ionization potential. Atoms are more easily ionized with decreasing ionization potential of the element. Table 2.2 shows ionization potentials of some elements. In the inert gas, helium has the highest ionization potential and it is not ionized even in high power CO_2 laser welding. Therefore, helium is sometimes used as a shielding gas of high power CO_2 laser welding to prevent attenuation of the incident laser energy by plasma absorption.

As the ionization potential of metals is about half that of argon, metals are more easily ionized than inert gases. Figure 2.7 shows the composition of Ar-50 vol% Fe plasma at various temperatures. Iron starts to ionize at lower temperature and shows the maximum ion density at around 10,000 K due to low ionization potential, whereas argon ion density has a peak at a higher temperature of around 17,000 K. Figure 2.8 shows the absorption

Table 2.2 Ionization potential of some elements

Element	H	He	Ar	N	O	Al	Ti	Fe	Ni	Cu
Ionization potential (V)	13.60	24.59	15.76	14.53	13.62	5.98	6.82	7.87	7.64	7.73

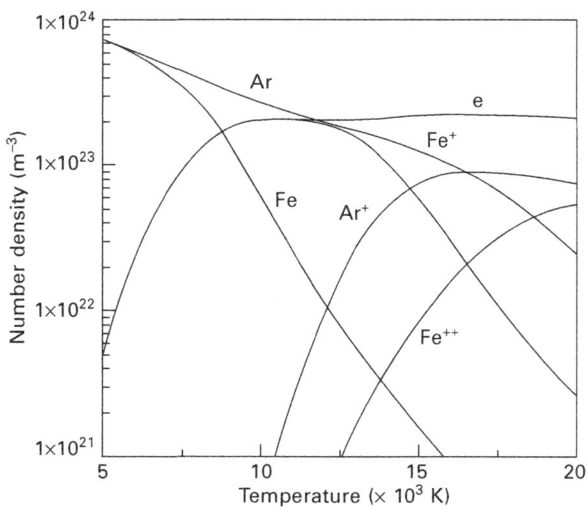

2.7 Composition of Ar-50 vol% Fe plasma as a function of temperature.

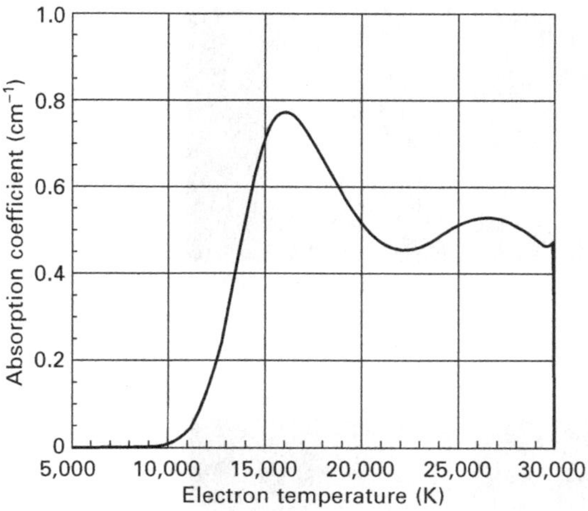

2.8 Absorption coefficient of Ar plasma.

coefficient of argon plasma as a function of the temperature. The absorption coefficient has a peak at around 17,000 K, because the argon ion and electron density has a maximum value at this temperature.

In low power CO_2 laser welding, only metals are ionized due to low plasma temperature, but the shielding gas is also ionized in high power laser welding. Figure 2.9(a) shows the laser-induced plasma formed during 5 kW CO_2 laser welding with a coaxial Ar shielding. Monochromatic images of some spectrum lines such as argon atom (Ar I: 696.5 nm), argon ion (Ar II: 480.7 nm) and iron atom (Fe I: 381.5 nm) are also shown in Fig. 2.9(b)–(d) (Tsukamoto *et al.*, 1999). Argon ion is clearly observed at the centre of the plasma, where the laser beam interacts directly with the plasma as shown in Fig. 2.9(c). It is surrounded by argon atoms as shown in Fig. 2.9(b). Thus, the gas plasma is normally formed above the workpiece in high power CO_2 laser welding. On the other hand, the metal plasma is mainly formed inside and just above the keyhole as shown in Fig. 2.9(d).

Formation of the plasma significantly affects the weld geometry. Figure 2.10 shows the penetration depth in 5.4 kW CO_2 laser welding with different shielding gases. The penetration depth is always lower in argon shielding. It is caused by attenuation of incident laser energy in argon plasma, whereas helium is not ionized during welding due to high ionization potential. Figure 2.11 shows the penetration depth as a function of the welding speed in 20 kW CO_2 laser welding with helium shielding. The penetration depth increases with decreasing welding speed, but it saturates below 50 cm/min. This phenomenon is not observed in solid-state laser welding. The salutation of the penetration depth in CO_2 laser welding is attributed to attenuation of laser energy by metal plasma formed inside the keyhole.

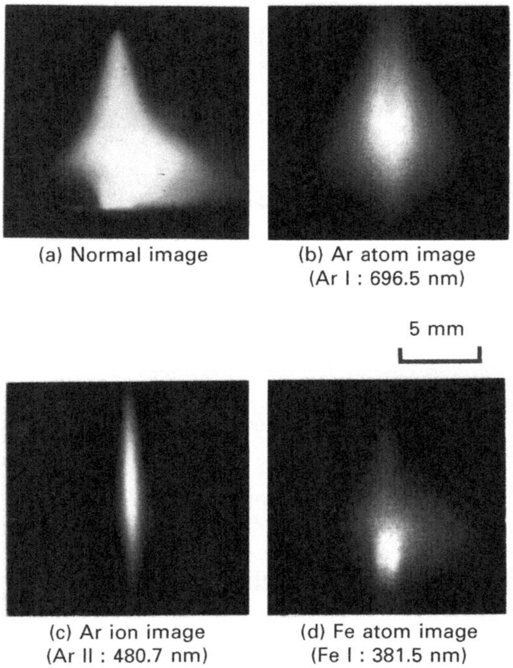

2.9 Monochromatic photographs of laser-induced plasma formed during CO_2 laser welding (5 kW, $f_d = -2$ mm, Ar shielding: 40 L/min): (a) normal image, (b) argon atom image (Ar I: 696.5 nm), (c) argon ion image (Ar II: 480.7 nm), (d) iron atom image (Fe I: 381.5 nm).

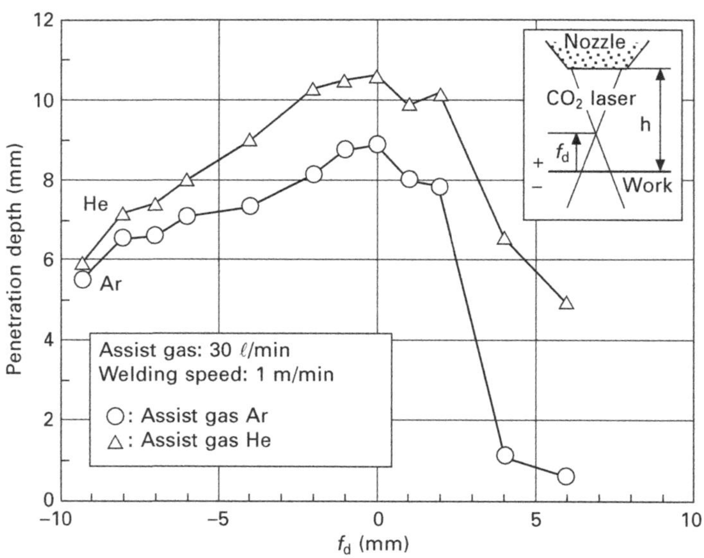

2.10 Effect of shielding gas on the penetration depth in CO_2 laser welding (C-Si-Mn steel).

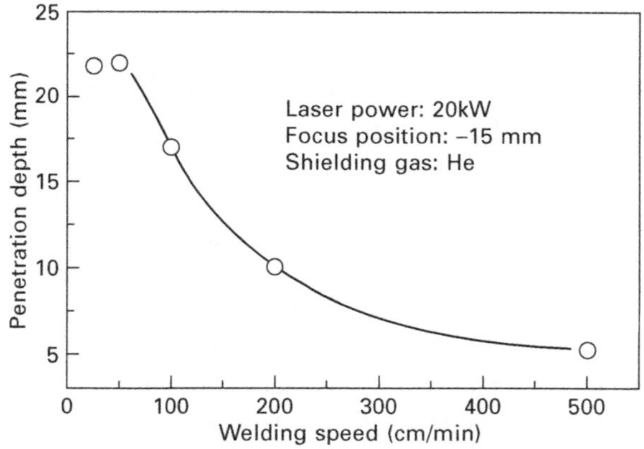

2.11 Penetration depth as a function of welding speed in CO_2 laser welding (C-Si-Mn steel).

Plasma diagnostics

From Eq. [2.4], the absorption coefficient in the plasma is a function of the electron density n_e and electron temperature T_e. n_e is determined by T_e with the assumption of LTE. Therefore, a lot of research work has been carried out to measure T_e during CO_2 laser welding by spectroscopic methods (Rockstroh and Mazumder, 1987; Mazumder *et al.*, 1987; Sokolowski *et al.*, 1988, 1989; Poueyo *et al.*, 1992; Miyamoto and Maruo, 1992; Verwaerde and Fabbro, 1995; Tsukamoto *et al.*, 1996). The intensity of a thermally excited spectral line ε_{nm} is given by:

$$\varepsilon_{nm} = \frac{hcA_{nm}n_0g_n}{4\pi\lambda\, Z(T)} \exp(-E_n/kT_e) \qquad [2.5]$$

where A_{nm} is Einstein transition probability of the line (s^{-1}), n_0 is the density of atom or ion (m^{-3}), g_n is statistical weight of the term n, $Z(T)$ is the partition function, E_n is the energy level (J). From Eq. [2.5], T_e can be determined by measuring the absolute value of spectral intensity ε_{nm} with the assumption of LTE. This is called the 'absolute spectral intensity method'. T_e can also be determined by the ratio of the spectral intensity ε_{nm} of same species in different spectral lines. This is called the 'relative intensity method'. In this case, assumption of LTE is not necessary. From Eq. [2.5],

$$\frac{\varepsilon_{nm}\lambda}{A_{nm}g_n} = K \exp(-E_n/kT_e) \qquad [2.6]$$

$$\ln\left(\frac{\varepsilon_{nm}\lambda}{A_{nm}g_n}\right) = -\frac{1}{kT_e}E_n + \ln K \qquad [2.7]$$

where K is a constant. If we measure some spectral intensities of the same species with different spectral lines and plot $\ln(\varepsilon_{mn}\lambda/A_{nm}g_n)$ as a function of E_n, we can know the electron temperature T_e from the gradient. This is called the the 'Boltzmann plot method' and is often used to determine the electron temperature in the plasma.

Figure 2.12 shows the temperature profile of the plasma formed during 5 kW CO_2 laser welding with argon shielding flowed from the coaxial nozzle at 30 L/min (Tsukamoto *et al.*, 1999). The focal point is located just on the sample surface. A peak temperature of around 22,000 K can be seen at the centre of the plasma, which corresponds to the argon ion enriched region in Fig. 2.9(c) due to direct interaction of the laser beam with the plasma. This temperature is similar to the peak temperature of argon TIG arc.

If we know the electron temperature distribution as shown in Fig. 2.12, we can calculate the electron density with the assumption of LTE and the absorption coefficient using Eq. [2.4] at any part of the plasma. As a result, we can estimate the absorptivity of the plasma under various welding conditions.

Figures 2.13–2.15 show the absorptivity for various welding conditions (Tsukamoto *et al.*, 1999). The absorptivity gradually increases with the laser power as shown in Fig. 2.13. The flow rate of the shielding gas does not exert a great influence on the absorptivity. The maximum temperature slightly decreases from 22,000 to 20,000 K with increasing the flow rate from 20 to

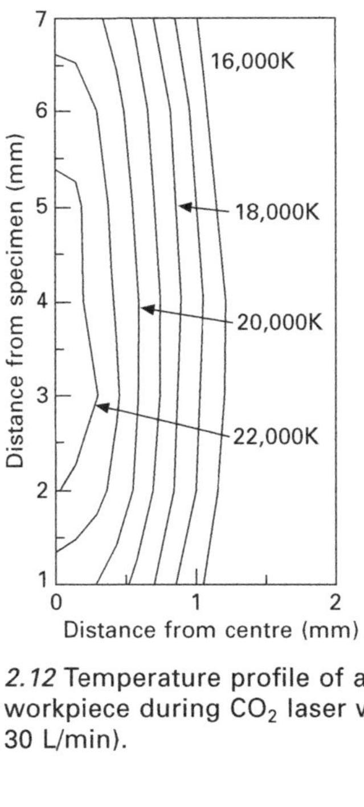

2.12 Temperature profile of argon plasma formed above the workpiece during CO_2 laser welding (5 kW, f_d = 0 mm, Ar shielding: 30 L/min).

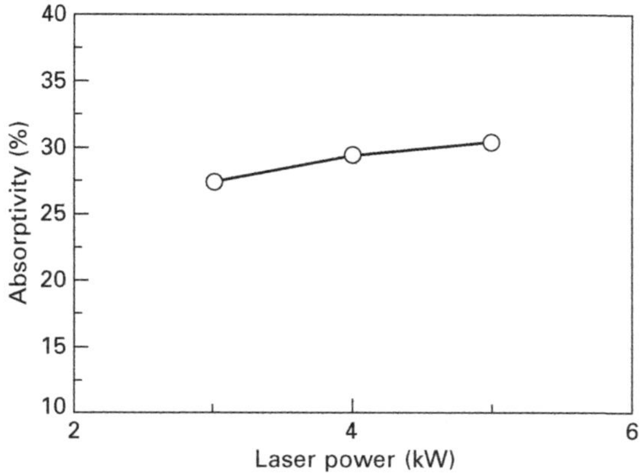

2.13 Effect of laser power on absorptivity.

60 L/min as shown in Fig. 2.14(a). In this temperature range, the absorption coefficient decreases with temperature as shown in Fig. 2.8, but the plasma size also decreases with the flow rate as shown in Fig. 2.14(b). As a result, the absorptivity shows approximately a constant value over a wide range of flow rates. The absorptivity is most significantly affected by the beam focusing condition. The maximum temperature is not affected significantly by the defocus distance f_d as shown in Fig. 2.15(a). The absorptivity, on the other hand, increases significantly as the focal point shifts upward from inside to above the sample as shown in Fig. 2.15(b). It is caused by expanding the plasma with f_d. The plasma size increases as the power density increases near the plasma-forming area.

To suppress attenuation of the incident laser energy by the plasma, flow of the inert gas with high ionization potential is effective. A helium or helium-argon mixture is normally used as a shielding gas in high power CO_2 laser welding. In the welding variables, the effect of beam focusing condition is most significant. To suppress the plasma absorption, power density just above the sample surface should be reduced by selecting the negative defocused distance.

Plasma monitoring

The plasma is generated and maintained by evaporation of materials and absorption of laser energy through inverse Bremsstrahlung. Therefore, the plasma emission includes a lot of information on the welding phenomena and it is sometimes used to monitor the welding conditions, welding phenomenon and formation of defects.

Figure 2.16(a) and (b) show one typical example of the in-process monitoring of penetration depth using the plasma signal (Mori and Miyamoto, 1997).

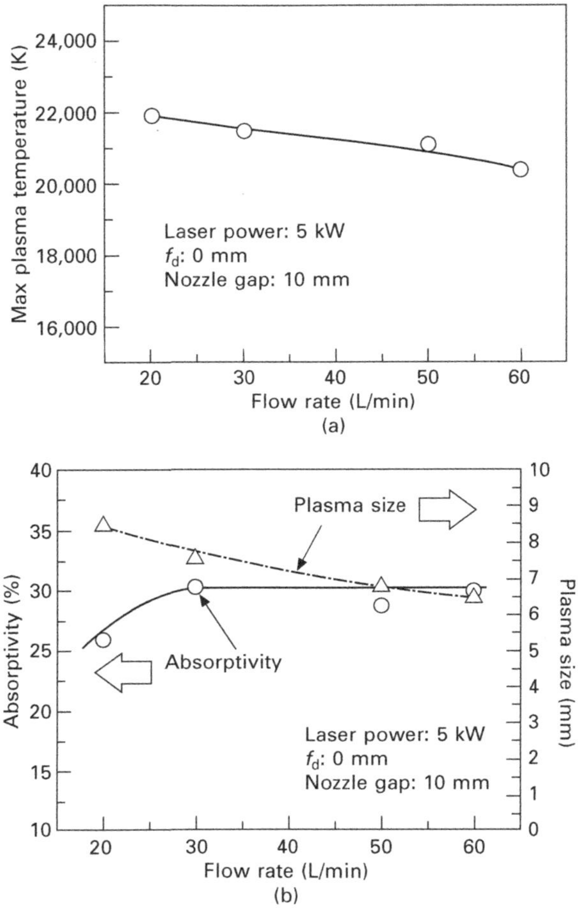

2.14 Maximum plasma temperature, plasma size and absorptivity for various flow rates of argon shielding gas: (a) maximum plasma temperature, (b) plasma size and absorptivity.

Partial and full penetration CO_2 laser welding was carried out on 1.6 and 4 mm thick mild steels. Plasma emission was detected by two photodiodes located at different angles of 75° and 5° as shown in Fig. 2.16(a). The 5° photodiode detects the emission from only the near surface plasma (A in Fig. 2.16(a)), whereas the 75° photodiode signal contains the emission from the keyhole plasma and near surface plasma (A+B in Fig. 2.16(a)). Figure 2.16(b) shows the mean square value of the plasma signals in a frequency range of 2–4 kHz as a function of the welding speed. The full penetration region is shaded in the graph. It is clear that full or partial penetration welding can be distinguished from the 75° photodiode signal. The mean square value of the plasma signal is much higher in full penetration than that in partial penetration.

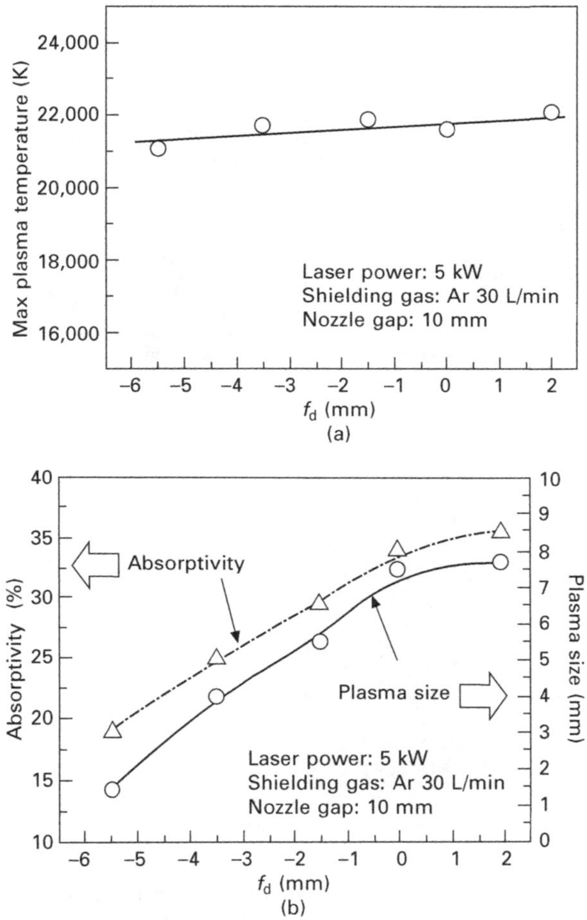

2.15 Maximum plasma temperature, plasma size and absorptivity for various defocus distances f_d: (a) maximum plasma temperature, (b) plasma size and absorptivity.

2.5 Welding phenomena and formation of defects

2.5.1 Pressure balance on keyhole wall

In keyhole mode laser welding, a narrow and deep keyhole is formed by evaporation of materials. The laser energy can be supplied inside the materials through the keyhole and then narrow and deep weld penetration is formed. However, the keyhole is basically unstable, if the length is larger than the circumference ($l > 2\pi r$, l: keyhole depth, r: keyhole radius). The keyhole tends to be closed at some parts along the penetration. It is called the Rayleigh instability (Rayleigh, 1945; Chandrasekhar, 1961). When the keyhole is being closed by instability, the laser beam strongly interacts with

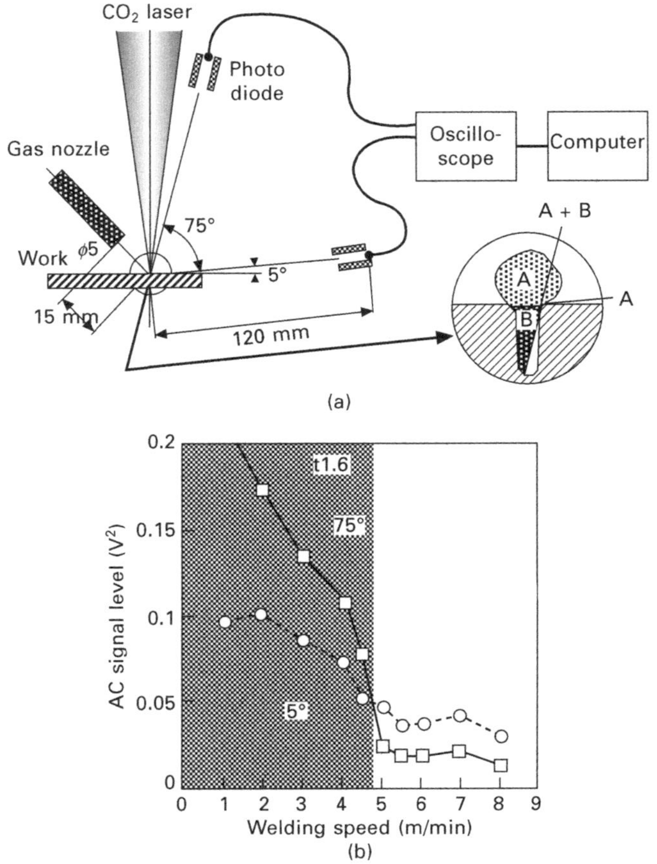

2.16 In-process monitoring of penetration depth by plasma emission signal (from Mori and Miyamoto, 1997): (a) schematic diagram of the experimental setup, (b) mean square value of plasma signal as a function of welding speed (plate thickness: 1.6 mm).

the molten metal at the closing part and opens the keyhole again by high recoil pressure as shown in Fig. 2.17. Thus, the keyhole oscillates in the radial direction while it is maintained during deep penetration laser welding and it sometimes causes weld defects.

Statically, the keyhole is maintained by the following pressure balance acting on the keyhole wall.

$$P_{abl} + \delta P_g = P_\sigma + P_h \qquad [2.8]$$

where P_{abl} is the recoil pressure (N/m^2), δP_g is the excess vapour pressure ($\delta P_g = P_v - P_0$, P_v: vapour pressure, P_0: atmospheric pressure) (N/m^2), P_σ is the surface tension pressure ($P_\sigma = \sigma/r$, σ: surface tension (N/m), r: keyhole radius (m)) (N/m^2) and P_h is the hydrostatic pressure (N/m^2). Pressure terms

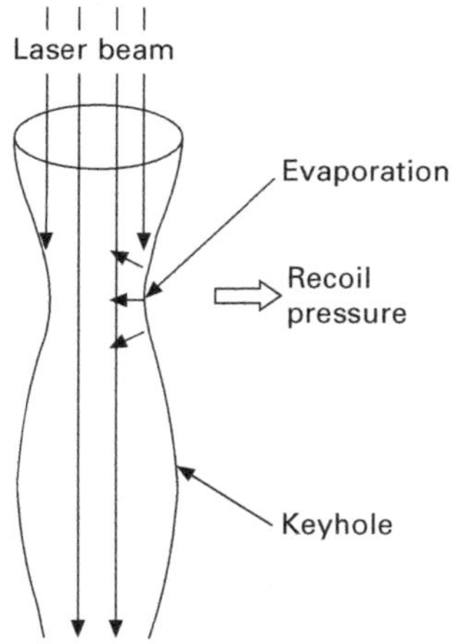

2.17 Keyhole behaviour during laser welding. Closure of the keyhole by Rayleigh instability is prevented by direct irradiation of the laser beam.

on the left side act to open the keyhole and those of the right side act to close the keyhole. During welding, these pressures are balanced at any parts on the keyhole wall. In CO_2 laser welding, the metal vapour is partially ionized and forms the plasma inside the keyhole. The plasma temperature is higher than that in the solid-state laser welding due to the high absorption coefficient of the CO_2 laser beam. High temperature plasma formed inside the keyhole may increase δP_g and the keyhole radius may increase in CO_2 laser welding. This may be one of the reasons why CO_2 laser welding is stable.

2.5.2 Defect formation in partial penetration CO_2 laser welding

Stability of the keyhole significantly affects the formation of defects. In heavy section laser welding, formation of the porosity is one of the major problems for practical applications. Especially in partial penetration welding, the porosity tends to be formed more frequently as shown in the longitudinal section of Fig. 2.18. Figure 2.19 shows an *in-situ* x-ray transmission keyhole image during partial penetration CO_2 laser welding of C-Si-Mn steel. The image is taken by a high-speed video camera while a micro-focused x-ray is irradiated from the side of the sample. Bubble formation is clearly observed at the keyhole tip (Matsunawa *et al.*, 1997; Matsunawa, 2001; Katayama *et al.*, 1997; Tsukamoto *et al.*, 2001, 2002). The bubbles formed at the keyhole tip combine with each other and most of them remain as porosity after solidification. Figure 2.20 shows the bubble formation process near the keyhole tip. During welding, a neck is formed near the keyhole tip (Fig. 2.20

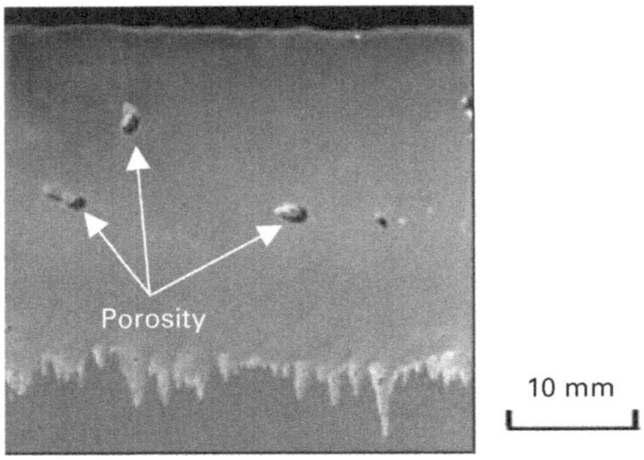

2.18 Longitudinal section of CW CO_2 laser welds (C-Si-Mn steel, 20 kW, f_d = 1.0, 1 m/min, shielding gas: He, 50 L/min).

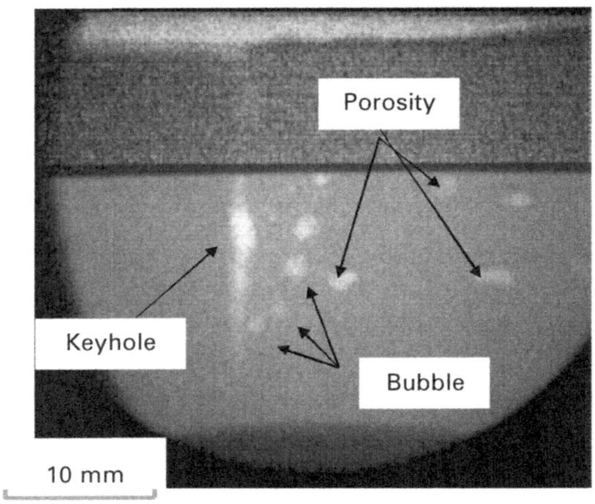

2.19 In-situ x-ray transmission image of keyhole (same condition as Fig. 2.18).

(c)) and then the keyhole is closed there by surrounding molten metal (Fig. 2.20(d)). The closure of the keyhole is normally prevented by irradiation of the laser beam to the closing molten metal as described above. However, if the keyhole fluctuates and laser/molten metal interaction is significant in the keyhole, a lot of laser energy is consumed by the interaction and it reduces the laser power density near the root to less than the critical value to maintain the keyhole. As a result, the keyhole is closed near the root by Rayleigh instability. This result indicates that stabilization of the keyhole should be effective in preventing the porosity.

Nitrogen shielding is normally used to prevent the porosity in high Cr stainless steel welding (Ishide *et al.*, 1992), because the solubility of nitrogen increases with Cr concentration. Bubbles are not formed during welding by dissolving nitrogen in the molten pool. However, nitrogen shielding is not

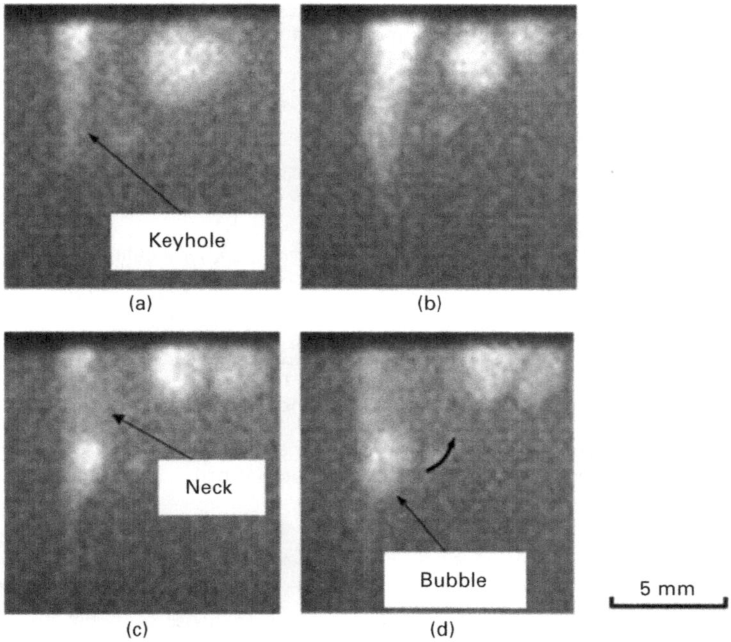

2.20 In-situ x-ray transmission images near the keyhole tip. A bubble is formed by closure of keyhole tip: (a) 0 ms, (b) 7 ms, (c) 16 ms, (d) 17 ms.

possible in very high power CO_2 laser welding due to significant attenuation of laser energy by large nitrogen plasma. Also nitrogen shielding is not effective in laser welding of carbon steels, because the solubility of nitrogen is quite low in carbon steels.

The other porosity prevention method is stabilization of the keyhole by laser power modulation (Tsukamoto *et al.*, 2002). The porosity is effectively suppressed by optimizing the modulation conditions. Figure 2.21 shows the porosity formation ratio Pr (ratio(%) of the sum of porosity area to the weld metal area) in CO_2 laser welding of C-Si-Mn steel as a function of the power modulation frequency. Laser power is modulated with square wave at the peak and base powers of 20 and 12 kW, respectively. The porosity is clearly suppressed at 16 Hz in this condition. This optimum modulation frequency coincides well with the natural frequency of the molten pool oscillation. Stabilization of the keyhole was confirmed under the optimum frequency by an *in-situ* x-ray transmission imaging system. The porosity suppression effect is further enhanced by optimizing the waveform. Almost all pores are prevented by optimization of the waveform and frequency as shown in Fig. 2.22 (a) and (b).

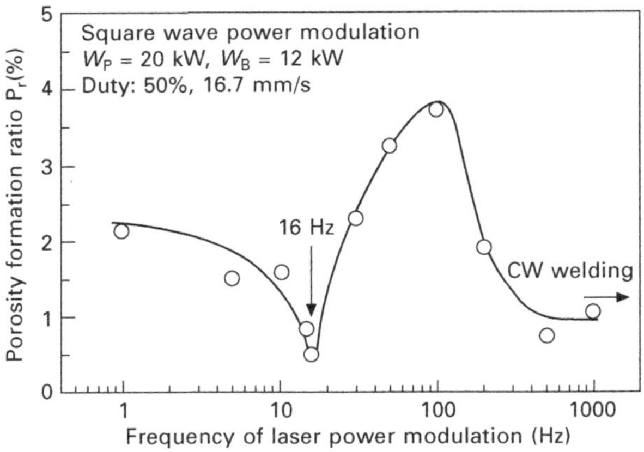

2.21 Porosity formation ratio Pr as a function of power modulation frequency (square wave power modulation).

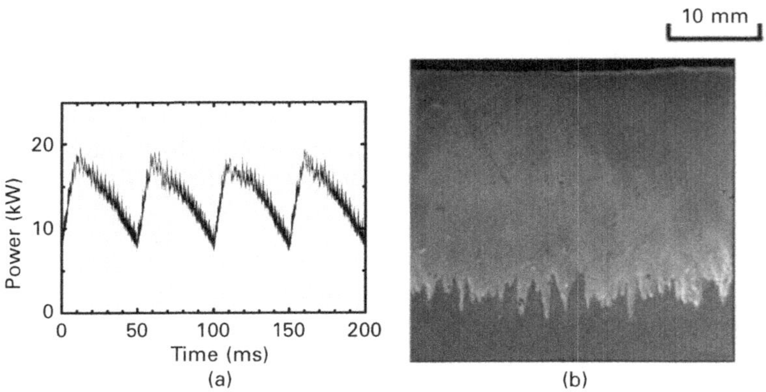

2.22 Prevention of porosity under optimum power modulation condition: (a) waveform of optimum power modulation, (b) longitudinal section of laser welds under optimum power modulation condition.

2.5.3 Defect formation in single pass full penetration CO_2 laser welding

In single pass full penetration laser welding, the porosity caused by keyhole instability is less likely to occur, but it is formed via a different mechanism. A typical example is shown in Fig. 2.23, where full penetration laser welding was carried out on C-Si-Mn steel without back shielding (Tsukamoto *et al.*, 2003). Figure 2.24(a) shows the *in-situ* x-ray transmission image for full penetration CO_2 laser welding of 15 mm thick C-Si-Mn steel without back shielding. Bubbles are generated in the molten metal just behind the keyhole. This is caused by supersaturated nitrogen in the molten pool. During

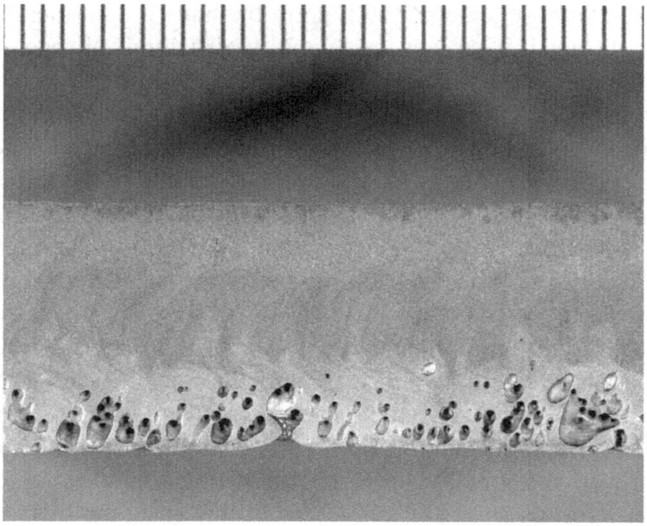

2.23 Longitudinal section of CO_2 laser welds of 11 mm thick C-Si-Mn steel without back shielding (11 kW, f_d = 0 mm, 1 m/min).

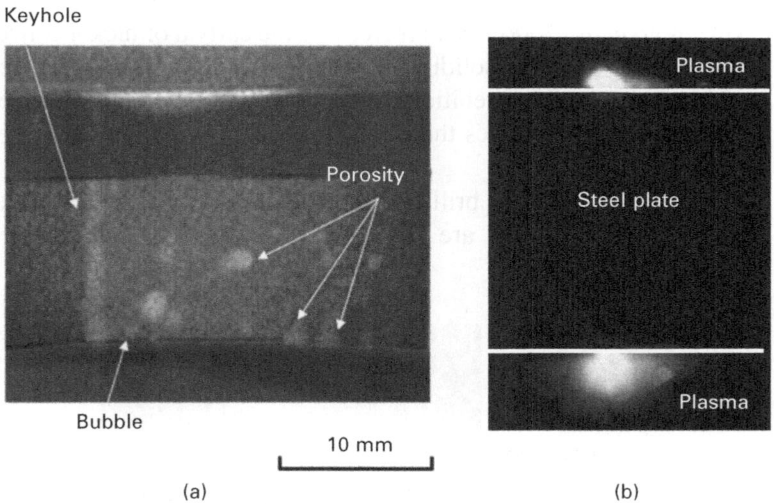

(a) (b)

2.24 Single pass full penetration CO_2 laser welding behaviour of 15 mm thick C-Si-Mn steel (20 kW, f_d = 0 mm, 1 m/min): (a) *in-situ* x-ray transmission image; bubbles are generated in the molten pool just behind the keyhole, (b) formation of air plasma just below the back surface.

full penetration laser welding, the excessive laser energy, which penetrates through the steel plate, forms air plasma on the back surface as shown in Fig. 2.24(b). The plasma contains a lot of monatomic nitrogen, which can much more easily dissolve in the molten steel than diatomic nitrogen (Dong *et al.*, 2004). The nitrogen concentration in the molten pool then increases near the bottom, and if it exceeds the solubility of diatomic nitrogen, bubbles are formed in the molten metal.

The inert gas back shielding can effectively prevent this type of porosity. However, the solidification crack is easily formed accompanied by the locally expanded weld cross section, which is very susceptible to the solidification crack, as shown in Fig. 2.25(a) (Tsukamoto *et al.*, 2003). This abnormal weld cross section is caused by a keyhole perturbation. During welding, the inert gas enters the keyhole from the back surface and expands in the keyhole as shown in the x-ray transmission image of Fig. 2.25(b). This local expansion of the keyhole forms the abnormal cross section, leading to occurrence of the solidification crack. Addition of a small amount of denitrification elements, such as aluminium, is effective to prevent the porosity without forming the solidification crack (Tsukamoto *et al.*, 2004). 0.8 mass% of aluminium is enough to prevent the porosity during the full penetration welding of 15 mm thick carbon steels as shown in Fig. 2.26. In this case, the inert gas back shielding is not necessary. The solidification crack then does not occur in the weld metal.

Addition of nickel and carbon is also effective in suppressing the porosity (Tsukamoto *et al.*, 2006). With increasing the carbon or nickel concentration in the base metal, the solidification mode changes from fully ferritic to ferritic-austenitic or austenitic-ferritic. Higher solubility of nitrogen in the austenite phase suppresses the porosity formation in high carbon or nickel steels.

Recently, high power brilliant lasers such as fibre and disk lasers were developed. These lasers are very attractive tools to weld heavy section

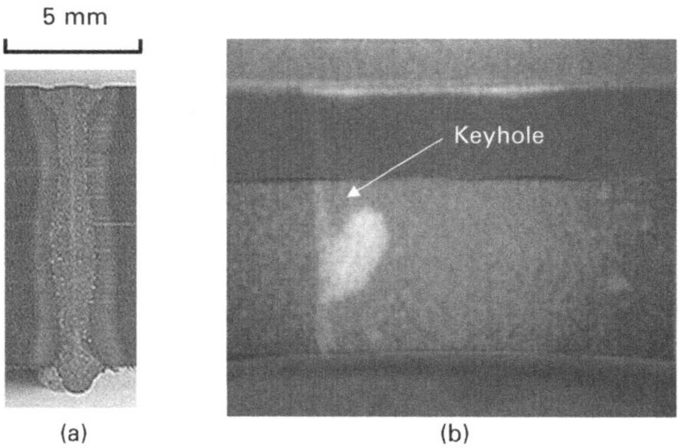

2.25 Formation of hot cracking susceptible cross section and keyhole behaviour during CO_2 laser full penetration welding of 15 mm thick C-Si-Mn steel with argon back shield (16 kW, f_d = 0 mm, 1 m/min): (a) transverse cross section of laser welds, (b) *in-situ* x-ray transmission image of keyhole.

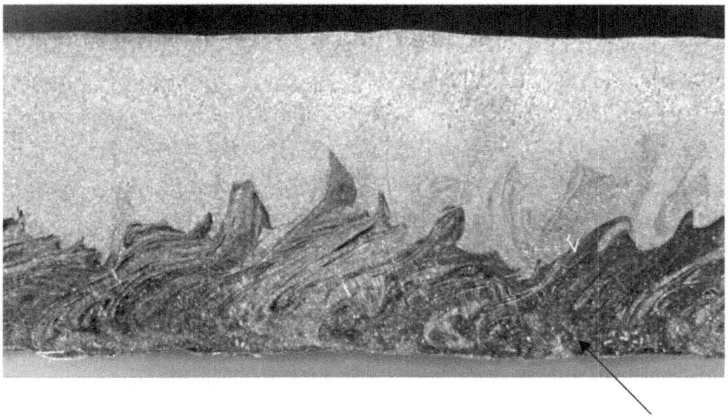

2.26 Longitudinal section of laser welds. Porosity formation is prevented by coating 200 μm thick aluminium on the back surface of the base metal prior to the welding.

plates. However, there are some problems to be solved (Rominger, 2011). Spatter generation, melt-through and underfill are more likely to occur and the optimum welding conditions are narrower in some cases compared with CO_2 laser welding. The mechanism of the different welding phenomenon between CO_2 and brilliant laser welding is not yet well understood. High temperature keyhole plasma formed in CO_2 laser welding may be one of the factors to stabilize the keyhole due to increase in δPg in Eq. [2.8].

2.6 Industrial applications of CO_2 laser welding

Multi-kW class high power CO_2 lasers, which are capable of keyhole mode welding, were developed in the mid-1970s. Since then, CO_2 laser welding has been applied in a variety of industries. In this section, some examples of the industrial applications are introduced.

2.6.1 Automotive industry

From the late 1970s to 1980s, application of CO_2 laser welding started in the automotive industry. Various parts of powertrains including transmission gears, hubs and shafts were welded by CO_2 lasers (Petring, 2004). High productivity, low heat input and low distortion are the main advantages of laser welding compared with the other processes. CO_2 lasers are still being used in these applications.

Laser welded tailored blanks are widely used in manufacturing parts of the car body such as doors, front and side panels, side beams and wheel arches.

As shown in Fig. 2.27, flat metal sheets with different gauge, strength and coating are joined by laser welding prior to forming and trimming. Mixture of different sheet metals allows a reduction in weight, number of components and total manufacturing cost and an improvement in crash energy management. This technique has been used since the mid-1980s. Most linear welding lines are equipped with 5–10 kW CO_2 lasers. On the other hand, Nd:YAG, fibre or disk lasers are used mainly in non-linear welding.

Application of laser welding in the body shop started in the mid-1980s. 5 kW class CO_2 lasers were installed by some automobile companies to weld body in white such as joining of the roof with the side frames. However, usage of CO_2 lasers decreased in the late 1990s due to installation of high power Nd:YAG lasers with robots, which are capable of three-dimensional laser welding.

Installation of remote laser welding started in the early 2000s to produce car body components. In this system, the laser spot quickly moves by high-speed scanning mirrors to weld some parts very quickly. A high beam quality laser is required to cover the wide welding area with a long focal length of normally greater than 1 m. A slab CO_2 laser was first used to attain a high beam quality laser beam. Reduction of the tact time and high productivity are the main advantages. Due to progress of high power disk and fibre lasers with high beam quality, a new remote welding system combined with a robot was developed. More flexible and faster welding was achieved using fibre and disk lasers. This system is called 'remote welding on the fly' and is used widely.

2.6.2 Aircraft industry

Airbus applied CO_2 laser welding of the stringer to skin sheets for aircraft fuselage panels in the early 2000s (Schumacher *et al.*, 2002; Vollertsen *et*

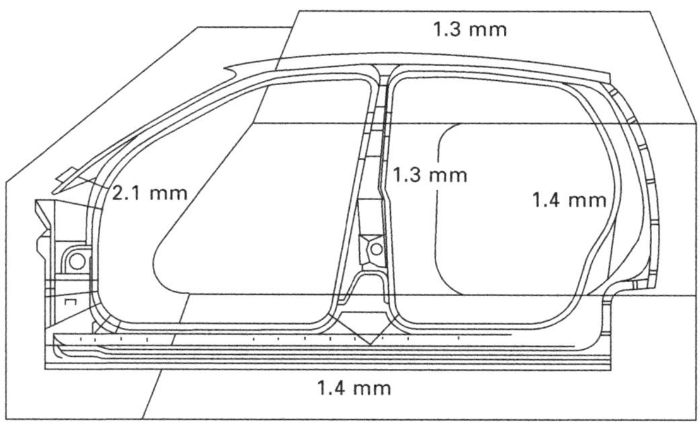

2.27 Laser welded tailored blanks.

al., 2004). This was achieved by developments in laser welding systems and weldable Al-Mg-Si-Cu alloy. Instead of conventional riveting, the stringers are joined to the skin sheet by simultaneous laser welding from both sides as shown in Fig. 2.28. Al-12% Si filler wire is used to avoid hot cracking. Weight reduction, high productivity, improvement of corrosion resistance and high cost performance are the main advantages of replacing the conventional riveting process with laser welding.

2.6.3 Shipbuilding industry

The first laser welding application in the shipbuilding industry was the manufacture of sandwich panels by Meyer Werft in the mid-1990s (Roland *et al.*, 2004). It was produced by stake welding of flat panels up to 10 mm in thickness to webs by 12 kW CO_2 laser as shown in Fig. 2.29. Weight reduction with high strength and stiffness and low distortion can be attained.

In Europe, guidelines for approval of CO_2 laser welding was laid down following a number of national and EC projects in the mid-1990s. This accelerated the applications of laser welding in the shipbuilding industry. Long straight butt and fillet welding lines of thick steel plates were installed at Odense Steel Shipyard (Sellerup, 1999), Fincantiaeri (Manzon, 1998) and Blohm & Boss (Kahl, 1998) in the late 1990s. The line consists of a gantry equipped with high power CO_2 lasers from 12 to 18 kW. The line includes monitoring and feedback control systems, edge preparation and machining and/or wire feeding systems. A most attractive advantage of installing high power laser is reduced distortion, which significantly reduces total man-hour consumption for additional straightening operations. Improvement of

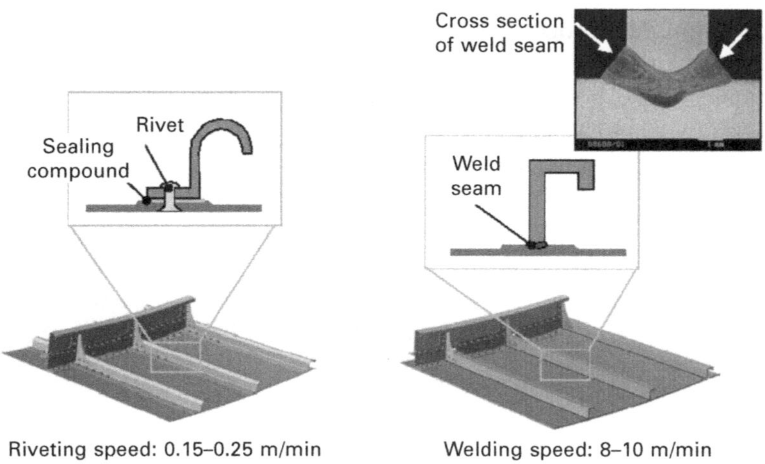

2.28 Comparison of riveting and laser welding for the skin to stringer joint (from Vollertsen *et al.*, 2004).

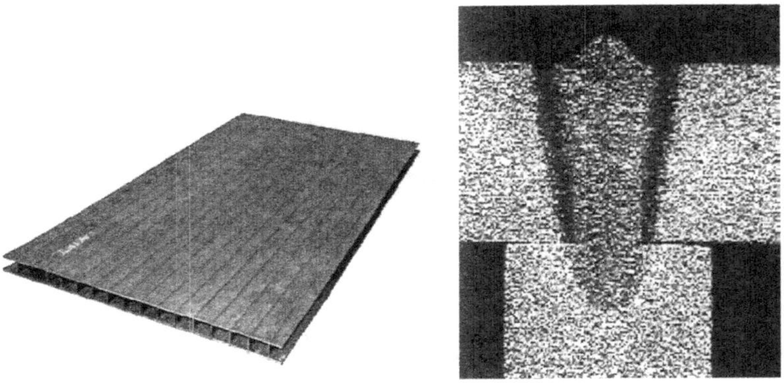

2.29 Laser welded sandwich panel.

mechanical accuracy enables automation of the production line and high productivity was realized. Meyer Werft installed four laser/GMA hybrid welding stations from the late 1990s to early 2000s (Roland *et al.*, 2004) and now the amount of laser hybrid welds in a large cruise ship is close to 50% of the total weld length.

Recently, fibre and disk lasers have been installed by some shipbuilding companies due to flexibility of fibre delivery systems. However, CO_2 laser has also been installed in new lines due to stability of the welding phenomena.

2.6.4 Steel industry

From the early 1980s, the steel industry installed 5–10 kW CO_2 lasers for welding of hot or cold rolled steel sheets in continuous coil welding lines. Joint interface property was much improved by replacing flash welding with laser welding. In the early 2000s, two 45 kW class CO_2 lasers were applied to welding of hot steel bars of 30 mm thickness at continuous finish rolling lines (Minamida, 2002).

CO_2 lasers were also installed for manufacturing small diameter stainless steel welded pipes to improve productivity in the early 1980s. CO_2 lasers are still being used in this application field. For manufacturing pipes with medium diameter and heavy wall thickness, 25 kW CO_2 laser welding combined with high frequency induction coil preheating was applied in the mid-1990s (Hayashi *et al.*, 1996; Ono *et al.*, 2001).

2.6.5 Other applications

High power CO_2 lasers have been used to fabricate parts of some chemical plants such as the plant for petroleum refining. A large skirt made by thick stainless steels is fabricated by CO_2 lasers as shown in Fig. 2.30(a). T-beams

(a)

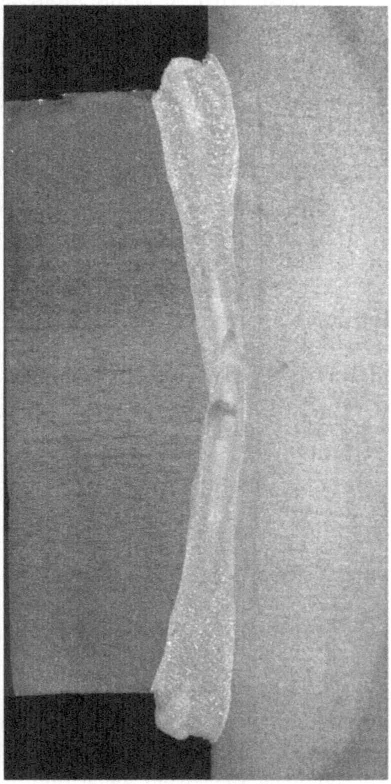

(b)

2.30 CO_2 laser welding for chemical plant: (a) a skirt 4 m in diameter fabricated by CO_2 laser welding, (b) CO_2 laser welding of 25 mm thick stringer to the web.

are also fabricated by CO_2 lasers as shown in Fig. 2.30(b), where 25 mm thick stringer is welded to the web. Components of plasma reactor used for production of compound, semiconductor and nanomaterials such as carbon nanotube are manufactured by CO_2 laser welding with low distortion and no weld defects. CO_2 lasers are also used to weld a pipe to a blade for skate shoe runners.

2.7 Future trends

CO_2 laser welding has been applied in a variety of industries since the mid-1970s. In the early 1990s, multi-kW class Nd:YAG lasers were developed. It has a great advantage in its fibre-optic delivery system, although the beam quality of rod-type YAG lasers is worse than that of CO_2 lasers. Therefore, applications of YAG laser welding immediately expanded mainly in the field of thin plate welding.

In the mid-2000s, high power brilliant lasers such as disk and fibre lasers were developed. As the beam quality of these lasers is very high, application fields of these lasers have been expanding not only in thin plate welding but also in heavy section welding. The development of these lasers significantly reduced the usage of CO_2 laser welding. However, some problems such as melt through, spattering and narrow optimum conditions still remain to be solved in heavy section fibre and disk laser welding. CO_2 laser welding will continue to be used in some areas unless these problems are not solved.

2.8 Acknowledgements

The author is grateful to Mr M. Dahmen, Fraunhofer Institute for Laser Technology, for his useful information on practical applications and CO_2 welding phenomena.

2.9 References

Allmen M. von and Blatter A. (1995), *Laser Beam Interaction with Materials*, 2nd edn. Berlin: Springer.

Chandrasekhar S. (1961), *Hydrodynamic and Hydromagnetic Stability*. (1961), Oxford: Clarendon Press.

Davis J. (2001), 'Slab lasers', in *LIA Handbook of Laser Materials Processing*, edited by Ready J. F. and Farson D. F., Orlando, FL: Laser Institute of America.

Dong W., Kokawa Y., Tsukamoto S., Sato Y. S. and Ogawa M. (2004), 'Mechanism governing nitrogen absorption by steel weld metal during laser welding', *Metall. Mater. Trans. B*, 35B, 331–338.

Hayashi T., Inaba Y., Matuhito Y., Yamada T. and Kudo T. (1996), 'Development of high power welding process for pipe', *Proc. ICALEO 1996*, D, 132–140.

Hughes T. P. (1975), *Plasma and Laser Light*. The Institute of Physics, Bristol: Adam Hilger, 39–54.

Ishide T., Nagura Y., Matsumoto O., Nagashima T., Kidera T. and Yokoyama A. (1992), 'High power YAG laser welded sleeving technology for steam generator tubes in nuclear power plants', *Proc. LAMP 1992*, 957–962.

Kahl A. (1998), 'Compact shipyard', *Proc. Intel. Conf. IIW'98*, DVS195, 60–63.

Katayama S. (2009), 'Laser welding for manufacturing innovation', *J. Japan Welding Society*, 78, 682–692 (in Japanese).

Katayama S., Seto N., Kim J. D. and Matsunawa A. (1997), 'Formation mechanism and reduction method of porosity in laser welding of stainless steel', *Proc. ICALEO'97*, G-83–92.

Manzon L. (1998), 'Welding in shipbuilding industry', *Welding Review International*, May, 6–8.

Matsunawa A. (2001), 'Problem and solutions in deep penetration laser welding', *Sci. Technol. & Weld Joining*, 6, 351–354.

Matsunawa A., Kim J. D. and Katayama S. (1997), 'Porosity formation in laser welding', *Proc. ICALEO'97*, G-73–82.

Mazumder J., Rockstroh T. J. and Krier H. (1987), 'Spectroscopic studies of plasma during CW laser gas heating in flowing argon', *J. Appl. Phys.*, 62, 4712–4718.

Minamida K. (2002), 'High power laser applications in Nippon Steel Corporation', *Proc. SPIE*, 4831, 402–410.

Miyamoto I. and Maruo H. (1992), 'Spacial and temporal characteristics of laser-induced plasma in CO_2 laser welding', *Proc. LAMP'92*, 311–316.

Modest M. F. (2001), 'Reflectivity and absorptivity of opaque surface', in *LIA Handbook of Laser Materials Processing*, edited by Ready J. F. and Farson D. F., Orlando, FL: Laser Institute of America, 182.

Mori K. and Miyamoto I. (1997), 'In-process monitoring of laser welding by the analysis of ripples in the plasma emission', *J. Laser Application*, 9, 155–159.

Ono M., Shinozaki T., Shinbo Y., Sekine Y., Iwasaki K. and Takahashi M. (2001), 'Development of high power pipe welding process', *Quarterly J. Japan Welding Society*, 19, 233–240.

Petring D (2004), 'Laser applications in European automotive manufacturing: Historical review and recent trends', *J. Japan Welding Society*, 73, 7–14.

Poueyo A., Deshors D., Fabbro R., de Frutos A. M. and Orza J. M. (1992), 'Study of laser induced plasma in welding conditions with continuous high power CO_2 laser', *Proc. LAMP'92*, 323–328.

Rayleigh J. W. S. (1945), *The Theory of Sound, Vol. 2*. New York: Dover Publications.

Rockstroh T. J. and Mazumder J. (1987), 'Spectroscopic studies of plasma during CW laser materials interaction', *J. Appl. Phys.*, 61, 917–923.

Roland F., Manzon L., Kujala P., Brede M. and Weitzenbock J. (2004), 'Advanced joining techniques in European shipbuilding', *J. Ship Production*, 20, 200–210.

Rominger V. (2011), 'High-performance laser welding – a comparison of CO_2 lasers and high brilliance solid-state lasers', *Laser Tech. J.*, 8, 32–35.

Schumacher J., Zerner I., Neye G. and Thormann K. (2002), 'Laser beam welding of aircraft fuselage panels', *Proc. ICALEO 2002*.

Sellerup M. (1999), 'Laser welded ship are afloat', *Industrial Laser Solution*, April, 20–22.

Sokolowski W., Herziger G. and Beyer E. (1988), 'Spectral plasma diagnostics in welding with CO_2 lasers', *Proc. SPIE'88*, 1020, 96–102.

Sokolowski W., Herziger G. and Beyer E. (1989), 'Spectroscopic study of laser induced plasma in the welding process of steel and aluminium', *Proc. SPIE'89*, 1132, 288–295.

Tsukamoto S., Hiraoka K., Asai Y., Irie H. and Yoshino M. (1996), 'Characteristics of stably induced laser plasma', *Proc. ICALEO'96*, 81, B76–B85.

Tsukamoto S., Asai Y., Tanaka H. and Shida T. (1999), 'High-speed measurement of plasma temperature in CO_2 laser welding', *Proc. ICALEO '99*, 87, D73–D82.

Tsukamoto S., Kawaguchi I., Arakane G. and Honda H. (2001), 'Suppression of porosity using pulse modulation of laser power in 20 kW CO_2 laser welding', *Proc. ICALEO 2001*, C1702.

Tsukamoto S., Kawaguchi I., Arakane G. and Honda H. (2002), 'Keyhole behaviour in high power laser welding', *Proc. SPIE, 1st Intl. Symp. on High Power Laser Macroprocessing*, 4831, 251–256.

Tsukamoto S., Arakane G., Kawaguchi I. and Honda H. (2003), 'Keyhole behaviour in high power laser welding of thick steel plates, *Proc. ICALEO 2003*, A176–A183.

Tsukamoto S., Arakane G., Honda H. and Kuroda S. (2004), 'Formation mechanism and prevention of weld defects in full penetration laser welding of thick steel plates', *Proc. ICALEO 2004*, 104–110.

Tsukamoto S., Arakane G., Kojima K., Otani T. and Ohkita S. (2006), 'Effect of alloying elements on porosity formation in laser welding of heavy section steel plates', *Proc. ICALEO 2006*, 26–33.

Verwaerde A. and Fabbro R. (1995), 'Experimental study of continuous CO_2 laser welding at subatmospheric pressure', *J. Appl. Phys.*, 78, 2981–2984.

Vollertsen F., Schumacher J., Schneider K. and Seefield T. (2004), 'Innovative welding strategies for the manufacture of large aircraft', *Proc. IIW Intl. Conf. 2004*.

3
Developments in Nd:YAG laser welding

R. FABBRO, Laboratoire PIMM (Arts et Métiers ParisTech – CNRS), France

DOI: 10.1533/9780857098771.1.47

Abstract: Laser welding in keyhole (KH) mode using solid-state laser emitting in near infrared at nearly 1 micron wavelength is discussed in this chapter. The main physical processes involved in laser welding are presented and the reasons for using this laser wavelength are shown. Section 3.2 describes the KH geometry and related physical mechanisms controlling its stability. The role of the main operating parameters is also presented. Section 3.3 shows examples of the evolution of keyhole and weld pool behaviour for various welding speeds illustrating the mechanisms discussed in Section 3.2. Finally in the conclusion, expected diagnostics improvements necessary for supporting adapted numerical simulations of this laser welding process are discussed.

Key words: welding with disk or fibre laser; keyhole, vapour plume and melt pool dynamics.

3.1 Introduction

The recent advent of the new generation of solid-state lasers based on Nd-YAG, disk or fibre technology, emitting in the near infrared region in the range of 1.03–1.07 micron, has renewed and strongly amplified the interest in using these types of lasers for the welding process. Compared to CO_2 lasers, their greater ability to transport this wavelength with a fibre offers an incomparable advantage for industrial applications particularly when anthropomorphic robots are used. Also, these lasers now have a very high level of reliability, with high efficiencies and presently achieve CW powers up to several tens of kW, with an outstanding beam quality reaching quite near diffraction limit. So, for these reasons they have become a very popular tool for joining technology where high performance and high quality with good flexibility are required for the welding of many different materials.

One usually defines two different types of welding. The first one is called the conduction mode: in that case, the absorbed intensity is lower than a given threshold, which depends on the material and the welding speed, where evaporation inside the focal spot is not sufficienthy intense to generate a keyhole. The size of the corresponding melt pool is then mainly controlled by the heat conduction from the focal spot and welding speed. Keyhole (KH) welding or deep penetration welding is the second type of laser welding, which occurs when the laser intensity inside the focal spot is high for generating a high evaporation rate, and so the corresponding recoil pressure depresses the melt pool surface and allows the laser beam to penetrate deeply inside the material. KH generation is therefore the basic mechanism of this mode of welding characterized by a combination of complex physical mechanisms such as laser absorption, conductive and convective heat transfer, a complex hydrodynamic flow around the KH and inside the melt pool, and a strong coupling of the vapour plume with this melt pool.

If conductive welding appears at first glance rather simple to describe, the dominant mechanisms being laser absorption at the melt pool surface, and conductive heat flow inside the melt whose hydrodynamics is controlled by Marangoni surface effect, this is not at all the case with KH welding. One can easily understand that there are many unknowns concerning these very different complex mechanisms that occur during that process and great efforts to understand them are still being undertaken in many laboratories in order to master this important laser process, to make it more reliable and also to improve it in order to achieve better performance. In the first part of this chapter, we will review the main phenomena that are believed to control this process: they concern basically the keyhole (KH) formation and its dynamics, and the resulting hydrodynamics of the melt pool that will define the final weld seam quality. The effect of operating parameters, such as welding speed, laser intensity or power and of course the nature of the material used are also of primary importance for understanding this global behaviour. In the second part, a typical example of evolution of the resulting weld seam is presented when two of the main operating parameters, the welding speed and the incident power, are modified. The chapter concludes by presenting possible directions for improvement of the present understanding necessary for future adaptations to numerical simulations of that process.

3.2 Basics of laser welding in keyhole (KH) mode with solid-state lasers

3.2.1 Main geometrical KH characteristics: wall inclination and KH depth

The basic phenomenon that is at the origin of the KH formation is the recoil pressure that pushes the liquid generated during the surface laser heating, on the sides of the focal spot and allows further penetration of the laser beam inside the material. This 'piston effect', which has been described by Semak and Matsunawa (1997) for welding conditions, can be easily understood in the case of pulsed laser welding, or of course during drilling processes that occur at much higher intensities than for welding, i.e. when there is no displacement of material during the laser pulse: the vapour/liquid interface moves into the laser direction, inside the material with a 'drilling velocity' V_d and the melt is ejected laterally with a velocity V_m mainly controlled by the recoil pressure P_{evap} resulting from the strong surface evaporation. When there is a relative displacement between the laser beam and material, defined by the welding speed V_w, the previous scheme is greatly modified: the KH geometry no longer has an axial symmetry: it is then defined by a keyhole front wall (KFW) and a keyhole rear wall (KRW), characterized by their respective inclinations along the welding direction and the side walls can be considered as quite vertical (see Fig. 3.1(a)).

For this configuration, the main process of laser energy deposition results from the incident laser beam that impinges the KFW under a large angle of incidence. Depending on the partial or full penetration regime of the KH, its aspect ratio L/D (L: KH depth, D: KH diameter), and its KFW inclination α, the local piston effect on the KFW ejects the melt sideways around the KH and downwards along the KFW. This KFW inclination angle α is a very important parameter: first, it defines the main absorptivity $A(\alpha)$ of the incoming laser beam on this surface. As the local angle of incidence is rather high, the absorptivity can be important if one considers the Fresnel equation behaviours and their dependence with the beam polarization (Dausinger and Shen, 1993) that can be modified by local surface roughness (Bergström *et al.*, 2008). Also, macroscopic or microscopic surface deformations resulting from surface instabilities generating humps or ripples (Golubev, 2004) have a non-negligible effect on the level of local absorptivity. Direct measurements of KFW absorptivity by using full penetration welding experiments have shown levels of absorptivity in the range of 60–70% (Fabbro *et al.*, 2005) for non-polarized beams. As a consequence, one should expect that multiple reflections of the laser beam inside the KH are not important for the process of energy distribution inside the KH when medium range incident laser intensities are used.

The second consequence that implies the KFW inclination is that it controls the penetration depth. From geometrical considerations (see Fig. 3.1(b)), it is easy to see that the contribution on the penetration depth L of the impingement of the incident beam with a diameter D, when the entire beam is intercepted, is given by $L = D/tg\alpha$. Physical analysis of the equilibrium of the KFW has allowed us to determine the main basic parameters of the KH geometry, for stationary conditions (Fabbro, 2010b). One can find that the KFW inclination α is given by a rather simple relation:

$$tg\alpha \approx V_w/(kI_0 A_0) \qquad [3.1]$$

In Eq. [3.1], the parameter k is a proportionality factor between the drilling speed V_d and the absorbed intensity: $V_d \approx k \, . \, I_{abs}$. This factor k is representative of some energy balance of the process that depends mainly on the workpiece material. It can be determined experimentally for similar

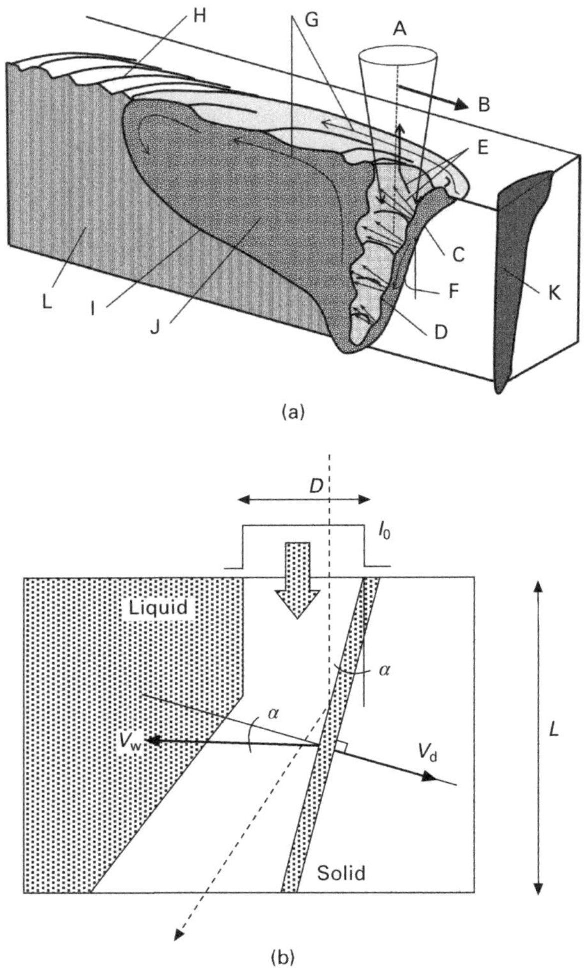

3.1 (a) General skech of laser welding: A: Laser beam; B: Welding speed; C: KH front; D: Humps; E: Vapour jets; F: FKW inclination; G: Melt flow; H: Chevron structure; I: Solidification front; J: Melt pool; K: Weld seam cross section; L: Resolidified material. (b) Scheme of the longitudinal section of a keyhole in a case of full penetration. Incident beam has a uniform intensity I_0, with a diameter D. Material is moving from right to left with the welding speed V_w. Under these conditions, the bottom of the KH can be enlarged by the reflected beam.

conditions of incident laser intensity (typically $k \approx 3 \times 10^{-11}$ m^3/J for 304 stainless steel) or estimated from the results of the 'piston model' of Semak and Matsunawa (1997). So the dependence on penetration depth L with the main operating parameters will be:

$$L \approx k\, I_0\, A_0\, D/V_w = (4A_0k/\pi) \cdot P/(D \cdot V_w) \qquad [3.2]$$

where P is the incident laser power. It is interesting to notice that this scaling dependence on the penetration depth L with the ratio $P/(D \cdot V_w)$ has been verified in many experiments (Beyer, 2008; Dausinger *et al.*, 2002; Verhaege and Hilton, 2005), particularly at high welding speeds V_w.

At low welding speeds, because of the small inclination of the KFW, it is easy to show that the resulting absorbed laser intensity on the KFW becomes very low and only depends on the welding speed, with the relation:

$$I_{abs} = V_w/k \qquad [3.3]$$

Equation [3.3] is interesting because it shows that the absorbed intensity depends only on the welding speed, and not on the incident intensity. This is a very simple self-regulating mechanism: for a given welding speed V_w, as the incident laser intensity I_0 increases, the KH depth L also increases (Eq. [3.2]), but the absorbed intensity on FKW is constant, due to the decrease of the FKW inclination (Eq. [3.1]).

Now, let's consider the KH stability at low welding speed. The KH will be maintained open, if the excess of recoil pressure P_{evap} over the ambient pressure P_a counterbalances the surface tension pressure $P_s = 2\sigma/D$ inside the KH (Hirano *et al.*, 2011); as this opening pressure is increasing with the absorbed intensity, Eq. [3.3] shows that for low welding speeds, this excess of recoil pressure should become lower than the characteristic KH closing pressure controlled by surface tension (Fabbro, 2010b). So, for these conditions, the KH cannot be stationary; it closes itself quite periodically, and is continuously re-opened, as in a drilling process, by the continuous incoming incident laser beam. From these considerations, one can estimate that the maximum penetration depth L_{max} for achieving a stable KH regime is given by:

$$L_{max} \approx (2CA_0/\pi\sigma)P \qquad [3.4]$$

where C is a constant defined by the relation between the evaporation pressure P_{evap} with the absorbed intensity: $P_{evap} - P_a \approx CI_{abs}$, where C is in the order of 1 bar/(MW/cm^2) for SUS304 (Fabbro *et al.*, 2006a). For greater penetration depths, one can expect strong fluctuations that may lead to some defects such as trapped porosity from the KH bottom inside the weld seam, strong spatter generation and, as a final consequence, some limitation of the penetration depth. Typically, for SUS304, one can estimate that (Fabbro, 2010b):

$$L_{max}(\text{mm}) \approx 1.5P\ (\text{kW}) \qquad [3.5]$$

The last important effect of the KFW inclination concerns the direction and the level of the dynamic pressure of the vapour emitted by the ablation process on the KFW. We have seen that for the range of incident intensities characteristic of KH generation, the process generates a recoil pressure that can be estimated to be linearly dependent on the absorbed intensity on the KFW. So, as the welding speed increases, the KFW inclination also increases as well as the absorbed intensity on the KFW. Consequently, the recoil pressure also increases and typically can reach a few bars for an absorbed intensity of about 1 MW/cm^2 on steel material (Fabbro *et al.*, 2006a). As the direction of expansion of the vapour is perpendicular to the local surface of the KFW, this vapour jet impinges on the KRW with a dynamic pressure P_d that can be similar to the recoil pressure, if the distance between the KFW and KRW is small, which is usually the case. It is the impact pressure P_d of this vapour on the KRW that maintains the KH opened, balancing the sum $P_s + P_m$ of the 'closing' pressures of the KRW that result from the surface tension pressure $P_s = 2\sigma/D$ and from the dynamic pressure $P_m \approx 0.5\, \rho_m\, V_m^2$ due to the liquid melt flowing laterally around the KH with the speed V_m, as a result of the recoil pressure applied on the KFW. (One must mention that if the welding velocity is not too low, the laser beam does not hit the KRW; so we do not have the term P_{evap} for the previous pressure balance on KRW.)

For these conditions, one can show that the equilibrium of the KRW cannot be stable (Fabbro and Chouf, 2000a, b), and so the KRW shows many fluctuations that are likely to generate some defects inside the melt pool. Moreover, these fluctuations can be even more easily triggered, if one considers that surface instabilities such as humps or ripples are present on the KFW surface, this effect being reinforced at low welding speeds (Golubev, 2004). In that case, important localized evaporation on these humps located on the KFW produces strong jets of vapour that can easily destabilize the KRW. As a result, bubbles of vapour can be generated inside the melt pool. Depending on the fraction content of the trapped shielding gas inside the KH (which may not be negligible, if one considers the very unstable behaviour of the KH in this regime), their position along the KH and the velocity field inside the melt pool, these bubbles can be evacuated at the melt pool surface, or if not, may result as porosity whose final size depends mainly on their shielding gas content when they are initially generated. The dynamic behaviour of these bubbles, whether or not it leads to porosity formation, has been remarkably demonstrated by the various X-ray shadowgraphy experiments realized by the Osaka team (Katayama, 2010) for the last 20 years and also more recently at IFSW in Stuttgart (Abt *et al.*, 2011). Also, some rather recent 2D and 3D numerical simulations have been able to reproduce the complex mechanisms of the interaction of the vapour plume with the melt pool (Amara *et al.*, 2006; Amara and Fabbro, 2008; L. Zhang *et al.*, 2011).

The previous considerations about the KH geometry (and more precisely its KFW inclination defined by the welding speed and the incident laser intensity), and the corresponding direction and dynamic pressure of the emitted vapour allow us to have a general view of the behaviour of the coupling between the KH and the surrounding melt pool:

- At low welding speed, the KH is rather deep, quite vertical and very unstable, leading to irregular collapses. The laser beam impinges on the collapsed liquid wall and pushes it downwards generating bubbles inside the melt pool that are not easily evacuated (with possible pore generation). Upward expansion of the vapour plume induces droplets and spatter generation.
- When the welding speed increases, KFW inclination increases and correspondingly the KH depth decreases. The evaporation on the KFW generates vapour jets that impinge on the KRW, producing different types of defects that may or may not be evacuated, depending on their size and the local induced upward flow of the liquid behind the KRW. So, bubbles leading to pores can be created and also droplets are emitted from the top part of the KRW and/or from the rear rim of the KH aperture. For some range of welding speeds, adapted conditions may lead to quite a stable configuration where the upper part of the KH is opened enough to allow a correct vapour ejection with a quasi-stationary velocity field inside the melt pool. In that case, porosity formation can be suppressed.
- At high welding speed, the KH depth still decreases and becomes rather small. The KFW inclination and therefore the absorbed laser intensity are important, so the resulting vapour jet is intense but it is emitted quite vertically; so, it no longer interacts with the KRW, and the previous defects cannot appear. However, as the melt flow expelled sideways from the KFW has a high velocity under these conditions, the resulting melt pool may present strong undercuts or even humping defects at welding speeds. This evolution of the KH and melt pool behaviour as a function of the welding speed will be detailed in Section 3.3.

3.2.2 Spatter formation

In addition to the porosity formation mechanism, the vapour jet issued from the KFW also plays a very important role in the spatter generation mechanism. From Eq. [3.1], we are able to know the mean direction of the ejected metal vapour, because it is emitted perpendicularly to the KFW whose inclination α increases when the welding speed increases or the incident laser intensity decreases. But when α increases, the vapour jet impinges on the upper part of the KRW, with an increase of its dynamic pressure due to higher absorbed intensity. Therefore, under these conditions, spatters can be generated and are ejected rearwards, in the direction of the vapour jet. By using high-speed video camera, the analysis of the maximum ejection angle of the trajectory of droplets emitted from the rear part of the KH rim confirms the linear relation between the KFW inclination and this maximum angle of droplet trajectory (Weberpals and Dausinger, 2007). In fact, the direction of this spatter emission is deviated from the direction of the vapour jet. This is because the final velocity of the spatters is given by a sum of momentum of the vapour jet emitted perpendicularly from the KFW and an upward directed melt flow occurring on the KRW (Weberpals, 2010). This deviation is more important in partial penetration than in full penetration. This spatter ejection can even be partly controlled and reduced: for similar operating conditions, by inclining the laser axis in the forward direction, the resulting reduction of the KFW inclination makes the vapour jet rather collide with the central part of the KRW, instead of its upper part, which is more sensitive to droplet ejection. In that case, the KH becomes more elongated, and spatter formation or induced defects inside the melt pool are minimized.

There is also another method for reducing spatter emission and for improving the final weld seam quality. It is usually observed that when the laser focal spot is located inside the workpiece, at a distance from the workpiece surface of about 1–3 mm (depending on the corresponding beam waist and beam intensity profile), the weld seam quality is significantly improved and even a greater penetration depth results. This effect is likely to result from the localization of the most intense zone of evaporation in the beam, which corresponds to the minimum focal spot, which is then deep inside the workpiece. At this depth, the KRW and behind it the melt pool, can sustain more easily the impact of the vapour jet, than in the wider upper part of the KRW. So, one can understand that the vapour jet is the main mechanism that can perturb and efficiently modify the hydrodynamics of the melt pool. Its control, if possible, is of primary importance during laser welding. Additional examples will be shown in Section 3.3.

A similar, and even stronger, perturbation is observed when Zn-coated sheets are welded in overlap joint configuration without a gap between the sheets. In that case, hot zinc vapours can only emerge inside the keyhole through a channel that is localized at the sheet interface and emerging at the KFW. Due to their high dynamic pressure (roughly corresponding to the saturation pressure of Zn at steel melting temperature), Zn vapours expel violently the liquid and large amounts of the rear melt pool can be blown away leading to cavities and blowholes in the weld seam. It has been shown experimentally that, depending on the relative position of this sheet interface with the upper surface, this liquid perturbation can be minimized in order to favour the Zn vapour expansion (see Fig. 3.2). The same effect is also obtained when a top-hat intensity distribution is used compared to a 'peaked' intensity distribution, because a clear aperture is similarly generated at the bottom of the KRW. Another possible solution consists in using an elongated focal spot along the welding direction that maintains the KRW at

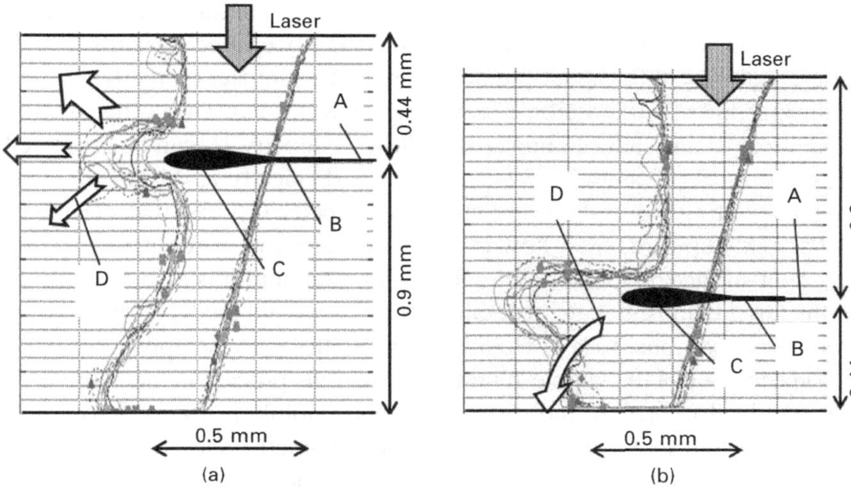

3.2 Computed keyhole profiles when two Zn-coated sheets of different thicknesses are welded in overlap configuration. The RKW is deformed by the Zn vapour jet escaping from a channel at the sheet interface: (a) 0.44 mm/0.90 mm and (b) 0.90 mm/0.44 mm. Experimentally, one observes that case (b) is much more stable than case (a) because the greater enlargement of the KH bottom allows the Zn vapour to escape more easily. A: Sheet interface; B: Zn vapour channel; C: Zn vapour jet; D: Flow followed by the Zn vapours (incident laser power: 4 kW, V_w: 3.5 m/min, top-hat intensity distribution, focal spot diameter: 0.45 mm, 5 ms time interval between each profile for visualizing their fluctuations) (Fabbro *et al.*, 2006b).

enough distance, at about 1 mm, in order to allow the dynamic pressure of Zn vapour to decrease. However, this solution requires a rather large incident laser power (Kielwasser *et al.*, 2000; Fabbro *et al.*, 2006b).

It is also well known that an efficient solution to this problem is to maintain a gap of about 0.1–0.2 mm depending on the thickness of the sheet material and the coatings, which allows Zn vapour venting between the two sheets. Some recent simulations of this effect show the resulting modification of the melt flow field (Geiger *et al.*, 2009). For generating such a gap between the two sheets, many techniques have been designed, but an elegant solution was recently proposed by pre-processing dimples on the upper surface of the second sheet before the final assembly (Gu and Shulkin, 2010). These dimples, with controlled height, can be easily generated using a laser weld seam realized in the humping regime, i.e. when the weld seam is obtained at very high welding speed, typically 60–80 m/min. This operation is then very rapidly obtained on these parts, using high-speed remote scanning optics and high beam quality lasers.

At low welding speed, the inclination angle of the KFW is very small, so the keyhole is quite vertical. It is also rather unstable, continuously re-opened and consequently the metal vapour plume is mainly ejected upwards. Under these conditions, this vapour plume has much less impact on the KRW, but interacts mainly by its friction effect along the KH walls. The corresponding shear stress τ_g and induced upwards melt flow velocity U_0 can be estimated using $\tau_g \approx 8\eta_g V_g/D$ and $U_0 \approx 3 (\tau_g L/\eta_m \rho_m)^{1/3}$. Typically $\tau_g \approx 100$–200 N/m^2 and the resulting induced upward melt flow velocity $U_0 \approx 5$–10 m/s (Fabbro, 2010a). This upwards melt flow is very important and, when it reaches the melt pool surface, it spreads along the radial direction. It is also responsible for the large bumps that are generally observed around the KH rim for these operating conditions. Kelvin-Helmholtz instability can also appear for these high speeds of ejected vapour plume; the previous process is then amplified and may lead to spatter and droplet ejection (Golubev, 1995). It is likely that high beam quality lasers, which generate small focal spot diameters, easily induce these effects. On the other hand, one can also understand the quality improvement generally observed when large keyhole sections such as those obtained with large or elongated focal spots are used (Hohenberger *et al.*, 1999; Luft, 2009). The use of large focal spot diameters is a general rule that can be applied for improving the KH stability.

3.2.3 KH opening stabilization using a side gas jet

We have seen that for low welding speeds, in stationary conditions, the evaporation pressure is not sufficiently intense to maintain an opened keyhole. It is then possible to maintain this opening by using a neutral gas jet delivered inside the keyhole by a small side nozzle (Kamikuki *et al.*, 2002; Fabbro *et al.*, 2006c) (see Fig. 3.3(a)). If the gas is correctly delivered, the dynamical pressure of the gas jet can be adjusted in order to balance the closing pressure controlled by its surface tension effects.

Figures 3.4 (a) and (b) show characteristic images of the melt pool with and without using the gas jet. The effect of this additional gas flow is very interesting: It maintains the KRW wall at a large distance from the KFW and so the laser beam is no longer intercepted by the KRW fluctuations. Therefore the laser beam can penetrate deeper inside the material. Also, the induced vapour plume is much less important and is more stable compared to conventional welding. Finally the hydrodynamic flow of the melt pool is very different from what is usually observed without the gas jet and it can be controlled: melt flow is ejected rearwards in a continuous, stationary and quite laminar flow. As a result of the use of the side gas jet, a gain of 40–50% in the KH penetration is observed, without any porosity. Also, the wine-cup-shaped weld seam usually observed at these low welding speeds is suppressed (see Fig. 3.3(b, c)) because the flow is no longer directed only upwards but also rearwards by the gas jet. This modified flow also explains

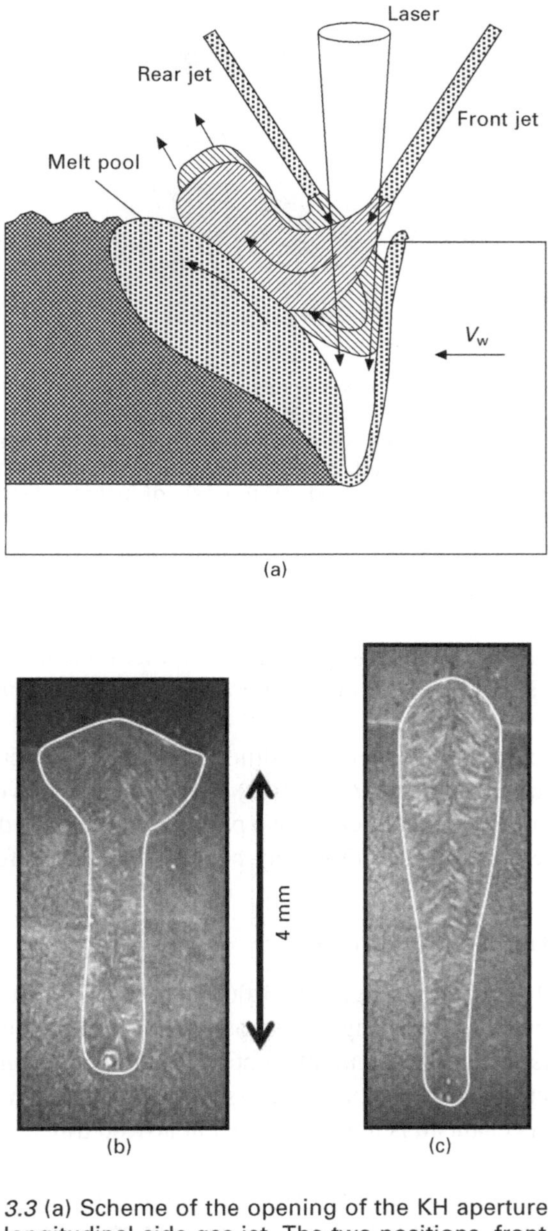

3.3 (a) Scheme of the opening of the KH aperture using a longitudinal side gas jet. The two positions, front or rear, can be used. Weld seam without (b) and with (c) a side gas jet. The weld seam contours have been highlighted (P_{laser}: 4 kW, V_w: 1 m/min, focal spot diameter: 0.6 mm). A flow rate of 20 l/min of Argon gas was delivered through a 2 mm diameter nozzle. It generates a dynamical pressure of 20 kPa at 5 mm from the nozzle aperture on its axis. By adjusting the argon flow rate, any dynamical pressure can be obtained inside the keyhole in order to avoid its closure.

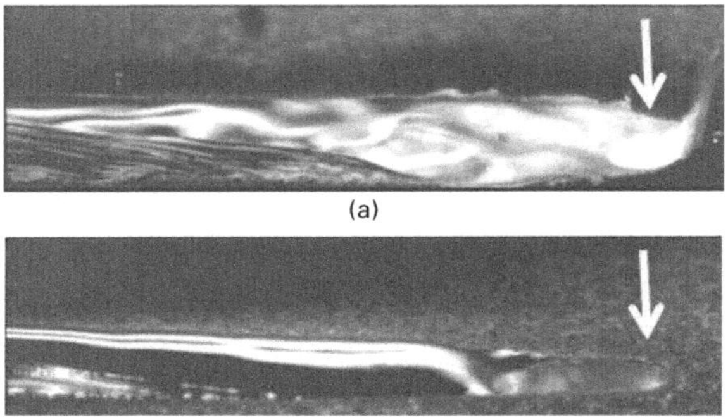

(a)

(b)

3.4 Melt pool behaviour and KH aperture without (a) and with (b) a side gas jet that delivers a dynamical pressure of 8 kPa inside the KH. The resulting elongated KH obtained in that case has a length of about 2 mm. (Welding speed: 3 m/min; P_{laser}: 3 kW; Spot diameter: 0.6 mm). White vertical arrows show the axis of incident laser beam location.

the increase of the height of the observed weld seam reinforcement (see Fig. 3.3(c)). 3D modelling of the interaction of a side gas jet with the melt pool was also realized (L. Zhang *et al.*, 2011): it was concluded that an improvement in the stability of the molten pool and a reduction in spatters and pores could be obtained.

This technique can be used for severe conditions: using this gas jet-assisted laser welding technique, X. Zhang *et al.* (2011) have been able to weld 316L stainless steel plates up to 40 mm thick, in two passes with an 8 kW disk laser and at 0.3 m/min welding speed without any porosity or other defects.

3.2.4 Vapour plume behaviour

Compared to the CO_2 laser wavelength, the 1.06 μm of the Nd-Yag laser has a very different behaviour concerning its interaction with the vapour plume. Firstly, as the Inverse Bremsstrahlung absorption coefficient of an ionized gas with a given electronic density follows a λ^2 scaling law, one can consider that this absorption mechanism is not at all relevant at 1.06 µm, for incident laser intensities characteristic of welding processes. As a consequence, the temperature of the vapour plume follows the temperature of the surface where the evaporation process occurs; in fact, this plume temperature may be even lower, due to its expansion into the ambient atmosphere and because its re-heating cannot occur due to the very low absorption coefficient, as discussed above. So these plume temperatures are rather low, in the range and above the evaporation temperature T_v of the material at atmospheric pressure. Spectroscopic measurements have confirmed this rather low temperature of the vapour plume and correspondingly its very low ionization state (Greses,

2003; Greses *et al.*, 2004; Katayama, 2010). This explains that, unlike CO_2 laser welding where the plume temperature in the range of 6,000–10,000 K results from its reheating by the Inverse Bremsstrahlung absorption, the nature of the shielding gas used, such as He, N_2 or Ar, has no effect on the final result of welding when 1 μm laser wavelength is used.

The second aspect of the laser–plume interaction concerns the important role of particles ejected back towards the laser along the direction of the beam. These particles have a very wide range of radius r that can vary from a few nanometres when they result from clusters of atoms, aggregates or ultra-fine particles generated by some complex mechanism of local condensation of these metallic atoms, to micron size or above, when they result from liquid droplets detached from the liquid wall by this violent expanding vapour jet as discussed in Section 3.2.2. Because this range of radius particles is typically smaller or about the 1.06 μm wavelength, it is expected that scattering and absorption by these small particles may play a much more important role than with the CO_2 10.6 μm laser wavelength. It is known that Mie scattering becomes effective when the ratio r/λ is about one and Rayleigh scattering scales with $(r/\lambda)^4$, so these mechanisms must be very efficient with short laser wavelengths. The analysis of the relative contribution of the beam absorption (that produces heating of these particles) and of the scattering terms (that induces a significant defocusing effect of the laser beam) shows that absorption is dominant for small particles, while scattering becomes important for larger ones ($2\pi r/\lambda > 1$).

Nevertheless, the application of these theories to experimental conditions is not satisfying due to the lack of precise knowledge of experimental parameters such as the volume ratio (ratio of particle volume to plume volume) or particle radius distribution along the vapour plume. However, Greses *et al.* (2004) has shown that for welding conditions at 1.06 μm, the particle average size inside the vapour plume varied from 20 to 50 nm when He or Ar shielding gas was used, while for CO_2 laser welding, this size was typically 10 times smaller; this likely results from the strong difference of the plume temperature between these two types of welding. Moreover, attenuation and scattering of a probe beam, of different wavelengths, transverse to the vapour plume, confirm the importance of this mechanism occurring at shorter laser wavelengths (Katayama, 2010). In fact, the real importance of the perturbing effect of these scattering mechanisms can really be emphasized when the vapour plume is blown away by an intense transverse gas jet located very near the workpiece surface. Typically, the penetration depth and weld seam area are improved by 15–20% when the vapour plume is correctly 'cut'. It is likely that the use of a side gas jet, as discussed in Section 3.2.3, also induces such effect by dispersing the vapour plume and therefore reducing its scattering and/or absorbing effect. One must also add that, apart from these localized scattering mechanisms, radial index variation on the shielding gas above the surface induced by thermal heating from the workpiece or from the vapour plume itself, can perturb beam propagation and modify the focal spot intensity distribution, particularly for fibre remote welding (Mitzutani and Katayama, 2009). These results clearly show that the control of the vapour plume and more particularly its size reduction or its suppression by adapted techniques, or of the shielding gas environment, improves the quality and the performance of the welding process. Moreover, due to the large relative variation of the focal spot by these perturbing mechanisms when small focal spots (and/or small apertures of focusing optics) are used, one may expect that these improvements should be very effective when high beam quality lasers are operated.

3.2.5 Welding under vacuum conditions

Welding under vacuum conditions is also another way of reducing the perturbing effects induced by the vapour plume previously discussed. This improvement has already been analysed for high power CO_2 laser welding (Poueyo-Verwaerde *et al.*, 1993). At low ambient pressure, the Inverse Bremsstrahlung process of laser absorption by the plasma plume is less efficient due to the decrease of its electronic density as a consequence of the reduction of ambient pressure. Therefore, plume temperature decreases and more laser energy can be transmitted into the KH; typically, on 35NCD16 steel, at a low welding speed of 0.3 m/min and P_{laser} = 7.5 kW, the KH penetration depth was increased by about 40% when ambient pressure was reduced to about 0.6 kPa. Simultaneously, the characteristic 'wine cup' shape of the weld seam top disappeared and large bulging of the melt pool was observed behind the KH.

At 1.06 μm laser welding, if very similar resulting effects are also observed (Abe *et al.*, 2010), the physical mechanisms are rather different. We have seen in Section 3.2.4 that the Inverse Bremsstrahlung absorption process is not effective at this wavelength; therefore the only mechanism that is likely to occur when the ambient pressure is reduced is the strong decrease of the vapour plume density during its expansion outside the KH into this reduced atmosphere. The size of the plume is efficiently reduced due to this 3D expansion just outside the KH aperture and the strong decrease of the observed plume luminosity and its steady behaviour confirm the reduction of perturbing effects. One can also expect that inside the KH, the vapour plume density is also decreased due to this easy and rapid expansion into vacuum. This point also explains the corresponding suppression of spatter emission and the much more stable behaviour of the KH and the melt pool. At low welding speeds, the KH being quite vertical, the vapour plume is then ejected vertically along the KH axis and so perturbs more efficiently laser beam propagation. Improvements observed for welding in vacuum conditions are therefore much more effective at low welding speeds. The consequence of the strong reduction of the absorption and scattering mechanisms in low vacuum conditions, with a more stable KH, is that more laser energy can be injected inside the KH, which then has a greater depth with a narrower size of the weld seam. Abe *et al.* (2010) obtained an increase of KH depth typically from 20 to 43 mm when the ambient pressure was reduced to about 1 kPa, on 304 stainless steel, at a welding speed of 0.3 m/min (with an incident laser power of 16 kW on a 0.4 mm focal spot diameter). For higher welding speeds, 3–6 m/min, the corresponding increase was only of 15–20%.

One must add that evaporation under vacuum conditions also modifies the temperature threshold for KH generation. As a general condition of KH equilibrium, we have previously seen that it is the excess of the recoil pressure P_{evap} due to evaporation, over the ambient pressure P_a, which must counterbalance the KH closure pressure P_s due to surface tension, at low welding speeds. Therefore, for vacuum conditions, this condition is realized for a surface temperature of the KH walls with a lower temperature than the usual evaporation temperature T_v. As this reduction of surface temperature can be about 10% of T_v (Hirano *et al.*, 2011), by considering the energy balance of the process, a significantly greater penetration depth has to be expected for similar operating parameters.

3.3 Examples of weld speed variation on global behaviour of keyhole (KH) and melt pool

The previous discussion will be illustrated now by some examples of evolution of the melt dynamics, analysed with a high-speed video camera, for variable welding speeds. Incident laser power and laser spot diameter were kept constant at 4 kW and 0.6 mm, respectively. 304 Stainless steel samples were used, in partial penetration mode. This analysis allows the definition of five main characteristic contiguous regimes, which are only controlled by the welding speed and where the action of the vapour plume on the melt pool is emphasized. This action is controlled by the geometry of the keyhole front, itself defined by the welding speed and incident laser intensity. Although the following observations are done for specific conditions of $P = 4$ kW and 0.6 mm focal spot diameter, it must be emphasized that these results are general and can be classified as a function of the FKW inclination angle α.

3.3.1 Welding speeds below 5 m/min: 'Rosenthal' regime

This regime, occurring for welding speeds lower than 5 m/min, is characterized by a rather large melt pool, even in front of the keyhole, because of the rather low welding speed. The melt pool surface shows many chaotic surface fluctuations and large swellings of liquids fluctuating around the keyhole aperture, which is rather well defined and remains circular (see Fig. 3.5(a)).

Because of these fluctuations, there is no clear laminar flow around the keyhole and the Marangoni effect is unlikely to occur for these conditions. Many spatters are emitted mainly from the keyhole rim, and particularly on its front side. On this range of welding speeds, one can see a rapid decrease of the penetration depth as well as large fluctuations of ejection direction of the vapour plume. These fluctuations can be correlated with the previously described melt pool swelling fluctuations. Macrographs of cross sections of the weld seam show a gradual transition of the shape from the well-known 'wine cup shape' characteristic of a low welding speed (typically about 1–3 m/min) to a more slender shape for 5 m/min. From the analysis of the keyhole tilting for these low welding speeds, discussed in Section 3.2.1, we know that the keyhole is quite vertical. Therefore, despite these fluctuations, one can schematize this regime by a vertical cylindrical keyhole surrounded by a large melt pool with very limited hydrodynamics. So the description of this situation as a kind of 'Rosenthal' heat flow regime, where the keyhole surface is uniformly heated at some temperature close to the evaporation temperature, could be used.

3.3.2 Welding speeds between 6 and 8 m/min: 'single wave' regime

This regime is observed for welding speeds ranging from 6 to 8 m/min. It is characterized by the presence of a rather large single swelling generated near the top of the rear keyhole wall (see Fig. 3.5(b)). It is only from this region that large melt droplets are emitted. This large wave is ejected rearwards quite periodically due to back and forth oscillations of the melt pool, leading to quite periodic closures of the keyhole. It is the vapour plume, which is emitted rather deeply inside the keyhole and which collides with the melt pool that triggers these oscillations.

One can observe that only the tilted KFW is clearly heated by the incident laser beam (unlike the 'Rosenthal' regime where the luminosity characterizing the laser heated surface was more uniformly and randomly distributed all around the keyhole surface). More interestingly, when the ejected vapour plume collides with the melt pool and lifts it, a corresponding local heating of the liquid surface results from the impact of this energetic heated vapour plume. So the vapour plume not only transfers impulse momentum, but also a non-negligible amount of energy due to its rather high temperature.

It is interesting to note that when high beam quality lasers are used, i.e. with smaller focal spots and correspondingly rather greater penetration depths

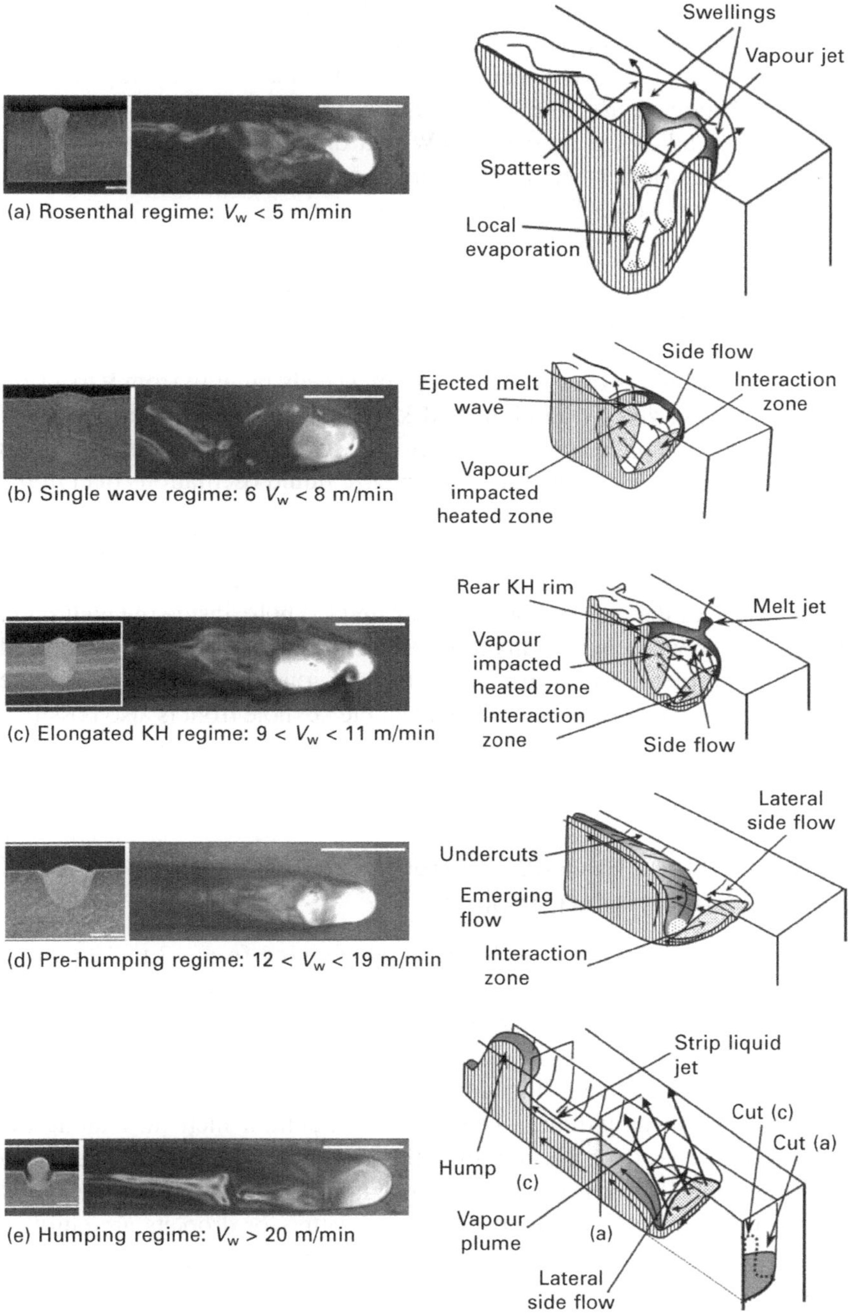

3.5 Characteristic views of KH with surrounding melt pool with corresponding weld seam cross section, and corresponding sketches of the five described regimes obtained for various welding speeds. (Focal spot diameter: 0.6 mm, stainless steel material.) Scale of melt pool: 1 mm.

obtained for this range of rather low welding speeds, one also observes a 'humping regime' characterized by very periodic humps generated at the rear of the KH on the weld seam (Behler and Schäfer, 2005; Miyamoto *et al.*, 2004; Thomy *et al.*, 2006). This results from the friction of the ejected vapour acting all along the deep KRW that drags and lifts this part of the melt pool, and whose effect is dominant compared with its colliding effect that occurs when larger focal spots are used.

3.3.3 Welding speeds between 9 and 11 m/min: 'elongated keyhole' regime

This regime, which is observed for welding speeds ranging from 9 to 11 m/min, is characterized by an elongated keyhole whose maximum length of about 2 mm, is obtained at 11 m/min (see Fig. 3.5(c)). This regime also shows some fluctuations, but with much less intense liquid oscillations than in the previous regime, and smaller induced swelling heights. This elongated keyhole shows two characteristic zones that are heated: the first one corresponds, of course, to the more inclined KFW, common to all regimes, and the second one is located at the rear end of this elongated keyhole, inside the melt pool. Moreover, vapour plume seems to be also emitted from the second heated spot and so directed frontward. As in the previous regime, heating by the collision of the vapour plume emitted from the keyhole front is also possible. One could also add that weld seams obtained with this regime have rather good quality, and do not show characteristic defects.

3.3.4 Welding speeds between 12 and 19 m/min: 'pre-humping' regime

This regime is observed for welding speeds ranging from 12 to 19 m/min. It is characterized by a rather strong sideward melt flow that contours the KFW, and an important tilting angle. The KH aperture is elongated (but with a shorter length than in the previous regime) and the surface of the following melt pool shows fluctuations characterized by only surface waves with rather small amplitudes. There is also a central melt flow along the KFW that emerges from the bottom of this very stable KFW and is deflected rearwards with its surface melt pool reaching a level close to the initial sample surface (see Fig. 3.5(d)). No more spatters or droplets are emitted, even from the rear rim of the keyhole as in the previous regime. Melt pool surface reaches the level of the surface sample and wets both sides. But above 15–16 m/min welding speeds, undercuts are observed on the sides of the seam. These undercuts begin to be important because the two side flows emitted from the KFW collide with the central emerging flow and press it towards its centre, slightly lifting it. Finally, the vapour plume is very stable;

its direction no longer fluctuates and is precisely ejected perpendicularly to the KFW interacting with the top part of the KRW.

3.3.5 Welding speeds above 20 m/min: 'humping' regime

For welding speeds above 20 m/min, a very characteristic melt flow appears which corresponds to the humping regime, where the weld seam has very strong undercuts, with solidified large swellings of quite ellipsoidal shape, separated by smaller valleys.

The main central flow always emerges from the bottom of the KFW (see Fig. 3.5(e)); it is strongly deflected rearwards and it rises up to a level that is much lower than the surface sample. So the mean melt velocity inside it is very high. It stays attached to the bottom of the resulting groove along a distance of about 2 mm and then a liquid jet is detached from the central part of this flow and forms a thin vertical strip propagating rearwards at high velocity. The hump appears at a certain distance from the detachment point along this vertical strip of liquid jet (Fabbro *et al.*, 2007). It is the shrinkage of this melt jet strip, due to the Rayleigh instability driven by surface tension, at about 2 mm from the detachment point that generates the instability. This shrinkage cools locally this strip of melt jet and so attaches it to the sample. The fluid flowing inside this strip of melt jet is then stopped at this point, and the hump can grow. The growth of this hump stops when a next shrinkage occurs at about a similar distance from the detachment point that allows the growth of the next hump. Of course, as this thin melt jet strip is located at the centre of section of the groove, very severe undercuts are generated even near the humps. For higher welding speeds, this scheme is not modified.

3.3.6 Analysis of the transition thresholds between these different regimes

The five previous regimes have been observed for a given incident laser power of 4 kW and focal spot of 0.6 mm, for different welding speeds. In fact, one can say that this is a very general result: any set of operating parameters will produce behaviour that corresponds to one of these five regimes. These experiments have been reproduced by using incident laser powers varying between 2.5 and 4 kW and in order to have a greater range of incident intensities, the focal spot was reduced to 0.45 mm. The same regimes were observed but with modified welding speed thresholds. These results are reported in Fig. 3.6 on a laser power–welding speed diagram. Welding is obtained with a KH mode in the region labelled K. Conduction mode is only observed in the region labelled C. Typically, for a welding speed of 25 m/min, KH regime is obtained for incident laser power greater than 0.5 kW. Of course, this power increases with the welding speed.

For incident power varying from 2.5 to 4 kW, we have reported the welding speed thresholds between the 'Rosenthal'/'Single-wave'/'Elongated'/'Pre-humping'/'Humping regimes (these five corresponding regimes are labelled, respectively R, S, E, P, H). Except for the P/H transition, all the other characteristic thresholds increase with the incident laser power. Best-fit lines passing through the origin of these three series of experimental points of these consecutive regimes have also been drawn.

For the P/H transition, it is interesting to note that for incident laser powers smaller than 2.5 kW, no humping instability is observed, even at very high welding speed. Also, when the incident power increases, the threshold for the P/H transition decreases; this behaviour is similar to previous experimental results obtained for different conditions (Thomy *et al.*, 2006). The different behaviour of this P/H transition, compared to the three previous ones, results from the very different mechanisms driving this humping instability. It is also interesting to notice that if we extrapolate the P/H transition to higher incident powers, the pre-humping regime should disappear.

From our previous analysis of the keyhole front equilibrium, one can relate the slope of the three best-fit lines plotted in Fig. 3.6 through the

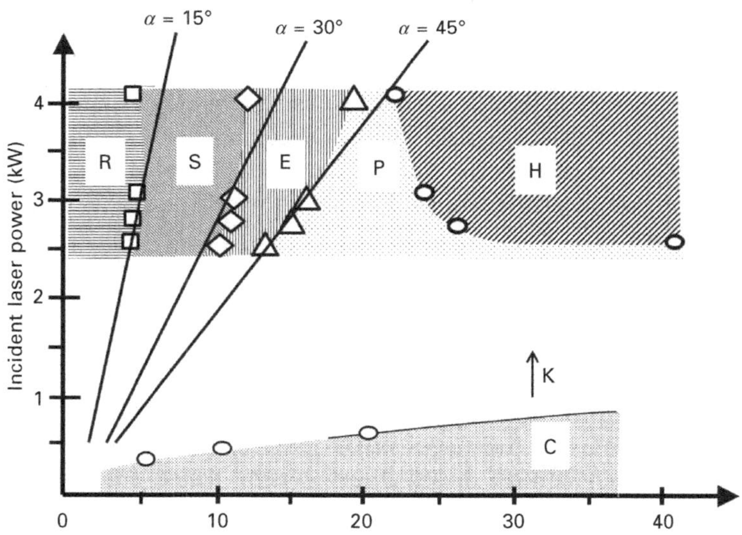

3.6 Location of different regimes respectively labelled R, S, E, P, H, inside the incident laser power–welding speed plane. Regions corresponding to keyhole (K) and conduction (C) regimes are also indicated. Linear best-fits for the R/S, S/E and E/P transitions have been drawn, with corresponding FKW tilting angle α. (Top-hat intensity distribution; Focal spot diameter: 0.45 mm; Stainless steel material.)

experimental thresholds of the corresponding transitions, with the mean inclination of the KFW. The combination of Eqs [3.1] and [3.2] gives a linear relation between incident laser power P and welding speed V_w:

$$P = B \cdot V_w, \tag{3.6}$$

where the slope B of this linear relation between P and V_w, is given by:

$$B = \frac{\pi LD}{4kA_0} = \frac{\pi D^2}{4kA_0 \tan\alpha} \tag{3.7}$$

Using Eq. [3.7], the KFW inclination angles α that would reproduce the slopes of these three best-fit lines in Fig. 3.6 are 15°, 30° and 45° for respectively the R/S, S/E and E/P transitions. In fact, these angles are in fair agreement with the corresponding experimental measurements obtained by using the high speed videos.

This model shows that this slope B is proportional to the penetration depth L (Eq. [3.7]). So, for a given transition (R/S, S/E or E/P), all the welding speed thresholds corresponding to that transition will occur at the same penetration depth or KFW inclination. Of course, as expected, when one goes from R to P regimes, the corresponding penetration depths decrease as the welding speeds increase.

So these results indicate that the transition between the first four regimes is basically controlled by a characteristic tilting angle of the KFW. As we have seen that this angle also defines the direction and the dynamic pressure of the emitted vapour plume (this vapour plume being directed perpendicularly to its emitting surface), these results clearly confirm that it is the level of interaction of the vapour plume with the rear melt pool that defines the type of hydrodynamic regime. At low welding speeds, the KFW is slightly tilted; the absorbed intensity is then rather small, as well as the dynamic pressure of vapour. The vapour plume is directed quite horizontally inside the melt pool and so perturbs the melt pool from its bottom. This would correspond to the occurrence of the single wave regime. As the welding speed increases, the tilting angle increases, as well as the absorbed intensity and correspondingly the dynamic pressure of the vapour plume. But this vapour plume is more emitted quite vertically and therefore perturbs more efficiently the top part of the melt pool as in the pre-humping regime.

3.4 Conclusion and future trends

We have described in this chapter the main physical mechanisms that control the quality of the process of laser welding in KH mode. The main conclusion that results is the very important role played by vapour plume generated by the evaporation process that leads to the KH formation. As important is the direction of this vapour plume defined by the KFW inclination. In fact, two main operating parameters, the incident laser intensity and the welding speed, control both this direction and the dynamic pressure of the vapour plume. We have seen that this vapour plume can perturb the welding process in two ways: first by perturbing the beam propagation, by scattering and/or absorbing the incident incoming beam, particularly when small focal spots or low aperture optics are used. As a general rule, it is recommended to blow away the vapour jet each time it is possible, in order to get reproducible, constant and non-perturbed conditions of irradiations. The vapour jet can also strongly perturb the melt pool geometry and its hydrodynamics by its impact on the KRW. The final quality of the weld seam is a sensitive function of the method used to attenuate these perturbations. We have seen that enlarging the focal spot, for example by using a side gas jet that repels the KRW to larger distances from the KFW, or finding an adapted defocused position, or inclining the incident laser beam, or even welding in vacuum conditions may be possible solutions that should improve the final quality, but of course their possible use must be balanced with the induced constraints.

However, there are still many unknowns related to different mechanisms and future experiments should be focused on them. One can mention several points that should be improved in the near future. A precise knowledge of absorption mechanisms on the KFW is necessary. This could be helped by determining the KFW surface geometry by adapted experiments with high time and space resolution, improving, for example, present high-speed video imaging and/or X-ray shadowgraphy techniques. This problem of surface instabilities generated by non-linear coupling of thermal field and hydrodynamic response should also be more deeply analysed on a theoretical basis. The knowledge of resulting recoil pressure is also rather poor; as we have seen, this parameter is very important because it controls melt side-flow velocity and dynamic pressure of the vapour jet. Presently, very few experiments have been done to determine this recoil pressure for this range of absorbed intensity. Moreover, the effect of several parameters such as the ambient pressure and the nature of alloying elements of the workpiece make the determination of recoil pressure rather complex. Also, the thermodynamic state of metal vapour (mean density, temperature, cluster or particles distribution, flow field, etc.) is completely unknown at present. We have seen that this point is very important for the beam propagation into the vapour plume outside the KH; but for inside the KH, beam scattering should control its propagation and so the resulting beam homogenization should interfere with the previous surface instabilities and absorption mechanisms.

Concerning the melt pool, its geometry, and particularly the shape of its KRW, and also its dynamics flow field, should be analysed with greater spatial and temporal precision than presently. High-speed X-ray shadowgraphy, possibly using tracers, will be an indispensable tool. One should also add that a better knowledge of the surface tension or the dynamic viscosity of the workpiece material dependence with temperature would be useful. Finally, one could also remark that there is an important thermodynamic parameter, which is the temperature, which is usually poorly measured during experiments and which is always a direct output of numerical simulations. Having diagnostics able to give time and space resolved temperature distribution in a sub-melting to evaporation temperature range is still a challenge.

So, the precise determination of all these very different parameters should be important data that must be compared with corresponding results of numerical simulations. It is interesting to note that already, several publications begin to show the possibility to realize 3D numerical simulations of laser welding where these main various mechanisms can be described self-consistently, for non-stationary conditions, and where the hydrodynamics of the melt pool and vapour/free surface interaction are also taken into account (Ki *et al.*, 2002a, b; Dasgupta *et al.*, 2007; Otto *et al.*, 2011; L. Zhang *et al.*, 2011; Pang *et al.*, 2011). Depending on their initial various hypotheses and also computational facility used, the main phenomena involved in the coupling between the KH and the melt pool dynamics discussed in that chapter, have already been correctly reproduced (Pang *et al.*, 2011). So, the numerical results of simulations are presently very encouraging. We are convinced that with continuous efforts on improvements to these numerical simulations on one hand, and, on the other, with the corresponding experimental progress on several experimental diagnostics as mentioned previously, one should be able to have, within a few years, very powerful simulation tools that are able not only to reproduce precisely the main observed characteristics, but also to optimize operating parameter selection for a targeted goal, by avoiding long series of experiments.

3.5 References

Abe Y, Mizutani M, Kawahito Y and Katayama S (2010), 'Deep penetration welding with high power laser under vacuum', *Proceedings of the ICALEO 2010 Conference*, Anaheim, CA, 648–653.

Abt F, Boley M, Weber R, Grag T, Popko G and Nau S (2011), 'Novel X-ray diagnostic for *in-situ* diagnostics of laser based processes – first approach', *Physics Procedia*, 12, 761–770.

Amara E H and Fabbro R (2008), 'Modelling of gas jet effect on melt pool movements during deep penetration laser welding', *J. Phys. D: Appl. Phys.*, 41, 055503.

Amara E H, Fabbro R and Hamadi F (2006), 'Modelling of the melted bath movement induced by the vapour flow in deep penetration laser welding', *J. Laser Appl.*, 18, 2–11.

Behler K and Schäfer P (2005), 'Melt pool dynamics in high speed welding with modern high power solid state lasers', *Proceedings of the ICALEO 2005 Conference*, Miami, FL, Oct. 31–Nov. 3, 1026–1031.

Bergström D, Powell J and Kaplan A (2008), 'The absorption of light by rough metal surfaces – a three-dimensional ray-tracing analysis', *J. Appl. Phys.*, 103(10), 103515.

Beyer E (2008), 'High power laser materials processing – new developments and trends', *Proceedings of the 3rd PICALO 2008 Conference*, Beijing, China, 5–9.

Dasgupta A K, Mazumder J and Li P (2007), 'Physics of zinc vaporization and plasma absorption during CO_2 laser welding', *J. Appl. Phys.*, 102, 053108.

Dausinger F and Shen J (1993), 'Energy coupling efficiency in laser surface treatment', *ISIJ International*, 33, 925–933.

Dausinger F, Berger P and Hügel H (2002), 'Laser welding of aluminium alloys: problems, approaches for improvement and applications', *Proceedings of the ICALEO 2002 Conference*, Scottsdale, AZ, Oct. 14–17.

Fabbro R (2010a), 'Physical mechanisms controlling keyhole and melt pool dynamics during laser welding', in *Advances in Laser Material Processing*, edited by J. Lawrence, J. Pou, D. Low and E. Toyserkani. Cambridge: Woodhead Publishing pp. 211–241.

Fabbro R (2010b), 'Melt pool and keyhole behaviour analysis for deep penetration laser welding', *J. Phys. D: Appl. Phys.*, 43, 445501.

Fabbro R and Chouf K (2000a), 'Dynamical description of the keyhole in deep penetration laser welding', *J. Laser Appl.*, 12, 142–148,

Fabbro R and Chouf K (2000b), 'Keyhole modelling during laser welding', *J. of Appl. Phys.*, 87, 4075–4083.

Fabbro R, Slimani S, Coste F and Briand F (2005), 'Study of keyhole behavior for full penetration Nd-Yag CW laser welding', *J. Phys. D.: Appl. Phys.*, 38, 1–7.

Fabbro R, Slimani S, Coste F, Briand F, Dlubak B and Loisel G (2006a), 'Analysis of basic processes inside the keyhole during deep penetration Nd-Yag CW laser', *Proceedings of the ICALEO 2006 Conference*, Scottsdale, AZ, Oct. 30–Nov. 2, 1–79.

Fabbro R, Coste F, Goebels D and Kielwasser M (2006b), 'Study of CW Nd-Yag laser welding of Zn-coated steel sheets', *J. Phys. D.: Appl. Phys.*, 39, 401–409.

Fabbro R, Slimani S, Doudet I, Coste F and Briand F (2006c), 'Experimental study of the dynamical coupling between the induced vapour plume and the melt pool for Nd-Yag laser welding', *J. Phys. D.: Appl. Phys.*, 39, 394–400.

Fabbro R, Slimani S, Coste F and Briand F (2007), 'Experimental study of the humping process during Nd:Yag CW laser welding', *Proceedings of the LIM Conf*, Munich, Germany, June 18–21.

Geiger M, Leitz K H, Koch H and Otto A (2009), 'A 3D transient model of keyhole and melt pool dynamics in laser beam welding applied to the joining of zinc coated sheets', *Prod. Eng. Res. Devel.*, 3, 127–136.

Golubev V S (1995), 'On possible models of hydrodynamical nonstationary phenomena in the processes of laser beam deep penetration into materials', *Proc. SPIE*, 2713, 219–230.

Golubev V S (2004), 'Melt removal mechanisms in gas-assisted laser cutting of materials' Eprint No. 3. Shatura: ILIT RAS.

Greses J., (2003), 'Plasma/plume effects in CO_2 and Nd:Yag laser welding', Dissertation, University of Cambridge, Cambridge.

Greses J, Hilton P A, Barlow C Y and Steen W M (2004), 'Plume attenuation under high power Nd:Yttritium–Aluminum–Garnet laser welding', *J. Laser Applications*, 16(1), 9–15.

Gu H and Shulkin B (2010), 'Remote laser welding of Zn-coated sheet metal component in a lap configuration utilizing humping effect', *Proceedings of the ICALEO 2010 Conference*, Anaheim, CA, 380–385.

Hirano K, Fabbro R and Muller M (2011), 'Experimental determination of temperature threshold for melt surface deformation during laser interaction on iron at atmospheric pressure', *J. Phys. D: Appl. Phys.*, 44, 435402.

Hohenberger B, Chang C, Schinzel C, Dausinger F and Hügel H (1999), 'Laser welding with Nd:Yag multi-beam technique', *Proceedings the ICALEO 1999 Conference*, San Diego, CA, 87, D167–176.

Kamikuki K, Inoue T, Yasuda K, Muro M, Nakabayashi T and Matsunawa A (2002), 'Prevention of welding defect by side gas flow and its monitoring method in continuous wave ND:Yag laser welding', *J. Laser Applications*, 14, 136–145.

Katayama S (2010), 'Understanding and improving process control in pulsed and continuous wave laser welding', in *Advances in Laser Material Processing*, edited by J. Lawrence, J. Pou, D. Low and E. Toyserkani. Cambridge: Woodhead Publishing, pp. 181–210.

Ki H, Mohanty P S and Mazumder J (2002a), 'Modelling of laser keyhole welding: Part I. Mathematical modeling, numerical methodology, role of recoil pressure, multiple reflections and free surface evolution', *Metall. Mater. Trans.*, A33, 1817–1830.

Ki H, Mohanty P S and Mazumder J (2002b), 'Modelling of laser keyhole welding: Part II. Simulation of keyhole evolution, velocity, temperature profile, and experimental verification', *Metall. Mater. Trans.*, A33, 1831–1842.

Kielwasser M, Fabbro R, Petring D and Poprawe R (2000), 'Physical processes during pulsed Nd:YAG laser and CO_2 laser welding of zinc coated steel', in *Laser Application to Auto Industry – Proceedings of ICALEO 2000 Conference*, Detroit, MI, 89, A10–19.

Luft A (2009), 'New applications for welding with diode lasers', *Proceedings of the 5th International WLT-Conference on Lasers in Manufacturing*, Munich, June.

Mitzutani M and Katayama S (2009), 'Numerical calculation of laser beam path influenced by high temperature gas above specimen during laser welding', *Proceedings of ICALEO 2009 Conference*, Orlando, FL, 600–609.

Miyamoto I, Park S J and Ooie T (2004), 'High-speed microwelding by single-mode fiber laser', *Proceedings of the 4th LANE 2004 Conf.*, Erlangen, Germany, 55–66.

Otto A, Koch H, Leitz K H and Schmidt M (2011), 'Numerical simulations – a versatile approach for better understanding dynamics in laser material processing', *Physics Procedia*, 12, 11–20.

Pang S, Chen C, Zhou J, Yin Y and Chen T (2011), 'A three-dimensional sharp interface model for self-consistent keyhole and weld pool dynamics in deep penetration laser welding', *J. Phys. D: Appl. Phys.*, 44, 025301.

Poueyo-Verwaerde A, Fabbro R, Deshors G, De Frutos A M and Orza J M (1993), 'Experimental study of laser induced plasmas in welding conditions with CW CO_2 laser', *J. Appl. Phys.*, 74, 5773.

Semak V and Matsunawa A (1997), 'The role of recoil pressure in energy balance during laser materials processing', *J. Phys. D: Appl. Phys.*, 30, 2541–2552.

Thomy C, Seefeld T, Wagner F and Vollertsen F (2006), 'Humping in welding with single mode fiber lasers', *Proceedings of the ICALEO 2006 Conference*, Scottsdale, AZ, Oct. 30–Nov. 2, 543–552.

Verhaeghe G and Hilton P (2005), 'The effect of spot size and beam quality on welding performance when using high-power continuous wave solid-state lasers', *Proceedings of the ICALEO 2005 Conference*, Miami, FL, 264–271.

Weberpals J (2010), 'Nutzen und Grenzen gutter Fokussierbarkeit beim Laserscheweinssen', Dissertation, University of Stuttgart, Herbert Utz Verlag GmbH.

Weberpals J and Dausinger F (2007), 'Influence of inclination angle on spatter behaviour at welding with lasers of strong focusability', *Proceedings of the ICALEO 2007 Conference*, Orlando, FL, 858–865.

Zhang L, Zhang J, Zhang G, Bo W and Gong S (2011), 'An investigation on the effects of side assisting gas flow and metallic vapour jet on the stability of keyhole and molten pool during laser full-penetration welding', *J. Phys. D.: Appl. Phys.*, 44, 135201.

Zhang X, Ashida E, Tarasawa S, Anma Y, Okada M, Katayama S and Mizutani M (2011), 'Welding of thick stainless steel plates up to 50 mm with high brightness lasers', *J. Laser Applications*, 23(2), 9–15.

3.6 Appendix: list of symbols

$A(\alpha)$	angular dependence of absorptivity
A_0	absorptivity under normal incidence
B	slope of relation between laser power and welding speed (J/m)
C	slope of relation between evaporation pressure and absorbed intensity (s/m)
D	spot diameter (m)
I_{abs}	absorbed laser intensity (W/m^2)
I_{cond}	conduction loss inside solid (W/m^2)
I_0	incident laser intensity (W/m^2)
k	constant (m^{3}J^{-1})
L	keyhole depth (m)
L_{max}	maximum depth of a quite stable keyhole (m)
P	incident laser power (W)
P_a	ambient atmospheric pressure (N/m^2)
P_d	dynamic pressure of ejected vapour plume (N/m^2)
P_{evap}	evaporation pressure (N/m^2)
P_s	keyhole closing pressure due to surface tension (N/m^2)
r	particle radius inside the vapour plume (m)
T_v	workpiece evaporation temperature (K)
V_d	drilling velocity (m/s)
V_g	vapour plume velocity (m/s)
V_m	melt velocity (m/s)
V_w	welding speed (m/s)
α	inclination of the front keyhole wall (rd)
λ	laser wavelength (m)
ρ_m	melt density (kg/m^3)
σ	surface tension of metal liquid (N/m)
τ_g	shear stress (N/m^2)
η_g, η_m	dynamic viscosity of metal vapour and melt (Pa.s)

4
Developments in disk laser welding

K. LÖFFLER, TRUMPF Laser and Systems GmbH, Germany

DOI: 10.1533/9780857098771.1.73

Abstract: The chapter presents different applications in welding with disk lasers. This includes the general concept of disk lasers and the advantages of disk laser welding in comparison to other laser concepts. Presented applications include laser welding of sheet metal, welding in precision machining, laser metal deposition, laser scanner welding and laser hybrid welding.

Key words: disk laser, welding in precision machining, laser metal deposition, laser scanner welding, laser hybrid welding.

4.1 Introduction: key principles of disk lasers

While the disk laser concept was invented in the early 1990s, the first industrial products were available in 2003. Before the disk laser was invented, the rod laser was dominating the solid-state laser market. But the relatively thick rod leads to a parabolic temperature profile (higher temperature in the inner part of the rod, lower temperature at the outer part) and limited the output power and beam quality respectively. The next step in laser development was a laser with a thin disk. The laser gain medium of such a laser is a wafer-thin crystalline disk made of Yb:YAG, about 15 mm in diameter and 0.2 mm thick. The disk rests on a heat sink, which dissipates heat and cools the disk, effectively eliminating the disadvantages of a rod and allowing highest output power at very good beam quality (see Fig. 4.1). The cooled rear side of the disk has a reflective surface, which reflects the laser beam and pump light and acts as a folding mirror. The disk laser is pumped by efficient diode lasers. Their pump laser beam is focused to a diameter of a few millimeters and then aimed at the disk. The disk is so thin that it absorbs only a fraction of the pump radiation that passes through. For this reason, the pump beam is passed through the disk several times, by means of a pair of folding mirrors and a parabolic mirror in the interior of the cavity (see Fig. 4.2).

The laser beam is produced in the disk and exits the cavity through a hole in the center of the parabolic mirror. The output mirror of the resonator is located outside the cavity. The laser power can be modified in different ways: on the one hand via the excitation power, in other words the power of the pump laser beam; and on the other hand via the irradiated area of the disk. Laser power of up to several kilowatts can be generated with one

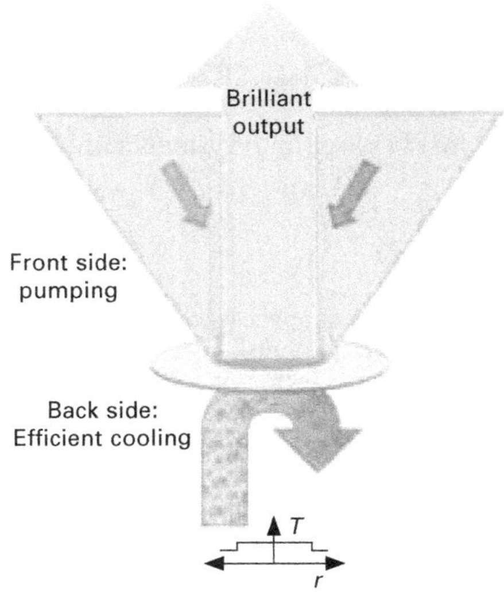

4.1 Basic principle of disk lasers.

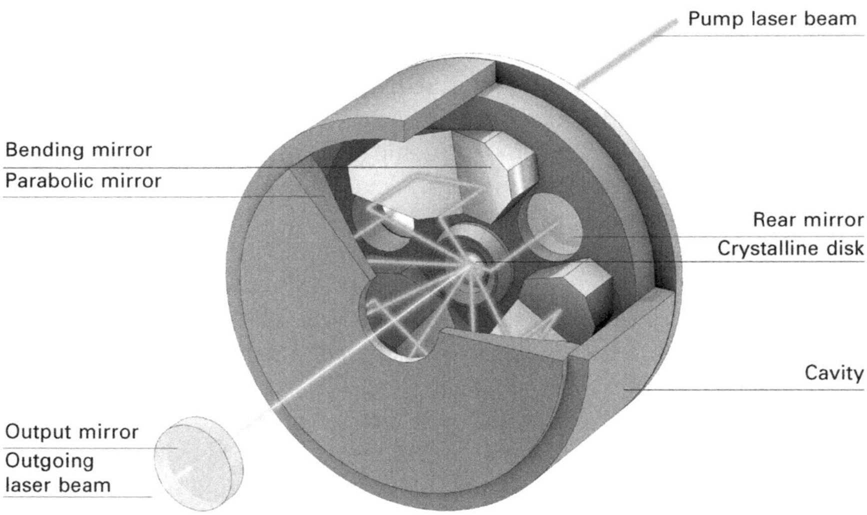

4.2 Cavity of a disk laser setup.

cavity. For even greater power levels, up to four cavities can be optically coupled. Folding mirrors (also called bending mirrors) guide the laser beam from cavity to cavity, connecting the cavities optically (see Fig. 4.3). Here, too, the end mirror and output mirror of the resonator are located outside the cavities.

The disk laser design benefits from unique properties:

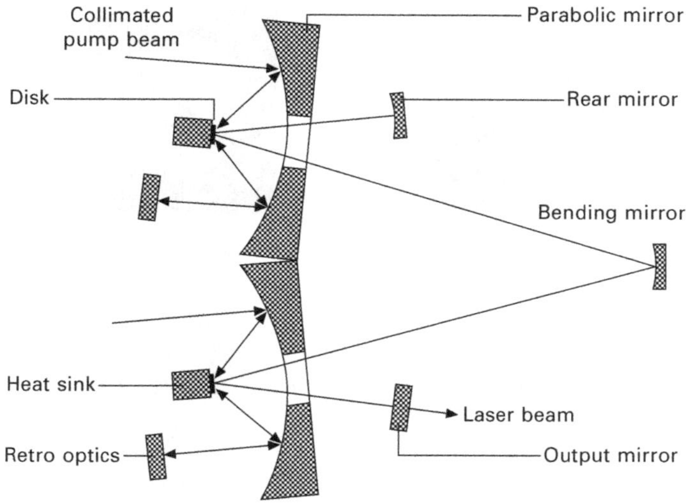

4.3 For higher laser power, two or more cavities are optically connected by mirrors.

- Virtually no thermal lensing (flat temperature profile) due to axial heat flow enables high brightness of the disk laser.
- Low power densities on the active material because of large pumping diameters.
- Low brightness requirements of the pump diodes enable cost-effective lasers with high electrical to optical conversion efficiency – especially in the high average power regime.
- Area scaling of the beam cross section enables power scaling while keeping constant internal intensities.
- Deep gain saturation eliminates harmful back reflection problems.
- The modal cross sections are generally large compared with the longitudinal extension of the gain medium. Therefore high peak power sources are possible without facing problems due to nonlinearities.
- Optimally suited for frequency conversion.

Figure 4.4 shows the option to adapt the pump spot on the disk, which is advantageous for the scalability of the output power per disk. The output power from a single disk can be increased while still keeping the pump power density constant. Only the pump spot diameter increases with the square root of the desired output power. If the pump spot diameter is doubled, one can increase the pump power by a factor of four without changing either the power density or the thermal conditions inside the disk. Contrary to other concepts, such as the rod or fiber laser, the power density on the disk itself is uncritical, even at the highest output power.

With this concept it was very easy to increase the output power of one disk

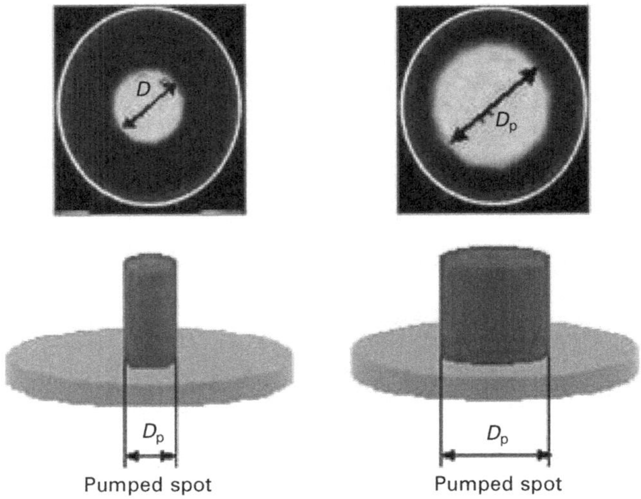

4.4 A certain laser power can be extracted per unit of area → $P_L \sim D_p^2$.

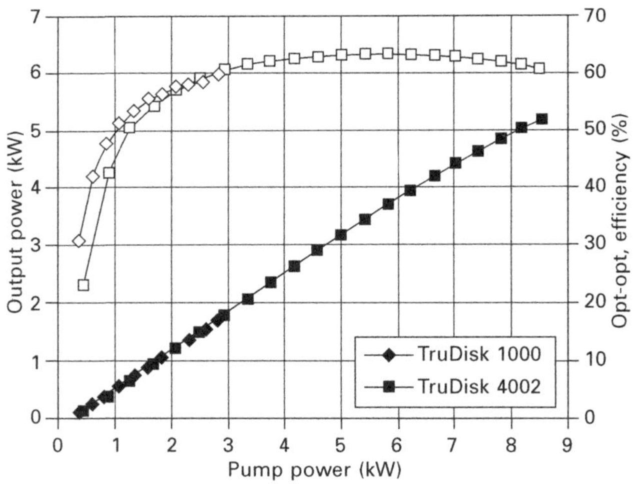

4.5 Output power and efficiency of a typical high power single disk laser oscillator for two beam qualities, BPP = 2 mm mrad (1 kW laser) and BPP = 8 mm mrad (4 kW laser).

without changing the beam quality. A change in beam quality has virtually no effect on the obtainable output power of the beam source. Therefore the beam quality of the disk laser can be adapted to the needs of the targeted applications.

In addition, Fig. 4.5 shows an output power out of one disk of about 5.5 kW achieved for industrial environments. This is the basis of the TruDisk 4002, which guaranties 4 kW at workpiece. The reserves of the laser can be used to realize a power feedback control and to compensate power losses in optical components such as beam guidance, laser light cables and focusing optics. As the achieved welding depth is directly related to the laser power, the end user cannot accept any variation of this parameter. Unlike other laser concepts, neither the environmental conditions nor internal parameters (like cooling water temperature of the laser) have any influence on the output power, which remains constant at all times in the case of the TruDisk. As a result of the increased output power per disk, the number of components and therefore the size of the laser and the cost of the system were dramatically reduced.

Besides the high potential for CW lasers, the mentioned advantages of the disk laser concept lead to a new generation of lasers: short-pulsed lasers (ns pulse duration) and ultra-short-pulsed lasers (ps pulse duration). The disk as a regenerative amplifier is used to amplify low power pulses and hence to achieve pulses with highest peak power and high pulse energy (up to 50 MW peak power (ps pulses) and up to 150 mJ (ns pulses) demonstrated at TRUMPF). These lasers are mainly used in micro applications such as cutting of mobile display glass, various applications in the photovoltaic industry, drilling holes, e.g., in injector nozzles or scribing materials like ceramics.

4.2 Technological trends and developments

The first available disk lasers have been operated with approximately 500 W output power per disk. Improving the technology in different steps lead to 1 kW per disk, 2 kW per disk, 4 kW per disk and nowadays even to 6 kW per disk output power. The result was a price decline, a dramatic reduction in footprint and an increase in overall efficiencies of up to 30% which is equal to a reduction in electrical power consumption by a factor of ten compared to lamp pumped rod lasers. The highest CW power at almost diffraction limited beam quality was demonstrated by Boeing with 27 kW laser power. Combining these results with the results achieved in ultra-short pulsed ps laser development (50 MW, 300 μJ, > 100 W average power), one can imagine the immense potential of the disk laser setup for high power CW lasers as well as for high peak power short pulse lasers. As mentioned, the theoretical limit calculated by the University of Stuttgart indicates that 30 kW from one disk could be possible. Hence, the future development of disk lasers will continue in three main areas:

- increase the output power per disk to reduce investment, operating costs and footprint at very good beam qualities suited for cutting, welding and brazing
- increase the peak power or pulse energy respectively for short-pulsed and ultra-short pulsed lasers for micro applications

- convert the wavelength: the features of the disk laser (e.g., modularity) makes it quite simple to convert the infrared wavelength to green or ultraviolet.

In summary, the disk laser principle will continue to play a decisive role for macro and micro material processing due to its high average power, high peak power, possibility for frequency conversion, truly scalability at constant internal intensities and its economical benefits in high efficiency, in short an ideal and versatile platform for industrial material processing.

4.3 Applications

4.3.1 Laser technology as part of the e-mobility mega-trend

E-mobility has generated a lot of interest in recent times. A big challenge to bring this trend into reality lies in the battery technology. Lithium-ion batteries seem most promising. The production process of lithium-ion batteries is already well known from the mobile phone industry. For e-mobility battery manufacturers now have to adapt the manufacturing process to their specific needs and to an even higher output. Laser technology is a key element. Besides laser marking, laser welding is indispensable for battery cell manufacturing and packaging of these. Remote welding with high-brightness disk lasers allows for the highest productivity.

Motivation

In the development of low emission technology for electric cars such as hybrid, plug-in hybrid or battery vehicles, the battery is the key component. The most preferred battery concept is the lithium-ion battery. A high power density and a very good reload behavior make lithium-ion batteries very attractive in comparison to other battery types. Within the production process of the lithium-ion battery, there are many applications for laser technology, from welding of the housing via welding of contacts to marking of the final battery.

Industrial implementation

Lithium-ion batteries for mobile phones, hand-held devices and laptop computers are more compact than those for electrical vehicles but the manufacturing tasks are the same. A major task is the hermetic sealing of the aluminum housings. In the electronics industry, laser welding has been used for many years now to close the battery housings by welding a lid to the can. It is a state-of-the-art production technique and known as a well-established reliable manufacturing process. Other processes, such as micro plasma welding, do not meet the demands of the industry.

Pulsed Nd:YAG lasers have the biggest installed base in welding of lithium-ion battery housings. They create homogeneous, perfect and hermetically sealed weld seams without holes and cracks (see Fig. 4.6). A precise control of the laser power is needed in order to guarantee the quality of the seam. It also prevents damage to the inner parts of the battery by assuring a constant welding depth. Therefore, the pulsed Nd:YAG lasers TruPulse from TRUMPF have a real-time power control. Although the heat-input is low in laser welding, specially designed fixtures made of copper help to dissipate excess heat from the battery and to improve the weld seam quality as well.

With an average power of several hundred watts, welding speeds of a few meters per minute are possible with pulsed Nd:YAG lasers.

Lithium-ion battery manufacturers for electrical vehicles now make use of the broad experience of the electronics industry. Most manufacturers are evolving from electronics into the fast growing field of e-mobility, so they have the process knowledge already in-house. The key exercise for them is to increase the productivity and therewith lower the total costs for battery manufacture. The reason for that is the large number of single battery cells needed for one electrical vehicle to drive for several hours and several hundred kilometers. Millions of batteries per year are expected to be produced for electrical vehicles in the near future and the numbers will even grow.

One step here is to use continuous-wave solid-state lasers with an output power of 1–6 kW. High-brightness disk lasers of the TruDisk series are

4.6 Lithium-ion battery for a mobile phone with a pulsed weld seam at front and laser marking.

the ideal tool for welding of high-reflecting materials like aluminum and copper (Fig. 4.7). They are insensitive against back-reflections and allow higher welding speeds than pulsed Nd:YAG lasers. The welding speed of disk lasers of several meters per minute is a prerequisite to limit the heat input and the formation of spatters. Using a scanner focusing optics makes the process often more profitable, because it allows guiding the laser beam a lot faster.

In many installations image processing helps to optimize the welding process by exactly positioning the laser beam on the weld seam. It can compensate tolerances in the parts and fixture and additionally can be used for quality inspection, too, if needed. For traceability reasons, all batteries are marked with a serial number and often with even more production-relevant data. Marking lasers are ideally suited for this permanent mark. They can be integrated directly into the production line.

4.3.2 Laser welding in sheet metal

The joining process has always played a crucial role in sheet metal fabrication. Alongside component marking, it comes at the end of the sheet metal process chain and therefore has to meet particularly high demands in terms of delivering a high-quality weld seam with little need for post-weld work (see Fig. 4.8).

The current machines for laser cutting, punching and bending sheet metal parts are high-precision systems that produce metal components to tight tolerances. Up to now the subsequent joining process has largely consisted of arc welding, such as tungsten inert gas (TIG) or gas metal arc (GMA) welding. Despite new developments to reduce the heat input per unit length of weld – using low-energy arc welding techniques – considerable distortion

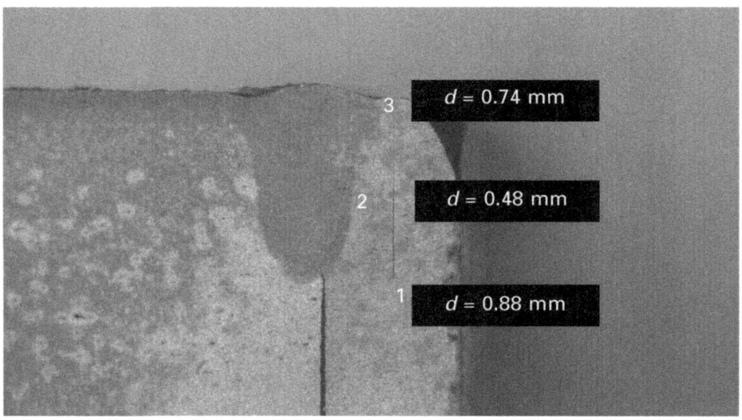

4.7 Cross section of a welded I-seam (aluminum).

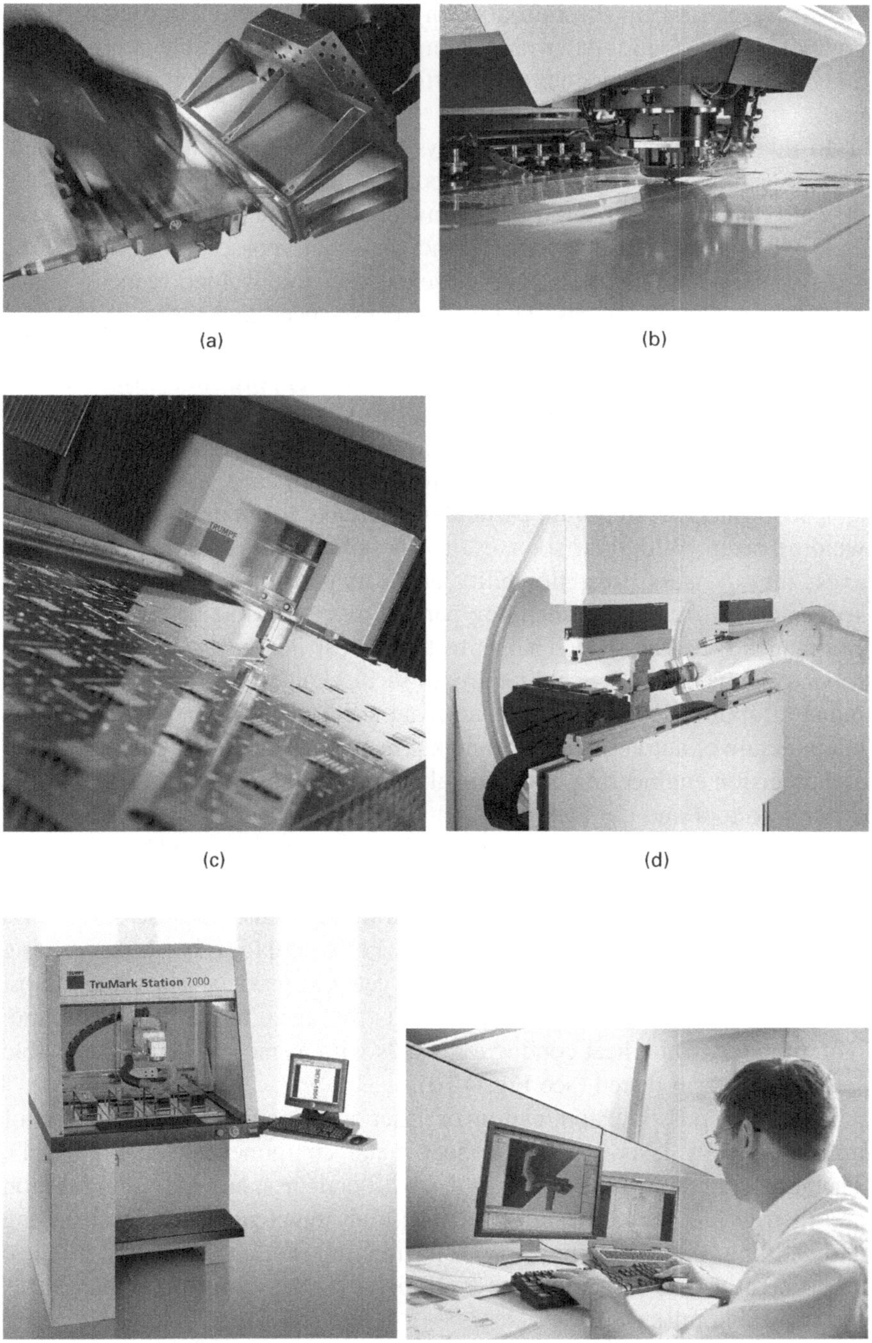

4.8 (a)–(f) Sheet metal process chain.

is usually caused by the introduction of energy over a large area of the component by the arc. After the joining process, considerable straightening and other post-weld work has to be performed on the precision-manufactured part.

In this respect laser welding enjoys an immense advantage. The small melt pool and low amount of heat introduced significantly reduce the amount of weld distortion compared with conventional methods. Hardly any post-weld work is therefore necessary, and the seams possess very high static and dynamic strength. Narrow, precise, contact-free welded seams produce smooth, fault-free surfaces. The high process speed and reproducibility of laser welding reduce costs, while delivering higher quality. As sheet metal fabricators increasingly individualize their product portfolios, laser welding yields decisive competitive advantages: thanks to the advantages described, the investment in a laser welding system is recouped even from small batch sizes when all cost-relevant factors are considered.

Robot-guided systems are particularly suitable for manipulating the laser welding beam. With their six axes, and possibly also additional positioning axes, they possess great flexibility and can perform complex movement sequences on frequently changing parts in response to standardized travel commands (see Fig. 4.9). Thanks to their great flexibility, fully automatic robot systems for laser welding are an ideal choice for an extremely wide range of applications. These extend from sheet metal users engaged in the manufacture of automotive parts and kitchens, to those involved in electrical and precision engineering, mechanical engineering and medical technology, through to tool and die making.

Possible applications in sheet metal fabrication are presented below, taking medical engineering components as an exemplary reference. Workpieces in this sector must meet the most exacting demands, because safety and hygiene have top priority in patient care. For example, robot laser welding systems have been used successfully for years to produce operating tables. Deep penetration welds are made on parts that have to meet high strength requirements, while heat conduction welds with a smooth and dense visible seam are also produced (see Fig. 4.10).

A further possible application of laser welding is the substitution of milled parts by laser-processed sheet metal components (see Fig. 4.11). The laser-cut sheet metal parts are positioned in sandwich configuration and then welded by the laser beam. The high travel speeds in laser cutting and welding, the low component distortion and the high strengths produce an efficient, low-cost component that possesses the same functionalities as a comparable milled part.

Laser welding with flexible robot systems has been established for years in many areas of industry. Ongoing optimization of the control technology has made it progressively easier to operate the robot systems, which has

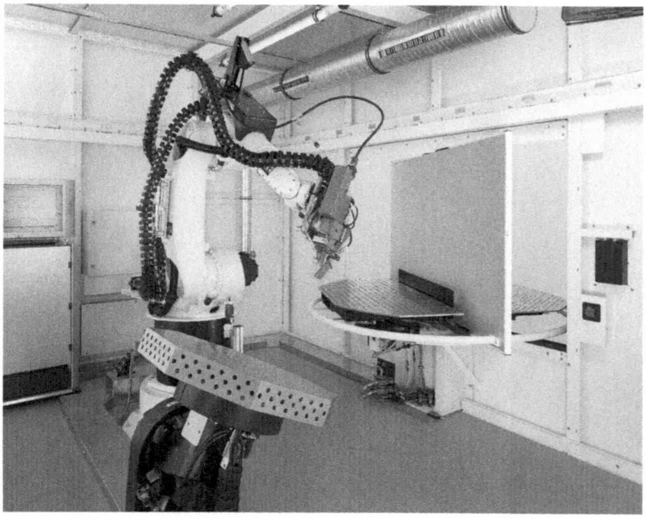

(a)

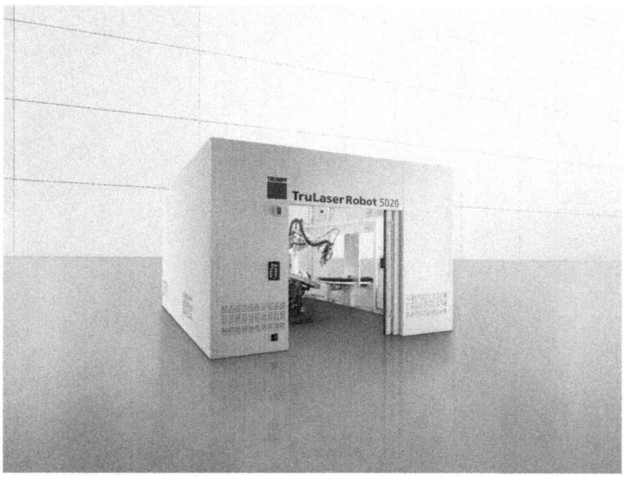

(b)

4.9 (a), (b) TruLaser Robot 5020 robot system for laser welding.

encouraged the more widespread industrial use of this technology. In addition, most robot installations today incorporate measuring systems which make them easier to use both in production planning as well as in the actual production process. As a result, robot laser welding systems can meet the major and diverse challenges that exist in sheet metal fabrication. This is true not only in respect of rising standards in component design, weld preparation and related process sensor technology, but also in the face of increased individualization, as batch sizes get smaller and change more frequently.

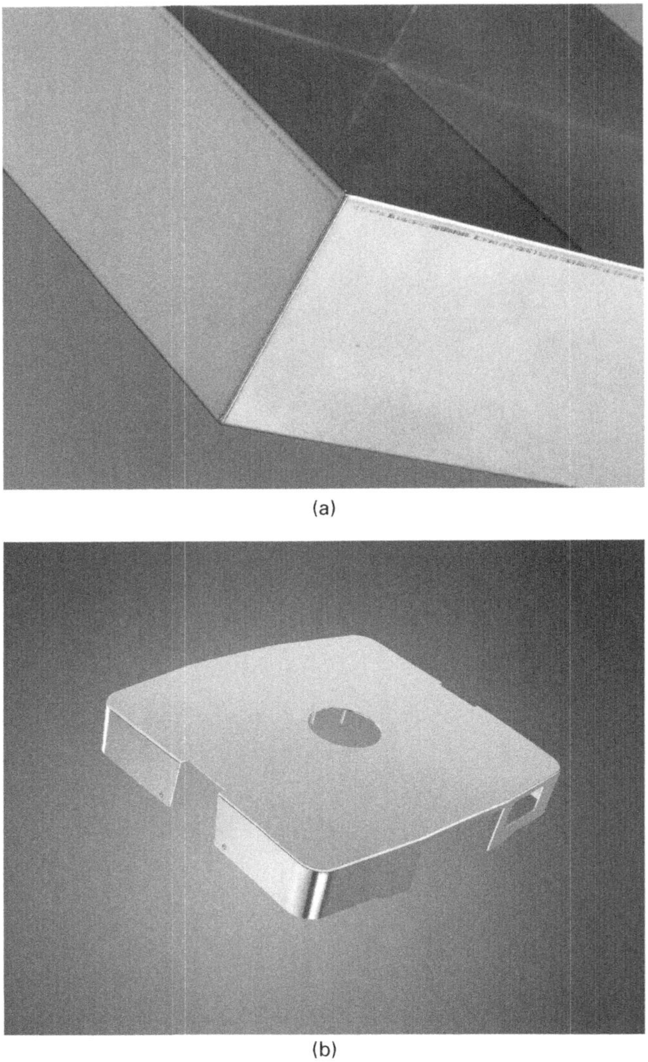

4.10 (a), (b) Heat conduction-welded top plate (source: TRUMPF Medical Systems).

4.3.3 Laser metal deposition by powder injection

Laser metal deposition (LMD) is used to produce coatings with defined properties as well as to repair areas where material is missing due to wear or damage. This technology can also be applied to add material on the component surface in order to change the part design.

By means of this technique, a deposit material is melted by the laser heat source. Part of the energy is absorbed by the substrate causing controlled

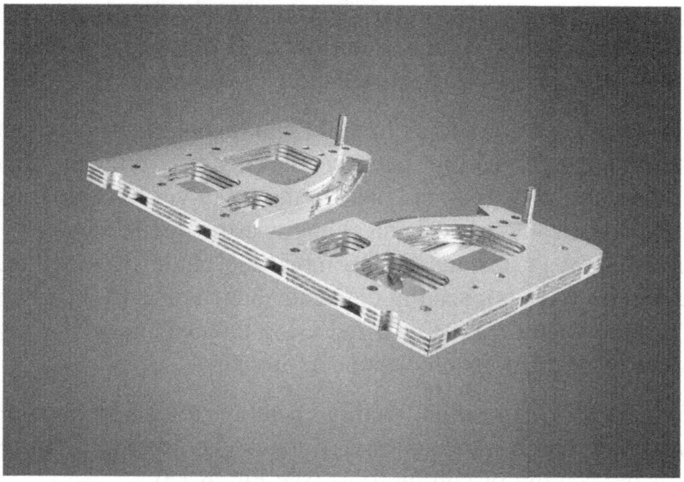

4.11 Laser-welded sandwich component for use in medical engineering.

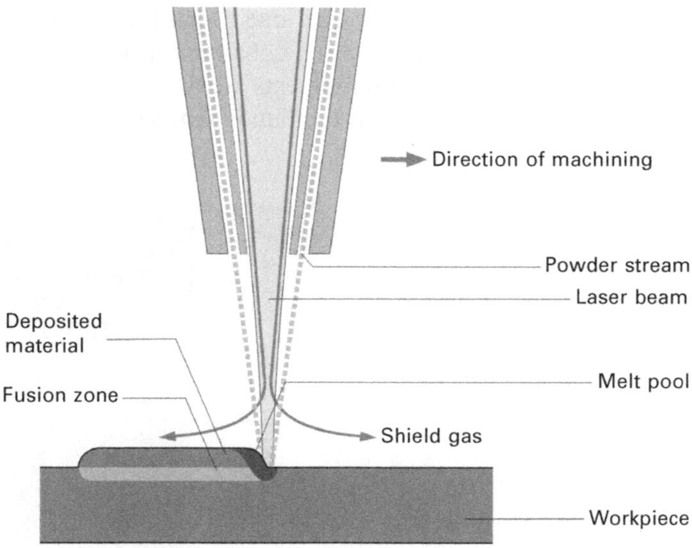

4.12 Principle of laser metal deposition, transfer of deposit material by powder injection.

melting of a thin layer of the base material. This ensures a metallurgical bonding between the coating and the base material. The deposit material can be transferred to the substrate by powder injection or wire feeding. If using deposit materials in powder form, the powder particles are fed into the melt pool by appropriate delivery nozzles (see Fig. 4.12). Single layers with a coating thickness between ca. 0.3 mm and 2 mm can be produced. Several layers can be deposited one upon the other in order to increase the total coating thickness. Track widths as narrow as 0.4 mm can be obtained.

Examples of application of this technology include:

- repairing of injection molds in mold making and tool manufacturing.
- reconditioning of bearing seats (bearing contact surfaces), shafts, rods and seals, sliding valves and discs, exhaust valves in engines as well as cylinders and rolls in mechanical engineering applications.
- refurbishment of turbine and compressor blades, Z-notches, nozzle guide vanes (NGVs), sealing elements and casings in jet engines; also gas turbine blades can be repaired using this technique.
- production of coatings on orthopedic implants.

Since, by means of deposition, welding coatings with improved resistance against abrasive, erosive and adhesive wear can be produced, this technique is used for the coating of cutting tools (see Fig. 4.13).

Also a major application of laser metal deposition is the repair and refurbishment of high-value components such as piston ring grooves in modern four-stroke engines. The surfaces of the piston ring grooves are affected by chemical corrosion and wear due to hard particles in the fuel. For reparation by laser metal deposition, cobalt and iron-based alloys can be used (see Fig. 4.14). Deposit requirements are not only appropriate wear resistance but also resistance against oscillating temperature profiles.

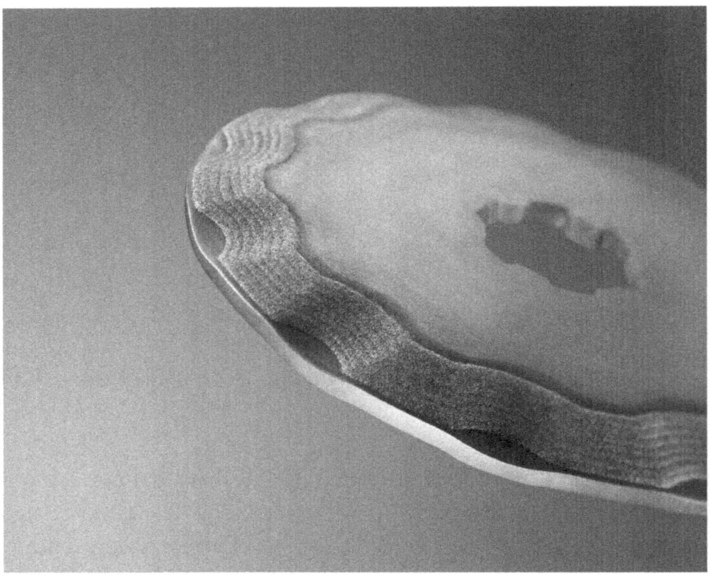

4.13 Laser metal deposition for coating production on agricultural cutting discs.

(a)

(b)

4.14 (a, b) Piston ring groove after laser metal deposition and machining, by courtesy of Gall & Seitz Systems GmbH (Hamburg, Germany). (c, d, e) Refurbishment of piston ring grooves by laser metal deposition.

(c)

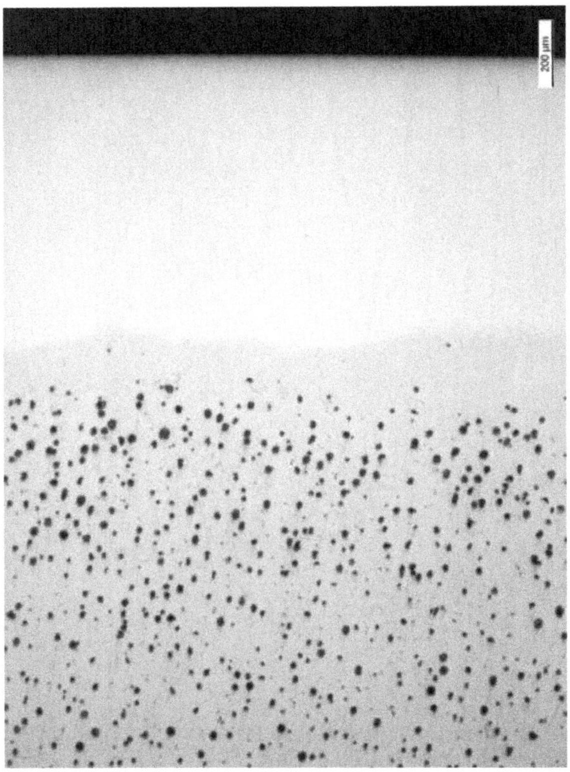

(d)

4.14 Continued

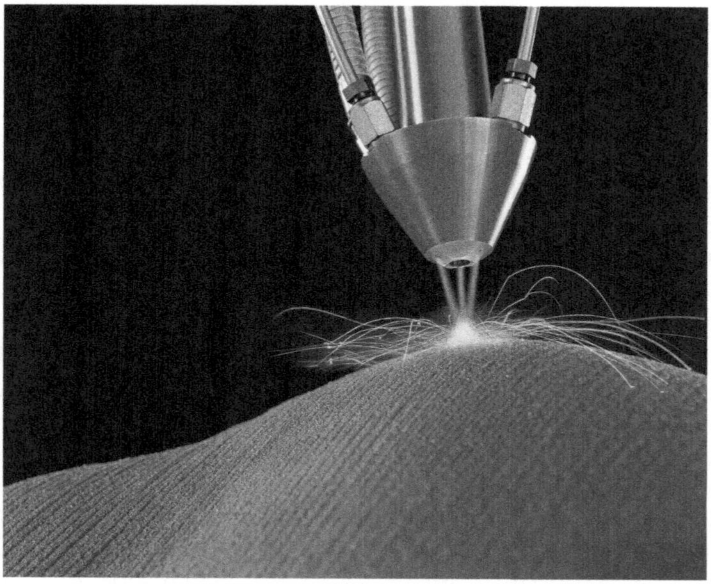

(e)

4.14 Continued

The laser technologies offer many advantages when compared with conventional processes such as TIG deposition welding: the thermal load produced by the laser beam affects only the area immediately surrounding the weld itself, making it possible to reduce dilution, distortion and thermal stress formation. In addition to this, by means of the laser, it is possible to coat complex 3D structures and to access hard-to-reach areas like inside edges or sunken holes. Disk or direct diode lasers (e.g., TruDisk, TruDiode) are suitable for automated deposition welding, the relative movement between the substrate and the laser optic being provided, depending on the required working envelope and positioning accuracy, by robotic manipulators or Cartesian machines.

4.3.4 Laser scanner welding

Remote laser welding or laser scanner welding is one of the most common and successful global applications of the disk laser. Many automotive companies as well as suppliers (e.g., seat industry) rely on the disk laser technology and have integrated remote welding in their production lines. The controllers of the TRUMPF programmable focusing optics (PFO) scanner systems can be linked together with the motion controllers of robotic systems in order to achieve full synchronization between the motion of the robot axes and the activity of the scanner optics. This so-called 'welding-on-the-fly' allows for the fastest material processing while the scanner optics is being moved in space by the robot which enlarges the scanner's processing space and, in addition, which provides three-dimensional access to the processed part, if needed.

In this way it only takes a few seconds to weld components like car doors precisely and firmly. Doing this, the robot arm carries the scanner optics along a smooth path at a constant working distance of about half a meter over the workpiece. Deflection mirrors within the PFO direct the focal point in fractions of a second from weld seam to weld seam. A TruDisk disk laser far away from the processing station generates the laser beam. Up to 100 m long, flexible laser light cables conduct it to the programmable focusing optics. The PFO at the end of the laser light cable is the central element for the precise positioning of the laser's focus point on the car body component. Inside the PFO, two very rapid and precisely pivotable deflection mirrors direct the beam through a so-called flat field objective. Thus, in fractions of a second they adjust the created focus point remotely in X and Y directions within the total work field, without any need to move the entire scanner optics itself. If in addition the PFO is also equipped with a moveable lens inside, this lens can also control the position of the focal point along the Z axis (Figs 4.15 and 4.16). The repositioning of the focused laser beam from one end to the other of one of these three positioning axes takes a mere 30 ms.

There are three basic preconditions for welding on the fly: first, a solid-state laser is needed as the beam source. Only flexible fiber optic cables that can follow the movement of the robot can conduct the laser beam to the scanner optics. The second precondition is high laser beam quality so that in spite of the long focal length of the used flat field objective, it is still possible to create a sufficiently small focus point. This is necessary because for the intended deep penetration welding process, the diameter of the focal point should not exceed 0.6 mm. Modern diode-pumped solid-state lasers like the disk laser generate the usual laser output powers for remote laser welding in car body production of 4–6 kW with beam qualities of 4–8 mm mrad. This allows for focusing distances three to six times the focusing distance being possible with a lamp pumped laser at the same laser power level. The third condition for the positioning of weld seams on the component exactly to the millimetre is the online synchronization between the robot and the scanner control. The scanner control can only take into account possible deviations of the robot from the originally programmed pathway, and compensate for them at any time of the on-the-fly laser welding process, if the robot control communicates to it the actual position of the scanner optics over the component in real time. The biggest advantage of the remote welding process is the extremely high productivity. Inefficient and unproductive times between welding steps can essentially be eliminated. Therefore, the laser,

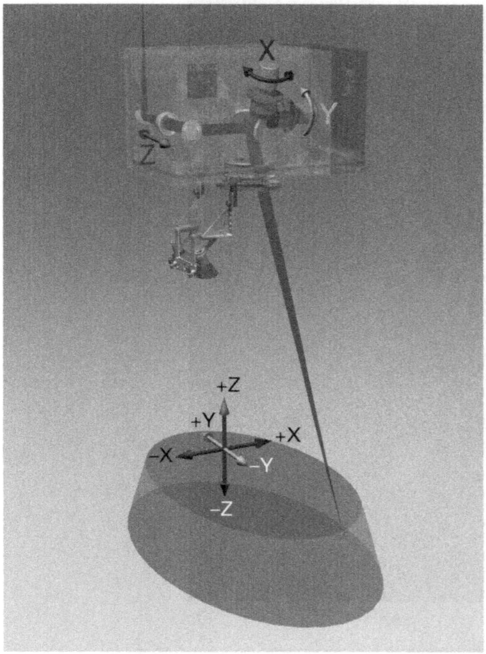

4.15 Principle of the programmable focusing optics (PFO) 3D. Two rotatable mirrors are positioning the beam in the X and Y directions. The movement of an additional lens is positioning the focus location in the Z direction.

as a welding tool, can be used in the most efficient way possible. While the laser-on time of conventional laser welding applications is about 30–40%, the remote laser process enables laser-on times of up to 90%. In comparison to conventional resistant spot welding, the process speed can be increased by a factor of three to ten! Another advantage of laser remote welding is a high flexibility. The geometry of the seam can be adapted to the actual stresses at the weld joint, which enables a continuously optimized laser suitable design. In addition, the seam geometry can be adapted to the local situation. While a slim flange leads to slim C-shape or linear welding geometries, circular C-shapes can be chosen, if the space enables such geometries. Figures 4.17 and 4.18 show a selection of different remote welding applications.

Welding patterns in any selected shape and size can be achieved with a PFO by means of the freely programmable built-in deflection mirrors with maximal welding speed and without any loss of precision, even when it is guided over the car body component by the rather clumsy and imprecise robot. The principal patterns used here are short lines, circles, C-shaped brackets or S-shaped welds. The free configuration, arrangement and orientation of the laser weld seams permit optimal adaptation of a welded joint to the

(a)

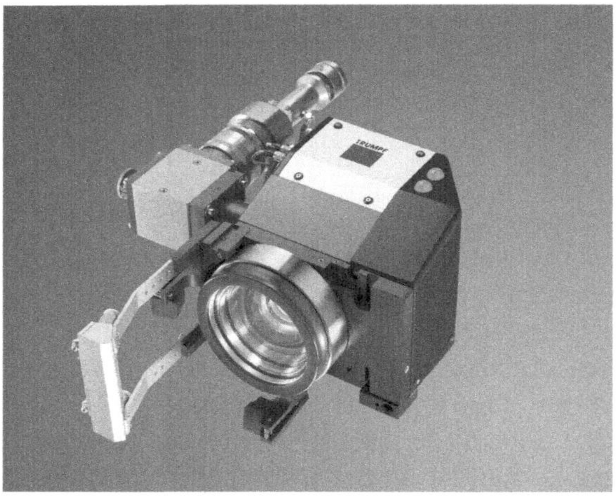

(b)

4.16 Rapid welding sequence performed by the programmable focusing optics (PFO) 33.

subsequent load in daily use. In addition, linear laser weld seams require less space on the component than the circular weld spots produced with the resistance spot welding process. As a result, the width of the flange can be reduced. Smaller, lighter, and more economical components can be produced in this way.

The advantages are not only quantitative; there are also advantages in

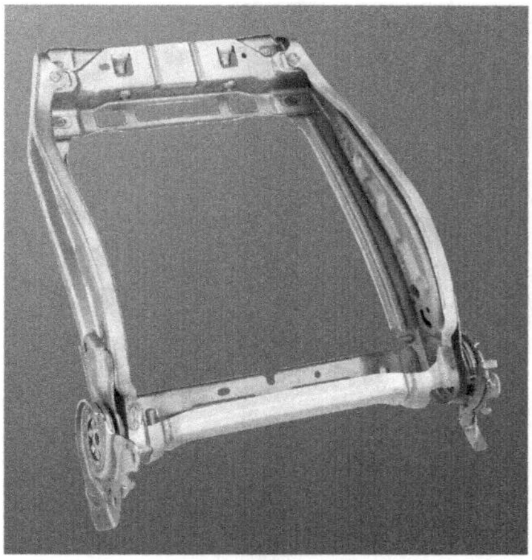

4.17 Laser remote welded car seat (courtesy of Faurecia).

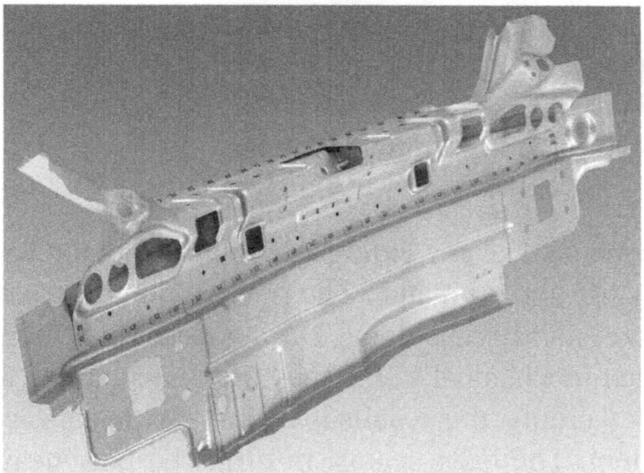

4.18 Laser remote welding in body in white: rear section (courtesy of Daimler).

quality of parts because remote laser welding decreases production tolerances. On the one hand, it subjects the parts to less heat input and mechanical stress than the usual welding processes. The parts are not distorted as much and their structural geometry is not affected. On the other hand, the lower number of positioning and clamping procedures contributes to an even greater dimensional stability of the welded components as well as the entire car body.

4.3.5 Laser application in powertrain production

'Powertrain parts' refers to the group of components that generate power and deliver it to the road surface, water, or air (e.g. crankshaft, clutches, torque converters, gear parts, drive shafts, wheels, etc.) (Fig. 4.19).

4.19 Laser welding of a truck differential gear.

Powertrain components and laser welding as a joining method found each other early on and they are a good match. The powertrain components have small tolerances and are mainly machined. During assembly, the parts are often additionally joined mechanically, i.e. they are attached, pressed in or shrunk. Ideally, in laser welding, this results in a joint gap that is very narrow or practically nonexistent. The laser seams, in turn, can bear heavy loads. The welding depth can be controlled more uniformly and precisely than with other welding processes. This contributes to high strength and quality of the welded components. The laser beam welds with very little or almost no distortion and the seam has a high finish quality. This frequently makes it possible to skip the refinishing step. Precision in laser welding also makes it possible to further reduce the tolerances in the transmission parts. The results are more durable transmissions that run quieter and are more comfortable for the driver and passengers. However, two additional factors are driving the current wave of investment in powertrain applications. There is political pressure to manufacture lighter-weight cars and, to help achieve that goal, find unused potentials in the powertrain and create processes to leverage them. For the example of a dual clutch, laser welding in production already achieves direct savings in terms of material and process steps. In addition, it can be automated and is suitable for manufacturing large production volumes (Figs 4.20 and 4.21).

In addition to the classic CO_2 lasers, solid-state lasers are also being used

4.20 Differential gear: old bolted version (left), new laser welded version (right).

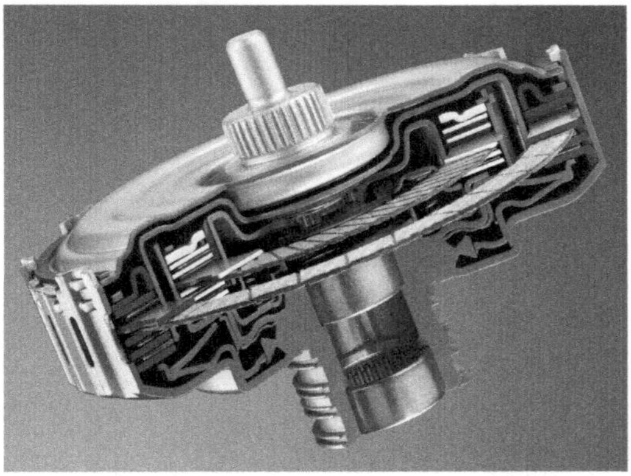

4.21 Laser welded dual clutch.

more often. The deep, subtle, highly stressed seams of many powertrain applications require a weld quality that for a long time primarily only CO_2 lasers were able to deliver. But now even the current disk lasers meet the high requirements and offer the additional advantage that their light can be distributed by means of a laser network and fast changing beam switches on the processing stations. Additionally, solid-state lasers can be smoothly integrated into the production line since laser light can be easily delivered to the work station via fiber optic cable. The length of this fiber optic cable and thus the distance between the beam generator and the welding machine can be up to 100 m (Figs 4.22 and 4.23).

All processing steps can be monitored and controlled using modern sensor and camera technology. This is especially true for process-critical positioning and the welding process itself. In order to achieve a perfect seam, the joining partners must be perfectly aligned and the center of the focus spot – measured to only a few tenths – aimed precisely at the center of the joint gap. CCD

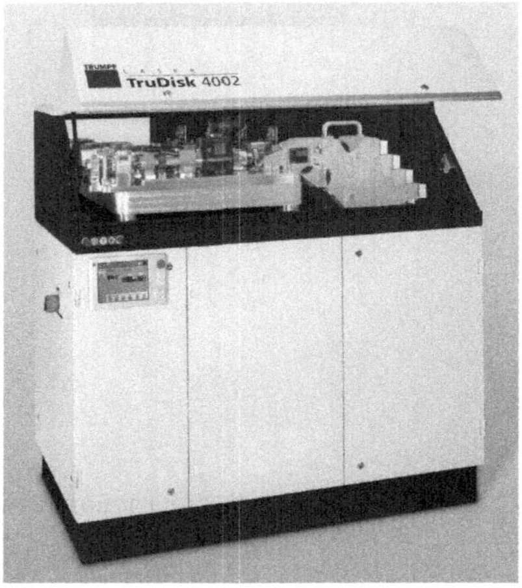

4.22 TRUMPF TruDisk laser.

4.23 TRUMPF TruDisk cavity.

Charge-coupled device (CCD) camera systems have proven reliable for this purpose, in that CCD systems are also able to identify different joining configurations or record and measure workpiece edges, openings and gaps and even an offset in the Z axis. If the components are suitably prepared, even a through-weld, for example in an annular duct, can be identified. The image processing system first gives the exact coordinates for positioning and determines, via a camera, whether the joint gap is optimal for welding. If errors are found, such as a joint gap that is too large, the control system reports this to the overriding controller and discharges the component as defective. The component can then be refinished and fed back into the process, which, in turn, reduces defects and makes the overall process more economical (Figs 4.24 and 4.25).

4.3.6 Hybrid laser welding with high laser power

Laser hybrid welding emerged in the 1980s, as CO_2 lasers were reaching the necessary performance yield for deep welding, often called keyhole welding. In this process, the laser beam partially vaporizes the work material. In this way, a deep, narrow vapor capillary is formed in the fused metal around the beam. This effect is both the strength and the weakness of the process. On the one hand, even very thick sheets can be welded in one pass over the entire face, something that is generally economically or technically not feasible with other processes. On the other hand, the beam requires a sophisticated

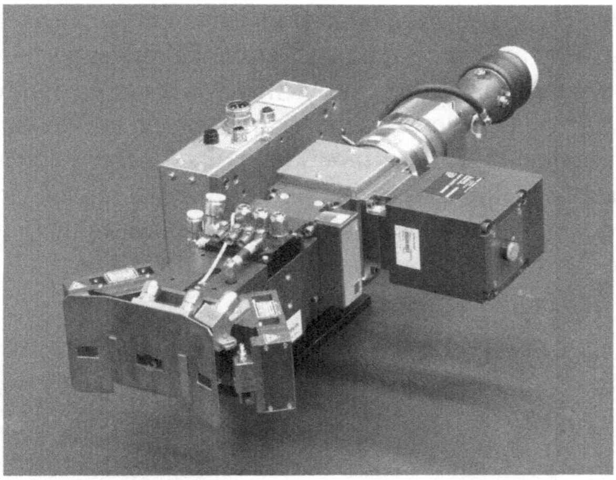

4.24 TRUMPF SeamLine Pro: process sensoric system including welding optic made by TRUMPF.

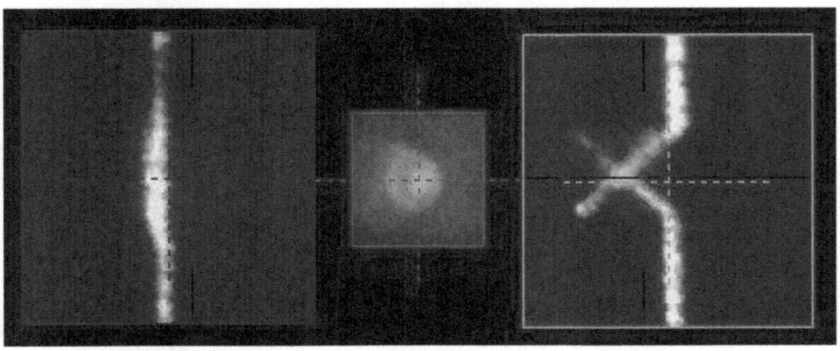

4.25 TRUMPF SeamLine Pro: pre, in, and post view of TRUMPF SeamLine Pro sensoric system.

focus in order to achieve the necessary power density. This means that the joining gap has to be so precisely prepared and the components so exactly clamped that the tolerances of the faces on the entire workpiece length measure in the tenths of a millimeter range. This is only worthwhile when the components are handy to work with and the sheets are thin or if the work or equipment required are justified by an increase in productivity or demands that will be made on the final product. To overcome these limitations, the idea arose that an electrical arc could be combined with the laser beam, because the metal gas welding melts more material into the surface, applies filler materials to the seam and can therefore better fill in the gaps. However, the weld penetration depth is comparatively small. If the sheets become thicker, nothing is left to do for weld preparation other than milling the offsets in a V shape and closing the 'valley' that develops with numerous welding beads in multiple layers. But arc welding introduces a large amount of energy into the workpieces, even when there are not many layers and on cooling, the beads shrink, which leads to deformations and internal stresses. And then there are the costs for fabrication time and additional materials.

Laser hybrid welding solves these problems. While the arc improves the filling in of the gap, the laser increases the weld depth and the welding speed, thus considerably reducing the heat input. In spite of larger tolerances, the seam can be made in a single pass. In addition, with a higher processing speed, the melted welding volume is considerably lower so that much less heat is applied to the surrounding material (Fig. 4.26).

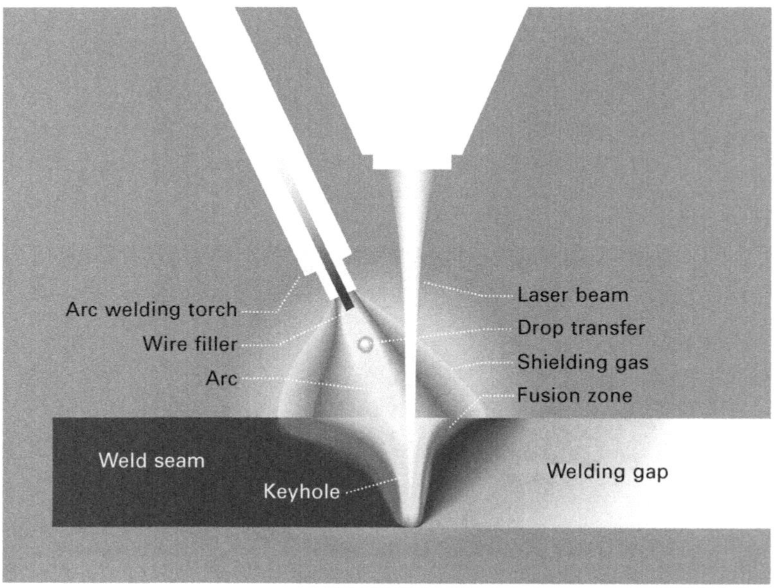

4.26 Principle of laser hybrid welding.

The breakthrough finally came with high-performance solid-state lasers. And yet there was a catch that delayed the spread of the process. The arc requires the plasma to do its work. However, this is a hindrance for the CO_2 laser because the plasma absorbs its beams. Added to that is the fact that the CO_2 laser itself produces plasma; it ionizes the metal steam cloud from the keyhole. When welding with CO_2 lasers, this plasma is 'thinned down' with gases such as helium that are hard to ionize. One non-ionizable helium cloud, however, is the last thing a MAG weld burner needs between the electrode and the workpiece. In addition, the CO_2 laser cannot be transmitted in a laser light cable, which makes the processing of three-dimensional parts more difficult. But this is important, because laser hybrid welding is generally applied to components with unusual dimensions. That is why the traverse paths are large and controlling the beam in the optical path becomes a challenge.

All of that changed when the disk laser broke through into the high performance range. The light from the solid-state lasers ionizes the metal vapor cloud very little or not at all. It is not absorbed by the plasma, anyway, so the arc plasma is no longer a problem. Moreover, with laser light cables it can be deployed to many processing stations. As a result, the distance from the processing head to the beam source or travel paths no longer makes a difference.

Figure 4.27 shows a cross section of a hybrid weld of 12 mm thick ship panels. The T-joint was done in one pass. The welding speed was 1.3 m/min at a laser power of 8 kW. The left-hand side of the picture shows the advantage of energy sharing the laser beam, which has been used for years with the disk laser. A simultaneous weld from both sides (using an energy shared laser beam) decreases distortion of the welded part. Another big advantage of the laser hybrid welding is the additional heated material of the arc welding which leads to a smoother cooling of the laser weld seam and helps in total to reduce negative effects like cracks and low hardness.

Nowadays 10–15 mm thick plates in a single pass have been economically realized in industrial production. The limits for industrial production are more metallurgical effects like weld defects, cracks, dropping, hardness, etc., than the availability of the adapted laser system or laser power, respectively.

But the laser hybrid welding has already proven its high potential and hence there are different applications which will lead to future growth in different industries:

- Welding of truck axles
- Welding of poles
- Welding of panels in shipyard area
- Pipeline construction
- Train and rail construction
- Finned tubes

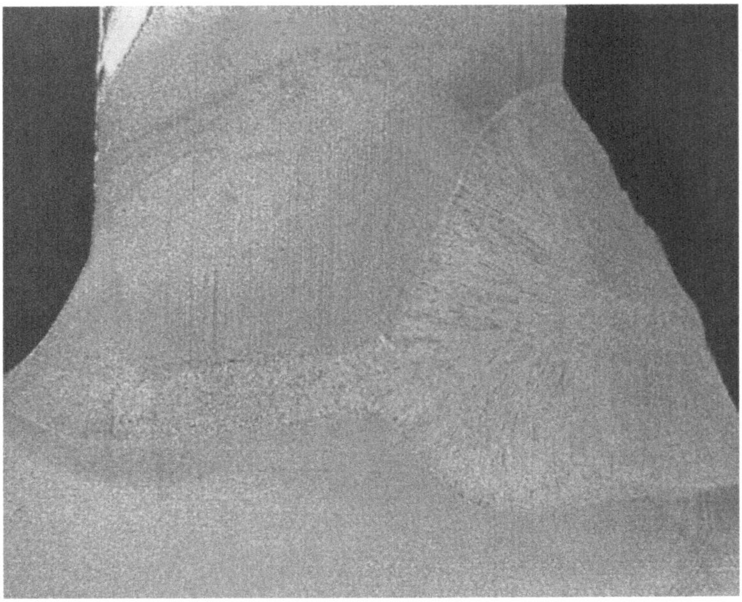

4.27 left: Laser hybrid welded T-joint; 8 kW laser power, welding depth 12 mm, welding speed 1.3 m/min right: double sided simultaneous welding of an 8 mm thick sheet (enabled due to energy sharing option of the disk laser).

- Heat exchanger/boiler
- Heavy industry (cranes, dredger, etc.).

In summary, the laser hybrid welding process offers a lot of advantages, like increased welding speed with high sheet thicknesses. On the one hand, process efficiency is increased; on the other, the process enables less heat input into the material. Therefore, thermal distortion can be minimized. In addition, the combination of laser and arc increases the size of the fit-up gap that can be bridged compared to welding with the laser beam only.

4.4 Future trends

TRUMPF has further developed its disk laser technology. In 2009 a new series of industrial TruDisk lasers was released, which is the third generation of TRUMPF disk lasers. Compared to the second generation of disk lasers, significant reduction in both investment and running cost is achieved. The actual TruDisk generation is more cost efficient than ever before. The footprint of a 4 kW laser was reduced by 60% without cutting any of the well-known functionalities of the former TruDisk generation. Due to the versatile concept of the disk, both the output power as well as the beam quality can be adapted to the needs of all applications. While the maximum output power is increased to 16 kW, a beam quality of 2–8 mm mrad is available.

The new generation of disk lasers has set new standards for industrial laser sources regarding price, running cost and reliability. Although the up-time availability of disk lasers is known to be >99.5%, the modular design allows a service and maintenance friendly system. Most of components can be exchanged by the user, without requiring a TRUMPF service technician.

The wide variety of different disk laser applications reflect the versatility of this laser concept. Applications in the micro area are served as well as in the thick sheet area. With the new 'Cut assist' option, the TruDisk possesses a superior intelligence which enables faster cutting on the one hand, as well as better cutting quality on the other.

Due to the excellent beam quality, the new TruDisk series is suitable for cutting and welding applications in the sheet thickness range of 0.05 mm to >15 mm. Remote processes belong to the most common processes of the TruDisk laser. The disk laser has been used for remote welding for years with several worldwide manufacturing automotive OEM companies and suppliers. The latest development with the multi-kW TruDisk lasers was the melt pressure induced remote cutting, which can be used with sheet thicknesses of up to 4 mm.

Further development of disk lasers aims mainly to:

- increase the output power per disk
- reduce the investment and operating costs
- make the laser even more compact
- convert the wavelength: disk lasers combine high efficiency, excellent beam quality, high average and/or peak power with moderate cost and high reliability at multiple wavelengths, ranging from the infrared over the green to the ultraviolet; therefore, they are the ideal platform for wavelength conversion in short-pulse lasers (e.g., for the processing of silicon) and on the other hand for CW lasers (advantages in the processing of copper).

4.5 Sources of further information and advice

Avizonis, P., Bossert, D., Curtin, M., Killi, A., Schmitz, C.: 'Physics of high performance Yb:YAG thin disk lasers', *Conference on Lasers and Electro-Optics* CLEO 2009, Baltimore, MD, 31 May 2009.

Bach, F. W., Möhwald K., Laarmann A., Wenz T.: *Moderne Beschichtungsverfahren*. Weinheim: Wiley–VCH Verlag, 2005.

Bernhardt, R., Schubert, H.: Laser welding applications in the new Mercedes-Benz C-class, *8th European Conference and Exhibition, European Automotive Laser Application*, Frankfurt, 30–31 January 2007.

Brockmann, R., Mann, K.: Optimierte Fertigungstechnologie bei Audi: Laserscannerschweißen an den Türen des neuen Audi A4, *Laser Technik Journal*, 4(2), pp. 27–28, 2007.

Contag, K., Karszewski, M., Stewen, C., Giesen, A., Hügel, H.: Theoretical modelling and experimental investigations of diode-pumped thin-disk Yb:YAG laser, *Quantum Electron*, 29, pp. 697–703, 1999.

Gasser, A.: Neue Entwicklungen im Laserauftragschweißen für industrielle Fertigungsverfahren. *Tagungsband des International Laser Technology Congress AKL'08*, 2008.

Gasser, A., Meiners, W., Weisheit, A., Willenborg, W., Stollenwerk, J., Wissenbach, K.: Maßgeschneiderte Oberflächen und Bauteile, *Laser Technik Journal*, 7(4), pp. 47–53, 2010.

Giesen, A.: Der Scheibenlaser – Potential und Perspektiven, Presentation at In-house 'Technologietag Scheibe', Trumpf Laser Schramberg, 26 April 2006.

Giesen, A., Speiser, J.: Fifteen years of work on thin-disk lasers: Results and scaling laws, *IEEE Journ. of Sel. Topics in Quantum Electron*, 13, pp. 598–609, 2007.

Giesen, A., Huegel, H., Voss, A., Wittig, K., Brauch, U., Opower, H.: Scalable concept for diode-pumped high-power solid-state lasers, *Appl. Phys. B*, 58, pp. 365–372, 1994.

Herwig, P.: Faserlaser zuverlässig vor Rückreflexen geschützt, *Photonic*, 5, 2007.

Jasnau, U., Hoffmann, J., Seyffarth, P.: Nd:YAG-Laser-MSG-Hybridschweißen von Aluminiumlegierungen im Schiffbau [Nd:YAG laser-GMAW hybrid welding of aluminium alloys in shipbuilding], *DVS-Berichte*, Vol. 225 (2003).

Kelbassa, I.: Qualifizieren des Laserstrahl-Auftragschweißens von Blisks aus Nickel- und Titanbasislegierungen. Dissertation, RWTH Aachen, 2006.

Klemm, P.: *Softwaretechnik für Prozessplanung und Leitsysteme*. Lecture notes, Universität Stuttgart.

Miebach, R., Lembeck, H.: Die neue Fertigung der Meyer-Werft: Laserhybridschweißen als Kerntechnologie [The new manufacturing process at the Meyer Werft shipyard: laser hybrid welding as key technology], *DVS-Berichte*, Vol. 225, pp. 187–192, 2003.

Ohnesorge, A.: Bestimmung des Aufmischungsgrades beim Laser-Pulver-Auftragschweißen mittels laserinduzierter Plasmaspektroskopie (LIPS). Dissertation, Technische Universität Dresden, 2008.

Peters, M.: Titan in Aufwind, *MB-Revue*, pp. 72–77, 2009.

Roland, F., Pethan, G.: Increasing efficiency and quality in shipbuilding and shiprepair by developing mobile laser equipment for the dock area. *4. Wismarer Fachtagung 'Maritime Technik'*, 2004.

Seyffarth, P.: Erfahrungen des Werftausrüsters IMG bei der Einsatzvorbereitung von Lasertechnologien für den Schiffbau [Experiences of shipyard supplier IMG in preparing laser technologies for shipyard applications], *5th Laser Users Forum*, Bremen, 2006.

Staufer, H., Reinthaler, G., Ennsbrunner, H.: Operation and visualisation of the LaserHybrid Twin welding process. Doc. 212-1174-10, XII-2003-10.

Wagner, F.: Laser-Pulver-Auftragschweißen zur Rekonditionierung von Schiffsmaschinenbauteilen. Thermisches Beschichten mit laserbasierten Fertigungsverfahren, Dresden, 2009.

5 Developments in pulsed and continuous wave laser welding technologies

J. ZHOU, Pennsylvania State University, USA and H. L. TSAI, Missouri University of Science and Technology, USA

DOI: 10.1533/9780857098771.1.103

Abstract: This chapter first describes the history and current status of the laser welding technology. Then, fundamentals of laser welding, such as keyhole formation and collapse, laser–plasma interaction, weld pool dynamics, porosity formation mechanisms, metal melting and solidification, etc., are elaborated. Finally, new developments in laser welding such as hybrid laser-arc welding, multi-beam laser welding, electromagnetically assisted laser welding, and future challenges and development of laser welding are discussed.

Key words: laser welding, hybrid laser-arc welding, multi-beam laser welding, keyhole formation and collapse, plasma, heat and mass transfer, droplet formation and impingement, electromagnetic force.

5.1 Introduction

Welding with lasers was first proposed in the early 1960s.[1] After that, especially in the last two decades, laser welding has been increasingly applied to various applications in industries like automotive, shipyard, oil, etc., due to its many advantages over traditional welding techniques, such as high energy intensity, deep penetration depth; high precision and good flexibility; less distortion, etc. Much of its success is based on the development and availability of enabling technologies, which include improvements in process understanding, enhancements in laser sources and systems, and continued development and progression in process technology in laser beam welding.[2]

In laser welding, high power lasers such as CO_2 lasers, Nd:YAG lasers, disk lasers and fiber lasers have been used or are under continuous development as a heating source.[1] According to the operation mode of these lasers, laser welding can be classified as pulsed laser welding and continuous wave laser welding. For the pulsed laser welding, the typical range of pulse durations is 1–20 msec. At the lower end of this range, pulse frequencies can approach 1 kHz. For the continuous wave mode laser welding, although penetration welding is possible at high average laser powers (>1 kW), this condition is not always obtainable at low average power. The CO_2 laser is the workhorse of industrial laser welding systems because it is simple and reliable with output powers up to 50 kW. It has higher output efficiency compared with other lasers, such as the Nd:YAG laser. The generation of high average power in Nd:YAG laser systems is accomplished by combining several individually pumped laser rods in a single resonator. One of the prime advantages of the Nd:YAG laser over the CO_2 laser is its ability to deliver laser radiation through optical fibers, which makes it very attractive in robotic or multiaxis laser welding applications. With fiber-optic coupling, it is also possible to combine the power from two or more lasers.

In recent years, many new developments and applications have occurred in laser welding, which includes the use of new laser source systems and laser welding process technologies.[2] In terms of the laser source, diode lasers and diode pumped solid-state lasers have been the subject of innovative developments in the last decade. Most recently high power diode pumped fiber lasers were developed with attractive characteristics for materials processing applications. High-power fiber lasers can be used for deep penetration welding of various materials since its characteristic low wavelength allows its absorption by almost all metals and alloys, and the fiber delivery system provides the necessary flexibility on the positioning of the beam.

The first use of fiber lasers dates back to the early 1960s, when low-power lasers were used in optical amplifiers. The first 100 W fiber laser was produced for materials processing in 2000.[3] Multi-kilowatt fiber lasers have now been introduced for materials processing and the maximum power from commercial fiber lasers can reach 100 kW. Multi-kilowatt power levels are achieved by combining the output from several single mode fiber lasers. The technology is under development to produce systems of even higher power. These new lasers have multiple advantages including high efficiency compared to lamp or diode pumped rod lasers; compact design; good beam quality due to the use of small diameter fibers, and thus small beam focus diameter; and a robust setup for mobile applications. The lifetime of the pumping diodes is expected to exceed the lifetime of other diode pumped lasers which leads to low cost.[3] In order to understand these new developments and continue to optimize the laser welding process for its current applications and expand its capacity into more applications, it is crucial to have a good understanding of the fundamental physics involved in laser welding. Then, the current developments and future trend in laser welding will be discussed.

5.2 Fundamentals of laser welding

Laser welding involves a lot of complicated physical processes including metal melting and solidification; keyhole formation and collapse; keyhole plasma formation and laser–plasma interaction, which results in very complex transport phenomena in the welding process.[4] As is well known, metallurgical structure and mechanical properties of the laser weld are strongly dependent on transport phenomena such as heat transfer, melt flow, and plasma flow in the welding process.[5–9] In the following section, major transport phenomena involved in laser welding will be discussed to understand the scientific fundamentals in the laser welding process. Special attention is given to discussions on: laser-induced keyhole plasma formation and energy transport in plasma; recoil pressure and mechanisms contributing to keyhole formation; keyhole collapse and solidification process; porosity formation; and metal melting and solidification process.

Due to the different nature of heat and mass transfer mechanisms in metal and plasma, different mathematical models have been developed to study the fundamental physics in laser welding. As shown in Fig. 5.1, one has been developed to study the transport phenomenon in the metal region containing base metal. The other is for the study of the keyhole region containing laser-induced plasma. There is a free surface (liquid/vapor interface) separating these two regions. For the metal region, continuum formation is used to calculate the energy and momentum transport.[10] For the keyhole plasma region, laser–plasma interaction and the laser energy absorption mechanism

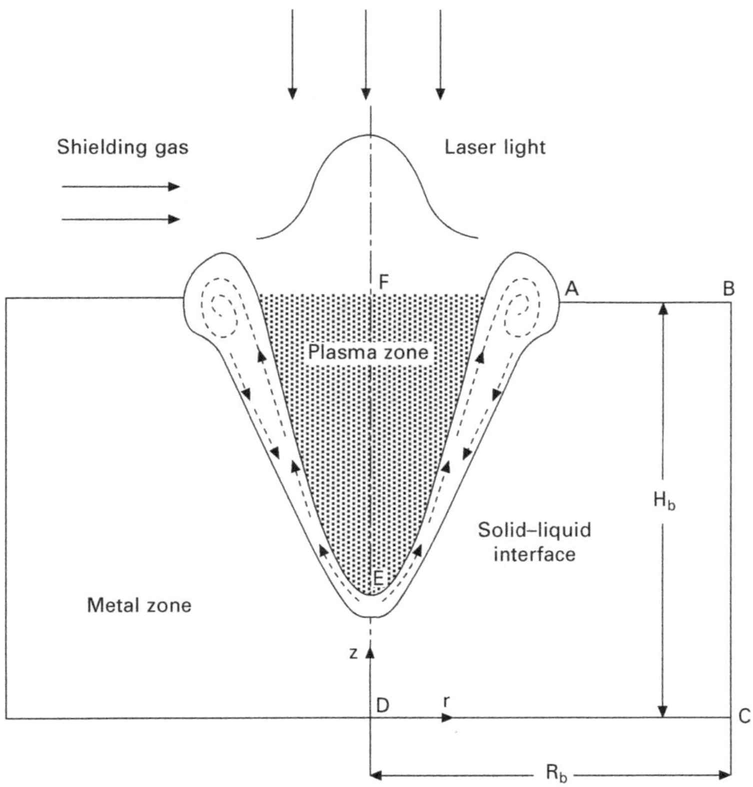

5.1 Schematic sketch of a pulsed laser keyhole welding process.

are discussed. These two regions are coupled together and the volume-of-fluid (VOF) technique is used to track the interface between these two regions.[10]

5.2.1 Transport phenomena in metal

Differential equations governing the conservation of mass, momentum and energy based on continuum formulation are given below.[10]

Conservation of mass

$$\frac{\partial}{\partial t}(\rho) + \nabla\cdot(\rho \boldsymbol{V}) = 0 \qquad [5.1]$$

where t is the time, ρ is the density, and $\boldsymbol{V}$ is the velocity vector.

Conservation of momentum

$$\begin{aligned}\frac{\partial}{\partial t}(\rho u) + \nabla\cdot(\rho \boldsymbol{V} u) &= \nabla\cdot\left(\mu l \frac{\rho}{\rho l}\nabla u\right) - \frac{\partial p}{\partial x} - \frac{ul}{K}\frac{\rho}{\rho l}(u - u_s) \\ &\quad - \frac{C\rho^2}{K^{0.5}\rho l}|u - u_s|(u - u_s) \\ &\quad - \nabla\cdot(\rho f_s f_l \boldsymbol{V} r u_r) + \nabla\cdot\left(\mu_s u \nabla\left(\frac{\rho}{\rho l}\right)\right)\end{aligned} \qquad [5.2]$$

$$\begin{aligned}\frac{\partial}{\partial t}(\rho v) + \nabla\cdot(\rho \boldsymbol{V} v) &= \nabla\cdot\left(\mu l \frac{\rho}{\rho l}\nabla v\right) - \frac{\partial p}{\partial y} - \frac{ul}{K}\frac{\rho}{\rho l}(v - v_s) \\ &\quad - \frac{C\rho^2}{K^{0.5}\rho l}|v - v_s|(v - v_s) \\ &\quad - \nabla\cdot(\rho f_s f_l \boldsymbol{Vr} v_r) + \nabla\cdot\left(\mu_s v \nabla\left(\frac{\rho}{\rho l}\right)\right)\end{aligned} \qquad [5.3]$$

$$\begin{aligned}\frac{\partial}{\partial t}(\rho w) + \nabla\cdot(\rho \boldsymbol{V} w) &= \rho g + \nabla\cdot\left(\mu l \frac{\rho}{\rho l}\nabla w\right) - \frac{\partial p}{\partial z} - \frac{ul}{K}\frac{\rho}{\rho l}(w - w_s) \\ &\quad - \frac{C\rho^2}{K^{0.5}\rho l}|w - w_s|(w - w_s) \\ &\quad - \nabla\cdot(\rho f_s f_l \boldsymbol{Vr} w_r) + \nabla\cdot\left(\mu_s w \nabla\left(\frac{\rho}{\rho l}\right)\right) \\ &\quad + \rho g \beta_T (T - T_0) + F_{\text{drag}}\end{aligned} \qquad [5.4]$$

where u, v and w are the velocities in the x-, y- and z-directions, respectively, and $\boldsymbol{V}_r$ is the relative velocity vector between the liquid phase and the solid phase. The subscripts s and l refer to the solid and liquid phases, respectively; Subscript 0 represents the reference conditions; p is the pressure; μ is the viscosity; f is the mass fraction; K, the permeability, is a measure of the ease with which fluid passes through the porous mushy zone; C is the inertial coefficient; β_T is the thermal expansion coefficient; g is the gravitational acceleration; and T is the temperature.

Conservation of energy

$$\frac{\partial}{\partial t}(\rho h) + \nabla\cdot(\rho \boldsymbol{V} h) = \nabla\cdot\left(\frac{k}{c_p}\nabla h\right) - \nabla\cdot\left(\frac{k}{c_p}\nabla(h_s - h)\right) - \nabla\cdot(\rho(\boldsymbol{V} - \boldsymbol{V}_s)(h_l - h)) \qquad [5.5]$$

where h is the enthalpy, k is the thermal conductivity, and c_p is the specific heat. The first two terms on the right-hand side of Eq. [5.5] represent the net Fourier diffusion flux. The third term represents the energy flux associated with the relative phase motion.

The third and fourth terms on the right-hand side of Eqs [5.2]–[5.4] represent the first- and second-order drag forces of the flow in the mushy zone. The fifth term represents an interaction between the solid and the liquid phases due to the relative velocity. The second term on the right-hand side of Eq. [5.5] represents the net Fourier diffusion flux. The third term represents the energy flux associated with the relative phase motion. All these aforementioned terms in this paragraph are zero except in the mushy zone. In addition, the solid phase is assumed to be stationary ($\boldsymbol{V}_S = 0$).

Conservation of species

$$\frac{\partial}{\partial t}(\rho f^{\alpha}) + \nabla\cdot(\rho \boldsymbol{V} f^{\alpha}) = \nabla\cdot(\rho D \nabla f^{\alpha}) - \nabla\cdot(\rho D \nabla(f_l^{\alpha} - f^{\alpha})) - \nabla\cdot(\rho(\boldsymbol{V} - \boldsymbol{V}_s)(f_l^{\alpha} - f^{\alpha})) \qquad [5.6]$$

where D is a mass diffusivity and f^{α} is a mass fraction of constitute. Subscripts l and s, represent liquid and solid phase, respectively.

5.2.2 Transport phenomena in laser-induced plasma

The vapor inside the keyhole is modeled as a compressible, inviscid ideal gas. No vapor flow is assumed in the keyhole and the energy equation is given in the following form:[11]

$$\frac{\partial}{\partial t}(\rho_v h_v) = \nabla \cdot \left(\frac{k_v}{c_v} \nabla h_v \right) + \nabla \cdot (-\boldsymbol{q_r}) + k_{pl} \cdot I_{\text{laser}} \cdot (1 - \alpha_{iB,1}) + \sum_{mr=1}^{n} k_{pl} I_{\text{laser}} \cdot (1 - \alpha_{iB,1}) \cdot (1 - \alpha_{Fr}) \cdot (1 - \alpha_{iB,mr}) \quad [5.7]$$

where h_v and ρ_v represent the enthalpy and density of the plasma; k_v and c_v represent the thermal conductivity and specific heat of the plasma. The first term on the right-hand side of Eq. [5.7] represents the heat conduction term. The second term represents the radiation heat term and $\boldsymbol{q_r}$ stands for the radiation heat flux vector. The fourth term represents energy input from the original laser beam. The last term represents the energy input from multiple reflections of the laser beam inside the keyhole.

5.2.3 Laser-induced recoil pressure and keyhole formation

In the laser welding process, when the laser beam is directed on the surface of the workpiece, it first melts the metal, forming a small molten pool in the workpiece. With the continuous heating from the laser beam, the liquid metal is then heated to high temperatures resulting in large evaporation rates. When the metal vapor escapes from the liquid metal, it exerts a pressure called recoil pressure to the liquid metal surface. This recoil pressure will depress the molten metal downward and then a cavity in the liquid metal known as keyhole will be formed. Many investigators believe that this recoil pressure combined with surface tension force is the main factor determining the shape of the keyhole. So, understanding the recoil pressure formation is critical for understanding the keyhole formation and thus for the control of the laser welding process. As known, the recoil pressure results from the rapid evaporation from a metal surface that has been heated to high temperatures. When the liquid–vapor interface temperature reaches boiling point, evaporation begins to occur. As shown in Fig. 5.2, there is a Knudsen layer adjacent to the liquid metal surface. In the Knudsen layer, the metal vapor escaping from the liquid surface is in a state of thermodynamic non-equilibrium, i.e., the vapor molecules do not have a Maxwellian velocity distribution. This occurs when the equilibrium vapor pressure (i.e., the saturation pressure) corresponding to the surface temperature is large compared to the ambient partial pressure of the vapor. Under these conditions the vapor adjacent to the surface is dominated by recently evaporated material that has not yet experienced the molecular collisions necessary to establish a Maxwellian velocity distribution. The Knudsen layer is estimated to be a few molecular mean free paths thick in order to allow for the molecular collisions to occur that bring the molecules into a state of translational equilibrium at the outer edge of the Knudsen layer.

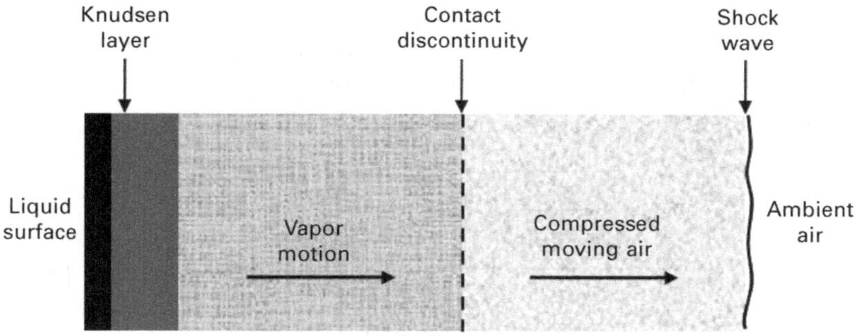

5.2 A schematic of the gas dynamic of vapor and air away from a liquid surface at elevated temperature.

Anisimov[12] and Knight[13] did the early investigations on the Knudsen layer. A kinetic theory approach[14] is used to study the behavior of this layer. The analysis proceeds by constructing an approximate molecular velocity distribution adjacent to the liquid surface. Equations describing the conservation of mass, momentum and energy across the Knudsen layer are developed in terms of this velocity distribution. This gives Eqs [5.8] and [5.9], as given below, for gas temperature, T_K, and density, ρ_K, outside of the Knudsen layer as functions of the liquid surface temperature and the corresponding saturation density, ρ_{sat}.

$$\frac{T_K}{T_L} = \left[\sqrt{1 + \pi\left(\frac{\gamma - 1}{\gamma + 1}\frac{m}{2}\right)^2} - \sqrt{\pi}\frac{\gamma - 1}{\gamma + 1}\frac{m}{2}\right]^2 \quad [5.8]$$

$$\frac{\rho_K}{\rho_{sat}} = \sqrt{\frac{T_L}{T_K}}\left[\left(m^2 + \frac{1}{2}\right)e^{m^2} erfc(m) - \frac{m}{\sqrt{\pi}}\right] + \frac{1}{2}\frac{T_L}{T_K}[1 - \sqrt{\pi}me^{m^2} erfc(m)] \quad [5.9]$$

The quantity, m, is closely related to the Mach number at the outer edge of the Knudsen layer, M_K, and is defined as $m = u_K/\sqrt{2R_V T_K} = M_K\sqrt{2/\gamma_V}$, where γ_V and R_V are the ratio of specific heats and the gas constant for the vapor, respectively. The value of m depends on the gas dynamics of the vapor flow away from the surface. The gas temperature, pressure and density throughout the vapor region (outside of the Knudsen layer) are uniform. The contact discontinuity, that is, the boundary between vapor and air, is an idealization that results due to the neglect of mass diffusion and heat conduction. The velocity and pressure are equal in these regions, $u_K = u_S$ and $P_K = P_S$, where the subscript, S, denotes properties behind the shock wave. Note that, in general, $T_K \neq T_S$ and $\rho_K \neq \rho_S$.

The thermodynamic state and velocity of the air on each side of the shock wave are related by the Rankine–Hugoniot relations, where the most convenient forms to this application are given by Eqs [5.10] and [5.11]. M_K is the Mach number in the vapor, $M_K = u_K / \sqrt{2\gamma_V R_V T_K}$.

$$\frac{P_S}{P_\infty} = 1 + \gamma M_K \sqrt{\frac{\gamma_V R_V T_K}{\gamma_\infty R_\infty T_\infty}} \left[\frac{\gamma_\infty + 1}{4} M_K \sqrt{\frac{\gamma_V R_V T_K}{\gamma_\infty R_\infty T_\infty}} + \sqrt{1 + \left(\frac{\gamma_\infty + 1}{4} M_K \sqrt{\frac{\gamma_V R_V T_K}{\gamma_\infty R_\infty T_\infty}} \right)^2} \right] \quad [5.10]$$

$$\frac{T_S}{T_\infty} = \frac{P_S}{P_\infty} \left(1 + \frac{\gamma + 1}{\gamma - 1} \frac{P_S}{P_\infty} \right) \bigg/ \left(\frac{\gamma + 1}{\gamma - 1} \frac{P_S}{P_\infty} \right) \quad [5.11]$$

The saturation pressure, P_{sat}, is obtained from Eq. [5.12], where A, B and C are constants which depend on the material. This is used to obtain the saturation density, $\rho_{sat} = P_{sat}/(R_V T_L)$, assuming an ideal gas.

$$\log(P_{sat}) = -\frac{A}{T_L} - B \log(T_L) + C \quad [5.12]$$

Eqs [5.8]–[5.12] are solved as a function of T_L using an iterative solution method. The vapor is assumed to be iron in the form of a monatomic gas with a molecular weight of 56, and $\gamma_V = 1.67$. Quantities of particular interest are the recoil pressure, P_r, and rate of energy loss due to evaporation, q_e, and they are given by:

$$P_r = P_K + \rho_K u_K^2, \quad q_e = H_V \rho_K u_K \quad [5.13]$$

The effect of recoil pressure and surface tension on the keyhole formation can be clearly seen in Plate I (between pages 328 and 329).[11] As shown, at the initial stage ($t < 1.8$ ms), the laser energy is mainly used to heat up the base metal. As the temperature of the base metal increases, a small weld pool starts to appear under the laser beam. When the laser power reaches its peak value (after 1.0 ms), the laser-induced recoil pressure starts to push down the molten metal under the laser beam. Since it takes time to accelerate the molten metal from a static condition and the weld pool is thin at this time, the surface of the weld pool is nearly flat. Only part of the laser energy is absorbed by the metal via Fresnel absorption (to be discussed later) and the rest of the laser energy is reflected away. Heat conduction is the major heat transfer mode at this stage. As the laser light continues to radiate the metal, the temperature of the weld pool surface increases. The large temperature difference along the free surface of the weld pool due to the Gaussian distribution of laser power leads to a high temperature-dependent Marangoni shear stress. Since the gradient of surface tension coefficient on temperature change is negative when T is above 2150 K, the temperature-dependent Marangoni shear stress is positive along the weld pool surface, which drives the fluid to flow outwards. In the surface area where temperature is below 2150 K, surface tension coefficient gradient will change to be a positive value. Since the temperature gradient is still negative, the Marangoni shear stress forces the liquid metal to flow inwards. A crater is then formed at some point of the weld pool where the temperature is above 2150 K, as shown in Plate II (between pages 328 and 329). Moreover, higher surface temperature leads to higher recoil pressure, which pushes the liquid metal under laser radiation downwards quickly. Although the liquid–solid interface keeps moving downwards, the liquid region of the weld pool under laser radiation remains only as a thin layer because the strong recoil pressure squeezes the liquid metal to flow outwards. Under the hydrodynamic force and the Marangoni shear force, the squeezed liquid metal flows upwards along the keyhole wall which facilitates the formation of a crater. As the liquid surface moves down and the crater level increases, a keyhole appears in the weld pool at $t = 6.0$ ms.

5.2.4 Laser–plasma interaction and multiple reflections of laser beam in keyhole

In the laser welding process, the laser beam is reflected several times inside the keyhole by the keyhole wall. Also, each time the laser beam travels inside the keyhole, it interacts with the keyhole plasma. Multiple reflection of the laser beam by the keyhole wall and absorption mechanism of the laser energy by the keyhole plasma (Inverse Bremsstrahlung absorption) and by the keyhole wall (Fresnel absorption) are critical in determining the energy distribution and are discussed below.

Inverse Bremsstrahlung (IB) absorption

As discussed above, with the continuous heating from the laser beam, the metal vapor inside the keyhole can be heated much above its evaporation temperature, resulting in strong ionization. Thus, keyhole plasma is produced. The resulting keyhole plasma can absorb a lot of laser power by the effect of Inverse Bremsstrahlung (IB) absorption, thus affecting the energy transport to the liquid metal. Equations [5.14] and [5.15] define the IB absorption fraction of laser beam energy in plasma by considering multiple reflection effects:[15]

$$\alpha_{iB,1} = 1 - \exp\left(-\int_0^{s_0} k_{pl}\, ds\right) \qquad [5.14a]$$

$$\alpha_{iB,mr} = 1 - \exp\left(-\int_0^{s_m} k_{pl} ds\right) \quad [5.14b]$$

where, $\alpha_{iB,1}$ is the absorption fraction in plasma due to the original laser beam; $\alpha_{iB,mr}$ is the absorption fraction due to the reflected laser beam. $\int_0^{s_0} k_{pl} ds$ and $\int_0^{s_m} k_{pl} ds$ are, respectively, the optical thickness of the laser transportation path for the first incident and multiple reflections, and k_{pl} is the plasma absorption coefficient due to Inverse Bremsstrahlung absorption:[16]

$$k_{pl} = \frac{n_e n_i Z^2 e^6 2\pi}{6\sqrt{3m}\varepsilon_0^3 ch\omega^3 m_e^2}\left(\frac{m_e}{2\pi kT_e}\right)^{0.5}\left[1 - \exp\left(-\frac{\omega}{kT_e}\right)\right]\bar{g} \quad [5.15]$$

where Z is the average ionic charge in the plasma, ω is the angular frequency of the laser radiation, ε_0 is the dielectric constant, k is the Boltzmann's constant, n_e and n_i are particle densities of electrons and ions, h is Planck's constant, m_e is the electron mass, T_e is the excitation temperature, c is the speed of light, and $\bar{g}$ is the quantum mechanical Gaunt factor. For the weakly ionized plasma in the keyhole, the Saha equation[16] can be used to calculate the densities of plasma species:

$$\frac{n_e n_i}{n_0} = \frac{g_e g_i}{g_0}\frac{(2\pi m_e kT_e)^{1.5}}{h^3}\exp\left(-\frac{E_i}{kT_e}\right) \quad [5.16]$$

Fresnel absorption

As discussed before, part of the laser energy will be absorbed by keyhole plasma and part of the laser energy can reach the keyhole wall directly. So, the energy input (q_{laser}) for the keyhole wall consists of two parts: Fresnel absorption of the incident intensity directly from the laser beam ($I_{\alpha,Fr}$) and Fresnel absorption due to multiple reflections of the beam inside the keyhole ($I_{\alpha,mr}$):

$$q_{\text{laser}} = I_{\alpha,Fr} + I_{\alpha,mr} \quad [5.17]$$

$$I_{\alpha,Fr} = I_{\text{laser}} \cdot (1 - \alpha_{iB,1}) \cdot \alpha_{\text{Fr}}(\varphi_1) \quad [5.18]$$

$$I_{\alpha,mr} = \sum_{mr=1}^{n} I_{\text{laser}} \cdot (1 - \alpha_{iB,1}) \cdot (1 - \alpha_{Fr}) \cdot (1 - \alpha_{iB,mr}) \cdot \alpha_{Fr}(\varphi_{mr}) \quad [5.19]$$

where I_{laser} is the incoming laser intensity and n is the total number incident light from multiple reflections. We assume the laser beam has in the simplest case a Gaussian-like distribution:

$$I_{laser}(x, y, z) = I_0 \left(\frac{r_f}{r_{f0}} \right)^2 \exp\left(-\frac{2r^2}{rf^2} \right) \quad [5.20]$$

where r_f is the beam radius and r_{f0} is the beam radius at the focal position; I_0 is the peak intensity.α_{Fr} is the Fresnel absorption coefficient and is defined by the following formula:[17]

$$\alpha_{Fr}(\varphi) = 1 - \frac{1}{2}\left(\frac{1 + (1 - \varepsilon \cos \varphi)^2}{1 + (1 + \varepsilon \cos \varphi)^2} + \frac{\varepsilon^2 - 2\varepsilon \cos\varphi + 2\cos^2\varphi}{\varepsilon^2 + 2\varepsilon \cos\varphi + 2\cos^2\varphi} \right) \quad [5.21]$$

where φ is the angle of incident light with the normal of keyhole surface, and ε is a material-dependent coefficient.

Plate II (between pages 328 and 329) shows a typical keyhole plasma formation process and the effect of keyhole plasma on energy transport in laser welding.[11] As shown, once the keyhole comes into being, some metal vapor will be trapped inside. The incoming laser beam interacts with this metal vapor and part of laser energy is absorbed by it. Since the heat capacity of the vapor is rather low, the temperature of the vapor increases very quickly. When the temperature of the metal vapor exceeds around 8,000 K, laser-induced plasma will be formed by the ionization process. The physical properties of plasma are quite different from that of the prior vapor, which absorbs more laser energy through the IB absorption process. Once the keyhole plasma comes into being, it plays two roles in keyhole formation. The coefficient of the IB absorption increases with the increase of plasma temperature[11]. Hence, once plasma forms, its temperature will increase very quickly. This hot plasma separates the keyhole wall from the cold shielding gas, which reduces the heat loss from the keyhole wall to the surroundings. Moreover, the hot plasma heats up the surrounding keyhole wall via radiation. This facilitates the temperature increase on the keyhole wall when the depth of keyhole is not large. As shown in Plate II (between pages 328 and 329), the temperature on the keyhole wall under laser radiation keeps increasing. Therefore, the recoil pressure becomes stronger accordingly, which helps the depth of keyhole grow very quickly. However, when the plasma temperature is above 10,000 K, the IB absorption coefficient decreases with the increasing plasma temperature. Although the coefficient of the IB absorption decreases with the small increase of plasma temperature, the travel length of laser light increases with the increasing keyhole depth, the overall effect of the plasma absorption becomes more and more remarkable, and it will block a greater percentage of the laser energy from reaching the keyhole bottom wall. Meanwhile, the temperature of plasma only increases a little. The heat radiation from the plasma also increases a little correspondingly.

Thus, the blocking effect on laser light becomes more and more dominant, which eventually will make the plasma become a negative factor for keyhole depth increase at a certain time.

The formation of the keyhole also enhances the laser light absorption through a mechanism called multiple reflections. As shown in Plate I (between pages 328 and 329), the surface of the weld pool is nearly flat at the beginning. Once laser light irradiates the flat surface, only part of the energy is absorbed by the surface via the Fresnel absorption mechanism. A large amount of the laser energy is reflected back. Since the direction of the reflected light is contrary to its incident direction, this part of the laser light will never reach the weld pool surface again. However, once the keyhole is formed, the surface along the keyhole wall is not flat. Part of the reflected light will be incident on the keyhole wall again and again. During each incidence, part of the laser energy is absorbed via the Fresnel absorption mechanism. This brings more laser energy input to the keyhole wall, which leads to higher recoil pressure on the keyhole wall. Moreover, as the keyhole deepens, the possibilities and times of multiple reflections increase. When the reflected laser light travels in the plasma, part of its energy will also be absorbed by the plasma to increase the plasma temperature.

5.2.5 Radiative heat transfer in laser-induced plasma

When an intense laser beam interacts with metal vapor, a significant amount of the laser radiation is absorbed by the ionized particles. The radiation absorption and emission by the vapor plume may strongly couple with the plume hydrodynamics. This coupling, shown on the right-hand side of Eq. [5.7], will affect the plasma laser light absorption and radiation cooling terms. The radiation source term $\nabla \cdot (-\mathbf{q_r})$ is defined by

$$\nabla \cdot \mathbf{q_r} = k_a(4\pi I_b - \int_{4\pi} I d\Omega) \qquad [5.22]$$

where k_a, I_b and Ω denote the Planck mean absorption coefficient, blackbody emission intensity and solid angle, respectively. For the laser-induced plasma inside the keyhole, the scattering effect is not significant compared with the absorbing and emitting effect. So it will not lead to large errors to assume the plasma is an absorbing-emitting medium. The radiation transport equation (RTE) has to be solved for the total directional radiative intensity I:[18]

$$(\mathbf{s} \cdot \nabla)\, I(\mathbf{r}, \mathbf{s}) = k_a(I_b - I(\mathbf{r}, \mathbf{s})) \qquad [5.23]$$

where $\boldsymbol{s}$ and $\boldsymbol{r}$ denote a unit vector along the direction of the radiation intensity and the local position vector. The Planck mean absorption coefficient is defined in the following:[18]

$$k_a = \left(\frac{128}{27}k\right)^{0.5}\left(\frac{\pi}{m_e}\right)^{1.5}\frac{Z^2e^6\overline{g}}{h\sigma c^3}\frac{n_e n_i}{T_v^{3.5}} \quad [5.24]$$

where n_i and n_e represent the particle density of ions and electrons, T_v is the temperature of the plasma, Z stands for the charge of ions, e is the proton charge and m_e is the mass of electrons.

5.2.6 Tracking of free surfaces

The algorithm of volume-of-fluid (VOF) is used to track the moving free surface.[10] The fluid configuration is defined by a volume of fluid function, *F(x,y,z,t)*, which is used to track the location of the free surface. This function represents the volume of fluid per unit volume and satisfies the following conservation equation:

$$\frac{dF}{dt} = \frac{\partial F}{\partial t} + (\boldsymbol{V} \cdot \nabla)F = 0 \quad [5.25]$$

When averaged over the cells of a computing mesh, the average value of F in a cell is equal to the fractional volume of the cell occupied by the fluid. A unit value of F means a cell full of fluid and a zero value indicates a cell containing no fluid. Cells with F values between zero and one are partially filled with fluid and identified as surface cells.

5.2.7 Melt flow and weld pool dynamics

Figure 5.3 shows the melt flow and weld pool dynamics in a typical laser keyhole welding process.[11] As known, in laser keyhole welding, the weld pool is subject to strong recoil pressure and Maragoni shear stress, which introduces complex melt flow in the weld pool. As shown, at the beginning of a laser welding process, since the weld pool is small and also it takes time to accelerate the molten metal, the melt flow is not significant. Once the keyhole is formed, due to the strong push from the laser-induced recoil pressure, some strong velocities appear in the weld pool. On the bottom of the keyhole wall, the strong push from the recoil pressure leaves a very thin layer of liquid there. Although the hydrostatic pressure and surface tension at the bottom of the keyhole become larger and larger as the keyhole deepens, which makes it difficult for the drilling process, the velocity of liquid along the bottom surface does not decrease, as shown in Fig. 5.3. Meanwhile, the hydrostatic pressure at the bottom of the keyhole increases as the keyhole deepens, which makes it more and more difficult for the recoil pressure to pump the liquid upwards. Moreover, the temperature difference along the sidewall of the keyhole decreases as the keyhole depth increases.

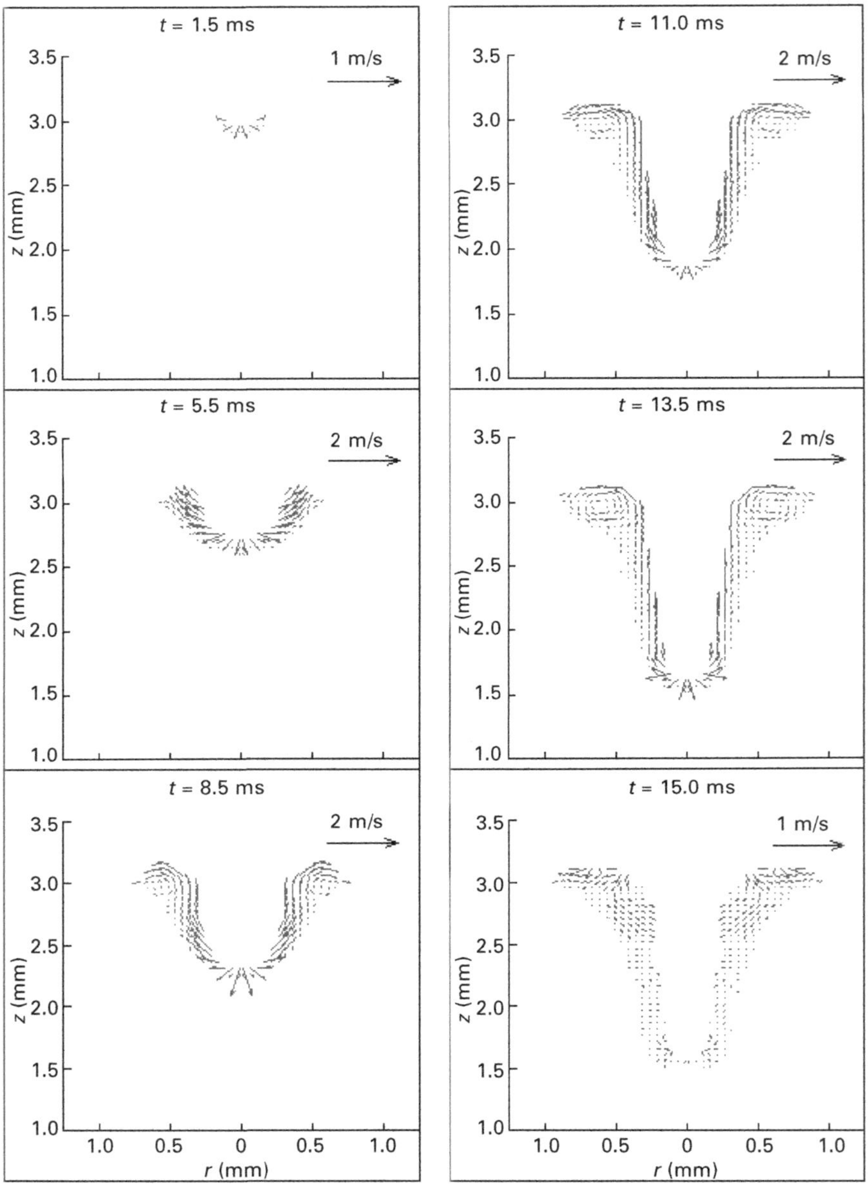

5.3 Weld pool dynamics and corresponding velocity distribution as shown in Plate I.

This decreases the Marangoni shear stress and makes it more difficult for the squeezed liquid metal to flow upwards. As shown in Fig. 5.3, both the downward velocities at the bottom keyhole and the upward velocities along the sidewall of the keyhole are smaller at $t = 13.5$ ms compared with those at t = 11.0 ms. At this time, since more laser energy is absorbed by the keyhole plasma, especially by that in the upper part of the keyhole, this hot plasma increases the heat input to the keyhole wall. Meanwhile, the multiple reflections also bring more and more energy to the sidewall of the keyhole as the keyhole deepens. As shown in Plate I (between pages 328 and 329), the increased heat input to the sidewall of keyhole causes the liquid–solid interface in the keyhole to move outwards, which enlarges the heat-affected zone in the weld substrate. Hence, the liquid region of the keyhole is getting bigger at t = 15.0 ms compared with that at t = 13.5 ms and the keyhole depth does not increase remarkably at this period. Also, the liquid metal on the bottom of the keyhole is getting thicker. This is because, with more and more laser energy being absorbed by the plasma, the recoil pressure on the bottom of the keyhole is decreasing at t = 15.0 ms, hence more liquid metal is accumulated there.

As also noticed in Fig. 5.3, there is a vortex in the upper part of the weld pool at t = 13.5 ms. The formation of the vortex is the result of the combined action of four driving forces: temperature-dependent Marangoni shear stress, surface tension, hydrodynamic force, and hydrostatic force. On the upper part of the vortex, the liquid is driven from the hot edge to the cold edge by the Marangoni shear stress force when the temperature of liquid is above 2,150 K. As the liquid metal moves to the cold edge, the temperature gradient becomes smaller and smaller which makes the Marangoni shear stress decrease. At the point where the liquid temperature is under 2,150 K, the Marangoni shear stress will change its direction. This change prevents the liquid from moving outwards. At the same time, the surface tension on the cold edge also blocks the liquid to flow outwards. Therefore, the velocity vector along the r-direction becomes smaller and smaller. Meanwhile, the hydrostatic force and surface tension cause the liquid to flow downwards, which makes the velocity along the z-direction become larger and larger. Finally, the liquid metal changes its flow direction to flow downwards at the upper of the cold edge. When this downward-flowing liquid metal meets the liquid–solid interface, it is blocked by this interface and flows along the interface. Meanwhile, the hydrodynamic force of the squeezed liquid from the lower part of the keyhole makes the liquid flow upwards. Thus, the liquid changes its direction to flow upwards at the lower part of the hot edge. As the liquid flows upwards, the hydrodynamic force is counteracted by the hydrostatic force and surface tension, which retards the upward tendency. Meanwhile, the Marangoni shear stress pushes the liquid to flow toward the cold edge. Thus, the liquid flow rotates clockwise on the shoulder of the keyhole and a vortex is formed as shown in the figure. This vortex enhances the heat transfer along the liquid–solid interface and enlarges the liquid region on the shoulder of the keyhole.

5.2.8 Keyhole collapse and porosity formation

Plate III (between pages 328 and 329) shows a typical keyhole collapse process in laser welding.[19] As shown, once the laser is shut off, the hot plasma radiation is the only heat input source for the keyhole wall. However, since there is no heat input to the plasma and the heat capacity of plasma is very small, the temperature of the plasma drops very quickly. Meanwhile, the heat conduction from the keyhole wall to the surrounding metal is very strong due to the high aspect ratio and high temperature gradient. Hence, the temperature of the keyhole wall drops very quickly, especially in the lower part of the keyhole since there is only a thin layer of the metal liquid there. Due to this quick temperature drop, the thin layer of liquid metal on the bottom of the keyhole completely solidifies. At this time, the temperature gradient along the sidewall surface of the keyhole is negative from the bottom to the top and the temperature of the liquid metal is above 2,150 K. Thus, the Marangoni shear stress is still positive, which helps the liquid to flow upwards. Since the temperature of the metal on the bottom surface drops much faster than that on the upper surface, the Marangoni shear stress there decreases quickly. Meanwhile, the hydrodynamic pressure of the squeezed liquid metal also decreases very quickly due to the removal of recoil pressure. The surface tension and hydrostatic pressure then make the liquid metal have a tendency to fill back the keyhole. However, the velocity of liquid metal needs time to change its direction, especially on the lower part of the keyhole wall due to their original high velocities inertia. On the keyhole shoulder, the Marangoni shear stress also decreases quickly with the decreasing temperature gradient. Hence, with the action of hydrostatic pressure and surface tension, the liquid metal on the upper part of the keyhole starts to flow inwards and downwards at $t = 15.5$ ms. Since there is a certain amount of liquid metal there and the flow friction along the liquid–solid interface is larger for the thin liquid layer, the backfilling velocities of the liquid metal on the upper part will be accelerated more easily than those on the lower part of keyhole. Thus, the keyhole is closed on the top first as shown in the figure. Meanwhile, the liquid metal continues to be accelerated by the hydrostatic force to flow downwards along the keyhole wall and finally the keyhole is refilled at $t = 30.0$ ms. The liquid region also shrinks as the liquid refills the keyhole, especially at the bottom because of fast conduction heat loss and low heat capacity. Finally, the liquid metal on the bottom completely solidifies. During the keyhole collapse process, the hot liquid metal cannot reach the far end of the top surface due to the disappearance of recoil pressure. Moreover, the velocity of the liquid metal flowing toward the edge decreases during the backfilling process. Thus, the liquid metal at the far end of the top surface cannot fill back and solidifies there, which results in a rough top surface on the final weld, as shown in Plate III (between pages 328 and 329) at $t = 30.0$ ms.

Compared with conventional welding processes, keyhole mode laser welding is more promising due to its large depth-to-width aspect ratio capability and high welding efficiency. However, in deep penetration laser welds, porosity is frequently observed.[20–23] Porosity deteriorates the strength of the welded part. Over the years, research has been conducted to study the porosity formation mechanisms in laser welding. Katayama *et al.*[24] reported that a lot of pores/voids were formed due to the entrainment of shielding gas or metal vapor and/or the dissolution of gas (for example, hydrogen) during the fast solidification process in high power YAG laser welding of A5083 alloy and 304 steel. Seto *et al.*[25] found that the rear side of the molten metal was depressed near the bottom when the keyhole front was inclined in the rear direction at the root. The reflected laser beam can superheat the molten metal near the bottom, resulting in a depression due to the high recoil pressure acting on the molten metal. They concluded that the shape of the keyhole front had a significant influence on porosity formation.

Katayama *et al.*[26] also found that the formation of porosity had a close correlation with the collapse of the keyhole in their studies of the fusion and solidification behavior of a molten puddle during laser spot welding of Type 316 stainless steel. Once the laser beam was terminated, the melt in the upper part of the keyhole flowed downward to fill the keyhole. Some gas seems to have been trapped in the lower part of the keyhole. At the same time, the upper part of the melt rapidly solidified, which prevented the melt from flowing to fill the keyhole. Thus porosity was formed in the weld. Zhou and Tsai.[19] found that porosity formation in laser welding is associated with the weld pool dynamics, keyhole collapse, and solidification processes, as shown in Plate III (between pages 328 and 329). More specifically, formation of porosity in pulsed laser welding was found to be due to two competing factors: one is the solidification rate of the molten metal and the other is the backfilling speed of the molten metal during the keyhole collapse process.[19] Porosity will be formed in the final weld if the solidification rate of the molten metal exceeds the backfilling speed of liquid metal during the keyhole collapse and solidification processes. As shown in Plate IV (between pages 328 and 329), porosity formation was found to be strongly related with the depth-to-width aspect ratio of the keyhole.[19] The larger the ratio, the easier porosity will be formed, and the larger the size of the voids.

5.3 New developments in laser welding

5.3.1 Hybrid laser-arc welding

In the last two decades, one of the major developments in laser welding has been the development of hybrid laser-arc welding technique and its applications in many areas in engineering.[27,28] As shown in Fig. 5.4, in

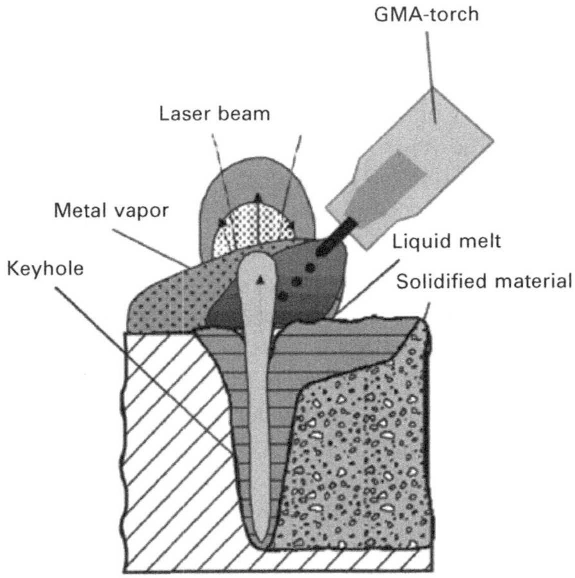

5.4 Schematic sketch of a hybrid laser-arc welding process.

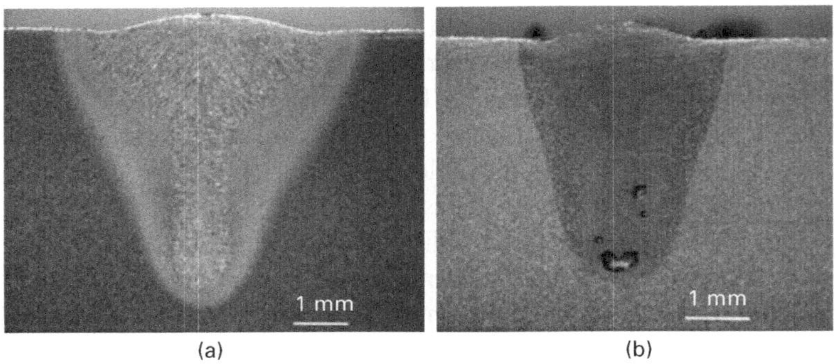

5.5 Comparison between (a) a laser welding and (b) a hybrid laser-arc weld in 250 grade mild steel.

hybrid laser-arc welding, the laser and the arc are integrated to provide primary and secondary heating sources for the joining process. Due to the synergistic action of the laser beam and welding arc, hybrid welding offers many advantages over laser welding and arc welding alone,[29,30] such as higher welding speed, deeper penetration,[31] better weld quality with reduced susceptibility to pores and cracks,[32–40] good gap bridging ability,[41–46] as well as process stability and efficiency, as shown in Fig. 5.5.[31]

The development of the hybrid laser-arc welding technique can be divided into three stages.[27] The concept of hybrid laser welding was first proposed by Steen *et al.*[47–49] in the late 1970s. In their studies, a CO_2 laser was combined with a tungsten inert gas (TIG) arc for welding and cutting applications. Their tests showed clear benefits of combining an arc and a laser beam, such as stabilized arc behavior under the influence of laser radiation; a dramatic increase in the speed of welding of thin sheets; and an increase in penetration depth compared with laser welding. Japanese researchers continued Steen's efforts and developed various methods and corresponding devices for laser-arc welding, cutting, and surface treatment. However, these efforts did not result in the introduction of this joining technique into engineering applications, particularly because laser welding itself was not yet an economic and viable joining technique at that time.[50]

In the second stage of the development of the hybrid laser welding technique, the observed influencing of the arc column behavior by laser radiation was used to improve the efficiency of arc welding processes, which leads to the laser enhanced arc welding technology.[27] A characteristic feature of this technology was that only a low-intensity laser beam was needed, i.e., the required laser power was small compared to the arc power. For TIG welding, Cui, Decker *et al.*,[51–53] demonstrated that a low-energy CO_2 laser beam with a power of merely 100 W could facilitate arc ignition; enhance arc stability; improve weld quality; and increase welding speed due to a reduced arc size and higher arc amperages. However, despite such reported improvements of the arc welding process by laser support, there were neither subsequent extensive investigations of this subject nor known industrial applications of the laser-enhanced arc welding technology.

The third stage of hybrid welding technology started in the early 1990s with the development of combined welding processes using a high-power laser beam as the primary and an additional electric arc as the secondary heating source.[54–62] At that time, although the continuous wave CO_2 laser welding process was already well established in industry, it had some known disadvantages, e.g., high requirements of edge preparation and clamping; high solidification rates leading to material-dependent pores and cracks; as well as the high investment and operating costs for the laser equipment. Additionally, some welding applications of highly practical interest could not be solved satisfactorily by the laser welding process alone, e.g., joining of tailored blanks in automotive engineering; welding of heavy plate under the conditions of the shipbuilding industry; as well as high-speed welding of crack-susceptible materials. In searching for suitable solutions, a hybrid welding technique was developed into a viable joining technique with significant industrial acceptance during the last decade.

According to the combination of various heating sources used, hybrid welding can be generally categorized as: (1) laser-gas tungsten arc (GTA) welding; (2) laser-gas metal arc (GMA) welding; and (3) laser-plasma welding.[25] Since laser welding offers deep penetration, primary heating sources commonly used in hybrid welding are CO_2, Nd:YAG, and fiber lasers.

The first two types of lasers are well established in practice and used for various hybrid welding process developments. While the fiber laser is still in development for industrial applications, it seems to be a future primary heating source for hybrid welding due to its high beam quality. The secondary heating sources used in hybrid welding are mainly electric arcs. Dedicated processes can be divided into GMA welding with consumable electrodes and GTA welding with non-consumable tungsten electrodes. In GMA welding, the arc is burning between a mechanically supplied wire electrode and the workpiece. The shielding gas used in GMA welding was found to have significant effects on arc shape and metal transfer.[63,64] Hence, GMA welding can be subdivided into metal inert-gas (MIG) and metal active-gas (MAG) welding according to the type of shielding gas used. In GTA welding, a chemically inert gas, such as argon or helium, is often used. A special form of this is the plasma arc welding (PAW), which produces a squeezed arc due to a special torch design and results in a more concentrated arc spot.

In hybrid welding, the laser and arc are arranged preferably in a way that they can compensate and benefit from each other during the welding process, which implies the creation of a common interaction zone with changed characteristics in comparison to laser welding and arc welding alone. In contrast to this is the arrangement in which the laser and arc are serving as two separate heating sources during the welding process. Several configurations have been proposed. In a parallel arrangement, there is a distance in either the vertical or horizontal direction along the path between both heating sources. In a serial arrangement, the primary and secondary heating sources are moved along the same welding path with a certain working distance, and the secondary heating source can either lead or follow the primary heating source.[27] The first one enables a preheating of the region to be welded. It can increase the efficiency of the laser welding process because the material to be welded is locally preheated and energy losses by heat conduction are reduced. In comparison, the second one often acts like a short-time post-heat treatment of the weld that can change the weld microstructure favorably. It is worth considering that there is an essential difference between parallel and serial process arrangements. In a serial arrangement, additional energy is dissipated within the weld seam region, whereas the parallel arrangement only reduces the heat flow across the weld seam. The option to move the working area temporally enables flexibility in influencing the cooling rates in order to avoid defects.

In recent years, hybrid laser-arc welding has been increasingly used to join aeronautical materials, the benefit mainly originates from the ability of this process to adjust filler metal additions, heat input, and post heat treatment processes. Although some initial successes have been obtained to join aeronautical materials, like stainless steels, titanium, magnesium, and aluminum alloys, in-depth research, rigorous characterization, and cost analysis are still needed. Hybrid welding of these alloys often involves combinations of different filler and base metals, which have to be determined for various hybrid welding conditions in order to obtain optimum weld properties. Detailed characterization of structure and properties for each alloy remains a major task and rigorous studies need to be carried out to understand the correlation between welding conditions and the resulting weldment structure and properties for these important engineering alloys. Hence, the applications of hybrid laser-arc welding technique can be widened in aerospace engineering.

Although hybrid laser-arc welding has been gaining increasing acceptance in recent years, good understanding of the underlying physics remains a challenge. For example, the interaction between the laser and the arc has been observed to enhance arc stability and push the arc towards the laser keyhole, resulting in a deeper penetration. However, the origin of this synergistic interaction between the arc and laser plasma is not well understood. Measuring the distributions of electron temperatures and densities in the plasma can provide a better understanding of the laser–arc interaction.[9] Porosity formation is believed to be strongly related to the keyhole collapse process. Hence, better understanding of keyhole stability and dynamics through experimental and theoretical studies would be beneficial. Hybrid welding is known to produce welds with desirable widths and depths, but the maximum gap tolerance and weld penetration for various welding conditions have not been quantified. In the future, advanced mathematical modeling of the heat transfer and fluid flow will enable accurate predictions of weld profile and cooling rates in the welding process, which is critical in understanding the evolution of weld microstructures and residual stress formation in welds. Thus, the hybrid welding process can be optimized to obtain quality welds with no cracking, no brittle phase and less thermal distortion. Better sensing and process control of the hybrid welding process would also be helpful in expanding its applications.

5.3.2 Multiple-beam laser welding

In laser keyhole welding, a number of defects, such as porosity, surfaces holes or pits, undercuts, humping and solidification cracking are frequently observed which deteriorates the quality of the welded parts.[20–23] Some researchers[65,66] believe these welding defects are related to the keyhole instabilities during the laser welding process. In order to overcome these defects to make high-quality and consistent laser welds, in recent years, a new welding technique combining two or more laser beam sources, called dual-/multiple-beam laser welding, has been proposed and is under development to become an enabling joining technique.[65–72] Studies have demonstrated that this new laser welding technique can provide benefits over conventional single-beam laser welding, such as increasing keyhole stability, slowing down cooling rate, and delaying the humping onset to a higher welding speed. It was reported that dual-beam laser welding can significantly improve weld quality. As shown in Fig. 5.6, two or more laser beams can be integrated in either traverse or longitudinal or mixed traverse and longitudinal arrangement in dual-/multiple-beam laser welding.

Although many experimental and numerical studies[73–89] have been conducted in recent years for single-beam laser beam welding processes, the investigations conducted for dual-/multiple-beam welding are limited. Most of the studies conducted to characterize the multi-spot laser welding process[65–72] are trial-and-error in nature. Xie[65,66] found that use of dual-beam laser processing can delay humping onset to higher speeds and slow down cooling rates. He also observed, in his experiments, for dual-beam laser welding of steel, surface quality had been improved with fewer surface defects such as undercut, surface roughness, spatter, and underfill. Weld hardness and centerline cracking susceptibility had also been reduced. For dual-beam laser welding of aluminum, quality improvements were in the form of smooth weld surfaces and fewer weld defects such as porosity, surface holes, and undercut. Through his experimental observations, Xie accredited the improved weld quality to better keyhole stability during the dual-beam laser welding process. Iwase *et al.*[67] have found a larger keyhole during dual beam Nd:YAG laser welding via real-time X-ray observations. An elongated keyhole was found to be the reason that enables the vapor to escape from the keyhole freely and thus results in a stable welding process. Coste *et al.*[68,69] found that deep penetration could be achieved up to 60 mm by combining two or three Nd:YAG lasers up to 12 kW. In the experiments conducted by Nakabayashi *et al.*,[70] a deep penetration up to 20 mm with a welding speed of 1 m/min could be achieved by combining a system of a 6 kW Nd:YAG laser and a 10 kW chemical oxygen-iodine laser. Gref *et al.*[71]

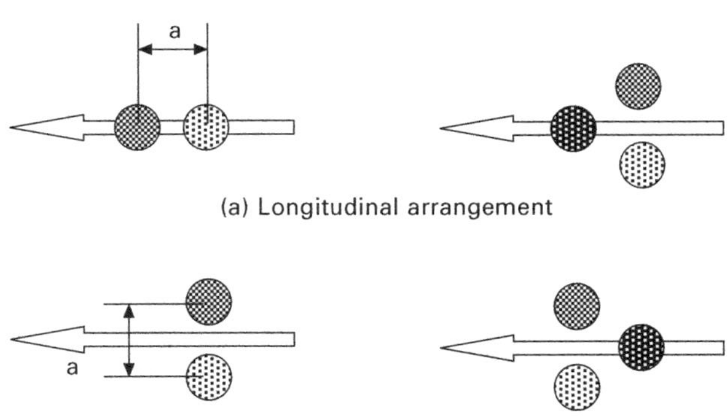

5.6 Typical beam arrangement in a dual-/multi-beam laser welding process.

experimentally studied the dual-beam laser welding process and concluded that the increased penetration depth is due to the improved energy coupling and process efficiency in the dual-beam welding process. They also pointed out that this efficiency is dependent on the configurations of two laser beam sources.

Although these experimental studies can provide some useful information on the dual-beam laser welding process, they cannot reveal the underlying physics of the process. For example, as shown in the experimental studies of Fabbro[90] and Coste *et al.*,[68,69] the shape of the weld pool in dual-beam welding is different from that observed for a single-beam welding. However, this difference cannot be well explained without the knowledge of weld pool dynamics during the welding process. Mathematical models have been proposed to understand the transport phenomena in this new welding process. Unfortunately, these studies are very limited. Chen and Kannatey-Asibu[91] studied the heat transfer during a conduction-mode dual beam laser welding process. Hou[92] and Capello *et al.*[93] simulated a welding process involving two laser beams as a system in which one beam served as a minor heat source and the other was a major heat source. The fluid flow was not considered in their models and the beam distance was much larger than the beam size. Hu and Tsai[94] simulated a stationary 3D dual-beam laser welding process; however, the keyhole plasma was not considered in the model.

Recently, Zhou *et al.*[73,74] extend their modeling work in a 3D moving single-beam laser welding process to investigate the heat transfer and fluid flow in the weld pool, the evolutions of two keyholes and their interaction in a 3D dual-beam laser welding process. As shown in Plate V (between pages 328 and 329), they found that the interaction of two keyholes caused the strong fluid flow in the direction perpendicular to the connection line of the two laser beam centers, leading to the oval-circle-oval shape change of the weld pool in the dual-beam welding process, as observed in some experiments.[90] Their study also clearly showed the meltdown and disappearance process of the liquid metal in the middle of the beam centers, which indicates that the process is greatly dependent on the distance between the two beam centers. Also, it was found that a large keyhole with a big weld pool in the rear of the keyhole was very helpful in improving the keyhole stability and delaying the cooling process and, hence, suppressing the humping.

5.3.3 Pulse control in laser welding for porosity prevention

Based on the studies of porosity formation mechanisms, in pulsed laser welding, controlling the laser pulse duration and power density has been proposed to delay the solidification process and increase the backfilling time laser welding, thus to prevent/eliminate porosity formation in laser welds.

Plate VI (between pages 328 and 329) shows the effect of pulse control on keyhole collapse and metal solidification in a medium depth-to-width ratio keyhole laser welding process.[19] As shown, when the main laser power is removed, the keyhole begins to collapse. However, with the pulse control, the laser is not completely shut down. Instead, the laser is still kept on for a while at a reduced power level. This reduced laser power helps to continue to heat the keyhole surface, which allows liquid metal at the bottom of the keyhole to remain as liquid for a while. As the liquid metal from the shoulder of the keyhole flows toward the center, it prevents part of the laser light from reaching the bottom surface of the keyhole, which leads to a temperature decrease on this part of the keyhole wall. However, the multiple reflections on the keyhole wall reflect some laser energy to the keyhole wall where laser light cannot radiate directly. This helps to delay the solidification process along the keyhole wall. When the keyhole is almost completely closed from the top, only a very small portion of the laser light can reach the bottom of the keyhole. However, the temperature on the bottom of the keyhole wall is still hot enough to allow the metal to stay liquid. At this time, most of the laser energy is used to heat the liquid metal on the top, which facilitates the liquid metal on the top to flow downward, as the downward momentum of the liquid metal is strong. Hence, within a short period of time, the downward flowing liquid metal almost completely fills up the keyhole and no pores will be found in the final welds. Although some gas is enclosed, the gas is surrounded by liquid metal, which can continue to flow into the gas region. During the backfilling process, the hydrostatic pressure continues to accelerate the melt flow, which facilitates the backfilling process. The downward flowing liquid metal makes the gas region smaller and smaller. In addition, since the gas is a mixture of shielding gas and metal vapor, some of it will be dissolved in the liquid metal during the compression process and eventually the gas region will disappear in the metal liquid, which is observed by Katayama and Matsunawa.[22]

The effectiveness of pulse control has also been investigated on a large depth-to-width ratio keyhole laser welding process. As shown in Plate VII (between pages 328 and 329) due to the additional heat input after the removal of the main laser power, the hot plasma tends to keep the liquid metal on the bottom of the keyhole and delay the solidification process of the liquid metal on the side of the keyhole. However, since the keyhole aspect ratio is larger in this case, more liquid metal is located on the shoulder, which leaves a larger crater on the top of the keyhole. Meanwhile, the reduced laser energy can keep the temperature of the liquid metal on the top high, which facilitates the melt flow there. Hence, after the removal of the main laser power, with the action of surface tension force and hydrostatic pressure, this large crater will drive the hot liquid metal on the top to move inward more quickly to close the keyhole and shorten the time for keyhole closure. When the liquid metal from the shoulder of the keyhole approaches the keyhole center, the amount of laser beam energy entering the keyhole becomes smaller and smaller. Although the multiple reflections can still bring some laser light to heat the keyhole wall where the laser light cannot radiate directly, the total area is relatively large compared with that in a medium aspect ratio keyhole, the heat input per unit area is reduced for a high aspect ratio keyhole. This facilitates the solidification process on the keyhole wall and bottom. As shown, since the keyhole collapsed and closed on the top very quickly, laser light cannot enter the keyhole after the close of the keyhole top and the liquid metal on the bottom of the keyhole quickly solidifies. Although, with the action of hydrostatic pressure, the liquid metal on the upper and middle part of the keyhole flows downward to fill the keyhole, due to the fast solidification rate, the bottom part of the downward flowing liquid metal completely solidifies before it can reach the keyhole bottom. After that, no liquid metal can reach the bottom of the keyhole and a pore/void is still found at the root of the keyhole. However, there is no pore or void found in the upper part of the keyhole.

5.3.4 Laser welding with electromagnetic force assistance

As discussed above, controlling the pulse shape of the laser irradiation was proposed to delay the solidification process, thus to reduce/eliminate the porosity formation in laser welding.[19] This method prolongs the solidification rate and has been demonstrated to be effective in preventing/eliminating porosity for a medium depth-to-width ratio keyhole laser welding process. However, it failed for a keyhole with large depth-to-width ratio. In recent years, a new method is proposed aiming to control the backfilling melt flow during the keyhole collapse process, thus to reduce/eliminate the porosity formation in laser welding.[95] This has been achieved by increasing the backfilling speed of the molten metal via the application of an electromagnetic force (Lorentz force) in laser welding. Electromagnetic force has been found to be able to affect melting and solidification processes.[96–98] It has been used in arc welding to change flow conditions.[99–102] In arc welding, when an electric current flows through the workpiece, it interacts with a self-introduced magnetic field, generating an electromagnetic force. This electromagnetic force can change the momentum of the melt flow and avoid weld slag or droplet detachment under the effect of gravitation when welding in a 'wall position'.[99,100] In gas metal arc welding, temperature and velocity evolutions, weld pool geometry and thermal cycles were also observed to be strongly influenced by electromagnetic force.[101,102] Recently, experimental studies have been conducted to investigate the usage of electromagnetic force in laser welding.[103,104] Electromagnetic force was found to be able to affect the melt flow and weld pool dynamics in laser welding to achieve good quality welds.

Plate VIII (between pages 328 and 329) shows the effectiveness of using electromagnetic forces on the prevention of porosity formation in laser welding.[95] As shown, an external current of 30 A and a constant magnetic flux $B_\theta = 300$ mT are applied as soon as the laser irradiation is shut off and they last for 5 ms. As shown, after the laser power is shut off, the recoil pressure is gone and the liquid metal near the top of the keyhole has a tendency to flow downward to fill the keyhole under the action of gravity and surface tension. As the electromagnetic force is downward and inward, it enhances the gravitational force and surface tension and shortens the time for the liquid metal to backfill the keyhole. As shown in Fig. 5.7, between $t = 15.0$ ms and 15.6 ms, the downward and inward velocities of the melt flow in the keyhole change more quickly than those without electromagnetic force. This is especially true for the liquid metal near the shoulder of the keyhole; the inward electromagnetic force 'sweeps' the molten metal surrounding the top of the keyhole into the keyhole. Thus, the keyhole is filled earlier with the applied electromagnetic force. As shown, the liquid–solid interface (i.e., solidification front) moves inward and upward from the keyhole in the metal after the laser irradiation is terminated. As the electromagnetic force is greater near the top of the keyhole and much less liquid metal exists near the bottom of the keyhole than that on the shoulder, the effect of electromagnetic force on the liquid metal near the bottom is not as remarkable as on that near the top of the keyhole. Hence, the application of electromagnetic force has little influence on the solidification rate of the liquid metal near the bottom of the keyhole. However, the downward velocity of liquid metal continues to be accelerated by electromagnetic force and the bottom of the keyhole at $t = 20.4$ ms is filled before complete solidification. Hence, there is no pore/void formed in the final weld.

As discussed above, since the formation of porosity in laser welding is strongly related to the depth-to-width ratio of the keyhole,[19] the larger the ratio, the more likely the porosity is easy to form. In order to prevent the porosity at the root of the keyhole in a large depth-to-width ratio laser keyhole welding process, the backfilling speed of the liquid metal must be accelerated faster than that in the medium ratio keyhole laser welding process. The effects of the strength of electromagnetic force and its duration on melt flow and porosity formation will be discussed in the following. In the following discussions, the electric current used in these study cases are all at 30 A, while the magnitude and duration of the external magnetic flux vary. Based on the relative magnitude of the magnetic flux, in the following discussions, small, medium, and large electromagnetic force will be used.

As shown in Plate IX (between pages 328 and 329), when a small electromagnetic force is applied, the liquid metal on the top flows inwards

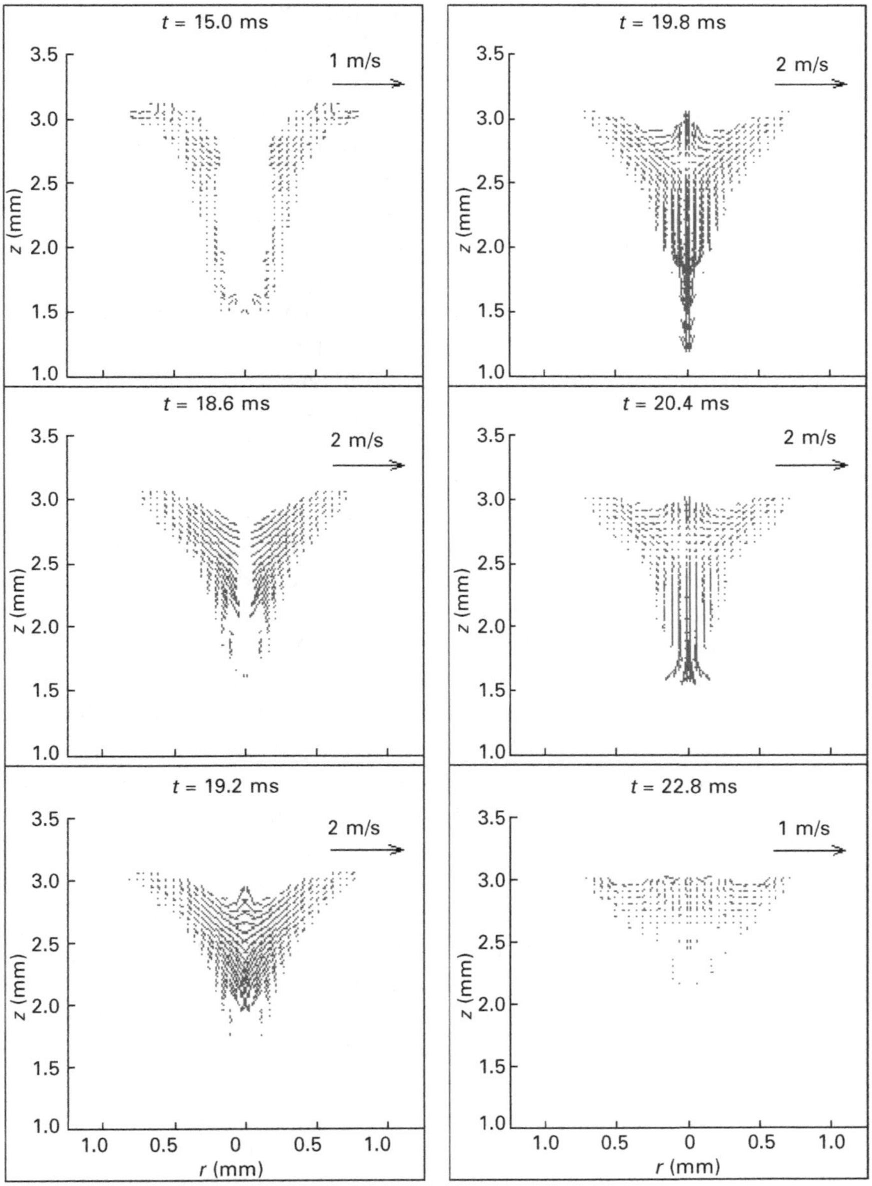

5.7 The corresponding velocity evolutions in weld pool as shown in Plate VIII.

and downwards to quickly refill the keyhole after the laser beam is shut off. In this case, the keyhole is deeper, which means it requires a longer time for the liquid metal from the top to reach the bottom of the keyhole. Since the magnitude of the electromagnetic force is relatively small, the liquid metal could not be accelerated fast enough to reach the bottom of the keyhole before its complete solidification. As shown, the bottom of the backfilling liquid metal solidifies at the depth of about 1.6 mm. After that, although there is still some liquid metal on the top, it was blocked by the solidified metal and could not flow farther to reach the bottom of the keyhole. Hence, a pore/void was found at the root of the keyhole, though the size of the pore/void is smaller than that in laser welding without using electromagnetic force.

A medium magnetic force is then applied with the magnetic flux B_θ increased to 300 mT and the duration of the magnetic field is 5.0 ms. Since the magnitude of the applied electromagnetic force is higher, the backfilling speed of the liquid metal from the top is faster. Due to the strong downward momentum induced by the downward electromagnetic force, the liquid metal on the top quickly flows downwards and reaches the bottom of the keyhole before complete solidification. Then, the keyhole was completely filled up by the liquid metal and the pore/void at the root of the keyhole was eliminated. A small extrusion was found in the center of the weld pool. This is due to the strong inward push of the liquid metal by the electromagnetic force. However, under the action of the hydrostatic force and surface tension, this squeezed liquid metal is able to flow back to the edge of the keyhole before complete solidification. Plate X (between pages 328 and 329) shows the effect of using a large electromagnetic force on porosity prevention. As shown, the magnetic flux B_θ is now increased to an even higher level at 400 mT. Since the magnitude of electromagnetic force is greater at t = 19.2 ms, the downward velocity of the liquid metal is greater. This strong electromagnetic force pushes the liquid metal from the top to flow downwards to fill up the keyhole much faster than those in the previous two cases. As shown, the liquid metal from the top almost completely fills up the keyhole in a very short time period. Hence, the pore/void at the bottom of the keyhole caused by rapid solidification after the shut-off of the laser power is eliminated. Although this large electromagnetic force can help prevent porosity formation at the root of the keyhole more easily, it can cause some other welding defects. As shown in Fig. 5.8, the inward velocity of the liquid metal becomes very strong due to the strong inward push from this large electromagnetic force that squeezes the liquid metal to move upwards. Even though the electromagnetic force is turned off at t = 21.0 ms, since the upward momentum of the liquid metal is so strong, it takes a relatively longer time for the hydrostatic force and surface tension to even the liquid metal. At this time, since the solidification proceeds quite quickly, the liquid metal flowing from the center toward the edge quickly solidifies. Undercuts at the edge of the final weld are clearly found as well as an extrusion in the center of the final weld. Hence, laser welding with electromagnetic force assistance must be optimized to achieve quality welds.

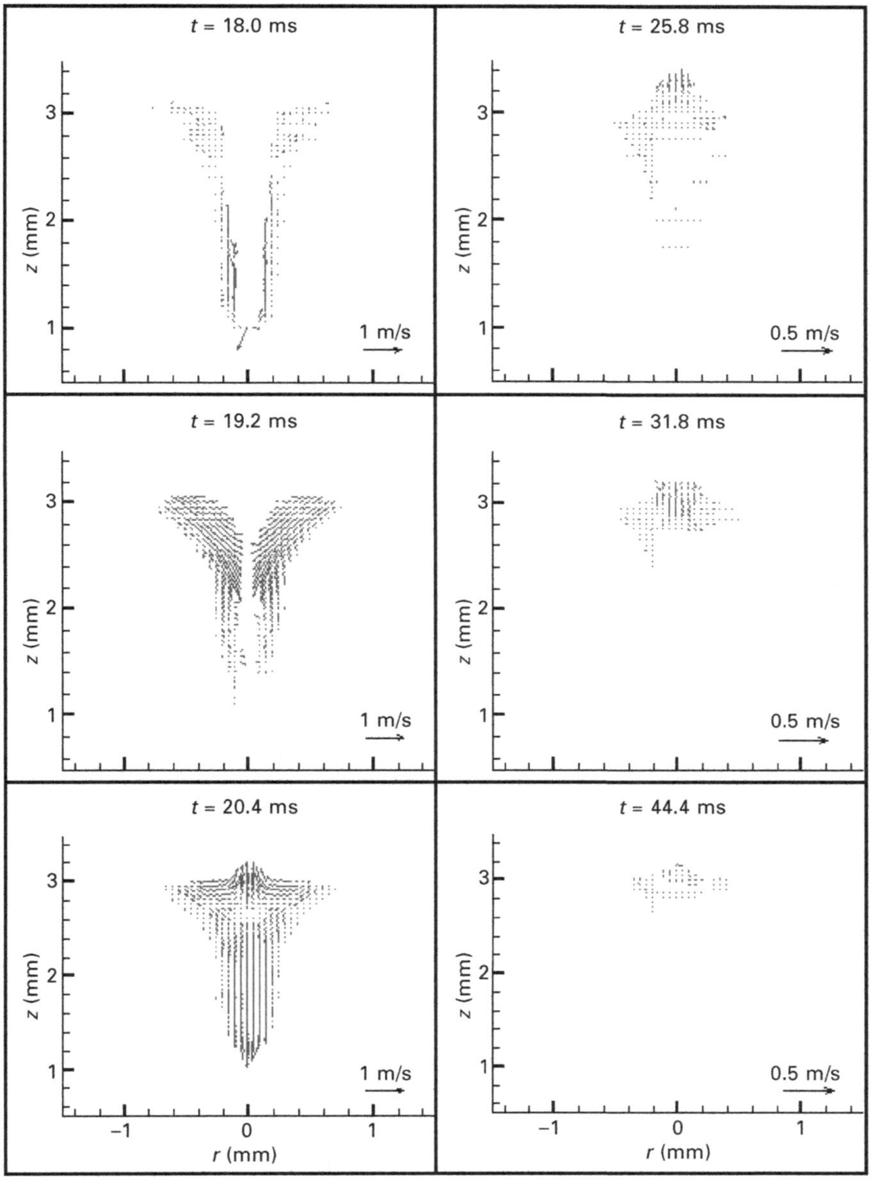

5.8 The corresponding velocity distributions in weld pool as shown in Plate X.

5.4 Future trends

Thanks to the improved understanding of the process, developments in laser source and process technology, the use of laser welding in industries has experienced tremendous growth over the last decade.[2] This will continuously be the trend to further develop and advance the laser welding technology in the future. Developments in laser source include the introduction of higher-power Nd:YAG lasers, utilizing diode pumped rods or disks, and fiber lasers, both providing the capability for fiber optic beam delivery. The use of higher power lasers with fiber optic beam delivery allows these systems to be easily integrated into manufacturing systems and provide dependable processing conditions. These systems will continue to gain attractiveness for a wide range of applications, with the potential for reduced operating costs and greater process capability. However, the application of these innovative sources will be related to advancements in process technology. A critical examination of the emerging laser process technology for welding must include the hybrid laser-arc welding processes, which are seeing considerable interest worldwide and are currently being evaluated for various applications within heavy industry and manufacturing.[2]

The benefit of the hybrid laser-arc process is the synergistic effect of improved processing rates and joint accommodation over either of the processes viewed separately. However, more studies need to be done to better understand the underlying physics in the process, such as keyhole plasma and laser plasma interaction. Other process technologies, such as the integration of dual or multiple laser beams in welding, are being developed to further improve the robustness of laser beam welding for industry. The most effective means of developing and implementing these advanced process techniques involves the use of improved process understanding refined through direct experimental observations and accurate theoretical simulations. Developments in mathematical modeling and numerical simulation techniques have provided and will continue to offer unprecedented opportunities to study the transient nature of the laser welding process. Accurate modeling of important phenomena in the welding process like laser–matter interaction, keyhole formation and dynamics, plasma formation, energy transport in plasma and metal; and their resultant effect on the weld quality will broaden the applications of laser welding in wider industries including shipbuilding, electronics, biology, automotive, oil and gas, and medicine. In the near future, laser welding will be continuously developed to become an enabling joining technology for assembling a broad range of materials like aluminum and magnesium alloys, plastics, biomaterials, dissimilar materials, etc., at both macro and micro scale.

5.5 References

1. Duley, W., *Laser Welding*, John Wiley & Sons, Chichester, 1999.
2. Martukanitz, R., 'A critical review of laser beam welding', *Proceedings of SPIE*, **5706**, 11–24, 2005.
3. Quintino, L., Costa, A., Miranda, R., Yapp, D., Kumar, V., and Kong, C.J.,

'Welding with high power fiber lasers – a preliminary study', *Materials & Design*, **28**, 1231–1237, 2007.

4. Zhou, J., and Tsai, H.L., 'Modeling of transport phenomena in hybrid laser-MIG keyhole welding', *International Journal of Heat and Mass Transfer*, **51**, 4353–4366, 2008.
5. Page, C.J., Devermann, T., Biffin, J., and Blundell, N., 'Plasma augmented laser welding and its applications ', *Science and Technology of Welding & Joining*, **7**(1), 1–10, 2002.
6. Joo, S.M., Kim, Y.P., Ro, C.S., Bang, H.S., and Park, J.U., 'Basic experiments by hybrid welding to steels for shipbuilding' *Adv. Nondestr. Eval.*, **270–273**, 2383–2388, 2004.
7. Graf, T., and Staufer, H., 'Laser-hybrid welding drives VW improvements', *Welding Journal*, **82**, 42–48, 2003.
8. Ono, M., Shinbo, Y., Yoshitake, A., and Ohmura, M., 'Welding properties of thin steel sheets by laser-arc hybrid welding: laser-focused arc welding', *Proc. 1st Int. Symp. on 'High-power laser macroprocessing'*, Osaka, Japan, 369–374, 2002.
9. Song, G., Liu, L.M., Chi, M.S., and Wang, J.F., 'Investigations of laser-TIG hybrid welding of magnesium alloys', *Materials Science Forum*, **488–489**, 371–375, 2005.
10. Chiang, K.C., and Tsai, H.L., 'Shrinkage-induced fluid flow and domain change in two-dimensional alloy solidification', *Int. J. Heat and Mass Transfer*, **35**, 1763–1769, 1992.
11. Zhou, J., Tsai, H.L., and Wang, P.C., 'Transport phenomena and keyhole dynamics during pulsed laser welding', *ASME Journal of Heat Transfer*, **128**(7), 680–690, 2006.
12. Anisimov, S.I., 'Vaporization of metal absorbing laser radiation', *Soviet Physics*, **JETP 27** (1), 182–183, 1968.
13. Knight, C.J., 'Theoretical modeling of rapid surface vaporization with back pressure', *AIAA J.*, **17**(5), 519–523, 1979.
14. Kanouff, M.P., Kassinos, A., Noble, D.R., and Schunk, P.R., 'Laser spot weld modeling using an ALE finite element method', Sandia National Laboratories Report.
15. Kaplan, A., 'A model of deep penetration laser welding based in calculation of the keyhole profile', *J. Phys. D: Appl. Phys.*, **27**, 1805–1814, 1994.
16. Duley, W., *Laser Welding*, John Wiley & Sons, Chichester 1999.
17. Ducharme, R., Williams, K., Kapadia, P., Dowden, J., Steen, B., and Glowacki, M., 'The laser welding of thin metal sheets: an integrated keyhole and weld pool model with supporting experiments', *J. Phys. D: Appl. Phys.*, **27**, 1619–1627, 1994.
18. Ho, J.R., Grigoropoulos, C.P. and Humphrey, J.A.C., 'Gas dynamics and radiation heat transfer in the vapor plume produced by pulsed laser irradiation of aluminum', *J. Appl. Phys.*, **79**, 7205–7215, 1996.
19. Zhou, J., and Tsai, H.L., 'Porosity formation and prevention in pulsed laser welding', *ASME Journal of Heat Transfer*, **129**, 1014–1024, 2007.
20. Ishide, T., Tsubota, S., Nayama, M., Shimokusu, Y., Nagashima, T., and Okimura, K., '10 kW class YAG laser application for heavy components', *SPIE High-Power Lasers in Manufacturing*, Osaka, **3888**, 543–550, 1999.
21. Katayama, S., Seto, N., Kim, J., and Matsunawa, A., 'Formation mechanism and reduction method of porosity in laser Welding of stainless steel', *Proceedings of ICALEO*, Sec. G, pp. 83–92, 1997.

22. Katayama, S., and Matsunawa, A., 'Formation mechanism and prevention of defects in laser welding of aluminum alloys', *Proceedings of CISFFEL 6*, 215–222, 1998.
23. Katayama, S., Seto, N., Kim, J., and Matsunawa, A., 'Formation mechanism and suppression procedure of porosity in high power laser welding of aluminum alloys', *Proceedings of ICALEO*, Sec. C, pp. 24–33, 1998.
24. Katayama, S., Seto, N., Mizutani, M., and Matsunawa, A., 'Formation mechanism of porosity in high power YAG laser welding', *Proceedings of ICALEO*, Sec. C, 16–25, 2000.
25. Seto, N., Katayama, S., and Matsunawa, A., 'A high-speed simultaneous observation of plasma and keyhole behavior during high power CO_2 laser welding', *Proceedings of ICALEO*, Sec. E, pp. 17–19, 1999.
26. Katayama, S., Kohsaka, S., Mizutani, M., Nishizawa, K., and Matsunawa, A., 'Pulse shape optimization for defect prevention in pulsed laser welding of stainless steels', *Proceedings of ICALEO*, pp. 487–497, 1993.
27. Tusek, J., and Suban, M., 'Hybrid welding with arc and laser beam', *Science and Technology of Welding and Joining*, **4**(5), 308–311, 1999.
28. Mahrle, A., and Beyer, E., 'Hybrid laser beam welding – classification, characteristics, and applications', *Journal of Laser Applications*, **18**(3), 169–180, 2006.
29. Casalino, G., 'Statistical analysis of MIG-laser CO_2 hybrid welding of Al–Mg alloy', *J. Mater. Process. Technol.*, **191**, 106–110, 2007.
30. Hu, B., and Richardson, I.M., 'Microstructure and mechanical properties of AA7075(T6) hybrid laser/GMA welds', *Material Science Engineering A*, **459**, 94–100, 2007.
31. Feng, Z.L., *Processes and Mechanisms of Welding Residual Stress and Distortion*, Woodhead Publishing, Cambridge, 2005.
32. Metzbower, E.A., Denney, P.E., Moon, D.W., Feng, C.R., and Lambrakos, S.G., 'Thermal analysis and microharness mapping in hybrid laser welds in a structural steel', *Materials Science Forum*, **426–432**, 4147–4152, 2003.
33. Ribic, B., Palmer, T.A., and Debroy, T., 'Problems and issues in laser-arc hybrid welding', *International Materials Reviews*, **54**(4), 223–244, 2009.
34. Song, G., Liu, L.M., and Wang, P.C., 'Overlap welding of magnesium AZ318 sheets using laser-arc hybrid process', *Material Science and Engineering A*, **429**, 312–319, 2006.
35. Nonn, A., Dahl, W., and Bleck, W., 'Numerical modelling of damage behaviour of laser-hybrid welds', *Engineering Fracture Mechanics*, **75**, 3251–3263, 2008.
36. Pinto, H., Pyzalla, A., Hackl, H., and Bruckner, J., 'A comparative study of microstructure and residual stresses of CMT-, MIG- and laser-hybrid welds', *Materials Science Forum*, **524–525**, 627–632, 2006.
37. Zhang, J.X., Xue, Y., and Gong, S.L., 'Residual welding stresses in laser beam and tungsten inert gas weldments of titanium alloy', *Science and Technology of Welding & Joining*, **10**, 643–646, 2005.
38. Liu, L.M., Song, G., and Wang, J.F., 'Investigations of laser-TIG hybrid welding of magnesium alloys', *Materials Science Forum*, **488–489**, 361–364, 2005.
39. Liu, L.M., Song, G., Liang, G.L., and Wang, J.F., 'Pore formation during hybrid laser tungsten inert gas arc welding of magnesium alloy AZ31B – mechanism and remedy', *Material Science and Engineering A*, **390**, 76–80, 2005.
40. Katayama, S., Uchiumi, S., Mizutani, M., Wang, J., and Fujii, K., 'Penetration and porosity prevention mechanism in YAG laser-MIG hybrid welding', *Welding International*, **21**, 25–31, 2007.

41. Fellman, A., and Kujanpaa, V., 'The effect of shielding gas composition on welding performance and weld properties in hybrid CO_2 laser-gas metal arc welding of carbon manganese steel', *Journal of Laser Applications*, **18**, 12–20, 2006.
42. Kim, T., Suga, Y., and Koike, T., 'Welding of thin steel plates by hybrid welding process combined TIG arc with YAG laser', *JSME International Journal Series A*, **46A**, 202–207, 2003.
43. Kim, Y.P., Alam, N., and Bang, H.S., 'Observation of hybrid (cw Nd:YAG laser + MIG) welding phenomenon in AA5083 butt joints with different gap conditions', *Science and Technology of Welding & Joining*, **11**(3), 295–307, 2006.
44. Murakami, K., Mitooka, Y., Hino, M., Iogawa, H., Ono, H., and Katayama, S., 'YAG laser-TIG hybrid welding of 800 MPa-class high tensile strength steel', *J. Japn. Inst. Met.*, **70**, 134–137, 2006.
45. Qin, G.L., Lei, Z., and Lin, S.Y., 'Effects of Nd:YAG laser + pulsed MAG arc hybrid welding parameters on its weld shape', *Science and Technology of Welding & Joining*, **12**(2), 79–86, 2007.
46. Swanson, P.T., Page, C.J., Read, E., and Wu, H.Z., 'Plasma augmented laser welding of 6 mm steel plate', *Science and Technology of Welding & Joining*, **12**(2), 153–160, 2007.
47. Steen, W.M., 'Arc-augmented laser processing of materials', *J. Appl. Phys.*, **51**, 5636–5641, 1980.
48. Eboo, M., Steen, W.M., and Clarke, J., 'Arc-augmented laser welding', *Advances in Welding Processes*, Proc. 4th Int. Conf., Harrogate, UK, 257–265, 1978.
49. Steen, W.M., and Eboo, M., 'Arc augmented laser welding', *Metal Construction*, **11**, 332–335, 1979.
50. Bagger, C., and Olson, F.O., 'Review of laser hybrid welding', *Journal of Laser Applications*, **17**(1), 2–14, 2005.
51. Cui, H., Decker, I., and Ruge, J., 'Wechselwirkungen zwischen WIGSchweißlichtbogen und fokussiertem Laserstrahl', *Proc. of the Conference Laser'89*, Springer, Berlin, 577–581, 1989.
52. Finke, B.R., Stern, F., and Decker, I., 'Auswirkungen eines unterstützenden Laserstrahls auf den WIG-Schweißprozess', *DVS-Ber.*, **135**, 149–152, 1991.
53. Cui, H., Decker, I., Pursch, H., Ruge, J., Wendelstorf, J., and Wohlfahrt, H., 'Laserinduziertes Fokussieren des WIG-Lichtbogens', *DVS-Ber.*, **146**, 139–143, 1992.
54. Dilthey, U., Lüder, F., and Wieschemann, A., 'Expanded capabilities in the welding of aluminium alloys with the laser-MIG hybrid process', *Aluminium*, **75**, 64–75, 1999.
55. Graf, T., and Staufer, H., 'Laser hybrid process at Volkswagen', *International Institute of Welding*, IIW-Doc. XII-1730-02, 1999.
56. Jokinen, T., Vihervä, T., Riikonen, H., and Kujanpaä, V., 'Welding of ship structural steel A36 using a Nd:YAG laser and gas metal arc welding', *Journal of Laser Applications*, **12**, 185–188, 2000.
57. Höfemann, M., Fersini, M., Szinyur, J., Haferkamp, H., and Cordini, P., 'Laser-GMA-hybrid welding of zinc-coated steel for hydroforming applications', *Proc. 1st Int. WLT-Conf. on Lasers in Manufacturing*, Munich, Germany, 466–472, 2001.
58. Dilthey, U., and Keller, H., 'Prospects in laser GMA hybrid welding of steel', *Proc. 1st Int. WLT-Conf. on Lasers in Manufacturing*, Munich, Germany, 453–465, 2001.
59. Waltz, C., Seefeld, T., and Sepold, G., 'Laser-GMA welding and its influence

on bead geometry and process stability', *Proc. 1st Int. WLT-Conf. on Lasers in Manufacturing*, Munich, Germany, 444–452, 2001.
60. Shibata, K., Sakamoto, H., and Iwase, T., 'Laser-MIG hybrid welding of aluminium alloys', *Proc. 1st Int. WLT-Conf. on Lasers in Manufacturing*, Munich, Germany, 436–443, 2001.
61. Jokinen, T., Jernström, P., Karhu, M., Vanttaja, I., and Kujanpää, V., 'Optimisation of parameters in hybrid welding of aluminium alloy', *Proc. 1st Int. Symp. on High Power Laser Macroprocessing*, Proc. SPIE 4831, 307–312, 2003.
62. Ishide, T., Tsubota, S., and Watanabe, M., 'Latest MIG, TIG Arc-YAG laser hybrid welding systems for various welding products', *Proc. 1st Int. Symp. on High Power Laser Macroprocessing*, Proc. SPIE 4831, 347–352, 2003.
63. Hyatt, C.V., Magee, K.H., Porter, J.F., Merchant, V.E., and Matthews, J.R., 'Laser-assisted gas metal arc welding of 25-mm-thick HY-80 plate', *Welding Journal Symposium*, Miami, FL, 163s–172s, 2001.
64. Fellman, A., Jernstrøm, P., and Kujanpä, V., 'The effect of shielding gas composition in hybrid welding of carbon steel', *Proceedings 9th NOLAMP, Conference on Laser Materials Processing in the Nordic Countries*, Trondheim, Norway, 103–112, 2003.
65. Xie, J., 'Weld morphology and thermal modeling in dual-beam laser welding', *Welding Journal*, pp. 283s–290s, 2002.
66. Xie, J., 'Dual beam laser welding', *Welding Journal*, 223s–230s, 2002.
67. Iwase, T., Shibata, K., Sakamoto, H., Dausinger, F., Hohenberger, B., Muller, M., Matsunawa, A., and Seto, N., 'Real time X-ray observation of dual focus beam welding of aluminum alloys', *ICALEO* 2000, Section C, pp. 26–34, 2000.
68. Coste, F., Janin, F., Jones, L., and Fabbro, V., 'Laser welding using Nd:Yag lasers up to 12 kW application to high thickness welding', *ICALEO 2002*, 131–137.
69. Coste, F., Janin, F., Hamadou, M., and Fabbro, R., 'Deep penetration laser welding with Nd:Yag lasers combination up to 11 kW laser power', *LAMP 2002*, 46–53.
70. Nakabayashi, T., Wani, F., Okado, H., Hayakawa, A., Suzuki, S., and Yasuda, K., 'Laser welding characteristics with high power combing beam', *ICALEO* 2001, Section A, 22–26, 2001.
71. Gref, W., Hohenberger, N., Dausinger, F., and Hugel, H., 'Energy coupling and process efficiency in double-focus welding with Nd:YAG', *ICALEO 2002*, 202–206.
72. Grupp, M., Seefeld, T., and Sepold, G., 'Laser beam welding of aluminum alloys with diode pumped Nd:YAG lasers', *ICALEO 2001*, 115–122.
73. Zhou, J., 'Modeling of three-dimensional laser welding process', PhD dissertation, University of Missouri-Rolla, 2003.
74. Zhou, J., Zhang, W.H., Tsai, H.L., Marin, S.P., Wang, P.C., and Menassa, R., 'Modeling the transport phenomena during hybrid laser-MIG welding process', *IMECE'03*, Washington, DC, November 16–21, 2003.
75. Semak, V.V., Hopkins, J.A., McCay, M.H., and McCay, T.D., 'A concept for a hydrodynamic model of keyhole formation and support during laser welding', *ICALEO 1994*, 641–650, 1994.
76. Semak, V.V., Hopkins, J.A., McCay, M.H., and McCay, T.D., 'Dynamics of penetration depth during laser welding', *ICALEO 1994*, 17–20, 1994.
77. Klemens, P.G., 'Heat balance and flow conditions for electron beam and laser welding', *J. Appl. Phys.* **47**, 2165–2174, 1976.
78. Allmen, M., and Blatter, A., *Laser-Beam Interaction with Material*, 2nd edn, Springer-Verlag, Berlin, Heidelberg, 1995.

79. Solana, P., Kapadia, P., and Dowden, J., 'Surface depression and ablation for a weld pool in material processing: a mathematical model', *ICALEO 1998*, Sec. F, 142–147, 1998.
80. Clucas, A., Ducharme, R., Kapadia, P., Dowden, J., and Steen, W., 'A mathematical model of laser keyhole welding using a pressure and energy balance at the keyhole walls', *ICALEO 1998*, Sec. F, 123–131, 1998.
81. Matsunawa, A., Kim, J., Seto, N., Mizutani, M., and Katayama, S., 'Dynamic of keyhole and molten pool in laser welding', *J. Laser Appl.* **10**, 247–254, 1998.
82. Chen, M.M., and Bos, J.A., 'Melt flow in deep penetration welding', *ICALEO 1998*, Sec. F, 187–196, 1998.
83. Ducharme, R., Williams, K., Kapadia, P., Dowden, J., Steen, B., and Glowacki, M., 'The laser welding of thin metal sheets: an integrated keyhole and weld pool model with supporting experiments', *J. Phys. D: Appl. Phys.* **27**, 1619–1627, 1994.
84. Sudnik, R., Rada, D., Breitschwerdt, S., and Erofeew, W., 'Numerical simulation of weld pool geometry in laser beam welding', *J. Phys. D: Appl. Phys.* **33**, 662–671, 2000.
85. Kapadia, P., Dowden, J., and Ducharme, R., 'A mathematical model of ablation in the keyhole and droplet formation in the plume in deep penetration laser welding', *ICALEO 1996*, Sec. B, 106–114, 1996.
86. Farson, D.F., and Kim, K.R., 'Simulation of laser evaporation and plume', *ICALEO 1998*, Sec. F, 197–206, 1998.
87. Miyamoto, I., Ohmura, E., and Maede, T., 'Dynamic behavior of plume and keyhole in CO_2 laser welding', *ICALEO 1997*, Sec. G, 210–218, 1997.
88. Solana, P., and Negro, G., 'A study of the effect of multiple reflections on the shape of the keyhole in the laser processing of material', *J. Phys. D: Appl. Phys.* **30**, 3216–3222, 1997.
89. Metzbower, E.A., 'Absorption in the keyhole', *ICALEO 1997*, Sec. G, 16–25, 1997.
90. Fabbro, R., 'Basic processes in deep penetration laser welding', *ICALEO 2002*, 182–187.
91. Chen, T.C., and Kannatey-Asibu, Jr., E., 'Convection pattern and weld pool shape during conduction-mode dual beam laser welding', *Proceedings of 24th North American Manufacturing Research Conference*, Ann Arbor, MI, 259–264, 1994.
92. Hou, C.A., 'Finite element simulation of dual-beam laser welding using ANSYS', *IASTE*, Pittsburgh, PA, 430–433, 1994.
93. Capello, E., Chiarello, P., Piccione, E., and Previtali, B., 'Analysis of high power CO_2 dual beam laser welding. Advanced manufacturing systems and technology', Courses and Lectures – International Center for Mechanical Sciences, No. 437, 2002.
94. Hu, J., and Tsai, H.L., 'Fluid flow and weld pool dynamics in dual-beam laser keyhole welding', *Proceeding of IMECE'03*, 353–360.
95. Zhou, J., and Tsai, H.L., 'Effects of electromagnetic force on melt flow and porosity prevention in pulsed laser keyhole welding', *International Journal of Heat and Mass Transfer* **50**, 2217–2235, 2007.
96. Sampath, R., and Zabaras, N., 'A joint variable method for the thermal design of eutectic directional solidification processes in an open-boat configuration', *Numer. Heat Transfer Part A: Appl.* **32**(6), 655–683, 2001.
97. Goncalves, E., Faghri, M., Asako, Y., and Charmchi, M., 'Numerical solution

of melting processes for unfixed phase-change material in the presence of electromagnetically simulated low gravity', *Numer. Heat Transfer Part A: Appl.* **46**(4), 343–365, 2004.

98. Goncalves, E., Faghri, M., Asako, Y., and Charmchi, M., 'Numerical solution of melting in side-heated rectangular enclosure under electromagnetically simulated low gravity', *Numer. Heat Transfer Part A: Appl.* **47**(4), 315–332, 2005.
99. Manabe, Y., Wada, H., Kondou, H., Hiromoto, Y., and Kobayashi, Y., 'Development of new GTAW process by magnetic control on molten pool', *International Institute of Welding*, IIW–XII, 1484–1497, 1997.
100. Shelenkov, G.M., 'Special features of the formation of joints in the arc welding of titanium with electromagnetic stirring', *Weld. Production* **3**, 24–25, 1997.
101. Kou, S., and Wang, Y.H., 'Computer simulation of convection in moving arc weld pools', *Metall. Trans. 17A,* 2271–2277, 1986.
102. Kim, W.H., Fang, H.G., and Na, S.J., 'Effect of various driving forces on heat and mass transfer in arc welding', *Numer. Heat Transfer Part A: Appl.* **32**(6), 633–652, 1997.
103. Kern, M., Berger, P., and Hugel, H., 'Magneto-fluid dynamic control of seam quality in CO_2 laser beam welding', *Weld. J.* **79**(3), 72s–78s, 2000.
104. Lindenau, D., Ambrosy, G., Berger, P., and Hugel, H., 'Effects of magnetically supported laser beam welding of aluminum alloys', *ICALEO,* Sec. A, 308–318, 2001.

6
Conduction laser welding

L. QUINTINO, Instituto Superior Técnico, Portugal and
E. ASSUNÇÃO, European Federation for Welding, Joining and Cutting, Portugal

DOI: 10.1533/9780857098771.1.139

Abstract: Laser welding has two different operational regimes: conduction and keyhole welding. The key difference between these two modes is the power density applied to the welding area. Conduction takes place when the intensity is not sufficient to cause boiling. In keyhole mode, the intensity used is high enough to cause vaporisation and create a keyhole in the melt pool. The current definition of conduction laser welding mode based on power density is discussed. Some studies focused on conduction laser welding are also discussed, emphasising the effect of beam diameter in this mode. Finally, an overview of some of the most relevant applications of conduction laser welding in industry is presented. This chapter not only emphasises how good and diverse the applications of this welding mode are, when compared to keyhole laser welding, but also helps in understanding the flexibility of this welding process.

Key words: conductive laser welding, power density, applications.

6.1 Introduction: comparison between keyhole and conduction laser welding

There are two main operational regimes in laser welding: keyhole and conduction laser mode.[1,2] The main difference between these two modes is the power density used. In this chapter the focus will be on the process differences between these two welding modes, such as productivity, stability, spatter and others. The characteristics of keyhole laser welding led industry to focus on this mode. The main characteristics of keyhole laser welding are deep penetration, the small heat affected zone and the high productivity of this process.[1] However, keyhole welding also possesses several problems that may lead to high levels of porosity, instabilities, spatters (Fig. 6.1) and other weld defects.[1,3–5] For example, keyhole laser welding of aluminium shows numerous problems. The high thermal conductivity, low boiling point elements and low viscosity are just some of the characteristics of aluminium alloys that lead to defects such as loss of alloy elements, degradation of mechanical properties, excessive porosity and blowholes, and to an unstable process when keyhole laser welding.[4,6–9] Ion showed in a review paper how different aluminium series behave when keyhole laser welded.[8] For example, when

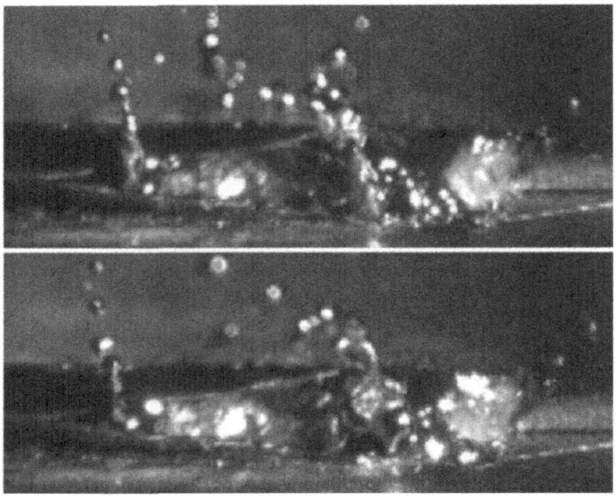

6.1 High-speed images of a blowout event that originates spatter in keyhole laser welding.[13]

laser welding a 5000 series there is significant vaporisation of magnesium. This vaporisation creates fluctuations in the keyhole mechanism, which affects the properties of the weld. This results in coarse porosity and in an irregular weld beam. Also, the magnesium loss to the vaporisation reduces solid solution hardening and degrades the tensile strength and ductility in comparison with the base material.[8] In terms of the laser system, keyhole laser welding requires a relatively good beam quality.[9]

On the other hand, conduction laser welding shows several advantages. This welding process can be a viable alternative to keyhole laser and to arc welding. It is a stable process which allows better control of the heat delivered to the workpiece. The use of large beams in this mode reduces fit-up problems and the laser system used does not require a high beam quality.[9] One of the disadvantages attributed to conduction mode welding was that the penetration depth of the welds was small, around 2 mm.[10] Recent work has shown that it is possible to obtain high penetration depth (see Fig. 6.2), using conduction mode welding and that this penetration is related to the parameters chosen.[9]

One of the disadvantages of conduction mode laser welding is the coupling efficiency of this process, especially when welding high reflective materials. Coupling is basically the percentage of energy that comes from the laser that is absorbed by the material. In keyhole laser welding, most of the beam is absorbed; however, in conduction the amount of laser beam absorbed by the material is smaller. For example, when laser welding stainless steel, the absorption of the laser beam in conduction mode is around 15% while in keyhole mode it jumps to 65%.[11] The increase of the absorption is a good indicator of the formation of a keyhole due to the increase of the coupling.

6.2 Conduction laser weld with 6.35 mm of penetration.[23]

Table 6.1 Advantages and disadvantages of conduction and keyhole laser welding

Process	Advantages	Disadvantages
Keyhole laser welding	Low heat input which originates low distortion Deep penetration welds with a big aspect ratio High productivity	Unstable process High levels of porosity Big amount of spatter Loss of alloy elements Degradation of mechanical properties Requires a laser system with a relatively good beam quality
Conduction laser welding	No vaporisation takes place – which means no porosity, no cracks and no undercut on the welds No spatter during welding Stable process with control of the heat input Good gap bridging ability due to the large beams used Does not require a laser system with a high quality beam	Low coupling efficiency Slow process High heat input which means higher distortion

However, the values of coupling depend on several factors, such as the wavelength of the laser, the surface finishing, the material and the thickness of the welded material.[11,12] Some studies have been made in measuring the coupling for different materials. While for steel the coupling was between 20 and 90%, in aluminium it varied from 10 to 80%.[12,13]

A brief comparison between conduction and keyhole mode is presented in Table 6.1. Due to the differences in the welds obtained and in the welding process itself, conduction laser welding tends to increase the applications of laser welding and not 'compete' with keyhole laser welding.

6.2 The transition between conduction and keyhole mode

The major difference between conduction and keyhole welding is the power density applied to the welding area. When the power density used is insufficient to cause boiling, the weld occurs in conduction mode. The alternative to this mode is keyhole mode welding, where the power density is sufficient to provide evaporation and open a hole in the melt pool. Despite different studies of these two welding modes, the boundaries between them are not clear in terms of processing parameters. Often it is considered that conduction welding occurs when the power density is lower than 10^6 W/cm^2.[14–16] This value is completely independent of the welding speed or the beam diameter used. It is also sometimes considered that conduction welding occurs when the value of the laser power is low, of the order of 1 kW; at this low value of power, the power density at the surface of the welded material is limited. Despite these different definitions for conduction laser welding, there is one fundamental point where there seems to be agreement, which is that conduction welding occurs when the vaporisation of the material is insignificant[1,15] or when the power density is not enough to cause boiling.[14] Another definition for conduction laser welding mode is that it shows an aspect ratio smaller than 0.5.[17] Nevertheless, these definitions have one thing in common: they do not take into account other welding parameters such as beam diameter and welding speed. A schematic of the current definition of conduction and keyhole laser welding is shown in Fig. 6.3.

There are not that many studies focused on the transition between conduction and keyhole laser welding mode. The following paragraphs review the relevant work done in this area.

Beyer *et al.* studied the absorption of a laser-induced plasma during laser

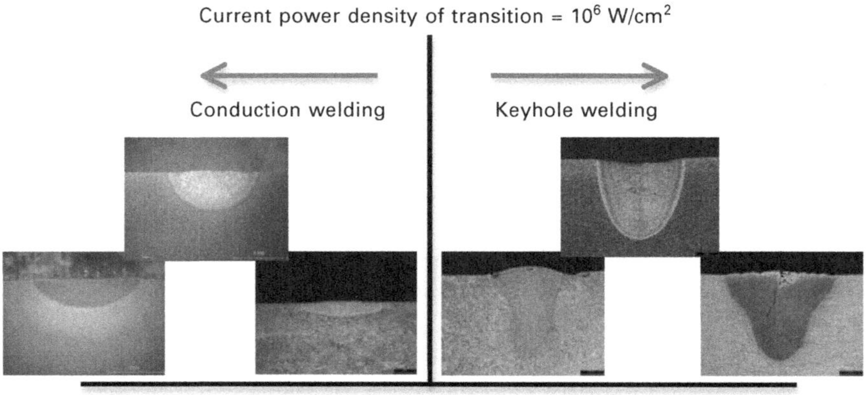

6.3 Current definition of the transition between conduction and keyhole welding.

processing.[18] The main focus of this paper was not the study of the transition between conduction and keyhole mode. However, a relation between power density, pulse duration and beam radius and the threshold power density for vaporisation was established, mainly in the region of laser processing (between 1 and 100 ms of interaction time) where the power density necessary for vaporisation is dependent on the beam radius and interaction time (Fig. 6.4). This part of the paper relied on an analytical equation that represents the surface at the beam centre for Gaussian beams.[19] However, no final conclusion was drawn regarding the transition between conduction and keyhole laser welding.

Liu *et al.* and Weckman *et al.* studied the effects of process variables on pulsed laser welding, for stainless steel and aluminium, respectively (Fig. 6.5). They concluded that the power density at which keyhole mode starts is independent of the pulse duration after a certain value of pulse time. However, there was a change in the power density at which a keyhole starts depending on the material. This was, in part, in agreement with what was previously considered to be true. However, the threshold power density for when the keyhole starts for stainless steel was 4 GW/m^2 and 10 GW/m^2 for aluminium (different from the 10^6 W/cm^2 mentioned previously) and so dependent on the material.[20,21]

Zhang *et al.* studied the transition between the different laser welding modes. In this work, it was concluded that laser power, focal position and travel speed will influence the welding mode. However, the double-U curve described is difficult to apply in a broader parameter range.[22]

Using a finite-element model, Buvanashekaran *et al.* evaluated the welding mode based on different energy levels. The authors concluded that

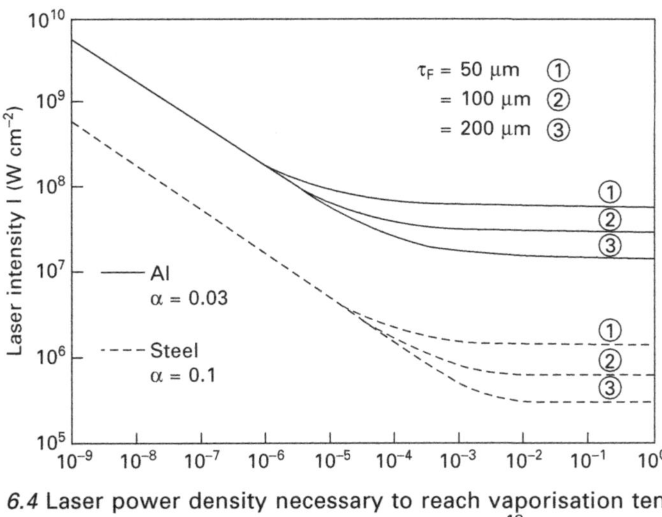

6.4 Laser power density necessary to reach vaporisation temperature at the centre of the beam as function of time.[18]

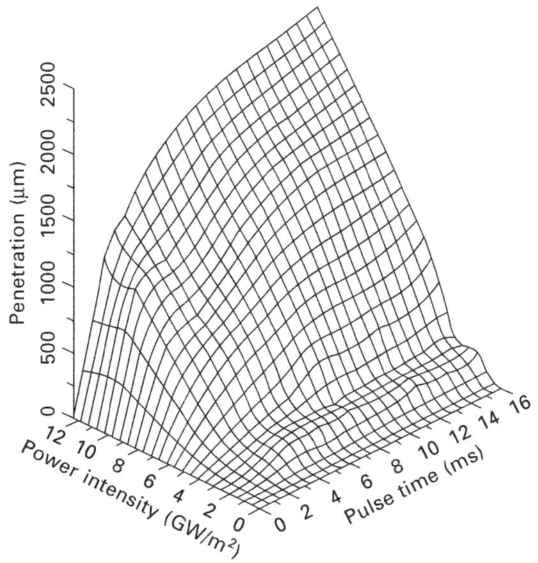

6.5 Variation of penetration, for aluminium, with the power density and the pulse time.[20,21]

there was a range of power densities where conduction mode takes place. At a welding speed of 1,000 mm/min, it was concluded that the transition from conduction to keyhole mode took place with 400 W of laser power.[23] In this work, despite concluding that there was a range of power densities where conduction mode takes place, no relation between the parameters and the welding mode was made and just a set of parameters was given as an example of where conduction would transit to keyhole mode.

Sibillano *et al.* used a real-time monitoring of the plasma, formed during the different welding regimes, to identify the welding mode.[24] However, this study showed no relation between the welding parameters and the welding mode. It also relied on the definition that the welding mode is defined based on the power density used ($>10^6$ W/cm^2 for keyhole and $<10^6$ W/cm^2 for conduction).

Nakamura *et al.* combined the optical and acoustic emissions from a laser irradiated spot. It was stated that, depending on the material, for 3 kW of laser power there was a welding speed range where conduction laser welding took place.[17] However, the identification of the welding mode was based on the ratio between penetration and width. Also no relation was established between the welding parameters and the welding mode.

All these works do not relate the transition from conduction to keyhole to the process parameters and how the process parameters influence the transition between conduction and keyhole. In some points, the works even show different conclusions. For example, while some studies show that there is a sharp transition between conduction and keyhole regime, other works identified a regime between conduction and keyhole. The presence of two or three welding modes in laser welding is also a common difference in some laser modelling papers. The following paragraphs present some examples of those differences.

Zhang *et al.* identified three kinds of laser welding process: stable heat conduction welding, stable deep penetration welding and unstable mode welding. The existence of this unstable mode welding or transition mode was attributed to thermal focusing or to an unsuitable selection of welding parameters.[22] The existence of this unstable mode, which was attributed to welding mode fluctuation, was previously identified by Chen *et al.*[25] and Zhang *et al.*[26]

The existence of just conduction mode regime and keyhole mode regime in PW laser welds was shown by the use of a model. The model shown by Semak *et al.*[27] relates metal displacement to the vaporisation recoil pressure (Fig. 6.6). In this model the melt displacement happens when the vaporisation recoil pressure accelerated the melt created in the fusion zone allowing this molten material to reach the edge of the molten pool. In this model it was possible to identify three different thresholds: one for when drilling starts to take place, another for when displacement starts to happen and another for melting. In this model, Semak *et al.* observed that at energy levels lower than the drilling threshold and higher than the displacement threshold was where the keyhole mode regime took place. On the other hand, it was assumed that

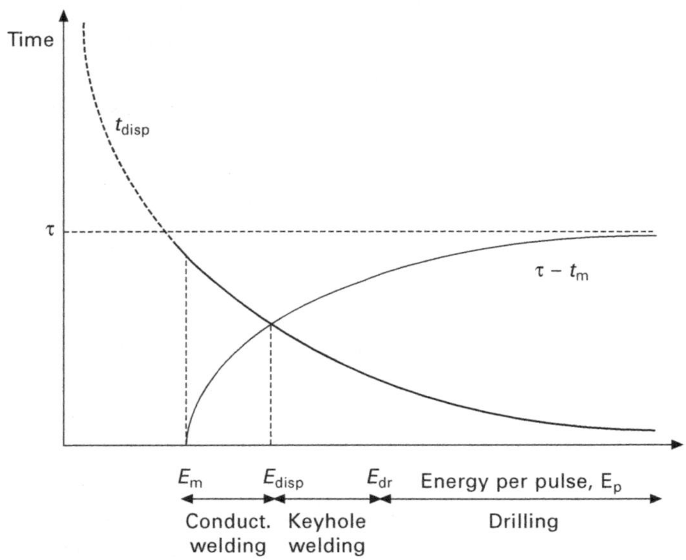

6.6 Identification of the different welding regimes, conduction and keyhole, in pulsed laser welding by Semak *et al.*[27]

the regime between the displacement threshold and the melting threshold was when the conduction mode regime took place.[27] In this work no regime was identified between conduction and keyhole regime. The model shown by Resch and Kaplan also considers two welding modes, conduction and keyhole laser welding.[28]

Based on the drilling model, Colegrove *et al.* identified three different regimes in pulsed laser welding.[29] They were conduction mode, mixed mode and keyhole mode regime (Figs 6.7 and 6.8). The presence of these three regimes had also been shown previously by Steen[14] and Ahmed.[30]

Buvanashekaran *et al.* identified three laser welding modes that are determined by the laser power and welding speed. Although, the authors established that the transition between conduction and keyhole for a 1,000 mm/min welding speed would occur at 400 W beam power, if a mode between conduction and keyhole was identified these values should represent the end of conduction or the beginning of keyhole mode and not the transition between these two modes (Fig. 6.9).[23]

Sibillano *et al.* also identified a mixed mode in laser welding. The authors defined this regime as having properties of both conduction and keyhole. This was noticed when the power density was just over the threshold for keyhole mode and so it was assumed to be caused by unstable keyhole formation.[24]

Based on the literature review, it is clear that the threshold of conduction laser welding mode is inaccurate and does not take into account the process parameters. An analytical approach to the transition between conduction and keyhole mode would also be possible. However, most of the analytical models found[19] are not dependent on the beam diameter or on the welding speed when the boundary conditions for determining the temperature at the centre of the laser beam are applied. The number of regimes and the transition

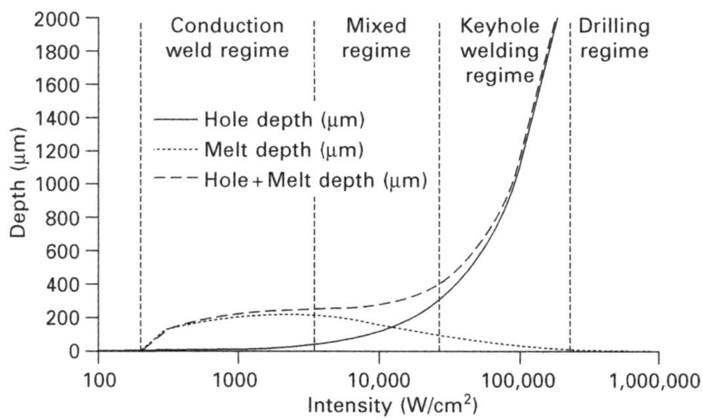

6.7 Identification of three welding modes in pulsed laser welding.[29]

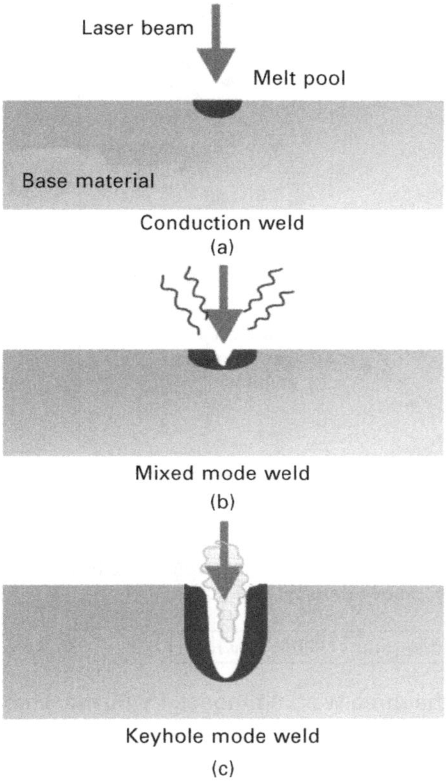

6.8 The three welding regimes in pulsed laser welding.[29]

between conduction and keyhole mode have been studied by several authors and progress made, though further research on this topic might clarify some of the aspects mentioned above.

6.3 Conduction laser welding

Conduction welding mode is mainly controlled by convection and conduction. In the specific case of laser welding at high intensities, it is normal to form a temperature gradient that itself will lead to surface tension also known as Marangoni convection. The literature about Marangoni convection is extensive; research by Zhao *et al.*,[31] Limmaneevichitr and Kou[32–34] and Chan *et al.*[35] are just a few examples. This convection is responsible for the geometry of the laser melt pool, e.g. its shape and aspect ratio. The composition of the melt pool during conduction laser welding is also influenced by convection, mainly due to the fact that convection is responsible for the mixing of material that happens during the laser welding process.[36] In this mode the

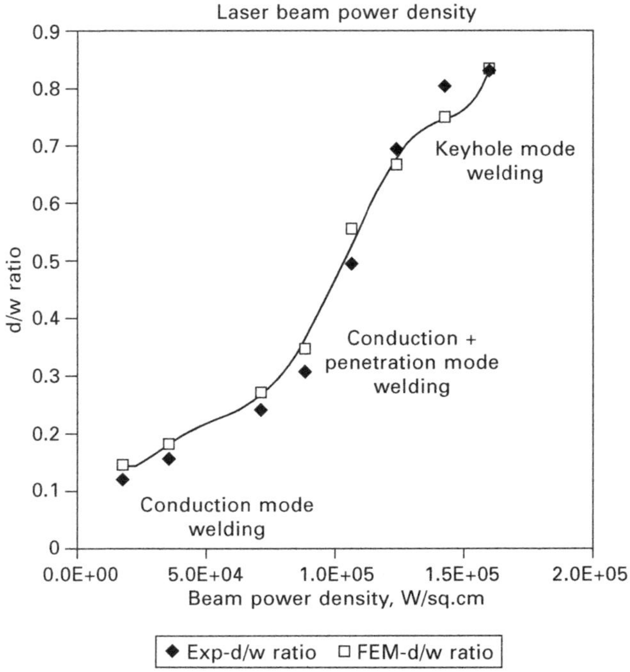

6.9 Identification of the three welding modes by Buvanashekaran *et al.*[23]

energy absorbed at the surface and by thermal conduction is transferred to the interior of the material (Fig. 6.10).[14]

Conduction welding can be a viable alternative to keyhole welding mainly because it is a very stable process and easier to obtain high quality welds free of pores and spatter.[37] This mode takes place when the vaporisation of the material is insignificant, in other words, when the thermal power density is not high enough to cause boiling.[38] Another advantage of conduction laser welding is that it can be made with a significantly low laser cost, because it does not require a high beam quality or very high power.[15]

Some studies, based on modelling and followed by experimental validation, have shown that there is an optimum beam diameter. This means that for a certain value of power and travel speed there is a specific beam diameter in which the penetration is maximum. This phenomenon occurs when, by increasing the beam diameter, the temperature at the surface remains at vaporisation temperature. In other words, when welding in conduction mode, a progressive increase in the spot size will result in an increase in penetration depth as long as the temperature at the surface of the workpiece is maintained at boiling temperature. However, when the optimum beam diameter is achieved, the increase of the spot size will result in a decrease

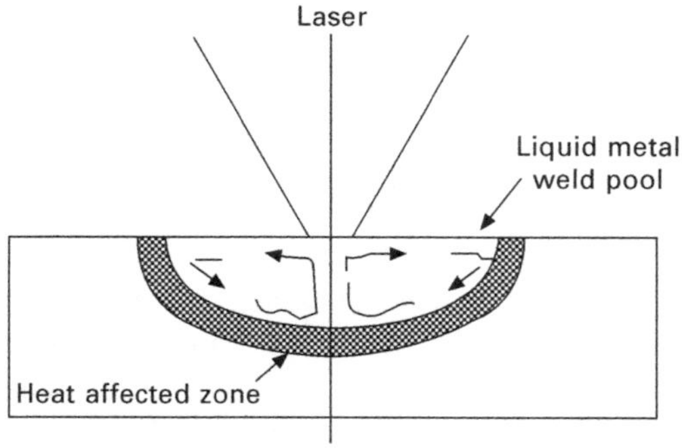

6.10 Schematic of conduction laser welding.[14]

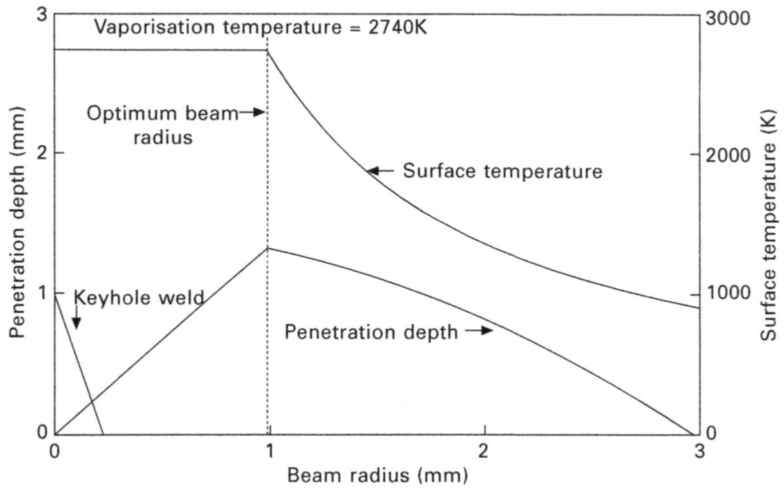

6.11 Variation of the surface temperature and the penetration, with beam radius, in aluminium.[9,38]

in the penetration depth because by increasing the spot size the temperature at the surface will decrease, which means it will be lower than the boiling temperature.[9,38]

Based on this principle, Bardin *et al.* were able to control the conduction mode welding process by thermal measurements (Fig. 6.11). Controlling the temperature of the welding process, keeping it just below the boiling point of the material, allowed the maximum penetration to be obtained. It was also noticed that when the temperature went over the boiling temperature, the presence of porosity was clear.[38]

Despite the fact that not many investigations have been carried out on conduction laser welding, some pioneer studies have been made in this welding mode and are reviewed in the following paragraphs.

One of the first works on conduction laser welding was by Esposito *et al.*[39] An evaluation of the welding parameters based on a theoretical model and experimental data was shown. One of the conclusions was that the temperature was controlled by the beam diameter and that the rise in temperature over the melting point would reduce the process efficiency. It was also concluded that the welding speed was related to the desired width of the weld and that the choice of the beam diameter was made by selecting a beam diameter that would provide the better coupling.[39] However, in this work the relation between welding speed, beam diameter and laser power was not fully discussed and evaluated. The principles behind this work were correct. The issue was how it related the choice of the beam diameter to the coupling and not to the control of the maximum temperature at the surface of the weld pool (as shown previously).

Paul and Debroy simulated the weld pool shape and the peak temperature based on numerical solutions of Navier–Stokes equation and the equation of energy conservation. The results showed good approximation to the experimental data; however, no evaluation of the welding parameters on the weld shape was made. They also showed that the temperature achieved in the laser–material interaction was close to the boiling point of the material. This emphasises that in this welding mode the vaporisation of material is insignificant.[40] Zhao and Debroy also evaluated the composition change in conduction mode laser welding of aluminium. They noticed that when vaporisation takes place in conduction mode, it is confined to a very small region under the laser beam. This vaporisation takes place due to a slight increase in the temperature over the boiling point of the aluminium. In this work, they evaluated that the welding speed and the laser power had no influence on the composition of the welded metal.[41] De and Debroy developed a model based on heat transfer and fluid flow for conduction mode spot welding. In this model they were able to optimise the laser beam absorptivity, effective thermal conductivity and the effective diffusivity based on a small number of experiments.[42]

More work has been carried out on modelling conduction laser welding. Bag *et al.* and Trivedi *et al.* studied the effect of power and on-time on the weld pool shape, using a finite element model with an adaptive heat source.[43, 44] Tsai and Kannatey-Asibu implemented a model suitable for feedback control based on a 2D heat flow model.[45] A numerical and experimental investigation was made to determine the importance of sulphur content and welding parameters on the fluid flow, heat transfer and shape of the welds obtained using a pulsed laser in conduction mode.[46] Pitscheneder *et al.* also developed a model to study the behaviour of solid precipitates.[47] The heat conduction in the weld pool and the convections phenomena were analysed by Bos and Chen, allowing a prediction of the pool geometry based on the parameters used during the welding process.[48] Conduction mode welds in different materials were carried out by Russo in order to allow a numerical and experimental investigation of the weld-pool surface phenomena.[49] Tobar *et al.* developed a three-dimensional numerical model to study aluminium laser welding in conduction mode and to evaluate the use of different surface treatments in order to increase the laser absorption. The authors concluded that in conduction mode the heat transfer is controlled by convection and not by conduction and that the flow of liquid material is caused by surface tension.[50] None of the works mentioned here related the influence of each process parameter and how they will influence conduction laser welding. Most of the work was actually based on modelling just using some welded samples as the starting point. This means that despite the models showing a good relation to the experimental data, there is no analysis over a range of parameters, which should include laser power, welding speed and beam diameter.

An interesting study has been carried out by Kell *et al.*[51] They evaluated the use of different beam profiles in conduction laser welding, maintaining the same energy density for the different beam profiles. The beam profiles used were: Gaussian, top-hat, rugby post and peaked edge line.[51] The conclusions were more related to the microstructure obtained by each beam profile and on the welded profile. However, based on Fig. 6.12, the assumption of the paper that all the welds were made in conduction mode does not seem to be completely correct. While the weld profile obtained using the peaked edge line is clearly in conduction mode, the weld profiles obtained using the other beams might not be. The reason might be related to the irradiance distribution obtained in all the profiles, which in the case of the peaked edge line might be more suitable for conduction laser welding. This paper is a good example of the lack of clarity in the definition of conduction laser welding.

6.4 Applications of conduction laser welding

The number of applications that use conduction laser welding is very low when compared to the number of applications that use keyhole laser welding. Even the applications that used conduction laser welding do not mention this welding mode. In these cases, a closer look at the welding parameters proves that this mode was used. One of the first applications of conduction laser welding goes back to 1970, when it was used to weld thermocouples to a sample container carried by the Apollo space shuttle.[51] Other applications included welding seals of electronic packages[53] and welding non-precious dental alloys.[54]

One of the applications where conduction laser welding has been most used is in welding aluminium alloys. The problems in using keyhole mode in laser welding of aluminium alloys can be overcome by using conduction laser welding. Some examples of the application of conduction laser welding

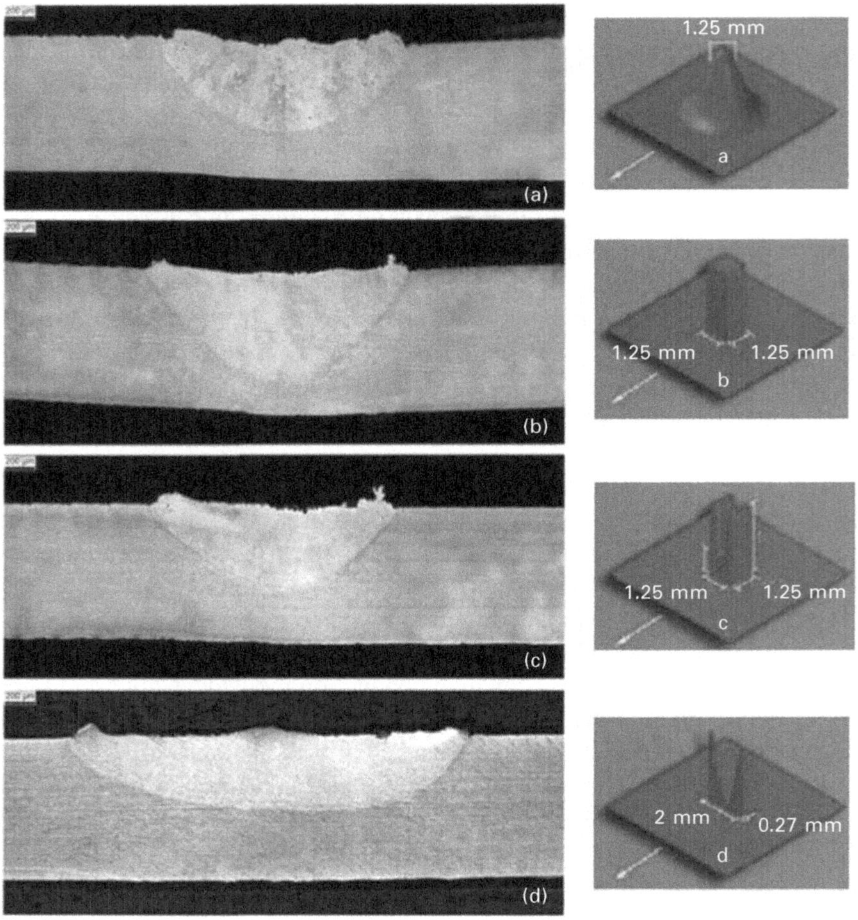

6.12 Micrographs and respective beam profiles of the welds obtained: (a) Gaussian, (b) top-hat, (c) rugby post and (d) peaked edged line.[51]

in aluminium are the work done by Morgan and Williams,[9] by Okon *et al.*[37] and by Sanchez-Amaya *et al.*[16,55] The results showed that by using conduction mode laser welding, it is possible to obtain welds free of pores and of cracks, two defects normally associated with laser welding of aluminium. In terms of the penetration achieved, Okon *et al.* were able to obtain full penetration in 3 mm thick AA5083,[37] Sanchez-Amaya *et al.* were also able to obtain full penetration on 3 mm thick AA5083 and 2.3 mm penetration on AA6082 (Fig. 6.13) while Morgan and Williams were able to achieve full penetration on 6.35 mm thick AA2024.[9]

Another application of conduction laser welding is in joining of dissimilar materials.[56–58] One of the most common combinations of materials is

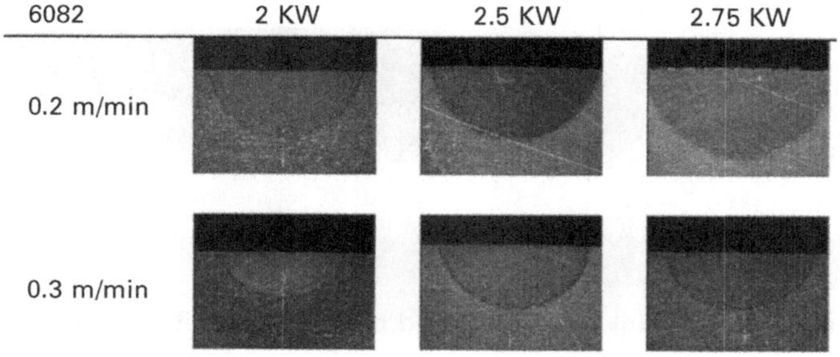

6.13 Some examples of conduction laser welding in AA6083 with different power and welding speeds.[55]

aluminium to steel weld. Due to the difficulty of this type of weld, several studies using resistance spot welding arc welding brazing and friction stir welding have been carried out. However, in recent years laser welding has been introduced as an alternative for welding steel to aluminium. Studies using keyhole welding in order to join aluminium to steel were made[59] but with limited success. Nevertheless, studies using conduction laser welding have shown very good results.[60–62] The advantage of using conduction laser welding in this application is related to the stability of the process that allows a better control of the temperature in the interaction area between the aluminium and the steel.[61]

The use of conduction mode welding on magnesium alloys has also been assessed.[63,64] It was possible, using conduction welds, to obtain a weld with less porosity, fewer cracks and with a smaller crater. However, in these works, it is unclear if the welds were made in 'pure' conduction mode. The profiles shown by the welds resemble a conduction mode. But the presence of a high vaporisation in the welding process raises some issues about the use of conduction laser welding.[64] Even so, the use of an 'almost' conduction mode weld has shown encouraging results in laser welding of magnesium alloys.

The introduction of new high power diode lasers will increase the use of conduction laser welding.[65] mainly because the use of this type of laser will solve one of the biggest issues in conduction laser welding, the low beam absorption, due to the wavelength of these lasers. Applications like aesthetical welds, brazing, cladding, soldering and polymer welding that are normally done in conduction mode will increase.[66] Figure 6.14 shows some examples of conduction laser welding using a diode laser.

The use of high power diode lasers allows the use of conduction laser welding in materials like aluminium and titanium[55,67] without any concerns related to the absorption of the laser beam by the material. Abe *et al.* showed some applications using this type of laser.[68] However, in no part of this work is conduction laser welding mode mentioned. A closer look at the welds

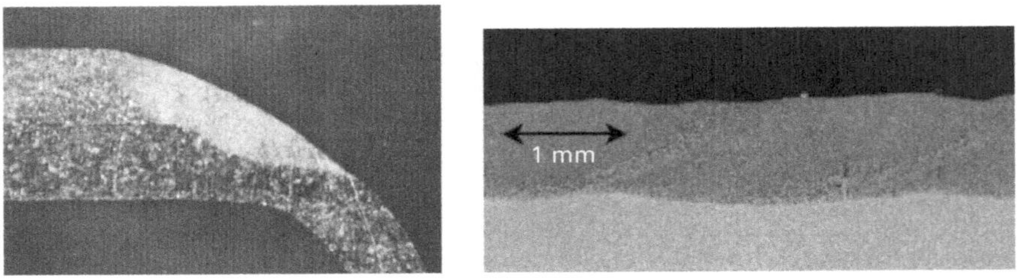

6.14 Kitchen sink weld (left) and powder deposition (right).[66]

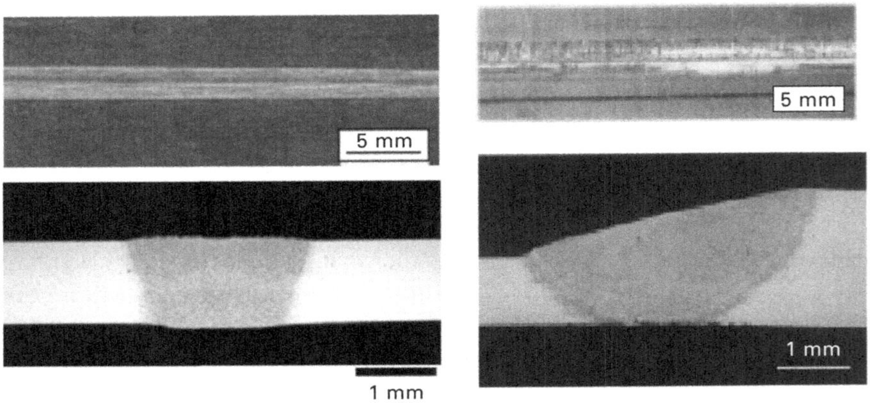

6.15 Examples of conduction laser welding using a high power diode laser.[68]

obtained (Fig. 6.15), and at the characteristics of the welding process (no spatter and no plasma), it was clear that the welds were made in conduction mode. Funada[69] and Abe[68] also used conduction laser welding on thin film with good results.[69–70]

Bassani *et al.* also used diode laser source to study the effect of process parameters on laser welding of an aluminium metal matrix composite reinforced with SiC particles. In this work a comparison was made between using a CO_2 laser to create a weld in keyhole mode, while a diode laser was used to produce a conduction mode weld (Fig. 6.16). The welds obtained in conduction mode showed a good metallurgical quality and the formation of Al_4C_3 was reduced considerably when compared to the welds made under keyhole mode.[71]

The application of a diode laser in welding titanium was also assessed by Block *et al.*[72] In this work no reference to conduction laser welding was made but the profile of the welds and the parameters used indicate that the welds were made in conduction mode (Fig. 6.17). Diode lasers have also been used in laser welding magnesium alloys with good results, which include the increased resistance to corrosion and wear.[73] Once again there

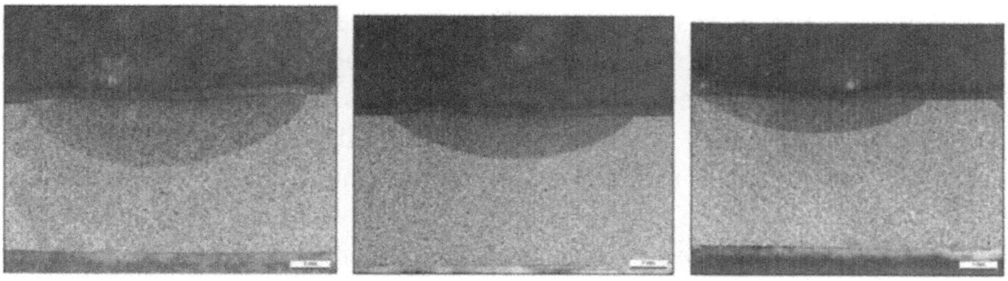

6.16 Some examples of diode laser welding, in conduction mode, of A359/SiC metal matrix composite.[71]

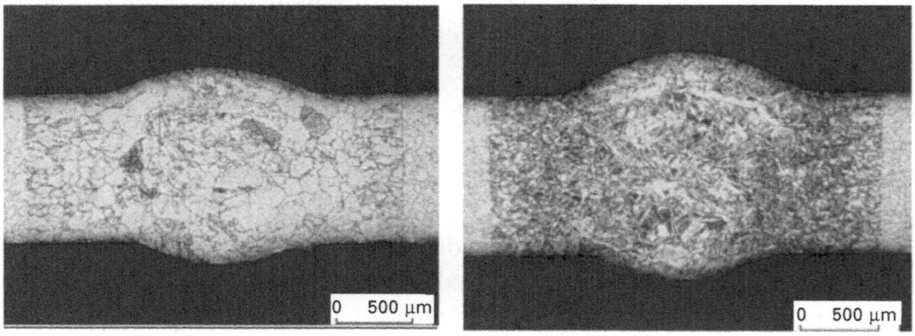

6.17 Laser filler weld seams on titanium.[72]

6.18 Macrographs of laser welded AZ61 magnesium alloy.[74]

is no reference to conduction mode laser welding, but the profiles shown in Fig. 6.18, mainly the low aspect ratio, indicate that conduction was the welding mode used.[74]

These last applications emphasise how the appearance of new high power diode lasers will increase the use of conduction mode laser welding. With the number of applications of conduction laser welding increasing, it is important to better understand the limits and the process of conduction laser welding.

However, conduction mode is not only used in laser welding. Another application is laser cladding, also known as laser metal deposition.[75–76] Despite in the literature related to laser cladding there being few references to conduction mode, based on the parameters used, mainly the low power densities, the properties of the cladding process (no spatter and stable process) and the properties of the clad (free of pores), it is possible to conclude that the process is normally done in conduction mode. Some studies mention that the power densities used are below the threshold for plasma formation, which means below the start of vaporisation,[76] emphasising the fact that conduction mode is used in laser cladding. Contrary to the applications of conduction mode in laser welding, the number of applications of laser cladding is extensive.[77–81] These types of applications show how, very often, conduction mode laser welding is mistaken for keyhole laser welding. As mentioned previously, the number of applications that use laser cladding is vast and it is common to see laser cladding parameterisation as if it was done in keyhole mode (Fig. 6.19).[81,82]

Conduction mode laser welding is also used in welding plastics. This process is sometimes referred to as laser transmission welding.[83,84] It requires a very controllable heat delivery to the workpiece and a very stable welding process.

An innovative application of conduction mode is in underwater laser beam welding. This technique is used in the nuclear industry for repair/maintenance applications. In this application, which is a laser cladding process and not a laser welding process, the heat input to the base material needs to be extremely well controlled. The principle behind this application is to deposit a cladding layer in order to avoid crack propagation in the base material (Fig. 6.20). It is the stability of the process, the control of the heat delivered to the workpiece and the quality of the welds obtained in this welding mode that trigger its increasing application.

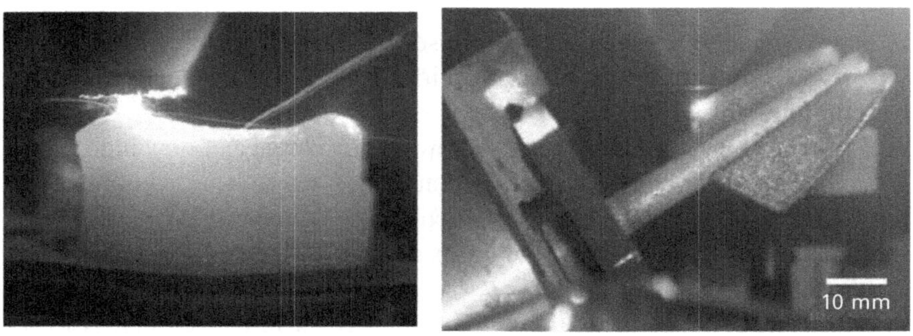

6.19 Manufacturing of a 3D block of a Ti-47Al-2Cr-2Nb (left)[82] and manufacturing of a Cp-Ti tube (right).[81]

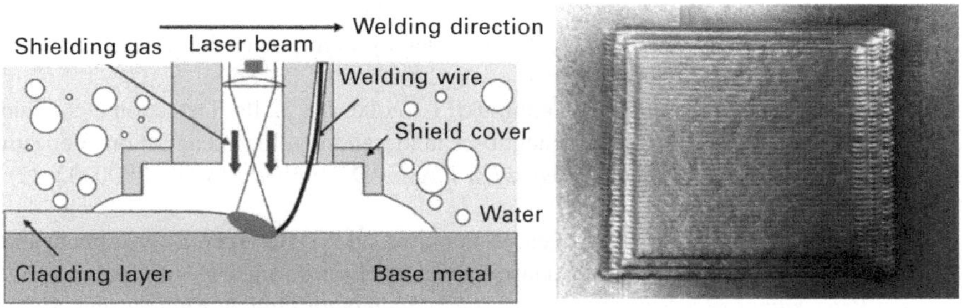

6.20 Schematic of underwater laser beam welding (left), underwater laser beam welding deposit (right).[85]

6.5 References

1. Ready, J.F. and D.F. Farson, *LIA handbook of laser materials processing*. 2001, Orlando, FL.: Laser Institute of America.
2. Bass, M., *Laser materials processing*. 1983, New York: Elsevier Science.
3. Lima, M.S.F., R. Riva, A.C. De Oliveira, and G.R. Siqueira, Laser beam welding aerospace aluminum using fiber lasers, in *Proceedings of SPIE – The International Society for Optical Engineering*. 2009. Lisbon.
4. Paleocrassas, A.G. and J.F. Tu, Inherent instability investigation for low speed laser welding of aluminum using a single-mode fiber laser. *Journal of Materials Processing Technology*. 210(10): 1411–1418.
5. Eriksson, I. and A.F.H. Kaplan, Evaluation of laser weld monitoring – A case study, in *ICALEO 2009 – 28th International Congress on Applications of Lasers and Electro-Optics, Congress Proceedings*. 2009. Orlando, FL.
6. Cao, X., W. Wallace, C. Poon, and J.P. Immarigeon, Research and progress in laser welding of wrought aluminum alloys. I. Laser welding processes. *Materials and Manufacturing Processes*, 2003. 18(1): 1–22.
7. Hu, B. and I.M. Richardson, Autogenous laser keyhole welding of aluminum alloy 2024. *Journal of Laser Applications*, 2005. 17(2): 70–80.
8. Ion, J.C., Laser beam welding of wrought aluminium alloys. *Science and Technology of Welding and Joining*, 2000. 5(5): 265–276.
9. Morgan, S. and S. Williams, Hybrid laser conduction welding, in *55th Annual Assembly of International Institute of Welding*. 2002. Copenhagen, Denmark.
10. Tam, S.C., R. Williams, L.J. Yang, S. Jana, L.E.N. Lim, and M.W.S. Lau, A review of the laser processing of aircraft components. *Journal of Materials Processing Tech.*, 1990. 23(2): 177–194.
11. Nath, A.K., R. Sridhar, P. Ganesh, and R. Kaul, Laser power coupling efficiency in conduction and keyhole welding of austenitic stainless steel. *Sadhana – Academy Proceedings in Engineering Sciences*, 2002. 27(Part 3): 383–392.
12. Weston, J.W., E.R. Wallach, Coupling for laser welds in aluminium alloys. Department of Materials Science and Metallurgy, University of Cambridge.
13. Mallory, L.C., R.F. Orr, and W. Wells, Effect of anodizing on laser welding of aluminium, in *Laser Materials Processing III*. 1989.
14. Steen, W.M., *Laser material processing*. 3rd edn. 2003, New York: Springer.
15. Ion, J.C., *Laser processing of engineering materials: principles, procedure and industrial application*. 2005, Oxford: Butterworth-Heinemann.

16. Sanchez-Amaya, J.M., T. Delgado, J.J. De Damborenea, V. Lopez, and F.J. Botana, Laser welding of AA 5083 samples by high power diode laser. *Science and Technology of Welding and Joining*, 2009. 14(1): 78–86.
17. Nakamura, S., M. Sakurai, K. Kamimuki, T. Inoue, and Y. Ito, Detection technique for transition between deep penetration mode and shallow penetration mode in CO_2 laser welding of metals. *Journal of Physics D: Applied Physics*, 2000. 33(22): 2941–2948.
18. Beyer, E., L. Bakowsky, P. Loosen, R. Poprawe, and G. Herziger, Development and optical absorption properties of a laser induced plasma during CO_2-laser processing, in *Proceedings of SPIE – The International Society for Optical Engineering*. 1984. Linz, Austria: SPIE.
19. Carlsaw, H.S. and J.C. Jaeger, *Conduction of heat in solids*. 2nd edn. 1959. Oxford: Clarendon Press.
20. Liu, J.T., D.C. Weckman, and H.W. Kerr, The effects of process variables on pulsed Nd:YAG laser spot welds: Part I. AISI 409 stainless steel. *Metallurgical Transactions B*, 1993. 24(6): 1065–1076.
21. Weckman, D.C., H.W. Kerr, and J.T. Liu, The effects of process variables on pulsed Nd:YAG laser spot welds: Part II. AA 1100 aluminum and comparison to AISI 409 stainless steel. *Metallurgical and Materials Transactions B: Process Metallurgy and Materials Processing Science*, 1997. 28(4): 687–700.
22. Zhang, X., W. Chen, J. Ren, G. Huang, and H. Zhang, Laser welding mode transition and influence of thermal focusing on mode transition, in *Proceedings of SPIE – The International Society for Optical Engineering*. 1996. Beijing, China.
23. Buvanashekaran, G., N. Siva Shanmugam, K. Sankaranarayanasamy, and R. Sabarikanth, A study of laser welding modes with varying beam energy levels. *Proceedings of the Institution of Mechanical Engineers, Part C: Journal of Mechanical Engineering Science*, 2009. 223(5): 1141–1156.
24. Sibillano, T., A. Ancona, V. Berardi, E. Schingaro, G. Basile, and P.M. LugarÃ, Optical detection of conduction/keyhole mode transition in laser welding. *Journal of Materials Processing Technology*, 2007. 191(1–3): 364–367.
25. Chen, W., X. Zhang, and J. Ren, Study of welding mode transition and stability of welding process in high laser welding. *Chinese Journal of Lasers*, 1996. 23(7).
26. Zhang, X., W. Chen, and J. Ren, Effects of processing parameters on mode and stability of laser welding, in *Photonics West'96*. 1996.
27. Semak, V.V., G.A. Knorovsky, and D.O. MacCallum, On the possibility of microwelding with laser beams. *Journal of Physics D: Applied Physics*, 2003. 36(17): 2170–2174.
28. Resch, M. and A.F.H. Kaplan, Heat conduction modelling of laser welding. *Lasers in Engineering*, 1998. 7(3–4): 229–240.
29. Colegrove, P., P.E. Simiand, A. Varughese, S. Williams, and D. Yapp. Evaluation of a drilling model approach to represent laser spot microwelding, in *ASM Proceedings of the International Conference: Trends in Welding Research*. 2009. Pine Mountain, GA.
30. Ahmed, N., *New developments in advanced welding*. 2005. Boca Raton, FL: CRC Press.
31. Zhao, C.X., C. Kwakernaak, Y. Pan, I.M. Richardson, Z. Saldi, S. Kenjeres, and C.R. Kleijn, The effect of oxygen on transitional Marangoni flow in laser spot welding. *Acta Materialia*. 2010. 58(19): 6345–6357.
32. Limmaneevichitr, C. and S. Kou, Visualization of Marangoni convection in simulated weld pools. *Welding Journal*, 2000. 79(5): 126–s.

33. Limmaneevichitr, C. and S. Kou, Experiments to simulate effect of Marangoni convection on weld pool shape. *Welding Journal*, 2000. 79(8).
34. Limmaneevichitr, C. and S. Kou, Visualization of Marangoni convection in simulated weld pools containing a surface-active agent. *Welding Journal*, 2000. 79(11): 324–s.
35. Chan, C.L., J. Mazumder, and M.M. Chen, Three-dimensional axisymmetric model for convection in laser-melted pools. *Materials Science and Technology*, 1987. 3(4): 306–311.
36. Cary, H.B., *Modern welding technology*. 2nd edn. 1989, Englewood Cliffs, NJ: Prentice Hall.
37. Okon, P., G. Dearden, K. Watkins, M. Sharp, and P. French, Laser welding of aluminium alloy 5083, in *ICALEO 2002 – 21st International Congress on Applications of Laser and Electro-Optics, Congress Proceedings*. 2002. Scottsdale, AZ.
38. Bardin, F., S. Morgan, S. Williams, R. McBride, A.J. Moore, J.D.C. Jones, and D.P. Hand, Process control of laser conduction welding by thermal imaging measurement with a color camera. *Applied Optics*, 2005. 44(32): 6841–6848.
39. Esposito, C., Duarelio, G. and A. Cingolani, On the conduction welding process of steels with CO_2 lasers. *Optics and Lasers in Engineering 3*, 1982: 139–151.
40. Paul, A. and T. Debroy, Free surface flow and heat transfer in conduction mode laser welding. *Metallurgical Transactions B*, 1988. 19(6): 851–858.
41. Zhao, H. and T. Debroy, Weld metal composition change during conduction mode laser welding of aluminum alloy 5182. *Metallurgical and Materials Transactions B: Process Metallurgy and Materials Processing Science*, 2001. 32(1): 163–172.
42. De, A. and T. Debroy, Improving reliability of heat and fluid flow calculation during conduction mode laser spot welding by multivariable optimisation. *Science and Technology of Welding and Joining*, 2006. 11(2): 143–153.
43. Bag, S., A. Trivedi, and A. De, Development of a finite element based heat transfer model for conduction mode laser spot welding process using an adaptive volumetric heat source. *International Journal of Thermal Sciences*, 2009. 48(10): 1923–1931.
44. Trivedi, A., S. Bag, and A. De, Three-dimensional transient heat conduction and thermomechanical analysis for laser spot welding using adaptive heat source. *Science and Technology of Welding and Joining*, 2007. 12(1): 24–31.
45. Tsai, F.R. and E. Kannatey-Asibu Jr, Modeling of conduction mode laser welding process for feedback control. *Journal of Manufacturing Science and Engineering, Transactions of the ASME*, 2000. 122(3): 420–428.
46. Pitscheneder, W., Numerical and experimental investigation of conduction-mode laser weld pools, in *Mathematical Modeling of Weld Phenomena 4*, H.E. Cerjak, Editor. 1998.
47. Pitscheneder, W., T. Hong, T. Debroy, R. Ebner, K. Mundra, and R. Benes, Experimental and numerical investigation of transport phenomena in conduction mode weld pools. *Welding in the World*, 2000. 44(6): 25–36.
48. Bos, J.A. and Chen, M.A., On the prediction of weld pool size and heat affected zone in Shallow-Pool Welding. *Transport Phenomena in Materials Processing and Manufacturing ASME*, 1996. 336.
49. Russo, A.J., Two-dimensional modelling of conduction-mode laser welding. L.I.A., 1984. 44 (ICALEO): 8–15.
50. Tobar, M.J., I.M. Lamas, A. Yáñez, J.M. Sánchez-Amaya, Z. Boukha, and F.J. Botana, Experimental and simulation studies on laser conduction welding of AA5083 aluminium alloys. *Physics Procedia*. 5(2): 299–308.

51. Kell, J., J. Tyrer, R. Higginson, R. Thomson, J. Jones, and S. Noden. Holographic diffractive optical elements allow improvements in conduction laser welding of steels, in *ICALEO 2006*. 2006.
52. Moorhead, A.J. and P.W. Turner, Welding a thermocouple gauge to Apollo lunar sample return containers. *Welding Journal*, 1970. 49: 15–21.
53. Emerson, W.F., Laser-beam welding seals electronic packages. *Welding Design and Fabrication*, 1996. 69(4): 43–44.
54. Bertrand, C., The laser welding technique applied to the non precious dental alloys procedure and results. *British Dental Journal*, 2001. 190(5): 255–257.
55. Sanchez-Amaya, J.M., T. Delgado, L. Gonzãlez-Rovira, and F.J. Botana, Laser welding of aluminium alloys 5083 and 6082 under conduction regime. *Applied Surface Science*, 2009. 255(23): 9512–9521.
56. Sepold, G., E. Schubert, and I. Zerner, Laser beam joining of dissimilar materials, in *IIW*. 1999. Lisbon.
57. Xiao, R., P. Dong, and K. Chen, Laser beam welding of dissimilar materials, in *ICALEO 2009 – 28th International Congress on Applications of Lasers and Electro-Optics, Congress Proceedings*. 2009. Orlando, FL.
58. Theron, M., C. Van Rooyen, and L.H. Ivanchev. CW Nd:YAG laser welding of dissimilar sheet metals, in *26th International Congress on Applications of Lasers and Electro-Optics, ICALEO 2007 – Congress Proceedings*. 2007. Orlando, FL.
59. Sierra, G., P. Peyre, F. Deschaux-Beaume, D. Stuart, and G. Fras, Steel to aluminium key-hole laser welding. *Materials Science and Engineering A*, 2007. 447(1–2): 197–208.
60. Sierra, G., P. Peyre, F. Deschaux Beaume, D. Stuart, and G. Fras, Galvanised steel to aluminium joining by laser and GTAW processes. *Materials Characterization*, 2008. 59(12): 1705–1715.
61. Kreimeyer, M., F. Wagner, and G. Sepold, Development of a combined joining-forming process for aluminum-steel joints, in *ICALEO 2004 – 23rd International Congress on Applications of Laser and Electro-Optics, Congress Proceedings*. 2004. San Francisco, CA.
62. Kreimeyer, M. and G. Sepold, Processing of laser joined aluminum-steel tailored blanks in overlap and butt joint configuration, in *ICALEO 2002 – 21st International Congress on Applications of Laser and Electro-Optics*. 2002. Scottsdale, AZ.
63. Zhu, J., L. Li, and Z. Liu, CO_2 and diode laser welding of AZ31 magnesium alloy. *Applied Surface Science*, 2005. 247: 300–306.
64. Min, D., J. Shen, S. Lai, J. Chen, N. Xu, and H. Liu, Effects of heat input on the low power Nd:YAG pulse laser conduction weldability of magnesium alloy AZ61. *Optics and Lasers in Engineering*. 49(1): 89–96.
65. Williams, S., G. Scott, and N. Calder, Direct diode laser welding of aerospace alloys. *Laser Opto*, 2001. 33: 50–54.
66. Bachmann, F., Industrial applications of high power diode lasers in materials processing. *Applied Surface Science*, 2003. 208–209: 125–136.
67. Caiazzo, F., F. Curcio, G. Daurelio, and F. Memola Capece Minutolo, Ti6Al4V sheets lap and butt joints carried out by CO_2 laser: mechanical and morphological characterization. *Journal of Materials Processing Technology*, 2004. 149: 546–552.
68. Abe, N., A. Morikawa, M. Koichi, and N. Keizo, Applicability of a high power diode laser to aluminum alloy welding, in *ICALEO 2002 – 21st International Congress on Applications of Laser and Electro-Optics, Congress Proceedings*. 2002. Scottsdale, AZ.

69. Funada, Y. and N. Abe, Lap welding with direct diode laser of thin foil on thick substrate, in *ICALEO 2004 – 23rd International Congress on Applications of Laser and Electro-Optics, Congress Proceedings*. 2004. San Francisco, CA.
70. Abe, N., N. Nakamura, Y. Funada, and M. Tsukamoto, The effect of direct diode laser beam size in heat conduction lap welding of a thin film on a thick substrate, in *Ceramic Transactions*. 2007: Kurashiki. 381–388.
71. Bassani, P., E. Capello, D. Colombo, B. Previtali, and M. Vedani, Effect of process parameters on bead properties of A359/SiC MMCs welded by laser. *Composites Part A: Applied Science and Manufacturing*, 2007. 38(4): 1089–1098.
72. Block, B., H. Haferkamp, A. Ostendorf, J. Bunte, J. Huang, T. Hassel, and C. Bruns, High-strength titanium laser welding with a subsequent combination of mechanical and laser heat treatment, in *ICALEO 2004 – 23rd International Congress on Applications of Laser and Electro-Optics, Congress Proceedings*. 2004. San Francisco, CA.
73. Abbas, G., L. Li, and Z. Liu, Effect of high power diode laser surface melting on wear resistance of magnesium alloys, in *15th International Conference on Wear of Materials*. 2005: San Diego, CA.
74. Abbas, G., L. Li, Z. Liu, and W.U.H. Syed, Diode laser surface treatment of magnesium alloy AZ61, in *ICALEO 2004 – 23rd International Congress on Applications of Laser and Electro-Optics, Congress Proceedings*. 2004. San Francisco, CA.
75. Grueninger, A., M. Hustedt, D. Herzog, M. Huse, D. Kracht, H. Haferkamp, and A. Ostendorf, Surface texturing by laser cladding. *Journal of Laser Applications*, 2011. 23(2).
76. Schubert, E., T. Seefeld, A. Rinn, and G. Sepold, Laser beam cladding: A flexible tool for local surface treatment and repair. *Journal of Thermal Spray Technology*, 1999. 8(4): 590–596.
77. Partes, K. and G. Sepold, Modulation of power density distribution in time and space for high speed laser cladding. *Journal of Materials Processing Technology*, 2008. 195(1–3): 27–33.
78. Del Val, J., R. Comesana, F. Lusquinos, A. Riveiro, F. Quintero, and J. Pou, Laser micro-cladding: A novel laser additive technique to produce hard micro-coatings, in *ICALEO 2010*. 2010.
79. Santhanakrishnan, S., F. Kong, and R. Kovacevic, Process parameters analysis of high power direct diode laser cladding, in *29th International Congress on Applications of Lasers and Electro-Optics, ICALEO 2010 – Congress Proceedings*. Anaheim, CA.
80. Emamian, A., S.F. Corbin, and A. Khajepour, Study on laser parameters effect on morphology of *in-situ* Fe-TiC particles deposition on mild steel using laser cladding process, in *29th International Congress on Applications of Lasers and Electro-Optics, ICALEO 2010 – Congress Proceedings*. Anaheim, CA.
81. Sankarã, S., D. Boisselier, T. Engel, and F. Hlawka, Additive laser manufacturing of small metallic components by laser micro-cladding, in *ICALEO 2009 – 28th International Congress on Applications of Lasers and Electro-Optics, Congress Proceedings*. 2009. Orlando, FL.
82. Aubry, P., T. Malot, P. Peyre, V. Ji, V.K. Rexerodt, M. Thomas, T. Vilaro, and C. Colin, Direct fabrication of a TI-47AL-2CR-2NB alloy by direct metal deposition, in *29th International Congress on Applications of Lasers and Electro-Optics, ICALEO 2010 – Congress Proceedings*. Anaheim, CA.
83. Binetruy, C., A. Bueter, P. Castaing, S. Clement, M. Deleglise, C. Franz, A. Giessl,

W. Knapp, M. Oumarou, J. Renard, and A. Roesner, Analysis of laser welding of long fiber reinforced composites, in *29th International Congress on Applications of Lasers and Electro-Optics, ICALEO 2010 – Congress Proceedings*. Anaheim, CA.
84. Franz, C., S. Mann, and S. Kaierle, Comparison of process monitoring strategies for laser transmission welding of plastics, in *26th International Congress on Applications of Lasers and Electro-Optics, ICALEO 2007 – Congress Proceedings*. 2007. Orlando, FL.
85. Bucurel, R. and G. Hlifka, Laser beam welding process automates underwater repairs. *Welding Journal*, 2010.

7
Developments in laser microwelding technology

M. NAEEM, JK Laser, GSI Group, UK

DOI: 10.1533/9780857098771.1.163

Abstract: The term microwelding, also known as precision welding or microjoining, is somewhat elusive and consensus on a precise definition is probably very difficult to achieve. For this chapter a working definition is suggested, i.e. laser microwelding concerns the joining of parts having a weld bead size <1 mm and a weld penetration depth of <1 mm. Laser microwelding is used for joining high value miniature components in a range of industries (i.e., electronics, telecom, automotive and medical). Typical applications include catheters or implants, orthodontic brackets, fibre optic couplings to laser diodes, hermetic seals, watch part components and other photonic fabrication devices. This chapter reviews the basics of laser microwelding, including the laser choice for microwelding, the microwelding process and the quality of microwelds. The chapter also examines applications of microwelding and covers future trends in microwelding technology.

Key words: fibre laser, Nd:YAG laser, beam quality, beam mode, power density, spot size, peak power, pulse energy, dissimilar materials, weld joints, weld defects, applications.

7.1 Introduction

A range of small components are increasingly being used in microelectronics and engineering devices. These miniature parts are made of complex shapes and material combinations which require joining techniques capable of producing good mechanically sound joints. This has increased the interest in the use of lasers for microwelding in a range of applications. The term microwelding, also known as precision welding or microjoining, is somewhat elusive and it is probably very difficult to reach consensus on a precise definition. For this chapter, a working definition is suggested, i.e. laser microwelding concerns the joining of parts having a weld bead size less than 1 mm and a weld penetration depth of less than 1 mm. Apart from the weld bead size and weld depth penetration, other laser microjoining requirements are:

- very fine welds (spot sizes could range from approx. 10um up to 100um)
- reduced distortion and heat affected zone

- process reliability and stability (pulse to pulse stability for micro spot welding)
- high process speed (welding with scanner)
- welding of reflective materials
- welding dissimilar materials.

Laser microwelding is used for joining high value miniature components in a range of industries (i.e., electronics, telecom, automotive and medical industry) (Table 7.1). Typical applications include catheters or implants, orthodontic brackets, fibre optic couplings to laser diodes, hermetic seals, watch part components and other photonic fabrication devices.[1–3]

Compared to conventional microwelding methods (i.e., resistance, flash, arc, TIG, MIG and plasma), laser microwelding offers a number of advantages including:

- low heat input: the weld energy is delivered only where it is needed and with exceptional control
- clean welds: in addition to the aesthetic benefits, clean welds result in products that are easier to sterilise or fit into other assemblies
- strong welds: lasers provide high strength with a minimum number of welds
- hermetic welds: unlike soldering or brazing, lasers can provide flawless hermetic welds essential for many micro-applications.

Some of the applications outlined in Table 7.1 require joining dissimilar metals such as stainless steel and copper alloys. Applications like electrical connectors and electromechanical components are normally made out of steel and copper-based alloys. The trends of joining dissimilar metals and

Table 7.1 Examples of microwelding applications

Market sector	Applications
Automotive	Air bag assemblies, fuel injectors, lighting assemblies, batteries, ignition controls
Medical	Hearing aids, prosthetics, pacemaker and defibrillator implants, implantable radioactive capsules, cochlear implants, surgical tools, ultrasound catheters, orthodontic appliances
Electronics	Leadframe assemblies, electrical connectors, relay terminal connections, mobile phone batteries, flexural arms for hard disc drives, mobile phones, stator spacers, fuel and solar cells
Other micro-products	Jewellery, razor blades, sensing devices, thermocouples, micro-pumps, ink-jet cartridges, micro-turbines, micro-motors, micro-gears, watch part components

the miniaturisation of components offer big challenges for welding methods and will be discussed in later sections.

7.2 Laser choices for microwelding

There are a number of laser sources available for laser microwelding applications, i.e. pulsed Nd:YAG, fibre, disk, diode and CO_2. The choice of laser will depend on the specific aspects of the application such as material type, cycle time, fit-up tolerance, integration requirement and, most importantly, the cost of the laser system. The choice of a laser source for a specific application is no longer straightforward and obvious. Laser users are now faced with additional questions of laser beam quality and brightness. The main advantage of high beam quality is that the laser beam can be transported via a small diameter optical fibre, which translates to a small spot size at the workpiece.

7.2.1 Beam quality

For the majority of microwelding applications, it is necessary to have a laser with good beam quality to produce a small spot size at the workpiece. A combination of good beam quality and small spot size offers a number of advantages during microwelding including:

- small focused spot size
- high power density at the workpiece
- reduced heat input
- reduced heat affected zone
- enhanced processing speeds.

The quality of a laser beam can be calculated by the beam parameter product (BPP),[4] which is defined as the product of a laser beam's divergence angle (half-angle) and the radius of the beam at its narrowest point (the beam waist):

$$\text{BPP} = \omega_0 \times \frac{\theta}{2} \qquad [7.1]$$

where ω_0 and θ denote the beam waste radius and the laser beam's divergence angle, respectively (Fig. 7.1) and the BPP quantifies how well the laser beam can be focused on to a small spot. A Gaussian beam has the lowest possible BPP, λ/π, where λ is the wavelength of the light.

The ratio of the BPP of an actual beam to that of an ideal Gaussian beam at the same wavelength is denoted M^2

$$M^2 = \frac{BPP}{BPP_q} \geq 1 \qquad [7.2]$$

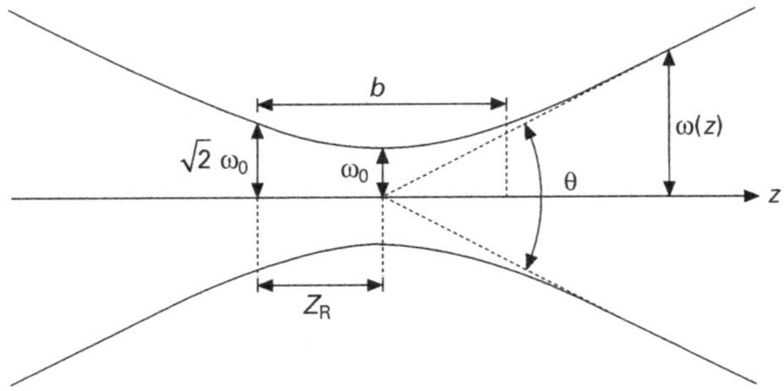

7.1 Gaussian beam width $\omega(z)$ as a function of the axial distance z. ω_0: beam waist; b: depth of focus; z_R: Rayleigh range; θ: total angular spread.

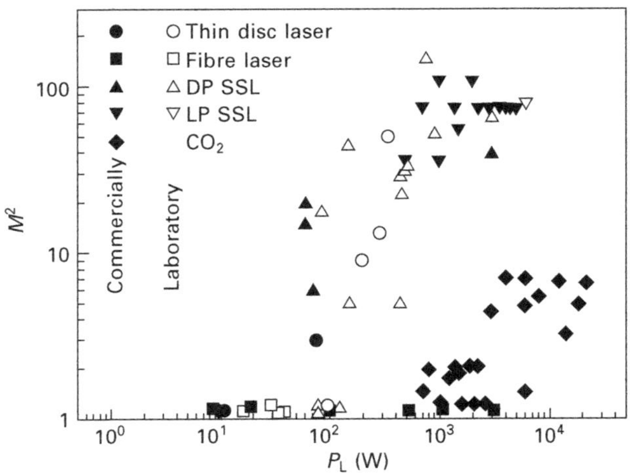

7.2 Beam quality (M^2) vs. laser output power (W) for selected laser sources used for laser material processing.[4]

The beam quality M^2 parameter is a wavelength-independent measure of beam quality. Figure 7.2 shows beam quality data for commercial and laboratory lasers.[4]

For a number of years, low power (up to 300 W) lamp-pumped pulsed Nd:YAG lasers with an adequate beam quality (M^2 = 15–20) have been the laser of choice for a number of microwelding applications. At a wavelength of around 1 μm, the focusing optics are smaller and simpler to enable smaller spot sizes than equivalent CO_2 lasers. As these devices are getting smaller and smaller, it is very important to have laser sources that have good beam quality to produce a small spot size and produce a very stable output for a consistent process.

The need for more efficient, compact and high beam quality lasers for very fine microwelding has fuelled the rapid growth in developing fibre lasers.

The following sections highlight the advantages and disadvantages for both pulsed lamp-pumped and fibre lasers for microwelding applications.

7.2.2 Laser structure

Figure 7.3 shows the basic components comprising a laser. The gain medium is capable of producing light when the excitation energy is applied to it. The front and rear mirrors define the resonator, which ensures the light reflects back and forth through the gain medium, thus gaining in intensity and reducing in angular spread. The fraction of the light which passes through the partially transmitting front mirror becomes the useful laser beam.

For a CO_2 laser, the gain medium is a mixture of gases (mainly helium, nitrogen and carbon dioxide) and the excitation energy comes from the electrical discharge through the gas. The laser beam wavelength is 10.6 μm. For a solid state laser, the gain medium is typically a rod of YAG (yttrium aluminium garnate) doped with about 1% Nd (neodymium) atoms, and the excitation energy is provided by the broadband light from a pulsed flashlamp of a continuous arc lamp. The laser beam wavelength is 1.064 μm.

For a fibre laser, the gain medium is an ytterbium-doped glass fibre, with the excitation energy being provided by laser diodes, operating around 950 nm, coupled by various schemes into the core of the doped fibre. The laser beam wavelength is typically in the range 1.07 μm to 1.09 μm and the physical dimensions of the gain medium for the fibre laser are very different from other laser types. A Nd:YAG rod might be 200 mm, a CO_2 discharge around 2 m, but the gain fibre in a fibre laser will be tens of metres long. The reflectors used in the fibre laser are physically very different from traditional lasers. Typically the mirror will be formed from a dielectric coating on substrate, which will be transmissive at the laser wavelength for the output coupler. For the fibre laser, Bragg gratings written into the core of a fibre are used. These fibre Bragg gratings (FBGs) consist of periodic refractive index variations. The longitudinal period of the grating determines the wavelength of the reflected light, and the magnitude of the variation controls the reflected percentage.

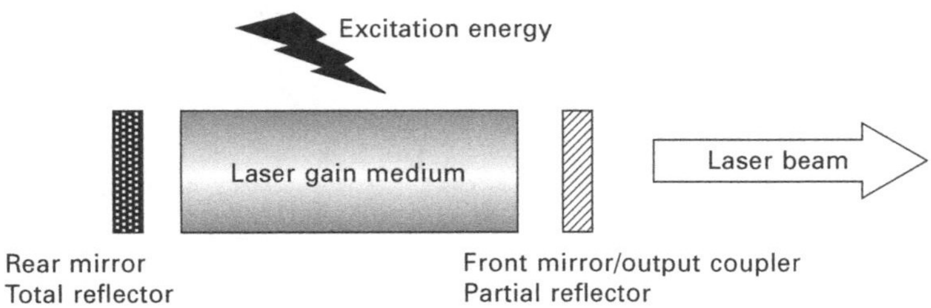

7.3 Basic components of a laser.

Nd:YAG laser

The Nd:YAG laser is a solid-state laser (Fig. 7.4) usually in the shape of a rod, operating at 1.06 μm.[5] The active species are neodymium ions present in small concentrations in the YAG crystal. Both continuous wave and pulsed laser outputs can be obtained at an overall efficiency in the 3–5% range. The laser is used in industry because of its efficiency, output power and reliability compared to other solid-state lasers. The crystal is grown using the Czochralski crystal growing technique[6] which involves slowly raising a seed Nd:YAG crystal from the molten crystal constituents to extract an Nd:YAG boule. A single boule typically yields several laser rods. The concentration of Nd ions in the boule is carefully controlled and is not greater than about 1.1%. Increasing the Nd doping further in order to increase the laser power can produce unacceptable strain in the crystal which can lead to a dramatic reduction in laser power.

Laser rods are typically 6 mm in diameter and 100 mm in length with the largest commercial size rods being 10 mm in diameter and 200 mm in length. Because of the crystal's small size, Nd:YAG lasers tend to be much more compact than CO_2 lasers. Illustrated in Fig. 7.4 are the main components of a single-rod Nd:YAG laser. Laser action is achieved by optically exciting the crystal by lamps placed in close proximity to it. The lamps have an emission spectrum, which overlaps the absorption bands of the Nd:YAG crystal at 700 nm and 800 nm. In order to couple the maximum amount of lamp light into the rod and extract the maximum laser power from it, the rod and the lamp are enclosed in specially designed and manufactured cavities.

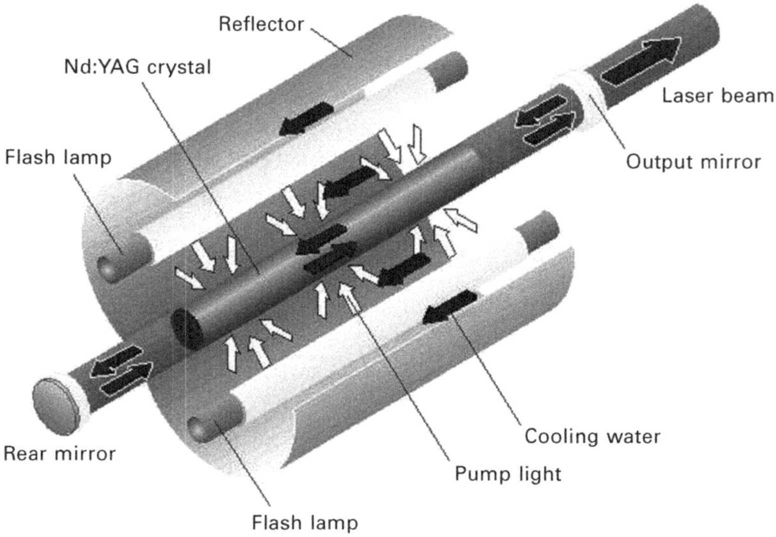

7.4 Schematic of Nd:YAG laser.

The two most common pump cavity configurations are elliptical and close-coupled. In the case of elliptical cross sections, the rod and the lamp are placed along the two foci, and in the case of close-coupled cavities, the rod and the lamp are placed close together at the axis. The inside surface of the cavity is normally coated with gold in order to maximise the coupling of lamp light into the rod. Some laser manufacturers also manufacture ceramic cavities which produce more uniform pumping of the rod but at the expense of lower efficiency (some 5% lower) compared to that of the gold-coated cavities.

For continuous operation, krypton arc lamps are most widely used, while for pulsed operation high-pressure xenon and krypton flashlamps are used. Lamp lifetime dominates the service requirement of modern Nd:YAG lasers. For arc lamps, the lifetime ranges between 400 and 1,000 hours, while for pulsed lasers it is about 20 to 30 million pulses depending on operating conditions.

As only a fraction of the emitted spectrum is absorbed by the laser crystal, the rest of the emitted light is dissipated as heat in the cavity and has to be removed for efficient laser operation. This is usually achieved by flowing deionised water around the rod and lamp in a closed loop cooling system. The loop is coupled to a heat exchanger for efficient heat removal.

Nd:YAG lasers are characterised by their power output over time, i.e. pulsed, continuous wave and super modulated[7,8] (a feature exclusive to JK continuous wave products).[9] Pulsed Nd:YAG lasers employ a power supply designed for delivering high peak powers during the laser pulse and do not have continuous wave (CW) capability. Pulsing implies that the laser's active medium is excited by a very quick response stimulus. This allows the laser to transmit a burst of energy for a brief length of time (generally in terms of milliseconds). Peak pulse powers for pulsing Nd:YAG lasers can reach values of over 30 times greater than the maximum average power levels. This allows low-to-medium power lasers to achieve enough energy to reach vaporisation temperatures for most materials. The basic laser pulse from the pulsed laser is a rectangular pulse as shown in Fig. 7.5. Often the single sector standard pulse is quite adequate when microwelding standard ferrous alloys without any coating or pulsed Nd:YAG microcutting applications. However, with most welding of reflective or dissimilar materials, pulse shaping has a measurable effect on the quality and consistency.[10] Most lasers are rated by their CW output, but pulsed lasers have pulse energy, peak power, pulse width and frequency terminology that must be understood.

Pulse energy

The volume of the melt puddle for each pulse is determined by pulse energy. There is a minimum pulse energy required for weld penetration to a certain

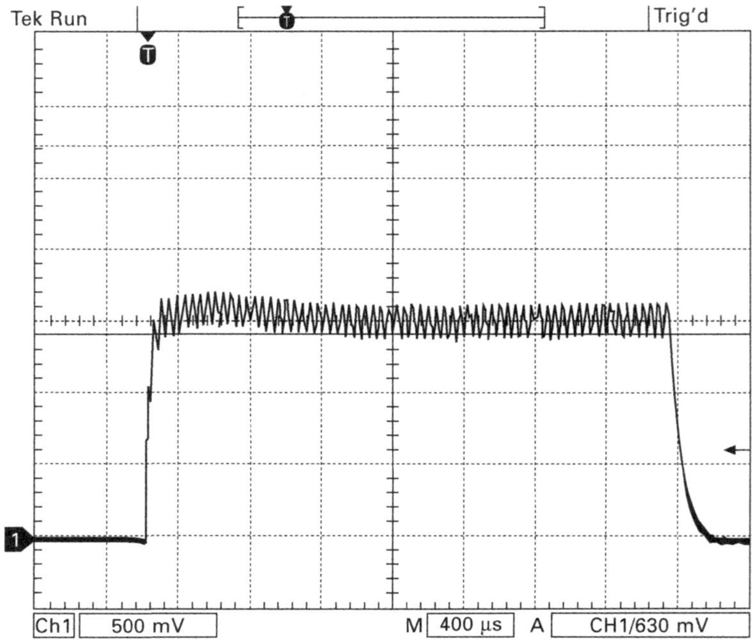

7.5 Basic temporal laser pulse shape from pulsed Nd:YAG laser.

weld depth for a given material. Energy per pulse in joules (E) is related to the average power (P) and the pulse frequency (f) by the following:

$$E \text{ (joules)} = \frac{\text{Power (watts)}}{\text{Frequency (Hz)}} \qquad [7.3]$$

Peak power

Height of the pulse is the peak power as shown in Fig. 7.6 and this peak power breaks down reflectivity and overcomes thermal diffusivity. High peak power is required for precious metal welding and a range of aluminium alloys. The peak power (P_p) can be calculated by the following:

$$P_p \text{ (kW)} = \frac{E\text{(joules)}}{t\text{(seconds)}} \qquad [7.4]$$

Pulse frequency and overlap

During pulsed Nd:YAG welding, seam welds are produced by a series of spot welds. The pulsing rate of the laser results in faster or slower seam welding as the rate is increased or decreased. To produce hermetic welds,

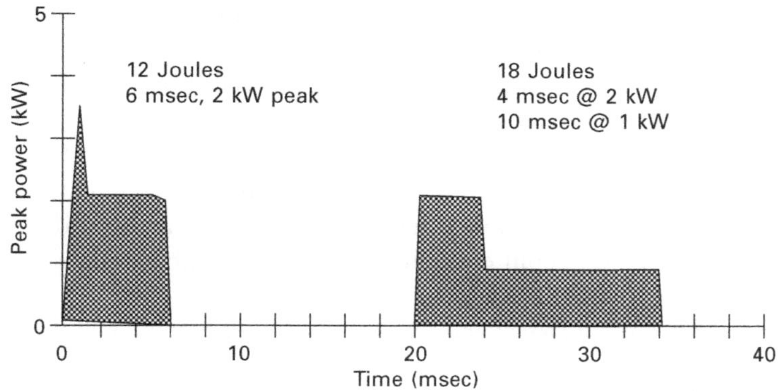

7.6 Temporal laser pulse with an initial spike and shaped pulse from pulsed Nd:YAG laser.

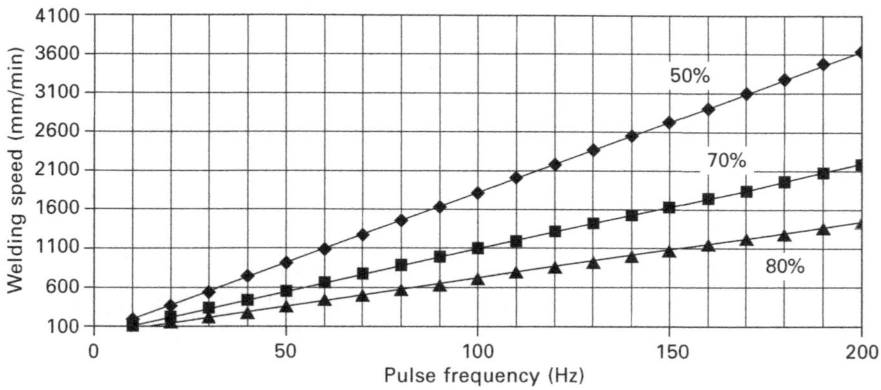

7.7 Welding speed vs. repetition rate for different overlap during pulsed laser welding (spot size = 0.60 mm).

pulse rate (*f*), spot diameter (*d*) and the weld speed (*v*) have to be matched to produce the required percent overlap (%OL). Generally speaking, typical values for hermetic welds are between 70 and 80%OL and for non-hermetic welds between 50 and 60%OL. The percent overlap can be calculated by the following:

$$\%OL = 10_0\{(d - v)|d\} \quad [7.5]$$

Figure 7.7 shows the relationship between welding speed and frequency for three different percentage overlaps.

Fibre laser

The low power fibre lasers are very compact and robust and have an edge over lamp pumped Nd:YAG lasers in terms of beam quality ($M^2 \approx 1.1$) and wall plug efficiency of approx. 25%.[11,12] Current investigations[13,14] have shown that a single-mode fibre laser is an efficient, reliable and compact solution for microcutting, microwelding and microdrilling. The diode-pumped technology offers low maintenance cycles and high conversion efficiency. Theoretical pump-light conversions of more than 80% are possible[15] but typical optical conversion efficiencies for ytterbium double-clad fibre lasers are 60–70%.[16] Average power levels up to 500 W are possible with air-cooling. Since the overall efficiency is high, most fibre lasers are powered by standard 110 V/230 V supplies.

In the design of a robust and reliable fibre laser system, a number of other fibre-based components enable the construction of monolithic 'all-fibre' laser cavities. The benefits of the all-fibre design have been widely discussed elsewhere,[17] with two of the principal advantages being the absence of optical alignment and exposed optical surfaces. Chief amongst these enabling components are the pump combiner and Bragg grating reflector. Figure 7.8 shows a schematic of such a fibre laser cavity design. The output fibre has a single-mode core with a diameter of less than 10 microns which ensures a high beam quality output. This architecture is designed to be capable of producing output powers of up to 500 W with a 100,000 hour diode ensemble lifetime in a water-cooled heatsink configuration and up to 120 W using forced air-cooling at ambient air temperatures of up to 35°C with the same level of diode reliability.

Central to the performance and reliability of the fibre laser system are the pump laser diodes. In the last ten years, multimode diode pump sources emitting many watts of output power in the 900–980 nm wavelength range have become commercially available. Some of these sources have extremely high levels of reliability, with mean time to failure of greater than 500,000 hours under normal operating conditions. By incorporating a suitable level of redundancy, ensembles of such pump sources having a mean time before failure (MTBF) of greater than 100,000 hours can be used to construct fibre lasers emitting several hundred watts of output power.

The advantages of the fibre laser for industrial applications can be summarised as follows:

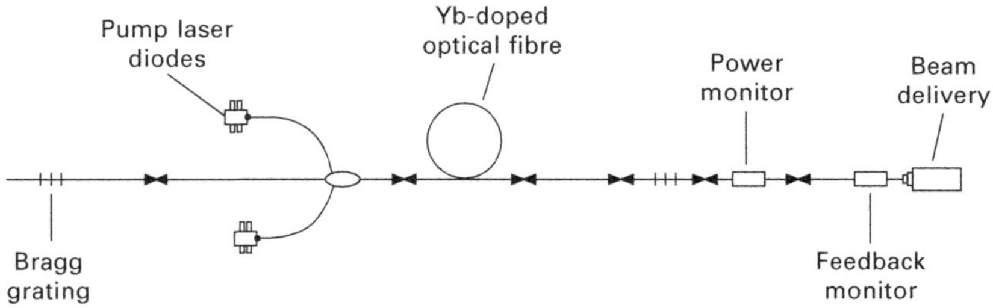

7.8 Fibre laser schematic.

Existing advantages of fibre lasers:

- Good reliability and lifetime
- High stability of laser output leading to consistency of processing
- Small size of overall unit
- Generally longer warranty than standard lasers
- Option of air-cooled or water-cooled up to a few hundred watts output power.
- Lower price than equivalent power traditional laser.

Advantages of emerging industrial fibre lasers:

- Integrated damage protection against backreflection issues
- Control software offering full functionality and ability to be integrated into system level controllers
- Fault diagnostics for improved warning or alarm identification
- End-of-life warning for tracking diode lifetimes
- Reliable, stable and linear power monitor integrated to laser
- Single sourcing for laser and process tools (cutting head, welding head or galvanometer based scanners)
- Ability to increase processing performance of reflective materials through periodic enhancements to laser peak power.

7.2.3 Beam delivery

From the laser microwelding point of view, the laser by itself is very much ineffectual. The laser must be used in conjunction with several different items of optical and mechanical equipment, which need to be integrated into a functional unit, in order to be able to weld materials. It comprises three key elements:

- laser source (Nd:YAG or fibre laser)
- beam delivery and focusing optics (fibre delivered)
- multi-axis, numerically controlled motion system or scanning head.

Fibre delivery

The delivery fibres used with 1.06 μm wavelength industrial lasers are of the 'step index' design. This means that they have a core with a high refractive index surrounded by a cladding with a lower refractive index (Fig. 7.9). Transmission of light occurs by total internal reflection at the core/cladding interface due to the refractive index difference between the core and the cladding. A property of these fibres is that the exiting beam has a relatively homogeneous intensity distribution over its diameter. Depending

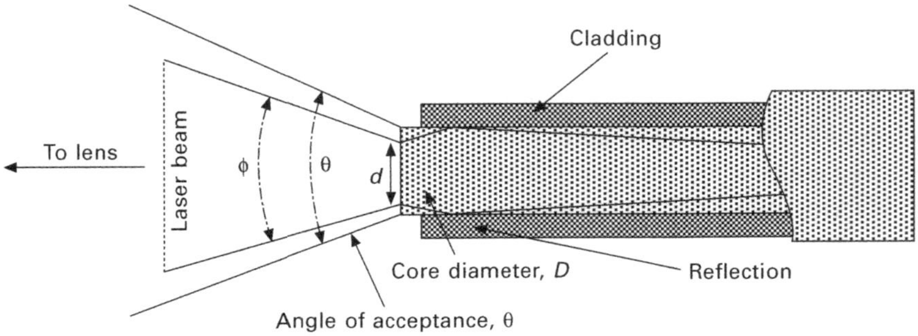

$d_{laser} < D_{fibre}$

$\sin(\phi/2) < \sin(\theta/2) = NA = \sqrt{(n_{core}^2 - n_{clad}^2)}$

7.9 The coupling of laser beam into an optical fibre.

on the power of the laser, the core diameters range in size from 100 to 600 μm for pulsed Nd:YAG lasers and from 10 to 20 μm for single-mode fibre lasers. The fibres are manufactured from high purity fused silica and possess minimal loss at the laser wavelength.

To launch a laser beam into a fibre, so that it experiences a minimal transmission loss as it propagates along it, requires that the diameter of the focused spot on the fibre face be smaller than or equal to the fibre core diameter. The focused spot size in commercial systems is normally 80–90% of the core diameter, which allows for easier adjustment of the fibre and any variation in the spot diameter due to laser parameters.[18] In addition, the divergence of the input laser beam must be less than the acceptance angle of the fibre defined by its numerical aperture, NA.

Delivery of the output beam to the workpiece is an important issue. A fibre laser beam delivery system must protect the fibre end-face from damage due to contamination or back-reflected light from the workpiece. Furthermore, to take advantage of the high beam quality in microwelding applications, high quality optics and 'through-the-lens' viewing are advantageous. GSI Group Laser Division manufactures fibre end-terminations, process and scanning heads.[19]

Figure 7.10 shows a schematic of a fibre termination (Luminator™ technology) designed for high-power YAG laser fibre beam delivery of the JK laser products. This patented technology makes use of a fibre mounting scheme and an angled capillary that is robust against back reflection. GSI are using a variation on this existing design for fibre laser beam delivery. By applying the patented Luminator™ technology to the beam delivery fibre on the GSI Group fibre lasers, an integrated protection system against back reflection issues has been provided that protects the fibre laser at a number of levels.

As the effective mode area at the end of the delivery fibre is very small

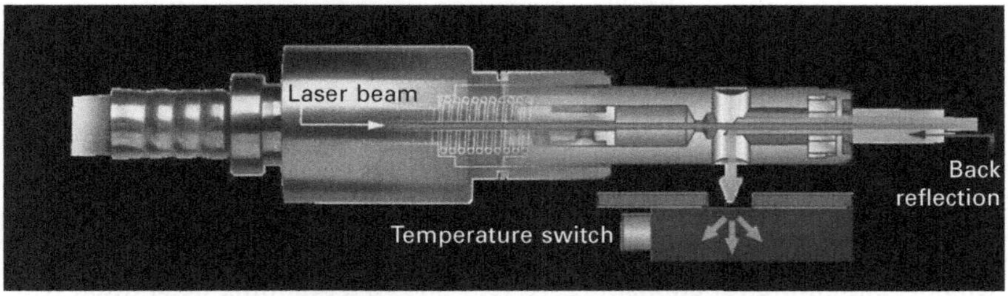

7.10 Fibre delivery incorporating backreflection protection. © JK Lasers, GSI Group.

Table 7.2 Range of scanning heads available for microwelding

Focus lens code	100	160	163	254	330	420
Scan head model	Scan dimensions (mm)					
LDS-7, LDS-10, LDS-14	62×62	99×99	115×115	157×157	217×217	291×291
LDS-20	N/S	N/S	110×110	157×157	217×217	283×291
LDS-25	N/S	N/S	84×84	130×130	170×170	216×216

for single-mode beams, so the power density on the fibre ends is extremely high, meaning that they can be easily damaged if contamination is present, especially in the presence of back-reflected light. By shaping the fibre end, so that the beam is allowed to expand in diameter before it exits the silica material of the delivery fibre, the occurrence of this damage mechanism can be greatly reduced.

Scanning heads

Laser microwelding with a scanning head provides significant advantages over the conventional process tool-based systems. Areas from fractions of square meters to a few square meters can be scanned depending on whether the galvanometer mirrors are in a fixed or movable arrangement (Table 7.2). This can be achieved at rates of many metres per millisecond, which significantly reduces processing cycle time and increases the productivity. In most welding applications the average welding time is 20–50% of the cycle time, which is relatively short compared to the positioning and loading time of the system. By reducing the non-productive times during the manufacturing process with a scanning head, the welding time can be increased by up to 90% or more. Additionally, the shape of the weld can be programmed in a variety of geometrical shapes and figures such as circle, sine wave, open circle and spiral, giving laser welding further structural advantage over conventional spot and MIG welding. As a leader in both the scanner and laser fields, GSI[19] has developed its own scanner and integrated software solution to enable laser-based scanner delivered processing. The system has been designed specifically for laser processing applications rather than the more traditional marking applications for which scan heads are predominantly used. The software application has been seamlessly integrated into its industry-leading LaserViewSE™ laser software.[19] The end result is a single user application that allows the operator to manage all laser and scan head programming functions. Figure 7.11 shows a scanning head fitted with delivery fibre from the laser and Fig. 7.12 shows spot welds made with the scanning head system.

Laser performance

An area where there is a significant difference between lamp-pumped YAG and fibre laser performance is pulsed operation. Lamp-pumped lasers are capable of producing long, multi-ms pulses with peak powers many times the rated average power of the laser, provided that the duty cycle is sufficiently low. This ability stems from the flash-lamp itself, which is often more constrained by the maximum average thermal load than the peak power output. Pulsed Nd:YAG lasers employ a power supply designed for delivering high peak powers during the laser pulse and do not have CW capabilities. These lasers are very good for welding highly reflective materials. High peak

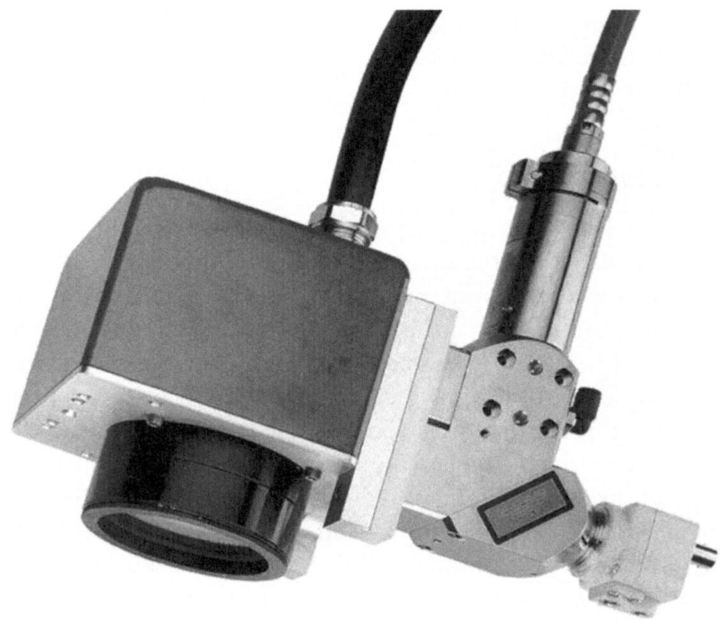

7.11 Scanning head fitted to delivery fibre from the laser.

7.12 Spot welds made with pulsed Nd:YAG laser fitted to a scanning head.

power overcomes the thermal diffusivity and reflectivity of precious metals, copper and aluminium alloys. Some materials, i.e. high carbon steels, casting alloys, and a range of aluminium alloys, are difficult to weld. These materials tend to either crack or produce welds with pores or other defects. It may be possible to produce defect-free welds in these materials by changing the laser pulse shape.[20] The basic laser pulse from a pulsed laser is a rectangular pulse which is adequate for microwelding most ferrous materials without any coating or plating. For crack-sensitive alloys, welding of dissimilar materials or materials with coatings, a metallurgical, or sometimes known as ramp-down, temporal pulse shape is used.[20]

By contrast, while the semiconductor laser diodes used to pump a fibre laser can be on-off modulated over a wide frequency range as shown in Fig. 7.13 (from DC to tens of kHz in most industrial applications), they cannot typically be over-driven for long periods (multi-ms), in the same way as a flash-lamp, without reducing the lifetime of the device to an unacceptable level.

7.2.4 Beam profiles

Apart from different temporal pulse shapes of both lasers, another factor to consider when choosing a laser for microwelding is the beam profile of the lasers. The beam profile of the laser can affect the welding performance, i.e. weld penetration depth, weld profile and weld quality. A top hat beam profile as for lamp-pumped pulsed Nd:YAG lasers (Fig. 7.14), for example,

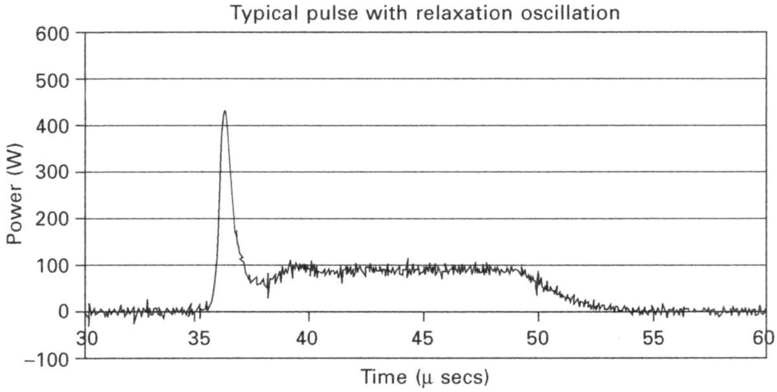

7.13 Typical output waveform with relaxation pulse, when turning the SM fibre laser on and off.

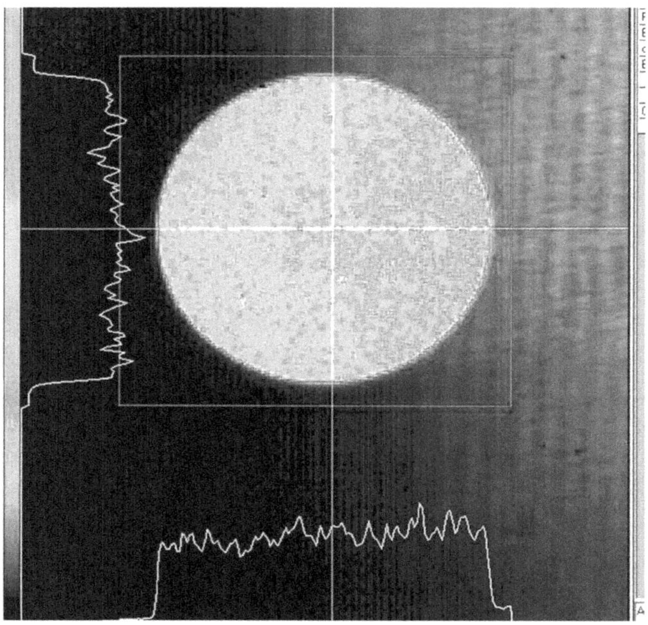

7.14 Nd:YAG pulsed laser beam profile (top hat).

is more forgiving since the power and energy are evenly distributed across the weld zone. Top hat energy distribution is achieved when the laser beam is delivered to the workpiece via a fibre optic cable. Increasing the pulse energy increases the weld depth (Fig. 7.15). Vaporisation does not occur if the power/energy density remains below the vaporisation point.

The beam profile from a single-mode fibre laser is Gaussian in profile (Fig. 7.16). This produces very deep and narrow welds (Fig. 7.17). For

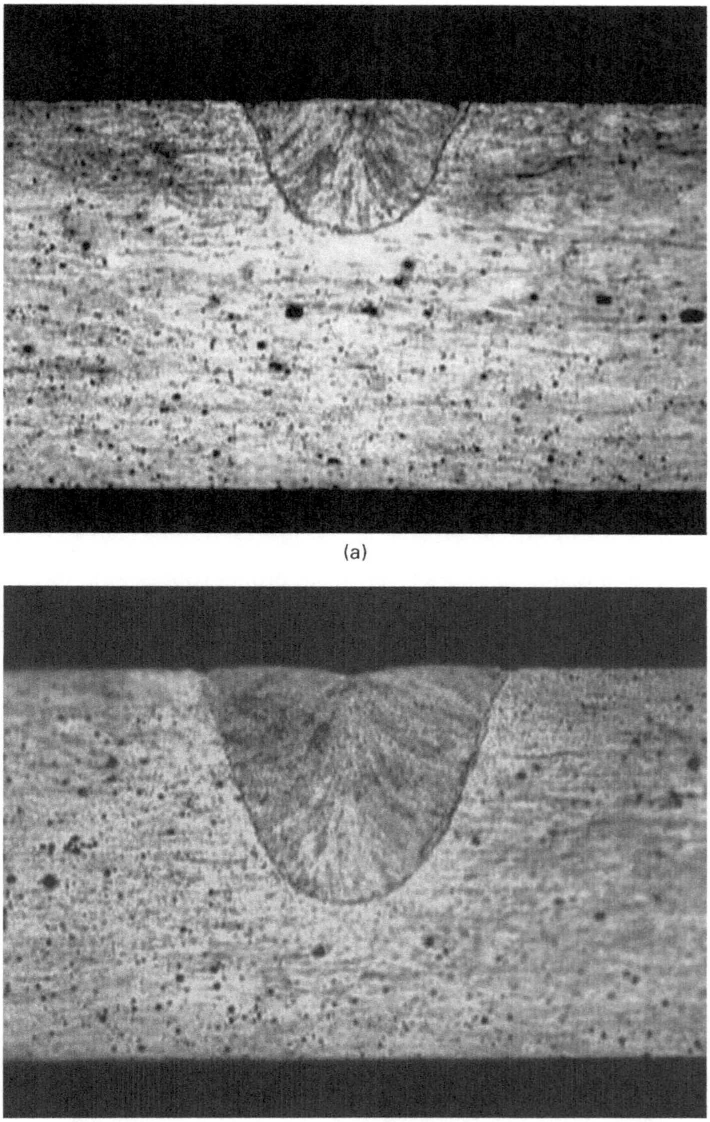

(a)

(b)

7.15 Cross sections of top hat welds at different pulse energies; *a* = 0.80 J, *b* = 1.96 J.

overlap joint welds, this weld shape is not so good in terms of mechanical properties because the interface is very small compared to the top hat beam profile. However, it is possible to have a low power fibre laser 100–500 W with a flat top beam profile (Fig. 7.18). This leads to a lower beam quality by introducing extra fibre modes and the delivery fibre is of large core

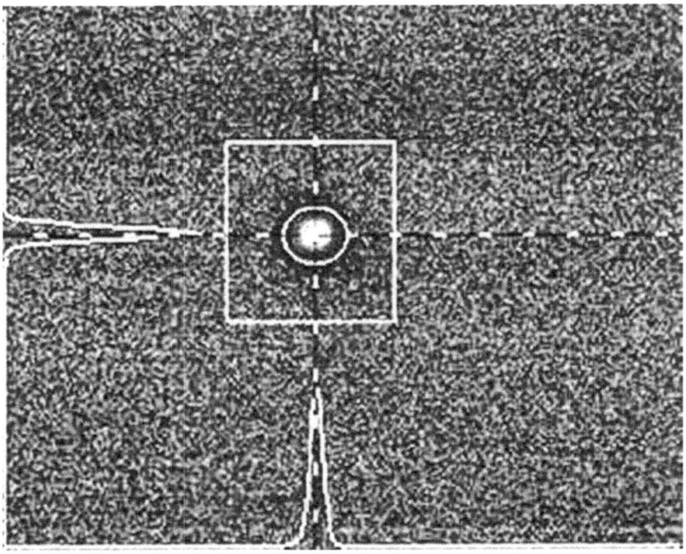

7.16 Single-mode fibre laser beam profile (Gaussian); fibre diameter size 10–12 μm.

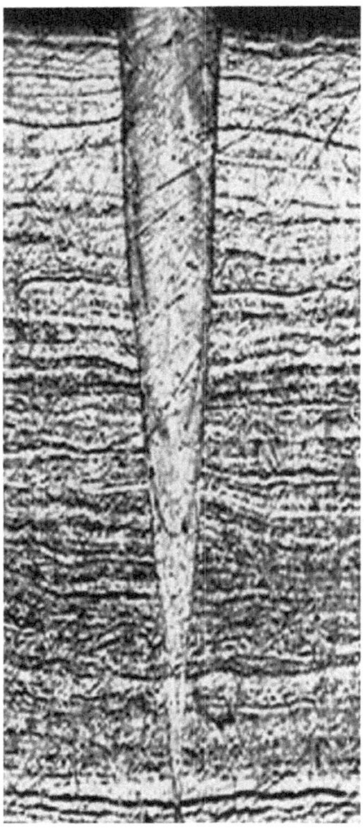

7.17 Cross section of Gaussian beam profile weld; 100 W average power.

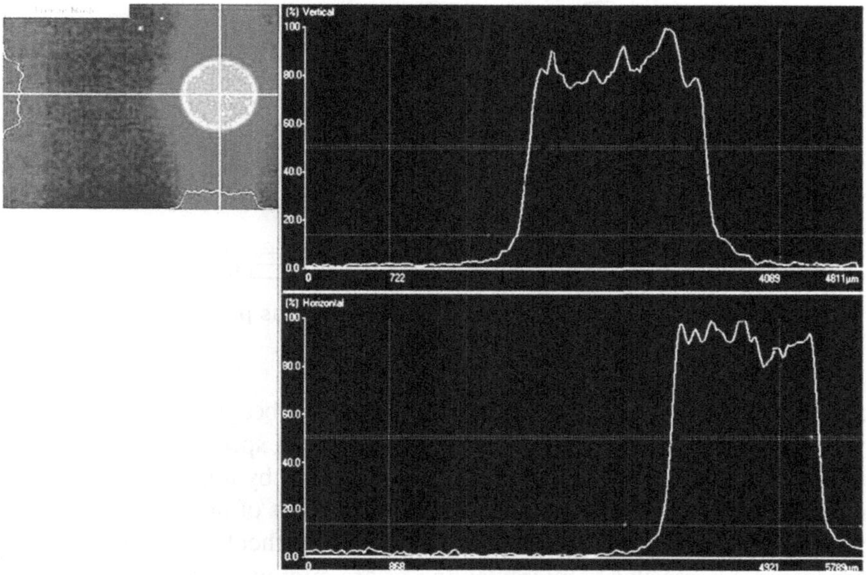

7.18 Top hat (flat top) beam profile of 300 W fibre laser; fibre diameter size 100 μm.

diameter, i.e. 100 μm compared to 10–20 μm diameter fibre for the Gaussian beam profile. Both of these beam profiles can provide advantages during microwelding.

7.2.5 Nd:YAG versus fibre

Considering the differences in beam quality and pulsed performance between the two types of laser, there are different operating regimes for the two types of laser. The lamp-pumped YAG laser is characterised by long high-energy pulses but poorer beam quality, and the fibre laser is associated with high repetition rate, on-off type modulation, single-mode beam quality but low pulse energy. From the perspective of microwelding applications, both of these regimes have their advantages and a few of the application examples are highlighted in Section 7.5.

7.3 Laser microwelding process

7.3.1 Welding modes

The heat transfer and molten metal flow experienced during laser microwelding can significantly influence the weld pool geometry, solidification structure,[21,22] and defect formation.[23] Laser beam microwelding is a process that melts and joins metals by heating with a focused laser beam. Figure 7.19 shows the

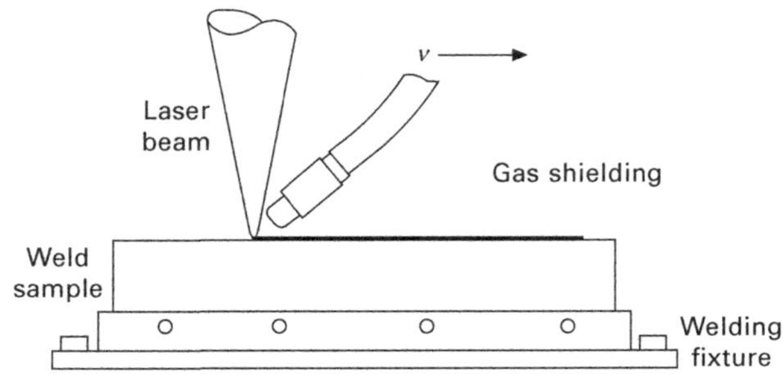

7.19 Schematic of laser welding, shielding gas prevents oxidation during welding.

schematic setup for laser welding. The focused laser beam is made to melt the workpiece or joint at given laser power and welding speed. A shielding gas, usually an inert gas, is supplied either coaxially or by a pipe to protect the molten weld pool from oxidation. The main features of laser microwelding, which make it an attractive alternative compared to other fusion microwelding processes (i.e. resistance microwelding or arc microwelding), are:

- precise narrow and deep welds can be produced with high metallurgical quality
- small heat affected zone which reduces metallurgical damage and also allows welds to be made close to heat-sensitive components
- the low heat input into the material obviating the need for complex jigging and allowing distortion-free welding of thick to thin sections
- high process speed (welding speeds in excess of 10 m/min can be achieved with materials of thickness about 1 mm)
- flexibility allowing one laser to be shared amongst a number of workstations
- post-weld treatment is not normally required
- welds can be performed in difficult geometries and dissimilar material thicknesses.

Laser microwelding of metals is classified into two welding modes i.e. thermal conduction and keyhole or sometimes known as deep penetration welding.[24,25] Both the laser beam power and the focused spot size with respect to the workpiece will govern the welding modes.

Conduction limited welding (Fig. 7.20) occurs when the laser beam is out of focus and power density is insufficient to cause boiling at the given welding speed. In this welding mode, overlapping spots from a pulsed laser or the beam from a continuous laser is absorbed by the surface of the material and the volume below the surface is heated by thermal conduction producing a semi-circular cross section. The minimum laser power needed to melt the material's surface can be evaluated by a simple equation.[26] When

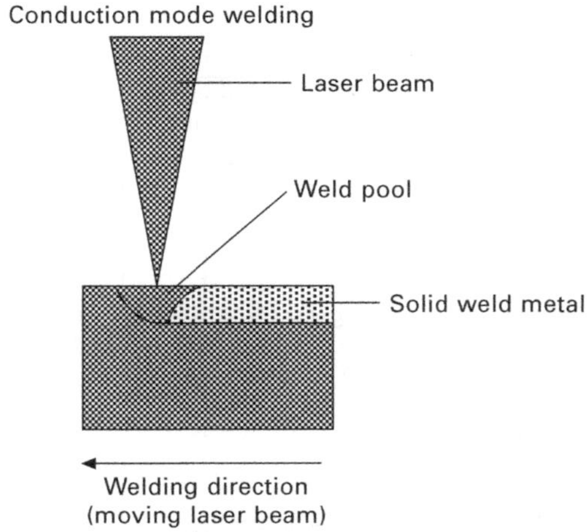

7.20 Principle of conduction mode laser welding.

a stationary Gaussian beam interacts with material surface, the temperature at the centre of the heat source $T(t)$ is given by:[27]

$$T(t) = \frac{2AW}{\frac{\pi 3}{2}} \tan^{-1}\left[\frac{4}{d}\sqrt{\alpha t}\right] \quad [7.6]$$

where d is focused laser spot diameter, W is the laser power, A is the absorptivity of the surface, K is the thermal conductivity and α is the thermal diffusivity. Thus the minimum laser power needed to melt a metal of melting temperature T_m is given by:

$$W_m = \frac{\sqrt{\pi K d T_m}}{A} \quad [7.7]$$

One of the most important parameters influencing the effects of laser–material interactions is the absorptivity of the material for laser radiation and the absorptivity of the material is greatly influenced by the wavelengths and temperature.[28] When the laser beam strikes the target material, part of the energy is reflected, the rest is absorbed, and the absorbed energy heats up the target material. The absorptivity is strongly dependent on the material being irradiated and on the laser wavelength. Figure 7.21 shows the absorptivity of some common metals which are routinely used in a range of microwelding applications with different laser wavelengths.[28] Figure 7.22[26] shows the minimum laser power needed with a given focused spot size to melt common microwelding metals with Nd:YAG wavelength (1.06 μm). It can be seen that silver is the most difficult material for conduction mode welding, requiring almost 900 W to melt. Aluminium-based alloys, steels and nickel are the easiest to weld. Copper is also difficult for conduction mode laser welding

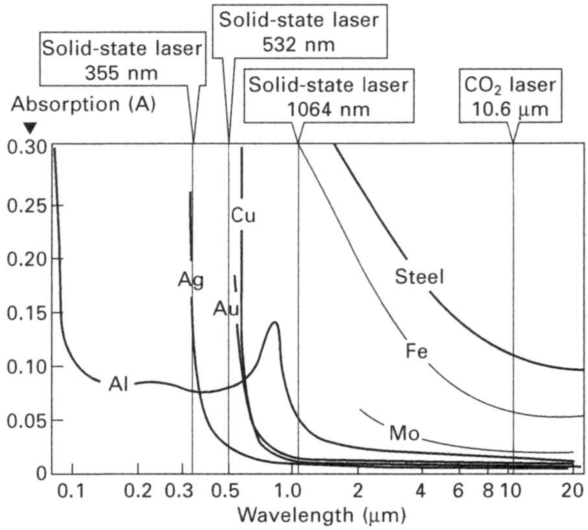

7.21 Absorptivity vs. wavelength of different metals for selected lasers.

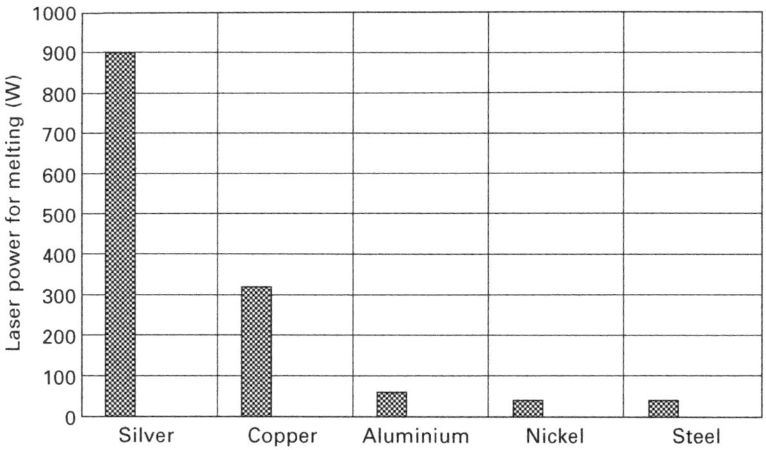

7.22 Minimum laser power needed to melt different metals with pulsed Nd:YAG laser.[26]

because it requires approximately 300 W to melt. Sometimes copper-based alloys are nickel plated to enhance laser absorption and reduce the laser power needed to weld this material.

In keyhole or deep penetration welding (Fig. 7.23), the metal surface is heated by a focused laser beam (increased power density) to a temperature where the surface evaporation creates a recoil force sufficient to depress the molten surface into a cavity or a small keyhole. The keyhole grows in

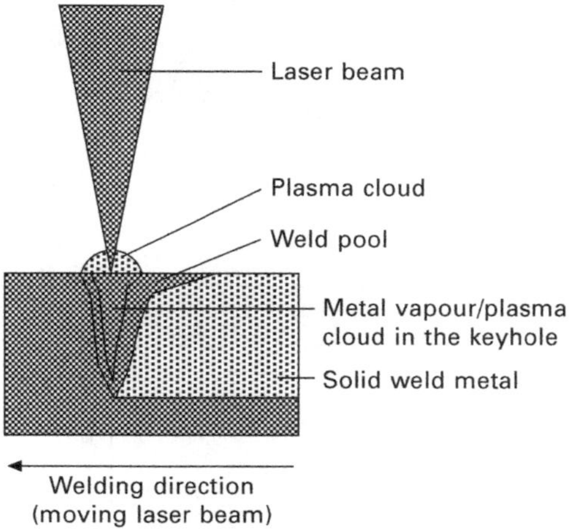

7.23 Principle of keyhole or deep penetration mode laser welding.

depth because of increased coupling of radiation into the workpiece, through multiple reflections of the laser beam off the keyhole walls, and material vaporisation.[25,29,30] The balance between the hydrostatic forces of the liquid metal surrounding the keyhole governs its existence and the pressure of vaporised and ionised material or plasma within it.

Figure 7.24 shows the transition in the weld bead shape and the weld penetration by changing the power density at the workpiece. The welds were made in stainless steel with a 400 W single-mode fibre laser.[31] The micrographs show that as the power density increases, the welding mode changes from conduction to keyhole. During these tests the welding speed was kept constant at 9 m/min.

7.3.2 Welding joints

When considering laser microwelding, the design of the joint and possibly the fabrication method of the whole product should be assessed in order to gain the maximum advantage from the process. Laser microwelding requires high tolerances in gap control and joint positioning (Fig. 7.25). For macrowelding, filler material can be used to widen the tolerance field, but in practice this is not very common because of the associated cost related with the process and also reduction in the weld speed. Laser microwelding is usually performed without any filler metals so gaps must be filled with metal from the adjacent area and bridging a gap requires extra laser energy. Generally, a gap no more than 10% of the thickness of the thinnest component is allowed. This can be relaxed for thicker materials greater than 1 or 2 mm but might need to be reduced to 5% or lower for materials less than 0.2 mm. Gap problems show themselves as very concave weld beads or failure to

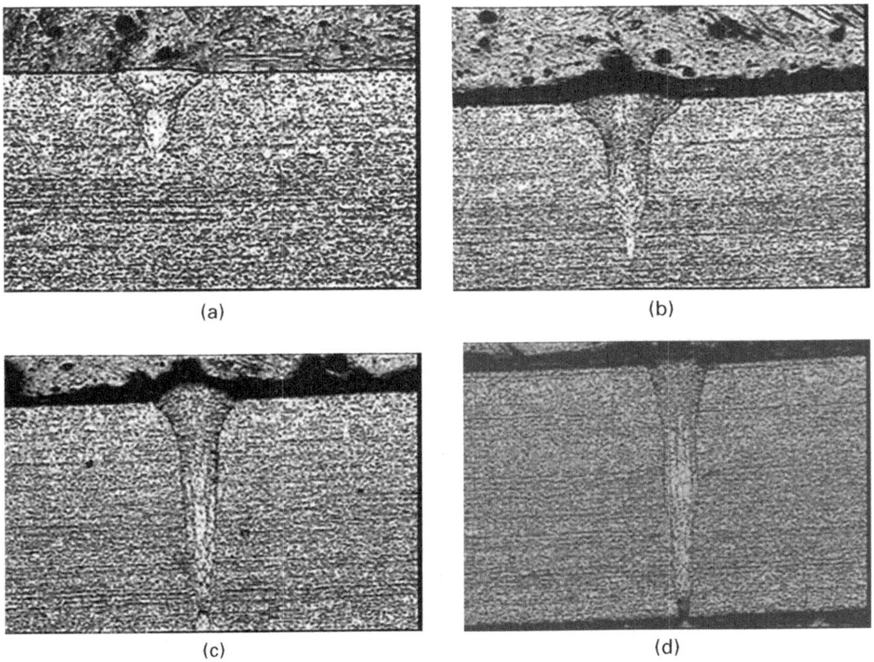

7.24 Welds made with 400 W (JK400FL) single-mode fibre laser; 304SS, 9 m/min, argon shield gas: (a) 4.98 MW/cm^2; (b) 5.98 MW/cm^2; (c) 6.98 MW/cm^2; (d) 7.98 MW/cm^2.[31]

bridge the gap between parts. It is also much more difficult to start a laser weld in a large-gap area compared to seam welds through a short section of high-gap area where bridging can sometimes be maintained. In addition to the suitable preparation and alignment of the joint faces, they should be free from contaminants such as grease, paint, dirt and oxide scales. Residue from chemical degreasing and cleaning agents should be carefully removed, since weld spatter and porosity (see below) can result if these substances are present on the workpiece.

7.3.3 Materials

The metallurgy of laser welding is not much different from other conventional welding techniques, but there are two very important features which make laser microwelding very challenging and the engineer/designer must be aware of these before designing the component and deciding on the material type. The first point to note is that laser welding is almost always an autogenous process, which means that no metal is added during the welding process. This was discussed previously in the consideration of fit-up tolerances. The second special feature of laser welding is the relatively rapid cooling rate

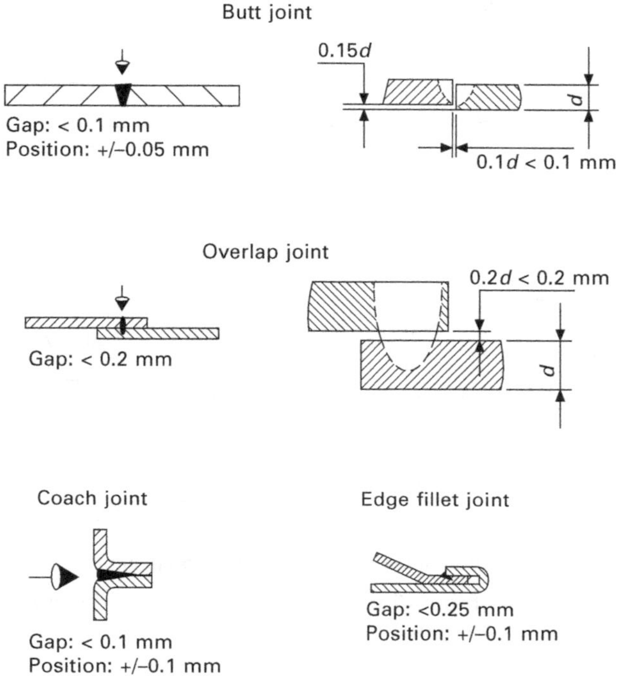

7.25 Maximum gap and positioning tolerances for different laser welded joints.

of the solidifying metal, which places some special constraints on a few metal choices.

When microwelding microwave packages, a designer requires a lightweight, corrosion-resistant, heat-dissipating, robust, and cheap material. Aluminium is usually the first choice. Aerospace packages for microwave circuits, sensor mounts, or small-ordinance initiators are the most common examples of aluminium components that need to be laser welded. Laser welding with penetrations up to 1.5 mm are common in these applications and 6061-T6 aluminium alloy (Al-Mg-Si) is the material of choice because of economics, rigidity and ease of machining. However, the material cannot be successfully laser welded to itself, because the partially solidified melt zone cannot withstand the stress of shrinkage upon solidifying, and cracks are formed, known as solidification cracking or hot cracking. Hot cracking is the cause of almost all cracking in aluminium weldments. Hot cracking is a high-temperature cracking mechanism and is mainly a function of how metal alloy systems solidify. This cracking mechanism is also known as hot shortness, hot fissuring, solidification cracking and liquation cracking.

Figure 7.26[32] shows the aluminium crack sensitivity curves and is a helpful tool in understanding why certain aluminium welds crack. The crack

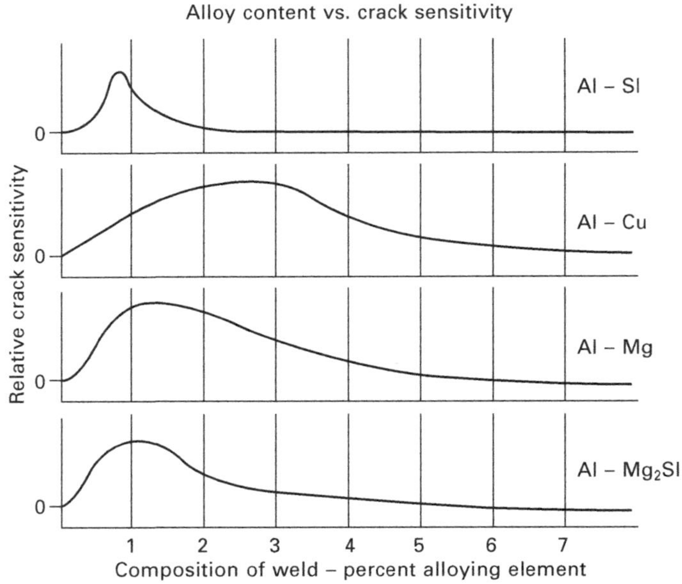

7.26 The effects of four different alloy additions on the crack sensitivity of aluminium.

sensitivity curves show that with the addition of small amounts of alloying elements, the crack sensitivity becomes more severe, reaches a maximum, and then falls off to relatively low levels. After studying the crack sensitivity curves, it is easy to recognise that most of the aluminium-based alloys are considered unweldable autogenously (without filler alloy addition). The solution to this problem is to improve the ductility of the weld metal by using aluminium with high silicon content, such as alloy 4047 (Al-12% Si). This alloy is very ductile as a solid and difficult to machine into complex shapes. Therefore, 6061 is usually employed as the package component with intricate features, and 4047 is used as a simple lid that is relatively thin (typically less than 1 mm). The only aluminium alloys that can be welded with low heat input and without the use of 4047 filler are the 1000 and 1100 alloys. These commercially-pure aluminium alloys have the metallurgical characteristics to avoid hot cracking, but their poor mechanical and machining properties usually prohibit use in most applications.

Kovar (Fe-29% Ni, 17% Co) is another material which is often chosen for microwelding because its thermal expansion coefficient matches that of other package constituents such as glass-to-metal seals. Plated Kovar offers good corrosion resistance and can be machined and drawn relatively easily. Kovar is denser and heavier than aluminium, but it presents few metallurgical problems compared to those of aluminium. In addition, it provides the benefit of a low coefficient of thermal expansion. Usually welded to itself, Kovar welds with ease up to 2 mm penetration. It is important to consider plating constituents when specifying Kovar package components.

Stainless steel with its excellent corrosion resistance and good metallurgical characteristics is used in a number of microwelding applications. Stainless steel is slightly more difficult to machine, and is heavier and more expensive than aluminium. Some aerospace packages employ stainless steel, but the majority of uses seem to be in the military, the medical field or in automotive airbag systems. The various types of stainless steel are identified, and guidance given on welding processes and techniques that can be employed in fabricating stainless steel components without impairing the corrosion, oxidation and mechanical properties of the material or introducing defects into the weld. The unique properties of stainless steels are derived from the addition of alloying elements, principally chromium and nickel, to steel. Typically, more than 12 or 13% chromium is required to produce a stainless iron.

Austenitic stainless steels

These steels are usually referred to as the 300 series and are generally suitable for laser microwelding. Slightly higher weld penetration depths or increased weld speeds can be achieved when compared with low carbon steels (Fig. 7.27)[33] due to the lower thermal conductivity of most stainless steel grades. The high speeds of laser welding are also an advantage in reducing susceptibility to corrosion caused by precipitation of chromium carbides at

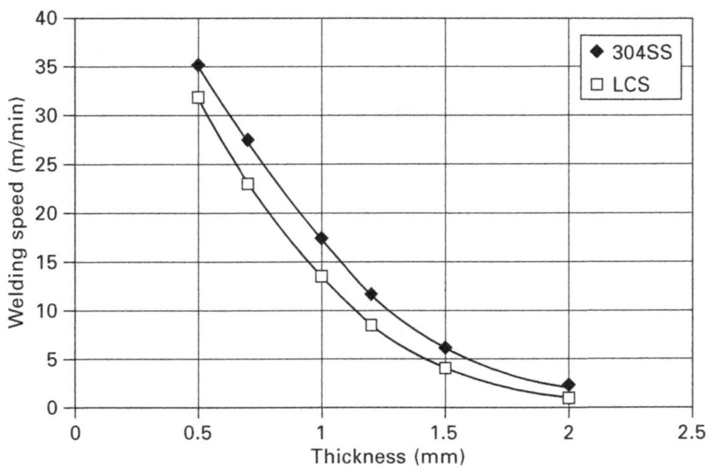

7.27 Welding performance of 304 stainless steel and low carbon steel. Welds made with 400 W single-mode fibre laser (JK400L), spot size 30 μm.[33]

the grain boundaries that can occur with high heat input welding processes. In addition, laser welding of these grades results in less thermal distortion and residual stresses compared with conventional welding techniques, especially as these steels have 50% greater thermal expansion than plain carbon steels. The use of free-machining grades should be avoided because these steels contain high sulphur and high phosphorous that can lead to hot cracking. These elements segregate to the weld centreline, causing a brittle zone that cracks under the stress of solidification (hot cracking).

Ferritic and martensitic stainless steels

These 400 series steels do not possess the good all-round weldability of the austenitic grades. Laser welding of the ferritic grades in some cases impairs joint toughness and corrosion resistance. The reduction in toughness is due in part to the formation of coarse grains in the heat-affected zone, and martensite formation, which occurs in the higher carbon grades. The heat-affected zone may have a higher hardness due to the fast cooling rate. Pre-heating can reduce martensite formation in 400 series alloys. Some 400 series alloys can be welded to 300 series alloys with good results, but again, results can vary from batch to batch, or with variations in heat input.

Titanium and its alloys is another material used extensively in microwelding applications, especially in the medical industry, i.e. for pacemakers and pacemaker battery packages. Commercially-pure titanium and Ti-6Al-4V weld extremely well, but nitrogen cannot be used as shielding gas because of the formation of titanium nitride. Argon or helium must be used to prevent oxidation.

Pure copper and its alloys are used mainly in microelectronic applications. Pure copper has good metallurgy for welding, but it is highly reflective to Nd:YAG laser energy and has high thermal conductivity, therefore making it difficult to achieve weld penetrations greater than 0.5 mm. The reflectivity of pure copper can be overcome by plating it with electroless nickel before welding. Beryllium copper (BeCu) has better weldability and can produce very good welds to pure copper.

Copper-tungsten is very heavy, but has very good heat conductivity, as well as a thermal expansion close to that of many electrical components. Copper-tungsten and copper-nickel alloys weld well. Brass alloys are not good candidates because of their zinc content. Zinc vaporises near the melting temperature of other metals, and vapour expansion tends to expel metal out of the weld pool. The little molten metal remaining solidifies, trapping the gas pockets in the joint, creating undercutting and porosity.

Both silver and gold can be laser welded but with penetration limited to less than 0.5 mm because of high reflectivity and thermal conductivity. These metals are used for special applications in aerospace, electronics and military ordnance, where corrosion resistance and electrical conductivity are paramount.

7.3.4 Microwelding of dissimilar materials

Many applications in various industrial sectors, i.e. electronics, medical, consumer goods, automotive applications, etc., require joining of dissimilar metals. The trends of joining dissimilar metals of components are creating big challenges for welding methods. The weldability of dissimilar metals depends on many different factors. The physical properties have a big influence on the amount of coupled-in energy and the heat transfer.[34] In principle, a laser can also weld any material that can be joined by conventional processes. Illustrated in Table 7.3 is the weldability of metal pairs. In the welding of dissimilar metals, good solid solubility is essential for sound joint properties. This is achieved only with metals having compatible melting temperature ranges. If the melting temperature of one material is near the vaporisation temperature of the other, poor weldability is obtained and often involves the formation of brittle intermetallics. When joining dissimilar materials, there may be certain advantages in using laser welding even though brittle intermetallics may tend to form. Since the weld itself is narrow, the volume of intermetallics may also be reduced to acceptable limits. Again, it may be possible to offset the beam in one direction or the other, thus allowing some control over composition of the resulting alloy. Although it may be possible to produce sound joints by these methods on a laboratory scale, it is more difficult to achieve similar control under production conditions. Mixing the molten metal in a laser weld seldom produces a chemically homogeneous

Table 7.3 Weldability of metal pairs

	Al	Ag	Au	Cu	Pt	Ni	Fe	Ti	W
Al	–	C	X	C	X	X	X	X	X
Ag	C	–	S	C	S	C	D	C	D
Au	X	S	–	S	S	S	C	X	N
Cu	C	C	S	–	S	S	C	X	D
Pt	X	S	S		–	S	S	X	X
Ni	X	C	S	S	S	–	C	X	X
Fe	X	D	C	C	S	C	–	X	X
Ti	X	C	X	X	X	X	X	–	X
W	X	D	N	D	X	X	X	X	–

Al: aluminium; Ag: silver; Au: gold; Cu: copper; Pt: platinum; Ni: nickel; Fe: iron; Ti: titanium; W: tungsten
C: complex structures may exist; X: intermetallic compounds formed, undesirable combination; S: solid solubility exists in all alloy combinations; D: insufficient data for proper evaluation; N: no data available

fused zone between the two dissimilar materials. Although the average chemical composition of the weld may be acceptable, local heterogeneity can be responsible for the presence of brittle zones. It will also be apparent that minor variations in the beam position can significantly influence the relative proportions of the two main constituents in the weld zone.

To date most microwelding of dissimilar metals has been carried out with pulsed lamp Nd:YAG lasers.[35,36] Lamp-pumped lasers are capable of producing long, multi-ms pulses with peak powers many times the rated average power of the laser, provided that the duty cycle is sufficiently low. This ability stems from the flash-lamp itself, which is often more constrained by the maximum average thermal load than the peak power output. High peak powers of pulsed lamp pumped Nd:YAG lasers coupled with pulse shaping make these lasers ideal for welding dissimilar materials. The shaping of pulses is of great importance since the temperature has to be controlled in the region where the two molten phases are mixed. Weld depth that is too deep – which can lead to defective joints – and also insufficient weld depths can be avoided by adjusting the high starting power and the correct decreasing power to the joint geometry and the material properties of the pulse shape. Figure 7.28 shows examples of dissimilar material welds made with a lamp pumped Nd:YAG laser. These welds were made with a ramp down pulse shape (Fig. 7.29).

Compared to conventional lamp pumped Nd:YAG lasers, fibre lasers offer a number of advantages when microwelding dissimilar materials. The big advantage of these new laser sources is the beam quality or focusability. Focusability is understood as the ability to achieve a small focus diameter with a given optical element.[37] The small focus diameter of the fibre lasers offers a number of advantages during laser microwelding, i.e.

- high power density at the workpiece
- reduced heat input
- reduced heat-affected zone
- reduced cycle time
- the volume of intermetallics may also be reduced to acceptable limits.

7.3.5 Plating and coatings

The majority of the materials used in microwelding applications are either plated or coated to improve corrosion resistance, solderability, or for better absorption of the laser energy. Nickel and gold are the most common plating choices. Nickel is used alone, and it is almost always used as under-plating for gold or tin. These metals are plated on top of the nickel, because they do not plate well directly to other base metals. Nickel can be plated by an electrolytic or electroless process. Ferrous alloys such as Kovar and

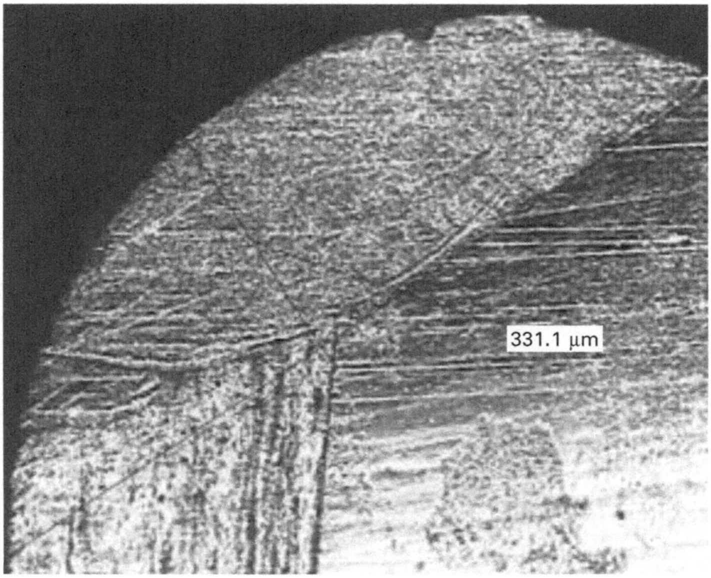

(a)

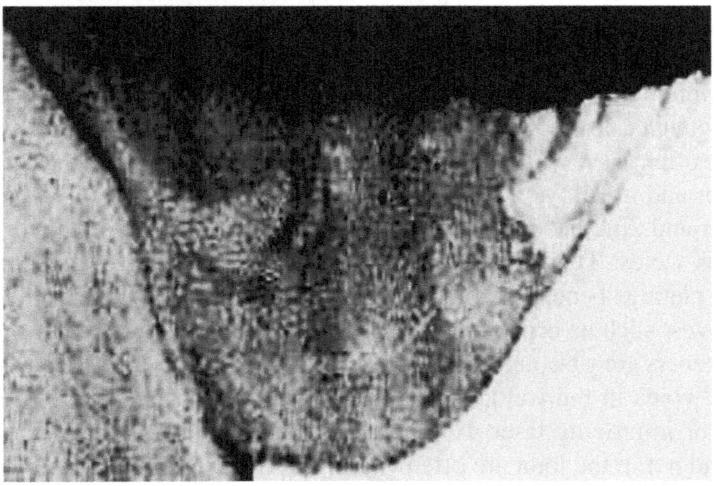

(b)

7.28 Examples of dissimilar weld joints; (a) Kovar to stainless steel, (b) 304 stainless steel and bronze. These welds were made with 125 W pulsed Nd:YAG laser (JK125); spot size 60 μm.

stainless steel must be plated with electrolytic nickel. Electroless nickel plating contains phosphorous, which produces intermetallics within the weld that cause porosity and cracking. Gold-plated Kovar packages weld well, as long as there is not under-plating of electroless nickel. Gold-plated mild steel produces weld voids due to intermetallic segregation in the weld.

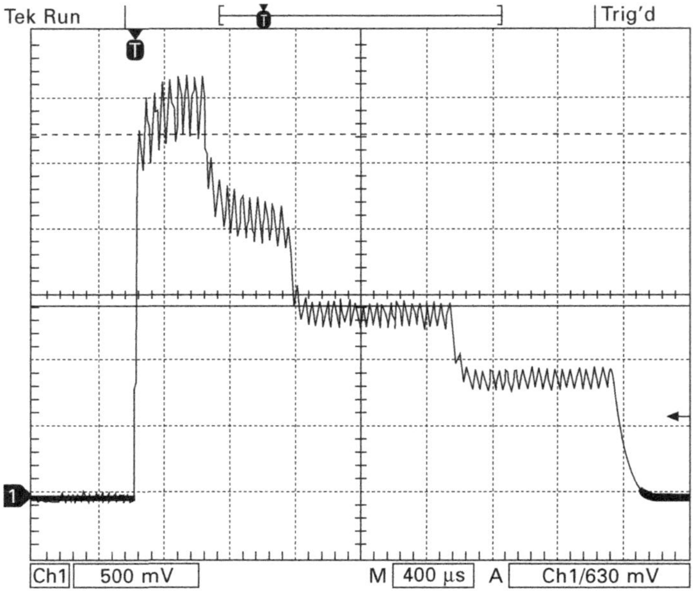

7.29 Example of ramp down temporal laser pulse shape used mainly for welding dissimilar or crack-sensitive materials.

Aluminium and copper can benefit from nickel plating, because the plating has better laser-energy absorbability than the base metals. This technique is not often used with aluminium but is an excellent aid in the welding of copper and its alloys.

Tin and zinc have low melting and vaporisation points, which result in porous welds. These platings should be absent from the weld joint. Even when platings beneficial to laser welding are chosen, there are other plating additives, such as organic brighteners, which can cause problems. Organic brighteners are gaseous when metals are liquid and, as discussed above, they create voids in the weld. Matte finishes with no organic brighteners work best for improving laser welding performance. Platings that only exist on top and not in the joint are often vaporised by the impinging laser beam and will not affect the weld quality.

7.4 Defects and evaluation of microweld joints

7.4.1 Weld defects

Welding defects such as porosity, lack of fusion, spatter, burn through, oxidation and different forms of cracking can occur during laser microwelding either due to material-related problems as outlined in Section 7.3.3 or during the welding process.

Weld cracking

There are a number of cracking mechanisms associated with the welding of metallic alloys. One of the most notorious is hydrogen cracking, also referred to as cold cracking. Hydrogen cracking is often a major concern when welding carbon steels and high strength low alloy steels. However, when welding aluminium alloys, hydrogen cracking cannot occur. Hot cracking is the cause of almost all cracking in aluminium weldments. Hot cracking is a high-temperature cracking mechanism and is mainly a function of how metal alloy systems solidify. This cracking mechanism is also known as hot shortness, hot fissuring, solidification cracking and liquation cracking. The cracking mechanisms that can occur during laser welding can generally be resolved by process parameter optimisation and fixturing. The combination of fixturing and weld configuration often introduces geometry defects that provide location of high stress concentration that can start a crack.

Figure 7.30 shows examples of microweld with excessive cracking and porosity, however, by just changing the temporal pulse shape of the pulsed laser, it is possible to produce defect-free cracks in the same materials.

Undercutting

Undercutting or weld blowout of the microweld is attributed to two main processing parameters. The first mechanism is due to high laser power which results in the expulsion and loss of liquid material from the weld pool (Fig. 7.31). The loss of the material comes from splattering and evaporation.[38] However, the loss of the material through splattering is just down to high laser power. A more valid criterion would be the laser power divided by the welding speed, as this governs the heat input and the rapid flow of the energy through the keyhole.[39] The other type of undercutting defect is observed when the gap is too big in overlap or butt joint welding (Fig. 7.32). This defect can be prevented by reducing the gap with good fixturing/clamping.

Weld spatter and porosity

Weld spatter is defined as the molten metal droplets that are thrown out of the fusion zone (Fig. 7.33). The formation of the spatter can be controlled by reducing the peak power and reducing the weld temperature. However, changing the peak power settings could also lead to a reduction in the weld penetration.

Rapid closing of the keyhole, thus preventing full penetration through the plate, can lead to the formation of porosity (Fig. 7.34). With the rapid cooling usually present with laser welding, pores are left in the weld when the rapidly solidifying pool traps pores before they can float to the surface.

However, with pulsed Nd:YAG welding, it is possible to tailor the pulse shape to allow the pores sufficient time to rise to the top. A long cooling segment of the pulse shape is normally utilised to reduce porosity.[40]

In the majority of cases, some sort of shielding gas is used during laser microwelding. The function of the shielding gas is to protect the weld surface

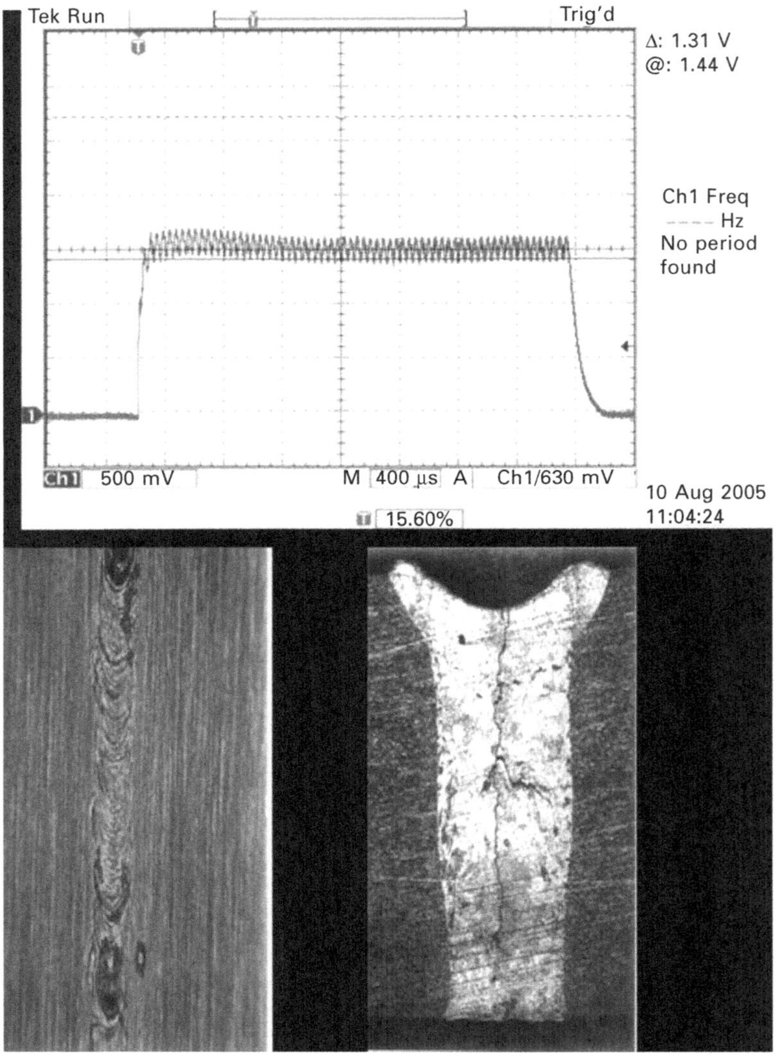

(a)

7.30 Pulsed laser welding of 6061 aluminium alloy: (a) the weld shows excessive porosity and centreline cracking with standard temporal laser pulse shape; (b) ramp down temporal laser pulse shape produces crack- and porosity-free weld.

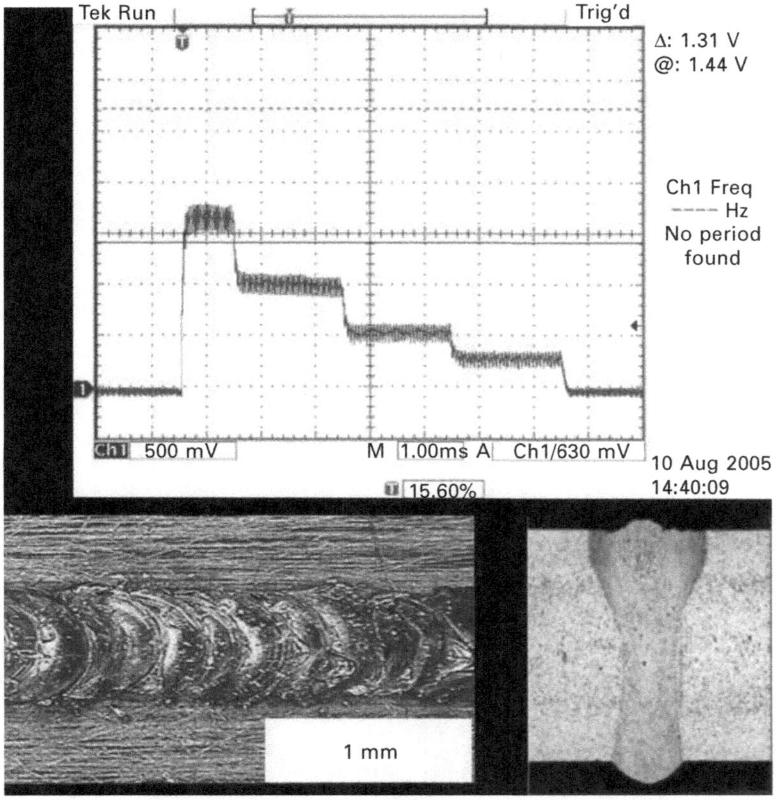

(b)

7.30 Continued

from oxidation and atmospheric contamination. A range of shielding gases can be used (i.e., N_2, Ar, He and CO_2, etc.). This will depend on the material being welded. The presence of a shielding gas can also cause porosity during laser welding. The gas is entrapped by the turbulent welding pool and keyhole.[41]

Microwelding of dissimilar materials can introduce a number defects caused by the difference in the mechanical and physical properties. One of the most serious involves the formation of intermetallic compounds in the fusion zone which can lead to microcracking. Figure 7.35 shows a microweld between copper and aluminium showing microcracks in the fusion zone.

7.4.2 Evaluation of welds

Once the microweld has been made, there are a number of different methods available (visual, metallographic sections, mechanical testing and leakage

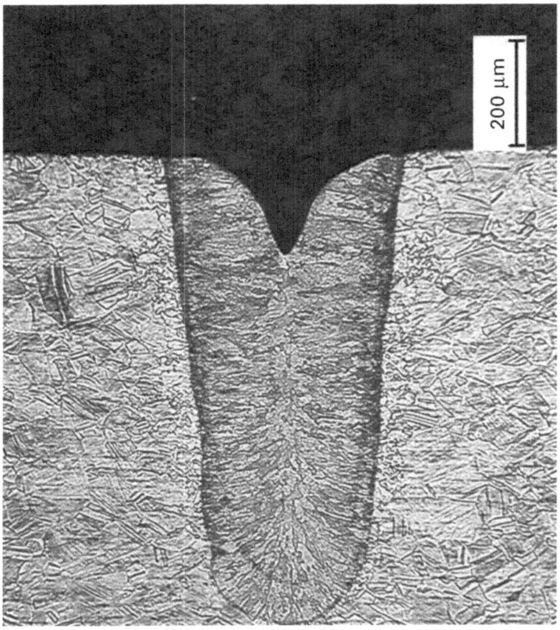

7.31 Undercutting or weld blowout caused by too much power/energy at the workpiece. This leads to vaporisation of material.

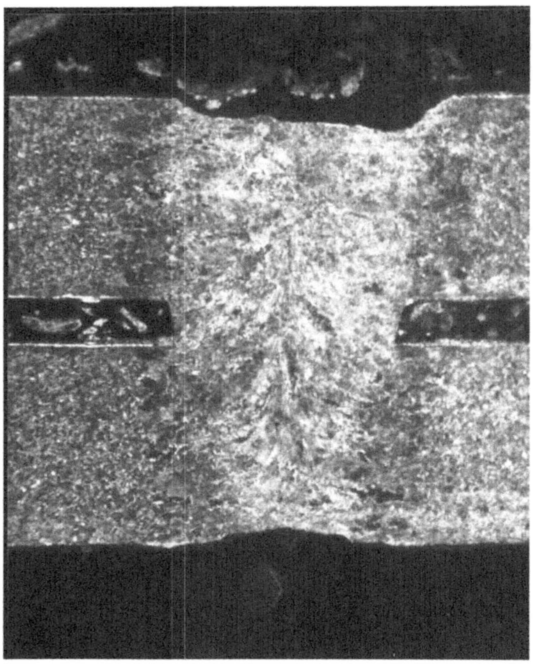

7.32 Excessive gap between two sheets during overlap welding leading to undercutting of both top and bottom weld beads.

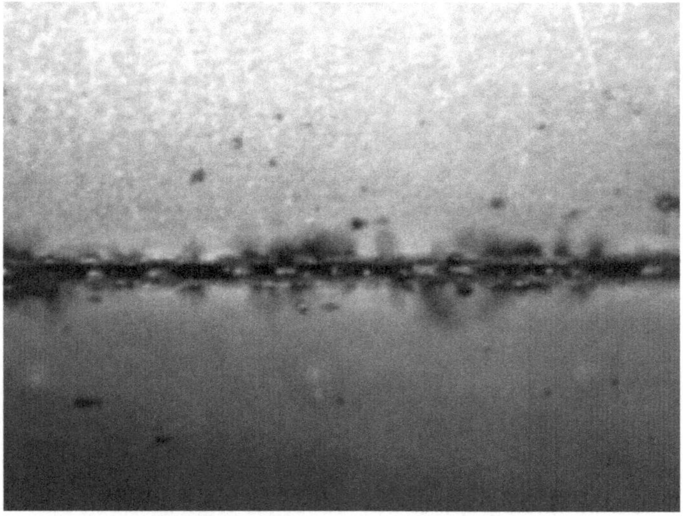

7.33 Combination of high power density at workpiece and low welding speed leading to porosity and spatter during laser welding.

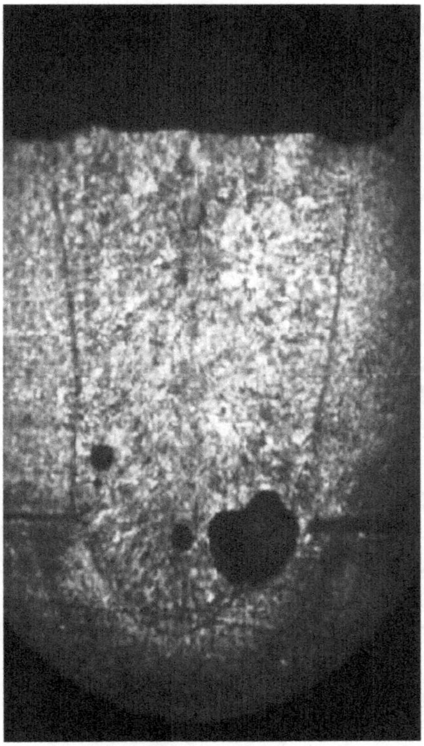

7.34 Rapid closing of the keyhole, thus preventing full penetration through the plate, can lead to the formation of porosity; 304 stainless steel overlap weld joint.

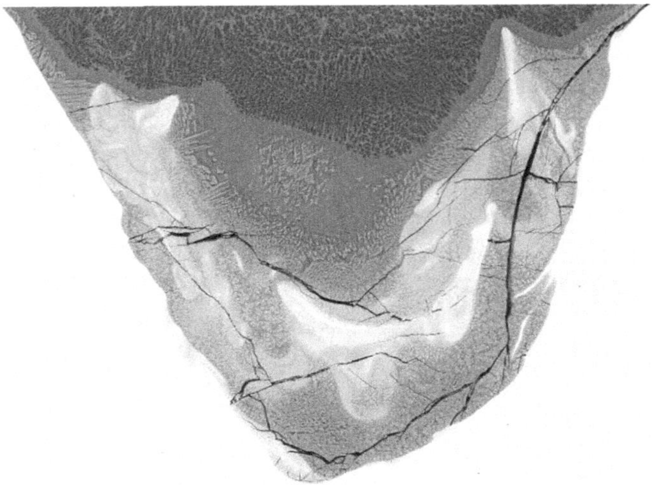

7.35 SEM micrograph of the root of the weld between aluminium alloy and pure copper. This shows formation of intermetallic compounds in the fusion zone leading to microcracking.

testing) to evaluate the quality of the welds, i.e. cosmetic appearance, porosity, cracking, hermeticity and strength.

The simplest and most widely used method is visual observation of the surface of the weld. This method is very useful for determining if the microwelds are free from spatter and surface oxidation, and for in-depth analysis (i.e. cracking, porosity weld penetration, etc.), metallographic methods are required. This method is very slow but very necessary to evaluate the microwelds, and extreme care should be taken when sectioning very small components. The metallurgical laboratory should be well equipped with the suitable equipment to cut and polish very small parts.

Leak testing is another method which is very common for checking hermeticity of the microwelds. Lithium ion batteries and microwave packages are two of the applications which are tested for leaks. Figure 7.36 shows a typical leak detection jig used for checking welded batteries. Leak tests are performed on the welded samples to check that battery case (can and cap) and electrolyte inlet (ball and cap) are hermetically sealed. The samples are tested with nitrogen gas at 15 kgf/cm^2 for 5 minutes in water. The schematic of the leak test is shown in Fig. 7.37. The welds are good, i.e. hermetically sealed, if no bubbles are observed along the length of the weld and around the ball and cap.

For tensile strength, small-scale tensile testing machines are used to evaluate the strength of the microwelds. During the pull test, the welded joint is often exposed to a complex stress state, resulting most often in tensile shear of the weld zone.[42] During laser microwelding it is important to consider the

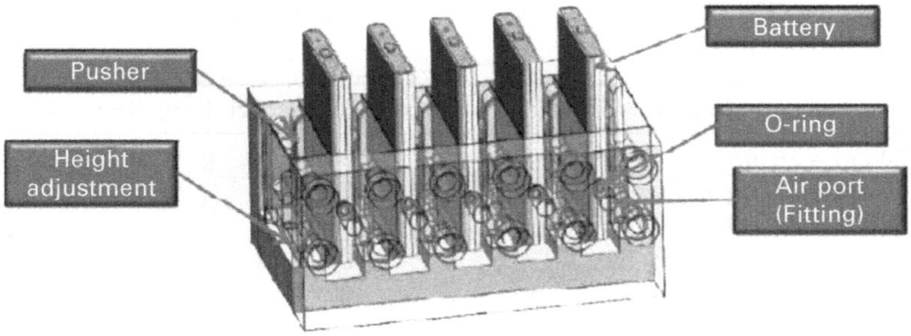

7.36 Leak testing jig mainly used for checking hermeticity of laser-welded lithium ion batteries (courtesy of Hana, South Korea).

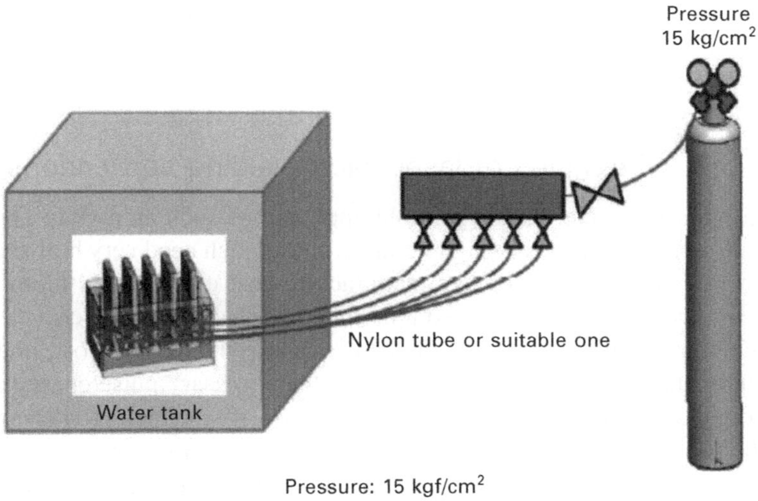

7.37 The schematic of the leak test. The welds are good, i.e. hermetically sealed, if no bubbles are observed along the length of the weld and around the ball and cap.

effect of the weld process on the raw parent material. The strength of the overlap weld joint is governed by power density and the spot size of the laser beam.

The mechanical properties of these joints are governed to an extent by the weld width at the interface. A weld width equal to or greater than the thinnest sheet thickness is a suitable target. Table 7.4 shows the effect of spot weld diameter and the nugget size on the shear strength of the spot welds. These spot welds were made with single-mode 200 W fibre laser in thin foils of stainless steel. At low power the weld had very little strength due to very small nugget size at the interface. As the power increases, the weld diameter also increases, which produces spots welds with good strength due to bigger nugget size.

Table 7.4 Effect of weld diameter on the nugget size and shear strength (overlap spot welding)

Material combination	Weld diameter (μm)	Nugget size (μm)	Shear strength (Psi)
20 μm + 25 μm	75	35	0.40
20 μm + 25 μm	115	65	1.39
20 μm + 25 μm	170	95	1.52
25 μm + 35 μm	95	55	0.57
25 μm + 35 μm	140	80	1.45
25 μm + 35 μm	195	110	1.60

7.5 Applications of laser microwelding

Laser microwelding plays a very important role in joining microparts for a number of industries, i.e. medical, electronics, automotive and consumer goods, and some applications are listed in Table 7.1.

7.5.1 Examples of laser microwelding applications

Lithium ion batteries for portable applications such as mobile phones and laptops are made in very high volumes and as such need very high throughput on production lines to make manufacture cost effective. Lithium cells are poisoned by water vapour, so a hermetic seal of the cell is critical for long life. Laser welding a lid to the can is fast, reliable, low cost, and easy to automate. Pulsed YAG lasers produce the necessary pulse energy and peak power to produce hermetically sealed welds with the required penetration for the mechanical strength. Galvo-mirrors or translating optics can be used to match the speed of the cells if a weld-on-the-fly application is needed. Figure 7.38 shows an example of a lithium ion battery welded with a JK300HP (300 W average power).

Disk drive flexures (Fig. 7.39) hold the ferrite reader that floats over the spinning magnetic media in a hard drive. The high speeds of positioning and the ultra-close proximity of the ferrite reader mean that this assembly must be accurately assembled from three or four pieces of stainless steel with thicknesses ranging from 20 μm to more than 200 μm. Their design must accommodate resonances, stiffness, and overall component accuracy in three planes. Assembly of these flexures must be highly automated for low-cost, repeatable manufacture.

Lasers have been the only joining solution available for this application since the early 1980s. Their ability to produce discreet energy levels and pulse shapes for each weld location and relative thickness means that each weld is tailored to its target and the welding rate can be over 150 welds/second. Most systems today use galvo-mirrors that direct the laser beam through a flat-field lens onto the part. Ultra-precise fixtures hold the parts together

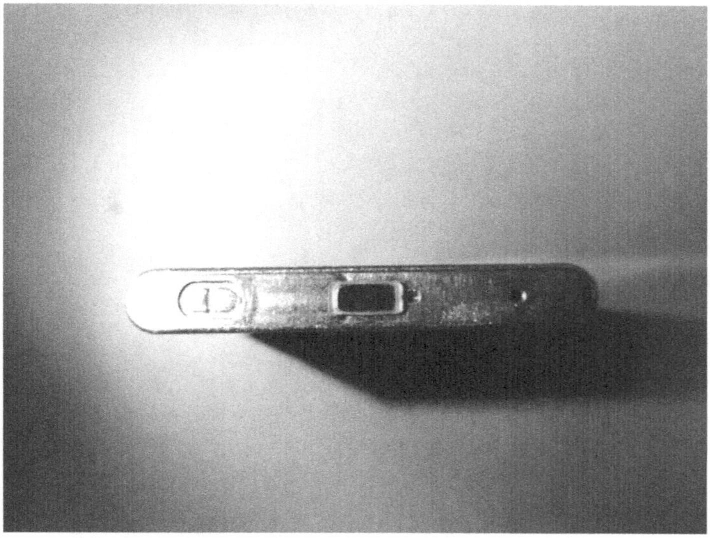

(a)

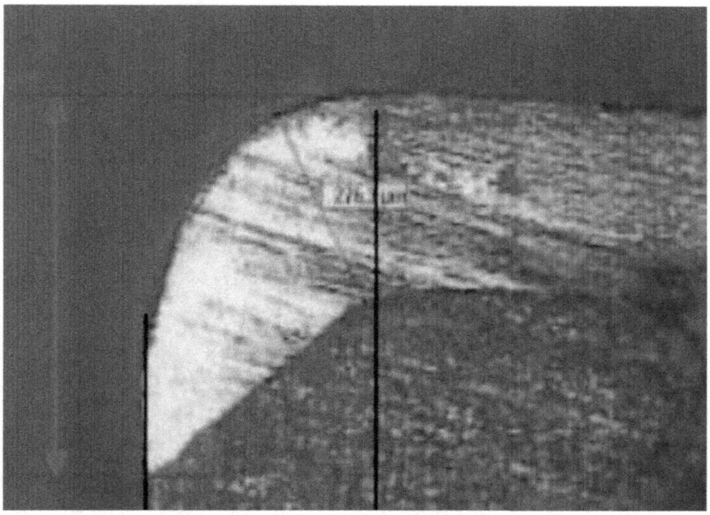

(b)

7.38 (a) Lithium ion battery welded with pulsed Nd:YAG laser at 50 mm/sec; weld penetration (b) 350 μm. The battery is made of aluminium alloy (0.2 mm thick case and 1 mm thick cap).

for laser welding and also feed inert gas to the weld area for clean, smooth, oxide-free welds. Any brittle oxide or melt spatter on the surface can flake off and contaminate the drive rendering it useless. For this application both low power Nd:YAG and fibre lasers can be used to produce welds. Figure 7.40 shows spot welds made with both lasers.

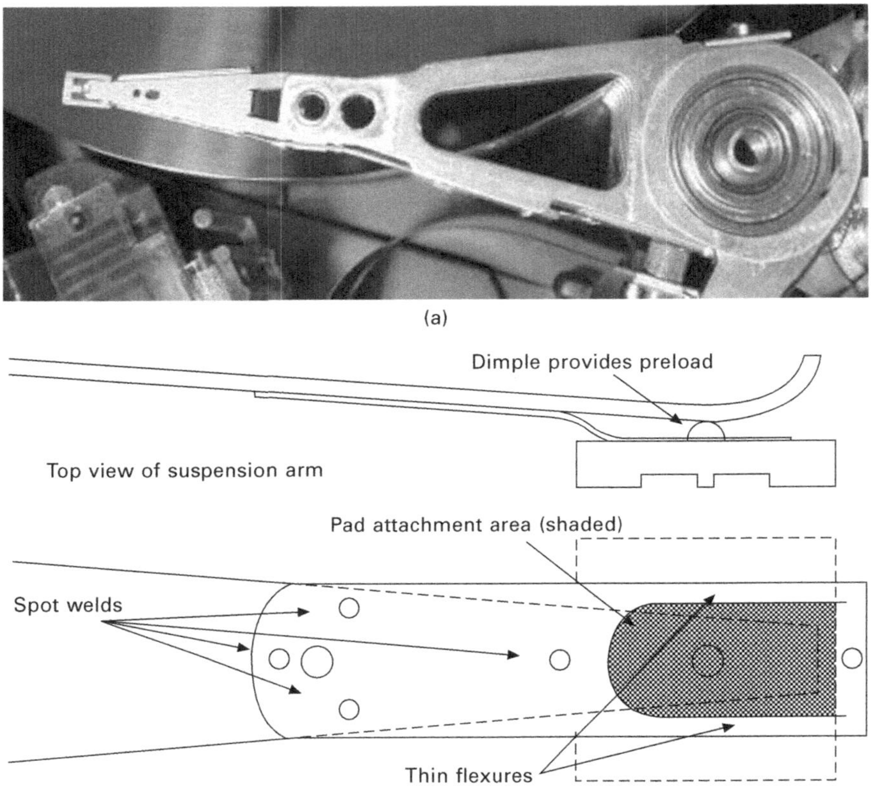

7.39 (a) Disk drive flexure; (b) flexure arm schematic, highlighting laser spot weld positions.

Medical devices – pacemakers, defibrillator and drug-pump – welding implanted medical devices that incorporate electronics and sensors, require process control and end-product quality that is second to none. These devices must not only function flawlessly, they must have an aesthetic quality that helps with sterilisation and does not irritate the tissue. The welds on these implantable devices must be a hermetic seal very close to heat-sensitive components and since the value of the component before welding is very high, the yield from the welding process must also be high. Pacermakers and defibrillators are electronic devices that are encased in a titanium package that holds the circuitry and battery. Most of the laser welding employed with these units is in the outer case weld. Drug pumps are more complex since at their core they are a mechanical pump and reservoir system along with control electronics and batteries. Laser welding of the different mechanical sub-assemblies and reservoirs means that dozens of laser welds can be part of these systems.

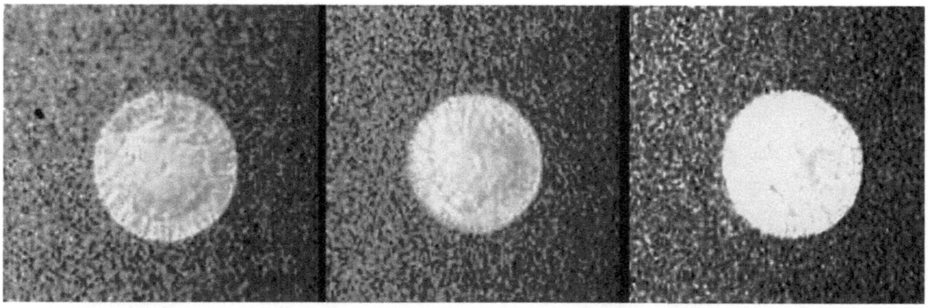

(a)

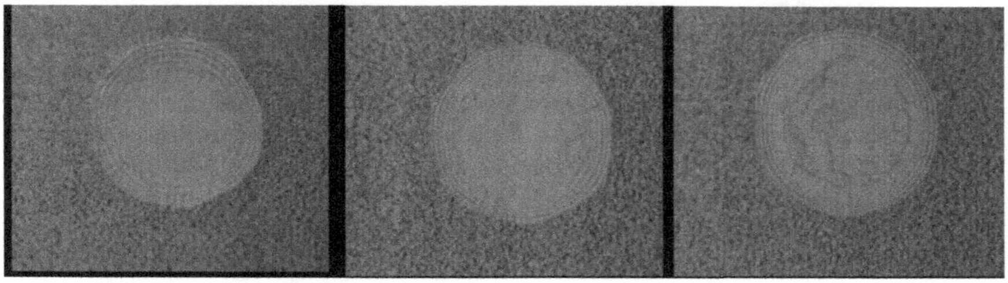

(b)

7.40 Laser spot welds, 304 stainless steel (51 + 102 μm); (a) pulsed Nd:YAG laser, (b) single-mode fibre laser.

Low-power pulsed YAG lasers (300 W) are the best choice for medical package assembly and sealing. Their high peak powers and low heat input can handle all the different alloys and create a very tolerant process while maintaining deep penetrations. Fibre optic beam delivery is most common for these parts because of the consistent focus spot size and top-hat energy distribution which helps with fillet and butt joints to fill any gaps and get good weld nugget size for maximum yield. It may be also possible to weld some of these medical devices with CW fibre lasers (200 W), but care is needed to optimise the laser parameters because pure CW power may cause distortion of the sensitive parts of the component. Figure 7.41 shows a titanium hearth pacemaker welded with a JK300HP (300 W). The weld penetration is approx. 200–250 μm.

Radioactive isotopes (Fig. 7.42) used for cancer therapy have a rather short half-life and must be produced to order and shipped immediately for implanting into the patient. The capsules or seeds containing radioactive elements are implanted into the body inside or near the cancer itself to produce a much localised radiation treatment. Other radioactive isotopes are used in heart procedures where the radioactivity helps reduce scarring around stents that begin to close the vessel sometime after the stent is inserted. In this case the radioactive material is laser sealed into a capsule and the capsule attached or laser welded to a lead for insertion into the body. It takes several minutes to locally radiate the artery around the stent.

7.41 The seam weld around the cardiac pacemaker demonstrates the precision that can be achieved with a laser process. A laser welding also offers a greater degree of control resulting in stability, excellent repeatability, and consistency. Pacemakers require high integrity hermetic seals at the same time as ensuring well-controlled depth of penetration and low heat input to prevent damage to internal components.

7.42 Laser welded radioactive isotopes. Pulsed YAG lasers are the best choice for capsule and seed assembly and sealing. Their high peak powers and low heat input create a very tolerant process while maintaining consistent shallow weld penetrations (200 μm).

For implantable medical applications, the alloy of choice is titanium which is very easy to weld due to its low thermal conductivity. Its reactivity with oxygen and nitrogen is the only process drawback. Weld shielding with argon is critical for bright silver welds. Pulsed YAG lasers are the best choice for capsule and seed assembly and sealing. Their high peak powers and low heat input create a very tolerant process while maintaining consistent shallow weld penetrations (200 μm).

In microwave packages welding (Fig. 7.43), the microwave circuitry is different from many other electronic devices in that the package around the

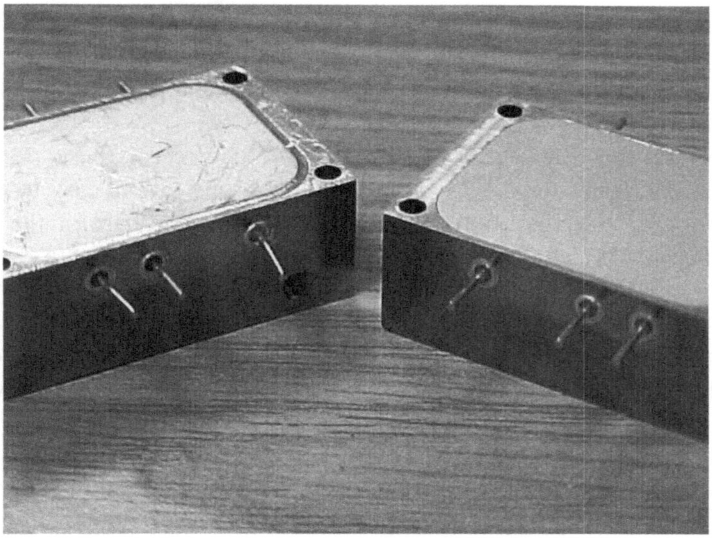

(a)

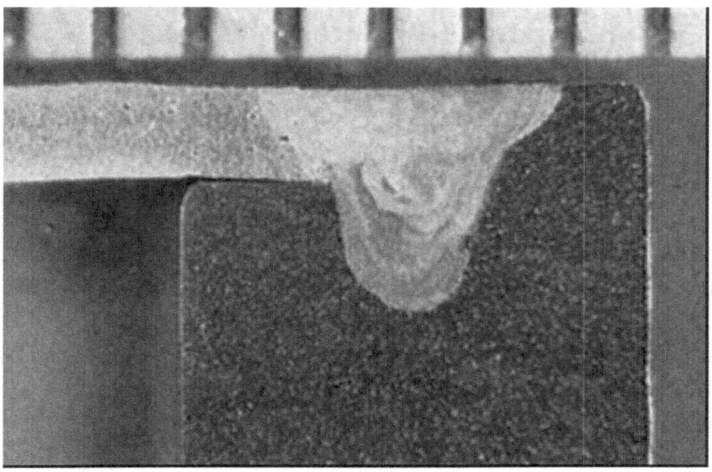

(b)

7.43 (a) Laser welding of microwave packages: a laser weld produces a hermetic seal so that no contaminants can reach the circuits and the weld also traps a very high quality inert atmosphere in the package. (b) Weld cross section of drop-in cover shown in (a). The cover is 0.50 mm thick 4047 aluminium alloy and the housing is 6061 aluminium alloy. Weld penetration is 1.20 mm.

active elements is also part of the microwave circuit. Since most microwave circuits involve expensive radar or communication systems, they require a very high reliability hermetic seal and a high yield. Therefore, their packaging must be robust, hermetically sealed, and capable of being opened and reworked when needed. Laser welding meets these requirements. A laser weld produces a hermetic seal so that no contaminants can reach the circuits and the weld also traps a very high quality inert atmosphere in the package. Laser welding has very low heat input and can be placed very close to polymer seals, glass-to-metal seals, soldered components and electronic circuits.

For aerospace applications, the alloy of choice for these packages is aluminium. It has the electrical and thermal properties, is easy to machine, and very light weight. 6061-T6 alloy is used for the base of the package and then a 4047 alloy lid is laser welded onto the base with a butt, lap or fillet joint. The 12% silicon-aluminium alloy 4047 can be fusion welded to the 6061 without cracking, creating a strong hermetic seal. For rework, an end mill will machine through the laser weld back to the same level as the original base. Once repaired, another lid can be located on the package and laser welded again. The number of repairs is virtually unlimited. In some instances, a laser weld is required through the lid into dividing labyrinth walls in the base to isolate certain parts of the circuit from others. Some microwave packages can use stainless steel or Kovar components.

7.6 Conclusion and future trends

Laser microwelding is an established technology that is becoming both a necessity and a valuable asset for manufacturing micro components for various industry sectors, and a range of industrial lasers with diffraction-limited spot sizes are available to meet the growing demand.

More and more of the microwelding applications will require joining dissimilar metals with different mechanical and physical properties which creates big challenges for welding. In the future, laser microwelding of dissimilar material combinations will benefit from a deeper understanding of welding methods and materials, so that good quality welds can be produced in a range of material combinations. The weldability data outlined in Table 7.3 need more detailed study in terms of welding technology, as most small-scale laser welds will cool very rapidly giving rise to many welding defects, and the process can violate what the handbook would indicate as a weldable material combination. There is also potential for using lasers to microweld plastics to metals or plastics to ceramics. Metals are strong and ductile, and plastics are light and easily formable, and a laser joining technology for direct bonding a metal and a plastic, called laser-assisted metal and plastic (LAMP) joining, has been developed.[43] LAMP joining is performed by melting the plastic near the joint interface and then forming small bubbles in the melted plastic by utilising a simple lap joint. The transmitted laser beam is absorbed to heat the metal surface, and the plastic near the joint can be melted to form bubbles by the heat conducted from the metal. The metal interface near the lap joint is not melted, but the plastic on the metal plate is melted to form small bubbles. The LAMP joining is applicable to bond thermoplastics, especially engineering plastics, to any metal. A strong LAMP joint is characterised by the formation of small bubbles in the plastic near the joint, which should be heated above the decomposition temperature of the plastic. This LAMP joining technology is applicable to many combinations of metals such as steels, titanium and aluminium alloys, and plastics such as PET, polyamide (PA) and polycarbonate (PC). It is also confirmed that LAMP joining is applicable to the joining of dissimilar metals with a plastic intermediate layer sheet in terms of the preventive means of local electric-physical corrosion in the metals.

Lastly, a key future trend in microwelding is a push towards more complete microwelding systems, i.e. good fixturing for good part fit-up, good gas shielding system, good vision system and some sort of in-process monitoring. It is very easy to produce sound welds in the laboratory environment by optimising laser and processing parameters. In production, however, it is not that easy to maintain optimal parameters of output and mode of laser beam, since the focusing optics or the cover glass slides can deteriorate in long-term operation. The weld quality can also be affected by non-laser conditions such as fit-up between the parts, variation in input materials, contamination (oil or dust), focus position and oxidation of the work surface. Thus it will be useful and cost saving to evaluate the weld quality during the welding process by in-process monitoring. The use of a complete microwelding system has the potential to improve the overall weld quality, reduce the cost and improve reliability by removing the human element and inherent variability of the manual process.

7.7 References

1. Olowinsky A.M., Kramer, T. and Durand, F., 'Laser beam micro welding in watch industry', *Proc. SPIE – The International Society for Optical Engineering,* **Vol. 4637**, 2002, Photon Processing in Microelectronics and Photonics, San Jose, CA, pp. 571–580.
2. Witte, R., Herfurth, H.J. and Bauer, I., 'Microjoining of dissimilar materials for optoelectronic and biomedical applications', *Proc. SPIE – The International Society for Optical Engineering*, **Vol. 4979**, 2003, Micromachining and Microfabrication Process Technology VIII, San Jose, CA, pp. 226–233.
3. Ready, J. and Farson, D. (eds), *Handbook of Laser Materials Processing*, Laser Institute of America, Orlando, FL, 2003.
4. Hugel, H., 'New solid-state lasers and their application potential', *Optics and Lasers in Engineering*, **34**, 2000, 213–229.

5. Koechner, W., *Solid-state Laser Engineering*, 2nd edn, Springer-Verlag, Berlin, 1988.
6. Dawes, D.G., 'Growing crystals for solid state lasers', *Lasers and Optronics*, December 1987.
7. Naeem, M., 'Material processing with super-modulation', *Proc. ICALEO 2002*, Laser Institute of America, Scottsdale, AZ, 2002.
8. Naeem, M., 'Welding performance of a 2kW continuous wave supermodulated Nd:YAG laser – increased weld speed, weld penetration and reduced porosity with supermodulated power', *Proc. PICALEO 2008,* Laser Institute of America, Beijing, China, 2008.
9. Graham, H., 'Throwing light on materials processing ... an addition to laser family', *TWI Bulletin*, May–June 2006.
10. Naeem, M., 'The influence of pulse shaping on laser material processing', *Proc. 9th NOLAMP Conference, Nordic Conference on Laser Material Processing*, Trondheim, Norway, 2003, pp. 239–249.
11. Norman, S., Zervas, M.N., Appleyard, A., Durkin, M.K., Horley, R., Varnham, M.P., Nilsson, J. and Jeong, Y., 'Latest development of high-power fibre lasers', *Proc. SPIE*, **Vol. 5335**, 2004, Fibre Lasers: Technology, Systems, and Applications, San Jose, CA, p. 229.
12. Dominic, V., MacCormack, S., Waarts, R., Sanders, S., Bicknese, S., Dohle, R., Wolak, E., Yeh, P.S. and Zucker, E., '110 W fibre laser', *Electronics Letters*, **35**, 1999, 1158.
13. Naeem, M., 'Micromachining with a single mode 100W fibre laser', *Proc. LAMP*, Kyoto Research Park, Kyoto, Japan, 2006.
14. Naeem, M. and Lewis, S., 'Pulsed Nd:YAG laser or continuous wave fibre laser for micromachining?' *Proc. WLT-Conference on Lasers in Manufacturing*, Munich, 2007.
15. Reichel, V., Unger, S., Hagemann, H. and Auerbach, M., '8W highly efficient Yb-doped fibre laser', 2000, *Proc. SPIE*, 3889.
16. Nilsson, J., Grudinin, A.B. and Turner, P.W., 'Advanced pulsed and CW high-power fibre laser', *Proc. CLEO*, 2000.
17. Schoenfelder, A., 'Fibre lasers address micromachining methods', *Laser Focus World*, June 1999.
18. Hunter, B.V., Leong, K.H., Miller, C.B., Golden, J.F., Glesias, R.D. and Laverty, P.J., 'Selecting a high-power fibre-optic laser beam delivery system', *Proc. ICALEO '96*, Laser Institute of America, Orlands, FL, 1996, Section E, pp. 173–182.
19. JK Lasers, GSI Company Group, Product Datasheets & Brochures, 2010, http://www.jklasers.com/downloads.
20. Naeem, M. and Collins, P., 'Micro-joining with pulsed Nd:YAG laser', *Proc. WLT-Conference on Lasers in Manufacturing*, Munich, 2005, pp. 643–649.
21. Zacharia, T., David, S.A., Vitek, J.M. and DebRoy, T., 'Weld pool development during GTA and laser beam welding of type 304 stainless steel, Part I – Theoretical analysis', *Welding Journal*, **68**, 1989, 499s–509s.
22. Zacharia, T., David, S.A., Vitek, J.M., and DebRoy, T., 'Heat transfer during Nd:YAG pulsed laser welding and its effect on solidification structure of austenitc stainless steels', *Metallurgical Transactions A*, **20A**, 1989, pp. 957–967.
23. Tomie, M., Abe, N., Yao, X. and Arata, Y., 1988, 'Tandem electron beam welding (Report – VIII)', *Transactions of JWRI*, **17** (1), 1988, 299–303.
24. Arata, Y and Miyamoto, I., 'Laser welding', *Technoctat*, **11** (5), 1978, 33–42.

25. Miyamoto, I., Maruo, H. and Arata, Y., 'Mechanism of bead-transition in laser welding', *Int. Conference Welding Research in the 1980s*, Osaka, The High Temperature Society, 1980, pp. 103–108.
26. Miyamoto, I. and Knorovsky, G.A., 'Laser microwelding', *Microjoining and Nanojoining* (Ed. Zhou, Y.), Woodhead Publishing, Cambridge, 2008, pp. 345–417.
27. Pittaway, L.G., 'The temperature distributions in thin foil and semi-infinite targets bombarded by an electron beam', *Brit. J. Appl. Phys.*, **15**, 1964, 967–982.
28. Ready, J.F., *Industrial Applications of Lasers*, New York, Academic Press, 1978.
29. Deininger, C., Mueller-Borhanian, J., Dausinger, F. and Huegel, H., 'Development of multi-detector systems for processing monitoring of laser beam welding capable for industrial use', *Proc. Lane 2004*, Erlangen, CIRP, 2004, pp. 107–117.
30. Seibold, G., Dausinger, F. and Huegel, H., 'Absorptivity of Nd:YAG laser radiation on ion and steel depending on temperature and surface condition', *Proc. ICALEO 2000*, Laser Institute of America, Dearborn, MI, 2000, pp. E125–E132.
31. Naeem, M., 'Modulation techniques for enhanced material processing using single mode fibre laser', *Proc. ICALEO 2008*, Laser Institute of America, Temecula, CA, 2008.
32. 'How to avoid cracking in aluminium alloys', ESAB cutting and welding data sheet.
33. Naeem, M., JK Lasers, GSI Group, internal report, 2007.
34. Klages, K., Ruettimann, C. and Olowinsky, A.M., 'Laser beam micro welding of dissimilar metals', *Proc. ICALEO 2003*, Laser Institute of America, Jacksonville, 2003.
35. Naeem, M., 'Microjoining of dissimilar metals with pulsed Nd:YAG laser', *Proc. PICALEO 2006*, Laser Institute of America, Melbourne, Australia, 2006.
36. Naeem, M., 'Microwelding performance comparison between a low power (125 W) pulsed Nd:YAG laser and a low power (100–200 W) single mode fibre laser', *Proc. PICALEO 2008*, Laser Institute of America, Beijing, China, 2008.
37. Dausinger, F., 'Benefit of enhanced focusability of new YAG-Lasers', *Proc. EALA, 2002*, Bad Nauheim/Frankfurt, Germany, 2002.
38. Boisselier, D., 'Comparison des performances laser YAG à fibre optique et CO_2 "Slab pour le soudage des alliages d'aluminium"', *Proc. CISFFEL 6*, Toulon, France, 1998.
39. Venkat, S., 'CO_2 laser beam welding of aluminium alloys', *Welding Journal*, **76**, 1976, 275s–282s.
40. Katayama, S., Kohsaka, S., Mizutani, M., Nishizawa, K. and Matsunawa, A., 'Pulse shape optimisation for defect prevention in pulsed laser welding of stainless steels', *12th International Conference Applications of Laser and Electro- optics (ICALEO' 93)*, Jacksonville, Laser Institute of America, 2001, pp. 487–497.
41. Katayama, S. and Matsunawa, A., 'Formation mechanism and prevention of defects in laser welding of aluminium alloys', *Proc. CISFFEL 6*, Toulon, France, 1998.
42. Dieter, G.E., *Mechanical Metallurgy*, McGraw-Hill, New York, 1986.
43. Katayama, S., 'Laser joining of metal and plastic', *Industrial Laser Solutions*, November 2010.

Part II

Laser welding technologies for various materials

8
Laser welding of light metal alloys: aluminium and titanium alloys

J. M. SÁNCHEZ AMAYA, Titania, Ensayos Y Proyectos Industriales S.L., Spain, M. R. AMAYA-VÁZQUEZ and F. J. BOTANA, University of Cadiz, Spain

DOI: 10.1533/9780857098771.2.215

Abstract: In this chapter, laser welding procedures employed in the literature to join aluminium and titanium alloys are reviewed. The chapter is divided into two main sections, comprising reviews of the literature on laser welding of aluminium alloys and titanium alloys, respectively. Each of these sections has an introductory part in which general concepts and specific properties influencing the weldability of alloys are described. Subsequently, the main experimental details of the reviewed papers are reported, including description of the laser equipment, welding modes, powers and processing rates, shielding gas, superficial pre-treatments of samples, etc. The microstructure, main defects, mechanical properties and corrosion behaviour of the weld beads are also reviewed. The main objective of this chapter is to introduce the reader to the latest results obtained in the literature regarding laser welding of these two light alloys, emphasising the most influential factors for each alloy.

Key words: Laser welding, aluminium alloys, titanium alloys.

8.1 Introduction to laser welding of aluminium alloys

Laser welding (or laser beam welding, LBW) presents some advantages in comparison with other conventional joining techniques, as reported in previous chapters of this book. However, there still exists the fundamental problem that the welding reliability of aluminium alloys is lower than that of other industrial metals, such as steels. The reason lies in the physical properties of aluminium alloys, especially their high thermal conductivity, high reflectivity and low viscosity (Kuo and Lin, 2006; Sánchez-Amaya *et al.*, 2009a, 2009b; Pierron *et al.*, 2007). The coefficient of thermal conductivity of aluminium alloys depends on different factors, such as the temperature, the amount and nature of the alloying elements and the previous heat treatment of the alloy (T3/T4 usually decrease the thermal conductivity in comparison with T0). Figure 8.1 shows the thermal conductivity at room temperature of some aluminium alloys in relation to their percentage of alloying elements. Data

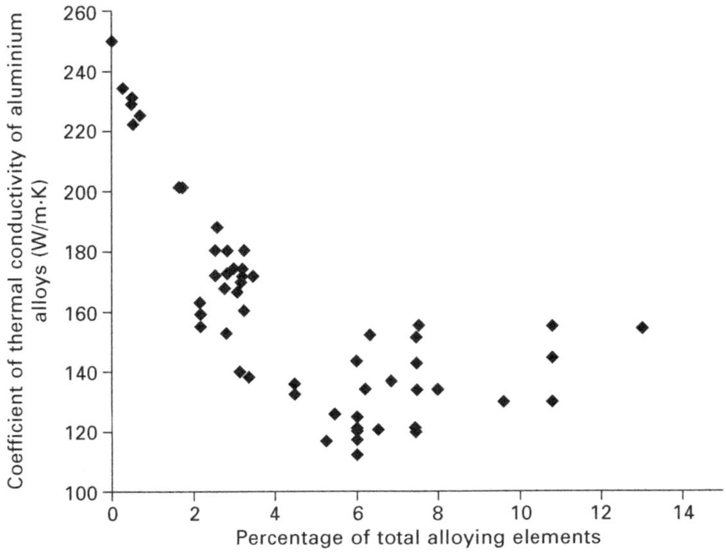

8.1 Thermal conductivity of some aluminium alloys at room temperature in relation to the total amount of their alloying elements.

shown in this figure highlight that thermal conductivity of aluminium alloys is usually higher the purer the alloy.

The relatively high thermal conductivity values provoke a rapid heat transfer when the aluminium alloys are subjected to a laser beam. This effect constitutes an important drawback for LBW, as the heat diffuses to the whole welding piece, avoiding the concentration of energy in the weld pool. Thus, aluminium alloys require higher laser energies than other alloys to be welded, as higher fluencies are needed to reach the melting temperature of the welding zones. It should be taken into account that each alloying element has a different influence on the coefficient of thermal conductivity. In the case of steels, it is claimed that molybdenum appreciably increases its thermal conductivity, nickel and copper increase it moderately, aluminium and silicon reduce it considerably, and chromium, vanadium, tin and antimony reduce it moderately (Rukadikar and Reddy, 1986). Unfortunately, such detailed information is not available for aluminium alloys. It is indicated in Sakamoto *et al.* (2003) that elements such as Si, Mg and Zn modify the thermal properties of aluminium alloys, therefore affecting their weldability. The effect of the Si content on the thermal conductivity of Al-Si diamond composites and Al-Si matrix has also been reported. In both materials, thermal conductivity decreases as Si content increases (Zhang *et al.*, 2009).

Some works demonstrate the influence of some alloying elements on other properties affecting the weldability of aluminium alloys. Thus, the

Mg and Zn content, which have lower boiling points than aluminium, are known to improve the weldability of aluminium alloys under the keyhole regime (Sakamoto *et al.*, 2003; Ancona *et al.*, 2007). In other work, taking into account the weld penetration and the susceptibility to solidification cracking, the weldability of six different aluminium alloys have been seen to depend on the magnesium content and, to a lesser extent, the zinc and silicon amount (Sánchez-Amaya *et al.*, 2012b). Authors conclude that these elements decrease the thermal conductivity and the melting temperature, therefore increasing the weldability of aluminium alloys.

The second important factor limiting the applicability of LBW to aluminium alloys is their high reflectivity. The radiation absorption of aluminium alloys depends on the input laser wavelength; Howard *et al.* (2006) indicate that the absorption of the high power diode laser (HPDL) wavelength by aluminium alloys is higher than the CO_2 and Nd:YAG wavelengths. Thus, the emission wavelength of the HPDL (808 nm) provokes a higher absorptivity in aluminium than the longer wavelengths of Nd:YAG (1,064 nm) and CO_2 (10,640 nm) lasers (Howard *et al.*, 2006). In any case, regardless of the laser source, the reflectivity of aluminium alloys can be higher than 80%, this value being higher as the aluminium alloy is purer. This high reflectivity makes the aluminium alloys absorb a small fraction of the incident radiation. For example, the absorptivity of pure polished aluminium at 800 nm is reported to be around 13% (Howard *et al.*, 2006). In practice, the actual absorptivity is influenced by the presence of an oxide layer, other surface contaminants and surface roughness (Howard *et al.*, 2006).

Different conventional surface treatments, such as sandblasting or the application of dark coatings are reported to decrease the reflectivity of aluminium alloys (Okon *et al.*, 2002; Duley, 1999; Howard *et al.*, 2006; Sánchez-Amaya *et al.*, 2009a). More sophisticated treatments based on laser texturisation treatments have recently been proved to increase the absorption of laser radiation of 5083 aluminium alloy surfaces, in comparison with the above commented conventional treatments (Sánchez-Amaya *et al.*, 2012a). These results indicate that laser texturisation treatments improve the absorption and weld penetration of aluminium alloys up to 20%. Further research is required in order to determine the precise influence of the experimental texturisation variables. In any case, superficial conditions are expected to play a larger role in conduction mode than in keyhole mode welding, as the formation of a keyhole increases the energy absorbed by multiple reflections and plasma interactions inside the keyhole (Howard *et al.*, 2006; Duley, 1999).

Finally, the third factor responsible for the low weldability of aluminium alloys is the low viscosity of the weld pool, which limits its expansion before solidification (Kuo and Lin, 2006; Sánchez-Amaya *et al.*, 2009a). To the authors' knowledge, there are no investigations that have succeeded at increasing the viscosity of the weld pool of aluminium alloys, this property being a constant drawback for laser welding of these alloys.

8.2 Laser welding technologies for aluminium alloys

As commented in previous chapters, there are two laser welding mechanisms: keyhole (usually involving density power higher than 10^6 W·cm^{-2}) and conduction (with density power normally lower than 10^6 W·cm^{-2}) (Okon *et al.*, 2002; Duley, 1999). The basic difference between the two modes depends on the characteristics of the generated weld pool. Thus, as the energy density of the laser beam is higher in keyhole welding than in conduction welding, the keyhole usually leads to narrower weld beads. This regime generates a cavity in the weld pool as a consequence of the metal evaporation. This cavity is stabilised by the pressure of the own vapour generated (Duley, 1999). The keyhole is associated with violent plasma generation which consists of metal vapour, ionised ions and electrons. The plasma resides both outside and inside the keyhole, known as the plasma plume and keyhole plasma, respectively (Jin *et al.*, 2012). Both plasma plumes and keyhole plasma have been studied with spectroscopic methods (Jin *et al.*, 2012; Cheng *et al.*, 2012). The stability of the keyhole depends on a balance between surface tension pressure and vapour pressure: surface tension pressure tends to close the keyhole while vaporisation tends to keep it open (Sibillano *et al.*, 2007). Keyhole welding is employed more than conduction welding, because it produces welds with high aspect ratios and narrow heat-affected zones (Dowden, 2009; Matsunawa *et al.*, 1998, 2000; Akhter *et al.*, 2007; Kuo and Lin, 2006; Shi *et al.*, 2007a; Bassani *et al.*, 2007). However, the keyhole welding mechanism can lead to some problems, such as instability, keyhole oscillation and intermittent closure of the keyhole that often produces porosity due to gas entrapment. Additionally, in some alloys, the high weld speed (and hence high rates of cooling) under the keyhole regime can lead to embrittlement in the weld or heat-affected zone (Okon *et al.*, 2002).

On the other hand, conduction welding is a more stable process, since metal vaporisation occurs at a lower level than in keyhole mode (Okon *et al.*, 2002; Sánchez-Amaya *et al.*, 2009a; Sibillano *et al.*, 2007). In the conduction regime, the energy density of the laser beam locally heats the material up to a temperature above its melting point and below its boiling point (Cao *et al.*, 2003). Conduction welding offers an alternative way of welding traditionally difficult materials such as aluminium alloys (Okon *et al.*, 2002). Different studies have been carried out concerning the laser welding of aluminium alloys by means of the conduction mechanism. Thus, the applicability of the diode laser to the conduction welding of AA2024 is analysed and interpreted by thermal modelling (Williams *et al.*, 2001). Zhao and DebRoy (2001) developed a model to study the vaporisation rate and weld metal composition change of automotive aluminium alloys, based on principles of transport phenomena, kinetics and thermodynamics. AA5083 welded under conduction has been reported to lead to magnesium loss of about 1–4% in the fusion zone when using a high power diode laser (Sánchez-Amaya *et al.*, 2009b) and around 8% when employing a high power CO_2 industrial laser (Sibillano *et al.*, 2007). This contrasts with the magnesium loss measured in 5083 welds produced under the keyhole regime, which range between 13% and 22% (Sibillano *et al.*, 2006a, 2007). As commented before, magnesium is the most influential alloying element on welding aluminium alloys; therefore, its evaporation must be reduced.

In addition to these well-known conduction and keyhole regimes, Assuncao *et al.* (2012) have just proved the existence of a third welding mode so-called transition regime, leading to welds with characteristics of both keyhole (undercut, not flat top profile and small depression at the surface) and conduction (low aspect ratio). These results also highlight that the welding mode (keyhole, transition or conduction) depends not only on the power density but also on the welding speed and beam diameter.

Examples of six aluminium butt welds using a high power diode laser under the conduction regime are shown in Fig. 8.2 (Sánchez-Amaya *et al.*, 2012b). The relatively high penetration of these welds, together with the low density of defects, demonstrate that laser welding of aluminium alloys under conduction regime is nowadays a reliable joining method.

In the literature, the possibility of changing the welding mode of aluminium

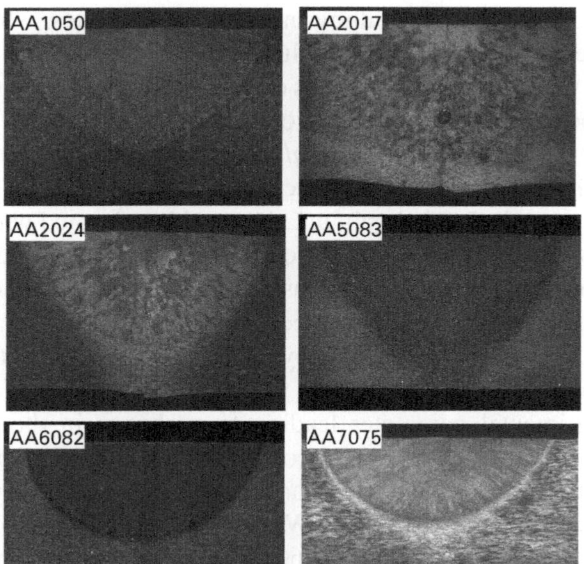

8.2 Examples of butt welds of aluminium alloys (1050, 2017, 2024, 5083, 6082, and 7075) generated with a HPDL under the conduction regime.

alloys is discussed (from the conduction to the keyhole regime or vice versa), by means of the variation of different laser experimental variables. For instance, using a CO_2 laser of 2.5 kW, Sibillano *et al.* (2007) showed that the flow of the shielding gas and/or the processing rate induced a change in the welding mode of 5083 samples. The results of this research indicate that low flows of shielding gas and high processing rates encourage conduction versus keyhole (Sibillano *et al.*, 2007). In a previous study employing a Nd:YAG laser of 1 kW, Hoult (1990) also observed that the processing rate controlled the welding mechanism and, therefore, the weld shape. Thus, at low rates (0.5 m/min) the welds were produced under the keyhole regime, while at higher processing rates (3 m/min), they were generated under the conduction regime (Hoult, 1990).

In more recent works, it is indicated that other variables can also induce a transition of keyhole to conduction, such as the laser beam defocusing (the laser beam defocusing provokes a spot size increase leading to a density energy decrease) (Okon *et al.*, 2002). In this context, Nakamura *et al.* (2000) proposed a technique to detect the transition between laser keyhole welding and laser heat conduction welding by monitoring optical and acoustic emissions from a laser-irradiated point. The study of the effect of defocusing high power laser beams was first restricted to distances between the focus and the workpieces ranging from –10 mm to +10 mm (Esposito *et al.*, 1982; Matsumura *et al.*, 1992). However, in Okon *et al.* (2002) the transition from keyhole to conduction mode is achieved when the distances of the workpiece range from +5 mm to + 100 mm below the focus (Okon *et al.*, 2002). This methodology offers an alternative method of welding resulting in lower cooling rates, which may enhance weld bead properties of aluminium alloys. Thus, full-penetration laser conduction welds produced by means of high-power defocused CO_2 and Nd:YAG laser beams could reduce or avoid defects related to the keyhole regime (Okon *et al.*, 2002). Sibillano *et al.* (2007) have recently developed a real-time monitoring technique to detect the transition between keyhole and conduction modes in 5083 aluminium alloy welds generated with a CO_2 laser. This methodology is based on the correlation analysis of the plasma plume optical spectra generated during laser welding. The principle of this technique, so-called covariance mapping technique (CMT), is that the keyhole mode provokes a higher evaporation of alloying elements, generating therefore higher plasma plume than the conduction mode.

Another experimental parameter leading to a change in the welding mechanism is working with continuous (continuous welding, CW) or pulsed laser beams (pulse welding, PW). Kuo and Lin (2006) have analysed the influence of the welding mode (CW or PW) of a Nd-YAG laser (2.5 kW) on the properties of butt welds of 5754 and 6022 aluminium alloys. This study shows that CW mode leads to a higher penetration than the PW mode. The alternating periods in the PW mode is thought to induce an unstable heat input, which would result in a dispersion of the energy in the specimen (Kuo and Lin,

2006). By contrast, the energy input remains uniform and stable when welding is performed in the CW mode, generating deeper penetration beads.

In order to give the readers a rough overview of the papers published in the literature, a summary of the most recent works dealing with laser welding of aluminium alloys is reported in Table 8.1. This table summarises the alloys welded, the applied superficial treatment, the welding mode (conduction or keyhole), weld type (butt welding, bead-on-plate, lap-fillet, T-joint, etc.), the laser source (Nd:YAG, CO_2, HPDL or Hybrid-GMAW), the applied laser power and welding rate, the shielding gas, and welding mode (pulse or continuous). Taking into account the data included in Table 8.1, it can be concluded that 5000 series alloys are most employed for research studies, providing their well-known good weldability. Concerning the laser devices, the most employed are the Nd:YAG and CO_2 lasers to obtain deep penetration under keyhole regime, although HPDL are also employed for morphological analyses of welds obtained under conduction mode. Additionally, Table 8.1 indicates whether or not the paper includes information related with the weld shape, microstructure, microhardness, defects, tensile strength, corrosion and simulation and/or modelling studies. A similar table has been put together in the subsequent section dealing with laser welding of titanium alloys.

Among the experimental variables detailed in Table 8.1, the shielding gas has been reported to have a strong influence on laser welding of aluminium alloys. Thus, the influence of the shielding gas composition (Ar, He, N_2, CO_2, or mixtures of them) on weld penetration has recently been analysed (Chung *et al.*, 1999; Sánchez-Amaya *et al.*, 2010; Katayama *et al.*, 2009; Matsunawa *et al.*, 1999). Concerning the nature of the shielding gas, contradictory results can be found in the literature. Thus, low density gases are reported in Tani *et al.* (2007a) to diffuse more easily in the welding, leading to worse protection than denser gases. However, Wang *et al.* (2007) claimed that He has a higher protection effect than Ar. High density gases (Ar) can be kept for longer times at the welding, but also have a higher absorption coefficient (leading to the absorption of higher laser radiation) than low density gases (He). These two effects can vary depending on the laser type and alloys, explaining the divergence of the results. Nitrogen gas has also been reported to reduce the porosity of 5083 weld beads, in comparison with Ar (Katayama *et al.*, 2009).

Optimisation of the shielding gas nature is a very active research line nowadays. Thus, mixtures of Ar-He (Akhter *et al.*, 2007), Ar-CO_2 or He, Ar and O_2 (Tani *et al.*, 2007b) have been demonstrated to improve the penetration of laser welds of aluminium alloys. The effect of Ar/He gas ratio on melting of different aluminium alloys (5052, 5083, 5182, 6061 and 7N01) has been investigated, revealing that the use of Ar is beneficial to welding more highly conductive and reflective alloys at lower power and higher welding speed, while a higher content of He is preferred for higher power and lower welding speed (Matsunawa *et al.*, 1999). In addition to the influence of the nature

Table 8.1 Summary of recent publications regarding laser welding of aluminium alloys

References	Alloys	Superficial pre-treatments	Regime (Conduction/Keyhole)	Weld type ((Butt, Bead-on-Plate, Lap-fillet, T-joint))	Shielding gas	Laser: Type	Laser: Power (kW)	Laser: Welding rate (m/min)	Laser: Mode (Pulse/Continuous)	Analysis of welds: Weld shape	Microstructure	Microhardness	Defects	Tensile strength	Fatigue	Corrosion	Simulation/ Modelling
Abe *et al.* (2006)	5052,6016		C	B, L	Ar	HPDL	4	2.4–12	C	X			X				
Akhter *et al.* (2007)	A356		K	B	Ar He	Nd-YAG	4.4	2.5–4.5	C	X	X	X		X			
Ancona *et al.* (2005)	5083		K	P, B	He	CO_2	2.5	2.7–7.2	C	X							
Ancona *et al.* (2007)	5083		K	B	He	CO_2	2–2.5	4.8–6	C			X	X	X			
Bassani *et al.* (2007)	A359/ SiC		K,C	P	N_2	HPDL, CO_2	2.4–4.0	0.9–3	C	X	X	X					
Campana *et al.* (2009)			K		He	Hybrid (GMAW)			C								X
Cheng *et al.* (2012)	6016		K			CO_2	2	1	C								X
Daneshpour *et al.* (2012)	6056					CO_2									X		X

El-Batahgy *et al.* (2009)	5052,5083 6061		K	P	Ar	CO_2	5	3–6		X		X	X	X		
Haboudou *et al.* (2003)	A356,5083	X	K	P	He	Nd-YAG	3.8	2.5–5	C	X	X	X	X			
Howard *et al.* (2006)	5182, 6013, 6111	X	C	L,P	Ar	HPDL	<4	1	C	X	X		X			
Jandaghi *et al.* (2009)	5754		K		Ar	Nd-YAG	0.4	0.54	C							X
Jin *et al.* (2012)	6016		K			CO_2	2.2	1	C, P							X
Katayama *et al.* (2009)	5083		K	P	Ar N_2	Fibre	6–10		C	X			X			
Katayama *et al.* (2011)	5052		K	P	Vacuum	Disk	10–16	0.3–6	C	X						
Kuo and Lin (2006)	6022 5754		K	B	Ar	Nd-YAG	2.5		P		X		X	X		
Malek Ghaini *et al.* (2009)	2024			P	Ar	Nd-YAG	0.4	0.24			X					
Matsunawa *et al.* (1999)	5052,5083, 5182,6061, 7N01				Ar/He	CO_2			C	X		X	X	X		
Mizutani *et al.* (2008)	5182			P	Ar	CO_2			C						X	
Okon *et al.* (2002)	5083	X	K,C	B	Ar	Nd-YAG, CO_2	4,5–1,8	0.36–0.72	P	X				X		
Pierron *et al.* (2007)	1050		K	P		Nd-YAG	0.5–6		C	X						

(Continued)

Table 8.1 Continued

Laser welding of aluminium alloys																	
References	Alloys					Laser				Analysis of welds							
		Superficial pre-treatments	Regime (Conduction/Keyhole)	Weld type ((Butt, Bead-on-Plate, Lap-fillet, T-joint))	Shielding gas	Type	Power (kW)	Welding rate (m/min)	Mode (Pulse/Continuous)	Weld shape	Microstructure	Microhardness	Defects	Tensile strength	Fatigue	Corrosion	Simulation/ Modelling
Sánchez-Amaya *et al.* (2009a)	5083	X	C	P	Ar	HPDL	1.5	1.5–3	P	X	X	X				X	
Sánchez-Amaya *et al.* (2009b)	5083-T0 6082-T6	X	C	B	N_2	HPDL	1.5–2.75	0.2–5	C	X	X	X				X	
Sánchez-Amaya *et al.* (2010)	5083-T0	X	C	B	Ar, N_2 CO_2	HPDL	2.8	0.3–1	C	X	X	X	X	X		X	
Sánchez-Amaya *et al.* (2012a)	5083-T0	X	C	P, B	Ar	HPDL	2–2.5	1–1.5	C	X							
Sánchez-Amaya *et al.* (2012b)	1050,2017 2024,5083 6082,7075	X	C	P, B	N_2	HPDL	1.5–2.75	0.2–6	C	X			X				
Sánchez-Amaya *et al.* (2012c)	2024,508 36082, 7075	X	C	B	N_2	HPDL	1.5–2.75	0.2–5	C	X	X	X					

Shi *et al.* (2007b)	1420	X	K	B		Nd-YAG	4.5	2.7	C		X	X	X	X	
Sibillano *et al.* (2005)	6082		C	P	N_2	CO_2	2.5	1.8–3.6	C				X		
Sibillano *et al.* (2006a)	5083		C	P	He	CO_2	2.5	4.8–7.2	C				X		
Sibillano *et al.* (2006b)	5083		K		He	CO_2	2.5	10	C	X			X		
Sibillano *et al.* (2007)	5083		C,K	P	He	CO_2	2.5	4.8–7.2	C	X					
Spina *et al.* (2007)	5083		K		He	CO_2	2.5	6–8.4	C						X
Zervaki *et al.* (2012)	2139		K	P	He, Ar	CO_2									X
Zhao and DebRoy (2001)	5182		C	P	He	Nd-YAG	1.5–3	5.72–7							X

of shielding gas, the effect of applying vacuum has also been analysed by Katayama *et al.* (2001, 2011). In Katayama *et al.* (2001), a reduction of the porosity is observed in the formation of AA5052 aluminium alloy weld beads when welding with high-power CW CO_2 and Nd:YAG laser under vacuum. More recently, the same authors (Katayama *et al.*, 2011) have developed a new vacuum chamber providing 0.1–10 kPa, with which a high power disk laser is successfully employed to obtain high penetration AA5052 welds. Weld penetration depth is seen to increase with decreasing ambient pressure (higher vacuum) and low welding speed. The morphology of the obtained welds is also seen to be modified by vacuum conditions (Katayama *et al.*, 2011). Other works have focused on the effect of the nozzle shape and design (Ancona *et al.*, 2005), or on the influence of shielding gas contamination, the inclination angle of the nozzle, the flow and the density of the protection gas (Tani *et al.*, 2007a, 2007b). Theoretical considerations analysed in Tani *et al.* (2007a) indicate that coaxial nozzles lead to a better protection against atmospheric contamination in the welding zone.

The effect of the welding rate and laser power (parameters influencing the laser fluence) on the weld shape have been also investigated. It is reported that for a specific alloy, the depth and width decreases as the welding speed is increased (Akhter *et al.*, 2007). When comparing the energy (fluence) needed to obtain similar penetration on 5000 and 6000 series alloys, it is found that the former require much less energy than the latter, for the 5754-O alloy being approximately 2.5 times less than that for the 6022-T4E29 alloy (Kuo and Lin, 2006). This means that the welding rate in 5000 series alloys can be increased by a factor of 2.5. This factor is seen to be 2.0 in Abe *et al.* (2006) for similar alloys (5000 and 6000 series). The welding rate and laser power have also been systematically analysed for butt welds generated under the conduction regime of six aluminium alloys (Sánchez-Amaya *et al.*, 2012b, 2012c). A linear relationship between the applied laser fluence and the penetration of the butt welds has been found in all alloys. Katayama *et al.* (2009) have employed a fibre laser to weld AA5083, at the power of 6 or 10 kW and several power densities from 0.4 kW/mm^2. The authors observe that the keyhole weld beads were narrower and deeper with the increasing laser power density.

8.3 Microstructure, defects, mechanical properties and corrosion behaviour of aluminium welds

8.3.1 Microstructure

The heat treatments produced during laser welding provoke significant microstructural changes on the processed materials. These changes have a strong influence on both the mechanical properties and corrosion behaviour of the joint pieces. The results obtained by different authors regarding the weld properties are reviewed subsequently.

Concerning the analysis of the microstructure, three different zones are generally identified at welding beads: fusion zone (FZ), heat-affected zone (HAZ) and base metal (BM), the extension of each depending on the alloy and the welding conditions (Nègre *et al.*, 2004; Sánchez-Amaya *et al.*, 2012c; Akhter *et al.*, 2007; Shi *et al.*, 2007a; Paleocrassas, 2005).

According to Paleocrassas (2005), the properties of the weld rely mainly on the microstructure of the FZ. Thus, depending on the alloy and laser processing conditions, the microstructure can consists of columnar dendrites and/or equiaxial grains. The latter microstructure is reported to be better than the former in 7000 series aluminium alloys, because of its isotropic properties, leading to an improvement in the mechanical properties of the weld, reducing its susceptibility to cracking during solidification (Paleocrassas, 2005).

The microstructure of butt welds of four aluminium alloys (2024, 5083, 6082 and 7075) generated with a HPDL under the conduction regime has been deeply investigated in Sánchez-Amaya *et al.* (2012c). The external zones of 5083 and 6082 weld beads, FZ close to the base metal, are characterised by showing dendritic growth, which correspond to the zones with higher solidification rates. Additionally, it can be seen that the microstructure of the inner part of the FZ consists of a fine precipitation of the second phases in a solid solution matrix of aluminium (Sánchez-Amaya *et al.*, 2009a). The fusion zone of A356 aluminium alloy welds analysed in Akhter *et al.* (2007) also consisted of a fine dendrite structure. The microstructure of the FZ in 2024 and 7075 alloys was seen to change gradually when approaching from the external part close to the HAZ to the centre of the bead. In 2024 and 7075 welds, the microstructure in the outer zone of the FZ is dendritic growth (oriented towards the bead centre), becoming a finer solidification as approaching the bead centre (Sánchez-Amaya *et al.*, 2012c). Figure 8.3 shows examples of the aforementioned typical microstructures described for 2024 and 5083 weld beads.

Figure 8.4 reports the microhardness measurements performed on the different zones of four aluminium alloy (2024, 5083, 6082 and 7075) welds obtained in similar experimental conditions (Sánchez-Amaya *et al.*, 2012c). AA5083 alloys showed FZ with higher microhardness values than the BM. Similar tendencies are found in other works, in which the fusion zone of 1420 Al–Li alloy (Shi *et al.*, 2007a), AA5083 (Ancona *et al.*, 2007) or A356 aluminium alloy (Akhter *et al.*, 2007) showed a microhardness increase of 20 HV in comparison with their base metal (microhardness increase between 10 and 16%; Ancona *et al.*, 2007). This microhardness evolution is due to the microstructural refinement obtained in the FZ. Thus, the increase in hardness in the weld metal of 5000 and 6000 series aluminium alloys is attributed to the fine dendrite structure of the weld fusion zone (Akhter *et al.*, 2007;

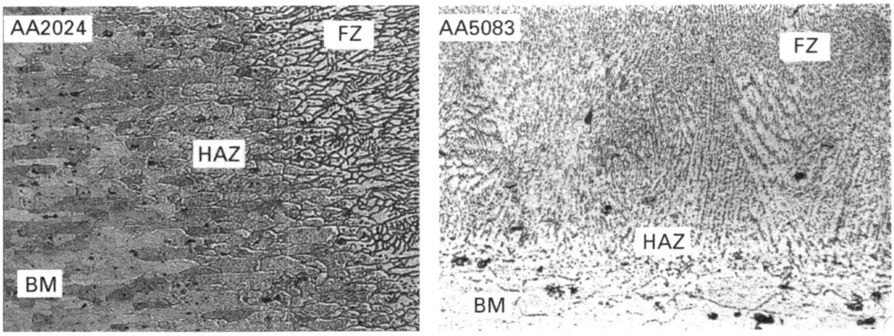

8.3 Microstructure of 2024 and 5083 weld beads.

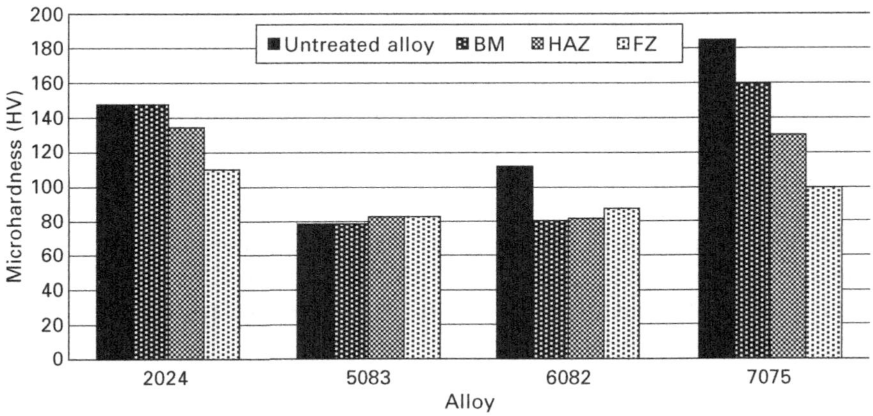

8.4 Microhardness of different zones of the aluminium alloy welds.

Ancona *et al.*, 2007) and to the fine precipitation of the second phases in the aluminium matrix (Sánchez-Amaya *et al.*, 2009b). Other aluminium alloys, such as 2024 and 7075 (Sánchez-Amaya *et al.*, 2009b) have presented the reverse tendency, FZ showing lower microhardness than BM. This microhardness decrease is related to the loss of hardening related to the original heat treatments (T3 in 2024 and T6 in 7075). These previous heat treatments are known to induce precipitation hardening processes in these alloys. Laser treatments are observed to provoke microstructural changes leading to microhardness decrease, especially in the FZ. This microhardness decrease associated to the loss of original heat treatment hardening in the FZ is also observed in 6000 series welds (Nègre *et al.*, 2004).

In order to minimise the differences in microstructures and properties between weld bead and base metal, a post-weld heat treatment (PWHT) is sometimes applied to the welds. This post treatment provokes a microhardness decrease in the bead, associated with coarsening and solution of strengthening precipitates (Shi *et al.*, 2007a).

8.3.2 Defects

Aluminium alloy welds can suffer from different types of defects, provoking an important decline in the quality of the welds (Duley, 1999; Paleocrassas, 2005; Matsunawa *et al.*, 1998; Seto *et al.*, 2000), such as lack of penetration (due to their poor absorption ability), weld oxidation (due to poor gas protection), porosities (due to gas entrapment during weld solidification), cracks (due to the high coefficient of thermal expansion of aluminium alloys, leading to a strong contraction during the solidification of the weld pool), inclusions (such as oxides, provoking a fluidity decrease of the melted aluminium before solidification) (Kwon and Lee, 2003; Di Sabatino *et al.*, 2005) or loss of alloying elements (Li, Mg, Zn) (Duley, 1999; Paleocrassas, 2005; Sanchez-Amaya *et al.*, 2009a). The meticulous control of the processing parameters (welding rate, laser power, shielding gas, etc.) is reported to minimise the appearance of these defects (Duley, 1999).

8.3.3 Mechanical properties

Regarding the analysis of mechanical properties, Shi *et al.* (2007a) have studied the effect of laser beam welding on tear toughness of a 1420 aluminium alloy. The tear energies of base metal are seen to be higher than those of weld metal and HAZ. The decrease in tear toughness of the weld metal is attributed to the transition of fracture mode from transgranular to intergranular (Shi *et al.*, 2007a). Akhter *et al.* have also analysed the mechanical properties (yield strength, YS and ultimate tensile strength, UTS) of welded A356 aluminium alloy samples, finding that welds reached around 80% of the mechanical properties measured for the BM (Akhter *et al.*, 2007). Kuo and Lin have also performed tensile strength tests on base metal and welds performed with different experimental conditions on AA5754 and AA6022 alloys (Kuo and Lin, 2006). Compared to the BM, the tensile strength of welds ranged between 83% and 90% for AA5754 (Kuo and Lin, 2006), around 80% for 5083 (Ancona *et al.*, 2007) and between 64% and 73% for AA6022 (Kuo and Lin, 2006). While in Akhter *et al.* (2007), the fracture in tensile tests always occurred in the BM, in Kuo and Lin (2006), all welds ruptured in the FZ. Micro-flat-tensile specimens taken from different zones of the weld joints of 6061 samples have been tested in Nègre *et al.* (2004). The results obtained show that yield strength and elongation at rupture of FZ are in the order of 70% of those of BM. Tensile strength tests performed in Okon *et al.* (2002) indicate that laser conduction welds were stronger than laser keyhole welds, providing fracture loads of 5.733 kN and 4.775 kN, respectively. These results are related to the higher alloy constituent depletion and to the smaller groove formed under the keyhole regime (Okon *et al.*, 2002). Ancona *et al.* (2007) observe a decrease of the mechanical properties (tensile strength) of the joints at lower welding speeds (probably because of the larger selective vaporisation of magnesium in the fused zone) and at low power laser conditions (leading to welds with high porosities) (Ancona *et al.*, 2007). Katayama *et al.* (1997) have also confirmed the influence of the porosity of the static mechanical properties (strength and elongation) of laser-welded A5182 joints. The authors conclude that large pores exerted a great influence on the reduction in strength and elongation but small ones hardly affected the tensile properties (Katayama *et al.*, 1997).

In summary, the tensile strength results reported in the literature indicate that, when welding under optimum experimental conditions, laser welds normally reach a minimum of 70% of the UTS obtained for the BM.

Very few studies regarding fatigue resistance of laser welds of aluminium alloys are available in the literature (Mizutani *et al.*, 2008). Among them, Mizutani *et al.* (2008) have studied the influence of weld defects, such as porosities, on the fatigue properties of AA5182 laser welded with a CO_2 laser. These authors have demonstrated the detrimental effect of porosities on the fatigue strength and life of these welds. Daneshpour *et al.* (2012) have studied the crack retardation mechanism due to overload to predict the crack growth rate of base metal and laser welded AA6056. Thus, using experimental and finite element methods, the authors determined the plastic deformation and resulting damage profile after overloading. Interestingly, welds presented significant crack retardation and longer fatigue life than base metal, since the application of a tensile overload induced a large crack tip plasticity confined to the weld zone, which elongated ahead of the crack tip.

8.3.4 Corrosion

Relatively few data have been found regarding the corrosion behaviour of laser welds of aluminium alloys. Recent results indicate that the weld bead generally behaves as a cathode, while the base metal surrounding the bead normally behaves as an anode (Mujibur *et al.*, 2007; Sanchez-Amaya *et al.*, 2009a). Therefore, it seems that the corrosion resistance of the fusion zone is slightly improved, due to the microstructural refinement. However, this different corrosion activity between BM and weld bead can lead to undesirable galvanic corrosion (Sanchez-Amaya *et al.*, 2009a). This galvanic corrosion is expected to provoke different degrees of intergranular corrosion in HAZ, especially in heat-treatable alloys, such as 2000 and 7000 series aluminium alloys. This topic, unfortunately, has not been explored systematically in the literature, and there is currently a lack of knowledge regarding the actual influence of the experimental laser conditions on the susceptibility to intergranular corrosion (IGC) of welds of the different aluminium alloys.

Post-weld treatment methodologies focused on reducing the IGC susceptibility of aluminium alloy welds have been little explored yet in the literature. In this context, conventional post-welding heat treatments have reported to improve the corrosion resistance of friction stir welds in aluminium alloys (Proton *et al.*, 2011; Paglia *et al.*, 2002). Post-welding laser treatments have also been employed for this purpose. Thus, laser surface melting (LSM) treatments performed with excimer laser have been applied after friction stir welding to AA2024 and AA7010 (Williams *et al.*, 2003) and to AA2024 and AA7449 (Padovani *et al.*, 2008, 2011, 2012), achieving significant corrosion resistance improvements. These LSM treatments lead to the formation of a thin layer (between 3 and 10 μm) where constituent particles are dissolved, the alloying elements being retained in solid solution (Williams *et al.*, 2003; Padovani *et al.*, 2008, 2011, 2012). These post-weld laser treatments reduce the intergranular corrosion of the HAZ and nugget (FZ) region. Diode laser surface treatments have also been applied after friction stir welding of AA2024 (Jyoti, 2011) and AA5083 (Kalita and Waldera, 2011). These latter laser treatments provoke thicker remelted layers (higher than 100 μm) and are proved to increase the pit nucleation resistance in comparison with the parent material. Unfortunately, as reported by Padovani *et al.* (2008, 2011, 2012), laser-treated layers can delaminate from the substrate as a consequence of corrosion propagation. Therefore, these treatments seem to require further research in order to provide better protective behaviour. Future work will be possibly directed towards the employment of unique laser devices to perform both the weld and the LSM in aluminium alloys.

8.4 Introduction to laser welding of titanium alloys

Attending to their microstructure, titanium alloys have been conventionally classified into three categories: α alloys (consisting of various grades of commercially pure titanium alloys), α + β alloys (presenting α phase and a small volume fraction of β phase in equilibrium, although martensite transformation can also occur upon fast cooling) and β alloys (which do not transform martensitically upon fast cooling from the β phase field) (Lütjering and Williams, 2007: 33–36). At room temperature, titanium alloys have a good corrosion resistance as a consequence of the formation of very stable, continuous, highly adherent and protective oxide films on metal surfaces formed when they are exposed to air (Lütjering and Williams, 2007: 16; Tsay *et al.*, 2006).

The weldability of titanium alloys strongly depends on the category previously described: Destefani (1992: 1771) reports that unalloyed titanium and all α titanium alloys have good weldability, toughness and strength. α + β alloys show good formability and weldability, their properties being highly influenced by heat treatments. β alloys are claimed to present bad weldability due to the degradation of strength after welding (Richter *et al.*, 2007).

Some interesting properties for welding of titanium alloys are their low thermal conductivity (lower than 22 $W \cdot cm^{-1} \cdot K^{-1}$), which prevents heat dissipation, and their low coefficient of thermal expansion (around 8.6 $\times 10^{-6}$ m/m K), which avoids the appearance of stresses during welding (Bertrand *et al.*, 2007). In other words, titanium alloys have low heat input, which minimises welding stresses and reduces distortion (Costa *et al.*, 2007). Titanium alloys also present a good rate of laser beam absorption (0.4%) and a high melting point (around 1670°C, i.e., 1943 K) (Bertrand *et al.*, 2007). These properties imply that high energy must be used to weld these alloys. Therefore, taking into account these considerations, laser welding technology is an adequate technology to join these alloys, as it allows a high localisation and low size of the melting pool, reducing the required energy considerably in comparison with other welding technologies.

Laser welding has some specific requirements for titanium alloys. Firstly, the cleaning procedure of samples before welding is of crucial importance. These strict measures are needed because of the high reactivity of titanium alloys at high temperature with gases such as oxygen and nitrogen, resulting in contamination by interstitial impurities in the weld. Titanium alloys also have high reactivity with grease, dirt, refractories and most other metals at high temperature, resulting in embrittlement of the weld (Costa *et al.*, 2007). Although titanium alloys are reported to have a limited reactivity with oxygen below 600°C (873 K), above this temperature an extensive growth of the oxide layer is formed due to fast diffusion of oxygen, leading to embrittlement of the adjacent oxygen-rich layer (Lütjering and Williams, 2007: 16). For this reason, the use of shielding gas is compulsory to prevent the oxidation and embrittlement during the formation of the weld molten pool. Under effective shielding, no significant oxygen contamination occurs in laser welding.

The most commonly used shielding gases to weld titanium alloys are argon, helium, or argon-helium mixtures (Caiazzao *et al.*, 2004). According to the results of Caiazzao *et al.* (2004), helium leads to greater penetration welds than argon, due to the lower ionisation energy of the latter gas. The results obtained in Wang *et al.* (2007) also show that helium provides better shielding efficiency than argon. Thus, the employment of argon means that part of the supplied laser energy is consumed by the argon ionisation, reducing therefore the actual energy transferred to the metal for melting (Caiazzao *et al.*, 2004). Due to their different density, argon is most used when applied from the top and helium when applied from the bottom of the welding zone.

The shielding gas used in laser welding to protect the molten pool can also provoke welding defects such as porosity and cracks. Porosity in the welding process is generated as a consequence of the entrapment of gas bubbles in the molten material before solidification. Employing a vacuum environment in the welding process is an easy way to reduce the porosity problem in titanium alloys (Akman *et al.*, 2009). The configuration of the gas delivery system must ensure that both the front and back of the weld zone is not in contact with atmospheric gases until the surface temperature is lower than 300°C. Several laser welding nozzles have been developed in the literature. Grevey *et al.* (2005) designed a conical coaxial nozzle system assuring protection against oxidation using minimal gas flow rates and increasing the welding penetration in the case of high-power Nd:YAG laser welding. By means of a lateral nozzle, Wang *et al.* (2007) have also analysed the influence of the lateral gas angle, shielding gas and flux rate on the shielding zone size and shielding efficiency. The authors conclude that the larger the inclination angle of the assist gas nozzle, the smaller the extent of the shielding zone.

8.5 Laser welding technologies for titanium alloys

Table 8.2 includes the main parameters employed by some recently published papers dealing with laser welding of titanium alloys. Note that, due to space limitations, only a few of the most recently published works in the field are reported in the table. This table specifically reports the laser beam parameters (laser type, laser power, frequency, etc.), the variables of the workpieces (surface treatments, weld type, etc.) and the experimental processing parameters (welding speed, shielding gas, etc.). Similarly to laser welding of other alloys, the laser fluence (energy density) employed in the treatments are modulated by experimental variables, such as laser source, laser power, welding speed, spot size, welding mode, etc. In addition, Table 8.2 indicates the properties of the welds analysed in each work: weld shape, microstructure, defects, mechanical or corrosion properties, and so on. As shown in the table, different types of laser have been used so far to weld titanium alloys: Nd:YAG, CO_2, fibre laser, diode laser and high power diode laser. The most widely employed types of laser equipment are Nd:YAG and CO_2 lasers. It can be observed in Table 8.2 that the titanium alloy most investigated is the $\alpha + \beta$ Ti6Al4V alloy (also called Ti64 or TiG5), because it is also the most employed in the industry. Butt welding is the most common configuration in the literature to investigate the laser welding of titanium alloys.

The experimental parameters mentioned above must be carefully optimised to obtain good quality welds. In this context, many studies focus on the optimisation of laser setting variables in order to monitor the morphology and mechanical properties of the welds obtained. Thus, Casalino *et al.* (2005) applied the analysis of variance (ANOVA) and Taguchi analysis

Table 8.2 Summary of recent publications regarding laser welding of titanium alloys

Laser welding of titanium alloys																	
References	Alloys					Laser				Analysis of welds							
		Superficial pre-treatments	Regime (Conduction/Keyhole)	Weld type (Butt, Bead-on-Plate, Lap, Overlap, Autogenous)	Shielding gas	Type	Power (kW)	Welding rate (m/min)	Mode (Pulse/Continuous)	Weld shape	Microstructure	Microhardness	Defects	Tensile strength	Fatigue	Simulation/ Modelling
Akman *et al.* (2009)	Ti6Al4V		K	B	He	Nd-YAG	1.12–2.68		P	X	X	X		X		
Balasubramanian *et al.* (2011a)	Ti6Al4V			B	He	CO_2	3.5	1.5	C	X	X			X	X	
Balasubramanian *et al.* (2011b)	Ti6Al4V		K	B	He	CO_2	3.5	1.5	C	X	X	X		X		
Bertrand *et al.* (2007)	TiG1	X			Ar	Nd-YAG	0.8–1		C	X	X	X				
Blackburn *et al.* (2010)	Ti6Al4V		K	B P	Ar	Nd-YAG	1–2	1.2–1.98	P C	X			X			
Buddery *et al.* (2011)	TiG1, TiG4		K		Ar	Nd-YAG	1.1–1.6		P		X	X				
Caiazzao *et al.* (2004)	Ti6Al4V		K	L B	He Ar	CO_2	2.5		C	X		X		X		

Cao and Jahazi (2009)	Ti6Al4V	K	B	He Ar	Nd-YAG	4	3–10.5	C	X	X	X	X	X		
Casalino *et al.* (2005)	Ti6Al4V	K	B O	He Ar	CO_2	0.5–2.5	3.6–7.2	C	X						
Casavola *et al.* (2009)	TiG2, Ti6Al4V		B	Ar	Nd-YAG	1.6, 2	3,1.5	C		X					X
Chen *et al.* (2012)	Ti6Al4V		B		CO_2	1.3	3	C		X			X		
Cheng *et al.* (2010)	Ti6Al4V			Ar	CO_2	1.3	3		X	X			X		
Cheng *et al.* (2011)	Ti6Al4V	K	L		CO_2	0.18–2.3	1.5–3	C	X				X		
Costa *et al.* (2007)	Ti6Al4V	K C		Ar	Fibre	4–8	1–3	C	X	X	X				
Gao *et al.* (2013)	Ti6Al4V		B	Ar	Nd-YAG		0.6	P	X	X	X	X	X		
Jianxun *et al.* (2012)	Ti6Al4V	K		Ar	CO_2	2.5	1.2–1.5	C		X	X		X		
Kwang Pan *et al.* (2007)	Ti6Al4V	K	B	Ar	Nd-YAG	0.2–0.3	0.65–0.75	P	X				X		
Lee *et al.* (2006)	CP Ti		T	He	Nd-YAG	0.076–0.128		P	X						
Li *et al.* (2005)	TiG1		P	Ar	Nd-YAG	0.23	0.024	P		X	X				
Liu *et al.* (2011)	TiG2		P	Ar	Fibre	8–10	2–8	C		X	X		X		
Liu *et al.* (2012)	TiG2	K	P	Ar	Fibre	10	2–8	C		X					
Nakai *et al.* (2012)	Ti-4.5Al-2.5Cr-1.2Fe-0.1C	K	B		Fibre					X	X		X	X	
Shi *et al.* (2007a)	Ti-6.5Al-2Zr-1Mo-1V			Ar	CO_2	0.26	2	C		X	X		X		

(Continued)

Table 8.2 Continued

Laser welding of titanium alloys																
References	Alloys					Laser				Analysis of welds						
		Superficial pre-treatments	Regime (Conduction/Keyhole)	Weld type (Butt, Bead-on-Plate, Lap, Overlap, Autogenous)	Shielding gas	Type	Power (kW)	Welding rate (m/min)	Mode (Pulse/Continuous)	Weld shape	Microstructure	Microhardness	Defects	Tensile strength	Fatigue	Simulation/Modelling
Squillace *et al.* (2012)	Ti6Al4V		K	A	He Ar	Nd-YAG	0.8–1.2	1.02–3.48		X	X	X		X	X	
Torkamany *et al.* (2012)	Ti Cp		K	O	Ar	Nd-YAG		0.12–0.6	P	X						
Tsay *et al.* (2006)	Ti6Al4V				He Ar	CO_2	0.8–2.5	48–72	C			X	X		X	
Tsay *et al.* (2010a)	Ti–6Al–6V–2Sn		K	P	Ar	CO_2	2.7	0.8			X	X				
Tsay *et al.* (2010b)	Ti–4.5Al–3V–2Fe–2Mo		K	P	Ar	CO_2	2.7	1	C		X	X				
Wang *et al.* (2003)	Ti6Al4V		K	P	He	CO_2	2.5	1.5	C		X	X		X		
Wu *et al.* (2002)	Ti–24Al–17Nb			P	He	CO_2	0.65–2.15	0.25–5.37		X				X		

to laser welding experimental data of Ti6Al4V (shape of welded area and hardness), leading to the forming of a suitable database on butt and overlap welds. Kwang Pan *et al.* (2007) have employed Taguchi method-based Grey analysis for optimising multiple quality characteristics in machining parameter settings of laser welded titanium alloys.

As can be observed in Table 8.2, the keyhole mode is more used to weld titanium alloys than conduction, because it leads to full penetration welds. The conduction mode is mostly employed in research work involving simulations and modelling of the weld. Thus, Cho *et al.* (2009) performed a simulation to understand the weld pool flow patterns of titanium. The results show that laser weld pool flow dynamics play a key role during the transition from conduction mode to keyhole. On the other hand, Du *et al.* (2004) presented a model for flow simulation of full penetration laser beam welding of titanium alloy, concluding that the molten pool becomes shorter and wider under the conduction regime, as a consequence of the Marangoni effect. More in-depth studies dealing with laser processing under the conduction mode are performed in Costa *et al.* (2007) and Amaya-Vázquez *et al.* (2012a), in which microstructure transformations and variations of mechanical properties are analysed.

Titanium alloys can be welded using pulsed or continuous wave mode laser (as can be observed in the table, continuous mode is the most employed). For example, Blackburn *et al.* (2010) conducted a study on the influence of different waveforms and frequencies on the quality of bead on plates (BOP) of Ti6Al4V produced with a Nd-YAG laser. The welds performed with square waveform at frequencies higher than 250 Hz considerably reduced the weld metal porosity when compared with BOP produced with a continuous wave output. The sinusoidal waveform increased the porosity formation compared with the continuous wave BOP tests. In a recent investigation regarding pulsed laser welding of Ti6Al4V, Akman *et al.* (2009) determined that the ratio between the pulse energy and pulse duration is the most important parameter in defining the penetration depth, while the variation of pulse duration at constant peak power was seen to not influence the penetration depth. These authors have also reviewed the relationship between welding mode (pulsed or continuous) and welding regime (conduction or keyhole). Thus, under pulsed mode, a small molten pool is formed by each laser pulse and within a few milliseconds it re-solidifies. When the peak power is low or the spot size is increased, welding occurs in conduction mode and a shallow, smooth weld pool is produced. On the other hand, when the peak power is increased or the spot size is reduced, a much deeper weld pool is obtained, leading to keyhole welding mode (Akman *et al.*, 2009).

In addition to avoiding surface oxidation, the application of surface pre-treatments to titanium alloys before welding to improve laser absorption has been analysed in the literature. Thus, Bertrand *et al.* (2007) carried out a study focused on improving laser beam absorption in TiG1 (α alloy) samples, concluding that sandblasted surfaces lead to the highest laser beam absorption, because they reduce the beam reflectivity. However, there is not a clear consensus about the real influence of surface treatments on laser weld penetration in titanium alloys, as preliminary works performed in our laboratory have shown that different pre-treatments lead to similar weld depths in Ti6Al4V (although resulting in different weld microstructure). In any case, superficial treatments in titanium alloys seem to be less influential than in aluminium alloys (Sánchez-Amaya *et al.*, 2009a), especially under conduction regime, as the laser reflectivity and the thermal conductivity of aluminium alloys are much larger than those of titanium alloys.

In the literature, not only previous superficial treatments have been applied, but also post-weld heat treatment (PWHT), with the objective of relieving stress (Shi *et al.*, 2007b). The authors observed that PWHT of Ti-6.5Al-2Zr-1Mo-1V weld leads to microstructural changes and a significant fracture toughness decrease. The welds of unalloyed titanium, however, do not seem to require PWHT unless the material is highly stressed in a strongly reducing atmosphere.

8.6 Microstructure, defects, mechanical properties and corrosion behaviour of titanium welds

8.6.1 Microstructure

When subjected to specific heat treatments, titanium alloys can undergo an allotropic phase transformation in which α(HCP) transform to β(BCC). The transformation temperature (so-called beta transus temperature) is modulated by the type and amount of impurities and/or by the alloying additions (Destefani, 1992). Thus, beta transus temperature is about 883°C (1156 K) for pure titanium (Destefani, 1992), and approximately 995°C (1268 K) for Ti6Al4V alloy (Akman *et al.*, 2009).

During laser welding, the joining zone of the samples undergoes microstructural changes due to the rapid cooling of the weld pool. The microstructure generated at the fusion zone (FZ) and heat affected zone (HAZ) strongly depends on the type of laser employed and on the energy density applied. Due to its great importance on the quality of welds, the weld microstructure is systematically studied by almost all the investigations reviewed (see Table 8.2).

The microstructure of Ti6Al4V base metal consists of two phases: intergranular beta phase and an equaixed alpha phase (Akman *et al.*, 2009; Amaya-Vázquez *et al.*, 2012a). Akman *et al.* (2009) described the microstructure of Ti6Al4V welds obtained with a Nd-YAG laser. Thus, the grain size is seen to increase with the average laser power in both HAZ and

FZ. The generated microstructure in FZ is described as acicular martensite (martensitic α') formed by columnar α and β grains. The average grain size in FZ is seen to range between 160 and 350 μm. HAZ microstructure is described as a mixture of martensitic α', acicular α and primary α. Similar microstructures in FZ and HAZ were reported by Wang *et al.* (2003) in Ti6Al4V laser beam welds obtained with a high power CO_2 laser. Balasubramanian *et al.* (2011a) depicted the microstructure of Ti6Al4V welds obtained with CO_2 laser as a fine lamellar and acicular α, although martensite is also present in the FZ. A similar microstructure is observed by Cheng *et al.* (2010) and Chen *et al.* (2012), together with coarse columnar β grains that grow opposite to the heat flow direction and then undergo martensitic transformation upon cooling. Similarly, Costa *et al.* (2007) have analysed the microstructure of Ti6Al4V welds performed with laser fibre. HAZ showed a fine cellular dendritic structure, while FZ presented martensitic structure, its morphology depending on the applied laser energy. Thus, martensite presented needle-like morphology at lower heat inputs (higher cooling rates) and plate-like morphology at higher heat inputs (lower cooling rates) (Costa *et al.*, 2007).

Martensitic phase can only be obtained when the temperature reaches the β transus (980°C, i.e., 1253 K, for Ti6Al4V) and the cooling rate is high enough (α' martensite is claimed to be fully formed in FZ when the cooling rate is higher than 410°C/s) (Cao and Jahazi, 2009; Squillace *et al.*, 2012; Ahmed and Rack, 1998). The HAZ of Ti6Al4V welds is reported to be a mixture of α' and primary α phases, corresponding to a structure quenched below the β transus (Cao and Jahazi, 2009; Squillace *et al.*, 2012).

Two different microstructural patterns have been identified in laser remelted Ti6Al4V samples, depending on the input laser fluence applied (Amaya-Vázquez *et al.*, 2012a). At high fluences (involving low cooling rates) only two zones are generated, melted zone (MZ) and heat affected zone (HAZ), both composed of clearly defined grains with microlaminar β in α matrix microstructure. The grains of MZ are bigger and more elongated than those generated in HAZ. At low and medium fluences (leading to high cooling rates), three zones are identified: MZ (martensite acicular α'), HAZ (mixture of $\alpha + \beta$) and base metal (equiaxed α with intergranular β). The formation of martensitic microstructure, generated in the FZ as a consequence of the high heating and cooling rates, has also been reported by other authors (Akman *et al.*, 2009; Amigó *et al.*, 2010; Candel *et al.*, 2010; Pasang *et al.*, 2011; Chen *et al.*, 2012). Some examples of metallographic images of the different microstructures reported for Ti6Al4V welds are shown in Fig. 8.5.

The microstructure of commercially pure titanium laser welds has also been described in the literature. The microstructure of TiG2 base metal (Fig. 8.6(a)) is elsewhere depicted as Ti-α type equiaxial grains with a relatively uniform grain size (Lampman, 1990; Polmear, 1995; Amaya-Vázquez *et al.*,

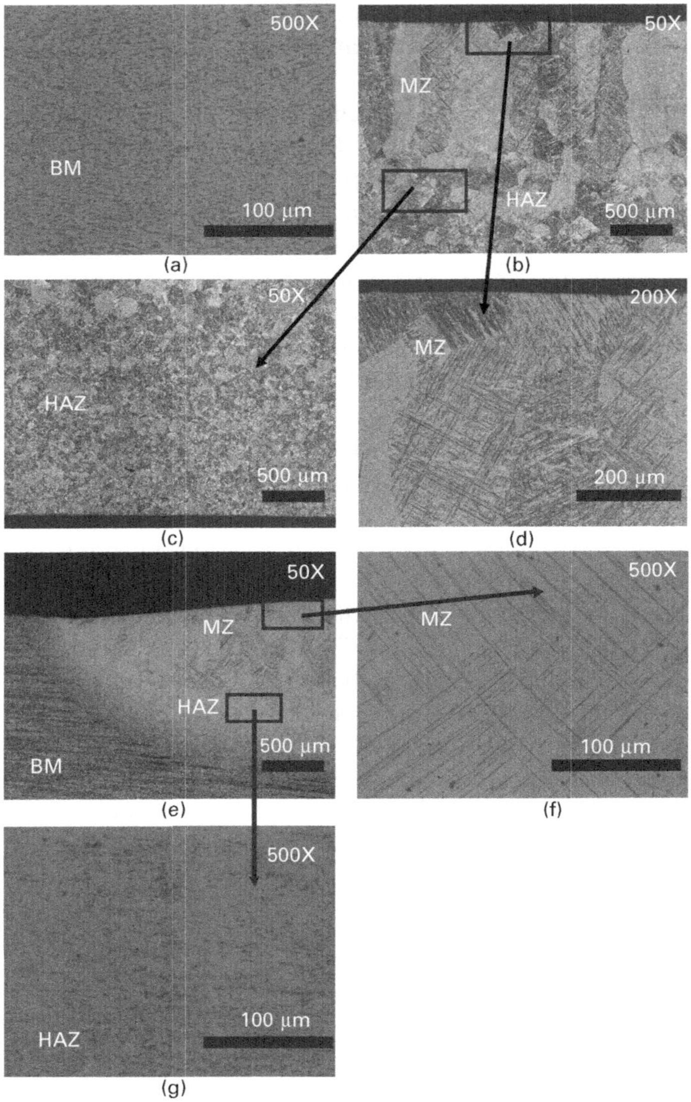

8.5 Examples of microstructures obtained in different zones of Ti6Al4V welds: (a) Ti6Al4V base metal; (b, c and d) melted zone (MZ) and heat-affected zone (HAZ) of Ti6Al4V weld obtained with low cooling rate (MZ with $\alpha + \beta$); (e, f and g) BM, HAZ and MZ of Ti6Al4V weld obtained with high cooling rate (MZ with α').

2012a). Li *et al.* (2005) and Amaya-Vázquez *et al.* (2012a) have reported that laser treatments with a pulse Nd:YAG laser and HPDL, respectively, modify the shape and size of the grains, the microstructure consisting of serrated Ti-α grains (S) with higher size than the base metal, together with some fine acicular α grains (A), as shown in Fig. 8.6(b). Clear evidence of martensite

8.6 Microstructure of (a) TiG2 base metal (with equiaxial α grains), (b) TiG2 fusion zone (with serrated alpha grains (S) and fine acicular α grains (A)).

has not been observed in TiG2 weld bead in either Li *et al.* (2005) or in Amaya-Vázquez *et al.* (2012a). These results contrast with those reported by Sun *et al.* (2003) and Zhang *et al.* (2007) in which martensite formation has been evidenced for CP titanium welded with a Nd:YAG pulsed laser (Sun *et al.*, 2003) and with a continuous wave CO_2 laser (Zhang *et al.*, 2007). These different results are related to the relatively higher cooling rates involved in these latter laser treatments.

Results obtained by different authors show that the weld microstructure of titanium alloys are strongly dependent on the laser processing parameters. Taking into account that weld microstructure is one of the main factors influencing both the mechanical properties and corrosion behaviour of titanium welds, a systematic study of the microstructure is always highly recommended when using laser welding.

8.6.2 Defects

The main defects reported in the literature for titanium alloy laser welds are described next. Regarding the superficial cleanliness, Buddery *et al.* (2011) have analysed the effect of contamination on the metallurgy of commercially pure titanium welded with a Nd:YAG pulsed laser beam. These authors conclude that Fe is a strong promoter of the martensitic phase, while O and N cause a significant amount of solid solution strengthening.

Unless the laser treatment is performed in a vacuum furnace or in an inert atmosphere, oxygen reacts with the titanium at the metal surface and produces an oxygen-enriched layer commonly called 'alpha case' (α case). This undesirable superficial microstructure is formed in titanium alloys when exposed to air at temperatures higher than 540°C (813 K) (Lampman, 1990; EN 3114-001:2006). α case is hard, brittle and considered detrimental because it results in marked lowering of fatigue properties (EN 3114-001:2006; AMS4945C, 2006). This brittle layer must be removed before the component is put into service. Standard practice is to remove α case by mechanical methods, chemical methods or by both (Lampman, 1990).

Contamination by interstitial elements such as O or N can decrease the tensile ductility of the titanium welds (Tsay *et al.*, 2006). Removal of external TiO_2 layer before welding has been reported to reduce weld metal porosity when CO_2 laser welding (Blackburn *et al.*, 2010; Chen *et al.*, 2006) because this oxide layer is hygroscopic and absorbs moisture from the atmosphere.

As mentioned above, one important defect to be taken into account is superficial oxidation, which can be due to poor cleaning procedures and/or an inadequate shielding atmosphere. Evidence of titanium oxide formation can be easily identified by a simple visual inspection of the surface colour. The standard (ANSI AWS D17.1:2001) specifies the discoloration of titanium oxides and the acceptance criteria of welds for aerospace applications. Acceptable colours of titanium welds are indicated to be: bright silver, silver, light straw, dark straw, bronze or brown. Meanwhile, unacceptable oxide colours for welds are: violet, green, blue, gray or white (Amigó *et al.*, 2010; ANSI AWS D17.1:2001).

Li *et al.* (2005) have also demonstrated the influence of adding oxygen to the shielding gas (argon) on the microstructure, discoloration and mechanical properties of CP Ti welds. Thus, as the oxygen content was increased, the weld surface colour changed from silver or straw to blue; the surface hardness increased; and the weld strength first increased and then decreased, because the microstructure changed from mainly serrated alpha to mainly acicular and platelet alpha (Li *et al.*, 2005).

In addition to the aforementioned superficial contamination effects and α case formation, different types of defects are reported for titanium welds, such as porosities, surface oxidation, microcracking and face/root undercut (underfill) (Blackburn *et al.*, 2010; Cao and Jahazi, 2009; Tsay *et al.*, 2006; Donachie, 2000; Bertrand *et al.*, 2007). Weld metal porosity is reported to form easily when keyhole laser welding (Blackburn *et al.*, 2010; Matsunawa, 2001). Pores are generated due to cavities formed by gas (metal vapour and/or inert shielding gases) trapped during solidification of weld metal. Porosities can break the surface, act as stress concentrators and reduce the fatigue resistance of the weld. Therefore, their formation should be minimised to fulfil the strict quality control standards (Blackburn *et al.*, 2010). Blackburn *et al.* (2010) have recently reported that the input waveform of Nd:YAG laser when applied to titanium alloy welding can influence the porosity formation. Thus, square waveform at frequencies >250 Hz reduced the weld metal porosity when compared with continuous wave output, while the sinusoidal waveform increased the porosity formation in the weld metal when compared with continuous wave. Similarly to the appropriate choice of the waveform, modulation frequency and modulation amplitude are reported to have importance in achieving excellent weld quality (Blackburn *et al.*, 2010). Micropores are also observed by Cao and Jahazi (2009) in butt joints of Ti6Al4V alloy welded using a high-power Nd:YAG laser, being more prevalent in thicker joints. Torkamany *et al.* (2012) have studied the spattering and weld porosity of thin CP Ti sheet welds performed with a Nd:YAG laser. The authors observed that increasing laser pulse peak power leads to a higher tendency of spattering and weld porosity formation. For this reason, titanium thin plates are not recommended to be welded under keyhole regime. Concerning underfilling, Cao and Jahazi (2009) reported that these defects decrease with increasing welding speed. At higher welding speed, less evaporation of molten metal leads to lower underfill area and depth. Therefore, high welding speed is preferred to minimise the underfill defects (also contributing to increase in productivity).

Other reported defects for titanium welds are microcracks, located at HAZ (Cao and Jahazi, 2009). Although titanium alloys are generally highly resistant to HAZ liquation cracking (Donachie, 2000; Cao and Jahazi, 2009), contamination cracking can still occur due to hydrogen delayed cracking or liquation of iron-rich eutectic (Donachie, 2000; Cao and Jahazi, 2009). Microcrackings are claimed to be caused by local melting of the titanium-iron eutectic (Cao and Jahazi, 2009).

8.6.3 Mechanical properties

Most investigations regarding laser beam welding of titanium alloys have focused on the study of the mechanical properties of the joints being normally desired to obtain welds with similar properties to the BM. Hardness and the ultimate tensile strength are the main mechanical properties analysed by the majority of the researchers (Table 8.2).

In order to characterise weld hardness, most research performs microhardness measurements on different zones of the welds. The melted zone (fusion zone) of the weld is normally harder than the BM. This is due to the very high cooling rates to which the melted zone is subjected following the laser welding process. This microhardness increase is attributed to the microstructure developed. For example, the FZ of α alloys, such as TiG2 samples, are reported to have microhardness values between 140 and 200 HV, while untreated base metal ranges between 110 and 140 HV (Liu *et al.*, 2011; Amaya-Vázquez *et al.*, 2012a). The welding heat input in this alloy led to grain coarsening in the HAZ and FZ. In this specific alloy, coarse grains resulted in increasing its hardness. The strengthening mechanism in the HAZ and FZ is attributed to substructure strengthening, solute solution strengthening (Liu *et al.*, 2011) and the formation of serrated/acicular α-grains (Amaya-Vázquez *et al.*, 2012a). Influencing factors on the microstructure obtained and, therefore, on the hardness of pure titanium welds, are the microelement contents and grain orientation due to its hexagonal close-packed structure (Liu *et al.*, 2011).

Similarly, in the case of laser welds of α + β alloys, such as Ti6Al4V, the zone with highest microhardness is also reported to be the melted zone, due to the formation of α′-martensite (Ahmed and Rack, 1998; Squillace *et al.*, 2012; Akman *et al.*, 2009; Costa *et al.*, 2007; Amaya-Vázquez *et al.*, 2012a, 2012b; Cao and Jahazi, 2009; Gao *et al.*, 2013). α′-martensitic microstructure in Ti6Al4V is reported to present microhardness values between 350 and 500 HV, while α + β base metal values ranged around 300–350 HV (Akman *et al.*, 2009; Squillace *et al.*, 2012; Costa *et al.*, 2007; Amaya-Vázquez *et al.*, 2012b; Cao and Jahazi, 2009; Gao *et al.*, 2013).

As can be seen in Table 8.2, most research works include studies of the tensile strength of titanium welds. Liu *et al.* (2011) have analysed the tensile strengths and elongations of the laser beam welds of TiG2, presenting the same values as those of BM. These results, together with the high microhardness measurements of the FZ and the fact that all tensile specimens fractured at the BM, permit the authors to conclude that the HAZ and FZ are stronger than the BM.

Regarding Ti6Al4V (Cao and Jahazi, 2009; Kwang Pan *et al.*, 2007; Wang *et al.*, 2003) and Ti24Al17Nb (Wu *et al.*, 2002) welds, some authors have also reported similar or slightly higher joint strength compared with metal base. Ti6Al4V BM usually has a yield tensile strength of 880–1045 MPa, tensile strength of 950–1081 MPa and elongation at fracture of 10–14% (Cao and Jahazi, 2009; Kwang Pan *et al.*, 2007). Meanwhile, Ti6Al4V butt welds are reported to have tensile strength from 975 to 1043 MPa, yield strength from 936 to 987 MPa and elongation at fracture of 6.5–12.4% (Cao and Jahazi, 2009). These results indicate that these laser weld joints have similar joint strength but lower ductility than base metal, the ductility decrease being associated with the presence of micropores and aluminium oxide inclusions. Similarly, Kwang Pan *et al.* (2007) obtained welds at optimum processing conditions with UTS values between 900 and 1000 MPa. On the other hand, Ti24Al17Nb (Wu *et al.*, 2002) and Ti6Al4V (Gao *et al.*, 2013) laser beam welds also presented similar tensile strength to the BM (around 1000 MPa), the fracture occurring in the base metal, the elongation of welds being slightly lower than that of BM.

The tensile strength and ductility are sensitive to temperature. Strength decreases and ductility increases when increasing temperature (Chen *et al.*, 2012). Wang *et al.* (2003) have analysed the influence of temperature (up to 450°C, i.e., 723 K) on the tensile properties of laser beam welds in Ti6Al4V. The authors observed that the UTS of the weldments was slightly superior to that of the BM; the yield stress of welds was inferior to that of the BM, especially above 150°C (423 K); and the elongation of welds was about 5% lower for the entire temperature range. At 300°C (573 K), the tensile yield strength of weldments decreases, due to the β precipitation in the tempered martensite of the fusion zone.

In contrast with previously commented results, other research work on Ti6Al4V welds reports tensile strength decrease in comparison with base metal, this mechanical detriment normally being related to the presence of defects in the welds, such as porosity, bad alignment or poor shielding procedure. For example, Akman *et al.* (2009) observed a tensile strength decrease in Ti6Al4V laser welds, this being associated with the microstructure change, growth in grain size and the presence of pores.

Among other properties, Caiazzao *et al.* (2004) have analysed the tensile strength of Ti6Al4V welds obtained with a CO_2 laser, observing UTS values for welds around 75% of that obtained for BM. Results obtained for different sheet thicknesses almost always highlighted a failure that began at the bead foot and spread towards the same bead, HAZ or base metal. The explanation given is the contamination of the join in the foot due to a low He protection at the lower side. The results showed lower UTS values attributed to a non-perfect alignment between the edges to be joined and an excessive air gap.

With the presence of sharp notches, the maintenance of high notched strength at elevated temperature is required for the material to resist cracking and/or embrittlement under tensile straining (Tsay *et al.*, 2010a). Tsay *et al.* (2010a, 2010b) conducted a study on the effect of test temperatures on the notched tensile strength (NTS) of Ti6Al6V2Sn and Ti4.5Al3V2Fe2Mo laser welds, with and without post-weld heat treatments (PWHT) at different temperatures. Results generally show that NTS decreases as temperature rises. In Ti6Al6V2Sn specimens, NTS values of welded specimens are normally much lower than those of the MB specimen at room temperature; however, the tendency is the reverse when testing at elevated temperatures (Tsay *et al.*, 2010a). These results are associated with the microstructure developed. The lack of deformation compatibility between the α and β phases enhanced interface separations in the MB specimen or induced voids in the welded specimens, resulting in lowered NTS at elevated temperatures. Ti4.5Al3V2Fe2Mo laser welds (Tsay *et al.*, 2010b) generally presented lower NTS than base metal at room temperature (except when performing PWHT). Thus, while as-welded and BM samples presented NTS and ductility decrease when increasing testing temperature, PWHT specimens preserved these properties even at 450°C (723 K). The fracture mode of the welds was strongly affected by the columnar grain size and the direction of the crack growth with respect to the columnar grain orientation (Tsay *et al.*, 2010b).

Casavola *et al.* (2009) pointed out the importance of investigating the fatigue behaviour of titanium welded joints, especially in light of the lack of official standards. This is a critical condition when the objective is to develop applications that must satisfy strict requirements on structural safety (Nakai *et al.*, 2012). Balasubramanian *et al.* (2011a, 2011b) have recently studied fatigue properties of Ti6Al4V welds obtained by gas tungsten arc

(GTAW), electron beam (EBW) and laser beam welding (LBW). Regardless of the joining technique, fatigue crack growth resistance of Ti6Al4V welds is reported to be much lower than that of base metal (Balasubramanian *et al.*, 2011a, 2011b; Tsay *et al.*, 2006). Fatigue crack growth of welds has been correlated with mechanical properties and microstructural characteristics of welds. Interestingly, the welds obtained by LBW exhibited higher fatigue crack growth resistance than those generated by EBW and GTAW, due to the presence of fine lamellar microstructure in the weld metal (Balasubramanian *et al.*, 2011a, 2011b). The formation of this microstructure using LBW, consequence of the lower heat input and associated faster cooling rate, also enhances the yield strength and ductility of the welds (Balasubramanian *et al.*, 2011a). However, of the three types of welded joints, the ones fabricated using EBW exhibit the highest strength values, due to the presence of fine serrate and regular plate-shaped α microstructures. Meanwhile, the joints obtained using GTAW exhibit higher impact toughness compared with the LBW and EBW joints, due to the presence of coarse serrate and acicular α structures of grain boundary α, massive α, and Widmanstätten $\alpha + \beta$ weld metal microstructure (Balasubramanian *et al.*, 2011b).

Squillace *et al.* (2012) have analysed the influence of welding parameters on morphology and mechanical properties of Ti6Al4V laser beam welded butt joints (including fatigue). The authors observed that the main defect is underfill, this causing a stress concentration at the weld edges. The fatigue life is, consequently, strongly influenced by the value of the underfill radius. The fatigue fracture initiates and propagates near the lower point of the underfill convexity at the interface between the FZ and the HAZ (Squillace *et al.*, 2012).

Casavola *et al.* (2009) have conducted an experimental and numerical study of static and fatigue behaviour of butt laser welded joints of CP Ti and Ti6Al4V. For almost all specimens, the fatigue final rupture in the welded titanium alloy is localised far from the seam, indicating that the welded zone is not the weakest part of the structure. The high fatigue performance observed is related to the microstructural transformations (Casavola *et al.*, 2009); the HAZ could result in a high resistance area for fatigue loads (Torster *et al.*, 1997).

Tsay *et al.* (2006) have investigated the effect of microstructures and weld porosity on the fatigue crack growth behaviour of Ti6Al4V laser welds. Fatigue crack growth rate (FCGR) of the fusion zone of welds was significantly lower than that of base metal. Randomly oriented martensite in the FZ led to local cleavage fracture, altering the crack growth direction out of the primary crack plane. The zigzag crack path in the FZ resulted in reduced FCGR. The porous weld showed a serration on the crack growth curve, the crack growth characteristics behaving similarly to the defect-free weld (Tsay *et al.*, 2006). Mechanical properties of Ti4.5Al2.5Cr1.2Fe0.1C laser welds have just been studied, presenting improved fatigue properties compared to the Ti6Al4V laser welds. Due to the presence of porosities, fatigue strength of Ti4.5Al2.5Cr1.2Fe0.1C welds is much lower than that of Ti4.5Al2.5Cr1.2Fe0.1C base metal (Nakai *et al.*, 2012).

8.6.4 Corrosion

Another interesting property of the titanium laser welds is their high corrosion resistance. Thus, Sun *et al.* (2003) have performed laser remelting (LR) treatments on TiG2 samples with a Nd:YAG laser, observing a microstructure change from α phase to acicular martensite, leading in improvements in the corrosion behaviour in 3.5 wt.% NaCl solutions. Amaya-Vázquez *et al.* (2012a, 2012b) have also recently reported that TiG2 and Ti6Al4V alloys subjected to LR treatments with a HPDL have similar corrosion behaviour in NaCl solutions to their BM. In agreement with the results obtained by Sun *et al.* (2003), the zones presenting martensitic microstructure have shown a slight corrosion resistance improvement in comparison with the base metal (Amaya-Vázquez *et al.*, 2012a, 2012b). In another study, Yue *et al.* (2002) applied surface treatments with an excimer laser to Ti6Al4V samples, achieving improvements in the pitting corrosion resistance of the alloy, especially when shielding with argon. More recently, Zaveri *et al.* (2008) have subjected Ti6Al4V samples to different processing conditions with a pulsed-wave Nd:YAG laser shielded with argon, to obtain surface oxide layers with high corrosion resistance. The authors reported that laser-treated specimens presented better corrosion resistance than the BM in three different simulated biofluids (SBFs): NaCl solution, Hank's solution (8 g/L NaCl, 0.14 g/L,$CaCl_2$, 0.4 g/L KCl, 0.35 g/L $NaHCO_3$, 1 g/L glucose, 0.1 g/L NaH_2PO_4, 0.1 g/L $MgCl_2{\cdot}6H_2O$, 0.06 g/L $Na_2HPO_4{\cdot}2H_2O$ and 0.06 g/L $MgSO_4{\cdot}7H_2O$), and Cigada solution (8.75 g/L NaCl, 0.35 g/L $NaHCO_3$, 0.06 g/L NaH_2PO_4, and 0.06 g/L $Na_2HPO_4{\cdot}2H_2O$).

8.7 References and further reading

Abe, N., Tsukamoto, M., Maeda, K., Namba, K. and Morimoto, J. (2006) 'Aluminum alloy welding by using a high power direct diode laser', *J Laser Appl*, 18(4), 289–293.

Ahmed, H., Wells, M.A., Maijer, D.M., Howes, B.J. and Van der Winden, M.R. (2005) 'Modelling of microstructure evolution during hot rolling of AA5083 using an internal state variable approach integrated into an FE model', *Mater Sci Eng A-Struct*, 390 (1–2), 278–290.

Ahmed, T. and Rack, H.J. (1998) 'Phase transformations during cooling in $\alpha + \beta$ titanium alloys', *Mater Sci Eng A*, 243, 206–211.

Akhter, R., Ivanchev, L. and Burger, H.P. (2007) 'Effect of pre/post T6 heat treatment on the mechanical properties of laser welded SSM cast A356 aluminium alloy', *Mater Sci Eng A-Struct*, 447, 192–196.

Akman, E., Demir, A., Canel, T. and Sınmazcelik, T. (2009) 'Laser welding of Ti6Al4V titanium alloys', *J Mater Process Tech*, 209, 3705–3713.

Amaya-Vázquez, M.R., Sánchez-Amaya, J.M., Boukha, Z. and Botana F.J. (2012a) 'Microstructure, microhardness and corrosion resistance of remelted TiG2 and Ti6Al4V by a high power diode laser', *Corros Sci*, 56, 36–48.

Amaya-Vázquez, M.R., Sánchez-Amaya, J.M., Boukha, Z., El Amrani, K. and Botana, F.J. (2012b) 'Application of laser treatments to improve the properties of Ti6Al4V', *Mater Sci Forum*, 713, 25–29.

Amigó, V., Candel, J.J., Amado, J.M. and Yañez, A. (2010) 'Evaluación de la oxidación superficial de recubrimientos de Ti6Al4V obtenidos por recubrimiento por láser', *Rev Metal Madrid*, 46, 13–18.

AMS 4945C (2006) 'Titanium Alloy Tubing, Seamless, Hydraulic Al-2.5V, Controlled Contractile Strain Ratio Cold Worked, Stress Relieved', Aerospace material specification.

Ancona, A., Sibillano, T., Tricarico, L., Spina, R., Lugarà, P.M., Basile, G. and Schiavone, S. (2005) 'Comparison of two different nozzles for laser beam welding of AA5083 aluminium alloy', *J Mater Process Tech*, 164–165, 971–977.

Ancona, A., Lugarà, P.M., Sorgente, D. and Tricarico, L. (2007) 'Mechanical characterization of CO_2 laser beam butt welds of AA5083', *J Mater Process Tech*, 191, 381–384.

ANSI AWS D17.1:2001 'Specification for fusion welding aerospace applications', American Welding Society.

Assuncao, E., Williams S. and Yapp. D. (2012) 'Interaction time and beam diameter effects on the conduction mode limit', *Opt Lasers Eng*, 50, 823–828.

Balasubramanian, T.S., Balasubramanian, V. and Muthu Manickam, M.A. (2011a) 'Fatigue crack growth behaviour of gas tungsten arc, electron beam and laser', *Mater Design*, 32, 4509–4520.

Balasubramanian, T.S., Balakrishnan, M., Balasubramanian, V. and Muthumanickan, M.A. (2011b) 'Influence of welding processes on microstructure, tensile and impact properties of Ti-6Al-4V alloy joints', *Trans Nonferrous Met Soc China*, 21, 1253–1262.

Bassani, P., Capello, E., Colombo, D., Previtali, B. and Vedani, M. (2007) 'Effect of process parameters on bead properties of A359/SiC MMCs welded by laser', *Compos Part A – Appl* S, 38, 1089–1098.

Bertrand, C., Laplanche, O., Rocca, J.P., Le Petitcorps, Y. and Nammour, S. (2007) 'Effect of the combination of different welding parameters on melting characteristics of grade 1 titanium with a pulsed Nd–Yag laser', *Lasers Med Sci*, 22, 237–244.

Blackburn, J.E., Allen, C.M., Hilton, P.A., Li, L., Hoque, M.I. and Khan, A.H. (2010) 'Modulated Nd:YAG laser welding of Ti–6Al–4V', *Sci Technol Weld Join*, 15(5), 433–439.

Buddery, A., Kelly, P., Drennan, J. and Dargusch, M. (2011) 'The effect of contamination on the metallurgy of commercially pure titanium welded with a pulsed laser beam', *J Mater Sci*, 46, 2726–2732.

Caiazzao, F., Curcio, F., Daurelio, G. and Memola Capece Minutolo, F. (2004) 'Ti6Al4V sheets lap and butt joints carried out by CO_2 laser: mechanical and morphological characterization', *J Mater Process Tech*, 149, 546–552.

Campana, G., Ascari, A., Fortunato, A. and Tani, G. (2009) 'Hybrid laser-MIG welding of aluminum alloys: the influence of shielding gases', *Appl Surf Sci*, 255(10), 5588–5590.

Candel, J.J., Amigó, V., Angarita, I. and Ramos, J.A. (2010) 'Microestructrura y comportamiento frente al desgaste de recubrimientos de Ti6Al4V sobre Ti obtenidos por laser cladding', *XI Congreso Nacional de Materiales*, Paper PO115. Zaragoza, Spain.

Cao, X. and Jahazi, M. (2009) 'Effect of welding speed on butt joint quality of Ti–6Al–4V alloy welded using a high-power Nd:YAG laser', *Opt Laser Eng*, 47, 1231–1241.

Cao, X., Wallace, W., Poon, C. and Immarigeon, J.P. (2003) 'Research and progress in laser welding of wrought aluminum alloys. I. Laser welding processes', *Mater Manuf Process*, 18(1), 1–22.

Casalino, G., Curcio, F. and Memola Capece Minutolo, F. (2005) 'Investigation on Ti6Al4V laser welding using statistical and Taguchi approaches', *J Mater Process Tech*, 167, 422–428.

Casavola, C., Pappalettere, C. and Tattoli, F. (2009) 'Experimental and numerical study of static and fatigue properties of titanium alloy welded joints', *Mech of Mater*, 41, 231–243.

Chen, B.K., Thomson, P.F. and Choi, S.K. (1992) 'Computer modelling of microstructure during hot flat rolling of aluminium', *Mater Sci Technol*, 8, 72–77.

Chen, L., Hu, L. and Gong, S. (2006) 'A study on the porosity of CO_2 laser welding of titanium alloy', *China Weld*, 15(1), 1–5.

Chen, S., Huang, J., Cheng, D., Zhang, H. and Zhao, X. (2012) 'Superplastic deformation mechanism and mechanical behavior of a laser-welded Ti–6Al–4V alloy joint', *Mater Sci Eng A*, 541, 110–119.

Cheng, D., Huang, J., Zhao, X. and Zhang H. (2010) 'Microstructure and superplasticity of laser welded Ti–6Al–4V alloy', *Mater Design*, 31, 620–623.

Cheng, D., Chen, Y., Hu, D. and Zhou, Y. (2011) 'Tensile shear strength of laser welded Ti–6Al–4V lap joints', *2nd International Conference on Mechanic Automation and Control Engineering, MACE 2011 – Proceedings*, art. no. 5987033, 741–744.

Cheng, Y., Jin, X., Li, S. and Zeng, L. (2012) 'Fresnel absorption and Inverse Bremsstrahlung absorption in an actual 3D keyhole during deep penetration CO_2 laser welding of aluminum 6016', *Opt Laser Technol*, 44, 1426–1436.

Cho, J.H., Farson, D.F., Milewski, J.O and Hollis, K.J. (2009) 'Weld pool flows during initial stages of keyhole formation in laser welding', *J Phys D: Appl Phys*, 42, 175502.

Chung, B.G., Rhee, S. and Lee, C.H. (1999) 'The effect of shielding gas types on CO_2 laser tailored blank weldability of low carbon automotive galvanized steel', *Mater Sci Eng A – Struct*, 272(2), 357–362.

Costa, A., Miranda, R., Quintino, L. and Yapp, D. (2007) 'Analysis of beam material interaction in welding of titanium with fiber lasers', *Mater Manuf Process*, 22, 798–803.

Daneshpour, S., Dyck, J., Ventzke, V. and Huber, N. (2012) 'Crack retardation mechanism due to overload in base material and laser welds of Al alloys', *Int J Fatigue*, 42, 95–103.

Destefani, J.D. (1992) 'Introduction to titanium and titanium alloys', in *Properties and Selection: Nonferrous Alloys and Special Purpose Materials*, Vol. 2, Materials Park, OH, ASM International.

Di Sabatino, M., Arnberg, L., Rorvik, S. and Prestmo, A. (2005) 'The influence of oxide inclusions on the fluidity of Al–7 wt.%Si alloy', *Mat Sci Eng A – Struct*, 413–414, 272–276.

Donachie, M.J. (2000) *Titanium: a Technical Guide*, 2nd edn, Materials Park, OH, ASM International.

Dowden, J. (ed.) (2009) *The Theory of Laser Materials Processing: Heat and Mass Transfer in Modern Technology*. Dordrecht Springer.

Du, H., Hu, L., Liu, J. and Hu, X. (2004) 'A study on the metal flow in full penetration laser beam welding for titanium alloy', *Compt Mater Sci*, 29, 419–427.

Duley, W.W. (1999) *Laser Welding*, New York, John Wiley and Sons.

El-Batahgy, A. and Kutsuna, M. (2009) 'Laser beam welding of AA5052, AA5083, and AA6061 aluminum alloys', *Advances in Materials Science and Engineering*, 2009, 974182.

EN 3114-001:2006, 'Test Method – Microstructure of ($\alpha + \beta$) titanium alloy wrought products – Part 001: General Requirements'.

Esposito, C., Duarelio, G. and Cingolani, A. (1982) 'On the conduction welding process of steels with CO_2 lasers', *Opt Laser Eng*, 3, 139–151.

Gao, X.L., Zhang, L.J., Liu, J. and Zhang, J.X. (2013) 'A comparative study of pulsed Nd:YAG laser welding and TIG welding of thin Ti6Al4V titanium alloy plate', *Mater Sci Eng A*, 559, 14–21.

Grevey, D., Sallamand, P., Cicala, E. and Ignat, S. (2005) 'Gas protection optimization during Nd:YAG laser welding', *Opt Laser Technol*, 37, 647–651.

Haboudou, A., Peyre, P., Vannes, A.B. and Peix, G. (2003) 'Reduction of porosity content generated during Nd:YAG laser welding of A356 and AA5083 aluminium alloys', *Mat Sci Eng A – Struct*, 363, 40–52.

Hoult, T. (1990) 'Welding with suprakilowatt solid state lasers', *Proceedings of SPIE – The International Society for Optical Engineering*, 1277, 209–216.

Howard, K., Lawson, S. and Zhou, Y. (2006) 'Welding aluminum sheet using a high-power diode laser', *Weld J*, 85(5), 101–110.

Jandaghi, M., Parvin, P., Torkamany, M.J. and Sabbaghzadeh, J. (2009) 'Measurement of the composition change in Al5754 alloy during long pulsed Nd:YAG laser welding based on LIBS', *J Phys D Appl Phys*, 42, 205301.

Jianxun, Z., Xu, S. and Li., Z. (2012) 'Investigation into plastic damage behavior of the CO_2 laser deep penetration welded joint for Ti–6Al–4V alloy', *Eng Fract Mech*, 83, 1–7.

Jin, X., Zeng, L. and Cheng, Y. (2012) 'Direct observation of keyhole plasma characteristics in deep penetration laser welding of aluminum alloy 6016', *J Phys D Appl Phys*, 45, 245205.

Jyoti Kalita, S. (2011) 'Microstructure and corrosion properties of diode laser melted friction stir weld of aluminum alloy 2024 T351'. *Appl Surf Sci*, 257, 3985–3997.

Kalita, S. and Waldera, B. (2011) 'Microstructure and corrosion investigation of a friction stir weld of AA 5083 H116 before and after diode laser surface treatment'. *Materials Science and Technology Conference and Exhibition, MS and T'11*, 2, 916–923.

Katayama, S., Yamaguchi, Y., Masami, M. and Matsunawa, A. (1997) 'Evaluation of mechanical properties of laser-welded aluminum alloy'. *LIA P*, 83(2), G93–G102.

Katayama, S., Kobayashi, Y., Mizutani, M. and Matsunawa, A. (2001) 'Effect of vacuum on penetration and defects in laser welding', *J Laser Appl*, 13(5), 187–192.

Katayama, S., Nagayama, H., Mizutani, M. and Kawahito, Y. (2009) 'Fibre laser welding of aluminium alloy', *Welding International*, 23(10), 744–752.

Katayama, S., Yohei, A., Mizutani, M. and Kawahito, Y. (2011) 'Development of deep penetration welding technology with high brightness laser under vacuum', *Physics Procedia*, 12, 75–80.

Kuo, T.Y. and Lin, H.C. (2006) 'Effects of pulse level of Nd-YAG laser on tensile properties and formability of laser weldments in automotive aluminum alloys', *Mat Sci Eng A – Struct*, 416, 281–289.

Kwang Pan, L., Chung Wang, C., Long Wei, S. and Feng Sher, H. (2007) 'Optimizing multiple quality characteristics via Taguchi method-based Grey analysis', *J Mater Process Tech*, 182, 107–116.

Kwon, Y.D. and Lee, Z.H. (2003) 'The effect of grain refining and oxide inclusion on the fluidity of Al–4.5Cu–0.6Mn and A356 alloys', *Mat Sci Eng A – Struct*, 360, 372–376.

Lampman, S. (1990) 'Wrought titanium and titanium alloys', in *Properties and Selection: Nonferrous Alloys and Special purpose Materials*, Vol. 2, Materials Park, OH, ASM International.

Lee, H.K., Han, H.S., Son, K.J. and Hong, S.B. (2006) 'Optimization of Nd-YAG laser welding parameters for sealing small titanium tube ends', *Mat Sci Eng A – Struct*, 415, 149–155.

Li, X., Xie, J. and Zhou, Y. (2005) 'Effects of oxygen contamination in the argon shielding gas in laser welding of commercially pure titanium thin sheet', *J Mater Sci*, 40, 3437–3443.

Liu, H., Nakata, K., Yamamoto, N. and Liao, J. (2011) 'Mechanical properties and strengthening mechanisms in laser beam welds of pure titanium', *Sci Technol Weld Join*, 16(7), 581–585.

Liu, H., Nakata, K., Zhang, J.X., Yamamoto, N. and Liao, J. (2012) 'Microstructural evolution of fusion zone in laser beam welds of pure titanium', *Mater Charact*, 65, 1–7.

Luijendijk, T. (2000) 'Welding of dissimilar aluminium alloys', *J Mater Process Tech*, 103, 29–35.

Lütjering, G. and Williams, J.C. (2007) *Titanium*, 2nd edn, New York, Springer.

Malek Ghaini, F., Sheikhi, M., Torkamany, M.J. and Sabbaghzadeh, J. (2009) 'The relation between liquation and solidification cracks in pulsed laser welding of 2024 aluminium alloy', *Mat Sci Eng A – Struct*, 519, 167–171.

Matsumura, H., Orihashi, T. and Nakayama, S. (1992) 'CO_2 laser welding characteristics of various aluminium alloys', *Proceedings of LAMP'92*, Nagaoka, Japan, June 1992.

Matsunawa, A. (2001) 'Problems and solutions in deep penetration laser welding', *Sci Technol Weld Join*, 6, 351–354.

Matsunawa, A., Kim, J.D., Seto, N., Mizutani, M. and Katayama, S. (1998) 'Dynamics of keyhole and molten pool in laser welding', *J Laser Appl*, 10(6), 247–254.

Matsunawa, A., Katayama, S. and Kojima, K. (1999) 'CO_2 laser weldability of aluminium alloys (Report 1): Effect of welding conditions on melting characteristics', *Welding Research Abroad*, 45(8), 2–11.

Matsunawa, A., Seto, N., Kim, J.D., Mizutani, M. and Katayama, S (2000) 'Dynamics of keyhole and molten pool in high-power CO_2 laser welding', *Proceedings of SPIE – The International Society for Optical Engineering*, 3888, 34–45.

Mizutani, M., Yamaguchi, Y. and Katayama, S. (2008) 'Fatigue properties of laser weld in aluminium alloys', *Welding International*, 22(10), 705–711.

Montero, M.J. (2006) 'Soldadura por conducción de aleaciones de aluminio con láser de diodos de alta potencia'. Diploma de Estudios Avanzados. Puerto Real, Cádiz, Septembee 2006.

Mujibur Rahman, A.B.M., Kumar, S. and Gerson, A.R. (2007) 'Galvanic corrosion of laser weldments of AA6061 aluminium alloy', *Corros Sci*, 49, 4339–4351.

Nakai, M., Niinomi, M., Akahori, T., Hayashi, K., Itsumi, Y., Murakami, S., Oyama, H. and Abe, W. (2012) 'Microstructural factors determining mechanical properties of laser-welded Ti–4.5Al–2.5Cr–1.2Fe–0.1C alloy for use in next-generation aircraft', *Mater Sci Eng A*, 550, 55–65.

Nakamura, S., Sakurai, M., Kamimuki, K., Inoue, T. and Ito, Y. (2000) 'Detection technique for transition between deep penetration mode and shallow penetration mode in CO_2 laser welding of metals', *J Phys D Appl Phys*, 33, 2941–2948.

Nègre, P., Steglich, D. and Brocks, W. (2004) 'Crack extension in aluminium welds: a numerical approach using the Gurson–Tvergaard–Needleman model', *Eng Fract Mech*, 71, 2365–2383.

Okon, P., Dearden, G., Watkins, K., Sharp, M. and French, P. (2002) 'Laser welding of aluminium alloy 5083'. *ICALEO 2002: 21st International Congress on Applications of Lasers and Electro-Optics*, Scottsdale, AZ, October 14–17.

Padovani, C.G., Davenport, A.J., Connolly, B.J., Williams, S.W., Groso, A., Stampanoni, M. and Bellucci, F. (2008) 'Corrosion and protection of friction strir welds in aerospace aluminium alloys', *Metall Ital*, 100(9), 29–42.

Padovani, C., Davenport, A.J., Connolly, B.J., Williams, S.W., Siggs, E., Groso, A. and Stampanoni, M. (2011) 'Corrosion protection of AA7449-T7951 friction stir welds by laser surface melting with an Excimer laser', *Corros Sci*, 53(12), 3956–3969.

Padovani, C., Davenport, A.J., Connolly, B.J., Williams, S.W., Siggs, E., Groso, A. and Stampanoni, M. (2012) 'Corrosion protection of AA2024-T351 friction stir welds by laser surface melting with Excimer laser', *Corros Eng Sci Techn*, 47(3), 188–202.

Paglia, C.S., Pitts, B.C., Carroll, M.C., Reynolds, A.P. and Buchheit, R.G. (2002) 'Investigating post-weld heat treatments to increase the corrosion and the environmental cracking behavior of 7075-T6 friction stir weld', *Trends in Welding Research: Proceedings of the 6th International Conference*, Phoenix, AZ, ASM International 15–19 April.

Paleocrassas, G.A. (2005) 'Feasibility investigation of laser welding aluminum alloy 7075-T6 through the use of a 300W, single-mode, ytterbium fiber optic laser', MSc thesis, North Carolina State University.

Pasang, T., Sabol, J.C., Misiolek, W.Z., Mitchell, R., Short, A.B. and Littleflair, G. (2011) 'Metallurgy and deformation of electron beam welded similar titanium alloys', *4th Manufacturing Engineering Society International Conference*. Cadiz, 21–23 September.

Pierron, N., Sallamand, P. and Matteï, S. (2007) 'Study of magnesium and aluminium alloys absorption coefficient during Nd:YAG laser interaction', *Appl Surf Sci*, 253(6), 3208–3214.

Polmear, I.J. (1995) *Light Alloys: Metallurgy of the Light Metals*, 3rd edn, New York, Halsted Press, imprint of John Wiley & Sons.

Proton, V., Alexis, J., Andrieu, E., Blanc, C., Delfosse, J., Lacroix., L. and Odemer, G. (2011) 'Influence of post-welding heat treatment on the corrosion behavior of a 2050-T3 aluminum–copper–lithium alloy friction stir welding joint', *J Electrochem Soc*, 158(5), C139–C147.

Ready, J.F. and Farson, D.F. (2001) *LIA Handbook of Laser Materials Processing*, Pineville, LA, Magnolia Publishing.

Richter, K., Behr, W. and Reisgen, U. (2007) 'Low heat welding of titanium materials with a pulsed Nd:YAG laser', *Material Wiss Werkst*, 38(1), 51–56.

Rukadikar, M.C. and Reddy, G.P. (1986) 'Influence of chemical composition and microstructure on thermal conductivity of alloyed pearlitic flake graphite cast irons', *J Mater Sci*, 21, 4403–4410.

Sakamoto, H., Shibata, K. and Dausinger, F. (2003) 'Effect of alloying elements on weld properties in CO_2 laser welding of aluminium alloys', *Welding International*, 17(7), 509–513.

Sánchez-Amaya, J.M., Delgado, T., De Damborenea, J.J., López, V. and Botana, F.J.

(2009a) 'Laser welding of AA 5083 samples by high power diode laser', *Sci Technol Weld Join*, 14(1), 78–86.

Sánchez-Amaya, J.M., Delgado, T., González-Rovira, L. and Botana, F.J. (2009b) 'Laser welding of aluminium alloys 5083 and 6082 under conduction regime', *Appl Surf Sci*, 255(23), 9512–9521.

Sánchez-Amaya, J.M., Boukha, Z., González-Rovira, L. and Botana, F.J. (2010) 'Influencia del gas de protección sobre la soldadura laser de AA5083', *XI Congreso nacional de materiales*. PO-095, Zaragoza, Spain, June 2010.

Sánchez-Amaya, J.M., Boukha, Z., González-Rovira, L., Navas, J., Martín-Calleja, J. and Botana, F.J. (2012a) 'Laser texturization to improve absorption and weld penetration of aluminium alloys', *J Laser Appl*, 24(1), 012002.

Sánchez-Amaya, J.M., Boukha, Z., Amaya-Vázquez, M.R. and Botana, F.J. (2012b) 'Laser weldability of aluminium alloys with high power diode laser', *Welding Journal*, 91(5), 155–161.

Sánchez-Amaya, J.M., Boukha, Z., Amaya-Vázquez, M.R., González-Rovira, L. and Botana, F.J. (2012c) 'Analysis of the laser weldability under conduction regime of 2024, 5083, 6082 and 7075 aluminium alloys', *Mater Sci Forum*, 713, 7–12.

Seto, N., Katayama, S. and Matsunawa, A. (2000) 'High-speed simultaneous observation of plasma and keyhole behavior during high power CO_2 laser welding: effect of shielding gas on porosity formation', *J Laser Appl*, 12(6), 245–250.

Shi, Y., Zhong, F., Li, X., Gong, S. and Chen, L. (2007a) 'Effect of laser beam welding on tear toughness of a 1420 aluminum alloy thin sheet', *Mat Sci Eng A – Struct*, 465, 153–159.

Shi, Y., Zhong, F., Xiaoyan, L., Gong, S. and Chen, L. (2007b) 'Effect of laser beam welding on fracture toughness of a Ti-6.5Al-2Zr-1Mo-1V alloy sheet', *J Mater Sci*, 42, 6651–6657.

Sibillano, T., Ancona, A., Berardi, V. and Lugara, P.M. (2005) 'Correlation analysis in laser welding plasma', *Opt Commun*, 251, 139–148.

Sibillano, T., Ancona, A., Berardi, V., Schingaro, E., Parente, P. and Lugarà, P.M. (2006a) 'Correlation spectroscopy as a tool for detecting losses of ligand elements in laser welding of aluminium alloys', *Opt Lasers Eng*, 44, 1324–1335.

Sibillano, T., Ancona, A., Berardia, V., Schingaro, E., Basilea, G. and Lugarà, P.M. (2006b) 'A study of the shielding gas influence on the laser beam welding of AA5083 aluminium alloys by in-process spectroscopic investigation', *Opt Lasers Eng*, 44, 1039–1051.

Sibillano, T., Ancona, A., Berardi, V., Schingaro, E., Basile, G. and Lugarà, P.M. (2007) 'Optical detection of conduction/keyhole mode transition in laser welding', *J Mater Process Tech*, 191, 364–367.

Spina, R., Tricarico, L., Basile, G. and Sibillano, T. (2007) 'Thermo-mechanical modeling of laser welding of AA5083 sheets', *J Mater Process Tech*, 191, 215–219.

Steen, W.M. (1991) *Laser Material Processing*, New York, Springer-Verlag.

Squillace, A., Prisco, U., Ciliberto, S. and Astarita, A. (2012) 'Effect of welding parameters on morphology and mechanical properties of Ti–6Al–4V laser beam welded butt joints', *J Materials Process Tech*, 212(2), 427–436.

Sun, Z., Annergren, I., Pan, D. and Mai, T.A. (2003) 'Effect of laser surface remelting on the corrosion behaviour of commercially pure titanium sheet', *Mater Sci Eng A*, 345(1–2), 293–300.

Tani, G., Ascari, A., Campana, G. and Fortunato, A. (2007a) 'A study on shielding gas contamination in laser welding of non-ferrous alloys', *Appl Surf Sci*, 254(4), 904–907.

Tani, G., Campana, G., Fortunato, A. and Ascari, A. (2007b) 'The influence of shielding gas in hybrid LASER–MIG welding', *Appl Surf Sci*, 253, 8050–8053.

Torkamany, M.J., Malek Ghaini, F., Papan, E. and Dadras, S. (2012) 'Process optimization in titanium welding with pulsed Nd-YAG laser', *Sci Adv Mater*, 4, 489–496.

Torster, F., dos Santos, J.F., Hutt, G. and Kocak, M. (1997) 'The use of titanium alloys for dynamic risers: a literature review', Report GKSS 97/E/38, GKSS Forschungszentrum Geesthacht GmbH, Geesthacht.

Tricarico, L., Spina, R., Sorgente, D., Ancona, A., Sibillano, T. and Basile, G. (2007) 'Experimental analysis of AA5083 butt joints welded by CO_2 laser', *Key Eng Mat*, 344, 745–750.

Tsay, L.W., Shan, Y.P., Chao, Y.H. and Shu, W.Y. (2006) 'The influence of porosity on the fatigue crack growth behaviour of Ti–6Al–4V laser welds', *J Mater Sci*, 41, 7498–7505.

Tsay, L.W., Hsu, C.L. and Chen, C. (2010a) 'The influence of microstructures on the notched tensile fracture of Ti–6Al–6V–2Sn welds at elevated temperatures', *ISIJ International*, 50(1), 128–132.

Tsay, L.W., Hsu, C.L. and Chen, C. (2010b) 'Notched tensile fracture of Ti–4.5Al–3V–2Fe–2Mo welds at elevated temperatures', *Mater Chem Phys*, 120, 715–721.

Wang, H., Shi, Y., Gong, S. and Duan, A. (2007) 'Effect of assist gas flow on the gas shielding during laser deep penetration welding', *J Mater Process Tech*, 184, 379–385.

Wang, S.H., Wei, M.D. and Tsay, L.W. (2003) 'Tensile properties of LBW welds in Ti–6Al–4V alloy at evaluated temperatures below 450°C', *Mater Lett*, 57, 1815–1823.

Williams, S.W., Scott, G. and Calder, N.J. (2001) 'Direct diode laser welding of aerospace alloys', *Laser Opto*, 33, 50–54.

Williams, S., Ambat, R., Price, D., Jariyaboon, M., Davenport, A. and Wescott, A. (2003) 'Laser treatment method for improvement of the corrosion resistance of friction stir welds', *Mater Sci Forum*, 426–432, 2855–2860.

Wu, A.P., Zou, G.S., Rena, J.L., Zhanga, H.J., Wangb, G.Q., Liub, X. and Xie, M.R. (2002) 'Microstructures and mechanical properties of Ti–24Al–17Nb (at.%) laser beam welding joints', *Intermetallics*, 10, 647–652.

Yue, T.M., Yu, J.K., Mei, Z. and Man, H.C. (2002) 'Excimer laser surface treatment of Ti–6Al–4V alloy for corrosion resistance enhancement', *Mater Lett*, 52(3), 206–212.

Zaveri, N., Mahapatra, M., Deceuster, A., Peng, Y., Li, L. and Zhou, A. (2008) 'Corrosion resistance of pulsed laser-treated Ti–6Al–4V implant in simulated biofluids', *Electrochim Acta*, 53, 5022–5032.

Zervaki, A.D., Haidemenopoulos, G.N., Vriami, D.P. and Lambrakos, S.G. (2012) 'Case-study inverse thermal analyses of Al2139 laser welds', *J Mater Eng Perform*, 21(6), 777–785.

Zhang, J., Fan, D., Sun, Y. and Zheng, Y. (2007) 'Microstructure and hardness of the laser surface treated titanium', *Key Eng Mater*, 353–358, 1745–1748.

Zhang, Y., Wang, X. and Wu, J. (2009) 'The influence of silicon content on the thermal conductivity of Al-Si/diamond composites', *International Conference on Electronic Packaging Technology & High Density Packaging* (ICEPT-HDP), 708–712.

Zhao, H. and DebRoy, T. (2001) 'Weld metal composition change during conduction mode laser welding of aluminium alloy 5182', *Metall Trans B*, 32B, 163–172.

Zhao, H., White, D.R. and Debroy, T. (1999) 'Current issues and problems in laser welding of automotive aluminium alloys', *Int Mater Rev*, 44(6), 238–266.

9
Laser welding and brazing of dissimilar materials

F. MÖLLER and C. THOMY, BIAS – Bremer Institut für angewandte Strahltechnik GmbH, Germany

DOI: 10.1533/9780857098771.2.255

Abstract: Mixed-material joints between, for example, aluminium and steel or aluminium and CFRP are increasingly used in modern automotive, airplane or ship lightweight structures. The chapter addresses the present state-of-the-art and future perspectives for laser-based joining of such material combinations, focusing on both joining processes and joint properties.

Key words: dissimilar materials, aluminium-steel joints, hybrid welding.

9.1 Introduction

Today's products cannot be thought of without joints between different materials, either between different metals or even between metals and plastics or composites. In those cases where the basic elements differ (e.g. aluminium and steel) or different classes of materials (e.g. aluminium and CFRP) are used, the joint is called a dissimilar or mixed-material joint.

Such mixed-material joints are increasingly used in modern lightweight structures, e.g. in cars or airplanes. The driving force behind this development towards multi-material lightweight designs is to reduce the ecological footprint of transportation by reducing fuel consumption. As an example, the European Commission has adopted a comprehensive strategy in 2007 to considerably reduce CO_2 emissions associated with new cars and vans in order to ensure that the EU complies with the greenhouse gas emission targets of the Kyoto Protocol. Both the production and the consumer (in use) side of CO_2 emissions are tackled, with the long-term objective of limiting average CO_2 emissions from new cars to 120 g/km by 2012 (a reduction of around 25% compared to 2006 levels) (Smokers, 2006).

One opportunity to reach these goals (aside from switching to alternative concepts such as electric vehicles) is to reduce weight. With the overall trend towards more and more amenities in passenger cars (air conditioning, electrical seats, and so on), the only option seems to be to reduce structural weight whilst maintaining or even improving structural performance (see survey by Thomy *et al.*, 2007a).

A recent example of such attempts is the European Super LIGHT-CAR project, where a wide variety of steel, aluminium, magnesium and CFRP materials are used to ensure maximum, locally optimised structural properties of the car body at minimised overall weight (N.N., 2009; Toelle, 2011) (Fig. 9.1). The predominant joint in this study was between aluminium and steel.

However, although the advantages of specifically adapted material combinations may be numerous, the challenges, especially considering an appropriate joining technology, are also significant. These challenges result from the differing properties of the materials, which have a considerable effect on both the joining process and the properties of the weld. With respect to thermal joining processes, these challenges are primarily associated with two properties:

1. Different coefficients of thermal expansion, potentially resulting in excessive distortion and residual stresses.
2. Limited solubility of the joining partners (in some important cases, such as aluminium and iron), potentially resulting in brittle intermetallic phases in the joining zone and inadequate joint properties.

Both effects are related to the heat input by the welding process and, in general, tend to be more detrimental with higher heat input. This makes it highly desirable to precisely control heat input and, if possible, limit it – an ideal playground for laser-based processes, which is the focus of this chapter on laser welding of dissimilar materials.

9.2 Special issues in joining of dissimilar materials

As illustrated above, the key challenges associated with joining of dissimilar metals are differences in (thermophysical) properties and potential limited

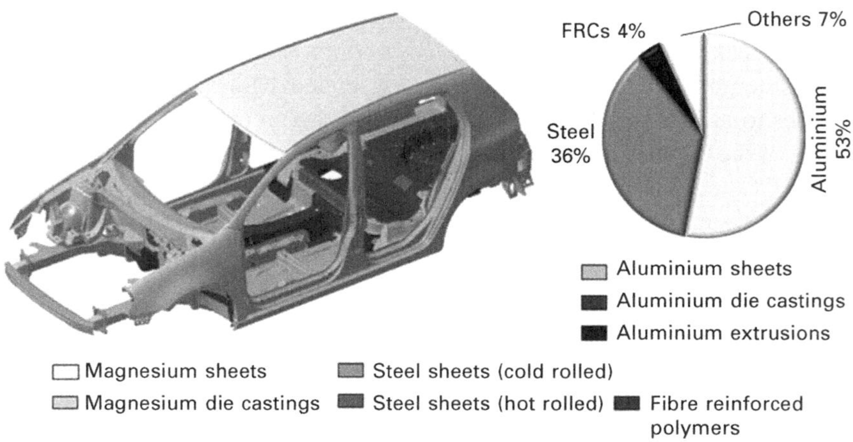

9.1 Mixed-material design in the automotive industry (N.N., 2009).

solubility of the materials with each other. Properties such as thermal conductivity, melting temperature and coefficient of thermal expansion can have a significant effect on the mechanical and micostructural properties of the resulting joint.

The main physical properties of exemplary (pure) metals considered as basic elements in alloys used in the transportation industry for either structure (aluminium, iron, titanium, magnesium) or (electrical) equipment components (aluminium, copper) are given in Table 9.1. As an example, for the case of aluminium-steel joints widely investigated (see survey in Thomy, 2009), the differences in thermal expansion in combination with the differences in the modulus of elasticity will result in thermally induced stresses, potentially leading to excessive distortion or residual stresss. This is often especially challenging in cases where complex parts requiring precise fit-up are to be joined.

The fundamental challenge to thermal joining of some (yet not all) dissimilar metals, however, arises from the limited solubility and, as a result, from the formation of intermetallic phases between the two metals. These are a potential location of brittle fracture under dynamic or even static load.

As an example, in the binary system Al-Fe, the solubility of Al in Fe is fairly limited, and exceeding an Al-content of 12 vol.% in Fe, intermetallic phase formation occurs. Conversely, the solubility of Fe in liquid Al is close to zero, which results in a very early interfacial phase formation when Fe is dissolved in Al. Depending on the aluminium content and the time–temperature cycle, the phases Fe_3Al, FeAl, $FeAl_2$ Fe_2Al_5 and $FeAl_3$ or even more complex structures of higher order may be formed (see Table 9.2). However, for typical laser-based joining processes, thermodynamic calculations show that the aluminium-rich phases will be formed preferably.

Table 9.1 Materials properties

Material	Aluminium Al	Iron Fe	Titanium Ti	Copper Cu	Magnesium Mg
Density (g/cm³)	2.7	7.85	4.5	8.92	1.738
Melting temperature (°C)	660	1563	1668	1357	650
Heat conductivity (W/mK)	238	75	22	400	160
Coefficient of thermal expansion (1/K)	23.8×10^{-6}	12.3×10^{-6}	10.8×10^{-6}	16.5×10^{-6}	26×10^{-6}
Modulus of elasticity *E* (N/mm²)	72,000	210,000	105,000	130,000	45,000
Tensile strength (N/mm²)	80	270–410	300–900	200–240	250

Source: Bargel and Schulze (2000); Klein (1997); Ostermann (1998).

This is also substantiated by experimental investigations (Wirth *et al.*, 2007). Unfortunately, as phase layer hardness increases with increasing aluminium content, these phases show a significant hardness in the range of 600–1100 HV (Table 9.2).

The binary system Ti-Al (Hansen, 1958) is also characterised by limited solubility, consequently showing comparable phase formation behaviour. Exceeding an aluminium content of approx. 10 at.% in the titanium, a composite of Ti and Ti_3Al will be formed. With rising aluminium content the phases TiAl, $TiAl_2$ and $TiAl_3$ are formed, respectively. On the aluminium-rich side, the low solubility of titanium leads to an early formation of the intermetallic phase $TiAl_3$ when the titanium content exceeds approx. 2 at.%. Based on thermodynamic assumptions, it is predicted that the $TiAl_3$ phase is formed preferably, which again is found in experimental studies on laser joining (Möller *et al.*, 2010b). Compared to the system Fe-Al, the intermetallic phases tend to show less hardness (Möller *et al.*, 2010b), which in turn makes them less detrimental to mechanical joint properties.

However, other systems, such as Fe-Cu, show excellent solubility with each other (ASM, 1993), consequently not producing any intermetallic compound in thermal joining. Yet, most technically relevant examples for dissimilar joints in structures require the joining of alloys with limited solubility, such as iron-based alloys with aluminium alloys. Because such joints typically require high strength and toughness, the formation of brittle intermetallic phases has to be limited to a minimum size (Radscheit, 1997; Zerner *et al.*, 1999).

Aside from alloying element distribution (Gebhardt and Obrowski, 1953; Rabkin and Rjabow, 1967), as an important driving mechanism for phase formation is diffusion (Achar *et al.*, 1980a), limiting the time where the joining zone is at elevated temperature is an effective means to reduce phase formation and potentially decrease joint brittleness (Achar *et al.*, 1980b). In other words, heat input has to be reduced and the time–temperature cycle has to be precisely controlled (Ozaki *et al.*, 2010).

In conclusion, both issues – distortion and phase layer formation – call for a reduction of the heat input by the joining process. Under these

Table 9.2 Properties of intermetallic phases for the binary system Fe-Al

Phase	Micro hardness HV
Fe_3Al	250–350
FeAl	400–1050
$FeAl_2$	1000–1050
Fe_2Al_5	1000–1100
$FeAl_3$	820–980

considerations, laser-based joining processes, as highly controlled processes typically characterised by low heat input and high speed, seem to be an appropriate solution to obtain optimised joints in dissimilar metals.

9.3 Laser joining processes and their applications

9.3.1 General considerations

With respect to laser-based joining processes for lightweight structures, the stage of the joining process in the process chain has to be considered in order to select the appropriate process for joining. Basically, the joining of dissimilar metals can be performed during the production of semi-finished products such as tailored blanks (Engelbrecht *et al.*, 2006) or during the production of the structure itself. The first option has the advantage of a mostly two-dimensional joining task and excellent fit-up tolerances. However, the tailored blank in general and the joint in particular will have to show a good formability for the subsequent stamping process, which is usually required to obtain three-dimensional structural elements. Conversely, joining of three-dimensional structural elements will not face the challenge of formability, but rather of poor fit-up tolerances (Thomy *et al.*, 2007a).

Table 9.3 gives a concise, yet not comprehensive overview of some recent research work in the field of laser joining of dissimilar materials, grouped by material combinations investigated. Although these combinations are among the most challenging, most research work was carried out on joining of aluminium to steel and aluminium to titanium, two combinations with significant relevance for the automotive industry (Al-Fe) and the aircraft industry (Al-Ti), respectively.

9.3.2 Laser brazing and welding

For material combinations with low solubility, most laser-based joining processes can generally be grouped into those designed to melt only the aluminium side of the joint with or without filler wire and laser brazing processes, trying to avoid the melting of both base materials.

In the latter processing approach for joining of aluminium to steel, mostly high-power diode lasers (Saida *et al.*, 2008) or Nd:YAG lasers are used for standard seam geometries such as raised edge welds and fillet welds in overlap configuration (Mathieu *et al.*, 2005). As filler materials, a wide range of alloy systems based on zinc, aluminium or silver were considered. Partly depending on the selection of filler materials as well as on the surface layer of the steel, the use of flux is suggested.

The other processing approach, i.e. melting only the lower melting material with or without filler wire, is called 'braze welding' by some authors (e.g.,

Table 9.3 Some recent research activities in the field of laser joining of dissimilar materials

Material 1	Material 2	Selection of publications
Aluminium	Steel	Dharmendra *et al.* (2011) Engelbrecht *et al.* (2006) Güngör and Gerritsen (2008) Katayama *et al.* (1998) Kreimeyer *et al.* (2001, 2002a, 2002b, 2004) Kreimeyer and Vollertsen (2006) Mathieu *et al.* (2005, 2007) Möller *et al.* (2009, 2010b) Ozaki *et al.* (2010) Radscheit (1997) Saida *et al.* (2004, 2008) Schubert and Zerner (1999) Thomy *et al.* (2007a, 2007b, 2007c, 2008); Thomy and Vollertsen (2009, 2011) Vollertsen and Thomy (2009, 2011) von der Haar *et al.* (2008) Vrenken *et al.* (2009) Zerner *et al.* (1999); Zerner (2002)
Aluminium	Titanium	Chen *et al.* (2010, 2011) Kreimeyer and Vollertsen (2005) Möller *et al.* (2010b) Ozaki *et al.* (2009)
Aluminium	Copper	Jokiel *et al.* (2005) Kraetzsch *et al.* (2011) Mai and Spowage (2004)
Steel	Copper	Mai and Spowage (2004)

Mathieu *et al.*, 2007). The bond is then obtained by brazing with the lower-melting melt onto the higher melting substrate material (e.g., steel).

A typical example for joining aluminium to steel in overlap configuration by this approach is provided by Mathieu *et al.* (2007), where heterogeneous steel–aluminium assemblies were joined by welding in the aluminium and brazing on the steel sheet. The filler material was a Zn-based alloy chosen for its compatibility with the aluminium alloys and its low melting temperature. Similar results are described in Vrenken *et al.* (2009), however at elevated brazing speed of up to 4 m/min at a laser power of 4.5 kW.

For joining of aluminium to steel (zinc-coated) in butt joint configuration, a laser welding process using either CO_2 or Nd:YAG laser for sheet thicknesses in the range of 1 mm was suggested (Fig. 9.2) (see, e.g., Thomy *et al.*, 2007a). The laser beam was positioned completely on the aluminium side of the joint to obtain a deep-penetration weld. The molten aluminium then wetted the steel sheet (to which flux was applied). Mainly for process stabilisation (and not predominantly for metallurgical reasons), filler wire was used.

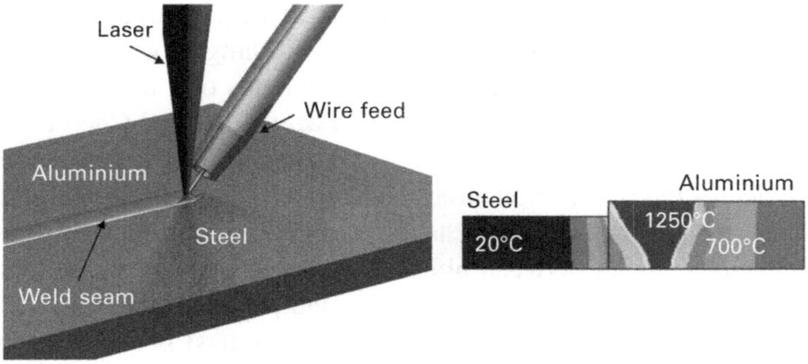

9.2 Laser welding of aluminium to steel.

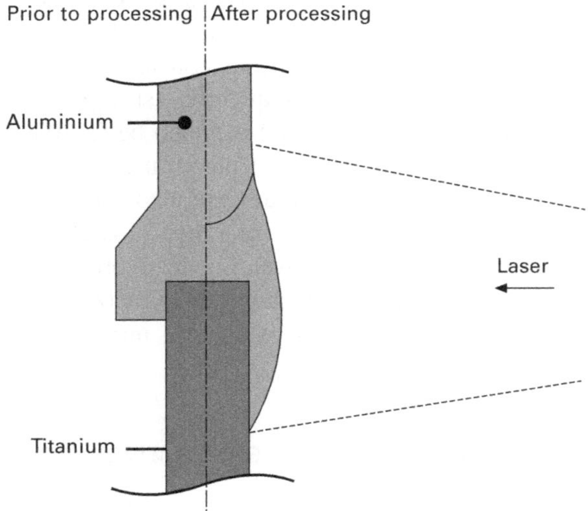

9.3 Process principle of the aluminium-titanium joining process.

Crucial factors of influence were the exact positioning of the beam relative to the interfacial region between aluminium and steel and the use of specially designed clamping devices to avoid the opening of a gap during the process. Using an Nd:YAG laser with 4 kW power, a joining speed of up to approx. 8 m/min was demonstrated, with the joints mostly failing in the aluminium.

A comparable approach was pursued for joining of aluminium to titanium for aircraft applications (see, e.g., Möller *et al.*, 2010a). In this process, again the aluminium is molten in order to wet the higher-melting joining partner, the titanium. However, contrary to the above example for joining of aluminium to steel, a heat conduction and not a deep-penetration laser welding process is carried out. The sheets are arranged in vertical position with aluminium on top to achieve a good wetting of the titanium sheet under the effect of gravitation (Fig. 9.3).

A typical joint appearance from Möller *et al.* (2010a) and a cross section are given in Fig. 9.4. By the simultaneous joining process, a reproducible, constant wetting of the titanium for joint lengths of 1 m was achieved. For the same material combination, Chen *et al.* (2010) and Chen *et al.* (2011) investigated a single-sided joining process using filler for good wetting.

Whereas the two combinations steel/aluminium and aluminium/titanium (which are the focus of this chapter) are mainly considered for structural applications, joints between aluminium and copper or stainless steel and copper are of importance in electronics and precision mechanics.

Welding of (stainless) steel to copper (or brass) is a task encountered, for example, in precision mechanics for joining heat conduction elements (Mai and Spowage, 2004). These materials, such as aluminium and iron, typically also show limited solubility. However, as part dimensions and the melt pools are typically much smaller, heat input is smaller and the cooling rate is much higher than for macro applications considered for aluminium-steel and aluminium-titanium joints. Consequently, using pulsed laser spot welding techniques (melting both materials) can also be feasible, resulting in at least sufficient mechanical properties of the joints.

Such pulsed laser spot welding techniques are also used for joining of aluminium to copper especially in the micro range (Jokiel *et al.*, 2005). For thicker sections, either direct laser welding in overlap configuration or laser joining using oscillating beam techniques (Kraetzsch *et al.*, 2011) was considered to join sheet thicknesses of up to 2 mm, for example for application in electrical conductors.

9.3.3 Combined and special processes involving lasers

In order to further improve joint properties, a variety of combined processes have been suggested:

9.4 Aluminium-titanium joint surface and cross section (laser power 1.75 kW each side, joining speed 0.22 m/min) (Möller *et al.*, 2010a).

- laser joining with superimposed force (Radscheit, 1997)
- laser-MIG hybrid welding (Kreimeyer and Vollertsen, 2006; Thomy and Vollertsen, 2009a)
- fluxless laser-plasma hybrid welding (Möller *et al.*, 2009).

In Radscheit (1997), a process for joining of aluminium to steel in overlap configuration with additional, superimposed clamping/forming force enacted by a specially designed welding head is presented. The joining zone is irradiated by either Nd:YAG or high-power diode laser, thus heating both materials, melting the aluminium and wetting the steel with the aluminium (Fig. 9.5). By optimising heat input, the formation of intermetallic phases was restricted significantly. The basic advantage of this process is that joints with one flat side (optimised for material flow in forming tools) can be obtained (Fig. 9.6), however at the limitation that only two-dimensional welds with limited curvature can be produced (Ozaki *et al.*, 2010). This seems to limit the process to tailored blank applications or aluminium.

More recently, a laser MIG hybrid welding process for dissimilar metals was suggested by Kreimeyer and Vollertsen (2006) and developed further by Thomy and Vollertsen (2009). In this process, a laser beam (CO_2, Nd:YAG, YB fibre or Yb:YAG) and a MIG-arc operated in the pulsed mode are used.

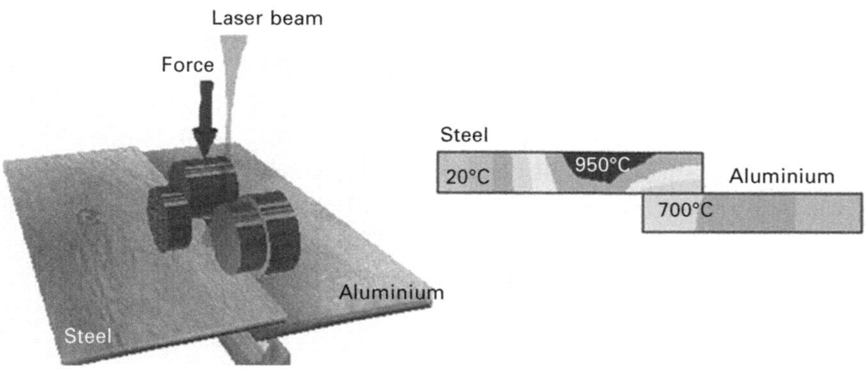

9.5 Laser joining with superimposed force – process principle and temperature distribution for overlap joints (see, e.g., Thomy, 2009).

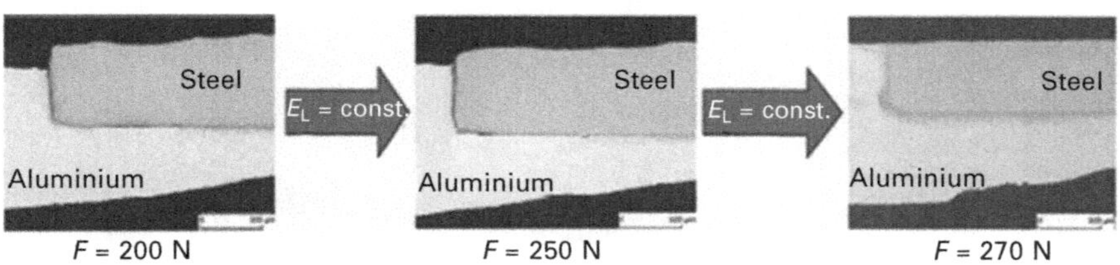

9.6 Laser joining with superimposed force – effect of force at constant heat input (see Thomy *et al.*, 2008).

The process principle is depicted in Fig. 9.7. The aluminium and steel sheet are arranged preferably in butt joint configuration, also allowing a gap. The laser beam is positioned on the aluminium side. During joining, the edge of the aluminium sheet is molten, and together with the molten wire, the gap between the aluminium and the steel is bridged and the steel is wetted by the aluminium melt. Figure 9.8 gives the top view and the cross section of a typical aluminium-steel joint obtained by laser MIG hybrid welding using optimised parameter settings. It is obvious that the weld seam is smooth and its width and the wetting length are fairly constant over the total seam length. In particular, no weld sagging, undercut or excessive spatters were observed. There was no melting of the steel sheet, and the weld showed only few small pores close to the face side of the steel sheet. This was attributed to the use of flux on the face side, which can contain humidity and, as a consequence, be a source of hydrogen to the melt pool. Moreover, the authors have suggested a parameter envelope giving basic rules for adjusting the parameters to obtain appropriate joint properties (i.e. with respect to seam appearance, tensile strength and phase layer thickness), which equal or surpass those reported in the literature, however at a significantly elevated speed of 6 mm/min or higher (Fig. 9.9).

Although this process yielded state-of-the-art properties at high productivity, it was still required to apply flux to the uncoated edge of the zinc-coated steel sheet to support wetting. In order to avoid the use of potentially corrosive flux, a laser-plasma hybrid welding was developed by Möller *et al.* (2009) as a coaxial combination of laser beam and plasma arc. The principle of the laser plasma hybrid joining process is depicted in Fig. 9.10. A ring-shaped plasma beam is used to heat the steel and the aluminium and activate and clean the surfaces, whereas an Nd:YAG laser beam increases heat input and

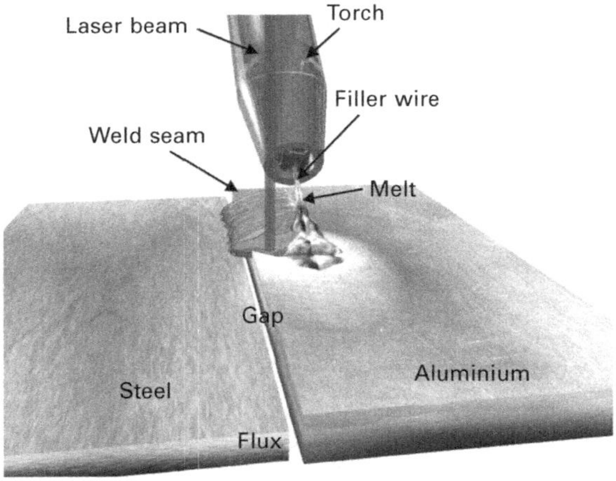

9.7 Principle of the laser MIG hybrid joining process (Thomy and Vollertsen, 2009).

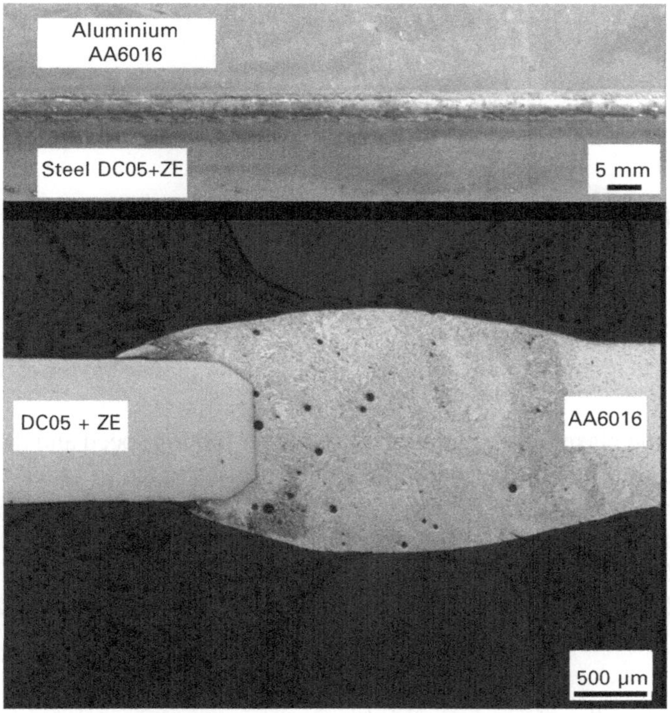

9.8 Top view and cross section of a laser MIG hybrid welded aluminium-steel specimen (Thomy and Vollertsen, 2009).

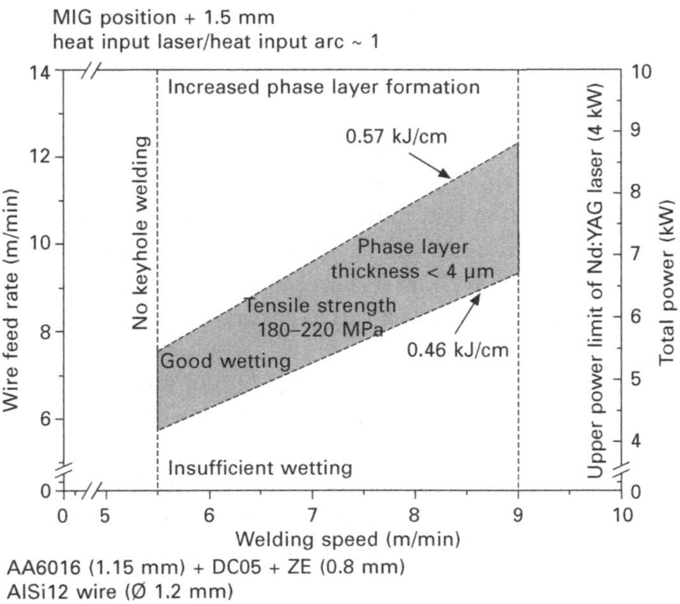

9.9 Process parameter envelope for laser MIG hybrid welding of aluminium to steel (Thomy and Vollertsen, 2009).

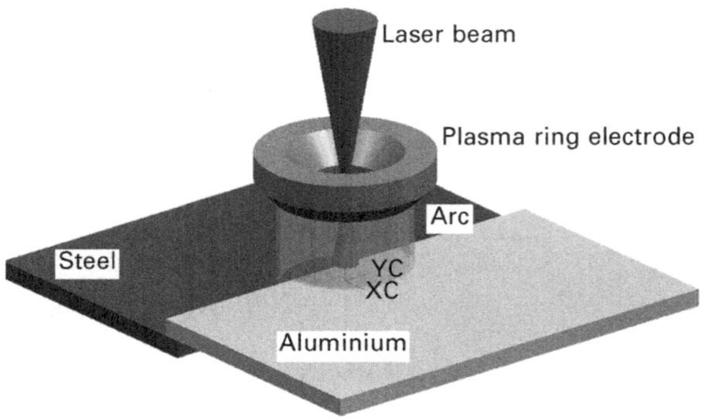

9.10 Principle of laser-plasma hybrid joining (after Möller *et al.*, 2009).

stabilises the plasma arc, thus allowing a higher joining speed and the joining of thicker aluminium sheets. The combined heat input led to a melting of the aluminium and a wetting of the steel by the aluminium, which was further promoted by the heating of the steel by the plasma arc. Overlap joints (EN-AW 6016, 2mm + DC 05+Z, 0.8 mm) were carried out using AlSi12 filler wire, and the authors found joining aluminium-steel tailored hybrid blanks without flux at an elevated speed of 2.6 m/min feasible.

9.3.4 Application potentials

As discussed above (and in detail in Thomy, 2009), one of the important application fields for laser joining of dissimilar materials is the transportation industry. The first significant area of application is the production of tailored hybrid blanks preferably from aluminium and steel, the focus of many application studies (see, e.g., Thomy and Vollertsen, 2009). An example of such a tailored hybrid blank (produced by Nd:YAG laser MIG hybrid welding) is shown in Fig. 9.11. The aluminium and steel sheet with a length of 400 mm were joined in butt joint configuration at a competitive speed of 66 m/min. Consequently, and taking into account the limits in formability in design of the parts as well as forming tools, such hybrid blanks may be used in lightweight structures in a way comparable to conventional tailored blanks.

The second field of application is the welding of tailored hybrid tubes or other three-dimensional structural elements (see, e.g., Thomy *et al.*, 2007a). As an example, the tube displayed in Fig. 9.12(a) was produced by Nd:YAG laser MIG hybrid welding (welding speed 4 m/min) from aluminium and zinc-coated steel tubes with a diameter of approx. 10 cm. Joint configuration was fillet welding in overlap configuration, with the aluminium tube on top. Comparable elements and structures may be used for structural elements (e.g. axles) in the automotive industry.

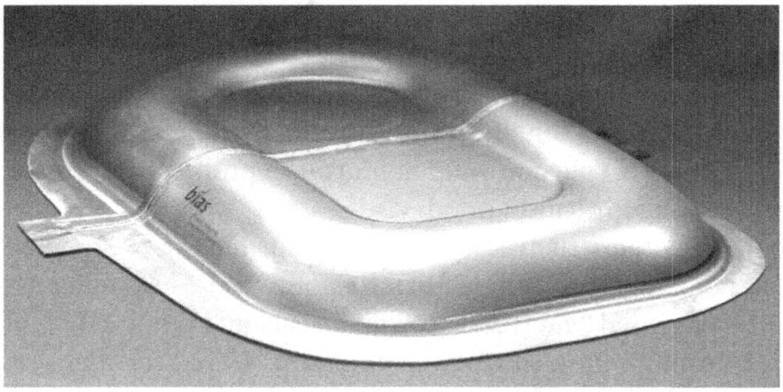

9.11 Aluminium-steel tailored hybrid blank after hydromechanical deep drawing (Courtesy: LFT, Chair of Manufacturing Technology, University Errlangen-Nuremberg and BIAS; Thomy and Vollertsen, 2009).

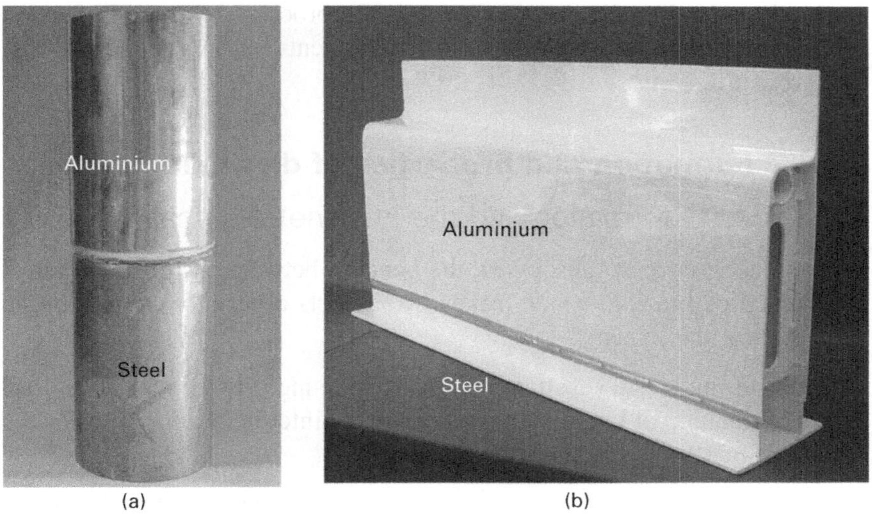

9.12 (a) Aluminium-steel tailored hybrid tube joined by laser-MIG process; (b) yacht deckhouse element, painted aluminium and steel (length: 1 m, material thickness: 33 mm) (after Thomy *et al.*, 2007a).

However, not only the automotive industry, but also shipbuilding may be a potential field of application of dissimilar joints in more complex structures. In modern shipbuilding (especially of yachts and military vessels), material combinations of aluminium and steel are already used in order to improve structural properties predominantly in view of stiffness, weight and position of the centre of gravity. Whereas the hull is normally manufactured from steel, the upper decks or the deckhouse are made from aluminium alloys. A structural part of a yacht deckhouse (steel on the lower part of the demonstrator and aluminium on the upper part, thickness 3 mm each, length 1 m) joined in butt joint configuration by laser MIG hybrid welding using a CO_2 laser is shown in Fig. 9.12(b) (see, e.g., Thomy *et al.*, 2007c). With this demonstrator produced in an industrial environment, a static strength of up to 140 MPa was achieved. This is a significant improvement compared to the so-called 'Triclad®' profiles (explosion welded transition profiles between aluminium and steel) used in shipbuilding today.

Comparably large structures from dissimilar materials are also found in aircraft. A potential application for laser joining of dissimilar materials is found in the so-called seat track, to which seats and other cabin interior are mounted. Currently, in most cases, aluminium is used for these seat tracks. However, to avoid defects from the presence of liquids in the passenger cabin, it is desired to use titanium instead of aluminium in the top area of the seat track. Since the remaining seat track part should still be made of aluminium (especially for cost reasons), laser joining for aluminium titanium seat tracks was developed (Möller *et al.*, 2010a). Figure 9.13 shows a demonstrator seat track with a length of 1 mm produced by the process presented in Section 9.3.3, fulfilling all relevant industrial requirements with respect to process and properties.

9.4 Formation and properties of dissimilar joints

9.4.1 Joint formation and the intermetallic phase layer

In most laser processes discussed, the bonding between the different joints is obtained by brazing, which makes two aspects especially crucial for the properties of the joints:

- the wetting of the substrate material by the melt (braze metal), and
- the formation of intermetallic phases in the interfacial zone between the braze metal and the substrate metal.

For the case of joining of aluminium to steel, several authors have identified the relevance of a zinc coating on the steel substrate (Kreimeyer *et al.*, 2004; Güngör and Gerritsen, 2008) for the wetting process in general and the resulting wetting length in particular. In case wetting occurs, typically

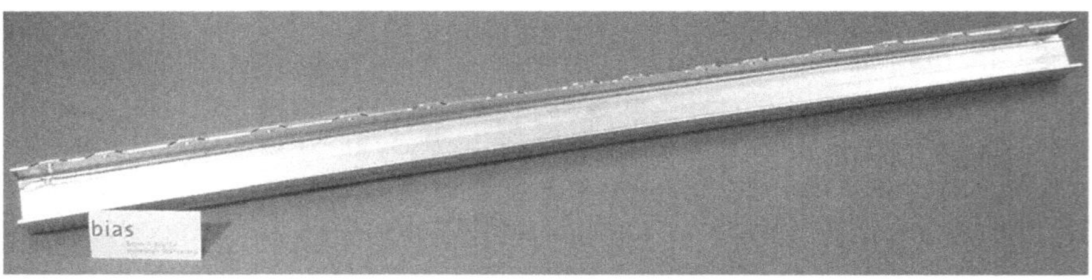

9.13 Demonstrator of titanium-aluminium aircraft seat tracks (Möller *et al.*, 2010a).

a marked zinc-rich region is observed in the interfacial zone, mostly close to the tip of the wetted zone (Fig. 9.14) (see, e.g., Mathieu *et al.*, 2007). Studies by Vollertsen and Thomy (2009, 2011) seem to suggest that the wetting length is essentially correlated to the melting isotherm of the zinc coating. However, the behaviour of the zinc coating during wetting is not completely elucidated yet.

Provided wetting has occurred, typically an intermetallic phase layer is formed. For the case of joining aluminium to steel, Wirth *et al.* (2007) has taken a closer look at a typical intermetallic phase layer (Fig. 9.15 right) and subjected it to EDX (Fig. 9.15 left) and EBSSD (Fig. 9.16) (see also Thomy, 2009). It was established that the intermetallic phase layer consisted of Al_5Fe_2 and Al_3Fe. The Al_5Fe_2 phase forms a continuous layer between the aluminium and the steel, whereas the Al_3Fe-phase forms needle-shaped

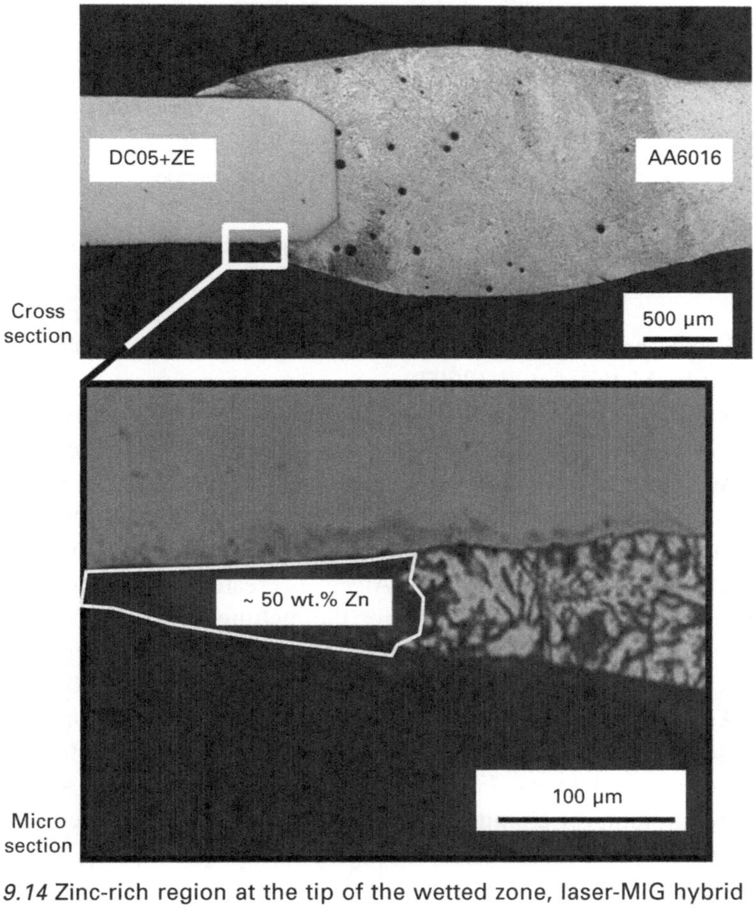

9.14 Zinc-rich region at the tip of the wetted zone, laser-MIG hybrid welding (after Vollertsen and Thomy, 2009).

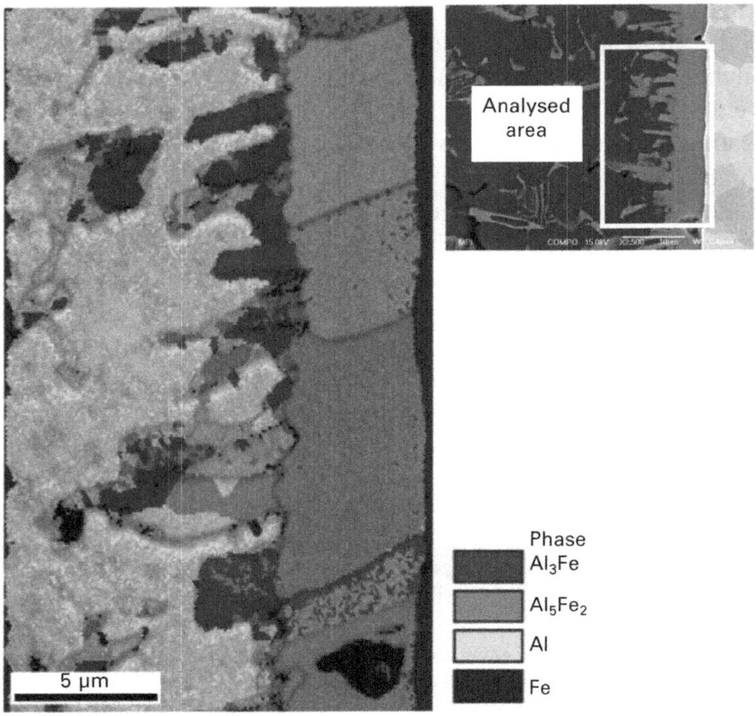

9.15 SEM (right) and EDX (left) of a typical phase layer in aluminium-steel joints (courtesy: MPIE Dusseldorf).

structures protruding into the aluminium matrix. Comparable results were found by Mathieu *et al.* (2007).

Regarding crystal orientation, there seems to be a preferred growth direction for Al_5Fe_2 perpendicular to the steel sheet surface, whereas the Al_3Fe needles show a more or less random orientation. This was interpreted by Wirth *et al.* (2007) as follows: initially, only the Al_3Fe-phase is created during heating by diffusion of Fe atoms into the aluminium as a layer between the two metal sheets. The continuous phase of Al_3Fe_2 is created by diffusion of Al atoms into the iron at temperatures above the melting temperature of aluminium (Radscheit, 1997). The formation of the needle-shaped Al_3Fe phase should then be connected to the cooling at temperatures between 50 and 350°C.

As is to be expected by the nature of phase formation, heat input and, consequently, process parameter variations can have a significant effect on phase layer thickness. For example, for the case of Nd:YAG laser MIG hybrid welding of aluminium to steel, Thomy and Vollertsen (2009) have found a slight trend towards an increase in intermetallic phase layer thickness with increasing laser power (Fig. 9.17), staying below approximately 4 μm, however, for most laser powers. For a laser power of 3400 W, a marked

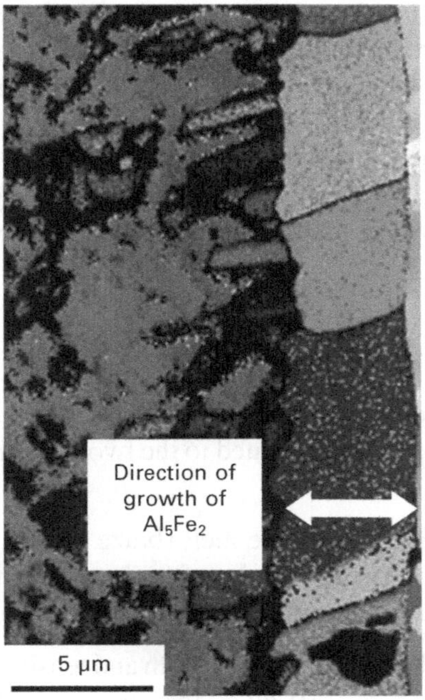

9.16 Orientation of the grains of a typical phase layer in aluminium-steel joints (courtesy: MPIE Dusseldorf).

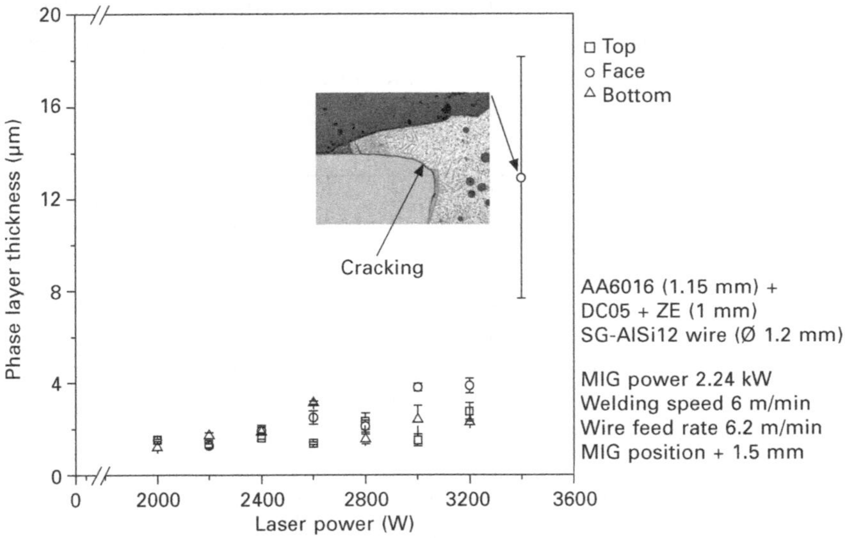

9.17 Effect of laser power on phase layer thickness for laser MIG hybrid welding of aluminium to steel (see, e.g., Thomy and Vollertsen, 2009).

increase in phase layer thickness to 12 μm on average at significant scatter was observed, and cracking occurred without external load. This substantiates the generally held belief that a phase layer thickness exceeding 10 μm should be avoided in all cases (see, e.g., Radscheit, 1997).

For the case of joining of aluminium to titanium, Möller *et al.* (2010a) have investigated phase layer thickness depending on heat input (Fig. 9.18). Again, it was found that with increasing heat input, phase seam thickness is also increased.

9.4.2 Mechanical properties and formability

The mechanical properties in general, and tensile as well as fatigue strength in particular, of dissimilar joints are closely related to the two crucial aspects of bonding:

- the wetting of the substrate material by the melt (braze metal), and
- the formation of intermetallic phases in the interfacial zone between the braze metal and the substrate metal.

Generally, a strong correlation between wetting length and tensile strength is found (Möller *et al.*, 2009). Figure 9.19 gives the correlation between wetting length and tensile strength for aluminium-steel overlap joints. For the same case, Mathieu *et al.* (2007) have found a roughly linear correlation between wetting length and tensile strength, despite significant scatter.

However, not only wetting length, but also the shape of the weld seam (which is, of course, often associated with wetting length; Möller *et al.*, 2009), have an influence on static strength. In general, it is found that a smoother transition between the braze metal and the substrate (i.e., a smaller wetting

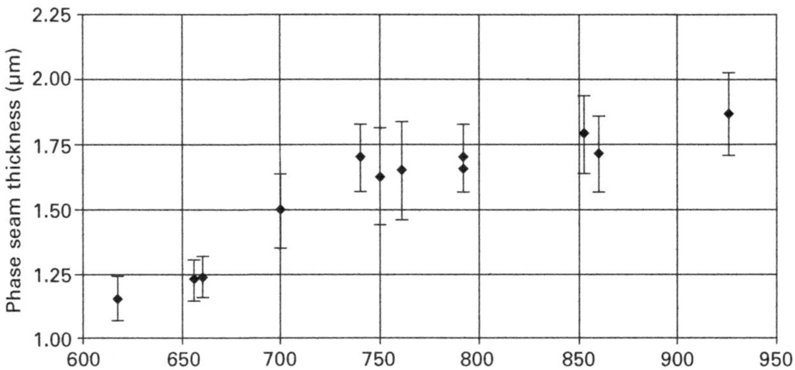

9.18 Effect of heat input per length on phase seam thickness for laser welding of aluminium to titanium (Möller *et al.*, 2010a).

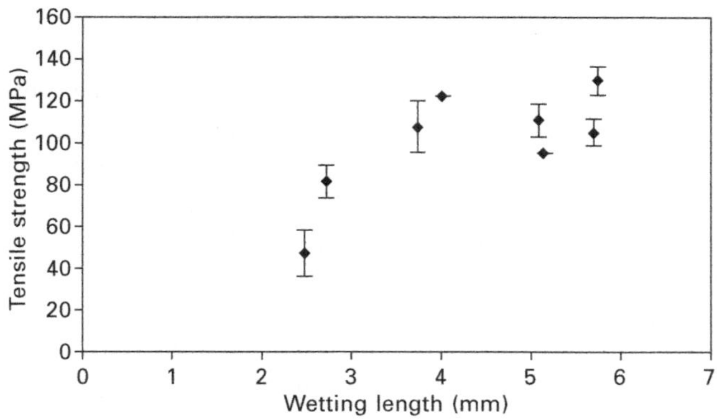

9.19 Relation between wetting length (i.e. the transverse length of the steel/seam interface) and tensile strength (Möller *et al.*, 2009).

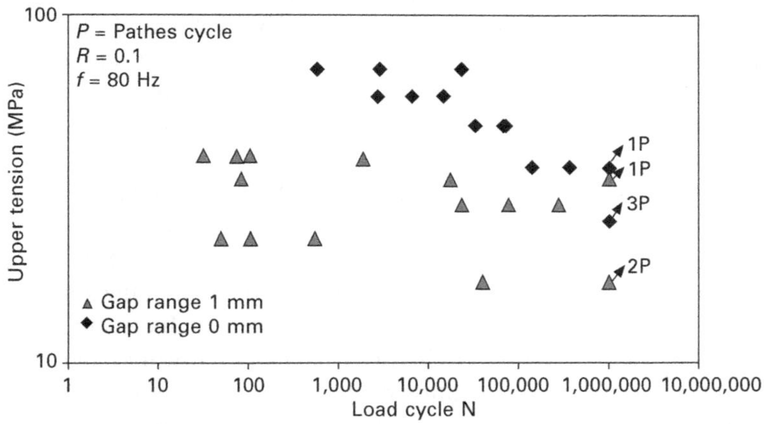

9.20 Fatigue behaviour for gap = 0 mm and 1 mm, respectively for laser MIG hybrid welded aluminium-steel specimens (Thomy *et al.*, 2007b).

angle, indicating better wetting) should improve tensile strength even under static conditions (Mathieu *et al.*, 2007).

In fatigue testing of laser MIG hybrid welded samples, an influence of several geometrical parameters of the seam on fatigue strength was identified (Thomy *et al.*, 2007b). In particular, with increasing gap and, consequently, decreased wetting length at otherwise unchanged parameters, the upper tension was reduced (Fig. 9.20).

Provided seam quality as such is adequate, the location of fracture not only depends on wetting length, but also on the cross section of the aluminium and the steel sheet. However, the interface zone with its steep property gradient is often associated with crack initiation or crack propagation under static load as well as in fatigue testing. Thomy and Vollertsen (2009) found that in those cases where the interface zone was involved in fracture, either a phase layer thickness in the range of 10 μm or above or a defect close to the interface zone was identified. In the latter case, fracture typically started at a small pore close to the interface zone and then deviated along the intermetallic phase layer (Fig. 9.21). A closer look at the fracture surface typically reveals remainders of the intermetallic phase layer on both the steel and the aluminium side (see, e.g., Mathieu *et al.*, 2007).

Comparable results were found for aluminium-titanium joints (Chen *et al.*, 2010). Whereas for lower heat input, fracture was in the fusion zone, higher heat input resulted in a steep decrease in tensile strength due to fracture in the interface zone, which was often associated with porosity occurring at higher heat input.

Of significant relevance, especially for tailored hybrid blanks, is formability. For aluminium-steel tailored hybrid blanks produced at adequate parameters, a bend test is typically passed (Thomy and Vollertsen, 2011; Fig. 9.22). In particular, transverse cracking normally does not occur, and delaminations or cracks between the aluminium weld metal and the steel sheet surface are not observed. Moreover, typical limit drawing ratios of at least 1.6 are achieved.

In hydromechanical deep-drawing (with a multi-axial stress state), the shape given in Fig. 9.12 (length approx. 250 mm, width approx. 2500 mm, height approx. 45 mm) was achieved applying a pressure of 65 bar. Aside from some slight transverse cracks in the weld metal in the zone of the edge radius of the die plate, no defects were observed. Increasing the pressure much further resulted in longitudinal cracks in the weld metal. Aside from the difference in stress condition compared to static tensile testing and bend testing, another reason is assumed to lie in the fact that both tool and part design were not yet fully optimised. In order to further improve formability,

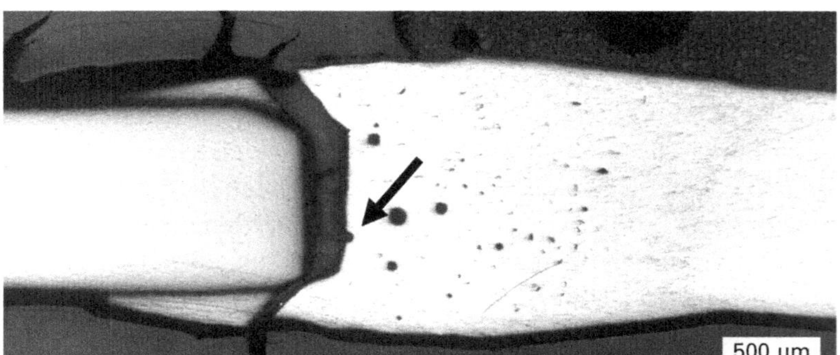

9.21 Cross section of a fractured laser MIG hybrid welded aluminium-steel specimen.

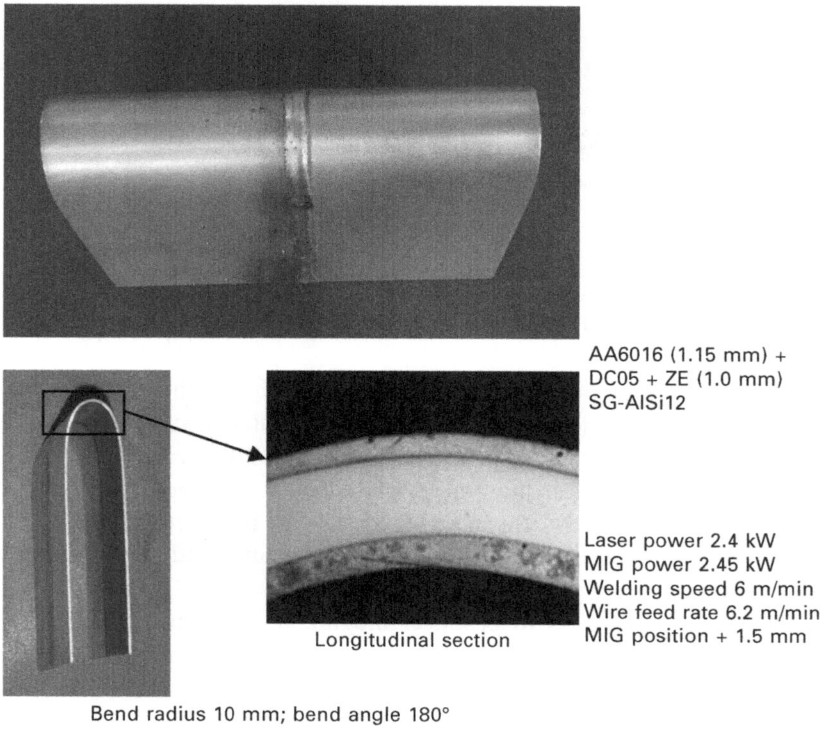

9.22 Result of bend test of laser MIG hybrid welded aluminium-steel sheets (Thomy and Vollertsen, 2011).

both the forming process and the joint properties of the tailored hybrid blanks before forming should be dealt with. One means to address the latter aspect should be connected to the selection of filler wire with the aim of obtaining a more ductile weld metal.

9.5 Future trends

Whereas until recently laser joining of dissimilar materials meant joining of dissimilar metals, the trend towards the use of composite materials such as CFRP in the aircraft industry, but also in the automotive industry (see Fig. 9.1) has called for appropriate joining processes using lasers also for these challenging combinations.

An approach to directly join plastics to metals or to use plastics as interface materials between dissimilar metals as a means of avoiding phase layer formation has been presented by the research group of Katayama (see, e.g., Katayama *et al.*, 2006, 2007; Niwa *et al.*, 2008) and also investigated by Wang *et al.* (2010). In this method (so-called LAMP joining), basically laser irradiation is used to promote bonding between the (partially) molten

9.23 Principle design approach for integral CFRP-aluminium structures (left); first prototype (right) (Möller *et al.*, 2010a).

and evaporating plastic (e.g. PET) and the metal. As this method can produce sound joints with considerable strength and is potentially suitable to solve the challenges of intermetallic phase layer formation and corrosion in joining of dissimilar metals such as aluminium and steel, it is to be expected that it will further attract significant interest. Moreover, this method should be of significant interest also for CFRP materials especially with thermoplastic matrix material.

A novel joining process to produce integral, load-bearing aluminium-CFRP-aluminium structures has been suggested by Möller *et al.* (2010a). The principle is based on the production of a fabric of carbon fibres and titanium wires, which is then thermally joined to the aluminium metallic structure (Fig. 9.23 left). As a thermal joining process, a heat conduction laser beam welding process is used. For the welding process, the aluminium and the titanium wire structure are arranged in a vertical position with the aluminium on top. The defocused laser beam is positioned directly onto the joining zone, simultaneously irradiating both sides of the specimen and creating a brazed bond between the aluminium and the titanium. After thermal joining, the structure is then immersed with the matrix material, e.g. a thermoplastic (Fig. 9.23 right).

9.6 References

Achar D R G, Ruge J and Sundaresan S (1980a), 'Verbinden von Aluminium mit Stahl, besonders durch Schweißen', *Aluminium*, 56(2), 174–193.

Achar D R G, Ruge J and Sundaresan S (1980b), 'Metallurgical and mechanical investigations of aluminium-steel fusion welds', *Aluminium*, 56(6), 391–397.

ASM (1993), *Alloy Phase Diagrams, Vol. 3*. ASM Metal Handbooks.

Bargel H-J and Schulze G (2000), *Werkstoffkunde*, 7th edn, Springer, Berlin.

Chen S, Li L Q and Chen Y B (2010), 'Interfacial reaction mode and its influence on tensile strength in laser joining Al alloy to Ti alloy', *Materials Science and Technology*, 26(2), 230–235.

Chen S, Li L, Chen Y and Huang J (2011), 'Joining mechanism of Ti/Al dissimilar alloys during laser welding-brazing process', *Journal of Alloys and Compounds*, 509, 891–898.

Dharmendra C, Rao K P, Wilden J and Reich S (2011), 'Study on laser welding-brazing of zinc coated steel to aluminum alloy with a zinc based filler', *Materials Science and Engineering A*, 528(3), 1497–1503.

Engelbrecht L, Meier O, Ostendorf A and Haferkamp H (2006), 'Laser beam brazing of steel-aluminium tailored hybrid blanks', in Ostendorf A (ed.), *Proc. 25th International Congress on Applications of Lasers & Electro-Optics ICALEO 2006*, 312–319.

Gebhardt E and Obrowski W (1953), 'Reaktionen von festem Eisen mit Schmelzen aus Aluminium und Aluminiumlegierungen', *Zeitschrift für Metallkunde*, 44(1), 12–17.

Güngör Ö E and Gerritsen C. (2008), 'Effect of filler wire composition and metallic coating on the joint performance of aluminium/steel braze-welds', *Welding and Cutting*, 7(5), 303–312.

Hansen M (1958), *Constitution of Binary Alloys*, McGraw-Hill, New York.

Jokiel M, Dürr U, Holtz R and Liebers R (2005), 'Laser welding of dissimilar metals: strategies with pulsed Nd: YAG lasers', *Proc. 3rd International WLT-Conference on Lasers in Manufacturing 2005*, Munich, 631–636.

Katayama S, Usui R and Matsunawa A (1998), 'YAG laser welding of steel to aluminium', *Proc. 5th International Conference on Trends in Welding Research*, 467–472.

Katayama S, Kawahito Y, Tange A and Kubota S (2006), 'Laser-assisted metal and plastic (LAMP) joining', *Online Proc. LAMP 2006*, JLPS, Kyoto, #06-7.

Katayama S, Kawahito Y, Niwa Y and Kubota S (2007), 'Laser-assisted metal and plastic joining', *Proc. 5th LANE 2007*, Erlangen, 41–51.

Klein B (1997), *Leichtbau-Konstruktion*, Friedr. Vieweg & Sohn, Braunschweig/Wiesbaden.

Kraetzsch M, Standfuss J, Klotzbach A, Kaspar J, Brenner B and Beyer E (2011), 'Laser beam welding with high-frequency beam oscillation: welding of dissimilar materials with brilliant fiber lasers', *Proc. ICALEO 2011*, 169–178.

Kreimeyer M and Vollertsen F (2005), 'Processing titanium-aluminium hybrid joints for aircraft applications', in Beyer F. (ed.), *Proc. 3rd International WLT-Conference on Lasers in Manufacturing LIM 2005*, 73–78.

Kreimeyer F and Vollertsen F (2006), 'Gap tolerant joining of aluminum with steel sheets using the hybrid technique', in Ostendorf A (ed.), *Proc. 25th International Congress on Applications of Lasers & Electro-Optics ICALEO 2006*, 947–952.

Kreimeyer M, Wagner F, Zerner I and Sepold G (2001), 'Laserstrahlfügen von Aluminium mit Titan unter Verwendung eines optimierten Arbeitskopfs', in *DVS-Berichte 212*, Düsseldorf, DVS-Verlag, 317–321.

Kreimeyer M, Wagner F, Sepold G, Criqui B and Joly M (2002a), 'Kombiniertes Schweiß-Lötverfahren zum Fügen von Aluminium-und Stahlblechen im Stumpfstoß mit Hochleistungslasern', *Proc. Schweißen und Schneiden 2002, DVS-Berichte 220*, Düsseldorf, DVS-Verlag, 256–261.

Kreimeyer M, Wagner F and Sepold G (2002b), 'Combined welding-brazing process for joining aluminium-steel tailored blanks with high power sources', in *Proc. IBEC 2002*, CD-ROM.

Kreimeyer M, Wagner F and Vollertsen F (2004), 'Einfluss von Zinkbeschichtungen beim Laserstrahlfügen von Aluminium-Stahl-Überlappverbindungen', in *DVS-Berichte 231*, Düsseldorf, DVS-Verlag, 223–227.

Mai T A and Spowage A C (2004), 'Characterisation of dissimilar joints in laser welding of steelkovar, copper-steel and copper-aluminium', *Materials Science and Engineering A*, 374, 224–233.

Mathieu A, Matteï S, Viala J-C and Grevey D (2005), 'Laser braze welding using hot

(88%-aluminium, 12%-silicon) filler material to join steel with aluminium', *ICALEO 2005 Congress Proceedings*, 985–991.

Mathieu A, Shabadi R, Deschamps A, Suery M, Matteï S, Grevey D and Cicala E (2007), 'Dissimilar material joining using laser (aluminium to steel using zinc-based filler wire)', *Optics & Laser Technology*, 39, 652–661.

Möller F, Thomy C and Vollertsen F (2009), 'Development of a plasma-laser brazing process for fluxless joining of aluminium to steel', in Ostendorf A, Graf T, Petring D, Otto A and Vollertsen F (eds), *Laser in Manufacturing (LIM09)*, AT-Fachverlag, Stuttgart, 83–88.

Möller F, Thomy C, Vollertsen F, Schiebel P, Hoffmeister C and Herrmann A S (2010a), 'Novel method for joining CFRP to aluminium', in Schmidt M, Vollertsen F and Geiger M (eds), *Physics Procedia*, 5/2, Meisenbach, Bamberg, 37–45.

Möller F, Thomy C and Vollertsen F (2010b), 'Joining of titanium-aluminium seat tracks for aircraft applications – system technology and joint properties', *Proc. IIW Assembly 2010 Com. XVII, Istanbul, Turkey*, IIW-Doc. XVII-0005-2010 (CD-Rom).

Niwa Y, Kawahito Y, Kubota S and Katayama S (2008), 'Evolution of LAMP joining to dissimilar metal welding', in Ostendorf A (ed.), *Proc. 27th International Congress on Applications of Lasers & Electro-Optics ICALEO 2008*, 311–317.

N.N. (2009), SuperLIGHT-CAR Project – Sustainable Production Technologies of Emission Reduced Light-weight Car Concepts, project brochure, http://www.superlightcar.com/public/docs/SuperLIGHTCar_project_brochure.pdf (accessed 30 September 2009).

Ostermann F (1998), *Anwendungstechnologie Aluminium*, Springer, Berlin.

Ozaki H, Ichioka R, Matsuura T and Kutsuna M (2009), 'Laser roll welding of dissimilar metal joint of titanium to aluminium alloy', *Welding International*, 23(7), 501–509.

Ozaki H, Kutsuna M, Nakagawa S and Miyamoto K (2010), 'Laser roll welding of dissimilar metal joint of zinc coated steel', *Journal of Laser Applications*, 22(1), 1–6.

Rabkin D M and Rjabow W R (1967), 'Das Schweißen von Aluminium mit Stahl', *Schweißtechnik*, 17(10), 448–454.

Radscheit C (1997), *Laserstrahlfügen von Aluminium mit Stahl*, BIAS Verlag, Bremen.

Saida K, Song W and Nishimoto K (2004), 'Laser brazing of aluminium alloys to steels with aluminium filler metals', *DVS-Berichte 231*, Düsseldorf, DVS-Verlag, 232–237.

Saida K, Ohnishi H and Nishimoto K (2008), 'Fluxless laser brazing of aluminium alloy to galvanized steel using a tandem beam – dissimilar laser brazing of aluminium alloy and steels', *Welding International*, 24(3), 161–186.

Schubert E and Zerner I (1999), 'Lasergestütztes Fügen von Werkstoffkombinationen', *Blech Rohre Profile*, 4, 76–81.

Smokers R (2006), 'Review and analysis of the reduction potential and costs of technological and other measures to reduce CO_2-emissions from passenger cars', Final report on European Commission Contract Nr. SI2.408212.

Thomy C (2009), 'Hybrid laser-arc welding of dissimilar metals', in Olsen F O (ed.), *Hybrid Laser-Arc Welding*, Woodhead Publishing, Cambridge, 270–295.

Thomy C and Vollertsen F (2009), 'Effect of process parameters on joint properties in laser MIG hybrid welding of aluminium to steel', in Ostendorf A, Graf T, Petring D, Otto A and Vollertsen F (eds), *Laser in Manufacturing (LIM09)*, AT-Fachverlag, Stuttgart, 59–64.

Thomy C and Vollertsen F (2011), 'Recent developments in laser joining of dissimilar materials', *Proc. ICALEO 2011*, 23–27 October 2011, Orlando, FL, 333–342.

Thomy C, Wirth A, Kreimeyer M, Wagner F and Vollertsen F (2007a), 'Joining of dissimilar materials – new perspectives for lightweight design in the transportation industries', *Proc. IW International Conference Welding & Materials*, Dubrovnik & Cavtat, 1–8 July, 311–326.

Thomy C, Wirth A, Kreimeyer M, Wagner F and Vollertsen F (2007b), 'Laser-MIG-Hybridfügen von Aluminium-Stahl Leichtbaustrukturen', *Laser Technik Journal*, 4, 36–40.

Thomy C, Wagner F, Vollertsen F and Metschkow B (2007c), 'Lasers in the shipyard – laser joining of aluminum to steel for modern yacht construction', *Industrial Laser Solutions*, 11, 9–13.

Thomy C, Walther R and Vollertsen F (2008), 'New developments when thermally joining hybrid materials in the automotive industry', *Proc. 11th German and 8th European Automotive Conference Joining in Automotive Engineering*, 24–25 April 2008, Automotive Circle International, Bad Nauheim, CD-ROM.

Toelle J (2011), 'A new path for lightweight designs by a systematic combination of materials', *Proc. 4th Nano and Material Symposium Niedersachsen*, CD-ROM.

Vollertsen F and Thomy C (2009), 'On the meaning of zinc coatings in hybrid welding of aluminium to steel', *Proc. ICALEO 2009*, 2–5 November, 2009, Orlando, FL, 89–97.

Vollertsen F and Thomy C (2011), 'Laser-MIG hybrid welding of aluminium to steel – a straightforward analytical model for wetting length', *Welding in the World*, 55(1/2), 58–66.

von der Haar C, Engelberecht L, Meier O, Ostendorf A and Haferkamp H (2008), 'Tailored hybrid blank production – new joining concepts using different solders', *Journal of Laser Applications*, 20(4), 224–229.

Vrenken J, Goos C, van der Veldt T and Braunschweig W (2009), 'Fluxless laser brazing of aluminium to steel', *Joining in Automotive Engineering*, CD-ROM.

Wang X, Li P, Xu Z, Song X and Liu H (2010), 'Laser transmission joint between PET and titanium for biomedical application', *Journal of Materials Processing Technology*, 210, 1767–1771.

Wirth A, Kreimeyer M, Gnauk J, Thomy C and Vollertsen F (2007), 'Analyses on the phase seam of a laser-MIG joined aluminum-steel sample', in Vollertsen F *et al.* (eds), *Proc. 4th International WLT-Conference on Lasers in Manufacturing LIM 2007*, 18–22 June 2007, Munich, Germany, 111–115.

Zerner I (2002), *Prozeßstabilisierung und Ergebnisse für das Laserstrahlfügen von AluminiumStahl-Verbindungen*, Bremen, BIAS-Verlag.

Zerner I, Schubert E and Sepold G (1999), 'Diode lasers join aluminium to steel', *Industrial Laser Solutions*, 5, 23–28.

10
Laser welding of plastics

I. JONES, TWI Ltd, UK

DOI: 10.1533/9780857098771.2.280

Abstract: Lasers may be used for rapid and reproducible welding of plastics. The principles of plastics welding and direct and transmission laser welding for plastics are described. The equipment variations and process parameter effects are discussed and guidance for practical application of the technique is provided. Typical application areas are summarised.

Key words: laser, welding, plastics, theory, modelling, equipment, process, parameter effects, monitoring, direct, transmission, applications.

10.1 Introduction

Welding of plastics is of practical importance in many automotive, medical and electronic packaging applications. There have also been developments in textile joining, and in joining dissimilar materials (e.g. plastics to metals or ceramics). The laser provides a heat source that is very controllable in terms of the amount of energy applied and the location or size of the applied heat. These properties are becoming ever more important for small and large devices alike where complex joint lines and products with thermally sensitive parts are made. Welds produced by lasers in plastics can be used for wide area laminating or can have a resolution of less than 100 microns enabling precise patterns, smaller scale and increased complexity in the joints made. In this chapter the principles of welding plastics and the use of lasers for this are introduced, and the methods available to the manufacturer of plastics components are described in detail.

10.2 History

Laser welding of plastics was first reported in the early 1970s (Silvers and Wachtell, 1970), when film materials were welded using a CO_2 laser. This was a 'direct' laser welding process, where the laser heated the films directly without the laser being transmitted through the upper part of the joint. The distinction between direct and transmission laser welding of plastics is

---

This chapter is a revised and updated version of Chapter 22 'Plastics Microwelding' by Ian Jones, originally published in *Microjoining and Nanojoining* edited by Norman Zhou, Woodhead Publishing Limited, 2008, ISBN 978-1-84569-179-0.

described in more detail below. In 1985 the first transmission laser welding process was demonstrated and patented by Toyota in Japan (Toyota, 1985) using carbon black additive to modify the absorption properties of the lower plastic. This was carried out using a near infrared Nd:YAG laser. Transmission laser welding of plastics is now carried out using diode and fibre lasers. Amongst the first mass production applications using a diode laser was a motor vehicle key fob, which was announced in 1997 (Puetz *et al.*, 1997). Further developments since then have included procedures using different laser wavelengths to modify the interaction and heating effect in the plastics. In this way it is possible to apply direct laser welding to thicker materials. Additionally, there have been developments of alternative absorbers, with minimal visible appearance unlike carbon black (Jones and Wise, 1998; Jones *et al.*, 1999). This has allowed application to parts and fabrics with almost any colour combination. Most recently, the use of high and low intensity parts of the beam has been applied to modify the absorption of the beam in the plastic and allow parts to be melted locally and internally only where the beam intensity is high enough (Roesner *et al.*, 2008).

10.3 Theory of welding plastics

10.3.1 Plastic materials and thermal effects

Thermoplastics may be divided into two groups, amorphous and semicrystalline, each with its own set of properties. Amorphous thermoplastics have no order beyond that on the level of the repeat units, which make up the long chain macromolecule. The linear chains exist often in random positions and orientations in the bulk polymer and are separated by a distance proportional to their dynamic state, which is governed by temperature.

If a linear thermoplastic has very regular chemistry within repeat units (stereoregular) and little or no chain branching, then it may exist in crystalline form (Stevens, 1953). Crystalline synthetic thermoplastics are invariably partly crystalline and partly amorphous and so the term crystalline polymer always implies partial crystallinity or semicrystalline (McCrum *et al.*, 1988).

For every amorphous thermoplastic homopolymer (one exhibiting no crystallinity) there exists a narrow temperature region in which it changes from a viscous or rubbery condition at temperatures above this region, to a hard and relatively brittle one (sometimes called glassy) below it (Wool, 1995). This temperature region is called the glass transition temperature T_g (Fig. 10.1) and is usually obtained from a volume versus temperature plot of observations taken on cooling (McCrum *et al.*, 1988). Since movement of whole polymer chain segments is a necessary prerequisite of welding, amorphous thermoplastics need to be above T_g before welding can take place.

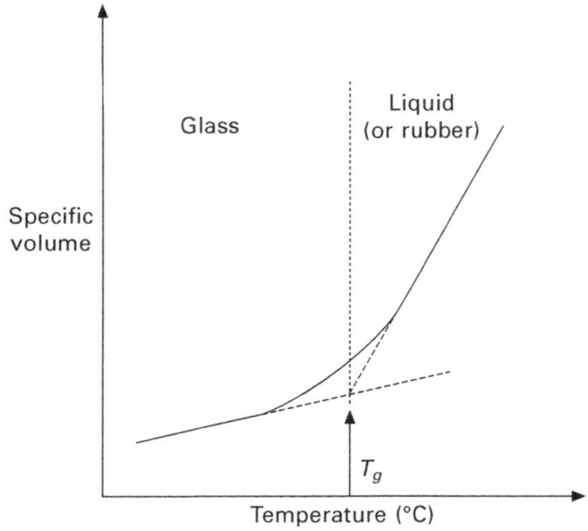

10.1 A typical graph of specific volume versus temperature for an amorphous thermoplastic (from Wise, 1999).

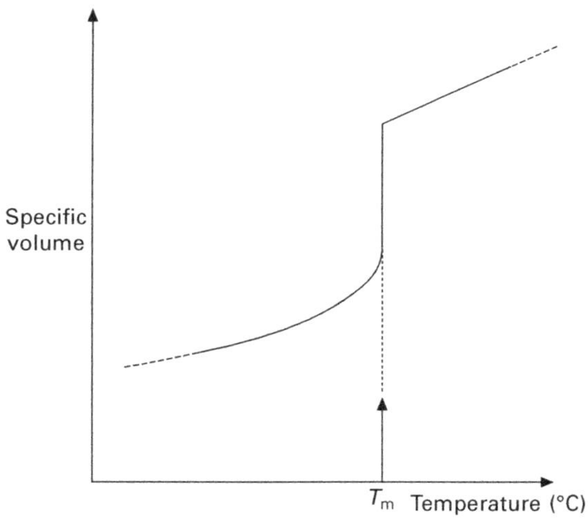

10.2 A typical graph of specific volume versus temperature for a semicrystalline thermoplastic (from Wise, 1999).

Semicrystalline thermoplastics are composed of crystalline regions and amorphous regions. For flow to occur in these polymers, their temperature must be above the crystalline melting point T_m, as shown in Fig. 10.2, which is the temperature at which all of the crystalline regions have disappeared (Stevens, 1953). Semicrystalline thermoplastics will generally have a T_g associated with the amorphous regions and T_m associated with the crystalline regions, with $T_m > T_g$. Viscous flow will generally only occur above T_m.

Thermal welding of thermoplastics relies on the temperature at the interface being above T_g for amorphous thermoplastics or above T_m for semicrystalline thermoplastics. Above T_m or T_g, there is increased chain mobility of the thermoplastic in the region of the weld. It is this increased mobility of the chains, which allows them to diffuse across the joint interface and entangle with chains on the other side of the interface. This is the mechanism for generation of strength at the joint and the formation of a weld between surfaces of similar thermoplastics.

10.3.2 Strength development at a joint: diffusion by reptation

A weld in a thermoplastic is generally formed when two pieces above their glass transition temperature or melting point are brought together for a period of time at temperature under sufficient pressure to ensure intimate contact at the joint interface. Based on this simple argument, early theories on welding were based around diffusion (Voyutskii, 1963). Molecular chains are said to diffuse across the interface and strength develops at the joint due to bridging by these chains (Fig. 10.3). Equations have been developed to account for the strength development as a function of time based on diffusion theory (Vasenin, 1961). The term 'tack' has been used to describe the build-up of mechanical strength at joints between similar and dissimilar materials. Tack at joints in similar materials is also sometimes called autohesion and was studied extensively by Voyutskii.

The main results of Vasenin's derivation are that autohesion force for separation (F) has the following dependencies:

$F \propto r$ (the rate of separation)

$F \propto M^{-2/3}$ (molecular weight)

$F \propto t^{1/4}$ (time of contact before breaking)

Or in combination, the force to separate two polymer surfaces is:

$$F \propto t^{1/4} M^{-2/3} e^{-1/T}$$

where t = time of contact, M = molecular weight, T = temp.

Following the work by Voyutskii and Vasenin, the next most significant advance in the theory of the welding of thermoplastics came from the development of reptation theory, which is important in explaining the development of weld strength.

The word reptation was originally coined by de Gennes (1971), from the Latin reptare (to creep), to describe the motion of polymer chains under certain circumstances. He first considered the snake-like movements of a linear polymer chain inside a strongly crosslinked polymeric gel. The gel provided a regular array of fixed obstacles through which the chain could not pass.

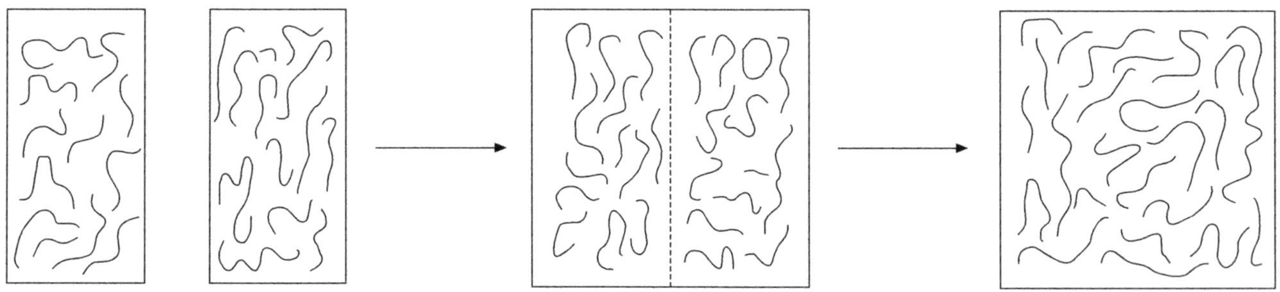

10.3 Scheme for polymer chains crossing an interface by diffusion leading to coalescence of two surfaces (from Voyutskii, 1963).

Instead the linear chain had to wriggle between the obstacles. The reason for this approach was that theories for the motion of polymer molecules in their molten state could not be made to agree with the experimental observations of viscosity and self-diffusion. De Gennes therefore simplified the problem to one linear chain moving in a crosslinked gel.

Amongst others, Wool developed the reptation theory into a model for welding. Wool's theory for the welding of polymers has evolved since the late 1970s (Wool and O'Connor, 1980) and one incarnation involves dividing the welding operation into five stages (Wool and O'Connor, 1981):

1. surface rearrangement
2. surface approach
3. wetting
4. diffusion
5. randomisation.

Using the five identified stages of welding, Wool next defined a welding function W, such that the mechanical energy G required to separate two welded polymeric parts, was given by:

$$G = W(t, T, P, M)$$

where W is a function of t = contact time, T = temperature, P = contact pressure and M = molecular weight. The general conclusions of the theory include:

- long molecules take longer to diffuse, increased weld time
- at least half a molecule is needed to crossweld for optimum properties
- high temperatures lead to rapid welds.

10.4 Effect of main welding parameters

From reptation theory and experimental observations, the inferences summarised in Table 10.1 can be drawn in relation to the effects of the main welding parameters.

10.5 Modelling of plastics welding

Molecular simulation of plastics welding has been studied using a Monte Carlo model at the mesoscale, i.e. with detail of polymer chain motion, but not with atomic scale detail (Haire and Windle, 2001). The models described the polymer chain interdiffusion and could predict completion times for welds when surfaces were brought into contact at a temperature greater than T_g. The models were successful for linear amorphous polymers and there is suggestion that the theory can be extended to semicrystalline and/or branched

Table 10.1 Effect of plastics welding processing parameters

	High	Low
Temperature	Fast weld Thermal degradation	Slow or no weld
Pressure	Chain diffusion inhibited	Insufficient surface wetting
Weld time	Oxidative degradation, flash	Insufficient diffusion, weak weld
Molecular weight	Very long weld times and high pressures required Slow diffusion weak parent materials and welds	Rapid weld times and low pressures required

polymers. More complex is the extension to all the commercial welding processes which can involve flow of the material and non-isothermal heating conditions in combination with chain interdiffusion. Figure 10.4 shows the output from a mesocale model of polymer welding after different periods of heating following contact of the two pieces.

Thermal modelling of plastics welding (i.e. using heat flow considerations rather than chain interdiffusion) is applied most easily when the process is predominantly one of diffusion at the surfaces and has minimal melt flow, which complicates the heat flow calculations. Transmission laser welding is one such process. A finite element thermal model for transmission laser welding has been developed (Jones and Olden, 2000) and used to simulate the welding of thermoplastic polymers. Cooling rate curves generated experimentally and using finite element analysis have shown good agreement as have the weld dimensions measured and calculated.

The model takes details of the materials' radiation absorption coefficient, thickness, reflectivity, melting point, density, heat capacity and thermal conductivity. It also takes details of the laser source, including the laser beam profile, laser power and welding speed. Through analysis of the heat flow, it is possible to generate the thermal history for all points in the joint during the welding cycle. Thus it can provide the weld dimensions, the peak temperature and, with details of the materials' mechanical properties, the residual stress remaining after welding can be estimated. Plate XI (between pages 328 and 329) shows a typical output from a finite element thermal model.

10.6 Introduction to plastics welding processes

Before considering the plastics laser welding methods, it is useful to review the other plastics welding methods available: techniques for thermally welding polymers can be broadly divided into three groups: those where heat is generated by mechanical movement of components to be joined;

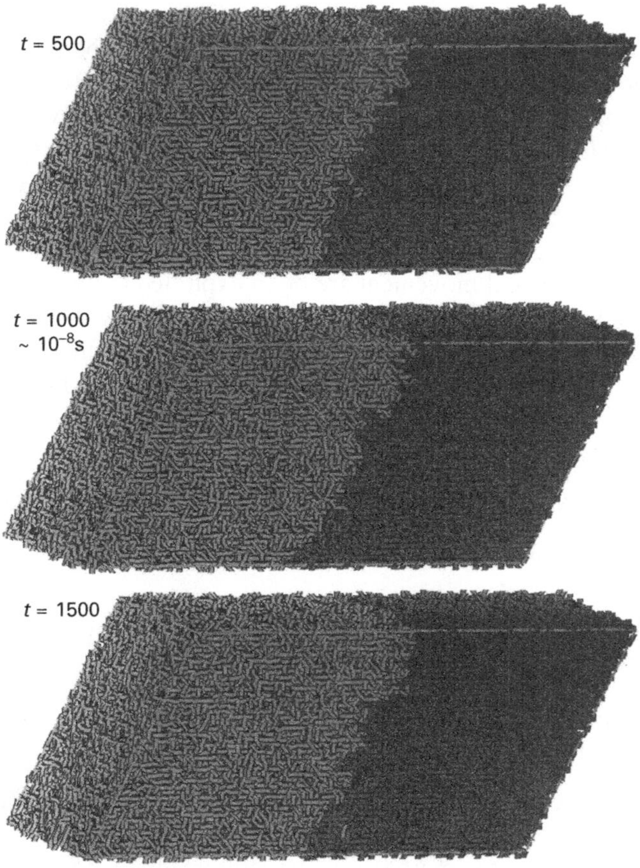

10.4 Simulation of polymer welding showing chain interdiffusion at a joint. (Haire and Windle, 2001).

those where heat is generated by an external source and this source heats the joint by thermal conduction; and those using electromagnetism directly (Wise, 1999).

10.6.1 Techniques in which heat is generated by mechanical movement

- Spin welding
- Vibration welding
- Ultrasonic welding
- Orbital welding
- Friction stir welding

When two surfaces are brought into contact and rubbed together under pressure, there is generally a temperature increase at the interface. The thermal welding of thermoplastics may most efficiently be accomplished by the generation of heat at the joint interface and for this reason welding techniques employing mechanical movement have found many commercial applications. The rate of temperature increase at the interface is generally related to the speed of the mechanical movement such that the greater the speed, the more rapid the temperature increase. High heating rates generally imply high welding rates and for this reason techniques where heat is generated by mechanical movement are often exploited in mass production. If the amplitude of vibration is small and correspondingly the frequency is high, then the methods become applicable to parts with small dimensions. Ultrasonic welding provides such conditions and is the only method in this group to be considered for microjoining applications.

10.6.2 Techniques employing a mechanical heat source

- Hot plate welding
- Hot bar welding
- Impulse welding
- Hot gas welding
- Extrusion welding
- Forced mixed extrusion welding
- Flash-free (BCF) welding

Thermoplastics generally have low coefficients of thermal conductivity particularly when compared to metals. This means that if heat can be applied to the joint area prior to joining, the temperature gradient normal to the surface may be high. However, in joining small parts, the loss of heat from the surface may be too rapid for the joint to be completed before cooling. Therefore methods where heating is carried out with the parts separated are normally only applied to large components. The only methods applicable to microjoining are hot bar or impulse welding, where the heat is applied externally to the parts as they are pressed together, and hot gas welding where a narrow gas stream may be used to provide a precise heating position to heat two small parts simultaneously.

10.6.3 Techniques which employ electromagnetism directly

- Resistive implant welding
- Induction welding
- Dielectric welding
- Microwave welding

- Infrared welding
- Laser welding

The welding techniques which directly employ electric or magnetic fields are listed above. Although most thermoplastics exhibit dielectric (non-electrically-conducting) properties, they generally produce some response when excited by an electric or magnetic field at a certain frequency.

Heat is generated at the joint line by an electric or magnetic field or both. In some cases, the field heats the thermoplastic directly, e.g. dielectric (high frequency) welding, and in other cases the field heats a different material at the joint which in turn heats the thermoplastic, e.g. resistive implant welding.

Of these methods, only laser welding has found application for joints at small scale. This is as a result of the precise control available on the position and amount of energy applied by the process at the joint.

10.7 Polymer combinations that can be welded

Only certain combinations of plastics are melt compatible and can be welded together. Generally the materials must have good miscibility, similar melting points and similar melt viscosities. Laser welding can provide a little more flexibility than other methods because the procedure can be completed rapidly without melting a large volume of material. Under these conditions, it becomes more feasible to combine materials with different melt properties. The combinations that can be welded are shown in Table 10.2.

10.8 Laser welding of plastics: process description

10.8.1 Introduction

The technique, suitable for joining both sheet film and moulded thermoplastics and textiles, uses a laser beam to melt the plastic in the joint region. Two general forms of laser welding exist: direct laser welding and transmission laser welding.

In direct laser welding, the materials are heated from the outer surface possibly to a depth of a few millimetres. Normally, no specific radiation absorber is added to the plastics. Laser sources from 2.0 to 10.6 μm wavelength are used. At 10.6 μm (CO_2 laser) radiation is strongly absorbed by plastic surfaces, allowing high-speed joints to be made in thin films (Jones and Taylor, 1994; Coelho *et al.*, 2000). At 2.0 μm, where the absorption is less strong, a fibre laser might be used to make welds in sheet a few millimetres thick. Direct laser welding is not widely applied for joining plastics. Alternatively, a beam with a large convergence angle can be used to make a weld where

Table 10.2 Compatibility in terms of welding performance between different thermoplastic materials.

	ABS	ABS/ PC	PMMA	Modified PPO	PA 12	PBT	PC	HDPE	PEI	PET	PSU	PP	PPO	PS	PVC	SAN–NAS–ASA	SBC	SBS
Acrylonitrile butadiene styrene, ABS	x	x	x	?		?	x		?					–	–	–		–
ABS/PC	x	x	–				x									–		
Acrylic, PMMA	x	–	x				–							–	?	–		x
Modified PPO	?			x			?						x	x		–		
Nylon 12, PA12					x		?											
Polybutylene terephthalate, PBT	?					x	?		?						?			
Polycarbonate, PC	x	x	–	?	?	?	x		–	?	–			–				
High-density polyethylene, HDPE								x				?					?	
Polyetherimide,PEI	?					?	–		x									
Polyester, PET							?			x								
Polysulphone, PSU							–				x							
Polypropylene, PP								?				x						
Polyphenylene oxide, PPO				x									x					
Polystyrene, PS	–		–	x			–							x		–		–
Polyvinyl chloride, PVC	–		?			?									x			
SAN-NAS-ASA	–	–	–	–										–		x		–
Styrene butadiene copolymer, SBC								?									x	
Styrene-butadiene-styrene, SBS	–		x											–		–		x

Blank cell = not weldable, x = weldable, – = some grades weldable, ? = non known the beam intensity is high enough to melt material, and otherwise leave the material unaffected (Roesner *et al.*, 2008).

Transmission laser welding is widely used, and the bulk of the following discussion refers to this technique. Laser sources from 0.8 to 1.1 μm wavelength such as diode, Nd:YAG and fibre lasers are applied. The radiation at this wavelength is less readily absorbed by natural plastics without absorbing additives. Laser absorbing additives are put into the lower part or applied as a thin surface coating at the joint, so the parts can be positioned together before welding and the laser beam passes through the upper part to heat the joint at the absorbing surface of the lower part (Fig. 10.5). The absorber in or on the lower plastic is typically carbon or an infrared absorber with minimal visible colour (Clearweld) (Jones and Wise, 1998). These and other variations allow a wide range of part colours and appearances to be welded. Transmission laser welding is capable of welding thicker parts than direct welding, and since the heat-affected zone is confined to the joint region, no marking of the outer surfaces occurs.

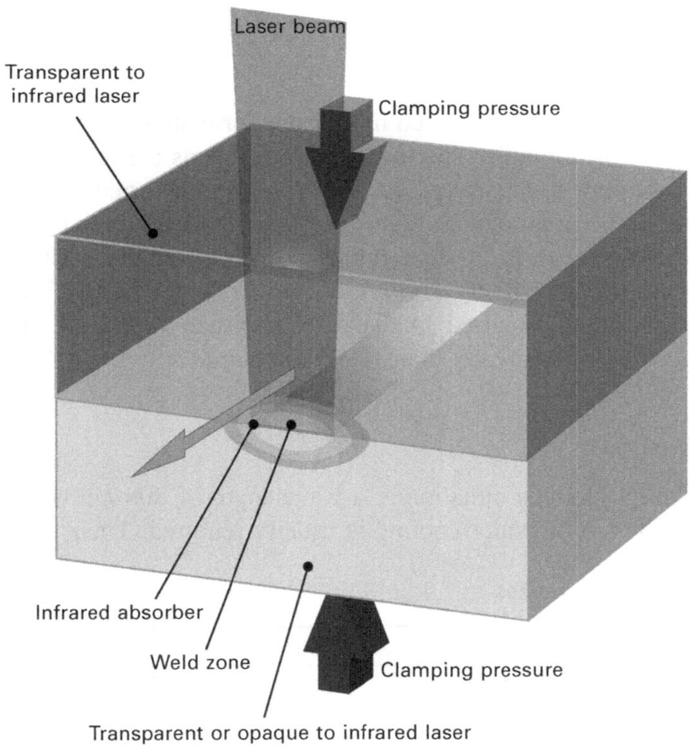

10.5 Diagram of transmission laser welding showing the movement of a beam over a workpiece. The lower part can be arranged to be an infrared absorber or the absorber may be placed at the joint surface.

10.8.2 Equipment and variations

The main elements of a laser welding system are:

- power supply including a chiller for higher power laser sources
- laser source
- beam delivery optics (lens, mirror or fibre-based)
- beam focusing or shaping optics including masks if required
- beam manipulation optics such as galvanometer controlled mirrors (if used)
- workpiece clamping and support
- workpiece manipulation
- monitoring and control.

10.8.3 Laser types for transmission laser welding

The main types of NIR laser used for through-transmission laser welding, and selected properties, are listed in Table 10.3. Additional details of each type of laser are given below.

Nd:YAG lasers

Nd:YAG lasers are widely used in industry for materials processing. High-power systems are bulky, but lower power systems are relatively compact. Water-cooling is usually required. The beam is transferred from the laser to the workpiece via an optical fibre. It is feasible to combine the beam from more than one laser to produce higher powers if required. The high beam quality allows a relatively small spot size to be produced if required for precision welding. Focused spot sizes of less than 50 μm are available for sources with power greater than 10 W if required.

Diode lasers

Diode lasers produce radiation at a wavelength of 808 nm (InGaAlAs) or 940 nm (InGaAs). Water-cooling is usually required. Their relatively low

Table 10.3 NIR laser types

	Nd:YAG	Diode	Fibre
Wavelength (nm)	1064	808 or 940	~1000 and higher
Efficiency (%)	3	30	20
Approximate cost for 100 W system (£k)	40	10	30
Beam quality	High	Low	High

beam quality means that they cannot be used to produce spot size as small as Nd:YAG or fibre lasers. However, this is rarely a problem for plastics laser welding, where the relatively low purchase and running costs have attracted a great deal of interest. The beam may be delivered by an optical fibre, but the diode laser is sufficiently small and light that it is often feasible to use a direct system, in which the diode is included with a lens system in a single unit, typically ~150 × 150 × 300 mm. This unit can be readily mounted on a gantry system or robot arm to manipulate the beam.

Fibre lasers

Rare-earth-doped fibre lasers typically supply a single wavelength in the range 1,000–2,100 nm. In the field of materials processing, much interest has focused on wavelengths around 1,100 nm to provide a direct replacement for Nd:YAG lasers, with equivalent beam quality, but greater efficiency. Systems are relatively compact and can be air-cooled. In the field of plastics welding, the use of fibre lasers has been demonstrated for a range of applications, including precision welding, films, textiles and larger moulded parts.

10.8.4 Manipulation equipment

The beam or workpiece manipulation equipment for laser welding will take one of the following forms as illustrated in Fig. 10.6 and described in more detail below.

Moving part

With the laser fixed, the part can be manipulated to form a continuous weld. This can be achieved, for example, with rollers, or a single- or two-axis moving table. This type of system is relatively simple to set up and program. It would not normally be used if three-dimensional welds are required.

Moving laser

The optical system for a fibre delivered laser, or the laser head for a direct diode laser, can be mounted on a variety of robotic systems. These range from simple two-axis gantry systems to multiple-axis robotic arms. The laser is then manipulated around the part to be welded, potentially allowing complex, three-dimensional welds to be produced. To facilitate automatic production, it is feasible to combine a moving laser with a moving part, for example by using a rotating table to present different faces of a component to a laser mounted on a robot arm. The resolution is limited by the size of the laser beam which could be as small as 10 microns.

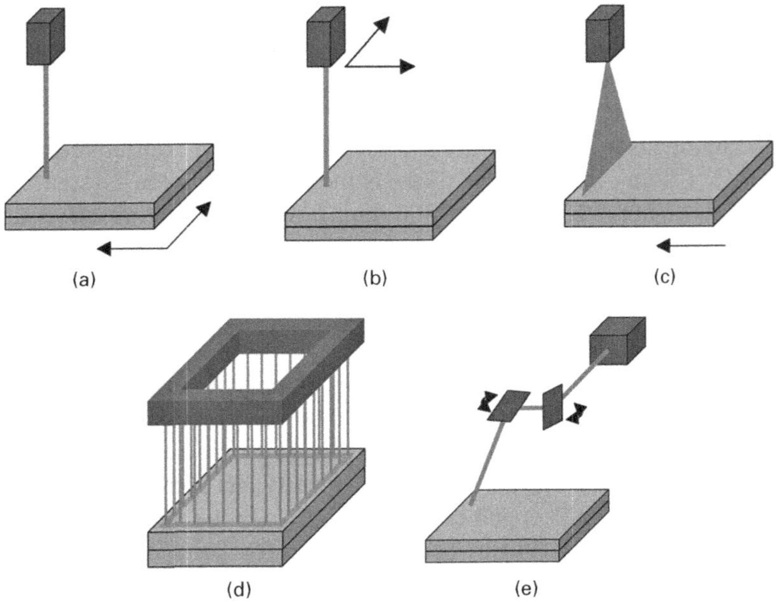

10.6 Welding methods (from Warwick and Gordon 2006): (a) moving workpiece; (b) moving laser; (c) curtain laser; (d) simultaneous welding; (e) scanning laser.

Curtain laser

The laser energy is spread into a line and then passed over the component, either by moving the laser or by moving the part. A mask is typically used to ensure that only the relevant areas of the component are exposed to the radiation. This is particularly suited to small components with a complex weld geometry. The process would usually be used only to produce two-dimensional welds. The welds are completed very quickly with a single sweep of a line source, but the resolution is limited to around 100 microns.

Simultaneous welding

If a large number of identical welds is required, then an array of diode lasers can be assembled in the shape of the weld. This is then used to irradiate the whole joint simultaneously, with a typical cycle time of 1–3 s. This approach is well suited to automated assembly. The equipment used is frequently based on ultrasonic welding equipment, and this process is typically used in place of ultrasonic welding where a good cosmetic appearance is required or for components that are sensitive to vibration. Two- and three-dimensional welds can be produced. The entire joint is welded at the same time, allowing more collapse of the polymer at the joint, and therefore allowing wider part tolerances.

Scanning laser

The laser radiation is manipulated by a pair of orthogonal rotating mirrors over an area that may range from 50 mm × 50 mm up to approximately 1000 mm × 1000 mm. In general, a larger working area implies a longer working distance and a larger spot size. It is possible to co-ordinate an assembly of a number of scanning systems to give a larger working area. In general, only two-dimensional welds can be produced.

Repeatedly scanning the laser at high speed over the same path can be used to give quasi-simultaneous welding. As for simultaneous welding, this welds the entire joint area at the same time, allowing more collapse of the material in the joint and potentially allowing wider tolerances.

10.8.5 Clamping systems

A wide variety of clamping systems have been used for through-transmission laser welding. They are mostly variants of the two systems illustrated in Fig. 10.7.

Variants of the fixed clamp include systems using mechanical fastenings, rather than an actuator, to apply a load. In the simplest variant, if the part design allows it, a bolt can be passed through the workpiece to apply the load. The transparent cover must be rigid enough to provide the clamping pressure. Thick acrylic or plain plate glass can be used. Borosilicate glass is less vulnerable to thermal shocks during welding, but more expensive. For welding of high-temperature polymers, quartz glass may be used. In all cases, it is important to ensure that suitable safety precautions are taken to avoid the risk of injury if the transparent cover breaks while it is under load.

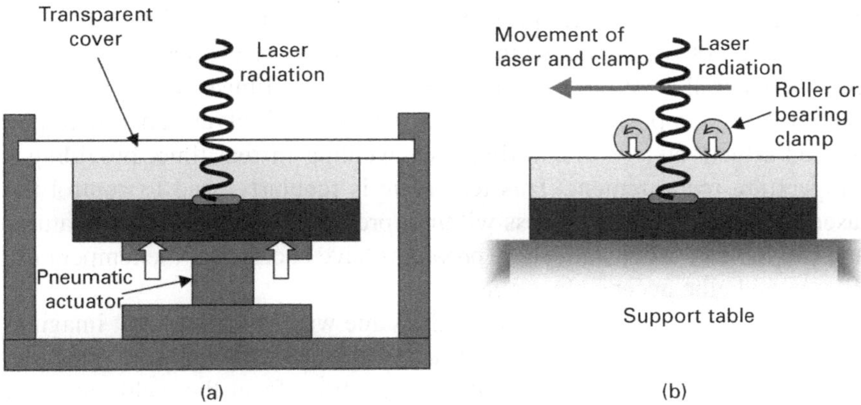

10.7 Clamping systems for through-transmission laser welding (from Warwick and Gordon, 2006): (a) fixed clamp; (b) moving clamp.

The moving clamp can use bearings, rollers, or a simple sliding shoe to apply a clamping load. Because the load is applied only at the point where the joint is irradiated, clamping loads may be much lower when a moving clamp is used. There is therefore less risk of distorting the workpiece, and equipment can be less bulky. This is particularly advantageous for large components, where application of a suitable clamping pressure to a large area can require large loads.

10.8.6 Monitoring and control methods

Although several commercial imaging techniques, such as infrared, ultraviolet, ultrasonic and visible light imaging are available (Jones and Rudlin, 2006) for viewing welding phenomena in metals and plastics, the transmission laser welding process poses a unique challenge, whereby the weld interface is typically behind two layers of polymer sheet: the pressure transmitter sheet, typically made from poly(methyl methacrylate) (PMMA), and the top layer of the part to be welded. Any top face monitored image of the weld region must therefore pass through these two layers. In addition, methods using IR absorbing coatings are designed to have minimal visible impact, making some monitoring methods difficult to use. Low colour IR absorbers can also be used in the resin of the lower part in a similar way to the normal use of carbon black absorber. Monitoring methods needed for this case are different again compared to use of IR coatings.

Visible light imaging is appropriate when transparent and translucent materials are used as the top layer. It is a low-cost option that uses off-the-shelf equipment. However, this method would be unsuitable when visibly opaque materials are used as the top layer.

Thermographic techniques remotely measure the radiation emitted from a hot surface by virtue of its temperature. The temperature of the surface can be calculated from the spectrum and intensity of the radiation emitted. The strength of the signal available is the product of the spectral radiance and the material's emissivity at a given wavelength. Emissivity values range from 1.0 for lampblack down to 0.02 for polished silver. Thermography is used effectively for monitoring laser welding in real time, providing a temperature measurement. This technique is regularly used to control the laser power to keep the process within a pre-defined range of temperatures and can indicate when and where problems have occurred if the temperature goes outside the acceptable range.

Infrared imaging is carried out in the same way as visible light imaging, with cameras sensitive to the near infrared range of wavelengths. In principle, if wavelengths of the thermal emission spectrum from the weld are used, then a thermal picture of the weld can be seen. A method of using an IR camera to image a complete weld line immediately after the welding process has finished has been described when welding clear to black polycarbonate parts. Regions of differing temperature could be identified and could indicate variations in weld quality along the joint.

Spectroscopic transmission measurements are used to measure the absorption properties of the IR absorber and substrate materials over a range of wavelengths. This procedure may be used before and after the coating deposition to check the position and amount of absorber present. As a result of the laser heating, some or all of the dye decomposes during the welding process into products that have a different absorption signature. Measurements made using a spectrophotometer before and after welding can therefore reveal information about the heat generated during welding.

Monitoring may be applied for a number of reasons, both assessing that various stages of the process have occurred, as well as indicating the quality of the weld:

- Check infrared absorber is applied correctly to parts before welding.
- View of weld during process.
- Indication that parts are in contact during process.
- Indication that weld heating is being carried out and control of temperature.
- View of weld after completion.
- Indication that parts are in contact after completion.
- Indication that weld has been achieved after completion.
- Indication that weld quality and strength are satisfactory.

Monitoring methods can give information for most of these points with varying degrees of confidence. However, it must be noted that a complete quality and weld strength assurance can only be achieved with destructive testing.

10.9 Welding parameters

As with all plastics welding processes there are three critical process parameters: temperature, time, and pressure.

The energy density used during welding combines the process parameters of temperature and time. It is determined by the laser power, the spot size at the joint, and the irradiation time (for fixed processes) or welding speed (for processes in which the part moves with respect to the laser):

$$\text{Energy density} = \frac{\text{Power} \times \text{Time}}{\text{Spot size}}$$

or

$$\text{Energy density} = \frac{\text{Power}}{\text{Spot width} \times \text{Speed}}$$

If the energy density is too low, then insufficient heating takes place and the material at the joint is not held at a high enough temperature for a sufficiently long time to form a strong weld. If the energy density is too high, then excess heating can degrade the polymer at the joint, resulting in porosity, or, in extreme cases, burning or charring of the polymer. Either case results in a weld of lower strength than the optimum. In practice, a relatively wide processing window can usually be found within which satisfactory welds can be produced. Typically, laser welding applications use an energy density within the range 0.1–2 J/mm^2, although this will vary depending on the depth of melt required to ensure a satisfactory joint. Although the energy density can be used to characterise the welding process, it should be treated with caution. The conduction of heat away from the joint during welding means that using the same energy density will not necessarily result in the same quality of weld. For example, with a constant spot width, doubling the power will usually allow the speed to be more than doubled, whilst retaining the same performance from the weld.

The pressure applied is controlled using the clamping system. If the workpieces are not clamped together during welding, or if the pressure at the joint is insufficient, then the joint faces will not be in intimate contact. This results in:

- poor conduction of heat to the upper workpiece
- limited interdiffusion of polymer chains on either side of the joint.

Both effects result in a weld of lower strength than the optimum. Care is needed to ensure that a clamping load actually provides pressure at the joint. Typically, clamping pressure in the range 0.1–1 N/mm^2 is used. If the workpieces bend under the clamping load in such a way that the joint is distorted, then a poor weld can result. For this reason, it is often useful to have some compliance, for example an elastomeric element, in the clamping system.

10.10 Advantages and disadvantages of transmission laser welding

10.10.1 Advantages

- Joint designs are simple flat-to-flat surfaces in general
- Hermetic seal possible
- Fast, <1 sec weld possible
- Non-contact
- No vibration
- No particulate generation
- Precise placement of welds

- No surface damage
- Low residual stresses
- Complex shapes possible
- Localised heating – no thermal damage to sensitive features close to weld
- Multiple layers can be welded simultaneously
- Thin or flexible substrates can be welded
- Little or no flash

10.10.2 Disadvantages

- Expensive compared to other plastics welding equipment
- Joint surfaces must be of good quality
- Part clamping must be designed carefully to ensure contact at the whole of the joint area during welding
- Laser absorbing material must be added to one of the plastics or at the joint surface
- The top part must transmit the laser radiation. This can limit the thickness of the top substrate when welding plastics with low transmission (e.g. PEEK or some filled plastics).

10.11 Applications

Transmission laser welding is used in a wide range of plastics joining application areas, including:

- Electronic packages
- Textiles and inflatable film products
- Biomedical devices including catheters and microfluidic chips
- Windows and signs
- Food and medical packaging
- Visual displays
- Automotive fluid and pressure vessels and brake components.

10.12 References

Coelho J P, Abrea M A, Pires M C (2000), 'High speed laser welding of plastic films', *Optics and Lasers in Engineering*, 385–395.

De Gennes R P (1971), 'Reptation of a polymer chain in the presence of fixed obstacles', *Journal of Chemical Physics*, 55(2), 572–579.

Haire K R, Windle A H (2001), 'Monte Carlo simulation of polymer welding', *Computational and Theoretical Polymer Science*, 11(3), 227–240.

Jones I A, Olden E (2000), 'A thermal model for transmission laser welding of thermoplastic polymers', TWI Members report 708/2000, July.

Jones I, Rudlin J (2006), 'Process monitoring methods in laser welding of plastics', *Conference Proceedings Joining Plastics 2006*, 25–26 April 2006, London.

Jones I A, Taylor N S (1994), 'High speed welding of plastics using lasers', *ANTEC '94 Conference Proceedings*, 1–5 May 1994, San Fransisco, CA.

Jones I A, Wise R W (1998), 'Welding method'. Patent application WO 00/201587, 1 October.

Jones I A, Hilton P A, Sallavanti R, Griffiths J (1999), 'Use of infrared dyes for transmission laser welding of plastics', *Proc. ICALEO*, November.

McCrum N G, Buckley C P, Bucknall C B (1988), *Principles of Polymer Engineering*, Oxford: Oxford University Press.

Puetz H, Haensch D, Treusch H, Pflueger S (1997), 'Laser welding offers array of assembly advantages', *Modern Plastics International*, September.

Roesner A, Abels P, Olowinsky A, Matsuo N, Hino A (2008), 'Absorber-free laser beam welding of transparent thermoplastics', *Conference Proceedings ICALEO 2008*, pp. 105–111.

Silvers H J Jr, Wachtell S (1970), 'Perforating, welding and cutting plastics films with a continuous CO_2 laser', PA State University, Eng. Proc, pp. 88–97, August.

Stevens M P (1953), *Polymer Chemistry*, Ithaca, NY: Cornell University Press.

Toyota, Jidosha K K (1985), 'Laser beam welding of plastic plates'. Patent application JP85213304, 26 September.

Vasenin R M (1961), 'The sticking pressure in the diffusion theory of adhesion to polymers', *Vys. Soed.* 3, 5, 679–685. (Rapra Translation 1010 by R Moseley, 1962.)

Voyutskii S S (1963), *Autohesion and Adhesion of High Polymers*, New York: John Wiley and Sons.

Warwick C M, Gordon M (2006), 'Application studies using through-transmission laser welding of polymers', *Proc. Joining Plastics 2006*, 25–26 April 2006, NPL, London.

Wise R J (1999), *Thermal Welding of Polymers*, Cambridge: Woodhead Publishing.

Wool R P (1995), *Polymer Interfaces: Structure and Strength*, Munich: Carl Hanser Verlag.

Wool R P, O'Connor K M (1980), 'Craze healing in polymer glasses', *ACS Polymer Preprints 21*, 40.

Wool R P, O'Connor K M (1981), 'A theory of crack healing in polymers', *Journal of Applied Physics*, 52(10), 5953–5963.

11
Laser welding of glass

I. MIYAMOTO, JWRI, Osaka University, Japan

DOI: 10.1533/9780857098771.2.301

Abstract: Laser welding of glass enables monolithic joint structure providing excellent mechanical strength, high temperature stability and space selectivity in comparison with existing joining procedures without pre- and post-heating. For welding of glass, CW CO_2 lasers and ultrashort pulse lasers (USPL) can be used. While CO_2 laser welding can be used for welding glass with a small coefficient of thermal expansion (CTE) based on linear absorption process similar to metal welding, USPL is applicable to any glass without restriction of CTE based on nonlinear absorption. This chapter discusses laser-matter interaction, welding characteristics and applications of laser welding of glass.

Key words: glass welding, nonlinear absorption, mechanical strength.

11.1 Introduction

Glass can be welded without pre- and post-heating by two different approaches using continuous wave (CW) laser and ultrashort pulse laser (USPL) to which glass is opaque and transparent, respectively. In the former approach, the laser energy is absorbed at the surface of the glass plate by a linear process following the same principles as the laser welding of metal discussed in the other chapters. Some applications of glass welding by CW laser are described here, which include observations on keyhole-welding, spot welding for optical assembly and welding of optical fibers to a substrate.

In the latter approach, USPL is absorbed within bulk glass by a nonlinear process due to multiphoton ionization followed by avalanche ionization, enabling internal melting of glass plate without giving any damage at the surface of the glass plate. The internal melting by USPL has been attracting much attention due to its unique absorption process and thereby an innovative welding technique for glass has been realized by the recent advancement of USPL technologies. Interaction between USPL and bulk glass during welding is still under investigation, since USPL with average power high enough for glass welding became commercially available only within the last five years. So, interaction between USPL and glass is discussed in some detail, and overlap-welding technology is introduced.

This chapter contains fundamentals and applications of CW-laser and USPL laser welding techniques to glass, which is followed by a conclusion and future trends.

11.2 Features of glass welding

There exist a variety of procedures for joining glass pieces other than fusion welding discussed in this chapter. They include adhesive bonding [1], soldering [2], optical contact [3] and direct bonding (diffusion bonding) [4]. Different joint performances are required depending on the application field. Table 11.1 compares joining procedures of glass in terms of the process performance (flexibility, process easiness and space selectivity) and the joint performance (mechanical strength, temperature stability, chemical resistance and optical indices matching). In the applications to mechanical structures, for instance, mechanical strength and temperature stability will be important. For joining optical components like optical fibers, for instance, the mechanical strength and the efficiency of the transmission of the optical signal can be major concerns.

While adhesive is most widely used in different industrial fields due to excellent joining performance such as flexibility in positioning and filling the space between the glass pieces, it has major disadvantages in terms of joint performance such as mechanical strength, temperature stability and chemical resistance. In soldering, solder glass, whose melting temperature is lower than that of the glass piece, is inserted at the interface of the glass pieces, and the solder glass is selectively melted by heating in furnace or laser irradiation. These two joining procedures provide a joint performance lower than that of the glass pieces themselves for mechanical construction, and also have disadvantages of absorption and back reflection of the optical

Table 11.1 Comparison of glass joining procedures and their performance

Evaluation		Adhesive	Soldering	Optical contact	Diffusion bonding	Fusion welding
Performance of joining process	Flexibility	◎	○	×	×	○
	Process easiness	◎	○	×	×	○
	Space selectivity	◎	×	×	×	◎
Performance of joint	Mechanical strength	×	○	○	◎	◎
	Temperature stability	×	○	○	◎	◎
	Chemical resistance	×	○	◎	◎	○
	Optical indices matching	×	×	◎	◎	◎

◎ = excellent; ○ = fair; X = poor.

signal at the joint due to mismatch of the refractive indices for optical applications.

The optical contact has disadvantages in that very high surface flatness ($\lambda/20$) and surface roughness are needed, and the strength of the optical contact is approximately two orders lower than that of the base material [5] since the joining is based on van der Waals interaction [6]. Although the joint strength of diffusion bonding is as high as that of the base material, this process is not practical, since the joining is based on diffusion of atoms near the joint interface, which requires not only high flatness and surface roughness of the glass pieces, but very long heating time at high temperature and high pressure. Optical contact and diffusion also do not have space selectivity of the joining area.

Fusion welding is based on local melting to provide excellent performance in terms of joining process and weld joint. While traditional glassblowing can be used for joining glass components, there is a desire for development of more precise procedures with automation, especially for the miniature part of the process. Advantages of glass welding include the monolithic joint structure with excellent spatial resolution of the joining area and its simple joining procedure in contrast to diffusion bonding. The monolithic joint structure provides advantages including not only thermal and chemical stability and mechanical strength comparable to glass itself, but no transmission and reflection loss at the weld joint due to no discontinuity in the refractive indices in the applications to joining transmission optics.

Glass can be welded using two contrastive lasers, CW-laser [7] and USPL [8–10], to which glass is opaque and transparent, respectively. In the former case, CW (or normal pulse) CO_2 laser ($\lambda = 10.6$ μm) is used where the laser energy is absorbed following the same principles as those of well-established laser welding technology of metal in the sense that the laser energy is absorbed at the surface of the material to be welded through linear absorption.

In the latter case, USPL with infrared wavelength (typically $\lambda = 1.06$ μm) is used, and the laser energy is absorbed only in the high intensity region in the focus volume through nonlinear absorption by multiphoton ionization followed by avalanche ionization without any absorbent. Since the laser energy is absorbed only in the focus volume and no laser energy is absorbed at the locations apart from the focus, the molten pool is embedded in the solid bulk glass. In overlap welding with USPL, the interface of the glass plates is selectively melted without giving any damage at the top and the bottom surfaces of the glass plates.

In laser welding of glass, prevention of cracks is one of the most important tasks, since local laser heating produces a steep temperature gradient so that cracks can be easily produced due to brittleness of glass at room temperature in contrast to metal. While tensile and compressive stress are developed by the local heating, cracks normally develop when the tensile stress exceeds the material strength, since the tensile strength of material is generally much smaller than the compressive strength.

11.3 Glass welding by continuous wave (CW) lasers

11.3.1 Fundamentals of CW laser welding

Welding modes

Glass is opaque to radiation in wavelength ranges of $\lambda < 250$ nm and $\lambda >$ 3500 nme. One of the most useful lasers in these wavelength regions is the CO_2 laser with a wavelength of $\lambda = 10.6$ μm. The Fresnel reflection at this wavelength is approximately 14%, and 86% is absorbed at the surface of the glass.

There are two types of glass welding using CO_2 laser: thermal conduction and deep penetration (keyhole welding) depending on the laser intensity as in the case of laser welding of metal, as schematically shown in Fig. 11.1 [11]. When the power density of the laser beam is not high enough to heat the surface above the boiling temperature of the glass, the interior of the bulk glass is heated by thermal conduction from the surface (Fig. 11.1a), providing

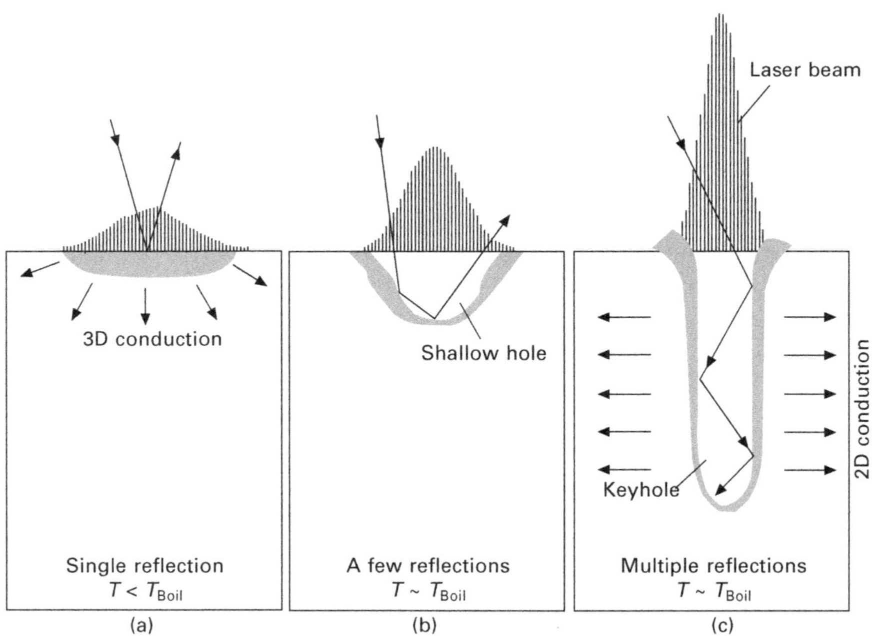

11.1 Schematic illustration showing different welding modes: (a) thermal conduction welding; (b) intermediate between thermal conduction and keyhole welding; (c) typical keyhole welding.

thermal conduction welding. Thermal conduction welding is useful in welding micro parts or thin sheets of glass, since the distance of thermal diffusion is limited due to low thermal diffusivity of glass. In thermal conduction welding, a rather large spot of laser beam is used at slow welding speeds. An application demonstrating hermetic sealing of a thin pressure-bearing quartz diaphragm to a thicker frame has been reported [12].

When the power density of the laser beam is high enough to heat the glass surface above the boiling temperature, the molten glass surface is depressed by the recoil pressure of the evaporation, and a shallow hole is produced. The keyhole grows deeper, when the laser beam is refocused to the bottom of the hole by the reflection at the wall of the hole (Fig. 11.1b). As the power density increases, the depth of the hole becomes deeper to produce a deep keyhole. Then the inside bulk glass is directly heated by the laser beam guided in the keyhole through multi-reflections (Fig. 11.1c).

Residual stress in melting containing free surface

It is not simple to evaluate the exact stress field in welding, and it has to be numerically simulated based on three-dimensional elasto-plastic analyses. However, the stress developed in fusion welding of glass can be qualitatively estimated by a simple three-bar model consisting of a center bar A and sidebars B as shown in Fig. 11.2. In this one-dimensional model, it is assumed that the three vertical bars are connected to rigid horizontal bars at their ends, and the laser energy is absorbed in A. In the early stage of the heating period (Fig. 11.2a: Phase I), compressive stress and tensile stress are elastically developed in A and B, respectively, since thermal expansion of A is constrained by B. When A is melted (or yielded), the resultant stress field depends on whether or not the molten region can be plastically deformed. In

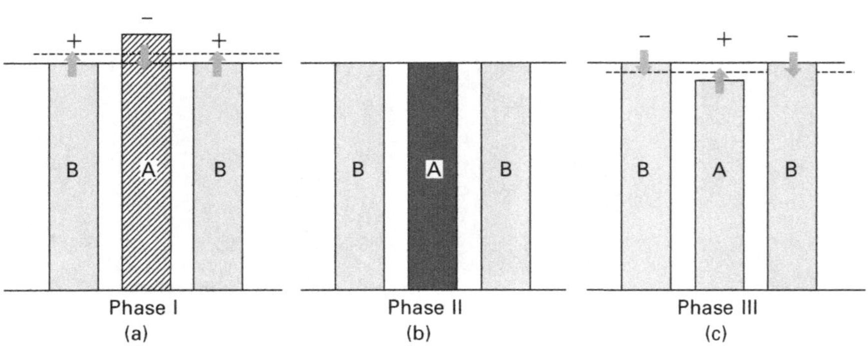

11.2 Three-bar model in CO_2 laser welding of glass, assuming laser energy is absorbed in A. (a) Heating period (elastic deformation). (b) Melting (stress is released). (c) Cooling periods. + and – represent tensile and compressive stress, respectively.

the case of CO_2 laser welding of glass, the molten region can be plastically deformed. Then the stress in A and B is completely released when A is melted, since the melted region contains free surface and hence the plastic deformation of the molten region can occur freely. Thus the lengths of A and B return to the original values on melting without producing any stress (Fig. 11.2b: Phase II).

In the cooling period after the solidification of the molten region, the shrinkage of A is constrained by B (Fig. 11.2c: Phase III). Therefore the tensile and compressive stresses are left in A and B, respectively, when A is cooled down to room temperature. Cracks are developed in A if the tensile stress (called shrinkage stress) exceeds the mechanical strength of the glass, since glass is brittle at room temperature. The residual tensile stress in A increases with increasing thermal expansion coefficient of the glass. Therefore in CO_2 laser welding of glass, crack-free welding is possible only in glass having a small thermal expansion coefficient like fused silica, as cracks are easily developed by the shrinkage stress. The shrinkage stress also increases with increasing absorbed laser energy. If the absorbed laser energy is not large enough to melt (or yield) A, no residual stress is left in Phase III, since no plastic deformation occurs and hence the shrinkage stress is compensated by the compressive stress produced in Phase I. The formation mechanism of the residual stress in glass welding is basically the same as metal welding. However, cracks can normally be prevented in metal welding, since metal has ductility even at room temperature in contrast to glass.

In the three-bar model, it is implicitly assumed that cracks due to the tensile stress in B in Phase I can be suppressed. This assumption is justified by the direct observation of glass welding by CO_2 laser as will be shown later. No cracks are observed during CO_2 laser welding of glass as will be shown below (see Figs 11.3–11.6). In the following sections, some applications of CO_2 laser welding of glass are introduced.

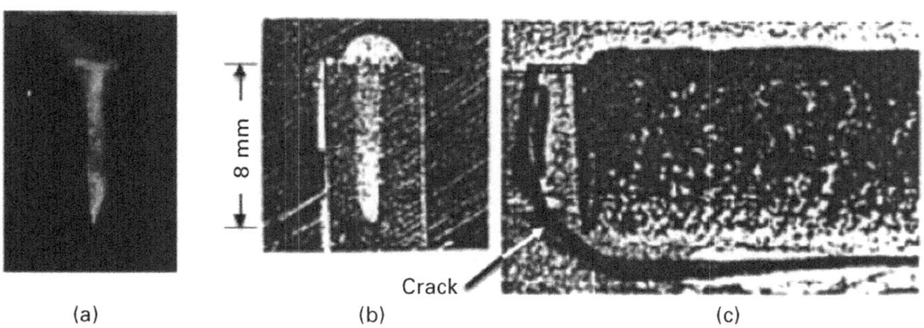

11.3 Keyhole welding in soda-lime glass using electron beam (150kV-0.5mA; 40mm/s). (a) Keyhole observed during welding. (b) Cross-section of weld bead. (c) Side view of weld bead, Cracks are developed after long delay at room temperature [7].

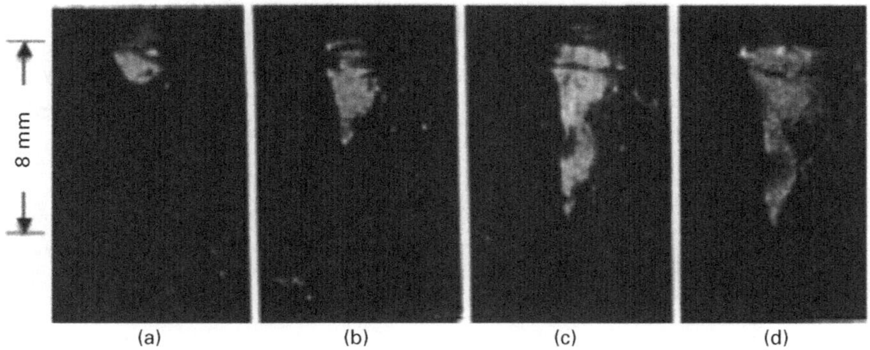

11.4 CO_2 laser welding of soda-lime glass at laser power of (a) 50 W, (b) 70 W, (c) 100 W and (d) 120 W at welding speed of $v = 70$ mm/s. Focus point was located 2 mm above the surface [7].

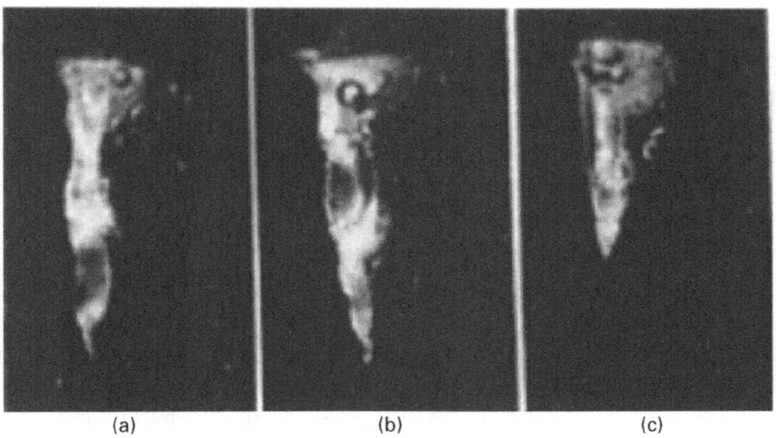

11.5 CO_2 laser welding of soda-lime glass at 150 W at welding speed of (a) 10 mm/s, (b) 60 mm/s and (c) 100 mm/s [7].

11.3.2 Applications of CO_2 laser welding

Observation of keyhole welding process

Keyhole welding is widely used for joining metal using a high power density source such as a laser beam and electron beam due to the advantage that thick sections can be welded in a single path with reduced heat-affected zone. The keyhole welding process, however, is not fully understood, since the dynamic behavior of the keyhole and the molten material is very complicated. Thus the keyhole welding process has been investigated experimentally by X-ray transmission imaging [13]. The X-ray transmission images, however, are not clear enough to investigate the dynamic behavior of the keyhole precisely.

It was found that a keyhole can be produced by focusing CO_2 laser in glass,

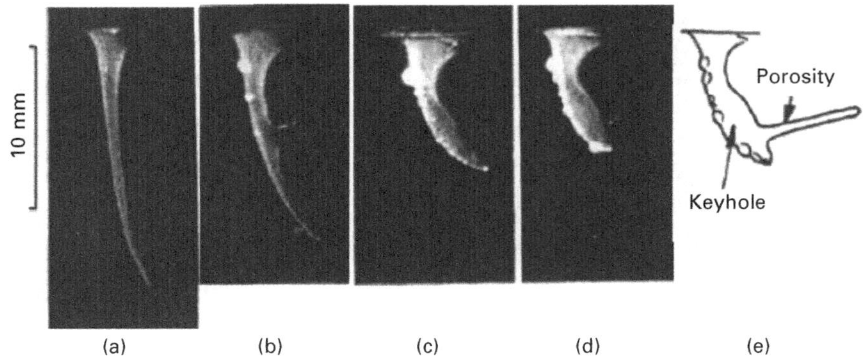

11.6 CO_2 laser welding of 96% silicate Vycol glass at 90 W at welding speeds of (a) 10 mm/s, (b) 20 mm/s, (c) 30 mm/s, and (d) 40 mm/s. (e) Schematics at 40 mm/s [7].

where the surface tension force is balanced with the recoil pressure of the evaporation at the keyhole wall [7]. When the glass sample moves forward, the keyhole is maintained stably by burying the backside of the keyhole with molten glass continuously in a similar way to the keyhole welding of metal. This means that the dynamic behavior of the keyhole welding process can be directly observed in glass with much higher spatial and temporal resolution than the X-ray transmission images in metal welding. This technique also provides a useful measure to investigate the effects of thermal properties such as viscosity and thermal conductivity on welding phenomena.

Figure 11.3(a) shows the side view of the keyhole welding in soda-lime glass by electron beam in vacuum. Interestingly, no influence of charge-up of electrons is found. High-speed photographs clearly show that as the glass sample moves forward at a constant speed, the backside of the keyhole is continuously buried by the molten glass to maintain a stable keyhole. The electron beam reaches directly the bottom of the vertical keyhole unlike the case of CO_2 laser welding of glass as will be described below. The cross-section of the molten zone is similar to that of keyhole welding of metal as shown in Fig. 11.3(b). This suggests that glass is a useful material for observing the dynamic behavior of the keyhole and the molten material.

It is noted that no cracks are developed during welding in spite of the steep temperature gradient developed during localized heating. This is considered to be because the heat-affected region gains ductility to prevent cracks in Phase I. However, cracks are always developed when the welded sample is cooled down to room temperature, sometimes after a very long delay as shown in Fig. 11.3(c). This indicates that the cracks are caused by shrinkage stress in Phase III, and that crack-free keyhole welding of soda-lime glass is not possible because of the large thermal expansion coefficient of soda-lime glass, $\alpha = 94 \times 10^{-7}/°C$ (T = 0–300°C). Crack-free welding is limited to glass having a thermal expansion coefficient one order smaller than soda-lime glass like fused silica or Vycol glass using beam sources to which glass is opaque as shown below.

Figure 11.4 shows the laser welding of soda-lime glass observed at different laser powers at a welding speed of 70 mm/s. In this experiment, a multimode CO_2 laser beam with a rather large focus diameter of 0.3 mm was used to make the observation easier, and the high-speed photographs were taken without lighting. A shallow keyhole is observed at 50 W (Fig. 11.4a). It grows quickly at 70 W by refocusing the reflected laser beam at the wall to the bottom of the keyhole. The inclination angle of the front wall of the keyhole in CO_2 laser welding is larger than that in electron welding, suggesting that CO_2 laser energy is absorbed by the multiple reflections at the keyhole wall. No humping and closure of the keyhole entrance were observed. The flow velocity of the molten glass was considerably slower than that of metal due to the larger viscosity of the molten glass than that of metal. This suggests that the instabilities of welding such as humping bead and closure of the keyhole entrance in metal welding are related to larger flow velocity of the molten metal. However, the dynamics of the keyhole process observed in glass is basically the same as that of metal.

Figure 11.5 shows the keyhole observed at different welding speeds at a laser power of 150 W. As the translation speed increases, the keyhole becomes shorter, and the front wall of the keyhole is somewhat bent as the bottom of the keyhole is approached.

Figure 11.6 shows the photographs observed in 96% silicate glass (Corning Vycol) at 90 W at different welding speeds. The keyhole is considerably thinner and longer than that of soda-lime glass. The thickness of the molten layer surrounding the keyhole appears to be much thinner (almost invisible near the bottom of the keyhole), since the viscosity of the molten glass in this material is much larger than that of soda-lime glass.

At a welding speed of 10 mm/s, a thin and long keyhole with large bending is produced indicating that at least three to four reflections are needed for the laser beam to reach the bottom of the keyhole. It is interesting to note that the keyhole swings back and forth near the bottom at a frequency of several tens Hz like a snake motion. Several inflection points are observed and the keyhole is bent even in the opposite direction with respect to the welding motion due to larger recoil pressure of evaporation at the reflecting points as shown in Fig. 11.7. This suggests that the laser intensity at the keyhole wall is not uniform. Such a stable keyhole cannot be produced in soda-lime glass, because the stability of the keyhole is prevented by the flow of the liquid glass.

At welding speeds faster than 20 mm/s, the inclination of the keyhole at the front wall becomes larger with accompanying increased diameter of the keyhole, and a long porosity is produced at the back of the keyhole,

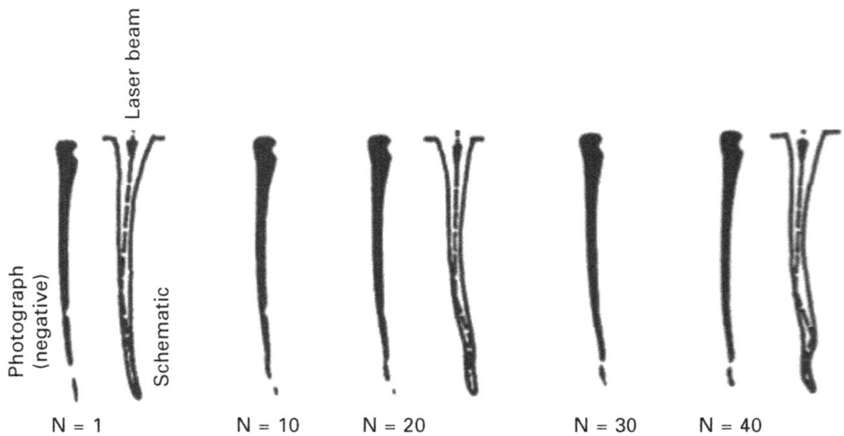

11.7 High-speed photographs of keyhole in Vycol glass (150 W, 10 mm/s, 1000 f/s). The keyhole swings back and forth in a snake-like motion. Beam propagation in the keyhole is drawn in schematics (N: frame number) [7].

suggesting the burial by the molten glass at the backside of the keyhole is only partially successful since the flow velocity of the molten glass cannot keep up with the welding speed. Such a porosity is not observed in soda-lime glass, suggesting higher welding speeds require larger fluidity of molten glass for obtaining sound keyhole welding.

It is known that the molten material is driven downward along the front wall of the keyhole by the recoil pressure of evaporation. If the melt flow stagnates due to larger viscosity of the molten material, the front wall has to be inclined so that the incident laser beam is absorbed at a larger incident angle to increase the evaporation recoil pressure. Figure 11.6 clearly shows that the inclination angle becomes larger with larger viscosity of the melt in Vycol. In laser welding of metal, bending of keyhole is much smaller, because the molten metal flows down quickly along the front wall due to smaller viscosity.

It is noted that no cracks were found in Vycol after cooling to room temperature unlike the case of soda-lime glass. This suggests that the CO_2 laser can be used for welding of Vycol. This is because the thermal expansion coefficient of this material is as small as 7.6×10^{-7}/°C, at least one order smaller than that of soda-lime glass, so that the shrinkage stress developed in Phase III can be suppressed below the tensile strength of the material.

Spot welding for optical assembly

A spot welding technique using a CO_2 laser was developed to join thick glass plates of fused silica with a small deformation, and was applied to assembly of the rooftop structure consisting of thick glass plates with a thickness of 10 mm [14]. Laser welding was adopted because it allows joining of glass pieces with an extremely high level of angle accuracy, while providing high stiffness and excellent stability over a large temperature range. This procedure was developed as a first step toward the fabrication of complete corner cubes.

A side of the glass piece with dimensions of 50 mm × 50 mm × 10 mm was set in contact with the second piece of 50 mm × 70 mm × 10 mm. The first piece was joined at a location 10 mm apart from the edge of the second piece to allow the formation of the welded spots located far enough from the edges of the glass pieces. For welding the thick glass plates, holes (diameter = 1.5 mm, depth = 9.2 mm, leaving a 800 μm layer) are drilled by an ultrasonic drill into the second glass piece to make the junction accessible, as shown in Fig. 11.8. The glass piece is deformed by the drilling process at the surface to form small bumps with a height less than 60 nm and a diameter of 2 mm. The glass pieces to be welded were pressed together by a mechanical holder. A Gaussian beam was focused into the bottom of the drilled hole to provide a focused waist of 300 μm. Average powers ranging between 20 and 70 W were irradiated at different exposure programs consisting of preheating by linearly increasing power, slowly decaying laser power for thermal relaxation and laser process duration. Total durations of the laser exposition recipes ranged from 110 to 900 s.

It is interesting to note that the glass pieces are pulled against each other by the shrinkage of the molten material during the cooling process, so that two glass pieces are brought into visual optical contact. The laser-welded rooftop assembly was coated with gold in a sputtering system after laser welding. This results in an interesting feature of the assembly that the junction between the two pieces cannot be distinguished when viewed from the gold-coated

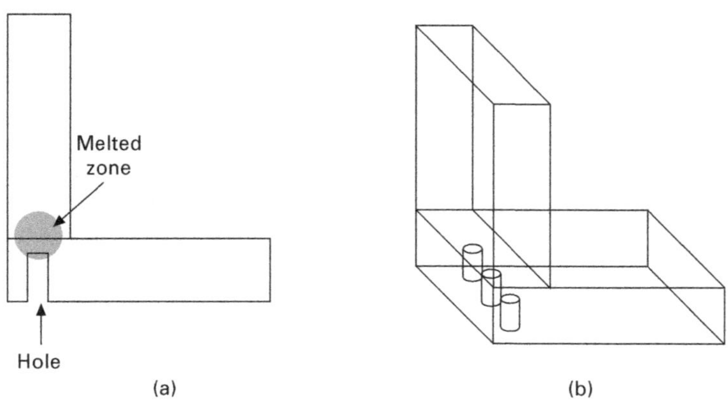

11.8 Spot welding of thick plates by CO_2 laser: (a) side view; (b) perspective view. [14]

side (Fig. 11.9). This is because the upper glass piece had no chamfer at the interface between the principal face and the contact face. This excellent optical quality junction can be an advantage for some applications.

The tensile strength of the weld spot was measured. The results show that the joint strength ranges from 10 kg to 30 kg per spot, which corresponds to approximately as large as 50–150 MPa, assuming the diameter of the spot welding is 1.5 mm. This indicates two welded spots were sufficient to provide a strong junction between the substrates.

The beam deviation error of the rooftop by the spot welding was measured with the Mark GPIxps interferometer from Zygo having an accuracy of ±0.05 arc second. The result indicates the angle deviation from 90° was 1.87 arc second. Considering that the width of the junction area between the two pieces of the assemblies is 10 mm, deviation of 1.87 arc second corresponds to 90 nm height error along 10 mm length. The measured deviation error indicates the welding process is very promising for the fabrication of high accuracy assembly.

The results of the wave front deformation errors also show the spot welding technique is very promising in keeping the quality of the initial optical faces. Melting and shrinking associated with the welding process provided negligible effect on the surface of the optical piece thanks to the distance separating the surface and the melted zones and the limited power range that minimized the side of the melted zone. One of the main advantages of laser welding is its stability over wide temperature ranges due to its monolithic structure. High temperature stability was verified by measuring the reflected wave front and angle over a temperature range between –60°C and +40°C. These tests indicated that the wave front distortion stayed below $\lambda/10$ and its

(a) (b)

11.9 Bare gold-coated laser-welded rooftop. (a) Photograph. The junction between two glass pieces cannot be seen on the photograph. (b) Drawing of the photograph. [14]

value varied less than 0.04 λ. In the same temperature range, the variations of the assembly angle were all within a 0.7 arc second interval, showing the spot welding of thick glass plates is promising.

Welding of optical fiber for fiber optical rotary joints (FORJ)

An interesting application of glass welding in the optical field is found in fiber optical rotary joints (FORJ), which are used for transmission at high data rates, for instance, between rotating radar antennas or medical sensors and stationary systems for analysis and further processing of the data. To establish an optical connection between the rotating and stationary fiber bundle, the optical fibers have to be attached to an array of collimator lenses, and the beam is focused onto the fiber core again and coupled into it on the opposite side, as schematically shown in Fig. 11.10.

The existing procedure for connecting fibers to collimator lenses in the production process of FORJ is adhesive bonding. This procedure, however, has disadvantages in that the laser energy absorbed in the adhesive can heat up the adhesive to reduce the mechanical stability. The temperature rise also changes the refractive index of the adhesive, resulting in an increase in the back reflection. As an alternative to adhesive bonding, a CO_2 laser welding technique has been developed. For heating the glass fiber at the end face and the glass substrate uniformly, a ring-shaped laser beam is ideal to heat the whole circumference of the glass fiber. In order to convert the Gaussian beam into the ring beam, the Schwarzschild objective was used, which consists of two spherical mirrors having concave and convex geometries. A laser welding system was designed to allow a minimum fiber pitch of approximately 1 mm [15].

The laser-irradiated region thus obtained is shown in Fig. 11.11, and the

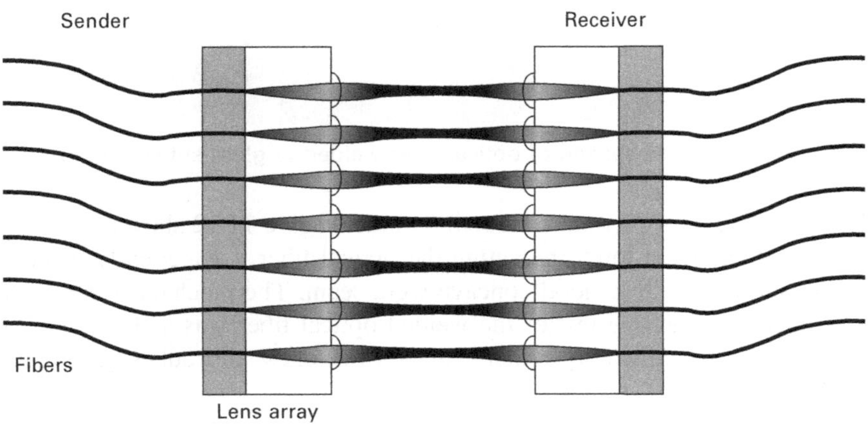

11.10 Coupling between two fiber arrays over freeboard.

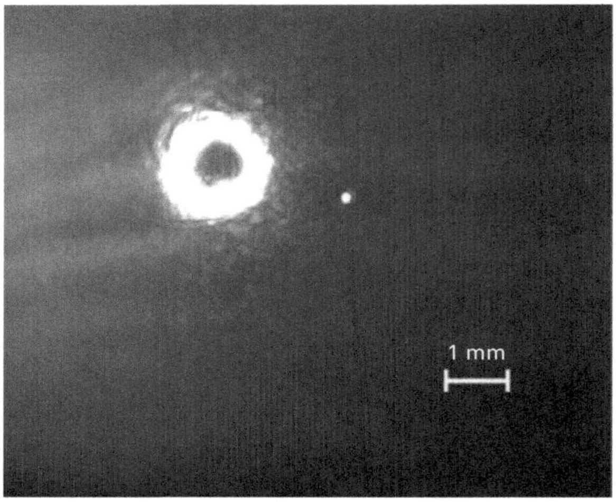

11.11 Profile of collimated beam provided by Schwarzschild objective [15].

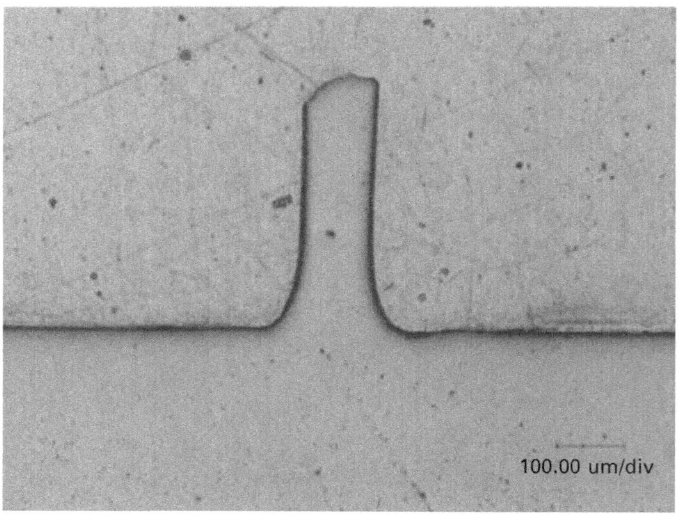

11.12 Cross section of optical fiber welded to glass substrate [15].

laser welding is of thermal conduction mode. Figure 11.12 shows the cross section of the weld joint, indicating the optical fiber is excellently welded to the substrate with a good concave weld seam. The mechanical strength evaluated by a pulling test of the welded optical fiber was as high as 326 MPa, which is significantly higher than the value of existing adhesive joining, 33 MPa [16].

Back reflection is an important factor for characterizing the quality of the optical data transmission system. The back reflection can be specified as an attenuation value (return loss). A return loss of around –40 dB is the best achievable value for adhesive joints [17]. The measured mean value of the weld joint is –49.4 dB with a deviation of 0.6 dB, which is in the range of the measurement accuracy. So there is an enhancement of the return loss of nearly 10 dB, corresponding to an absolute improvement of a factor of 10.

11.4 Glass welding by ultrashort pulse lasers (USPL)

11.4.1 Fundamentals of USPL welding of glass

Nonlinear absorption process

Ultrashort pulse laser (USPL) is a useful energy source for internal modification of glass due to its unique absorption process. When glass is irradiated by infrared CW laser with normal intensity, no laser energy is absorbed in the glass, since the band gap energy E_g is larger than the photon energy. When infrared USPL is tightly focused into bulk glass, the laser energy is absorbed into glass by a nonlinear absorption mechanism by multiphoton ionization to promote a valence electron to the conduction band as is schematically shown in Fig. 11.13 [18,19]. Then the free electron gains kinetic energy by absorbing the laser energy by Inverse Bremsstrahlung. When the kinetic energy of a free electron exceeds the conduction band minimum by more

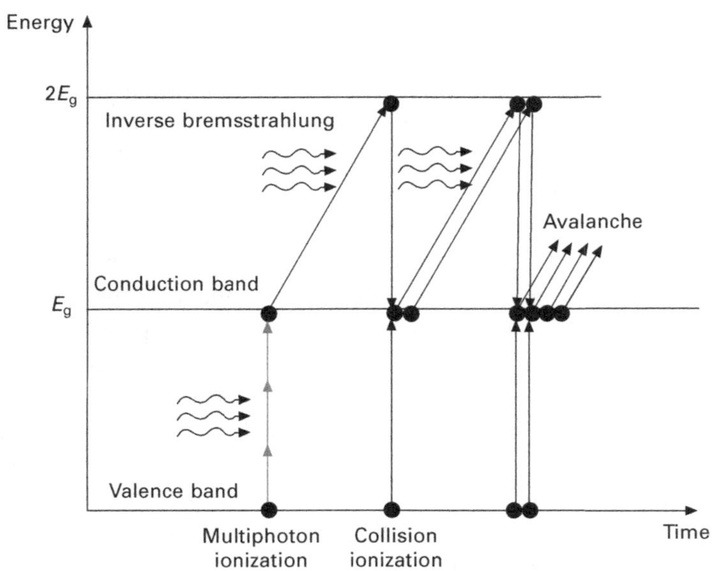

11.13 Nonlinear absorption process consisting of multiphoton ionization and avalanche ionization.

than E_g, the electron can then collisionally ionize another electron from the valence band (avalanche ionization). While the breakdown process is dominated by the multiphoton ionization in the early period of the laser pulse, the avalanche ionization starts to govern the breakdown dynamics to produce more free electrons thereafter.

Recent rapid advancement of USPL technology [20] has enabled attractive applications of internal modification of glass such as refractive index change [21], nano-grating [22], selective etching [23], three-dimensional memory [24] and fusion welding [5,8,10,25]. In particular, the advent of USPL with high-pulse repetition rates has made it possible to achieve internal modification at high throughput thanks to cumulative heating arising when the interval between laser pulses is less than the time required for the absorbed laser energy to diffuse out of the laser-irradiated region [10,26].

The laser energy absorbed by free electrons is transferred to lattice (atoms) to elevate the temperature of the glass after the laser pulse. Then the temperature field is developed in bulk glass by thermal diffusion to produce a molten region. The thermal conduction theory can be utilized to simulate the temperature distribution or the melt dimensions in the glass, assuming the intensity distribution of the absorbed laser energy is known [10].

Melting structure of glass

Since avalanche ionization requires seed electrons, which can be provided by multiphoton ionization, the laser-absorption region is limited only to the focus volume, enabling localized internal melting of bulk glass, if the focus is located inside the bulk glass. The glass molten pool can actually be observed by focusing USPL at the rear surface of the glass plate. Figure 11.14 shows a SEM photograph of the molten pools with a size of 3–4 μm produced by single laser pulses of 10 ps pulse duration [9].

Figure 11.15 shows a cross section of overlap welding of borosilicate glass plates (Corning 0211) [27]. Two advantages are found in the figure in comparison with CO_2 laser welding. First, selective melting is possible only at the interface of glass plates without any damage at the top and bottom surfaces of the glass plates. Second, crack-free overlap welding is possible with most commercially available glass materials even though their thermal expansion coefficient is 10 times larger than fused silica. The cross section of the weld joint consists of a teardrop-shaped inner region and an elliptical outer region as seen in Fig. 11.15(b). The glass plates are joined together within the outer region, indicating the glass is melted in the outer region. Since glass does not show a clear melting point, the characteristic temperature of the outer structure (T_{out}) is defined as the forming temperature of the glass, of which viscosity is $\eta = 10^4$ dPas. The inner structure corresponds to the region where the laser beam is absorbed as discussed later [10].

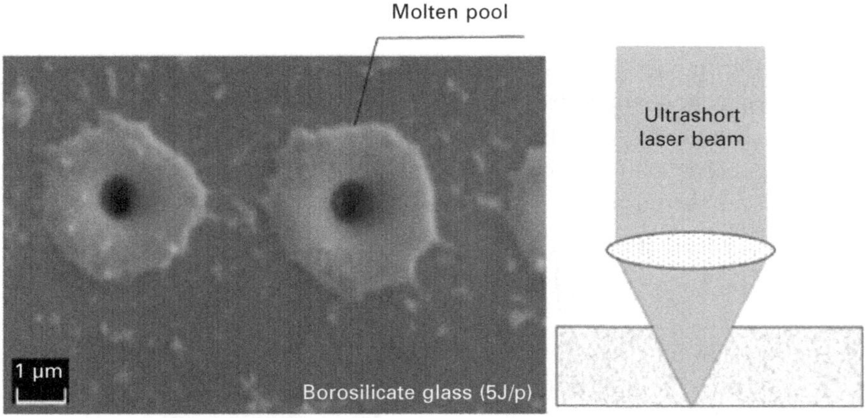

11.14 Local melting of glass plate (Schott D263) at the bottom surface using USPL (τ_p = 10 ps, λ = 1.064 μm) [9].

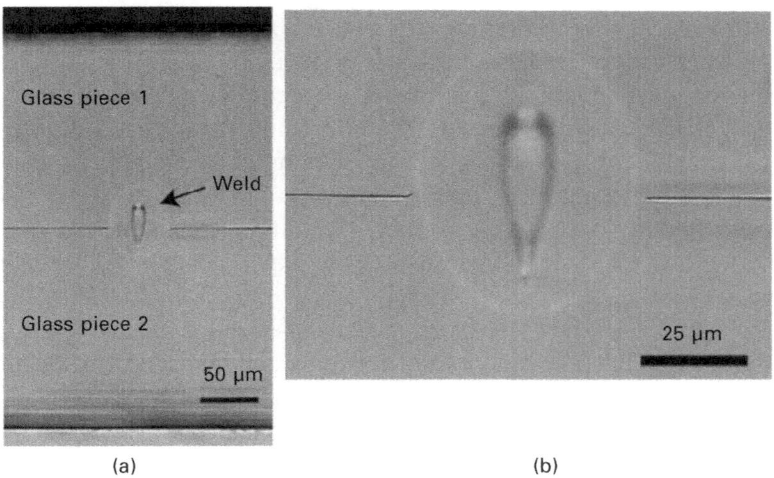

11.15 Example of overlapped welding of borosilicate glass (Corning 0211) using USPL (τ_p = 360 fs, λ = 1.045 μm, f = 1 MHz): (a) cross section of overlap welding; (b) magnified picture [27].

The stress field developed in USPL welding of glass is different from that in CO_2 laser welding. Although no difference is found between USPL and CO_2 laser in Phase I, the plastic deformation with the internal melting is prevented by surrounding solid glass in Phase II, since the molten region produced by USPL is embedded in the solid bulk glass. The embedded molten region is considered to behave like an elastic body, so that no residual stress is produced in Phase III. This is because the stresses in A and B induced during Phases II and III are elastic, although detailed study is needed to understand the exact mechanism.

Thermal conduction model for analyzing laser absorption

A simulation model has been developed to correlate between the intensity distribution of the average absorbed laser power and the temperature distribution in a glass sample by Miyamoto *et al.* [10], assuming the thermal properties are independent of temperature, for simplicity. The steady temperature distribution is given by:

$$T(x, y, x) = \frac{1}{4\pi K} \int_0^l \frac{w(z')}{s} \exp\left\{-\frac{v}{2\alpha}(x + s)\right\} dz' + T_0 \qquad [11.1]$$

where $w(z)$ is average absorbed laser power at z, l is the length of the absorption region, K is the thermal conductivity, α is thermal diffusivity given by $K/c\rho$ (c = specific heat and ρ = density), T_0 = room temperature and $s^2 = x^2 + y^2 + (z-z')^2$. In this model, it is assumed that a line heat source with continuous heat delivery [28] of $w(z)$ appears in an infinite solid moving at a constant speed of v along the x-axis, and $w(z)$ is given by $w(z) = Q(z)f$, where $Q(z)$ is pulse energy absorbed at location of z and f is pulse repetition rate. Despite the actual laser beam having finite spot size with pulsed energy delivery, the simple line heat source model with continuous heat delivery can be used to calculate the temperature at the contour of the molten zone, because the temperature rise at locations apart from the heat source is spatially and temporally averaged. Using the simulation model given by Eq. [11.1], one can determine $w(z)$ and its extent l by fitting the simulated isothermal line of the characteristic temperature to the experimental outer structure (melt region) [10].

Figure 11.16 shows the experimental cross section of borosilicate glass and the simulated isothermal lines of the inner and outer regions, assuming that $w(z)$ is given by a simple quadratic function. Excellent agreement is found between the simulated isothermal lines and the experimental contours of the outer structure of T_{out} = 1,051°C, which is the forming temperature of the material (Schott D263). The intensity distribution $w(z)$ is also plotted in the figure. The length l agrees with the vertical length of the tear-drop-shaped inner structure. This means that the inner structure is defined as the region where the laser energy is absorbed.

Nonlinear absorptivity

Figure 11.17(a) shows the cross sections of borosilicate glass at different pulse energies obtained at a constant welding speed of 20 mm/s. The geometrical focus was determined at the bottom tip of the inner structure from the modified region at lowest limit of internal melting assuming the self-focusing [29] is negligible. The inner structure, which corresponds to the laser-absorbed region [10], extends toward the incoming laser side as

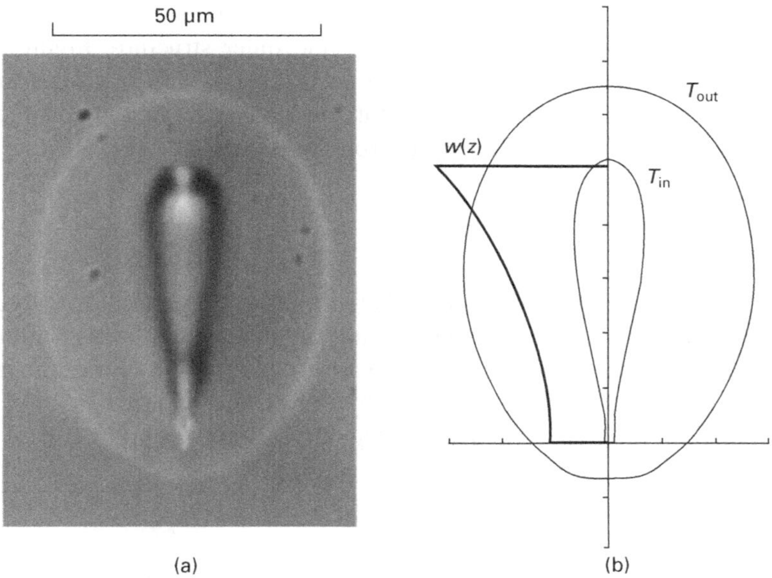

11.16 (a) Cross section of modified zone of borosilicate glass (Schott D263) using USPL (τ_p = 10 ps, λ = 1.06 µm, f = 500 kHz) at 20 mm/s. (b) Simulated isothermal lines at T_{out} = 1051°C and T_{in} = 3600°C [10].

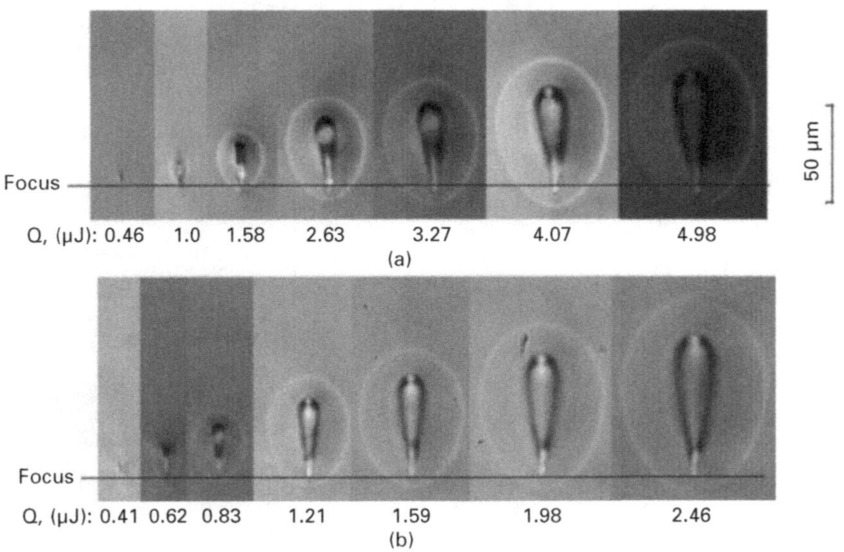

11.17 Cross section of modified zone of borosilicate glass (Schott 263) using USPL (t_p = 10 ps, l = 1.06 µm, v = 20 mm/s) at pulse repetition rate of (a) 200 kHz and (b) 500 kHz.

the pulse energy increases. The outer region (molten region) also extends in accordance with the extension of the inner structure. Figure 11.17(b) shows the cross sections at different pulse repetition rates at a constant pulse energy of 3.8 μJ. The inner structure also increases as the pulse repetition rate increases. The nonlinear absorptivity A_{Cal} is given by [10]:

$$A_{Cal} = \frac{1}{fQ_0} \int_0^l w(z)dz \qquad [11.2]$$

where f is pulse repetition rate and Q_0 is pulse energy. Nonlinear absorptivity can also be determined experimentally by measuring the transmitted pulse energy below the glass plate (see the inset of Fig. 11.18). Since the reflection and scattering of the laser beam by the plasma induced in bulk glass is negligible [30], the nonlinear absorptivity A is given by [9,10]:

$$A = 1 - \frac{Q_t}{Q_0} \frac{1}{(1 - R)^2} \qquad [11.3]$$

where Q_0 is incident pulse energy, Q_t is transmitted pulse energy and R is Fresnel reflectivity of the glass plate.

Figure 11.18 shows the simulated and the experimental nonlinear

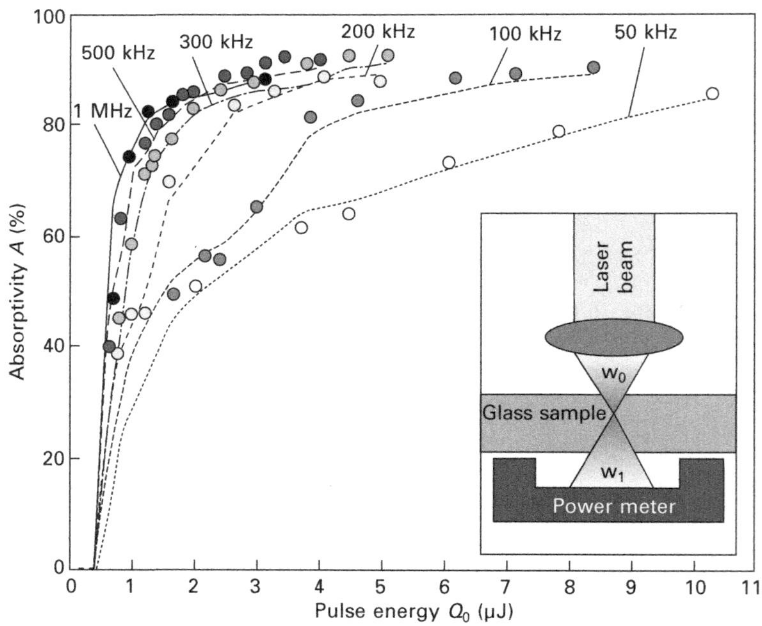

11.18 Nonlinear absorptivity of USPL (τ_p = 10 ps) at different pulse energies and pulse repetition rates. Data points and solid lines show simulated and experimental values, respectively [10].

absorptivity at different pulse energies and pulse repetition rates, showing excellent agreement between the simulated and the experimental values. The simulation model can predict the nonlinear absorptivity with an uncertainty less than ±3%. The nonlinear absorptivity increases with increasing pulse energy, since the multiphoton ionization rate increases with increasing pulse energy. The nonlinear absorptivity also increases with increasing pulse repetition rate at constant pulse energy. This can be attributed to the heat accumulation effect; the density of the thermally excited free electrons to the conduction band increases due to the heat accumulation, and contributes as seed electrons for avalanche ionization [10], as shown below.

Figure 11.19 shows the simulated absorption region l plotted vs. the average absorbed power W_{ab} (= $Q_{ab}f$: Q_{ab} = absorbed pulse energy) for pulse durations of 10 ps and 400 fs. The length of the laser absorption region l increases monotonically with increasing W_{ab} independently of the pulse repetition rate, the pulse energy and the duration of the laser pulse.

11.4.2 Overlap welding of glass plates

Mechanical strength of internally melted zone

Overlap welding technology has been developed with Foturan glass (Schott), which is widely used for three-dimensional structuring of components and is

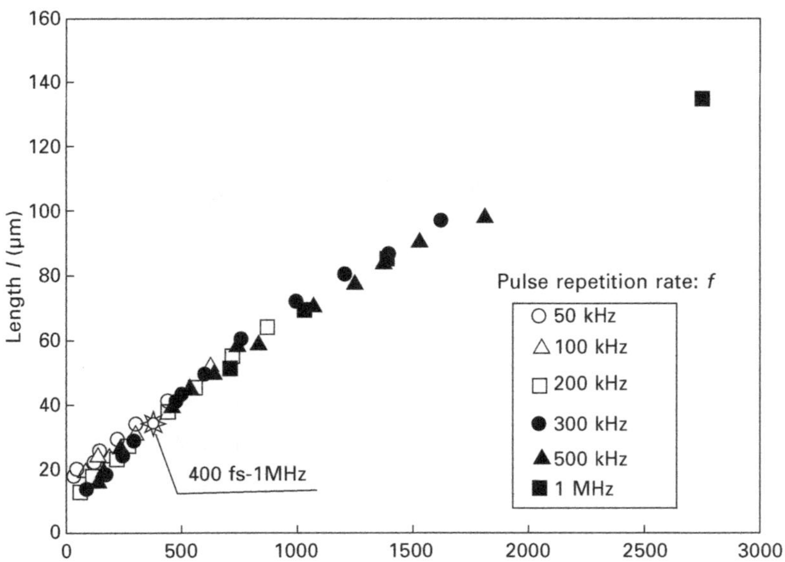

11.19 Length of laser absorbed region l vs. W_{ab} (=$A_{Cal}fQ_0$) at different energies Q_0 and repetition rates of laser pulse f at τ_p = 10 ps. Data for τ_p = 400 fs are also plotted [10].

finding increasing use in devices ranging in application from the medical to the development of miniaturized satellites [31–33]. Prior to evaluating the mechanical strength of the overlap weld joint, single glass samples of Foturan glass were internally melted using USPL, and the mechanical strength of the molten zone was tested. The laser-irradiated samples were lapped and polished parallel to the melt line to expose the maximum width of the melt line to the surface. Then the samples were cut perpendicularly to the melt line, and the mechanical strength was determined with a three-point bending test by applying the maximum stress at the molten region.

Figure 11.20 shows the mechanical strength determined at different pulse repetition rates *f* and translation speeds *v*. The strength of the virgin material is in the region of 115–200 MPa (average value: 152 MPa), and the average value of all the internally melted samples (60 data points in total) is approximately 145 MPa, which is equivalent to the virgin or base material, showing no reduction in the mechanical strength due to internal melting. This is because no shrinkage stress is left at room temperature as mentioned above. However, the strength of the internally melted single plate is approximately 30 MPa lower than that of the base material at the lowest pulse repetition rate of $f = 0.2$ MHz, suggesting that some slight damage is produced when the peak intensity of USPL is too high. There is a tendency for the strength to increase with increasing *v* and *f*, reaching values equivalent to or even higher than that of the base material.

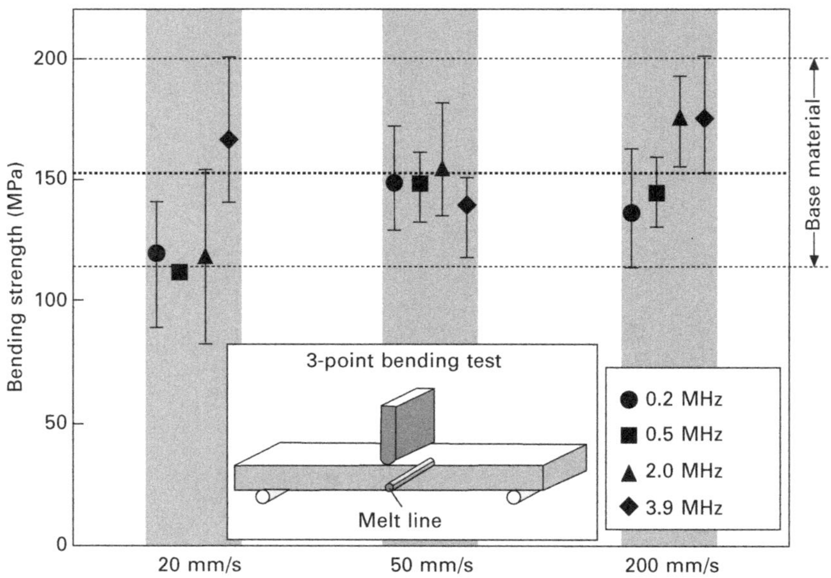

11.20 Mechanical strength of internally melted single Foturan sample determined by a three-point bending test at different *f* and *v* at a constant average laser power of 2.5 W. [5]

The bending strength data of the internally melted single sample is replotted as a function of the averaged absorbed laser power W_{ab} in Fig. 11.21. The shaded zone in the figure highlights the region where the measured data distributes. While the mechanical strength is lower than the base material for larger W_{ab} values, there is a net positive effect on the mechanical strength with decreasing W_{ab}. As W_{ab} decreases, the mechanical strength increases such that in the region $W_{ab} < 1.0$ W there is the possibility of strengthening the material beyond the strength of the base material. Similar results with the internally melted single plate of fused silica with 10 ps are also reported [34]. These results are in accordance with the results using *fs* laser pulses, showing the strengthening of bulk glass by the irradiation of USPL that were evaluated by different testing procedures including the four-point bending test [35], the double torsion test [36] and nano-indentation [37, 38].

Laser welding system

Figure 11.22 shows the laser welding system of glass using USPL. Optical contact is strongly recommended to prevent the molten glass from flowing out of the laser-irradiated region. The optical contact is marked by a lack of reflections from the joining surfaces, since the gap between the plates is small enough to excite evanescent waves by an impinging light wave at

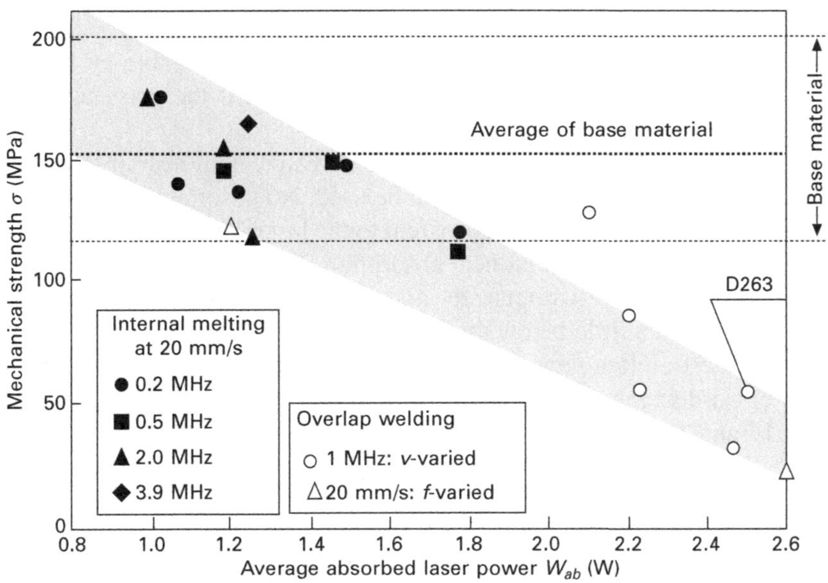

11.21 Mechanical strength of internally melted Foturan (single glass sample) and overlap-welded joint plotted in Foturan glass vs. average absorbed laser power W_{ab}. The result of overlap welding of D263 is also plotted [5].

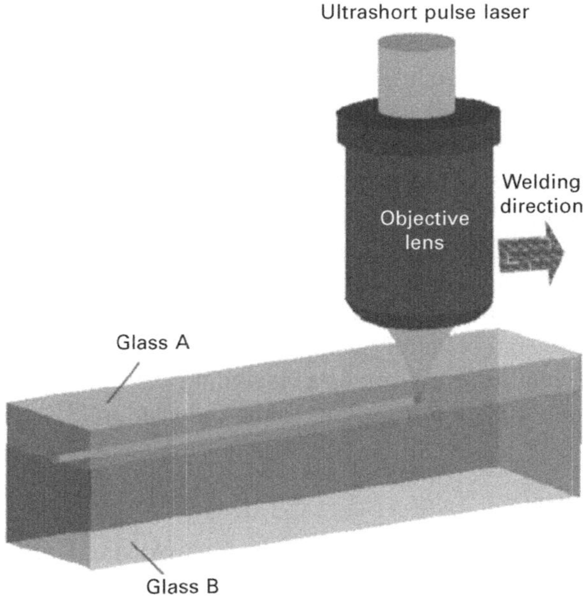

11.22 Schematic illustration of overlap welding of glass plates with optical contact. USPL is tightly focused by an objective lens for optical microscopy.

the surface of the first glass part, almost without attenuation, a propagating wave inside the other part [39]. The optical contact is available by facing the commercially available float glass plates together, if the glass surfaces are carefully cleaned.

For focusing USPL tightly into bulk glass to realize nonlinear absorption, an objective lens for the microscope can be used. No absorbent is needed in spite of the fact that the glass is transparent to the laser wavelength, since the laser energy is absorbed by nonlinear absorption. The welding can be carried out at room temperature without pre- and post-heating. The focus location should be adjusted a little below the interface of the glass plates in order to make the widest molten region at the interface, since the absorption region extends upward as the average absorbed power W_{ab} increases, as shown in Figs 11.17 and 11.19.

USPL welding of glass has a lot of advantages, which include

- welding is possible in a clean room due to no fume emission
- crack-free welding is possible independently of thermal expansion coefficient
- only the interface is selectively melted without absorbent
- high throughput welding is possible

- high mechanical strength of the weld joint is available with hermeticity.

Overlap welding

Figure 11.23 shows the peeled-off sample of overlap welding [9]. The fractured face does not show the original flat face but has an irregular appearance, suggesting that the two plates were strongly joined with vacuum tightness.

Figure 11.24 shows the cross section of multi-path welding with a separation of 15 μm for borosilicate glass [34]. No cracks were developed with overlaying the beads, showing excellent versatility of the USPL welding

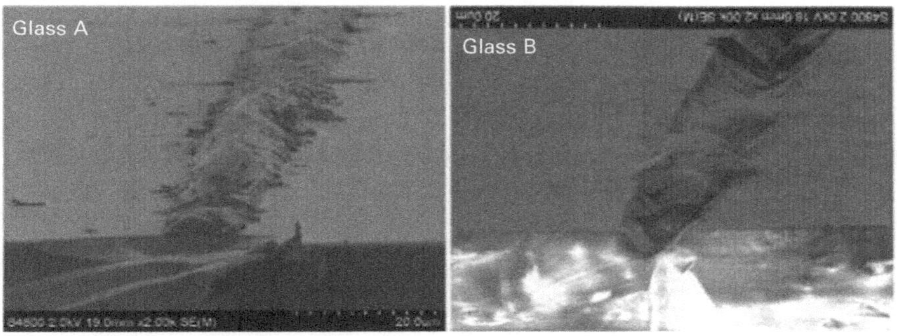

11.23 SEM photograph of the peeled-off weld joint in overlap welding of D263 [9].

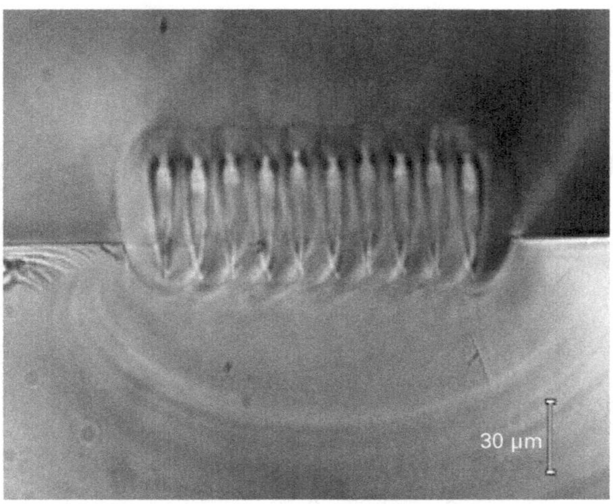

11.24 Multipath overlap welding of borosilicate glass of D263. The weld paths are overlapped with a separation of 15 μm (τ_p = 10 ps, f = 640 kHz, v = 20 mm/s) [34].

of glass and the possibility to increase the joint strength without restriction of the overlay number.

The mechanical strength of the overlap weld joint was evaluated by a shear test. In evaluating shear strength, the effect of the optical contact force F_{OC} has to be taken into account [5], since the attractive force caused by van der Waals interactions works at the interface [6]. Figure 11.25 shows the shear force F_{OC} required to break the optical contact between commercially available float glass plates without welding plotted as a function of optical contact area S_{OC}. The optical contact force per unit area σ_{OC} (= F_{OC}/S_{OC}) is approximately 1.5 MPa at L = 1 mm, and tends to decrease with increasing S_{OC}.

Overlap welding was performed with L = 10 mm corresponding to $F_{OC} \approx 88$N ($\sigma_{OC} \approx 0.6$ MPa), which is strong enough to keep optical contact in handling the sample for the welding operation [5]. While $\sigma_{OC} \approx 0.6$ MPa is approximately two orders smaller than the strength of the glass, L = 10 mm is also two orders larger than the typical width of the weld bead. Therefore the strength of the weld joint σ_W was determined by subtracting the optical contact force F_{OC} from the rupture load F_{RUP}, and then dividing by the cross-sectional area of the weld bead S_W as is given by the following equation [5, 39]:

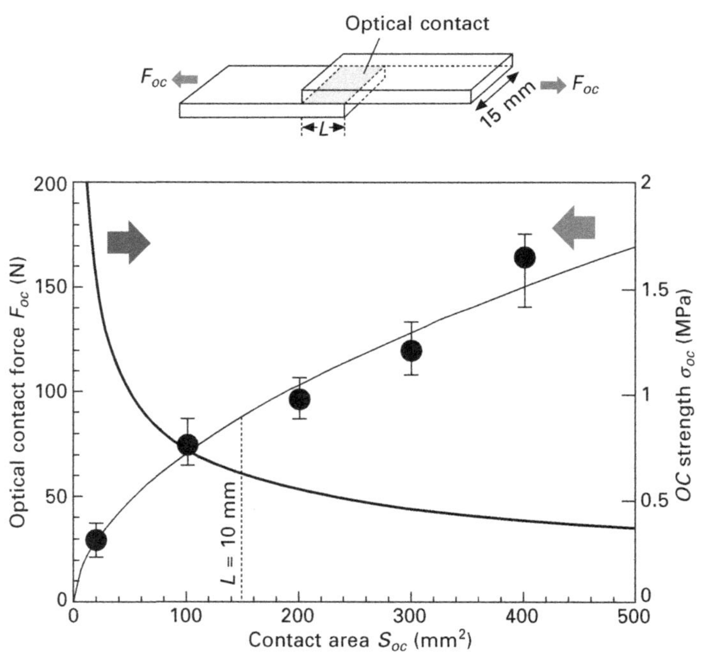

11.25 Shear force to break optical contact F_{OC} plotted vs. contact area S_{OC}. The optical contact force σ_{OC} is also plotted [5].

$$\sigma_W = \frac{F_{RUP} - F_{OC}}{S_W} \qquad [11.4]$$

Overlap welding was performed at different pulse repetition rates and translation speeds. Figure 11.26(a) shows the rupture load F_{RUP} and the welded area S_W plotted as a function of translation speed v at a constant pulse repetition rate of f = 1MHz. Figure 11.26(b) shows F_{RUP} and S_W plotted as a function of pulse repetition rate at a constant translation speed of v = 20 mm/s. It is noticeable that the rupture strength F_{RUP} is increasing in spite of the fact that the width of the weld bead decreases as the translation speed v or pulse repetition rate f increases.

The weld joint strength σ_W was calculated by Eq. [11.4] using the data plotted in Fig. 11.26, and is plotted vs. average absorbed laser power W_{ab} in Fig. 11.21. Although some data scattering is found, most data points of the overlap weld joint fall in the zone that highlights the data from the internal melt single plate studies, suggesting that higher joint strength can also be obtained at reduced W_{ab}.

11.5 Conclusion and future trends

Although glass is widely used in a variety of industrial fields due to its excellent physical and chemical properties, the welding technology of glass has not been used widely in industry, and the assembly of glass parts is actually dependent on technologies other than welding at the moment. This is because only CO_2 lasers have been used for long time, and crack-free welding is available only in glass having low thermal expansion coefficient such as fused silica or Vycol due to cracks caused by the shrinkage stress.

It is expected that such a situation will be changed dramatically by the recently developed novel welding technology of glass using ultrashort pulse laser (USPL). This welding technology has been attracting much attention, because of many of the advantages discussed above. Among these, most importantly cracks can be prevented, and hence most commercially available glass can be welded without pre- and post-heating independently of the thermal expansion coefficient. It is expected USPL welding technology will become popular in glass-made products in a variety of micro applications such as electronics packaging, MEMS and micro optics in different industrial fields, as is the case for laser welding technologies for metal. Fundamental research work is, however, still needed in parallel to developing industrial applications of USPL welding of glass, since the laser–matter interaction during the welding process has not yet been fully understood.

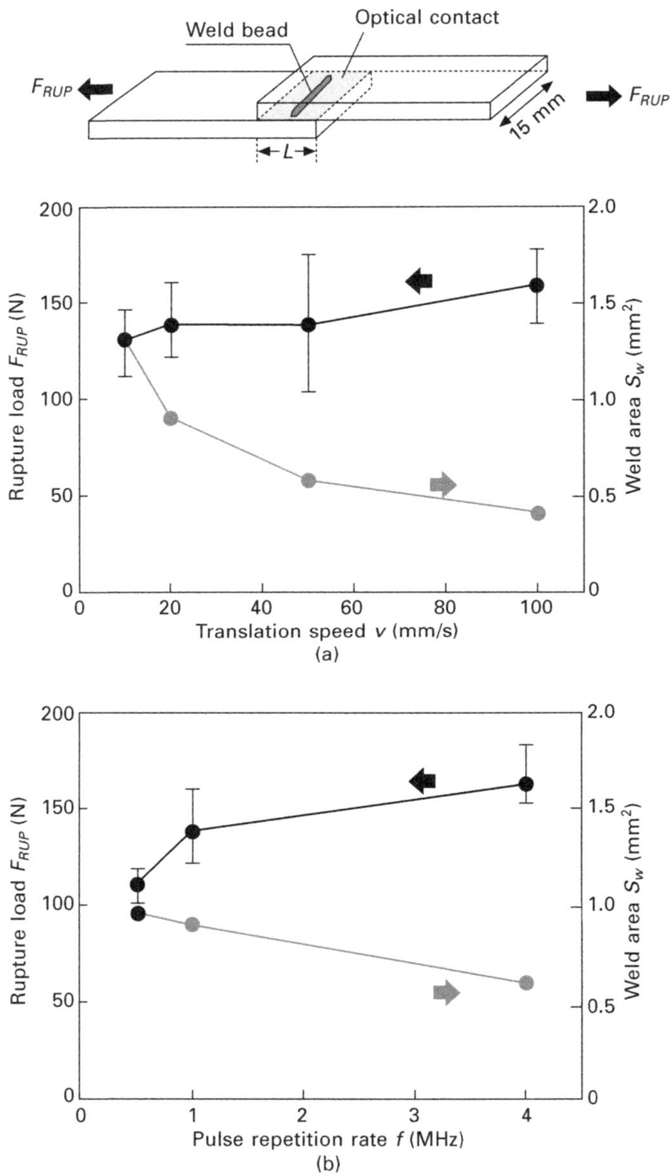

11.26 Effect of (a) translation speed at f =1 MHz and (b) pulse repetition rate at v = 20 mm/s on the rupture strength and weld area of the overlap-weld samples. (L = 10 mm, length of weld path = 8~9 mm) [5].

11.6 References

1. L. Tong, J.K. Speit, G. Fernlund, *Recent Advances in Structural Joints*, Dordrecht: Kluwer Academic Publishers, 27–66 (2003).

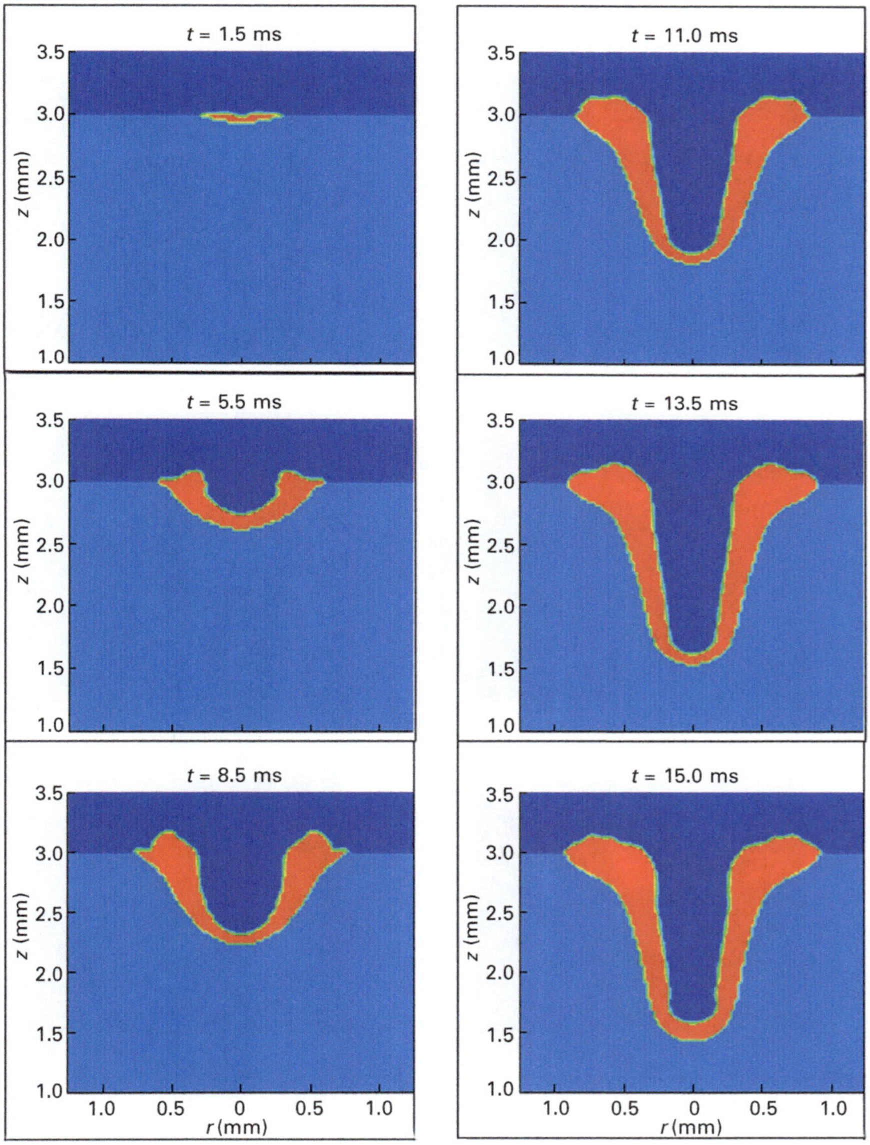

Plate I Keyhole formation in a pulsed laser welding process.

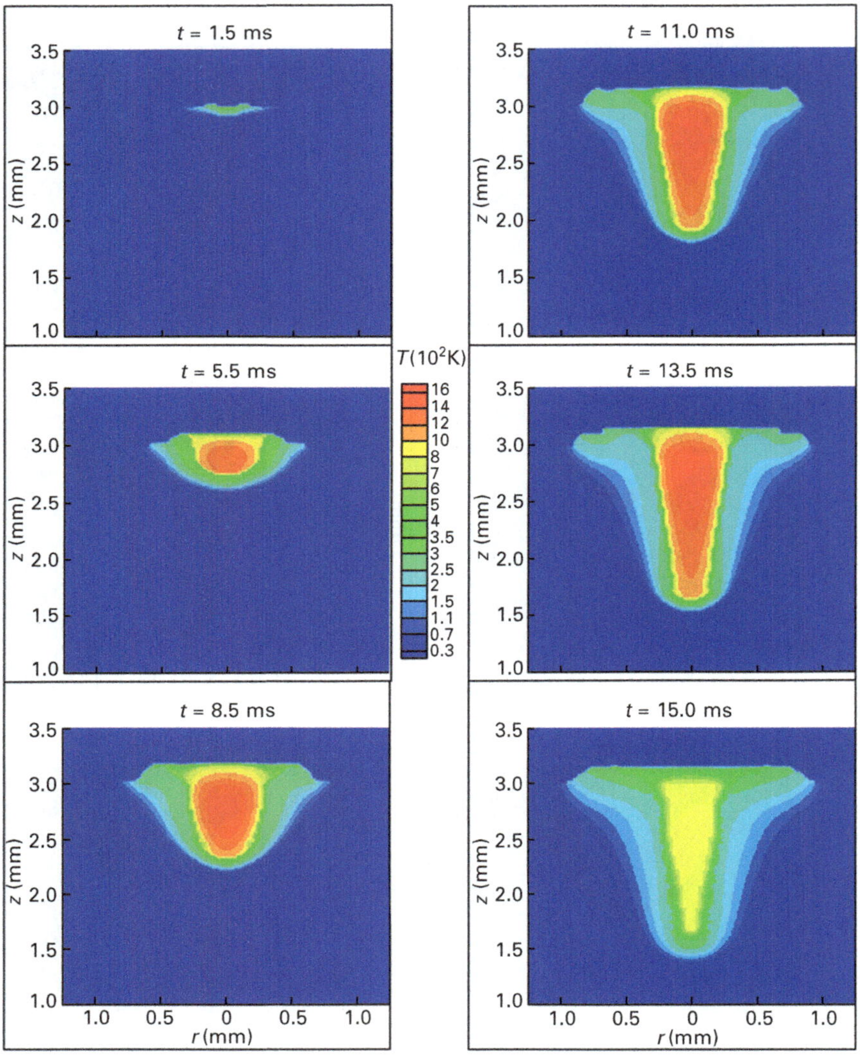

Plate II Keyhole plasma formation and corresponding temperature distributions as shown in Plate I.

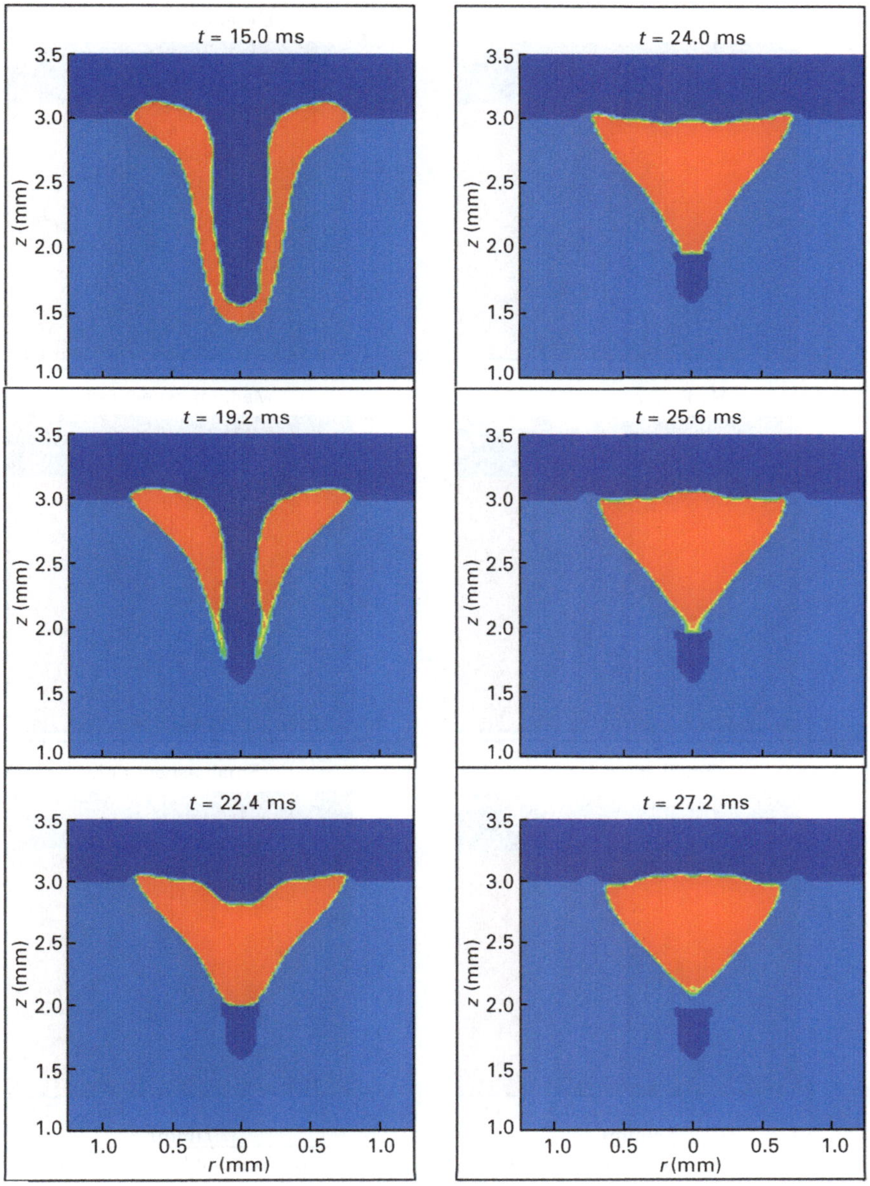

Plate III A typical keyhole collapse process in pulsed laser keyhole welding.

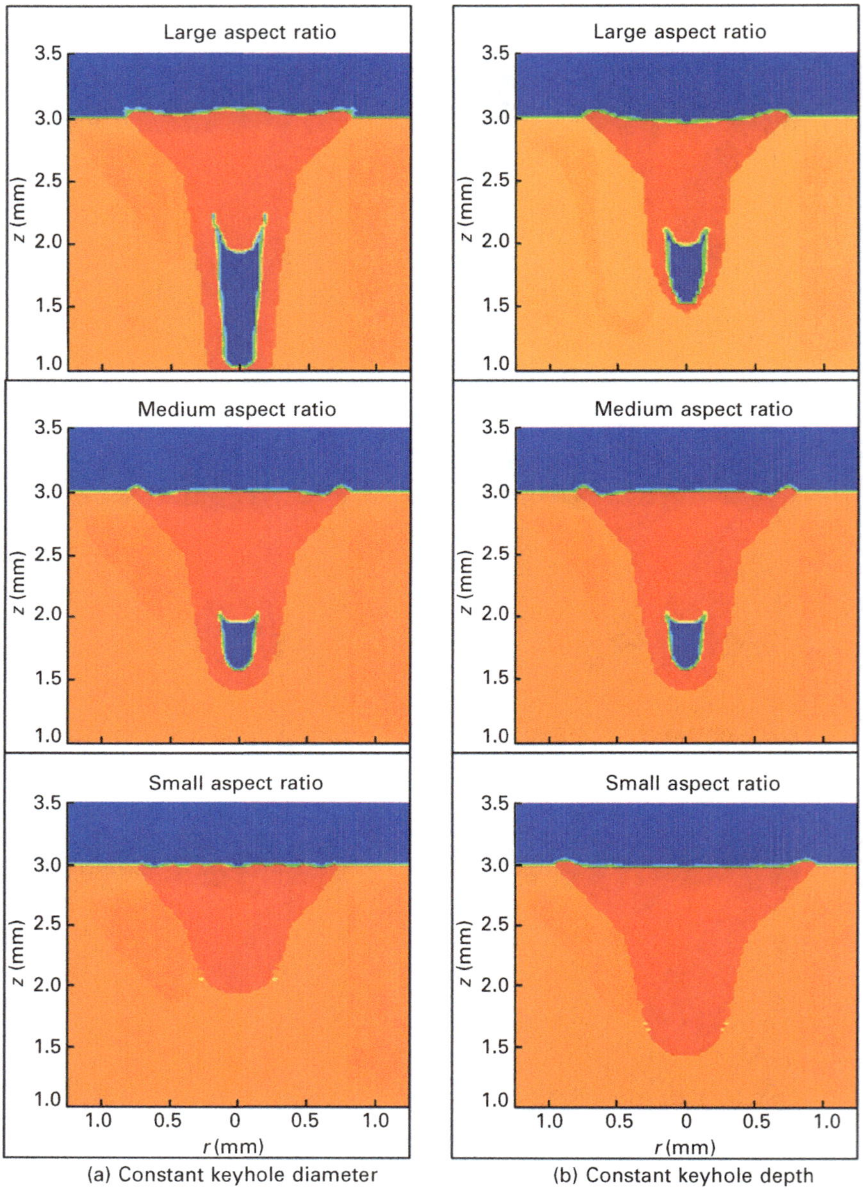

Plate IV Effect of depth-to-width ratio of the keyhole on porosity formation in laser keyhole welding.

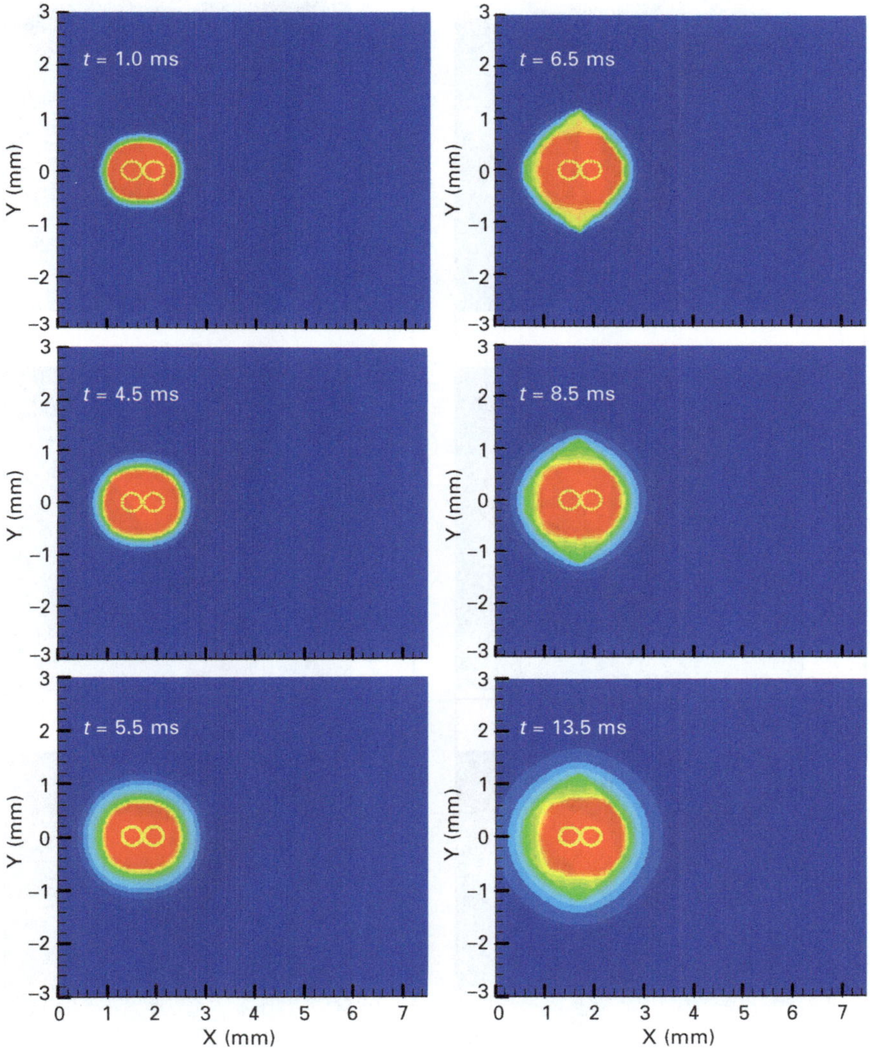

Plate V Weld pool dynamics and temperature evolutions in dual-beam laser keyhole welding.

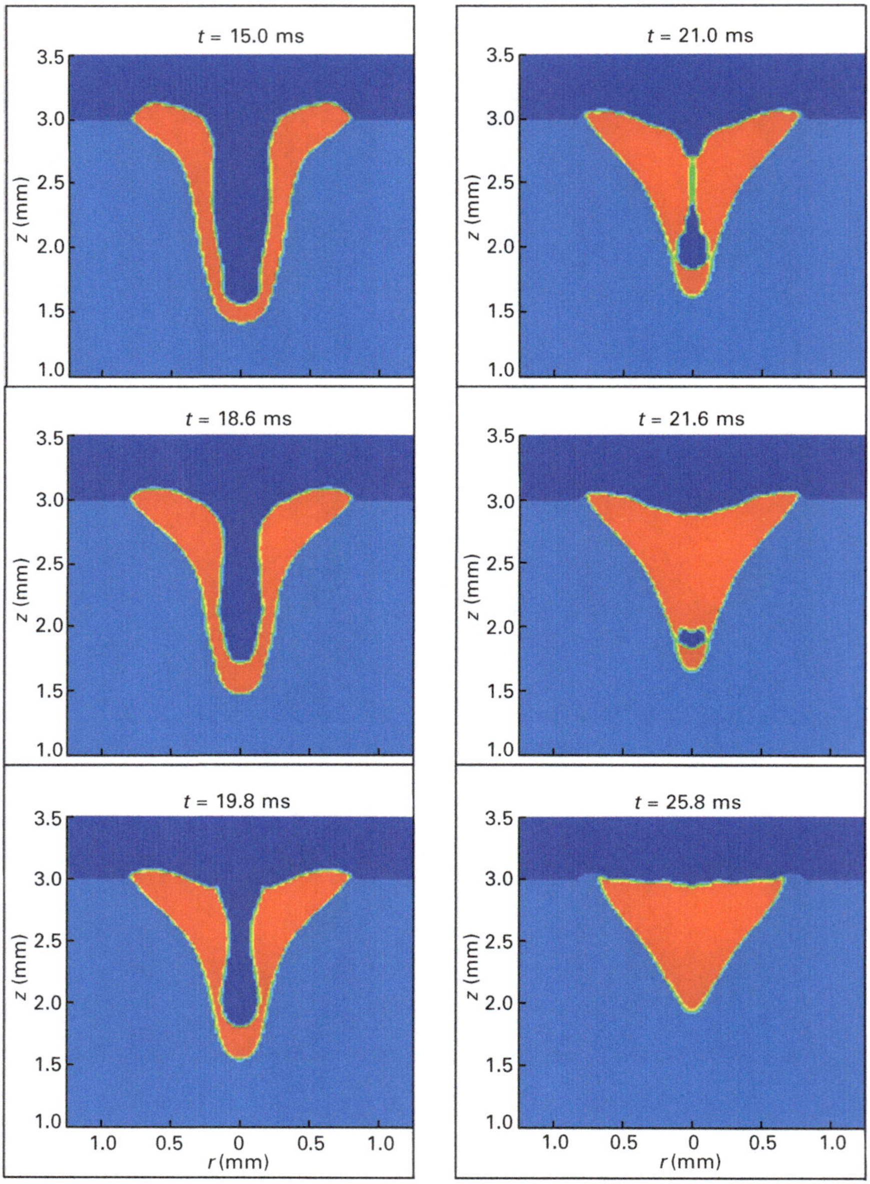

Plate VI Keyhole collapse and porosity elimination in a medium depth-to-width ratio keyhole laser welding with pulse control.

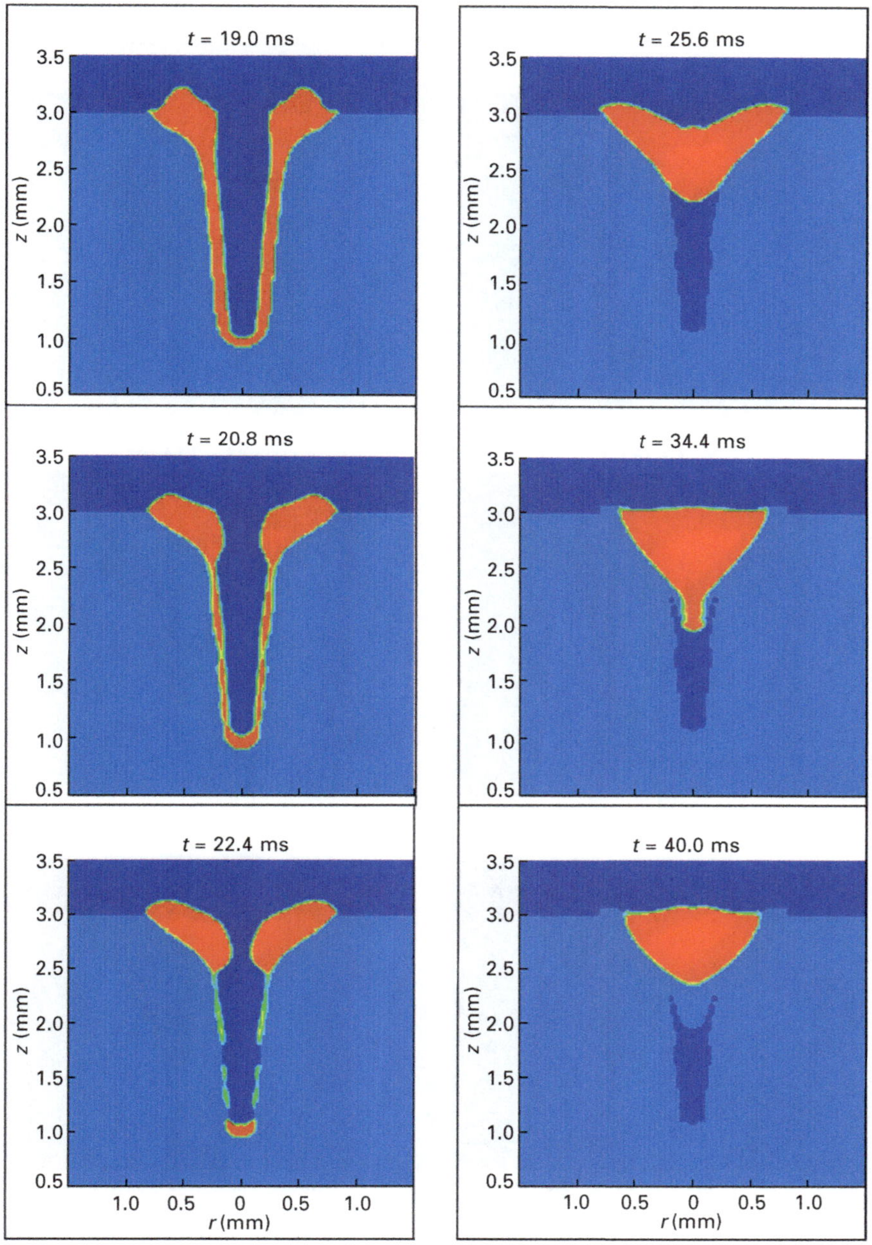

Plate VII Effect of pulse control on keyhole collapse and porosity prevention in a large depth-to-width ratio keyhole laser welding process.

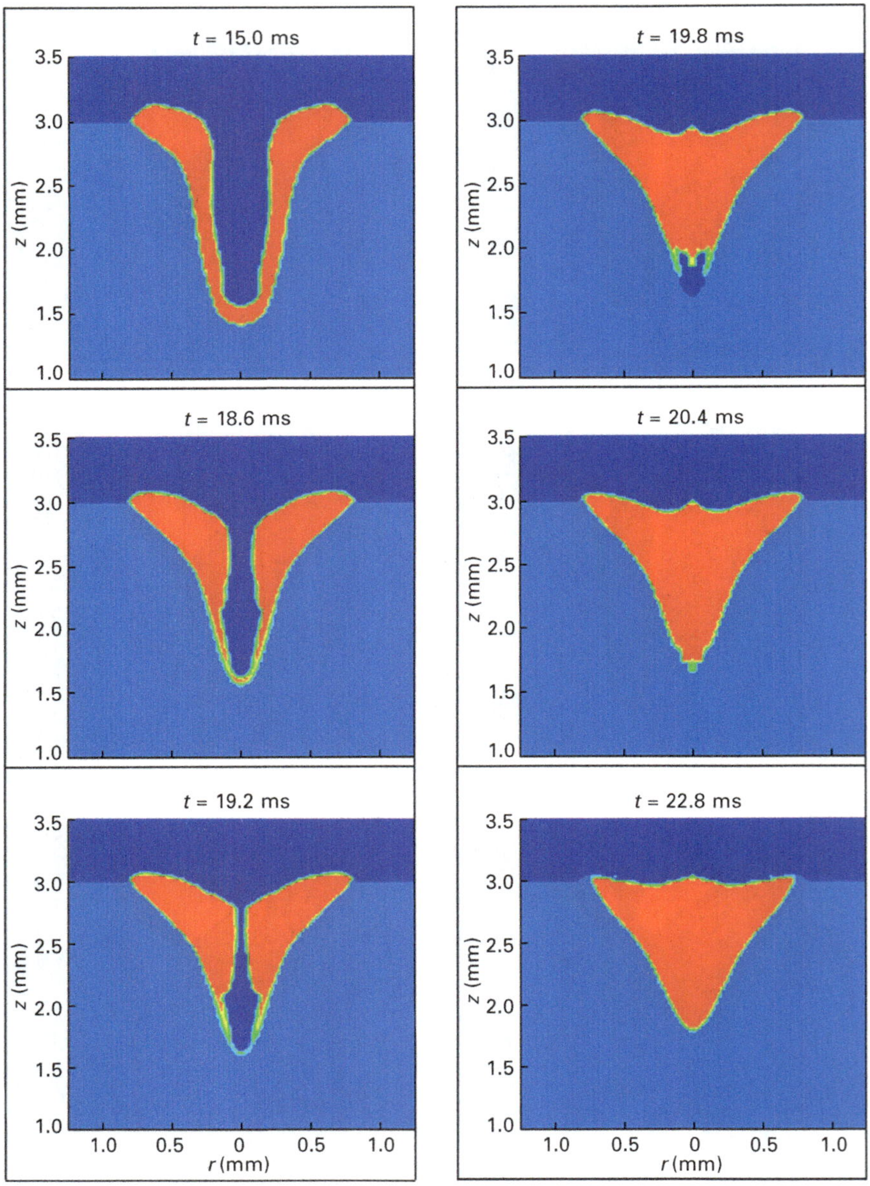

Plate VIII Effect of external electromagnetic force on keyhole collapse and porosity prevention in a medium depth-to-width ratio keyhole laser welding process.

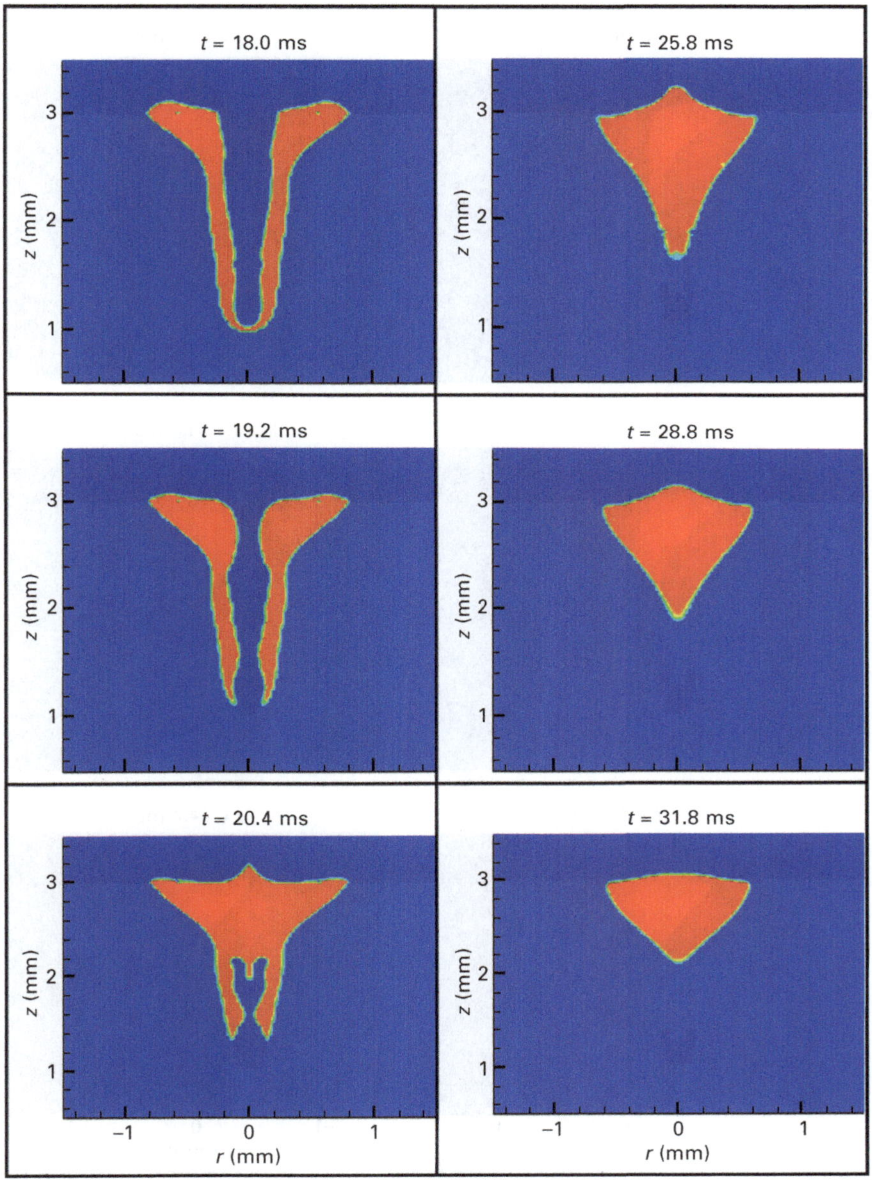

Plate IX Effect of applying a small electromagnetic force on porosity prevention in a large depth-to-width ratio keyhole laser welding process.

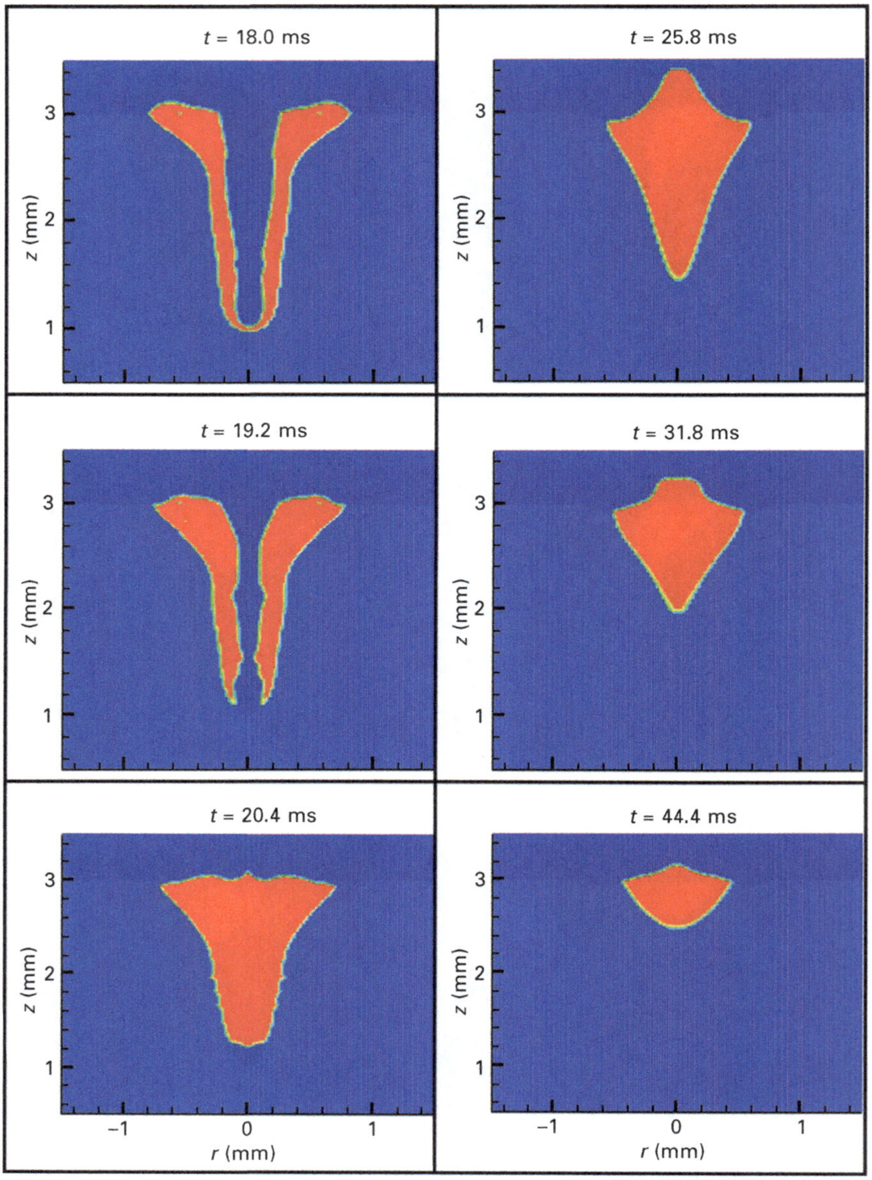

Plate X Effect of applying a large electromagnetic force on porosity prevention in a large depth-to-width ratio keyhole laser welding process.

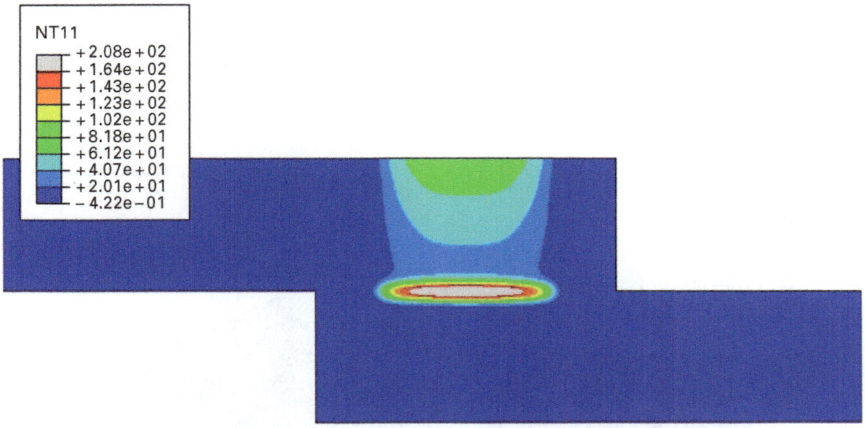

Plate XI Typical temperature distribution for a laser welded specimen resulting from the finite element analysis.

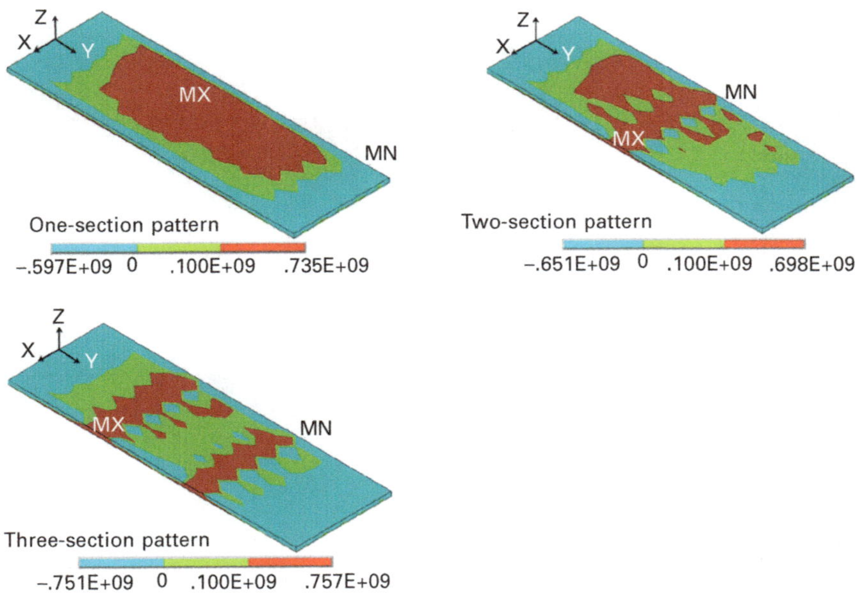

Plate XII Computed effect of deposition pattern on residual stress distribution (from Foroozmehr F and Kovacevic R (2009) 'Effect of path planning on residual stress distribution in laser powder deposition process', *Proceedings of ICALEO'09*, 830–839).

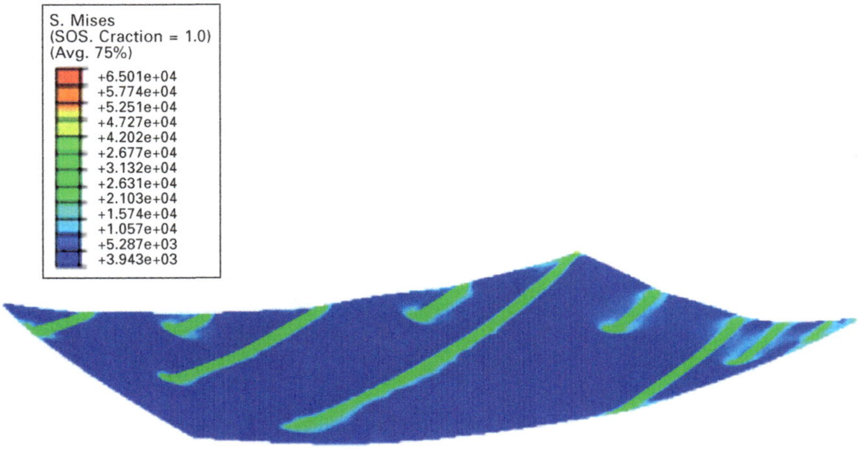

Plate XIII Computed deformation of plate after forming by CO_2 laser (from Romero P, Otero N, Pintos V, Arias R (2011) 'Experimental analysis of automated laser forming for shipbuilding industry', *Proceedings of ICALEO'II*, 1272–1279).

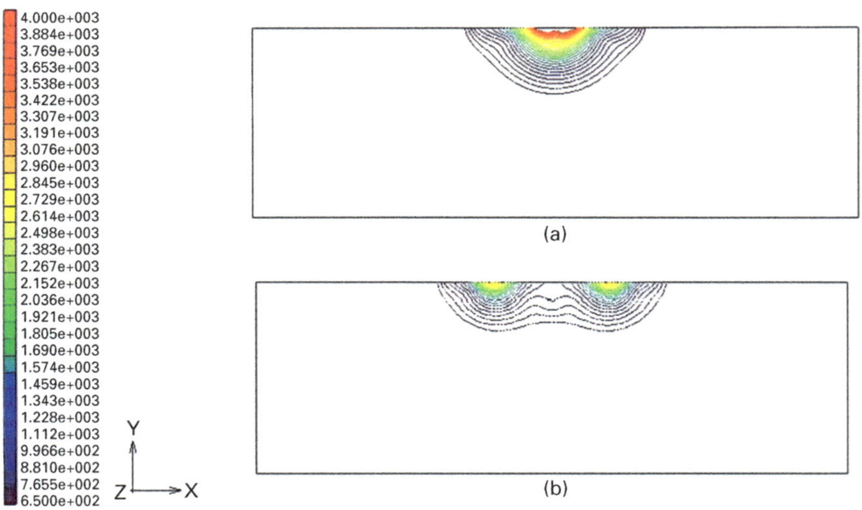

Plate XIV Weld pool shape in the cross section perpendicular to the laser scanning direction obtained by FEM analysis: (a) single-beam irradiation, (b) parallel twin-beam irradiation.

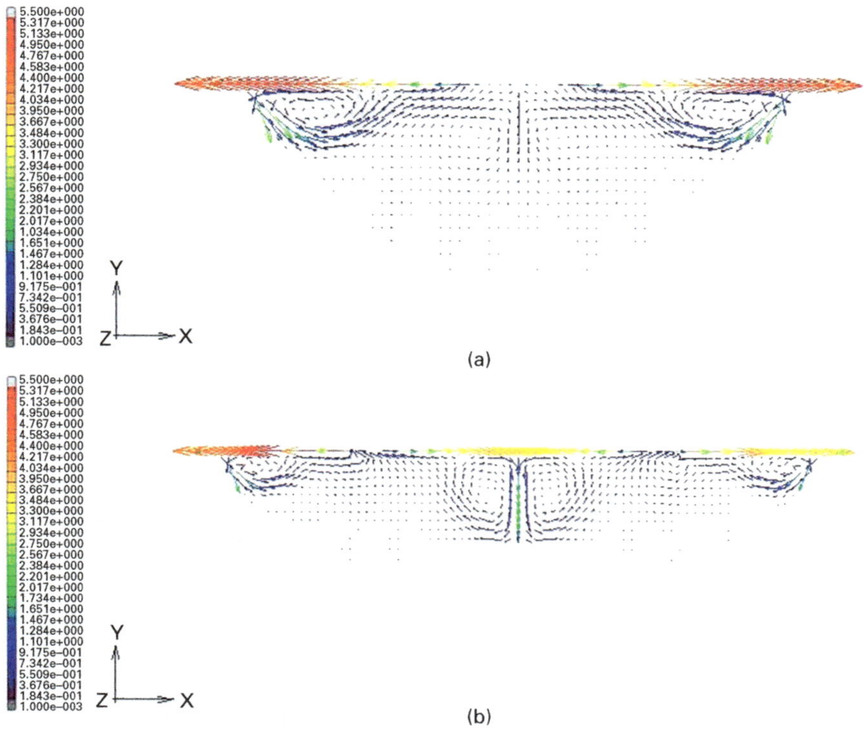

Plate XV Flow pattern in the cross section perpendicular to the laser scanning direction obtained by FEM analysis: (a) single-beam irradiation, (b) parallel twin-beam irradiation.

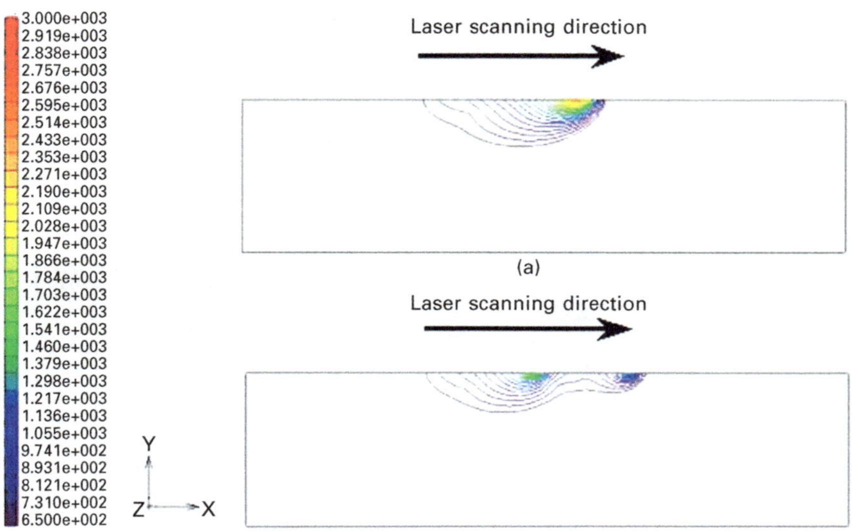

Plate XVI Weld pool shape in the cross section parallel to the laser scanning direction obtained by FEM analysis: (a) single-beam irradiation, (b) parallel twin-beam irradiation.

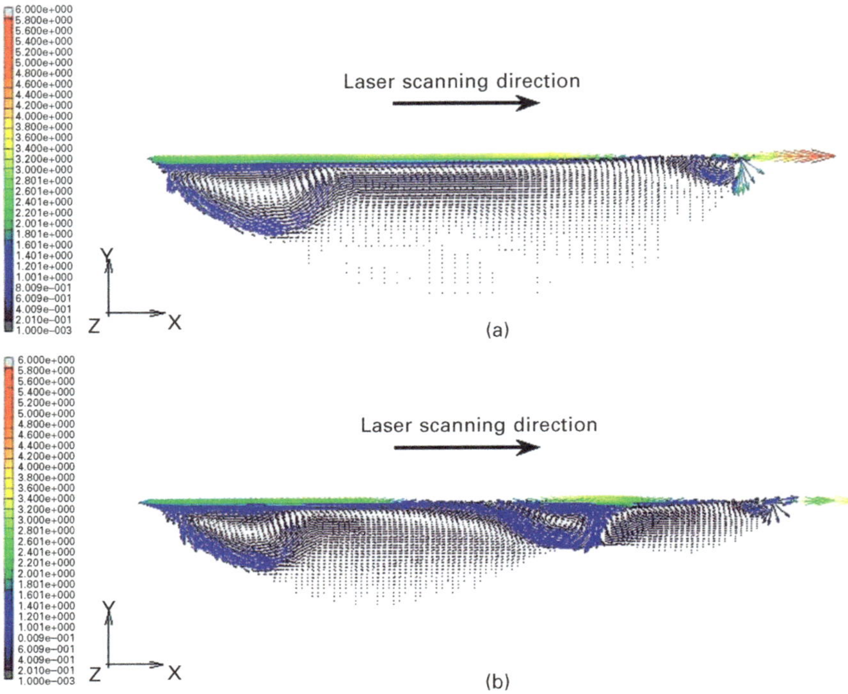

Plate XVII Flow pattern in the cross section parallel to the laser scanning direction obtained by FEM analysis: (a) single-beam irradiation, (b) parallel twin-beam irradiation.

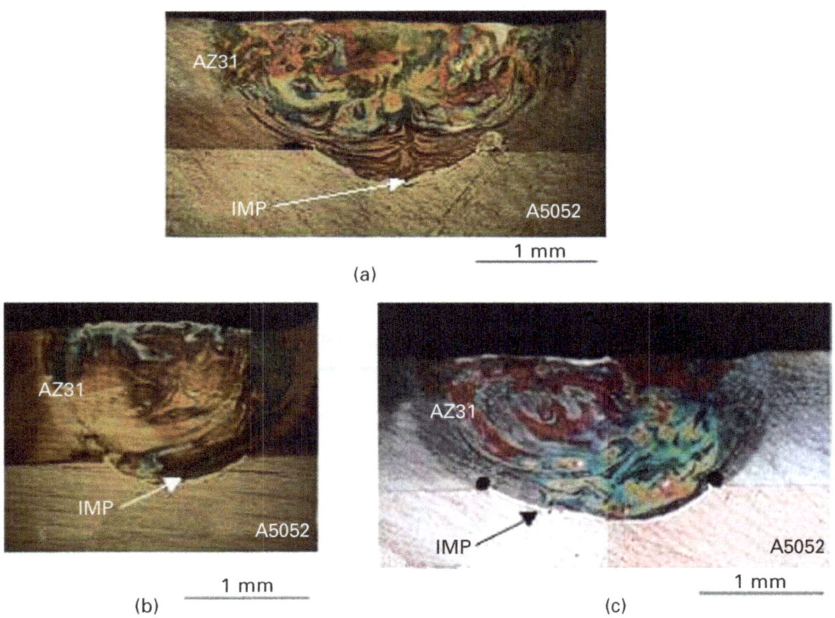

Plate XVIII Cross-sectional observations of AZ31/A5052 dissimilar materials direct joint: (a) single beam, (b) twin beam (wedge angle 2°) and (c) twin beam (wedge angle 4°).

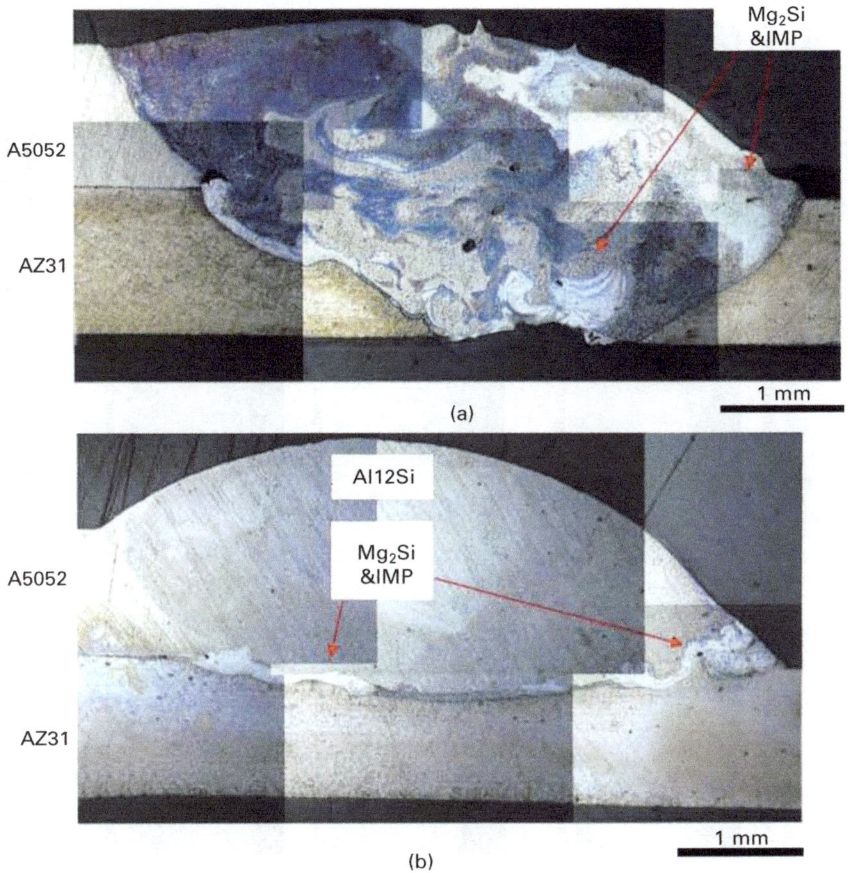

Plate XIX Cross-sectional observations of the AZ31/A50502 dissimilar materials joints welded with (a) single beam and (b) twin beam in the welding with filler wire.

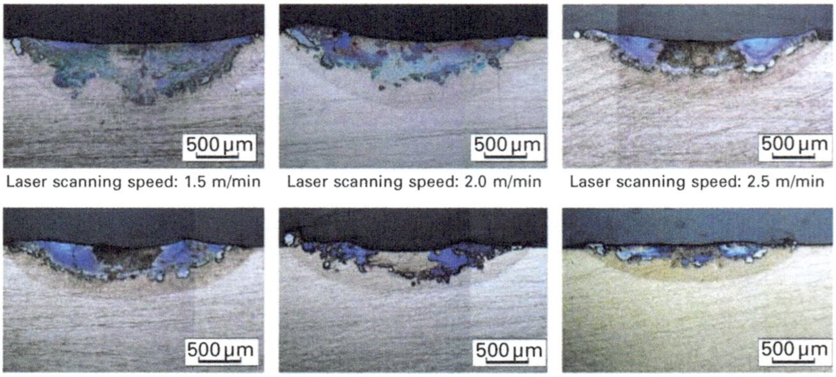

Laser scanning speed: 1.5 m/min | Laser scanning speed: 2.0 m/min | Laser scanning speed: 2.5 m/min

Laser scanning speed: 3.0 m/min | Laser scanning speed: 3.5 m/min | Laser scanning speed: 4.5 m/min

(a)

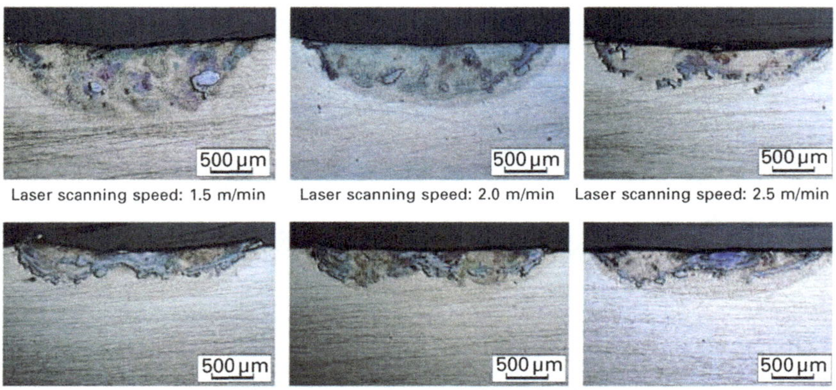

Laser scanning speed: 1.5 m/min | Laser scanning speed: 2.0 m/min | Laser scanning speed: 2.5 m/min

Laser scanning speed: 3.0 m/min | Laser scanning speed: 3.5 m/min | Laser scanning speed: 4.5 m/min

(b)

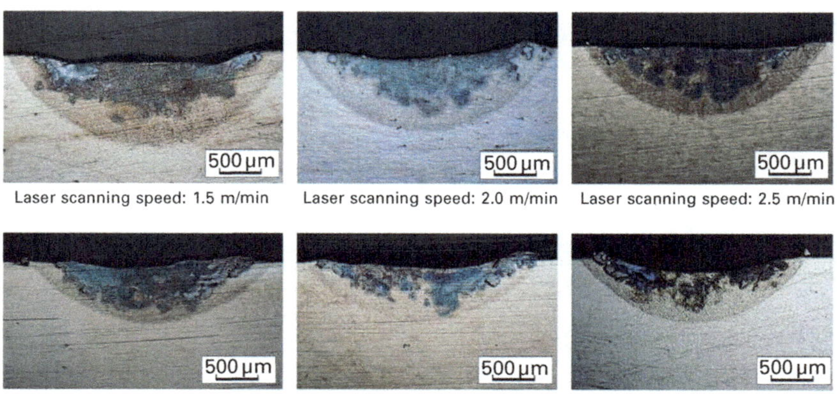

Laser scanning speed: 1.5 m/min | Laser scanning speed: 2.0 m/min | Laser scanning speed: 2.5 m/min

Laser scanning speed: 3.0 m/min | Laser scanning speed: 3.5 m/min | Laser scanning speed: 4.5 m/min

(c)

Plate XX Cross-sectional observations of cladded specimen: (a) single-beam irradiation, (b) parallel-beam irradiation and (c) tandem-beam irradiation.

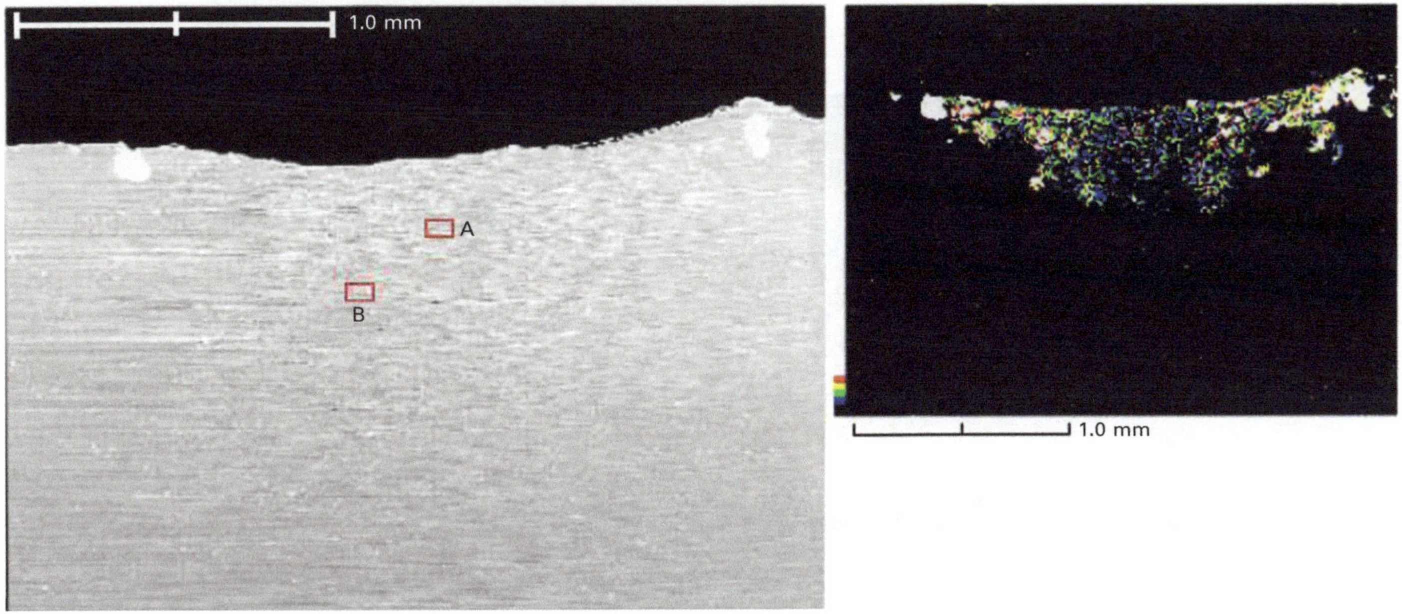

Plate XXI Scanning electron microscope image and Si distribution at molten zone obtained by the tandem-beam cladding.

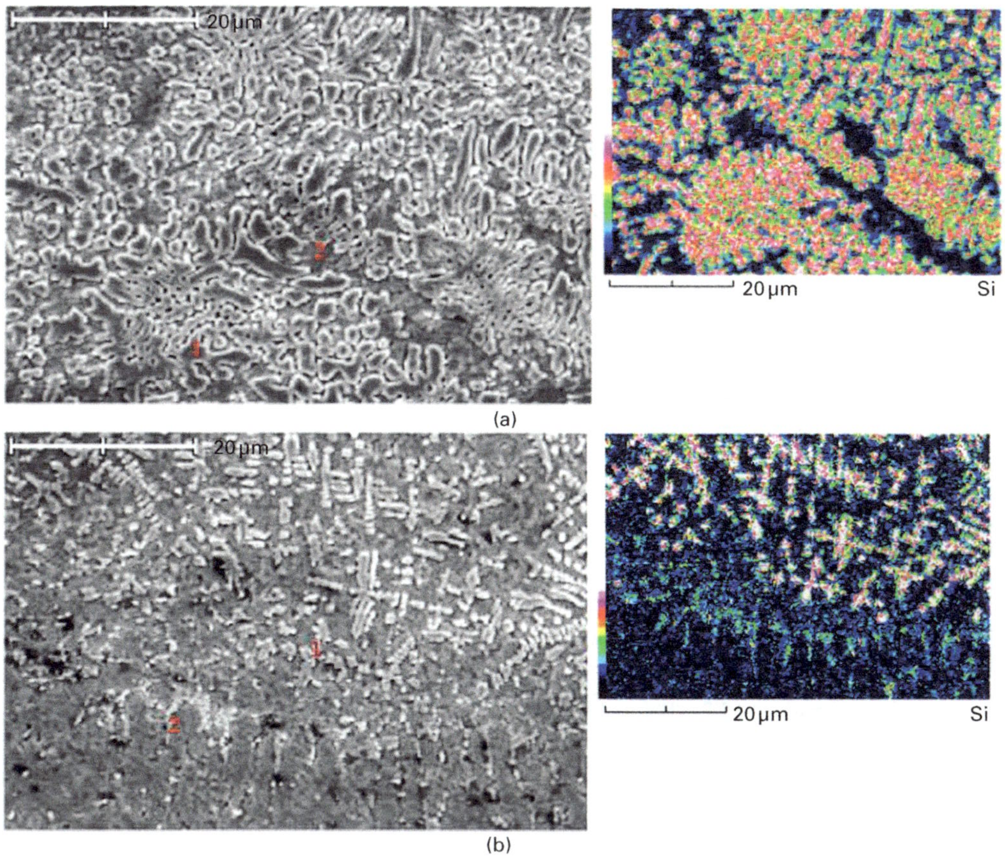

Plate XXII Enlarged images of Plate XXI showing Si distribution: (a) area 'A' and (b) area 'B'.

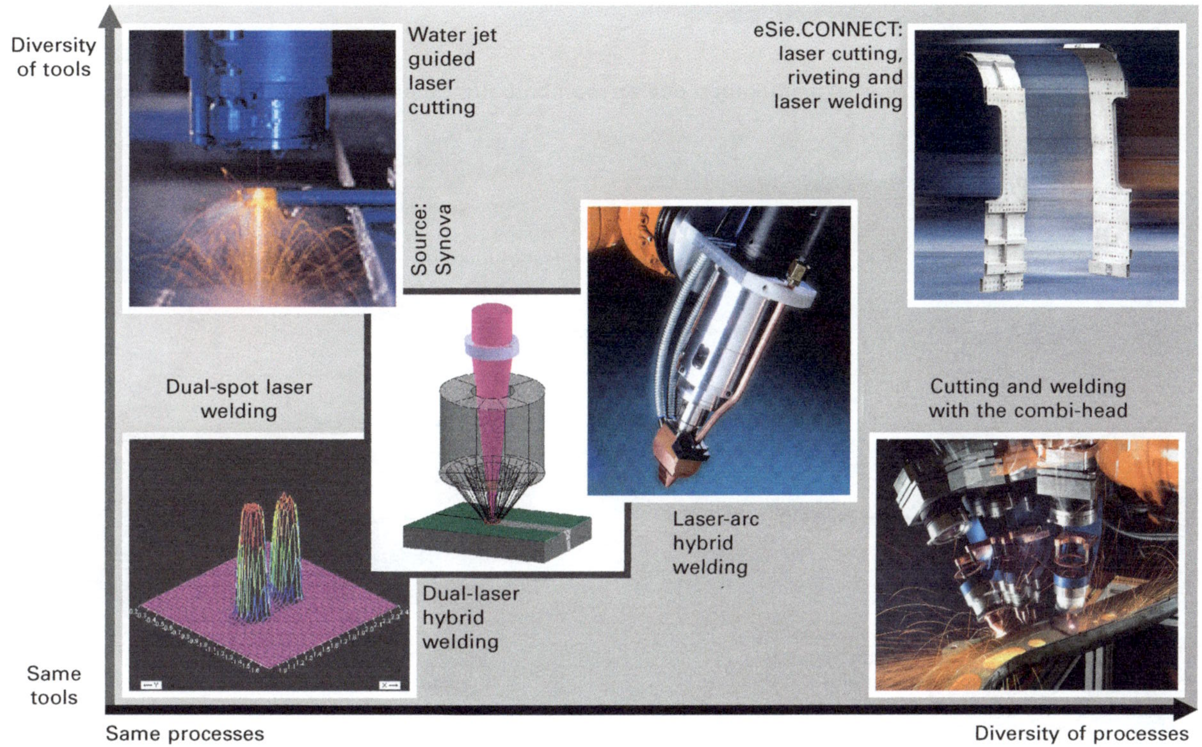

Plate XXIII Map of combined laser processes and tools.

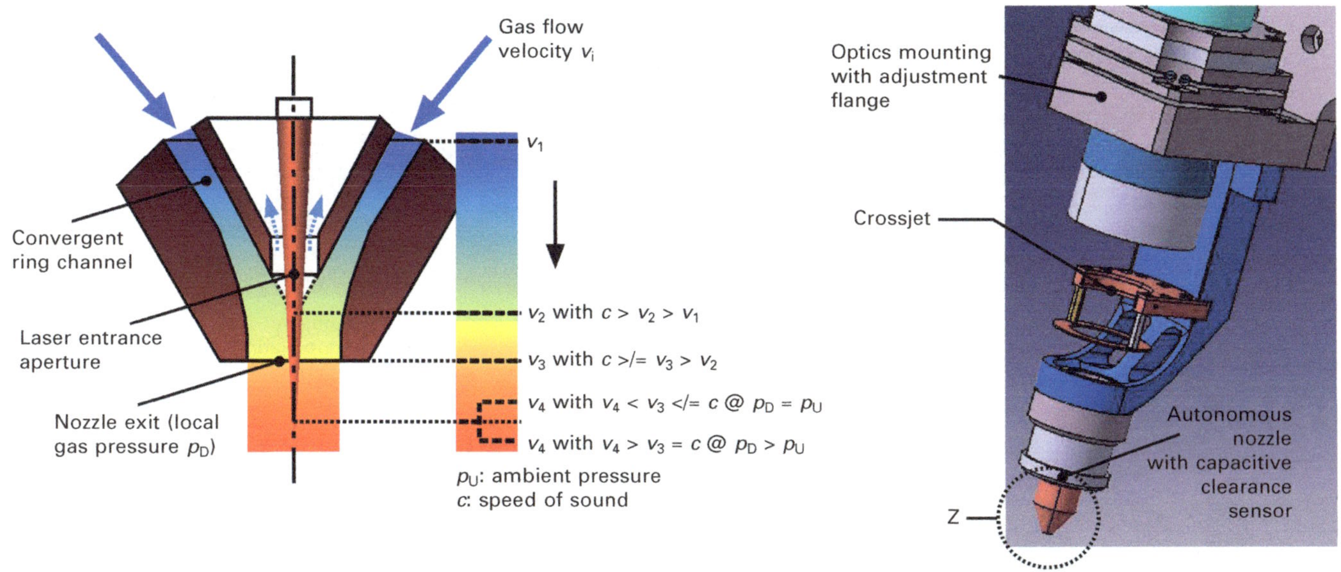

(a) (b)

Plate XXIV Scheme and principle of the autonomous nozzle and combi-head.

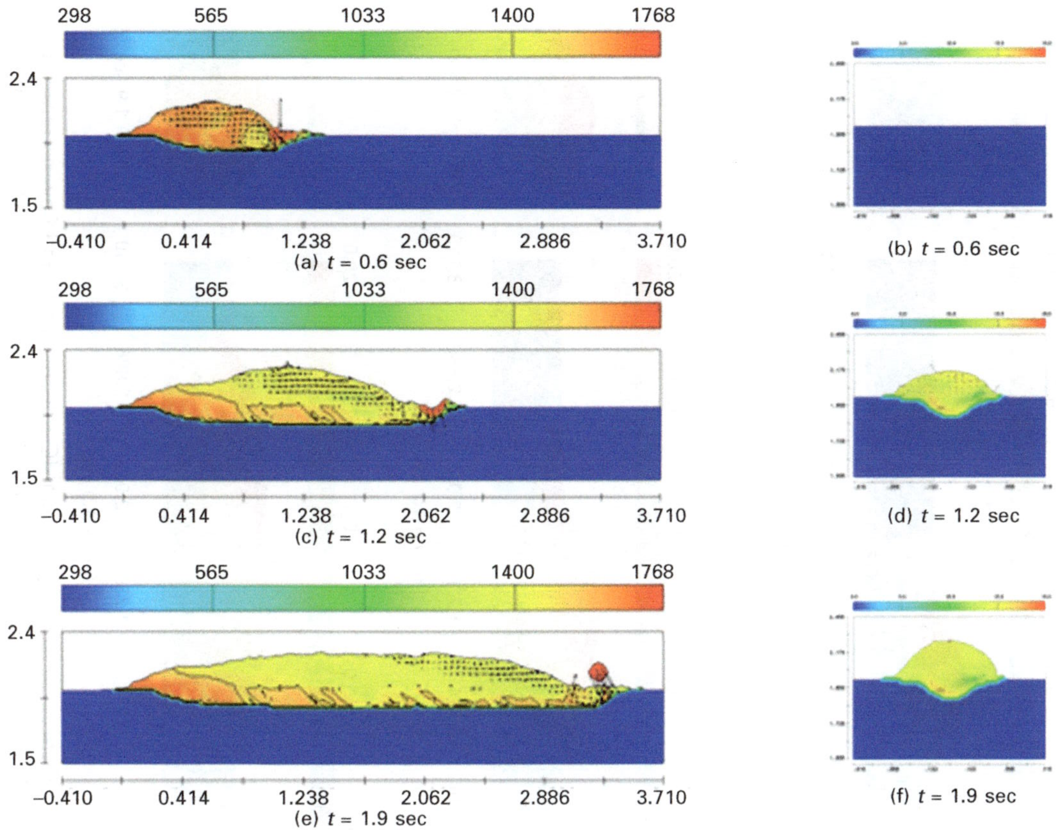

Plate XXV Temperature profiles and flow patterns in GMA welding (cross-sectional side view: a, c, e and cross-sectional front view: b, d, f).

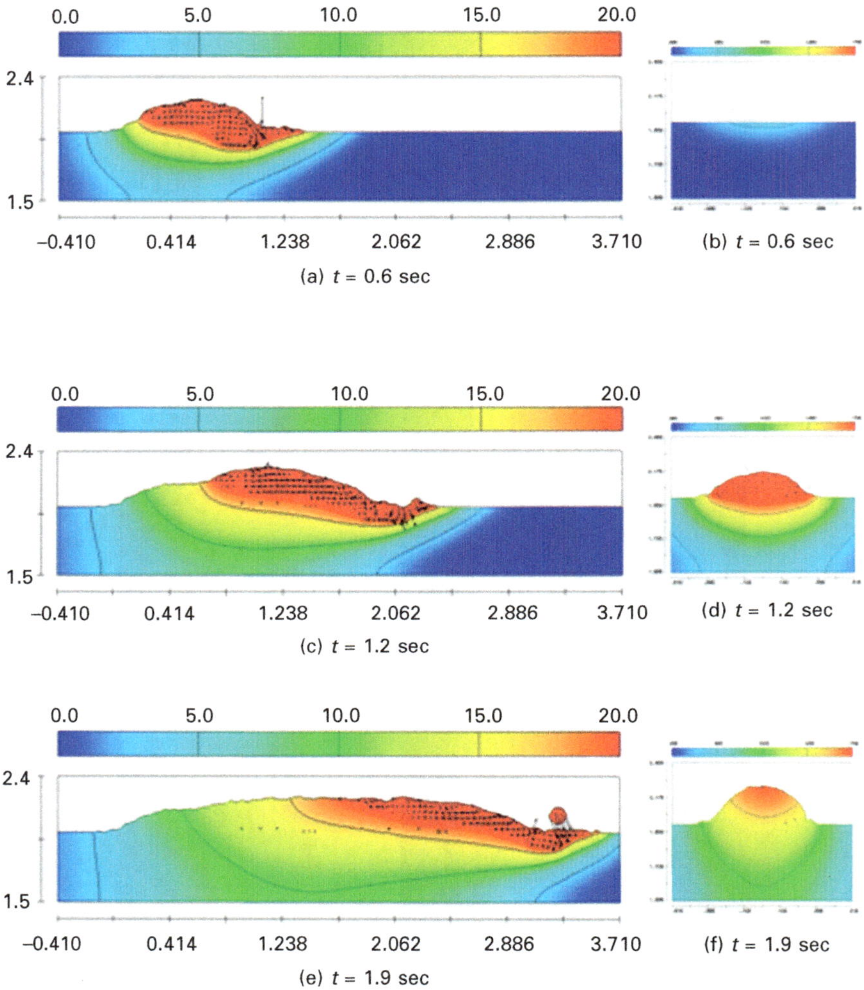

Plate XXVI Alloying element (Cr) distributions and flow patterns in GMA welding (cross-sectional side view: a, c, e and cross-sectional front view: b, d, f).

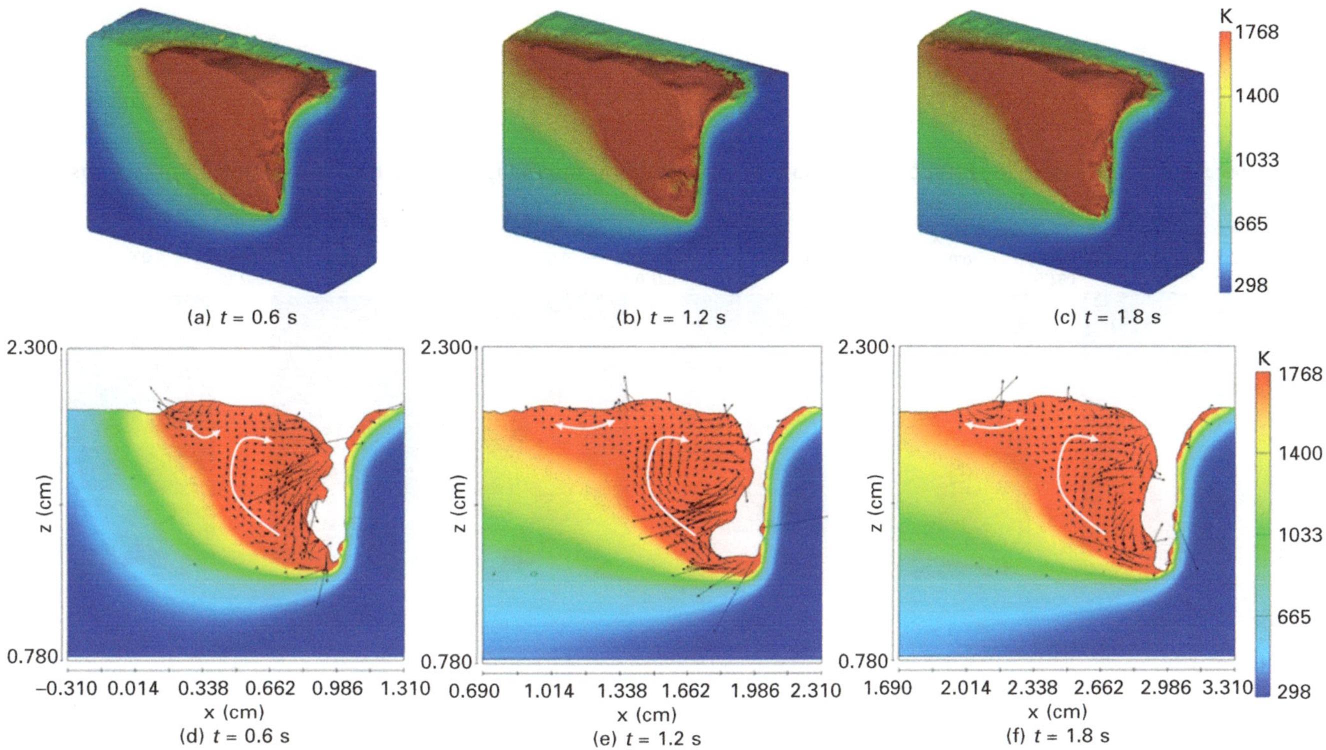

Plate XXVII (a)-(c) Calculated temperature profiles and flow patterns in a three-dimensional cross-sectional view; (d)–(f) cross-sectional side view during laser welding.

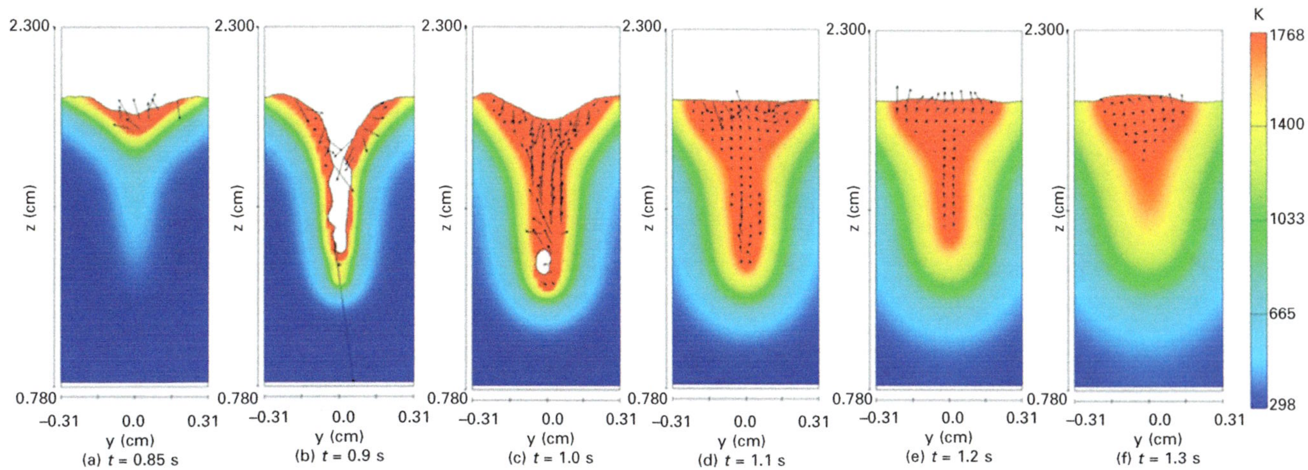

Plate XXVIII Calculated temperature profiles and flow patterns in a cross-sectional front view during laser welding.

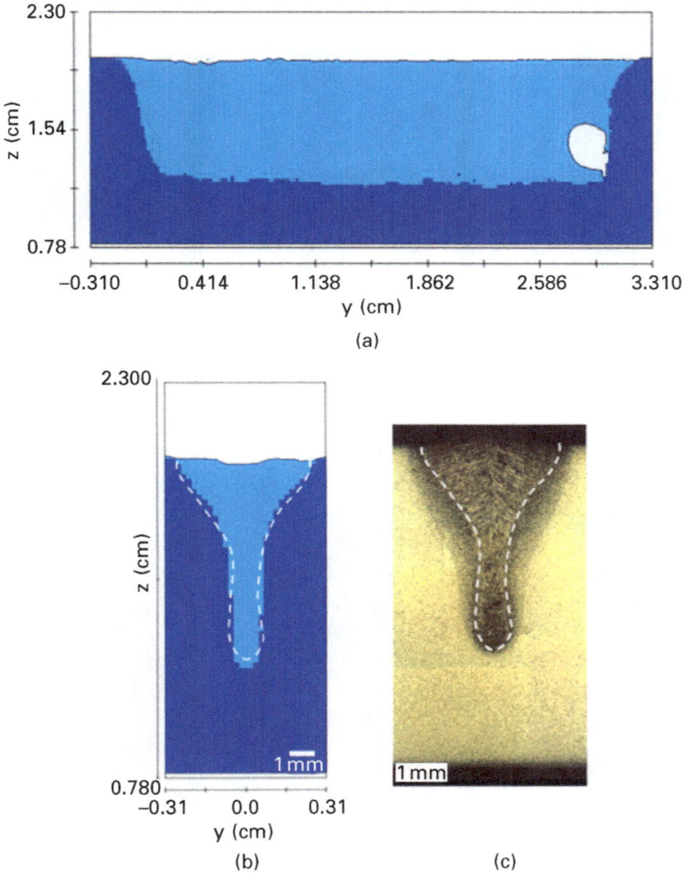

Plate XXIX (a) and (b) Calculated fusion zone profiles and (c) an experimental fusion zone profile.

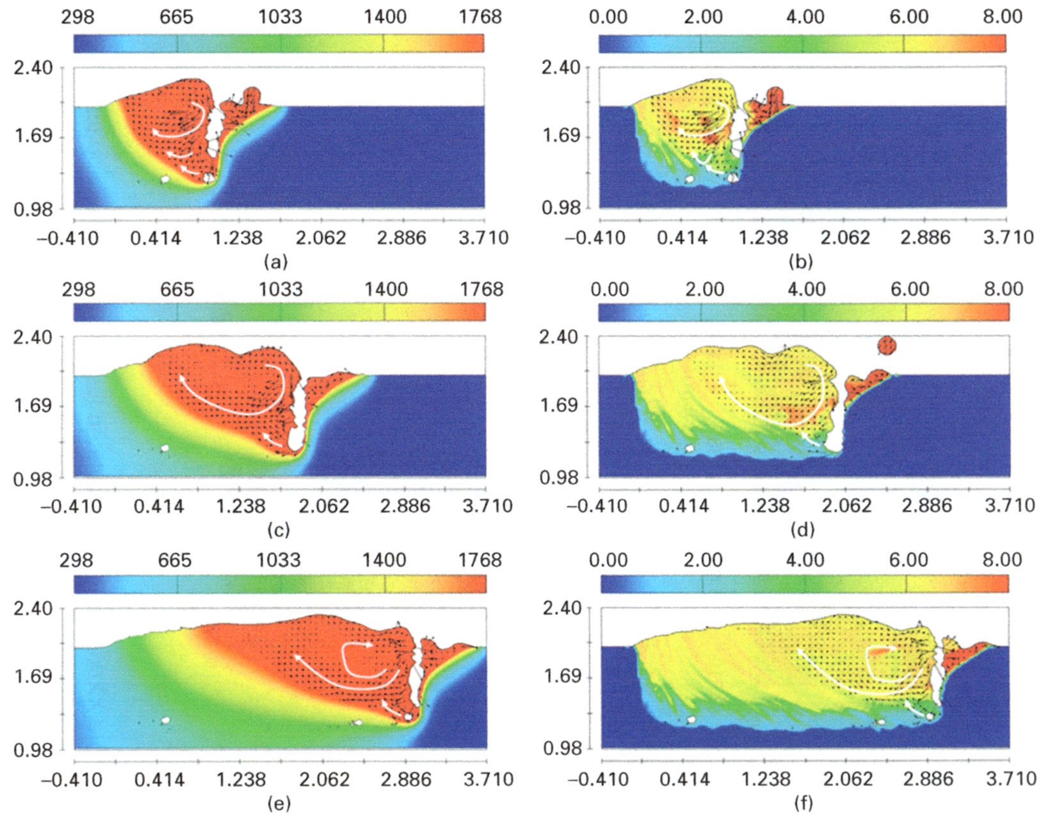

Plate XXX Simulation results in longitudinal cross sections during hybrid welding. (a) temperature profile at t = 0.6 sec, (b) Cr profile at t = 0.6 sec, (c) temperature profile at t = 1.2 sec, (d) Cr profile at t = 1.2 sec, (e) temperature profile at t = 1.8 sec and (f) Cr profile at t = 1.8 sec.

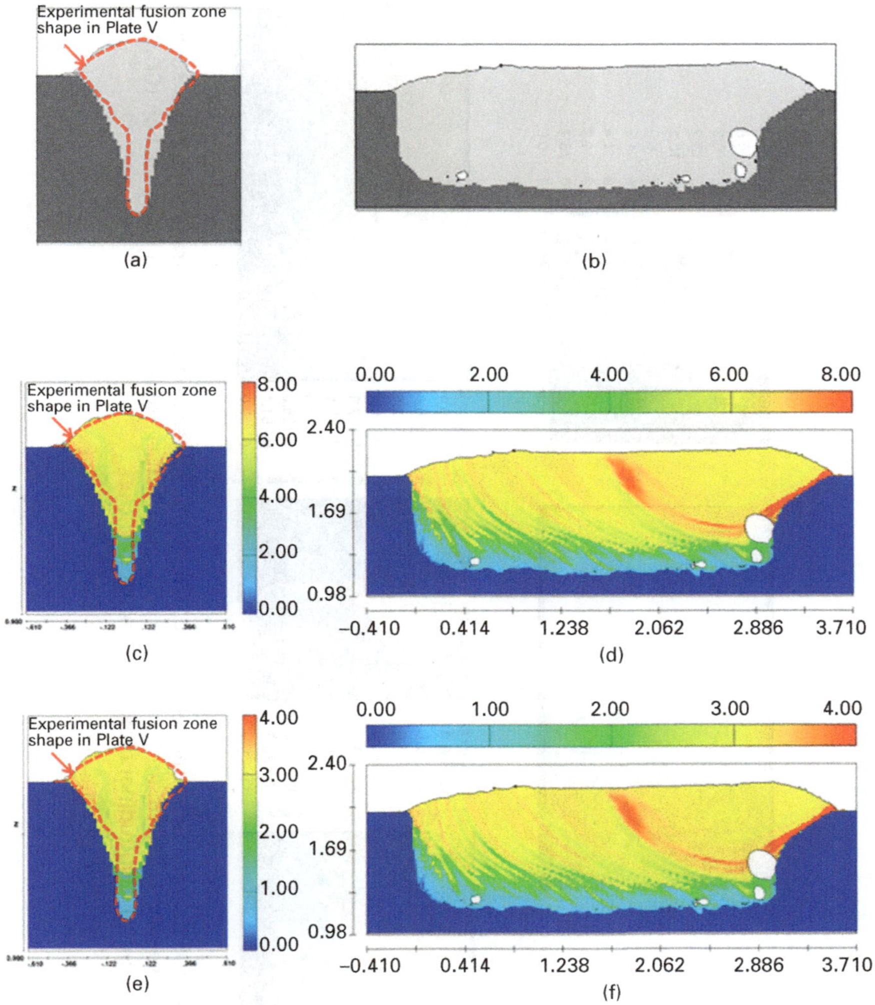

Plate XXXI Simulation results in transverse and longitudinal cross sections after solidification: (a) transverse cross section (fusion zone), (b) longitudinal cross section (fusion zone), (c) transverse cross section (Cr), (d) longitudinal cross section (Cr), (e) transverse cross section (Ni) and (f) longitudinal cross section (Ni).

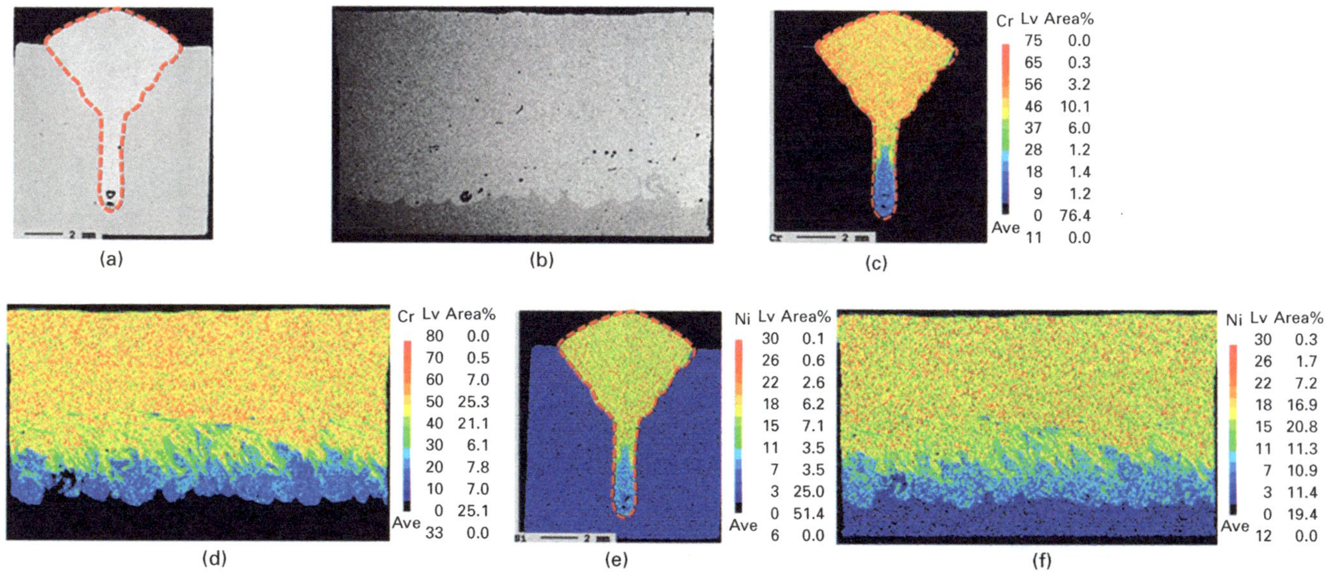

Plate XXXII Experimental results in transverse and longitudinal cross sections: (a) transverse cross section (etched), (b) longitudinal cross section (etched), (c) transverse cross section (Cr), (d) longitudinal cross section (Cr), (e) transverse cross section (Ni) and (f) longitudinal cross section (Ni).

2. C. Chaminade, A. Olowinsky, H. Kind, 'Laser-based glass soldering for MEMS packaging', *Proc. 26th Int Congress on Application of Lasers and Electro-Optics (ICALEO)*, 143–148 (2007).
3. S.S. Kachkin, Yu. V. Lisitsyn, 'Optical-contact bonding strength of glass components', *Sov. J. Opt. Technol*, **47**, 159–161 (1989).
4. M. Shimbo, K. Fukukawa, K. Fukuda, K. Tanzawa, 'Silicon-to-silicon direct bonding method', *J. Appl. Phys.*, **60**, 2987–2989 (1986).
5. I. Miyamoto, K. Cvecek, Y. Okamoto, M. Schmidt, H. Helvajian, 'Characteristics of laser absorption and welding in FOTURAN glass by ultrashort laser pulses', *Optics Express*, **19**, 22961–22973 (2011).
6. V. Greco, F. Marchesini, G. Molesini, 'Optical contact and van der Waals interactions: the role of the surface topography in determining the bonding strength of thick plates', *J. Opt. A: Pure Appl. Opt.*, **3**, 85–88 (2001).
7. Y. Arata, H. Maruo, I. Miyamoto, S. Takeuchi, 'Dynamic behavior of laser welding and cutting', *Proc. Symp. Electron and Ion Beam Science and Technology 7th Int. Conf.* 111–128 (1976)
8. T. Tamaki, W. Watanabe, J. Nishii, K. Itoh, 'Welding of transparent materials using femtosecond laser pulses', *Jpn. J. Appl. Phys.*, **44**, L687–L689 (2005).
9. I. Miyamoto, A. Horn, J. Gottmann, 'Local melting of glass material and its application to direct fusion welding by ps-laser pulses', *J. Laser Micro/Nanoeng.*, **2**, 7–14 (2007).
10. I. Miyamoto, K. Cvecek, M. Schmidt, 'Evaluation of nonlinear absorptivity in internal modification of bulk glass by ultrashort laser pulses', *Optics Express*, **19**, 10714–10727 (2011).
11. I. Miyamoto, G.A. Knorovsky, 'Laser microwelding', in *Microjoining and Nanojoining* (Y.N. Zhou, ed.). Cambridge: Woodhead Publishing (2008).
12. D.O. MacCallum, G.A. Knorovsky, S.T. Reed, 'CO_2 laser welding fused silica', *Proc. 24th Int. Congress on Application of Lasers and Electro-Optics (ICALEO)*, 687–695 (2005).
13. Y. Arata, 'Challenge to laser advanced materials processing', *Proc. 2nd-LAMP*, 3–12 (1987).
14. M. Levesque, B. Labrnche, R. Forest, E. Savard, S. Deshaies, A. Cournoyer, 'Welding of glass pieces', *Proc. 6th Int. Conf. on Photonic Technologies (LANE)*, 139–144 (2010).
15. L. Schaefer, M. Schmidt, 'Welding of glass fibers onto large-scale substrates with high mechanical stability and optical quality', *Proc. 6th Int. Conf. on Photonic Technologies (LANE)*, 145–152 (2010).
16. DELO Technical Information: DELO-PHOTOBOND®GB310 (09.12, Reversion 24). Available from: http://www.delo.de/fileadmin/datasheet/DELO-PHOTOBOND_GB310_(TIDB-GB).pdf (accessed 4 January 2013).
17. D. Du, X. Liu, G. Korn, J. Squier, G. Mourou, 'Laser-induced breakdown by impact ionization in SiO_2 with pulse widths from 7 ns to 150 fs', *Appl. Phys. Lett.*, **64**, 3071–3073 (1994).
18. B.C. Stuart, M.D. Feit, S. Herman, A.M. Rubenchik, B.W. Shore, M.D. Perry, 'Nanosecond-to-femtosecond laser-induced breakdown in dielectrics', *Phys. Rev.* B, **53**, 1749–1761 (1996).
19. A. Vogel, V. Venugopala, 'Mechanisms of pulsed laser ablation of biological tissues pulsed', *Chemical Review*, **103**, 577–644 (2003).
20. P. Russbueldt, T. Mans, G. Rotarius, J. Weitenberg, H.H. Hoffmann, R. Poprawe,

'400W Yb:YAG Innoslab fs-amplifier', *Optics Express*, **17**, 12230–12245 (2009).
21. K.M. Davis, K. Miura, N. Sugimoto, K. Hirao, 'Writing waveguides in glass with a femtosecond laser', *Opt. Lett.*, **21**, 1729–1731 (1996).
22. C. Hnatovsky, R.S. Taylor, P.P. Rajeev, E. Simova, V.R. Bhadwaj, D.M. Rayner, P.B. Corkum, 'Pulsed duration dependence of femtosecond-laser-fabricated nanogratings in fused silica', *Appl. Phys. Lett.*, **87**, 014104 (2005).
23. C. Hnatovsky, R.S. Taylor, E. Simova, V.R. Bhardwaj, D.M. Rayner, P.B. Corkum, 'Photoionization selective etching in femtosecond laser-assisted microfluidic channel fabrication in fused silica', *Optics Letts.*, **30**, 1867–1869 (2006).
24. M. Watanabe, H. Sun, S. Juodkazis, T. Takahashi, S. Matsuo, Y. Suzuki, J. Nishii, H. Misawa, 'Three-dimensional optical data storage in vitreous silica', *Jpn. J. Appl. Phys.*, **37** (Part 2, No. 12B), L1527–L1530 (1998).
25. I. Miyamoto, A. Horn, J. Gottmann, D. Wortmann, F. Yoshino, 'Fusion welding of glass using femtosecond laser pulses with high-repetition rates', *J. Laser Micro/Nanoengineering*, **2**, 57–63 (2007).
26. C.B. Schaffer, J.F. Garcia, E. Mazur, 'Bulk heating of transparent materials using a high-repetition-rate femtosecond laser', *Appl. Phys. A*, 351–354 (2003).
27. J. Bovastek, A. Arai, C.B. Schaffer, 'Three-dimensional micromachining inside transparent materials using femtosecond laser pulses: new applications', *Proc. CLEO/Europe – EQEC2005* (2005).
28. H.S. Carslaw, J.C. Jaeger, *Conduction of Heat in Solids*, Oxford: Clarendon Press (1959).
29. S. Tzortzakis, L. Sudrie, M. Franco, B. Prade, A. Mysyrowicz, A. Couairon, L. Bergé, 'Self-guided propagation of ultrashort IR laser pulses in fused silica', *Phys. Rev. Lett.*, **87**, 213902 (2001).
30. K. Nahen, A. Vogel, 'Plasma formation in water by picosecond and nanosecond Nd:YAG laser pulses – Part II: transmission, scattering, and reflection', *J. Sel. Top. Quant. Electron.*, **2**, 861–871 (1996).
31. B. Fisette, M. Meunier, 'Three-dimensional microfabrication inside photosensitive glasses by femtosecond laser', *JLMN-J. Laser Micro/Nanoeng.*, **1**, 7–11 (2006).
32. Y. Cheng, K. Sugioka, M. Masuda, K. Toyoda, M. Kawachi, K. Shihoyama, K. Midorikawa, '3D microstructuring inside Foturan glass by femtosecond laser', *RIKEN Review*, **50**, 101–106 (2003).
33. H. Helvajian, P.D. Fuqua, W.W. Hansen, S. Janson: 'Nanosatellites and MEMS fabrication by laser microprocessing', *Proc. 1st Int. Symp. on Laser Precision Microfabrication, LPM2000* (2000).
34. I. Miyamoto, A. Horn, J. Gottmann, D. Wortmann, I. Mingareev, F. Yoshino, M. Schmidt, P. Bechtold, Y. Okamoto, Y. Uno, T. Herrmann, 'Novel fusion welding technology of glass using ultrashort pulse lasers', *Proceedings of the Int. Cong. Appl. Laser and Electro Optics (ICALEO)* (2008).
35. K. Hirao, Y. Shimotsuma, J. Qiu, K. Miura, 'Femtosecond laser induced phenomena in gasses and photonic device application', *Mater. Res. Soc. Symp. Proc.*, 13–23 (2005).
36. N. Borrelli, J. Helfinstine, J. Price, J. Schroeder, A. Atreltsov, J. Westbrook, 'Glass strengthening with an ultrafast laser', *Proceedings of the Int. Cong. Appl. Laser and Electro Optics (ICALEO) (2008)*, pp 185–189.
37. Y. Bellouard, T. Colomb, C. Depeursinge, M. Dugan, A. A. Said, P. Bado, 'Nanoindentation and birefringence measurements on fused silica specimen

exposed to low-energy femtosecond pulses', *Opt. Express*, **14**, 8360–8366 (2006).

38. P. Kongsuwan, H. Wang, S. Vukelic, Y.L. Yao, 'Characterization of morphology and mechanical properties of glass interior irradiated by femtosecond laser', *J. Manuf. Sci. Eng.*, **132**, 041009 (2010).
39. K. Cvecek, I. Miyamoto, J. Strauss, M. Wolf, T. Frick, M. Schmidt, 'Sample preparation method for glass welding by ultrashort laser pulses yields higher seam strength', *Appl. Opt.*, **50**, 1941–1944 (2011).

12
Defect formation mechanisms and preventive procedures in laser welding

S. KATAYAMA, JWRI, Osaka University, Japan

DOI: 10.1533/9780857098771.2.332

Abstract: This chapter deals with the terminology, characteristics, formation conditions, causes, occurrence mechanisms and preventive procedures or measures of laser welding defects or imperfections. In particular, porosity and solidification cracking are described in detail in conjunction with laser welding phenomena.

Key words: laser welding defects, welding imperfection, weldability, porosity, solidification cracking, melt flows, keyhole behavior, bubble formation.

12.1 Introduction

Laser welding can join parts together at very fast speeds or produce deep-penetration welds efficiently in any environment such as air, shielding gas or vacuum. Consequently, it is known as a high-quality, high-precision, low-distortion, low-deformation, high-efficiency, high-performance, high-flexibility and high-speed joining process. It also facilitates robotization, automation, saving manpower, systematization, etc. Accordingly the applications of laser welding in many industrial fields are increasing together with the development of carbon dioxide (CO_2) laser, YAG laser, diode laser (LD), LD-pumped solid-state laser, disk laser and fiber laser of high power, high efficiency or high quality. However, welding imperfections or defects leading to a fracture or a disaster of manufactured goods or constructions may take place under improper conditions. In order to manufacture sound goods or constructions of high quality and high reliability with a laser beam, it is inevitably important to establish preventive procedures for welding defects and to take the necessary measures on the basis of understanding of welding phenomena as well as their formation conditions and mechanisms.[1]

In the following sections, the formation mechanisms and preventive measures of laser welding defects or imperfections are reviewed. In particular, the formation mechanisms and preventive procedures of porosity and solidification cracking will be described in detail together with laser welding phenomena.

12.2 Terminology, characteristics, causes and preventive procedures of laser welding imperfections and defects

In laser welding, a variety of defects or imperfections may occur depending on the kinds of material and their compositions, welding conditions, and so on. Laser welding defects are classified into three characteristic groups: geometrical or appearance defects, internal or invisible defects, and property or quality defects[1,2] (which are modified from arc welding defects[3]). The kinds, features, schematic representation, causes and suppression or prevention procedures of the main laser welding defects are summarized in Table 12.1.[1,2] In laser welding, porosity is easily formed in deeply penetrated welds, characterized by high-power laser welding. Solidification cracking and liquation cracking may occur in laser weld fusion zones and the heat-affected zone (HAZ), respectively, in aluminium alloys, Ni-base alloys, etc. Such internal defects result in degradation in mechanical properties or may cause a catastrophic-failure, and their detection is sometimes difficult. Therefore, porosity and solidification cracking, which should be prevented, will be described in further detail in the other sections.

12.2.1 Geometrical or appearance defects

Laser welding deformation or distortion

During laser welding, the metal or material is subjected to thermal cycles or histories, such as heating, melting, evaporation, solidification and cooling, and is inevitably accompanied by volume changes of thermal expansion and contraction. This causes distortion or deformation, residual stresses, or even cracking in the welded joint.

Welding deformation and/or distortion should always occur. In thin plates or sheets and various manufactured goods of the materials such as aluminum alloys and austenitic stainless steels with great thermal expansion coefficients, welding deformation or distortion takes place easily. It occurs more appreciably in the case of larger heat inputs, a larger welding molten pool, and a greater difference between the upper and lower bead widths. Angular deformation and/or lateral contraction distortion happens in thick plates, and warping such as angular deformation, longitudinal distortion and buckling deformation occurs in thin sheets.[4] Solidification cracking may take place due to rotary deformation during welding near the edge or the end of thin plate.[1]

As suppression procedures for welding deformation or distortion, the following measures are considered:

- the use of rigid fixture or jig,

Table 12.1 Kinds, features, schematic representation, causes and suppression or preventive procedures of main laser welding defects

Features	Kinds	Depiction	• Causes and ○ remedies for prevention
Geometrical defect	• Deformation (strain) (distortion)		• Great thermal effect • Incomplete jigging and restraint conditions ○ High speed, deep penetration welding ○ Utilization of rigid jig
	• Bad appearance (dirty surface) (rough surface)		• Improper welding conditions ○ Low power density welding ○ Removal (blow-away) of evaporated particles
	• Drop-through (burn-through)	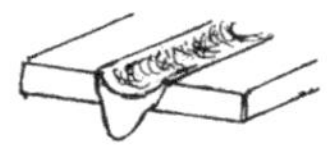	• Thin sheet • Wide bead width due to excessive energy ○ Optimization of welding conditions (lower heat)
	• Undercut		• Slow welding speed and high heat input • Properties of weld fusion zone (viscosity) ○ Proper welding conditions (gas, etc.)
	• Underfill	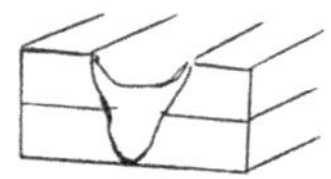	• Spattering due to excessively power density and small melt zone ○ Avoidance of low enegy and focused conditions ○ Optimization of pulsed laser welding
Internal defect	• Crack (hot crack) (solidification crack) (liquation crack) (crater crack) (bead crack) (cold crack)	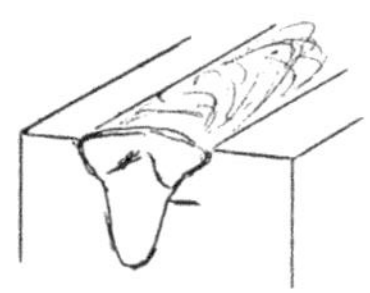	• Low melting point products and segregation • High stress/strain (rapid strain rate) ○ Proper selection of alloying compositions in weld metal and base metal ○ Optimization of pulsed laser welding ○ Avoidance of high speed welding ○ Prevention of rapidly augmented strain-rate
	• Porosity (blowhole) (pore) (pit)		○ Elimination of hydrogen and oxygen sources ○ Formation of stable keyhole ○ Proper selection of shielding gas and flow rate

	• Swell in lap joint		• High power welding of sheets (deformation) ○ Close adhesion (no gap in lap part) ○ High speed welding at low heat input ○ Utilization of pulsed laser welding
	• Incomplete penetration • Lack of fusion		• High reflectivity and effect of plasma ○ Proper conditions for deep penetration ○ Prevention of misalignment
	• Inclusion (oxides, etc.)		○ Surface polishing ○ Proper shielding conditions
	• Evaporation loss of alloying elements		• Volatile elements and low temperature of melt ○ High-speed, deep-penetration welding ○ Pulsed deep-penetration welding
	• Macrosegregation		• Incomplete mixing of dissimilar materials ○ Mixing due to beam oscillation or pulsation
	• Microsegregation		• Elements of low distribution coefficients ○ Welding at high speed and low heat-input ○ Solution heat treatment after welding
Property defect	• Mechanical properties	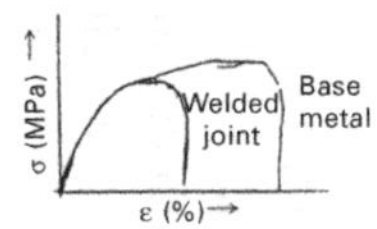	• Formation of soft zones (disappearance of age- or stain-hardening) and defects ○ High speed, deep-penetration welding ○ Welding conditions for prevention of defects ○ Optimum heat treatment
	• Chemical properties		• Stress corrosion cracking • Grain boundary corrosion due to precipitates ○ Low heat-input (deep-penetration) welding ○ Relief and relaxation of residual stress

- rapid cooling or quenching of the welded part and its vicinity just after the molten pool,[5]
- keeping the welded products cooled to room temperature in the fixture,
- formation of a parallel full-penetration weld bead (that is to say, almost the same weld bead widths in the penetration depth direction), and
- optimization of welding procedure and conditions.

In addition, the deformation and/or distortion is much smaller in laser welding than in any other fusion welding, since the heat input and bead width of a laser weld are generally smaller and narrower, respectively, than those of any other welding process.[6]

Poor surface appearance of laser welds

Poor surface appearance of laser or hybrid laser-arc welds is caused by oxidation, the deposition of ultrafine particles, fumes and debris, and/or rough surfaces due to spatters, pits, humps, underfills, undercuts, etc. (These technical terms will be described in detail in later section.[1]) The surface of a laser welded part and its vicinity is exposed to high temperatures during laser welding, and therefore a shielding gas is used to protect a molten pool and the surrounding part from oxidation. Nevertheless, the surface is oxidized or suffers from deposition of ultrafine particles produced by evaporation or spatters of melt particles. The color of the surface may be changed from metallic silver to purple, dark blue and black, golden or white depending on the oxidization levels in stainless steels or titanium alloys. Spattering results in the deposition of spatters on or near the weld beads and the formation of concave bead surfaces,[7] which are judged to be poor surface appearances. Pits or rough surfaces are observed in the weld beads of cast alloys enriched with segregated inclusions or hydrogen sources, Zn-coated steel lapped sheets or some alloys produced under improper conditions.

To counteract poor surface appearance of welds, the respective proper measures should be taken by considering their causes:

- In the case of the deposition of ultrafine particles or evaporated particles and oxidation, a proper amount of an inert shielding gas (for example, Ar, He or N_2) should be flowed from the optimum direction.[8]
- In the case of laser-arc hybrid welding of aluminum alloy, a weld bead made under the condition of laser-leading and arc-following exhibits a better surface appearance than that of arc-leading and laser-following because of the surface cleaning action of an arc.[9]

In order to suppress spattering in bead (melt-run) welding with CW (continuous wave) lasers, the following measures may be recommended:

- The adoption of beam-defocused conditions or heat-conduction welding.
- The adoption of an extremely small beam spot in comparison with the molten pool size.
- The enlargement of a keyhole inlet by a proper flow of a shielding gas.
- The utilization of pulse-modulated mode with power decrease before the occurrence of spattering instead of CW mode.
- The use of minus defocused conditions (setting a focal point below the specimen plate surface) in high power laser welding.
- The utilization of forward inclination of a laser beam.
- The use of vacuum environment.

In the case of laser welding of zinc(Zn)-coated steel sheets, a proper gap (e.g., 0.05–0.3 mm) between sheets should be employed to reduce the detrimental effect of the Zn vapor against the molten pool, although no gap at all is normally used in laser welding of other metals.

In spot welding with a PW (pulsed wave) laser, spattering takes place due to a small tolerance of a melt around a keyhole when the keyhole is formed in a small molten puddle. Therefore, to suppress spattering in spot welding, a relatively larger molten pool should be formed, and the (peak) laser power density should be increased gradually so as to slowly deepen a keyhole.

Burn-through or melt down

Burn-through is the phenomenon where the melt in a molten pool is dropped down during welding to form an underfilled bead with concave top surface and convex bottom surface. Burn-through occurs easily in the case of full penetration welding of a thin sheet or a thick plate with wide bottom surface of a molten pool at high heat input.[1] It happens more readily in materials with lower surface tension. In the case of a thick plate of steel or stainless steel with a higher surface tension at a lower temperature, noticeable humping on the bottom surface takes place if the laser power is not so high as to produce a full penetration weld with a low content of oxygen with a through-keyhole.[10,11] Figure 12.1 shows examples of a sound full penetration weld bead and a poor one with humps on the bottom surface in HT 780 steel plate of 12 mm thickness, observation results of a molten pool on the top and bottom surfaces during 10 kW fiber laser welding with a hot wire, and the mechanism of the formation of underfilling and bottom humps.[11] In the weld metal with an appropriately high content of oxygen, bottom humping is prevented. The melt at lower temperatures near the bottom rear end of the molten pool with a low content of oxygen has such a high surface tension to pull the higher temperature liquid near the keyhole tip, resulting in the

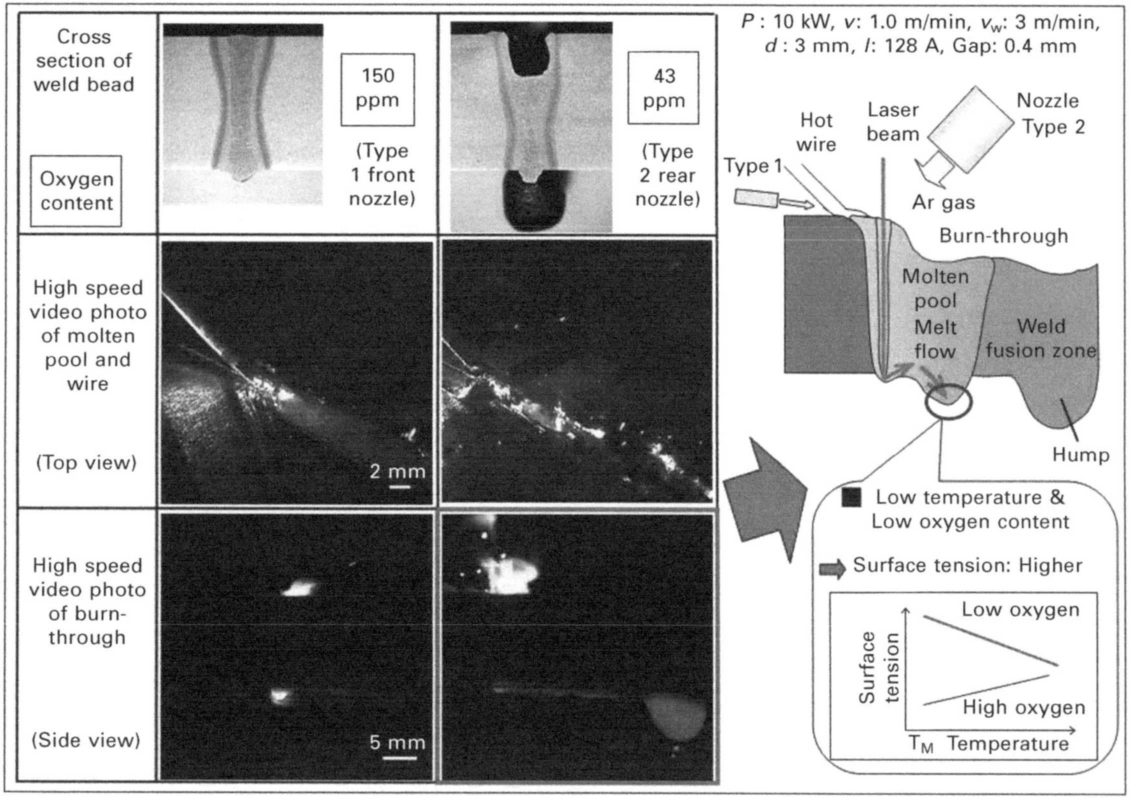

12.1 Cross sections of weld beads and video observation results of molten pool from back-upper and horizontal sides during welding of butt joints of 12 mm thick HT 780 steel plates with 0.4 mm gap subjected to fiber laser welding with hot wire at 10 kW and 3 m/min in Ar shielding gas by using Type 1 and Type 2 nozzle, showing sound full penetration weld bead and underfilled bead with humps on the bottom surface, and schematic formation mechanism of underfilling and bottom humps, showing melt flows due to effect of surface tension considering temperature and oxygen content.

formation of bottom humps.[11] It is noted that surface tension and its gradient may affect melt flows and the formation of welding defects.

To prevent the burn-through or melt down, the welding conditions should be optimized. The use of a hot wire and a proper content of oxygen in a molten pool should prevent bottom humping in laser welding of thick steel plates.[10,11] The use of a backing plate or a shielding gas to suspend the bottom molten pool surface may be required as a preventive measure against burn-through.

Undercutting

Undercut is the name for a groove along the toe of the weld bead. Examples of a sound laser weld bead and poor ones with underfill or undercut in 3 mm thick AZ31 alloy plates are shown in Fig. 12.2.[12] Undercutting is likely to occur in a wide bead made with a high power laser, at high pressure of an assist gas or under a large amount of flow of shielding gas from the front side of welding. In laser-arc hybrid welding, undercutting may occur easily in full penetration weld bead made at high arc current. A high arc current must induce shielding gas and plasma flows backward. In the case of fillet welding for a T-joint, an undercut is easily formed in the vertical plate. The cause is attributed to the effect of gravity which causes the upper melt in the molten pool to sag.

The optimization of welding conditions is needed as a measure to prevent undercutting. In hybrid welding, an additional filler wire may prevent undercutting.

Underfilling

Underfilling is the name for a concave surface of a weld bead, as shown in Fig. 12.2(b).[12] It occurs when welding the butt-joint with a wide gap with a shortage of filler wire, in the case of full penetration weld accompanied with burn-through,[10] or severe spattering of melts.[7,12]

Underfilling is suppressed or prevented by beam-scanning, utilization of a filler wire, or hybrid welding (with additional filler wire for a wider gap[13]) under the proper conditions.

In the case of lap welding of Zn-coated steel sheets without gap, spattering occurs very easily due to evaporation of the Zn layer, resulting in the formation of underfilling.[14] In this case, control of the gap between sheets is needed.[14,15] In spot welding, when a high power density laser is shot rapidly, spattering takes place due to a keyhole formation because of a small tolerance of melt in the molten pool, leading to a underfilled weld. In such a case, the optimization of welding conditions such as slow increase in laser power or energy is required.[16]

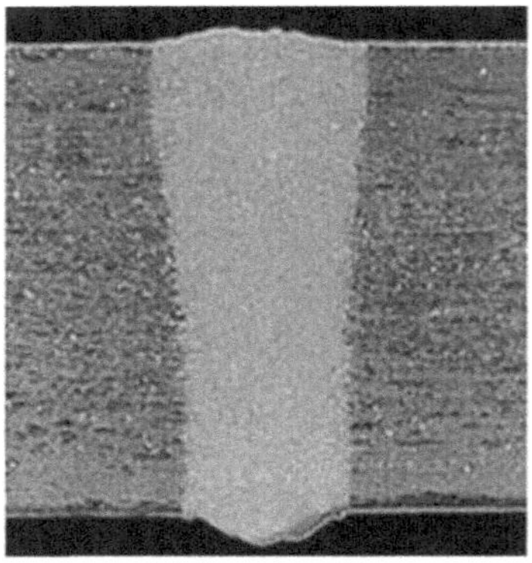

(a) Sound weld

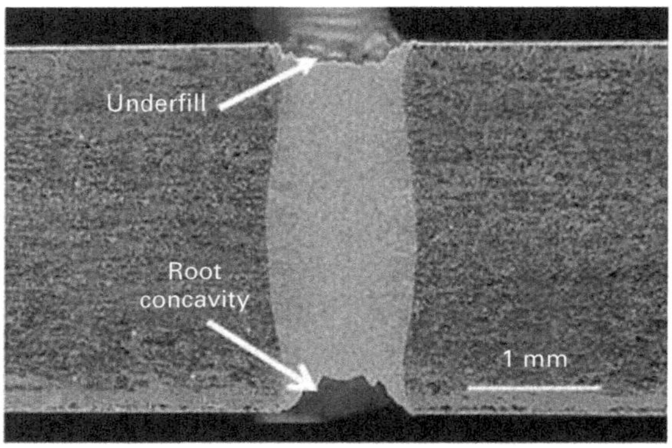

(b) Underfill

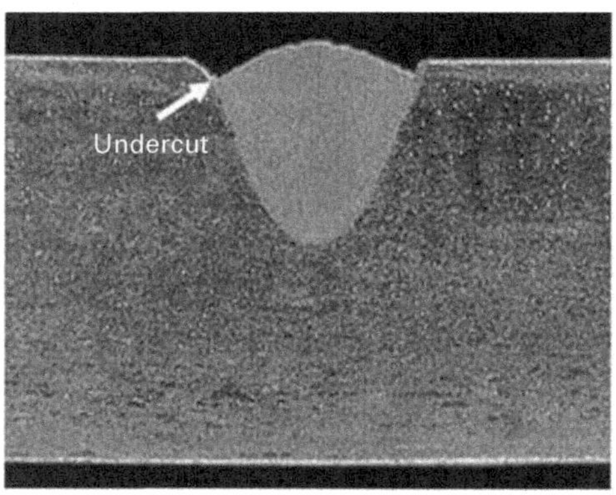

(c) Undercut

12.2 Examples of (a) sound weld bead, (b) underfilled weld bead and (c) weld bead with undercuts produced by disk laser welding of 3 mm thick AZ31 alloy plates in Ar shielding gas.

Reinforcement of weld or overlapping

Reinforcement of a weld may be formed in the case of welding with the excessive amount of a filler wire or with high power density laser with a small beam size at higher welding speed.[7] It is well known that fatigue properties of the toe of the weld are degraded. Therefore, the reinforcement of a weld may be removed to improve the fatigue properties of a welded joint.

Overlapping occurs due to the overflow of a melt on the base metal, and it may take place in fillet welding with a filler wire. The means of optimization of filler wire supply and sufficient melting of metal are needed.

Humping

Humping, which means the periodical formation of humps of weld metal on the bead surface, may occur in a narrow weld bead produced with a very small focused beam during high speed welding[7] or during low vacuum welding at a focal point.[17] Examples of humping and preventive conditions are shown in Fig. 12.3.[7] Humping is formed owing to the backward flow of a melt ejected by plume ejection and the high surface tension of the accumulated melt due to a narrow molten pool width.[7]

Such humping may be suppressed under the conditions of defocused distances (a focal point below the plate surface leading to wider molten pool surface) or by changing partial penetration to full, leading to the suppression of a melt ejected upward, as shown in the middle and lower parts of Fig. 12.3, respectively.[7]

12.2.2 Internal or invisible defects

Hot cracking (solidification cracking and liquation cracking)

Hot cracking (or high temperature cracking) in the weld fusion zone and the HAZ is called 'solidification cracking' and 'liquation cracking', respectively. Examples of such cracking in laser-welded A2090 aluminum alloy are shown in Fig. 12.4.[3,18] In aluminum alloys, fully austenitic stainless steels, Ni-based alloys, etc., solidification cracking and liquation cracking may occur along grain boundaries. In particular, solidification cracking can occur easily during spot welding with a pulsed laser or during high speed welding with continuous wave (CW) laser.[2] These causes are attributed to microsegregation and resultant formation of low solidification temperature liquid films along the grain boundaries, and thus the selection of proper materials and the process to reduce tensile load or strain during welding are important.[1,2] In the case of spot welding with a pulsed laser, the irradiation conditions of tailing laser power so as to narrow the area of a mushy zone (coexistence of solid and liquid) should be preferably selected to suppress solidification cracking.[16,19]

In mild or high tensile strength steels, hot cracking or pear-shaped cracking may take place near the middle or bottom part of a partially penetrated deep

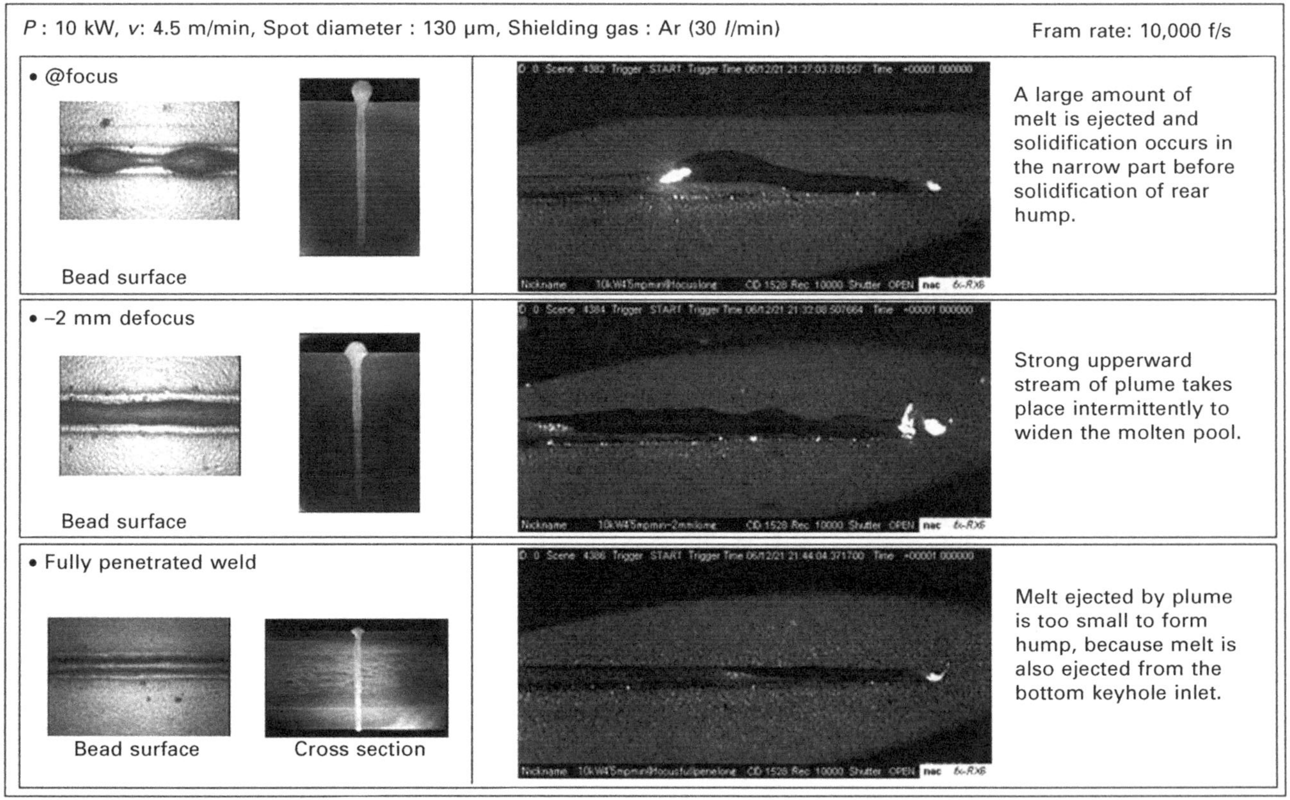

12.3 Surface appearances and cross sections of weld beads and video observation pictures during welding, showing examples of humping (upper), and suppressed humping due to defocused conditions (middle) and full penetration welding (lower).

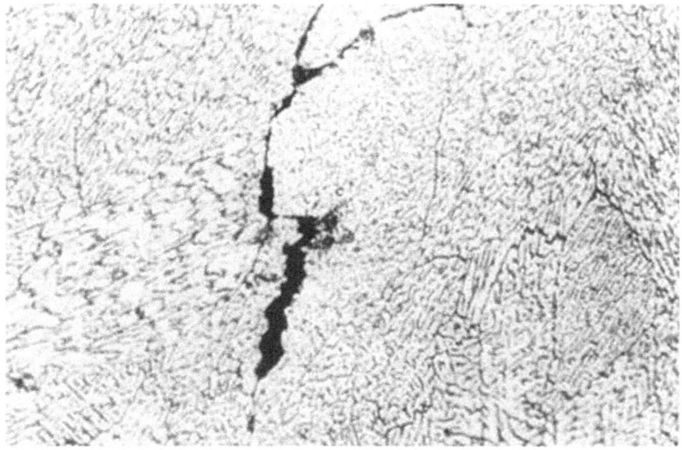

(a) Solidification cracks in weld fusion zone

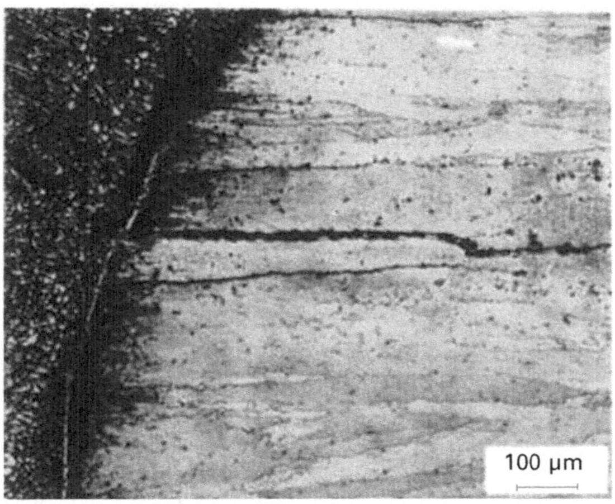

(b) Liquation crack in HAZ

12.4 Examples of solidification crack in laser weld fusion zone (a) and liquation crack in HAZ (b) of A2090 alloy plate.

weld bead made with high power laser[17] or laser-arc hybrid. The causes may be ascribed to the formation of retained liquid areas due to melt flows near the bottom part of the laser or hybrid weld beads.[17] The optimization of weld penetration depth or bead geometry or the selection of a filler wire is recommended.

In welding of dissimilar materials, hot cracking occurs very easily if a large amount of intermetallic compound is formed.[20] In this case, the area of the intermetallic compound should be minimized by controlling the mixing of a molten metal of dissimilar materials. It is important to suppress the melting area in the lower plate at high welding speeds.

Cold cracking

Cracking which occurs below 300°C is generally called 'cold cracking'.[3] In general structural steels and low-alloyed high tensile strength steels, cold cracking known as 'delayed cracking' or 'hydrogen-assisted cracking' may take place due to the harmful effect of hydrogen accumulated in the highly stress-concentrated zone,[21] while in the case of medium and high carbon steels, 'quenching cracking' or 'restraint cracking' may occur due to the formation of brittle martensite and/or cementite during welding. Cold cracking is affected by hydrogen content in the weld, restraint stress and the formation of hard martensite phase, and occurs in a shorter period under conditions of higher hydrogen content, higher restraint stress and/or higher hardness of the weld.

Cold cracking may be prevented by pre- or post-heating treatment similar to the other welding processes.[21] Nevertheless, laser welding or hybrid welding can be performed in the high tensile strength steels at room temperature without pre- or post-heating because of low hydrogen content in the weld metal.[22]

For high carbon steels and cast iron, quenching cracking is easily generated, and so the pre- or post-heating treatment, slow cooling of a weld or the use of a Ni-enriched wire is recommended to prevent cracking.[23]

Porosity

Porosity refers to blowhole, pore, wormhole, pit or bubble. In this chapter, porosity and bubbles are used in the 'solid' weld fusion zone and the 'liquid' molten pool, respectively.[1,2]

Porosity is very easily formed during laser welding. In the keyhole type of deep penetration welding, the majority of bubbles, which are generated from the tip of a keyhole during welding, lead to the formation of porosity in the weld fusion zone. Porosity formation is affected by the surface cleanness of the butt-joint, the kind of shielding gas, and welding conditions.[1,2]

The suppression or prevention procedures for porosity are:

- keyhole laser welding under the proper conditions (for full penetration),
- welding with a shallow, stable keyhole or heat-conduction type welding,
- low or high vacuum laser welding,
- forward welding with an inclined laser beam,
- pulse-modulated welding,
- twin laser beam welding,
- welding under the properly defocused conditions (for the formation of a conical, stable keyhole),
- the selection of a proper shielding gas for the material, and so on.[1,2]

In aluminum alloys, especially, the cleaning of butt-joint or lap-joint surfaces is needed to remove oxide films and hydrogen sources. In spot welding with pulsed laser, a part of a keyhole is retained as porosity in the middle or bottom part of the weld metal due to the rapid collapse of the keyhole.[2,16,24] To prevent such porosity, the control of laser power (or its density) or the selection of proper pulse waves in the final stages of laser irradiation is required.[2,16,24]

In cast or die-cast alloys, porosity is formed by the enlargement of the origins of bubbles in the base materials.[25,26] Thus it is difficult to prevent such bubbles. It may be slightly reduced by the use of an intermediate (insert) sheet or a filler wire.[26] In laser welding of Zn-coated steel sheets, a proper degree of gap between the sheets is required to prevent porosity.

Spiking

Saw-like penetration with periodical variation may be formed near the bottom part of a partially penetrated weld bead made with CO_2 laser, LD-pumped YAG laser, disk laser or fiber laser with a high power and high density under the defocused conditions of large minus values.[8,27,28] This is called 'spiking' or 'spiking phenomenon'. Welding defects called spiking porosity, root porosity and cold shut occur in deep weld metal parts of spiking. In order to prevent such defects, a defocused beam (of focal point above the plate surface) with reduced power density, beam scanning, and tandem beam welding can be applied.

Swelling of lap weld

Swelling may take place in thin aluminum alloy sheets subjected to low welding speed and high heat input. It is also observed in a pulsed YAG laser spot weld of 0.1 mm thick upper sheet onto 1 mm thick lower sheet of aluminium alloy.[2,29] It is especially present in the case of a larger amount of melting in the upper sheet in comparison with the lower plate. Some measures such as use of a strong restraint to prevent deformation or distortion of the upper sheet, improvement in thermal conductivity or heat transfer of the upper and lower sheets, the reduction in heat input, and re-melting treatment of the upper part of swelling are effective.

Incomplete penetration, incomplete fusion or lack of fusion

Incomplete penetration of a weld bead accompanied with the change from a keyhole type of deep penetration to a conduction type of shallow penetration may take place easily in the material with high reflectivity and high thermal conductivity such as aluminum alloys and copper alloys.[2,28,30] Such a change may be caused by argon (Ar) plasma in CO_2 laser welding in addition to the material properties of high light reflection and high heat conduction. To prevent this, the use of helium (He) shielding gas or a laser with about 1 μm wavelength is needed. The lack of fusion may be observed in the case of multi-pass deep welds in a narrow gap with a filler wire. To prevent this lack of fusion in multi-layers, beam scanning method, higher laser power or a smaller amount of deposited layer are considered to be effective.

Inclusions

Inclusions such as oxides or nitrides are formed in the case of oxidized plate surface or butt-joint surfaces, or welding with an active shielding gas such as O_2, N_2 and CO_2. Oxides are sometimes detrimental to ductility in the weld of martensitic steel and austenitic stainless steel, but have a beneficial effect on the ductility by forming embryos for ferrite phase formation in the steel weld fusion zone.[31] It is known that nitrides, which are formed at high temperatures in the molten pool, are effective in reducing solidification cracking by the presence across the grain boundaries in A6xxx aluminum alloy.[32] To reduce inclusions, the plate surfaces and butt-joint surfaces must be cleaned, and laser welding should be performed in the appropriate shielding gas conditions.

Evaporation loss of alloying elements

Evaporation always occurs during laser welding, and the contents of impurity or volatile alloying elements with low vaporization temperatures (e.g., Mn in Fe, and Mg, Zn, Li, etc. in Al) decrease by a higher amount of evaporation than the base metal.[2,30,33] Evaporation loss should be reduced by welding at high speed or compensated by using a filler wire with a higher content of alloying elements.

Macrosegregation

Macrosegregation is present in the weld fusion zone in the case of laser welding of dissimilar materials[20] or using a filler wire different in chemical composition from the base metal.[23] Mixing or steering of melt in the molten pool can reduce macrosegregation. The cross section and schematic of a laser weld, EDX analysis results of Fe and Ni elements and line analysis results of Fe and Ni along the bead centerline are shown in Fig. 12.5, where laser welding was performed for a butt-joint of 0.4 mm gap between cast iron and low carbon steel plates using a 97%Ni filler wire.[23] In laser welding, sufficient mixing of a filler wire into the base metal is difficult, and consequently the compositions of the bottom part of the weld fusion zone are approaching

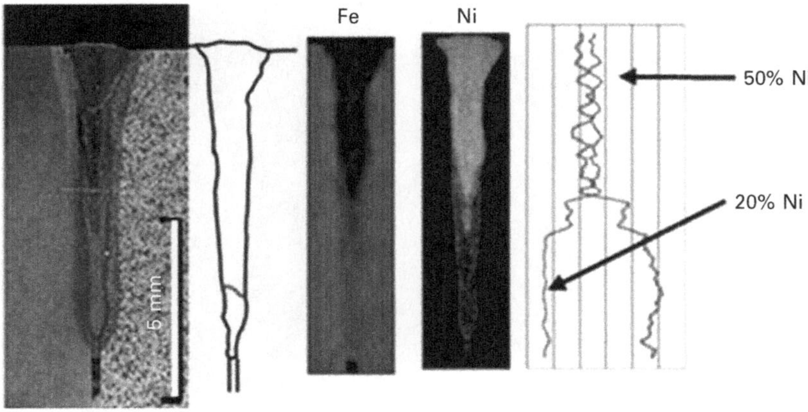

12.5 Cross section and schematic of laser weld made in butt-joint of cast iron and low carbon steel plate with 0.4 mm gap using 97%Ni filler wire, EDX analysis results of Fe and Ni elements and line analysis results of Fe and Ni along bead centerline, showing enrichment of Ni in upper part of weld fusion zone.

those of the base metal.[23,34] The wider the gap in the butt-joint, the more deeply the compositions of the filler wire reach.[23]

Microsegregation

Microsegregation necessarily occurs due to the solidification process in fusion welding (such as arc, plasma, laser and electron beam welding) of alloys.[1,2,35] An SEM photo of the laser weld fusion zone in A6061 aluminum alloy, and EPMA analysis results of Fe, Mg and Si of some areas are shown in Fig. 12.6.[33] EPMA and EDX analysis results can reveal microsegregation of Fe, Mg and Si along cellular dendritic boundaries and the same enrichment of Mg and Si at the partial cell boundaries. White phases in the SEM photo correspond to the microsegregation of Fe (of higher density). Thus the microsegregation takes place more easily for an alloying element or an impurity with a smaller distribution coefficient and at a higher content.[35] As a result, the solidification temperature of the melt falls during solidification, leading to the occurrence of solidification cracking and degradation in the strength of the weld joint.[36–38] In the case of age-hardenable alloys, aging recovery is insufficient due to lower contents of chief aging elements near the cell cores.[37] To prevent the reduction in hardness and strength due to microsegregation, solid solution heat treatment of the joint after welding is effective.

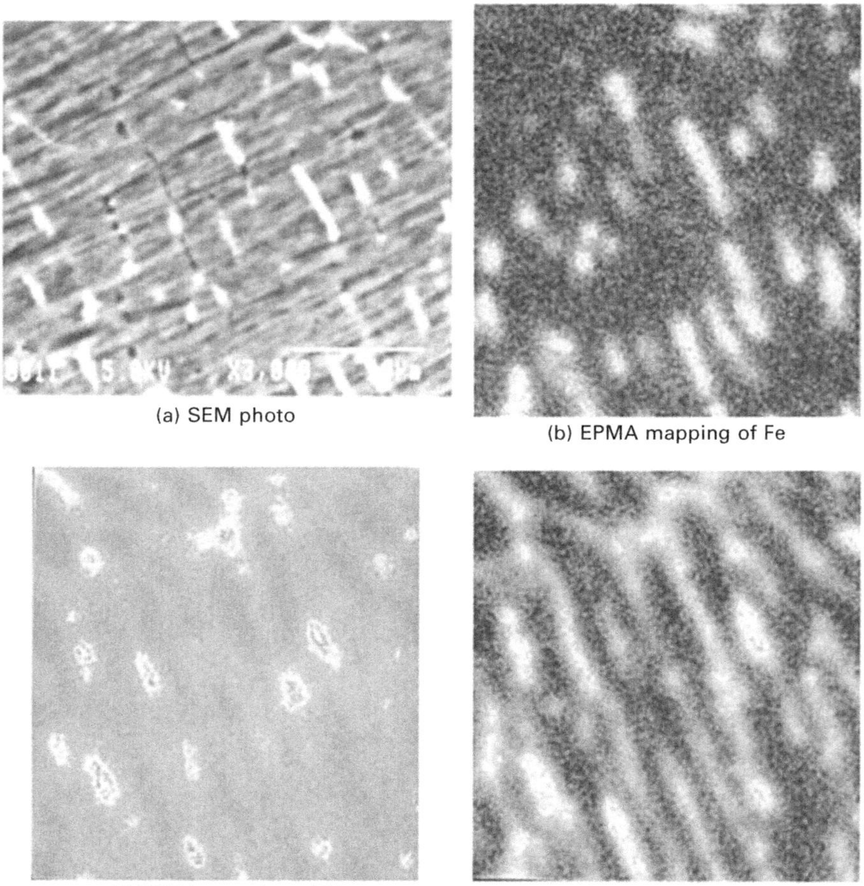

(a) SEM photo (b) EPMA mapping of Fe

(c) EPMA mapping of Mg (d) EPMA mapping of Si

12.6 SEM photo and EPMA analysis results of Fe, Mg and Si in laser weld fusion zone of A6061 aluminium alloy, showing formation of inclusions and microsegregation along cellular dendritic boundaries.

12.2.3 Quality or property defects

Reduction in mechanical properties

In the case of normal carbon steels, the welds are hardened due to the formation of harder phases such as lower bainite and martensite, and the strengths of the joints are generally higher than those of the base metals.[39] On the other hand, in the case of quenched and tempered high tensile strength steels, and age-hardenable or work-hardenable materials such as aluminum alloys, Ni-base alloys and copper alloys, the hardness or strength of the weld fusion zone and HAZ becomes lower than that of the base metal.[30,33,37] The tensile strength results of welded joints of various aluminum alloys are shown in Fig. 12.7. In the annealed base alloys such as A5052, A5083 and A5182, the strengths and strains of the joints are slightly lower and appreciably lower than those of the base alloys, respectively. The reduction in the strength and strain is attributed to the formation of porosity. The strength and the strain can be increased by a decrease in the porosity with an increase in the welding speed. On the other hand, the strength and the strain of age-hardenable alloys become lower than those of the base alloys because of the disappearance of age-hardenable phases by melting and the suppression of precipitation-hardenable phases due to microsegregation, as shown in the right column of Fig. 12.7. To fully improve the joint strength of the age-hardenable alloy, the heat treatment (i.e., solid-solutionizing and aging treatment) is required after welding.[37] To lessen the reduced strength of a work-hardenable alloy, higher welding speed may be employed.[33] Furthermore, the mechanical properties are degraded in the weld with some welding defects.[38] The prevention of welding defects is needed in terms of a sufficient welded joint. For quenched and tempered high tensile strength steels, rapid welding speed may be used.

The toughness of steels for pipelines, ships, and so on is evaluated by the V-notch Charpy impact test. Fracture pass may occur to apparently show high toughness in the test,[40] and then the Charpy test specimens with side notches should be used.[41,42]

The fatigue strengths of laser and hybrid weld joints are generally higher than those of other fusion welding processes.[43] The strengths are higher when the welds have higher hardness and strengths, while they are lower when the welds are weaker than the base metal. It is noted that the fatigue strength of the high tensile strength steel is not so high in comparison with that of the base metal, because the softer zones are formed in the HAZ. The reinforcement may sometimes be removed to improve the fatigue strength by suppressing the stress concentration.

Welding defects such as underfilling, cracking and porosity may remarkably reduce tensile strength and strain (elongation) as well as the fatigue strength by enhancing an actual applied load and by inducing stress concentration.[38,44,45] Therefore, to improve the mechanical properties, the prevention of welding defects is essential.

Degradation in chemical properties

In some steels and alloys, stress-corrosion cracking and grain boundary corrosion may take place in the HAZ[1]. The selection of a material, the use of low heat input welding, reduction or removal of residual stresses, good protection of a weld for preventing oxidation due to an inert shielding gas, etc., are required to improve chemical properties. Grain boundary corrosion due to the precipitation of carbides in the HAZ of austenitic stainless steel is

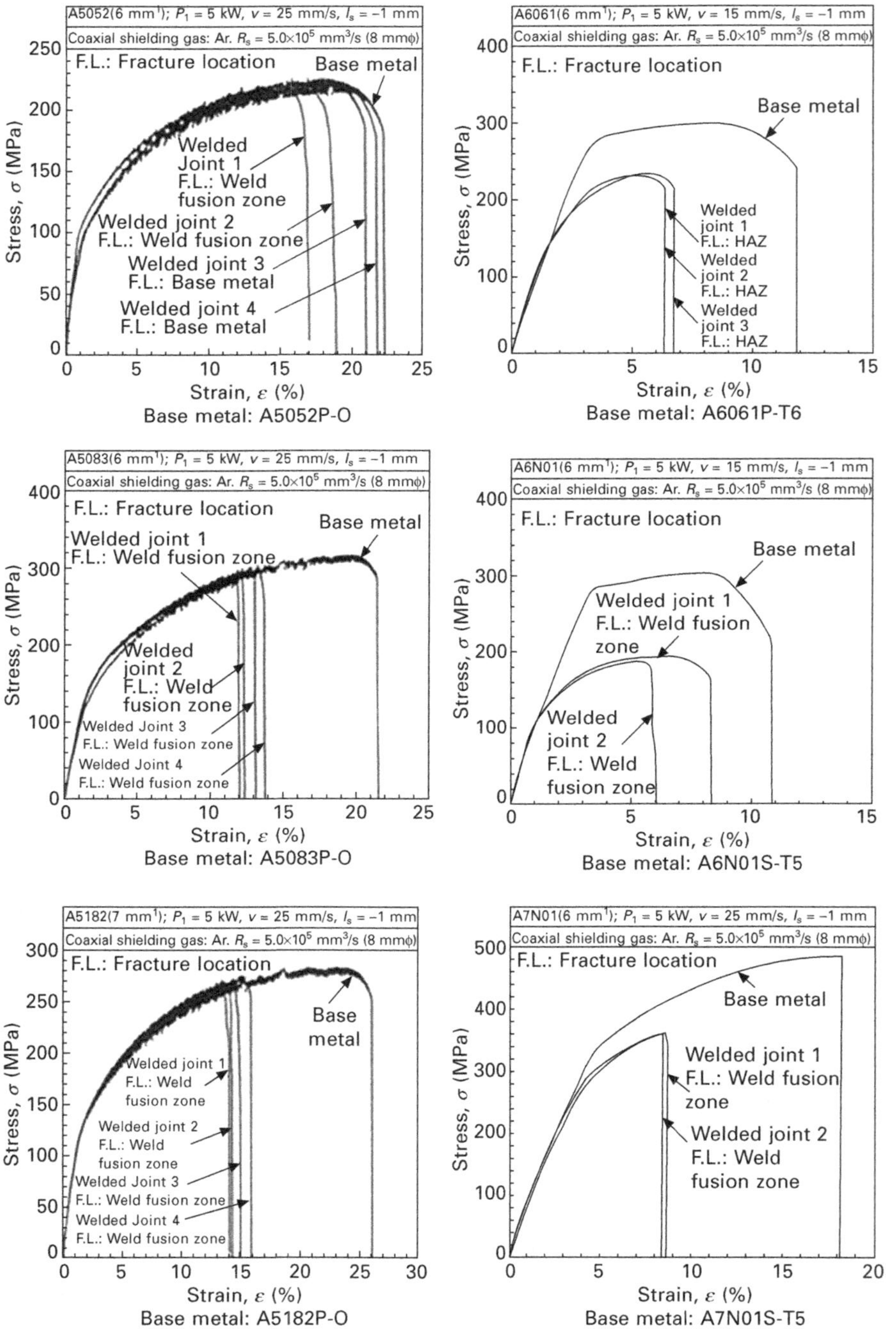

12.7 Tensile test results (stress–strain curves) of base metals and welded joints of A5052P-O, A5083P-O, A5182P-O, A6061P-T6, A6N01S-T5 and A7N01S-T5, showing respective reductions in strengths of welded joints.

a big problem in other fusion processes such as arc, but may be lessened by low heat input in laser welding. In welding or joining dissimilar metals or alloys, the problem of galvanic corrosion may occur.[20] The normal measures such as painting protection and proper applications without water or corrosive environment are required.

12.3 Formation mechanisms and preventive procedures of porosity

12.3.1 X-ray transmission apparatus for observation of bubble and porosity formation

Concerning the formation of porosity during laser welding, brazing and hybrid welding, keyhole behavior, bubble generation situation and melt flows in the molten pool were elucidated by utilizing the microfocused X-ray transmission *in-situ* observation system. The system is schematically shown in Fig. 12.8.[2,24,27,46,47] X-ray transmission observation results are described for spot welding with pulsed YAG laser, bead (melt-run, butt-joint or lap-joint) welding with CW laser, hybrid welding with laser and arc, brazing, welding of Zn-coated steel and cast alloys in the following subsections.

12.3.2 Formation mechanism and preventive procedure of porosity during spot welding with pulsed laser

When the metal or alloy is subjected to spot welding with pulsed YAG laser of rectangular shape, porosity is easily formed in the middle or bottom part of the weld fusion zone in the case of extremely short pulses or deep

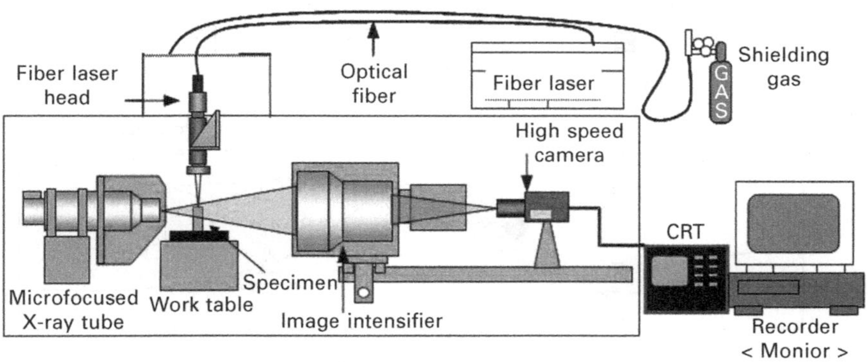

12.8 Schematic arrangement of microfocused X-ray transmission *in-situ* imaging system for observation of keyhole behavior, bubble generation resulting in porosity formation and melt flows in molten pool during laser welding.

penetration welds, as illustrated by the welding results of austenitic stainless steel in Fig. 12.9.[24,48] Examples of X-ray transmission observation results exhibiting keyhole behavior, and a schematic representation showing formation mechanisms of typical and root porosity are shown in Fig. 12.10.[1,24,47] The porosity was observed to form from a collapsed keyhole. The keyhole grows deeper together with some fluctuation. If the laser power is rapidly terminated when the keyhole tip reaches the bottom of a molten pool, the root porosity is formed. When the laser power is terminated by keyhole fluctuation, the porosity is formed near the middle of the spot weld. The gas inside the porosity is mainly air or environmental shielding gas in the case of air or shielding gas atmosphere, respectively. Such porosity can be reduced or prevented by controlling the pulsed laser power.[1,17,24,48] Therefore, in the final stage of laser irradiation, a keyhole becomes shallow slowly through the process whereby the melt flows into the bottom tip to bury the keyhole gradually due to the gradual decrease in the laser power. As shown in Fig. 12.11, the

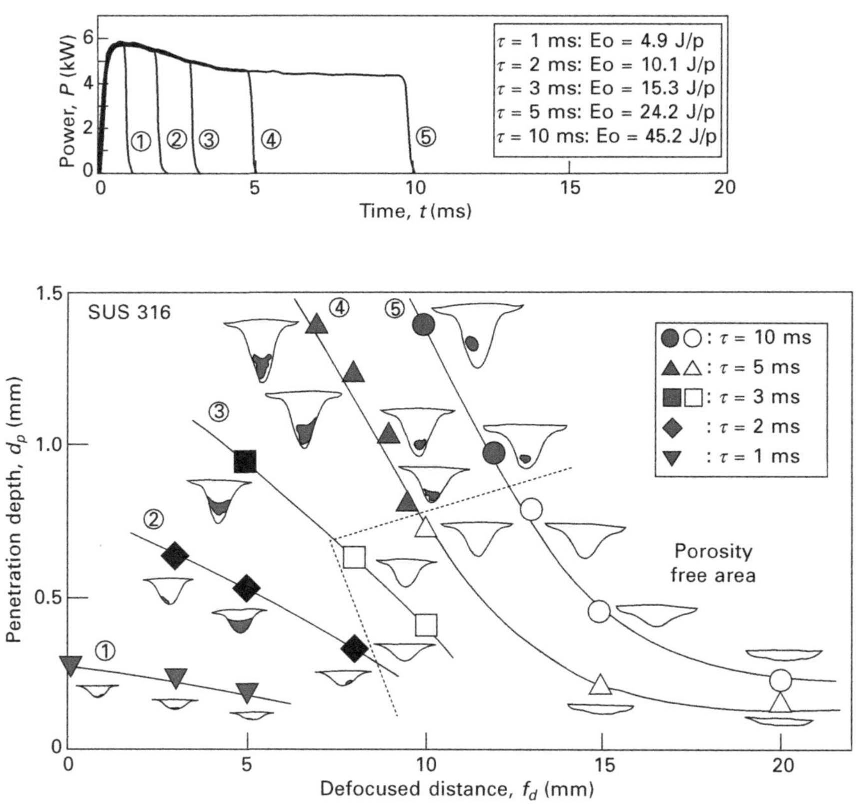

12.9 Effect of defocused distance and pulse width on penetration depths of spot welds with pulsed YAG laser of rectangular shapes, showing formation conditions of porosity.

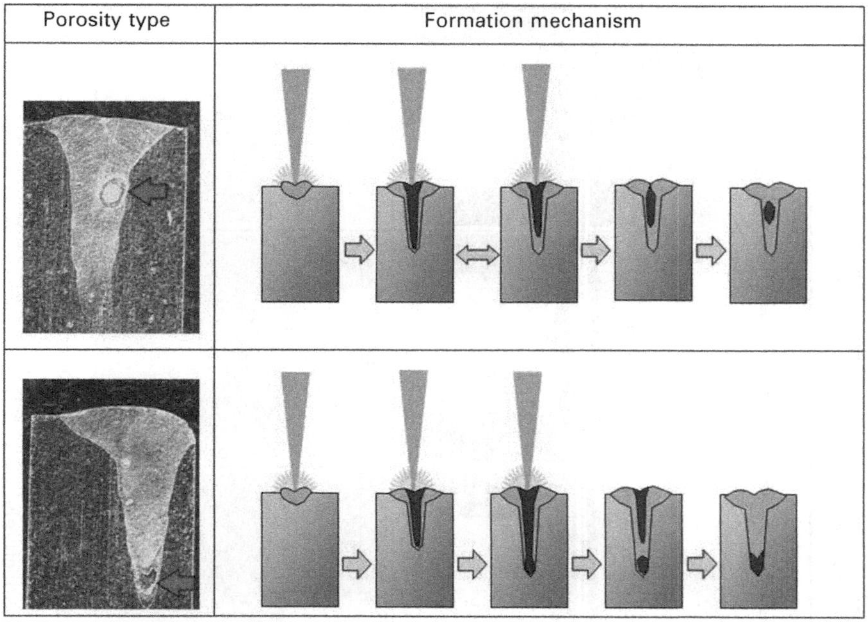

12.10 Examples of typical and root porosity in spot welds of stainless steel, showing schematic illustration of keyhole behavior and porosity formation mechanisms.

prevention of porosity is feasible by a saw-like pulse wave with a reduced power.[17,24] The first gradual increase in pulse laser power is used to suppress spattering, and the subsequent three pulses can produce a keyhole from deep to shallow depths depending on the power levels. Bubble generation from a shallow keyhole, leading to porosity formation, should be avoided by the last pulse wave shape.

12.3.3 Formation mechanisms and preventive procedures of porosity during bead welding with CW laser

In welding of various metals and alloys with high power CW laser, porosity such as a large pore of more than 1 mm size, root porosity, porosity near the fusion boundaries of the butt or lap joint, small porosity of about 0.1–0.4 mm or less than 0.1 mm size, and healed porosity filled with melt during solidification may be present in the weld fusion zones, as illustrated by the examples of CO_2 laser welds of aluminum alloys in Fig. 12.12.[2,28,46] Small or large pores are formed from the keyhole tip during welding, as examples of X-ray transmission observation photos during laser welding of A5083 aluminum alloy and Type 304 stainless steel.[8,24,27,46] Examples of observation photos during fiber laser welding are shown in Fig. 12.13.[8] A

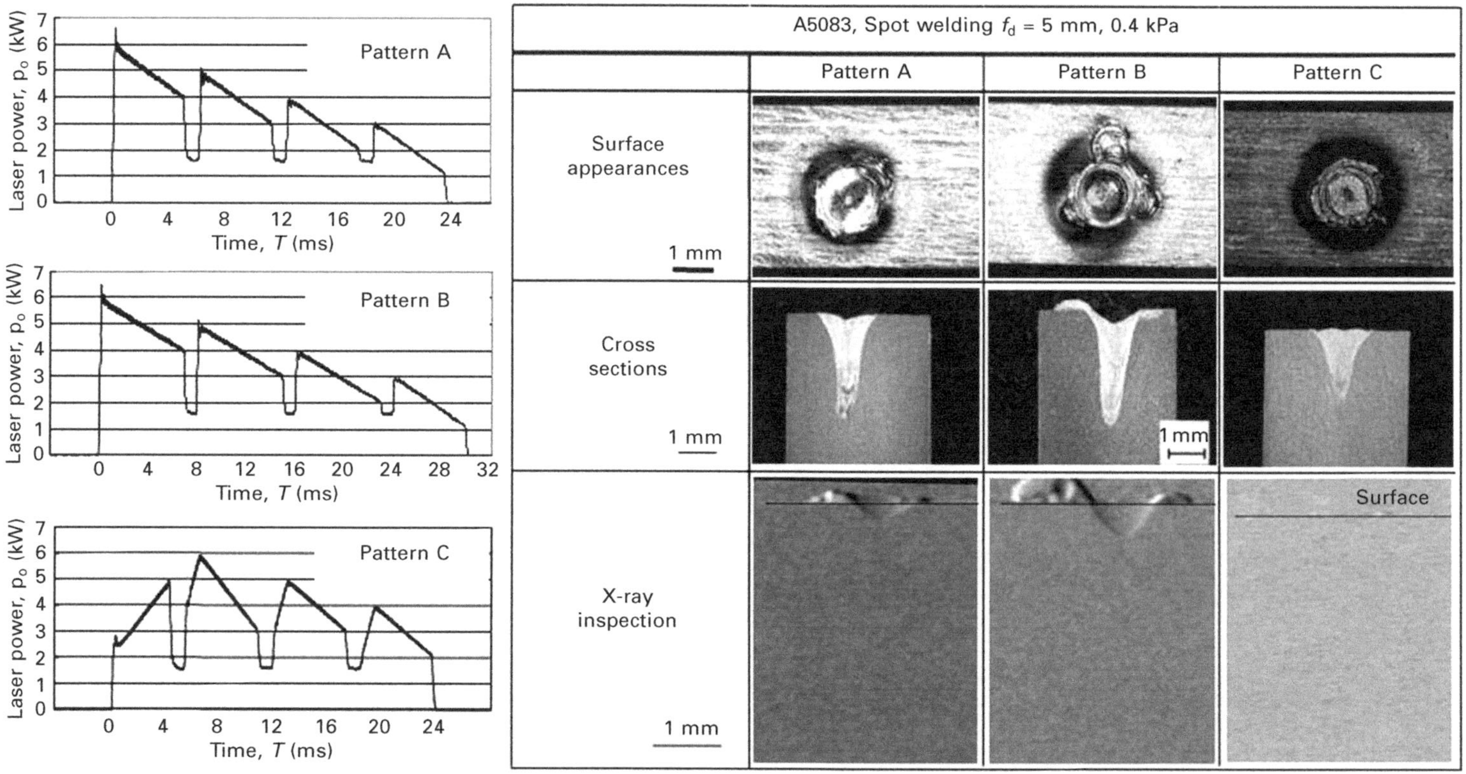

12.11 Examples of saw-like pulse waves, and surface appearances, cross sections and X-ray inspection photos of spot welds produced, showing beneficial effect of controlled shape and gradual increase in pulse laser power on prevention of porosity and suppression of spattering and concave surface, respectively.

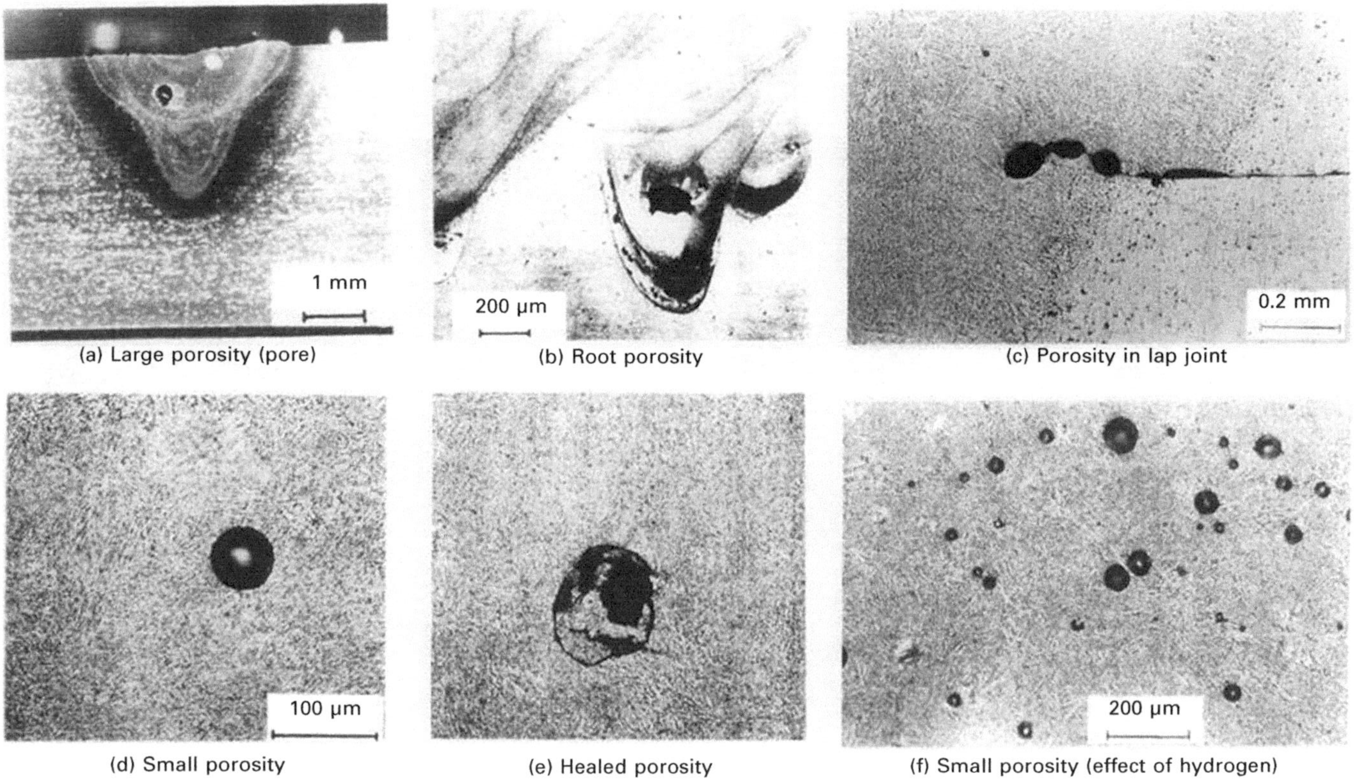

12.12 Examples of various types of porosity in laser weld fusion zones.

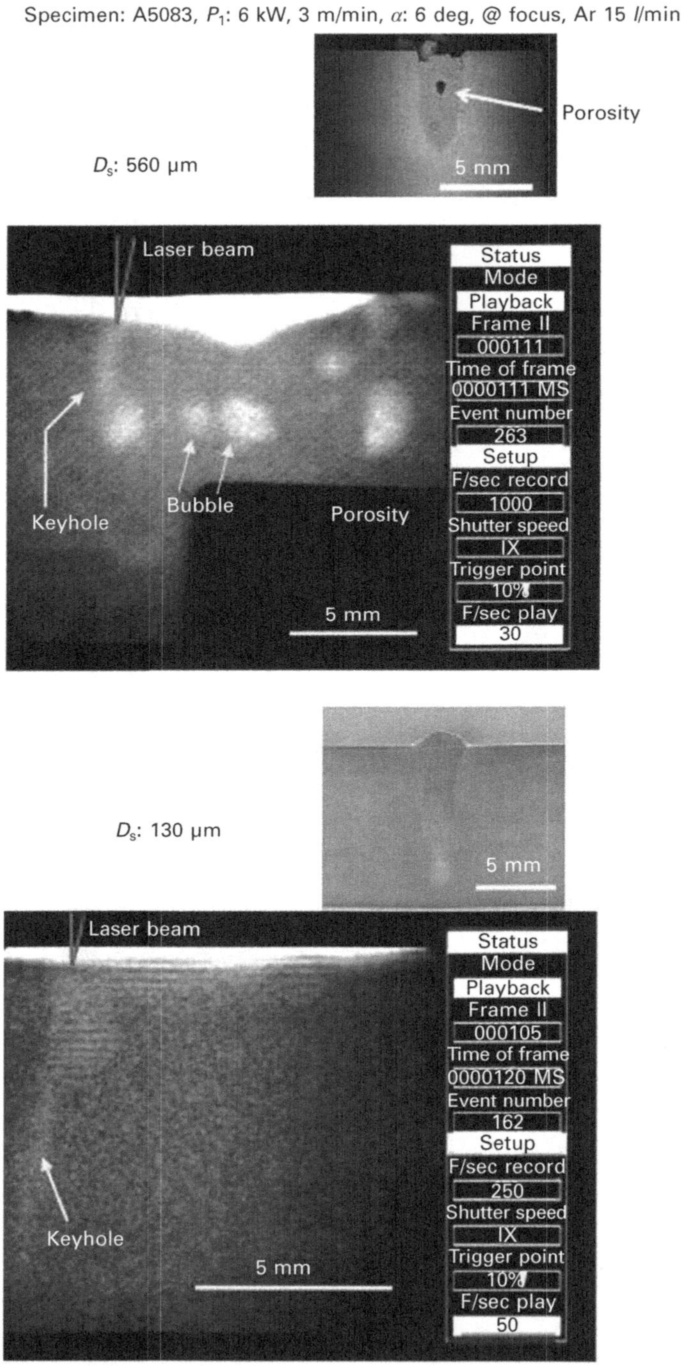

12.13 Cross-sectional weld beads and X-ray transmission *in-situ* imaging photos observed during welding with fiber laser of 560 and 130 μm in focused beam diameter, showing generation of bubbles leading to formation of porosity.

lot of bubbles are generated and result in porosity in the case of about 0.6 mm beam diameter, but bubble formation is hardly seen in the case of about 0.13 mm diameter. Very small porosity of less than 0.1 mm in size is formed due to the effect of hydrogen in CO_2 laser welding of aluminum alloys with Ar-1%H_2 mixture shielding gas. Porosity near the fusion boundaries of the butt or lap joint in aluminum alloys is also formed owing to the effects of a gap between the plates or hydrogen when hydrogen is enriched beneath the surface oxide film or the moisture is high on the surface. Root porosity may occur under defocused minus (–) conditions (of a focal point under the plate surface) or in the case of the existence of interaction between a laser beam and induced plume or plasma in high power CO_2, fiber or disk laser welding.

Based upon the X-ray transmission observation results, bubbles and porosity formation situations are shown together with keyhole behavior and melt flows in Fig. 12.14.[1,2,47] At very slow welding speed, an unstable keyhole is easily formed and leads to the formation of bubbles, since a laser beam is irradiated on the wall, which causes the melt downward flows to produce a bubble by closing the bottom keyhole. In this case, the bubbles sometimes disappear from the molten pool surface. Porosity is more easily formed by the generation of bubbles from the tip of the keyhole at slow welding speed. Such large bubbles, resulting in pores or porosity, are generated from evaporation of the keyhole (front) wall. Such porosity is prevented by full-penetration welding. Pulse modulation is sometimes effective. In the case of high power density due to a 0.2 mm small focal beam, bubbles leading to pores or porosity are formed from the middle part of a full-penetration keyhole (at about 5 mm below the focal point). In steels or austenitic stainless steels, porosity is absent due to the ejected plume upwards at high welding speed.

According to the gas analyses of porosity, a large amount of shielding gas, a small amount of hydrogen (H) and sometimes a small amount of nitrogen (N) are detected.[1,2,27,45] It is generally considered that the bubbles have the same compositions as a keyhole such as a large number of evaporated vapors and a small number of the other elements composing a shielding gas, and N (nitrogen) or O (oxygen) in air. Vapors and oxygen make oxide films on the porosity inside surface, and the remaining vapors also deposit on the porosity surface as the temperature falls. Consequently, a shielding gas is mainly left in the porosity, and H is contained because it can diffuse into the porosity during and after welding.

The procedures for the suppression or prevention of porosity in laser welding are as follows:[1,2]

- a heat-conduction type of shallow welding (but it should be noted that another type of porosity may be formed due to the effect of a gap or H in aluminum alloys),

- laser welding with a conical (V-geometry) stable keyhole under the properly defocused conditions,
- full-penetration welding due to a complete-through keyhole (but the laser power density with 0.2 mm focal spot size is too high to induce bubbles in thick plate, and so moderate power density is required),
- vacuum laser welding under the proper conditions,
- welding with a forward inclined laser beam,

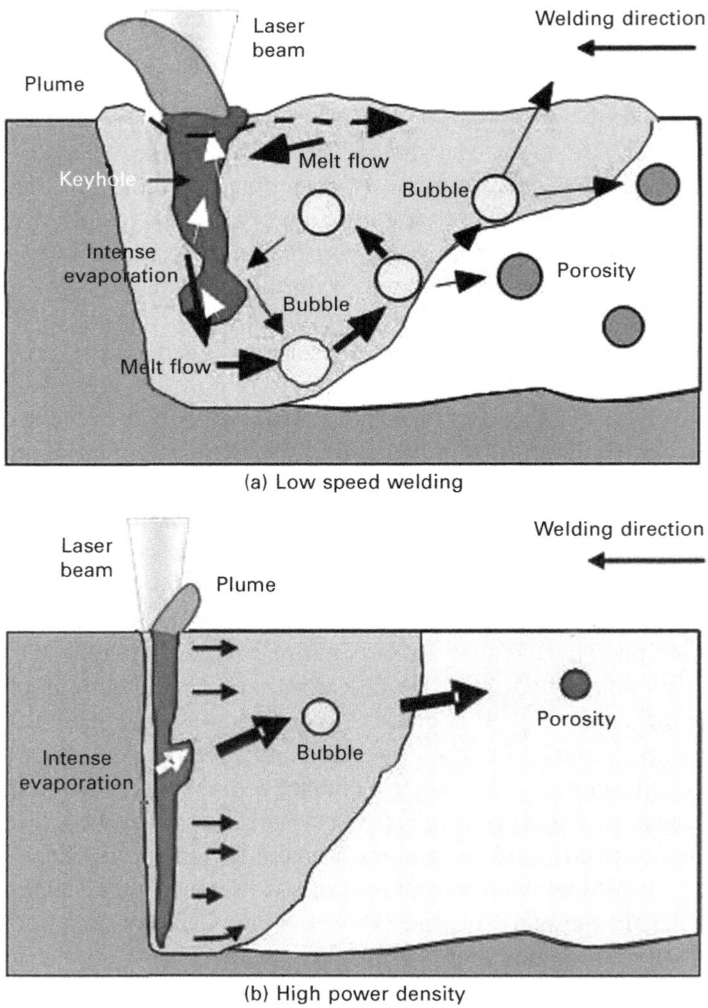

12.14 Schematic illustration of laser welding phenomena under various welding conditions, showing bubbles and porosity formation situations at low speeds and no porosity at high speed, together with plume ejection, keyhole behavior and melt flows in molten pool.

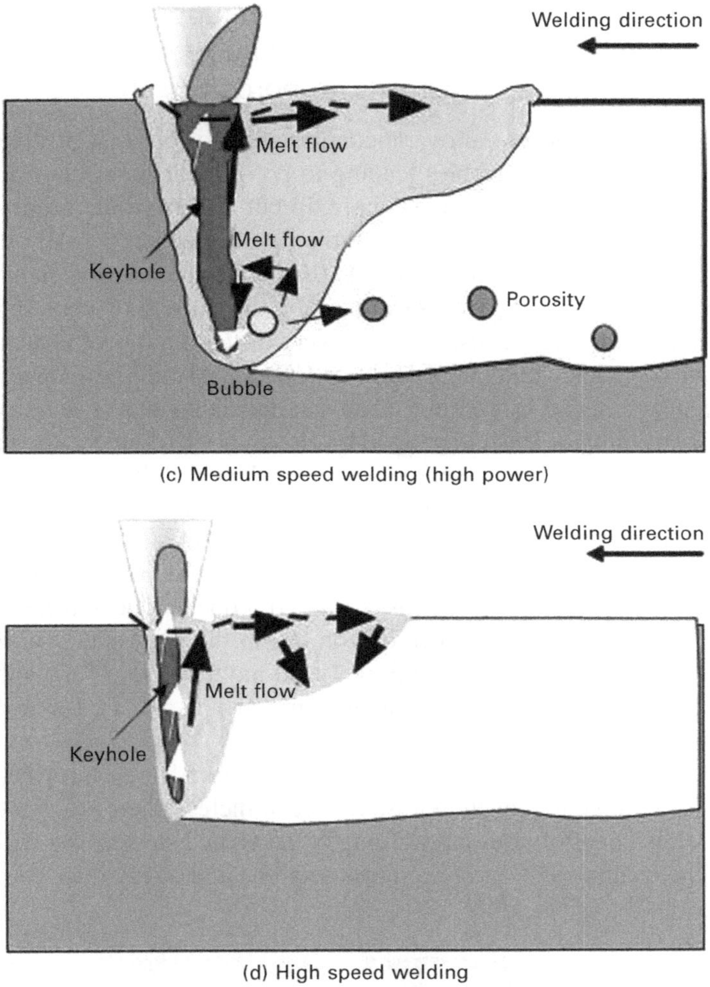

12.14 Continued

- a low welding speed at a high power, which induces melt flows upwards for bubbles to disappear from the molten pool surface, (it was especially observed in laser welding of aluminum alloy),
- a high welding speed, which can suppress or prevent bubble formation,
- pulse-modulated welding for moderate penetration,
- twin laser beam welding (not always effective but the location/arrangement of beams is important to produce a wide and stable keyhole),
- the selection of a proper shielding gas for the material (e.g., N_2 gas for austenitic stainless steel and CO_2 gas for some steels).

12.3.4 Formation mechanisms and suppression procedures of porosity during welding of materials with great sensitivity to porosity

In laser welding of cast alloys, including gases or die-cast or thixomolded alloys, small or large bubbles leading to porosity or pores of various sizes are formed not only from the keyhole tip but also from the origins of very small bubbles of H, N or CO_2 gas in the base materials.[25,26] An example of porosity in a laser weld of die-cast AZ91 magnesium alloy is shown in Fig. 12.15.[25] Such porosity is characterized by the growth of very small holes from the fusion boundaries. Thus it is very difficult to prevent such bubbles and porosity. The contents of H, N or CO_2 gas should be reduced. It may be slightly reduced by the use of an intermediate (insert) sheet or a filler wire. In welding of cast alloys, good welds may be formed if the parts with reduced macro- and microsegregation are selected to be welded.

There is a demand for laser butt-joint welding of specimens whose surfaces are prepared by laser cutting. In the case of steels prepared by laser cutting with pure O_2 gas, the cut surfaces are covered with oxide films, and consequently porosity may be easily formed probably due to rimming action (FeO + C > Fe + CO) between carbon (C) in the steel and O in the oxide films or a similar phenomenon in the weld molten pool.[1,49] To suppress or prevent this porosity, deoxidation elements such as Al, Ti, Mn and Si may be used effectively.[49]

In laser welding of Zn-coated steel sheets, a proper degree of gap between the sheets is required to prevent porosity.[15] Nevertheless, there is a great demand for establishment of laser lap welding of Zn-coated steel sheets without any gap. The examples of surfaces, cross sections and X-ray inspection results

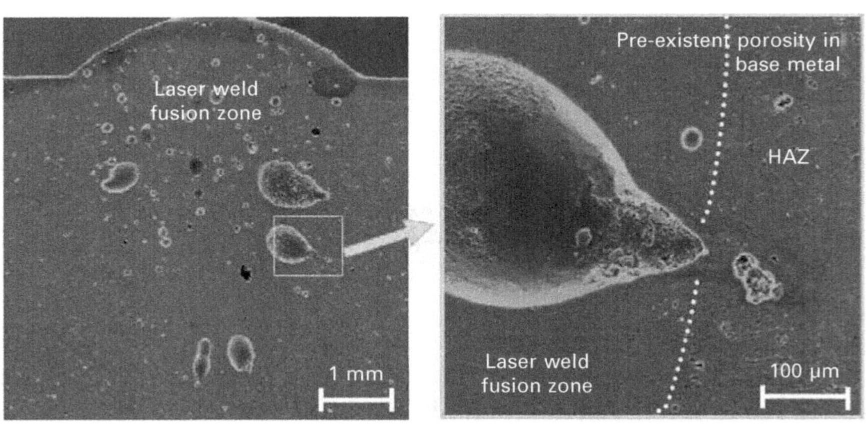

12.15 Example of porosity in laser weld fusion zone of die-cast AZ91 magnesium alloy.

of laser welds made at 1–5 mm/s under the conditions of pulse-modulation are shown in Fig. 12.16.[14] In the case of a 0 mm gap, worm-like porosity is formed from the fusion boundaries of the lap joint at low welding speed. On the other hand, at high welding speed, underfilled weld beads are formed by spattering of melts. Nevertheless, under the conditions of proper pulse modulation and adequate welding speeds, porosity and underfilling are reduced or suppressed by the procedures that laser power decreases before the occurrence of spattering and low heat input due to pulsation of laser power can prevent Zn vapors from invading into the molten pool by suppressing the amount of Zn evaporation.

Laser welding phenomena are schematically shown in Fig. 12.17.[14] In the case of a large heat input, the molten pool is so big that spattering is suppressed but porosity is formed. In the case of normal focal beam spot (of about 0.6 mm diameter), the smaller the molten pool, the more severely the spattering occurs. Such spattering leading to poor weld beads can be prevented by setting the gap to about 0.05–0.3 mm for lapped sheets. Spattering should be reduced with an increase in the thickness of the upper sheet. Porosity should also be reduced with an increase in the welding speed in thick Zn-coated plates.

In laser lap welding of Zn-coated steel sheets, the procedures recommended for the production of good weld beads are as follows:[1,14,15,50–53]

- control of the gap between sheets at about 0.05 to 0.3 mm depending on the Zn-coating layer thickness,
- device or modification of lap joint geometry (which can produce a proper gap),
- removal of Zn-coated layer at the lap joint part,
- backward inclination welding with high power laser beam,
- hybrid welding,
- the use of thin Cu sheet,
- laser brazing with Cu-Si alloy filler wire, etc.

It is probably considered that backward inclination welding and hybrid welding can produce a gap naturally during welding by inducing deformation of the upper sheet. Besides, laser butt-joint welding of Zn-coated steel sheets can produce welds without defects, since the Zn layer on the surface has little effect on the welding results.

12.4 Formation mechanisms and preventive procedures of hot cracking: solidification cracking and liquation cracking

Hot cracking generally occurs along grain boundaries in the welds of aluminum alloys, Ni-based alloys, etc., when the liquid films are formed along grain

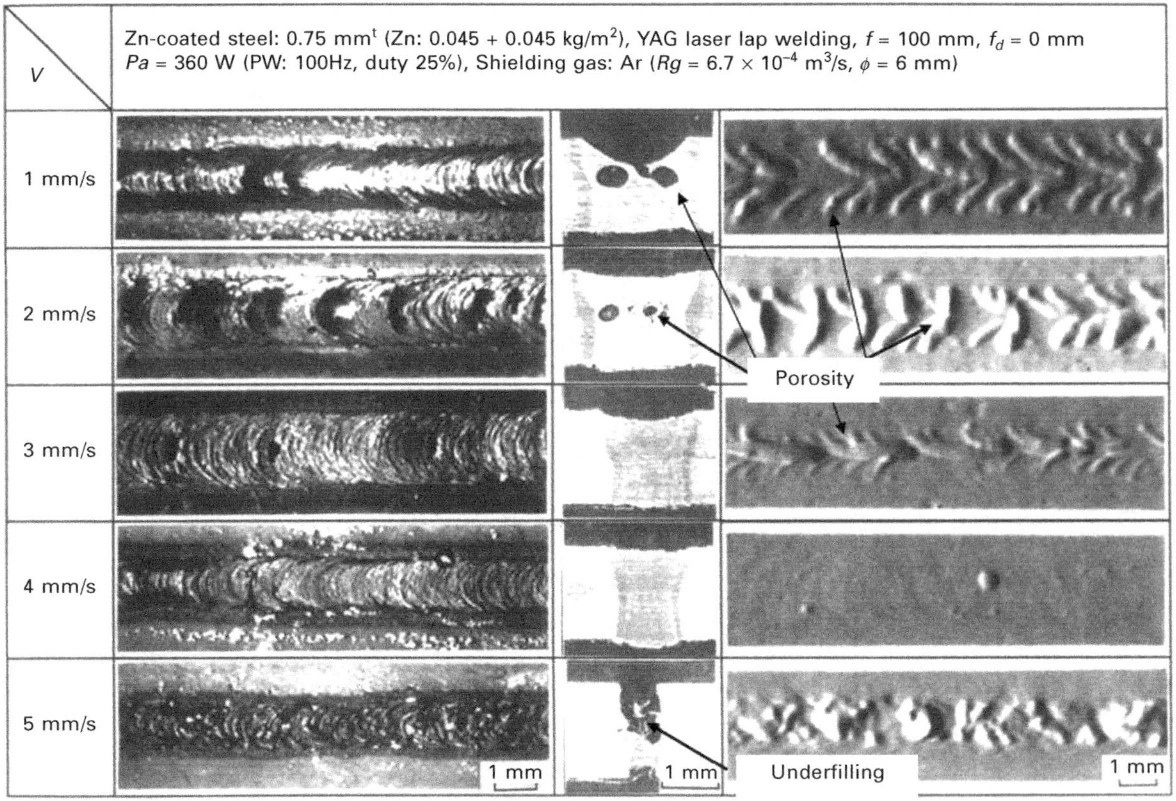

12.16 Surface appearance, cross sections and X-ray inspection results of laser welds produced at 1 – 5 mm/s under pulse-modulation conditions, showing effect of welding speed on formation of porosity and underfilling.

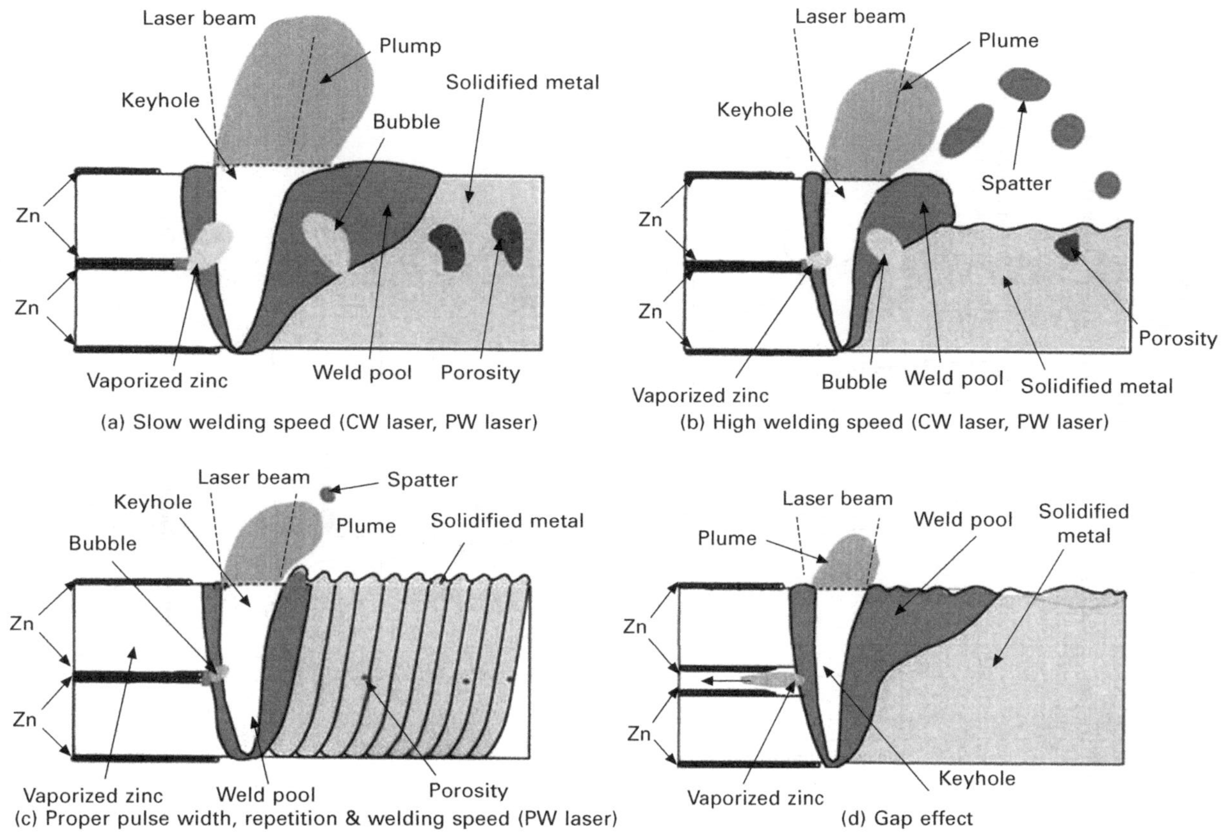

12.17 Phenomena during laser welding of Zn-coated steel lap sheets with or without gap, showing porosity (a), underfilled bead due to spattering (b), reduced porosity due to pulse modulated laser (c), and sound weld bead due to gap effect (d).

boundaries near the nominal solidus temperature, and tensile strains higher than the minimum ductility for cracking initiation are applied at a higher rate than the critical strain rate for cracking occurrence, as illustrated by the example of solidification cracking shown in Fig. 12.18.[2,36] The cracking takes place in the weld fusion zone and in the HAZ near the fusion boundary. The hot cracking in the weld fusion zone and the HAZ is called 'solidification cracking' and 'liquation cracking', respectively.[1,2] In Invar, hot cracking called 'ductility-dip cracking', which has no relationship with liquid films,

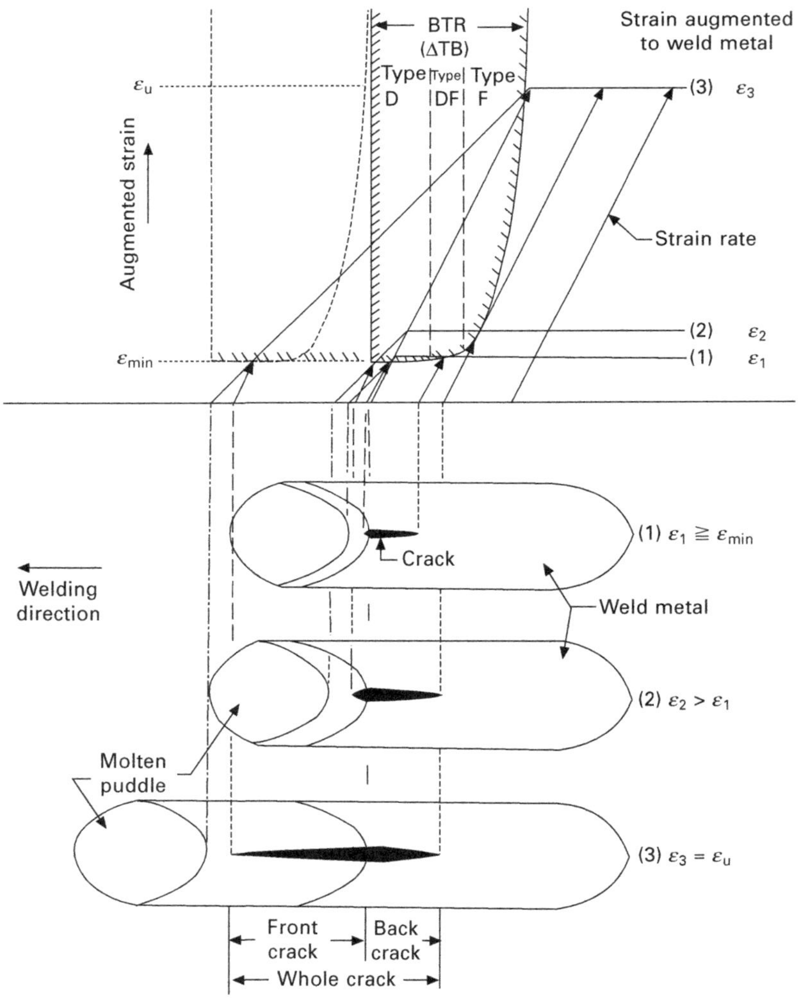

12.18 Strain levels applied at constant strain rate for ductility curve of BTR (brittleness temperature range) during weld solidification, showing crack initiation above the minimum strain and solidification length according to applied strain level.

may occur in the previous weld metal heated as HAZ by the subsequent welding process.[1]

In laser welding, cracking is most likely to occur as solidification cracking along grain boundaries due to the rapid growth of cellular dendrite tips and the consequent formation of low solidification temperature liquid films in the central part of the spot made with a pulsed laser. In spot welding, the tailing laser power is preferable if the area of mushy zone can be narrowed owing to the suppression of the rapid growth of the dendrite tips, as shown in Fig. 12.19.[19]

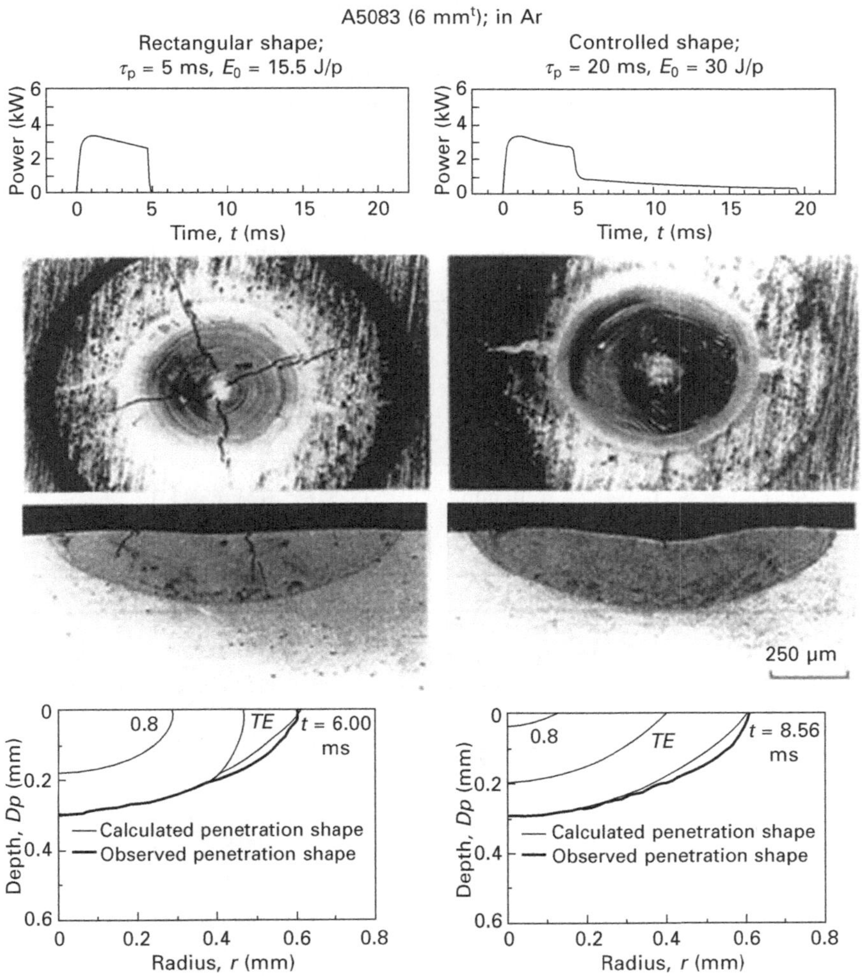

12.19 Surface appearance and cross sections of spot welds made with pulsed lasers of rectangular shape and controlled shape, and calculation result of solid–liquid ratio at initiation of solidification of spot center, showing cracking due to wide solid–liquid zone by rapid solidification and reduced cracking due to narrowed solid–liquid zone with additional level of tailing laser power.

Solidification cracking may take place in deep, partial-penetration weld beads made in thick steel plates with CW high power laser or by hybrid welding with high laser power and high arc current.[17,22] The causes may be attributed to the formation of wide retained molten areas due to melt flows near the bottom part of the laser or hybrid weld beads. The optimization of weld penetration depth or bead geometry or the selection of a filler wire is recommended.[22] In a thick plate of high tensile strength steel, cracking occurs in the weld bead made in a shielding gas (at 1 atm), but is prevented in the weld produced at 16 kW, 1.2 m/min and 0.1 kPa, as shown in Fig. 12.20.[17] Under low vacuum, a sound full-penetration weld bead can be produced in 23 mm thick plate by one-pass welding.[17]

The conditions for cracking in laser-welded A7N01 alloy are shown as functions of plate thickness and welding speed in Fig. 12.21.[2,3] It is considered

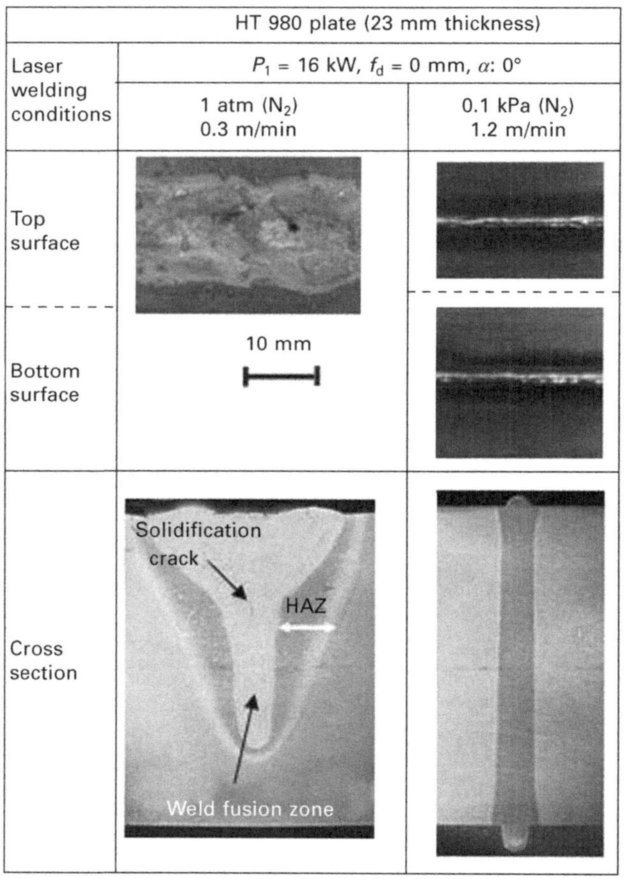

12.20 Partial-penetration and full-penetration welds in HT 980 with disk laser at 1 atm and 0.1 kPa, showing crack and no crack in the weld fusion zone, respectively.

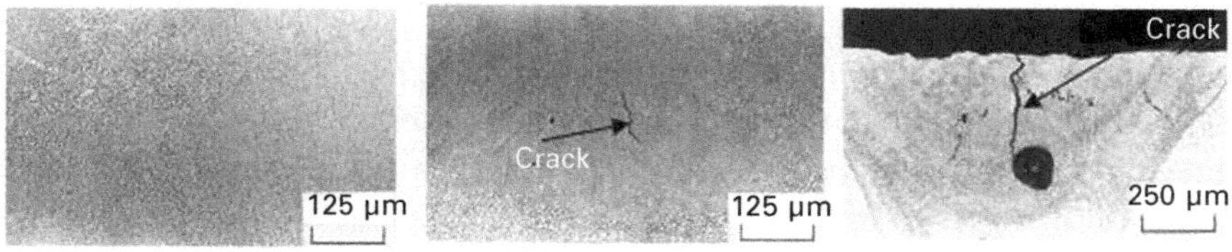

A7N01; P_1 = 5 kW(CW), f = 254 mm, fd = 0 mm, Ar (Rg = 5 × 10^5 mm^3/s, ϕ = 8 mm)

	25 mm/s	50 mm/s	75 mm/s	100 mm/s	125 mm/s	150 mm/s
1 mm^t			○	○	○	○
2 mm^t	○	○	○	○	○	○
3 mm^t	○	○	○	○	▲	▲
4 mm^t	○	○	▲	▲	●	●
5 mm^t	○	○	●	●	●	○
6 mm^t	○	▲	●	●	●	○

○: No cracking ▲: Crack in half sections observed ●: Crack in all sections

12.21 Conditions of cracking formation in laser weld fusion zones of A7N01 alloy, showing effects of welding speed and plate thickness on cracking susceptibility.

in aluminum alloys that solidification cracking is more likely to occur in a thicker plate at a higher welding speed.

Observation of cracking surfaces is important in understanding the occurrence mechanism of solidification cracking. Generally, long cellular dendrites are observed. If the surface is characterized by clear dendrites, cracking should occur at high temperatures near the liquidus. If both cellular-dendritic and flat features are observed, it is supposed that the strain should be rapidly applied to the boundary, and consequently cracking occurs at a dendritic area of higher temperature and propagates along the grain boundary to a lower temperature region. On the other hand, in the HAZ of Ni-based alloys, etc., as the heating speed is fast during laser welding, compositional liquation should occur along the grain boundary between the base metal and an inclusion, which results in a higher possibility of liquation cracking.

The hot cracking susceptibility of aluminum alloys is summarized as a function of alloying content in Fig. 12.22.[2,54–61] In each alloy system, cracking susceptibility increases with an increase in the alloying content to a certain level, and then the susceptibility decreases with the increase in the eutectic content. In normal welding, therefore, cracking prevention is tried by controlling compositions in the molten pool using a filler wire with a high content of Si or Mg.[2,32,54,62]

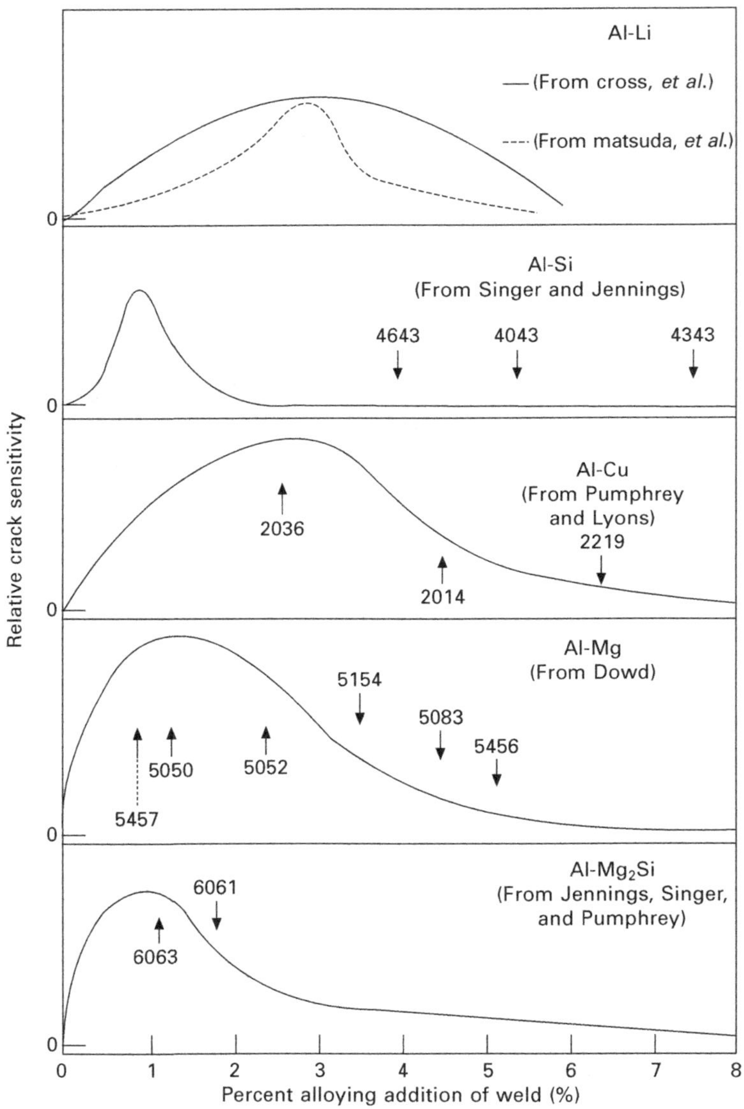

12.22 Effects of alloying element and content on hot cracking susceptibility of welds in aluminum alloys.

From the above results, the relationship between cracking susceptibility, laser welding process and the plate thickness is schematically summarized in Fig. 12.23.[1,2,36] Cracking susceptibility is the highest in spot welding with a pulsed laser, and is the lowest with CW laser welding of thin sheet (of about 1–3 mm thickness) under the proper welding conditions. However, it is noted that a rotational crack may occur when the full-penetration welding of thin sheets starts near the end.

In order to reduce or prevent weld hot cracking, it is necessary to take measures from the viewpoints of both material or metallurgical and mechanical factors. In laser welding, therefore, the following measures are considered:

- proper selection of the base metal, its thickness and weld fusion zone geometry,
- proper control of molten pool compositions by using a filler wire, and
- adoption of proper welding conditions for narrowing a mushy zone as well as suppression of rapid solidification and rapid tensile strain during solidification.

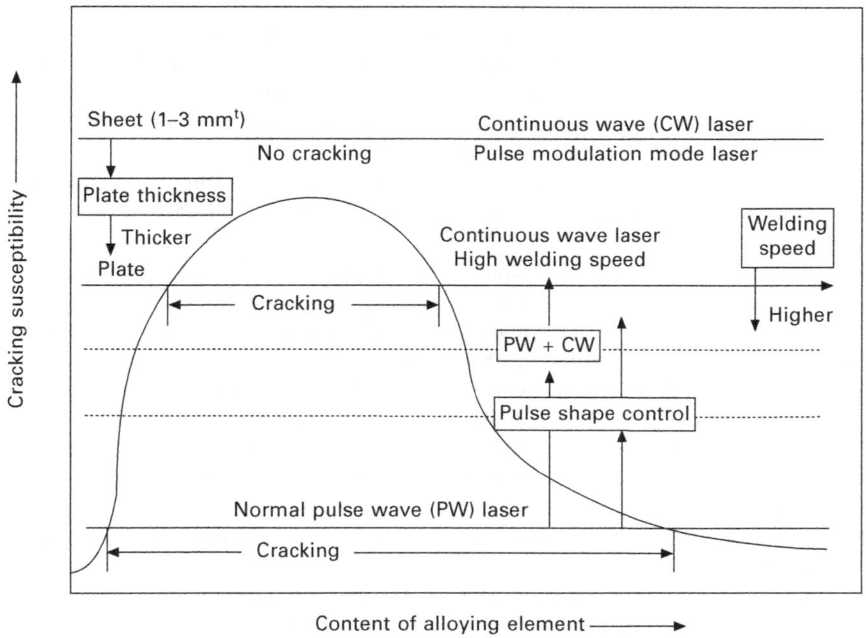

12.23 Schematic representation of effects of various factors on solidification cracking susceptibility of laser welds.

12.5 References

1 Katayama S (2005) 'Formation mechanisms and preventive procedures of laser welding defects', *2004 LMP Symposium on High Quality and New Trend of Laser Welding*, Laser Materials Processing (LMP) Research Committee of Japan Welding Engineering Society (JWES), January, 8-1–8-15 (in Japanese).

2 Katayama S and Matsunawa A (1998) 'Laser weldability of aluminum alloys', *Proc. 43rd Laser Materials Processing Conference*, Japan Laser Processing Society (JLPS), **43**, 33–52 (in Japanese).

3 Suzuki H (1971) *Recent Welding Engineering*, Tokyo: Corona Publishing, **142**, 206 (in Japanese).

4 Japan Welding Society Edition (1990) *Welding and Joining Handbook*, Tokyo: Maruzen, 13–41 (in Japanese).

5 Gabzdyl J, Johnson A, Williams S and Price D. (2003) 'Laser weld distortion control by cryogenic cooling', *Proc. SPIE (LAMP 2002)*, JLPS, Osaka, **4831** 269–274.

6 Mataumoto T. and Yoshida M (1998) 'Welding conditions and mechanical properties of laser welded butt joints of aluminum alloy', *Proc. 43rd Laser Materials Processing Conference*, Japan Laser Processing Society (JLPS), **43**, 71–82 (in Japanese).

7 Kawahito Y, Mizutani M and Katayama S (2009) 'High quality welding of stainless steel with 10 kW high power fibre laser', *Science and Technology of Welding and Joining*, **14**(4), 288–294.

8 Nagayama H, Mizutani M, Kawahito Y and Katayama S (2007) 'Laser welding phenomena of aluminum alloy with high-power fiber laser', *Proc. 10th International Conference on Joints in Aluminium (INALCO 2007)*, Osaka, 137–142.

9 Katayama S, Uchiumi S, Mizutani M, Wang J and Fujii K (2006) 'Penetration and porosity prevention in YAG laser-MIG arc hybrid welding of aluminum alloy', *J. Light Metal Welding & Construction*, **44**(3), 99–109 (in Japanese).

10 Onishi T H, Mizutani M, Kawahito Y and Katayama S (2009) 'High-power fiber laser butt welding of thick high-strength steel plate using sensing system with hot wire', *Proc. 28th International Congress on Applications of Lasers & Electro-Optics (ICALEO 2009)*, LIA, **102**, 773–779.

11 Katayama S, Onishi T H, Mizutani M and Kawahito Y (2010) 'High-power laser butt welding of high-strength steel thick plate', *IIW Annual Meeting 2010*, Commission IV, Turkey, IV-1031-10, 1–7.

12 Wahba M, Mizutani M, Kawahito Y and Katayama S (2010) 'Keyhole stability in disc laser welding of AZ31B and AZ61A magnesium alloys and weld metal properties', *Science and Technology of Welding and Joining*, **15**(7), 559–566.

13 Wang J, Nishimura H, Katayama S and Mizutani M (2006) 'Penetration and porosity prevention in YAG laser-MIG arc hybrid welding of aluminum alloy', *J. Light Metal Welding & Construction*, **44**(3), 99–109 (in Japanese).

14 Katayama S, Wu Y and Matsunawa A (2001) 'Laser weldability of Zn-coated steels', *Congress Proc. of ICALEO 2001 (Laser Materials Processing Conference)*, LIA, Jacksonville, Welding-P520 (CD).

15 Akhter R, Steen W M and Watokins K J (1991) 'Laser welding of Zn-coated steel sheets', *J Laser Applications*, **3**(2), 9–20.

16 Katayama S, Mizutani M, Ikeda H, Nishizawa K And Matsunawa A (1992) 'Effect of pulse shaping on defect reduction in pulsed laser welding', *Proc. ICALEO '92*, LIA, Orlando, FL, **75**, 547–556.

17 Katayama S, Abe Y, Ido R H, Mizutani M and Kawahito Y (2011) 'Deep penetration

welding with high power disk lasers in low vacuum', *Congress Proc. 30th ICALEO*, LIA, Orlando, FL, **104**, 669–678.

18 Katayama S and Lundin C (1991) 'Laser weldability of aluminum alloy 2090', *J. Light Metal Welding & Construction*, **29**(9), 403–415 (in Japanese).

19 Katayama S, Mizutani M and Matsunawa A (1997) 'Modelling of melting and solidification behaviour during laser spot welding', *Science and Technology of Welding and Joining*, **2**(1), 1–9.

20 Katayama S (2010) 'Laser welding of dissimilar materials', *The Review of Laser Engineering*, **38**(8), 594–602 (in Japanese).

21 Stout R D (1987) *Weldability of Steels*, 4th edn, New York: Welding Research Council.

22 Katayama S (2012) 'Laser welding', *Ferrum (Bulletin of The Iron and Steel Institute of Japan)*, **17**(1), 18–29 (in Japanese).

23 Vondrous P, Katayama S and Dunovsky J (2010) 'Specifics of laser welding of ductile iron, Cr-Mo steel and their dissimilar butt joints', *Proc. 74th Laser Materials Processing Conference*, JLPS, **74**, 89–98.

24 Katayama S, Seto N, Mizutani M and Matsunawa A (2001) 'X-ray transmission *in-situ* observation of keyhole during laser spot welding and pulse-shaping for prevention of porosity', *Congress Proc. of ICALEO 2001*, LIA, San Jose, **74**, Session C: Welding-804 (CD).

25 Morita S, Katayama S, Matsunawa A and Hino M (2001) 'Laser weldability of magnesium alloys', High Energy Density Research Committee of Japan Welding Society (JWS), (2001), EBW-495-01.

26 Wahba M, Mizutani M, Kawahito Y and Katayama S (2011) 'Laser welding of die-cast AZ91D magnesium alloy', *Materials and Design*, **33**, 569–576.

27 Katayama S, Seto N, Kim J D and Matsunawa A (1998) 'Formation mechanism and suppression procedures of porosity in high power laser welding aluminum alloys', *Proc. of ICALEO '98*, LIA, Orlando, FL, **85**, Sec. C, 24–33.

28 Katayama S and Lundin C (1991) 'Laser weldability of aluminum alloy 5456', *J. Light Metal Welding & Construction*, **29**(7), 295–307 (in Japanese).

29 Kawahito Y and Katayama S (2005) 'In-process monitoring and adaptive control for stable production of sound welds in laser micro-spot lap welding of aluminum alloy', *J. Laser Applications*, **17**(1), 30–37.

30 Katayama S and Lundin C (1991) 'Laser weldability of various commercially available aluminum alloys', *J. Light Metal Welding & Construction*, **29**(8), 349–360 (in Japanese).

31 Todate A, Ueno Y, Katsuki M, Katayama S and Matsunawa A (2001) 'Porosity formation in carbon steels welded with YAG laser in CO_2 shielding gas', *Pre-Prints of the National Meeting of JWS*, **69**, 46–47.

32 Katayama S, Kim J D and Matsunawa A (1999) 'Cracking susceptibility of laser welds in aluminum alloys, and effect of nitrogen shielding gas on cracking', Symposium on 'Present state and future trend of laser welding of aluminium alloys', Japan Light Metal Welding and Construction Association, Tokyo, 66–77.

33 Katayama S, Kojima K and Matsunawa A (1999) 'CO_2 laser weldability of aluminum alloys (Report 3) – Metallurgical characteristics of laser welds', *J. Light Metal Welding & Construction*, **37**(2), 47–58 (in Japanese).

34 Vollertsen F (2011) 'Laser joining of aluminium alloys – state of the art', *IIW Annual Meeting 2011*, Commission IV, Chennai, IV-1057-11.

35 Katayama S (2000) 'Solidification phenomena of weld metals (2) – Solidification

theory, solute redistribution and microsegregation', *J. Light Metal Welding & Construction*, **38**(4), 168–180 (in Japanese).

36 Katayama S (2000) 'Solidification phenomena of weld metals (3) – Mechanism and susceptibility of solidification cracking', *J. Light Metal Welding & Construction*, **38**(9), 417–427 (in Japanese).

37 Katayama S and Ogawa K (2010) 'Laser weldability and aging characteristics of welds – laser weldability of commercially available A7N01 alloy (1)', *J. Light Metal Welding & Construction*, **48**(12), 463–474 (in Japanese).

38 Katayama S, Kojima K, Kuroda S and Matsunawa A (1999) 'CO_2 laser weldability of aluminum alloys (Report 4) – Effect of welding defects on mechanical properties, deformation and fracture of laser welds', *J. Light Metal Welding & Construction*, **37**(3), 95–103 (in Japanese).

39 Yasuyama M, Uchihara M and Fukui K (2005) *Proc. 64th Laser Materials Processing Conference*, Japan Laser Processing Society (JLPS), **64**, 52–59 (in Japanese).

40 Arata Y, Matsuda F, Shibata Y, Hozomi S, Ono Y and Fujihira S (1975) 'Mechanical properties on electron beam welds of constructional high tension steel (Report III)', *Trans. JWRI*, **4**(2), 181–187.

41 Minami F (2011) 'Evaluation of fracture toughness and structural integrity of laser welds of structural steels', *J. Japan Laser Processing Society*, **18**(2), 94–102 (in Japanese).

42 Inose K (2011) 'Functional assessments of laser welded joint', *J. Japan Laser Processing Society*, **18**(2), 63–68 (in Japanese).

43 Inose K, Oowaki K and Kurata Y (2011) 'Application of laser welding to bridges', LMP Symposium 2011 on Latest Trend of Laser Processing Technology in Japan and Europe, Laser Materials Processing (LMP) Research Committee of Japan Welding Engineering Society (JWES), January 117–125 (in Japanese).

44 Katayama S, Yamaguchi Y, Mizutani M and Matsunawa A (1997) 'Evaluation of mechanical properties of laser-welded aluminum alloy', *Proc. ICALEO'97*, LIA, San Diego, CA, **83**, Part 2, Section G, 93–102.

45 Mizutani M, Yamaguchi Y and Katayama S (2008) 'Fatigue properties of laser-welded aluminum alloy', *J. Light Metal Welding & Construction*, **46**(2), 56–64 (in Japanese).

46 Katayama S, Seto N, Kim J D and Matsunawa A (1997) 'Formation mechanism and reduction method of porosity in laser welding of stainless steel', *Proc. ICALEO '97*, LIA, San Diego, CA, **83**, Part 2, Section G, 83–92.

47 Katayama S and Kawahito Y (2007) 'Visualization of welding phenomena with high speed video camera and X-ray transmission real-time observation system', *J. High Temperature Society*, **33**(3), 118–127 (in Japanese).

48 Katayama S, Kohsaka S, Mizutani M, Nishizawa K and Matsunawa A (1991) 'Pulse shape optimization for defect prevention in pulsed laser welding of stainless steels', *Proc. of ICALEO '93*, LIA, Orlando, FL, **77**, 487–497.

49 Ono K, Adachi H, Katada K, Miyoshi M, Miyamoto I and Inoue T (2002) 'Laser welding phenomena of steel plate with oxide film – Effect of oxide film in high power CO_2 laser welding (Report I)', *Quaternary J. of Japan Welding Society*, **20**(2), 213–219 (in Japanese).

50 Graham M P, Hirak D M, Kerr H W and Weckman D C (1993) 'Nd:YAG laser welding of coated sheet steel', *Proc. of ICALEO '93*, LIA, **77**, 651–660.

51 Gu H (2000) 'A new method of laser lap welding of zinc-coated steel sheet', *Proc of ICALEO*, LIA, Dearborn, **89**, Section C, 1–6.

52 Ono M, Shinbo Y, Yoshitake A and Ohmura M (2003) 'Welding properties of thin steel sheets by laser-arc hybrid welding – Laser focused arc welding', *Proc. SPIE (LAMP 2002)*, JLPS, **4831** Osaka, 369–374.
53 Dasgupta A, Mazumder J and Bembenk M (2000) 'Alloying based laser welding of galvanized steel', *Proc. of ICALEO 2000*, LIA, Dearborn, **91**, Section A, 38–45.
54 Martukanitz R P (1988) 'The metallurgy of aluminum welding', Product Manufacturing Technology Division Report Number 52–88–06 Alcoa Technical Center, PA.
55 Singer A R E and Jennings P H (1947) 'Hot-shortness of the aluminum–silicon alloys of commercial purity', *J. Inst. Metals*, **74**, 197–212.
56 Dowd J D (1952) 'Welding cracking of aluminum alloys', *Weld. J.*, **31**, 10448s–10456s.
57 Cross D E, Olson D L, Edowards G R and Capes J F (1984) 'Weldability of aluminum–lithium alloys', *Aluminum–Lithium Alloy II*, eds. T. H. Sanders and E. A. Starke, TMS/AIME, Warrendale, PA, 675–682.
58 Matsuda F, Nakata K and Tsukamoto K (1987) 'Effect of Li on solidification cracking susceptibility of aluminum alloy', *Preprints of the National Meeting of J.W.S.*, **41**, October, 308–309 (in Japanese).
59 Jennings P H, Singer A R E and Pumphrey W I (1948) 'Hot shortness of some high-purity alloys in the systems aluminum–copper–silicon and aluminum–magnesium–silicon', *J. Inst. Metals*, **74**, 15–52.
60 Pumphrey W I and Moore D C (1948) 'A consideration of the nature of brittleness at temperature below the solidus in castings and welds in aluminum alloys', *J. Inst. Metals*, **75**, 257–267.
61 Pumphrey W I and Moore D C (1948/49) 'Cracking during and after solidification of some Al–Cu–Mg alloys of high purity', *J. Inst. Metals*, **74**, 425–438.
62 Binroth C A, Zuo T C and Sepold G (1990) 'CO_2-laser welding with filler material of high strength aluminum alloys', *Proc. of Power Beam Technology*, September, 119–127.

13
Residual stress and distortion in laser welding

H. MURAKAWA, JWRI, Osaka University, Japan

DOI: 10.1533/9780857098771.2.374

Abstract: Typical examples of welding distortion and residual stress are presented first and the mechanism by which they are produced is explained using a simple mechanical model. Further, influential factors on various types of welding deformations and the distribution of residual stress are discussed based on the temperature distribution and the restraint. To understand the welding distortion and the residual stress associated with laser welding, typical examples selected from published articles are presented.

Key words: residual stress, distortion, transverse shrinkage, angular distortion, longitudinal shrinkage, buckling, thermal cycle, mismatch, inherent strain, inherent deformation, inherent force, heat input, simulation, prediction, control.

13.1 Introduction

Fusion weldings, such as arc, electron beam weldings and laser welding, are commonly used to assemble a wide range of metal structures, ships, automobiles and passenger trains for example, because of their high productivity and advantages in weight saving. However, distortion and residual stress are formed as unavoidable consequences. Figure 13.1 shows the typical forms of welding distortions. Transverse shrinkage, longitudinal shrinkage and angular distortions are the three fundamental modes of welding deformations. The welding distortions observed in real welded structures are a combination of these modes. When the plate is thin and the heat input is large relative to the plate thickness, buckling type distortion can be produced by welding. Welding distortion is one of the major causes of the dimensional error of welded structures and the deflection of panels reduces the buckling and the load carrying capacity of thin plate welded structures. If the welding distortion exceeds the tolerable limit, it may cause delay in production schedules and reduces the productivity due to additional correction works.

Figure 13.2 illustrates the distribution of welding residual stress on a butt welded plate. Figure 13.2(a) shows the distribution of longitudinal residual stress (component in welding direction) along the welding direction and the transverse section at the center. On the center section, the largest tensile

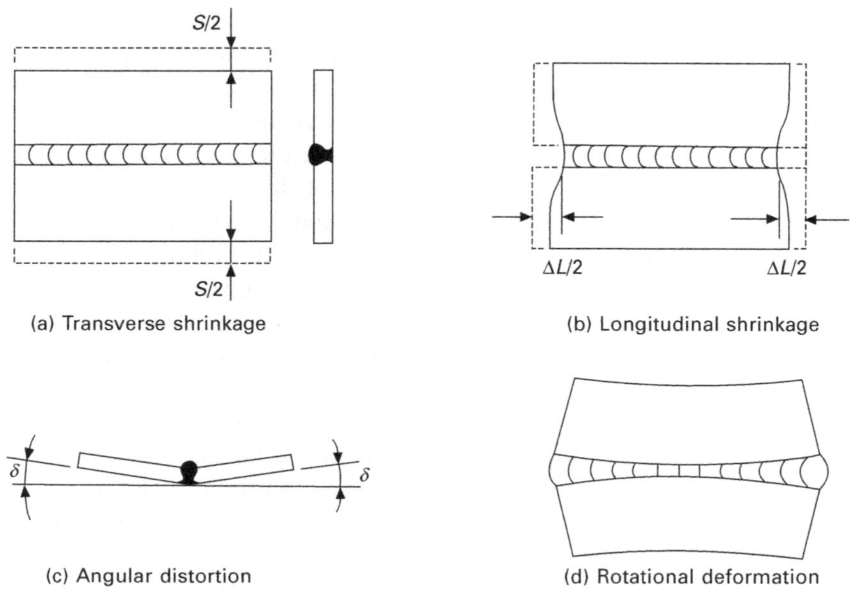

13.1 Typical welding distortions: (a) transverse shrinkage; (b) longitudinal shrinkage; (c) angular distortion; (d) rotational deformation.

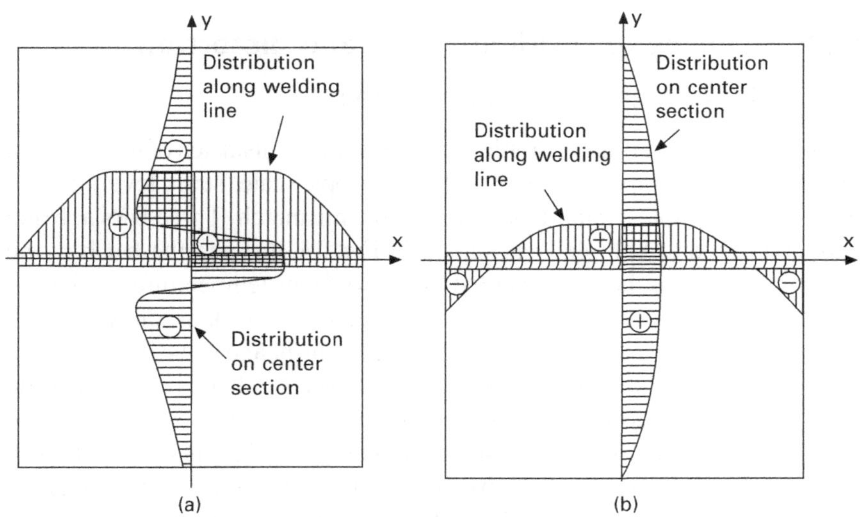

13.2 Schema of welding residual stress distribution: (a) stress in welding direction (b) stress in transverse direction.

residual stress is observed in the location around the welding line and the value reaches almost the yield stress when the material is mild steel. To keep the balance, small compressive stress is produced in the area away from the weld line. As for the distribution along the welding line, the longitudinal residual stresses at both ends become zero because of the free boundary. In the same manner, the distribution of the transverse residual stress is shown in Fig. 13.2(b). The transverse residual stress is tensile at the center and it is compressive at the start and finish ends of the welding line. The transverse residual stress becomes zero along both edges of the plate.

As will be discussed in this chapter, the magnitude of the welding distortion and the residual stress are strongly influenced by the welding heat input. Generally, the magnitude of the welding distortion and the residual stress increase with the heat input. In this sense, high energy density welding such as laser welding is advantageous because it requires smaller heat input compared to conventional arc welding.

The welding distortion and the residual stress are produced as an undesirable consequence of the thermal cycle generated by the heat concentrated in a small area. The same mechanism can be utilized to form or bend plate to a desired curved shape by controlling the heat given to the plate. Since laser is superior in controllability, it can be used to form flat plate to arbitrary 3D curved shapes for shipbuilding applications and it can be employed for fine adjustment of the geometry of machine parts. Also it can be employed to mitigate or improve the residual stress distribution of welded structures, such as the pressure vessels in power plant.

13.2 Causes of residual stress and distortion

13.2.1 Mismatch and inherent strain

Welding residual stress and distortion are produced by mismatch of macroscopic scale (structural component level) and microscopic scale (strain level). Figure 13.3 shows a simple example of macroscopic mismatch. Three bars a, b and c are connected at both ends as shown in Fig. 13.3(a) and the bar b in the center is melted by heat and cooled down to room temperature. Since bar b shrinks during cooling, the connected three bars shrink as shown in Fig. 13.3(b). The deformation corresponding to this shrinkage is the displacement u and compressive stresses σ_a and σ_c are produced in bars a and c. Tensile stress σ_b is produced in bar b. Figure 13.3(c) shows the lengths of three bars separated by cutting. Because of the thermal cycle, the length of bar b becomes shorter than the other two by ΔL. This difference of length ΔL is the mismatch and it produces the distortion u and the residual stresses σ_a, σ_b and σ_c.

At this stage, it is worth noting that the distortion and the residual stress can be estimated if the mismatch ΔL is known. Let A_a, A_b, A_c be the cross-sectional area of three bars and E is the Young's modulus. Then the deformation of the three bars joined together u and the residual stresses σ_a, σ_b, σ_c are obtained as:

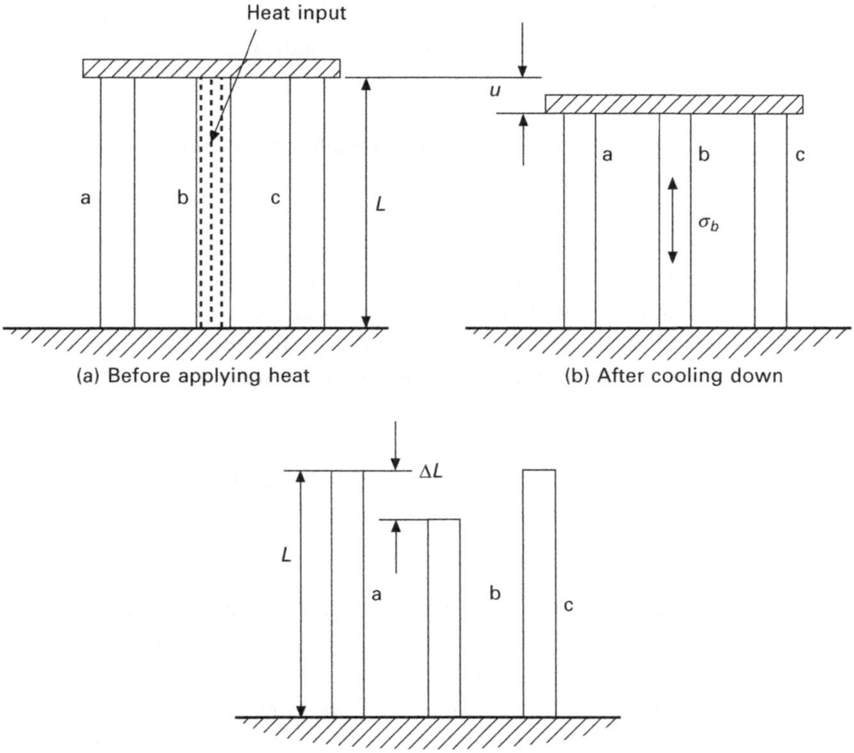

13.3 Deformation and force produced by mismatch.

$$u = \Delta L\, k(k + k^*) = (1 - \beta)\, \Delta L \quad [13.1]$$

$$\sigma_a = \sigma_c = -E(u/L) = -(1 - \beta) E(\Delta L/L) \quad [13.2]$$

$$\sigma_b = E(\Delta L - u)/L = \beta E(\Delta L/L) \quad [13.3]$$

In the above equations, k represents the stiffness of bar b and k^* represents the sum of the stiffnesses of bars a and c, i.e.:

$$k = EA_b/L \quad [13.4]$$

$$k^* = E(A_a + A_c)/L \quad [13.5]$$

Then the parameter β defined by the following equation means the degree of restraint from bars a and c on bar b.

$$\beta = k^*/(k + k^*) \quad [13.6]$$

When the restraint is small, in other words ($k^* << k$), $\beta \approx 0$. If the restraint is large ($k^* >> k$), $\beta \approx 1$. Applying these relations to Eq. [13.1] and Eq. [13.3], it is readily seen that the mismatch turns into the deformation, and stress is not produced when the restraint is extremely small, i.e:

$$u \approx \Delta L, \qquad \sigma_b \approx 0 \qquad \text{when } \beta \approx 0 \qquad [13.7]$$

If the restraint is extremely large, the mismatch turns into stress and no deformation is produced, i.e.:

$$u \approx 0, \qquad \sigma_b \approx E(\Delta L/L) \qquad \text{when } \beta \approx 1 \qquad [13.8]$$

The microscopic mismatch can be explained using decomposition of strain into its components. The total strain ε can be decomposed into the sum of the elastic strain ε^e, the plastic strain ε^p, the thermal strain ε^T, the creep strain ε^c and that produced through phase transformation ε^t, i.e:

$$\varepsilon = \varepsilon^e + \varepsilon^p + \varepsilon^T + \varepsilon^c + \varepsilon^t \qquad [13.9]$$

Noting that the deformation and the stress are produced by the total strain ε and the elastic strain ε^e, Eq. [13.9] can be rearranged to:

$$\varepsilon - \varepsilon^e = \varepsilon^p + \varepsilon^T + \varepsilon^c + \varepsilon^t = \varepsilon^* \qquad [13.10]$$

This equation means that the distortion ε and the residual stress ε^e are produced by the inherent strain ε^* which consists of the plastic, the thermal, the creep strains and that caused by the phase transformation. The inherent strain is sometimes called 'misfit strain'. Though the detail is not explained here, the distortion and the residual stress after welding can be computed by elastic analysis if the distribution of the inherent strain ε^* is known as in the case of the mismatch ΔL in the three-bar model.

13.2.2 Inherent strain and deformation in thin plate

Welding deformations in thin plate, such as the transverse shrinkage, longitudinal shrinkage and angular distortion shown in Fig. 13.1, are mostly produced by the longitudinal and the transverse inherent strains ε_x^* and ε_y^* where x, y and z are coordinates taken in the welding, the width and the thickness directions. By integrating the inherent strain over the cross section normal to the welding line and taking the average through the thickness h, the inherent deformations are obtained:

$$\text{Longitudinal shrinkage} \qquad \delta_L^* = \frac{1}{h}\int \varepsilon_x^* dydz \qquad [13.11]$$

$$\text{Transverse shrinkage} \qquad \delta_T^* = \frac{1}{h}\int \varepsilon_y^* dydz \qquad [13.12]$$

$$\text{Longitudinal bending} \qquad \theta_L^* = \frac{1}{l}\int z\varepsilon_x^* dydz \qquad [13.13]$$

Transverse bending $$\theta_T^* = \frac{1}{I}\int z\varepsilon_y^* dydz \quad [13.14]$$

where $$I = \frac{1}{12}\int z^2 dydz$$

Among the above four components of the inherent deformation, three components except the longitudinal bending are important components in the distortion of welded structures. The inherent deformations can be calculated from the computed results by finite element analysis according to Eqs [13.11]–[13.14]. They can be also measured by experiments and used to predict the welding distortion of large-scale complex real structures (Murakawa *et al.*, 1998).

13.2.3 Inherent stress and inherent force in thin plate

As the inherent deformation is defined based on the concept of the inherent strain, inherent stress and inherent force can be defined as relevant concepts. Through the constitutive relation, the inherent strains ε_x^*, ε_y^* are transformed into inherent stresses σ_x^*, σ_y^*:

Transverse inherent stress $$\sigma_x^* = \frac{E}{1-\nu^2}(\varepsilon_x^* + \nu\varepsilon_y^*) \quad [13.15]$$

Longitudinal inherent stress $$\sigma_y^* = \frac{E}{1-\nu^2}(\nu\varepsilon_x^* + \varepsilon_y^*) \quad [13.16]$$

Further, the inherent forces and the inherent moments can be defined by the following equations.

Longitudinal inherent force $$F_L^* = E\int \varepsilon_x^* dydz = Eh\delta_L^* \quad [13.17]$$

Transverse inherent force $$F_T^* = E\int \varepsilon_y^* dydz = Eh\delta_T^* \quad [13.18]$$

Longitudinal inherent moment $$M_L^* = E\int z\varepsilon_x^* dydz = EI\theta_L^* \quad [13.19]$$

Transverse inherent moment $$M_T^* = E\int z\varepsilon_y^* dydz = EI\theta_T^* \quad [13.20]$$

As explained, the mismatch (inherent deformation) turns into deformation when the restraint is small and it turns into force when the restraint is large. It is natural to use the inherent strain or deformation for the transverse direction in which the restraint is small and to use the inherent stress or force for the longitudinal direction in which the restraint is large.

13.3 Mechanism of formation of longitudinal and transverse shrinkage of welded joints

13.3.1 Inherent strain in three-bar model subjected to thermal cycle

As discussed, the welding distortion and the residual stress are caused by the mismatch in microscopic and macroscopic scales or the inherent strain and the inherent deformation. In this section, the mechanism by which the inherent strain and the inherent deformation are produced is explained using a simple model that consists of three bars, a, b and c as shown in Fig. 13.4. In this model, there is no mismatch in the beginning and the initial length of the three bars is L.

If the temperature of bar b is raised by T°C in the heating process and the thermal expansion ratio is α, the length of bar b in stress-free state is increased by αTL due to the thermal expansion. The stress σ_b acting on bar b can be obtained by substituting this value into Eq. [13.3] as the mismatch ΔL, i.e.:

$$\sigma_b = -\beta E\alpha T \tag{13.21}$$

Assuming E = 200 GPa, $\alpha = 10^{-5}$ 1/°C, the thermal stress induced in the steel bar which is fully restrained ($\beta = 1$) becomes

$$\sigma_b = -2.0T \text{ MPa} \tag{13.22}$$

If the yield stress is 300 MPa, the stress reaches the yield stress when the temperature becomes 150°C and plastic deformation is produced during the further increase in temperature.

As a general case, the case in which $0 < \beta < 1$ and the thermal cycle is 0°C → T_{max} °C → 0°C is considered. The time history of the stress during

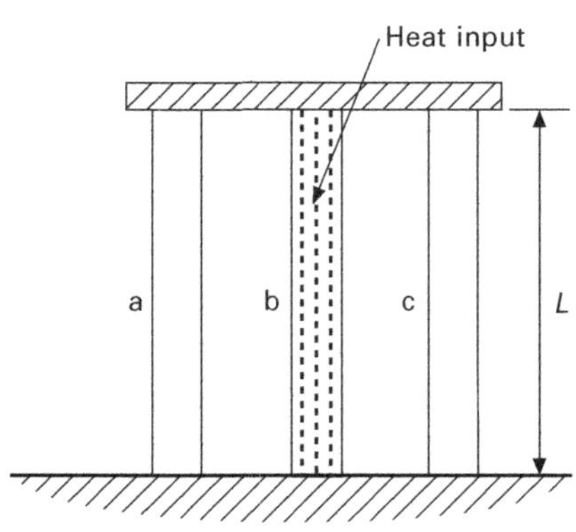

13.4 Three-bar model under thermal cycle.

the thermal cycle can be divided into three types depending on the maximum temperature reached T_{max} as shown in Fig. 13.5:

1. $T_{max} < T_1$: (O → A → O)

   Since the thermal stress remains below the yield stress, no plastic strain or inherent strain ε_b^* is produced.

2. $T_1 < T_{max} < T_2$ (O → B → C → D)

   Compressive plastic strain is produced only in the heating process.

3. $T_2 < T_{max}$ (O → B → G → H → F)

   Compressive plastic strain is produced in the heating process. Then tensile plastic strain is produced in the cooling process. As a result, compressive plastic strain remains as the inherent strain.

Though the details of the derivation are not shown, the relation between the highest temperature reached T_{max}, the inherent strain ε_b^* and the residual stress σ_b are summarized as follows:

1. $$T_{max} < T_1 \quad \varepsilon_b^* = 0 \qquad [13.23]$$

   $$\sigma_b = 0 \qquad [13.24]$$

2. $$T_1 < T_{max} < T_2 \quad \varepsilon_b^* = (T_1 - T_{max})\alpha \qquad [13.25]$$

   $$\sigma_b = \beta E\alpha(T_{max} - T_1) \qquad [13.26]$$

3. $$T_2 < T_{max} \quad \varepsilon_b^* = -\varepsilon_Y/\beta \qquad [13.27]$$

   $$\sigma_b = \sigma_Y \qquad [13.28]$$

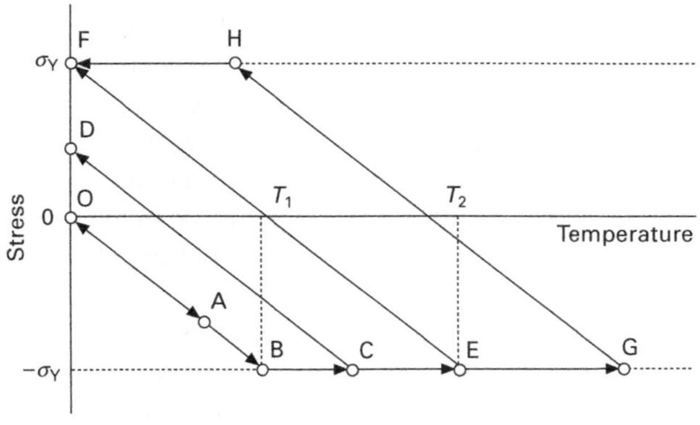

13.5 Stress in bar b under thermal cycle.

where

$$T_1 = \sigma_Y/\beta E\alpha \quad [13.29]$$

$$T_2 = 2T_1 \quad [13.30]$$

$$\varepsilon_Y = \sigma_Y/E \quad [13.31]$$

The above equations can be used to calculate the longitudinal shrinkage δ_L^* or the tendon force F_L^* as will be discussed in the next section. Also, Eqs [13.29] and [13.27] tell us that critical temperature T_1, which corresponds to the transition point of the inherent strain, becomes high and the maximum value of the inherent strain ε_b^* becomes large when the restraint β is small.

13.3.2 Welding deformations in longitudinal and transverse direction

If the temperature distribution in a structure is uniform and there is no external restraint, neither residual stress nor distortion is produced by the thermal cycle. However, the metal is melted and joined by applying concentrated heat to a small area in welding. The temperature distribution in welding is not uniform and it changes with time. Figure 13.6 illustrates the temperature distribution during butt welding. The highest temperature is observed at the position of the heat source and the equitherms near the heat source are ellipsoids close to circles. As they move away from the heat source, the equitherms become long ellipsoids. This means that the temperature gradient near the heat source is large and that in the welding direction and the transverse direction is roughly of the same order. At the point behind the heat source, the temperature gradient in the transverse direction is large while that in the welding direction is small.

The temperature gradient can be related to the restraint using the three-bar model. Figure 13.7 illustrates the temperature distribution and the restraint in the welding direction which produce longitudinal residual stress. Since the temperature gradient is large in the transverse direction, the area close to the weld can be considered as the heated bar b in the center. The remaining areas on both sides of the weld, corresponding to bars a and c, have large stiffness and their temperature is close to room temperature. The temperature distribution in each bar is assumed to be uniform for simplicity. It is readily understood that large tensile residual stress is produced in the weld zone due to the large restraint and the shrinkage in the welding direction is small for the same reason.

Similarly, the illustration for the transverse stress and deformation is given in Fig. 13.8. As shown in the figure, the temperature gradient in the heating and the cooling processes are different. The temperature gradient in the welding direction in the heating process is large while that in the cooling

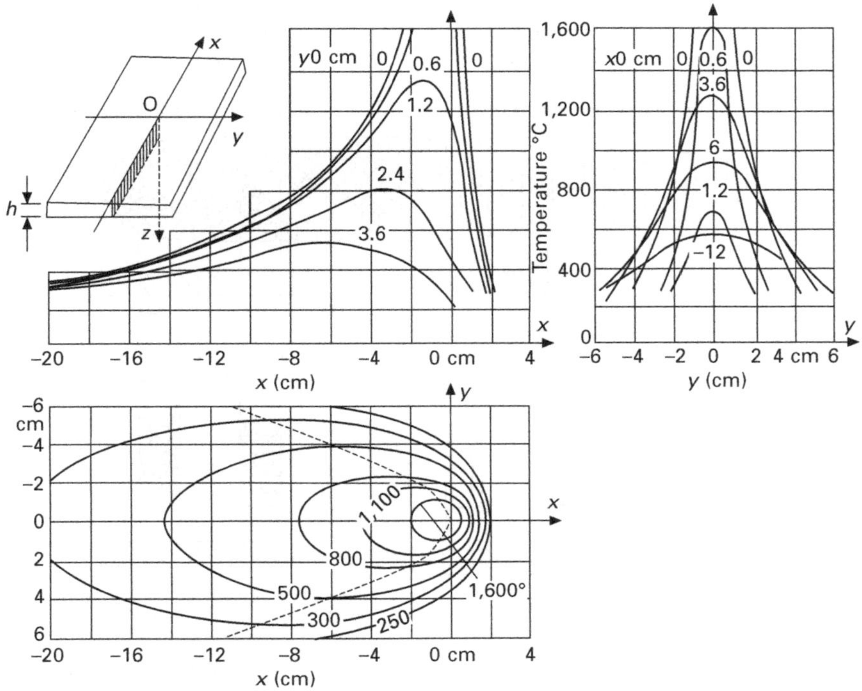

13.6 Schema of transient temperature distribution during butt welding.

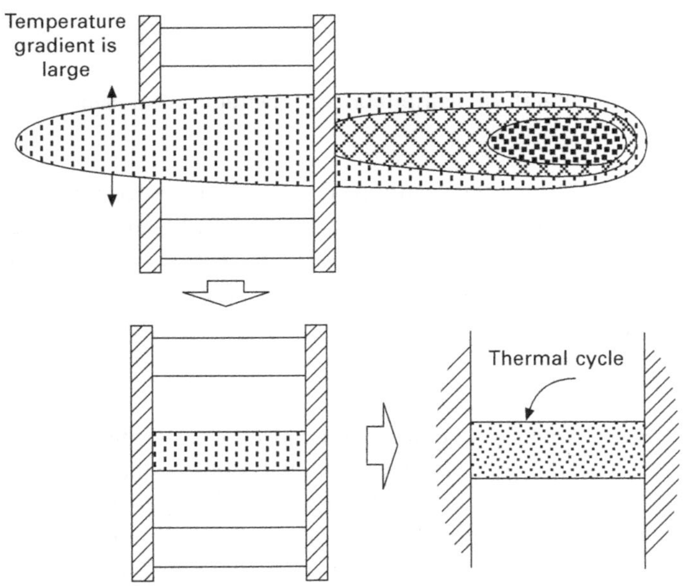

13.7 Temperature distribution and restraint in welding direction.

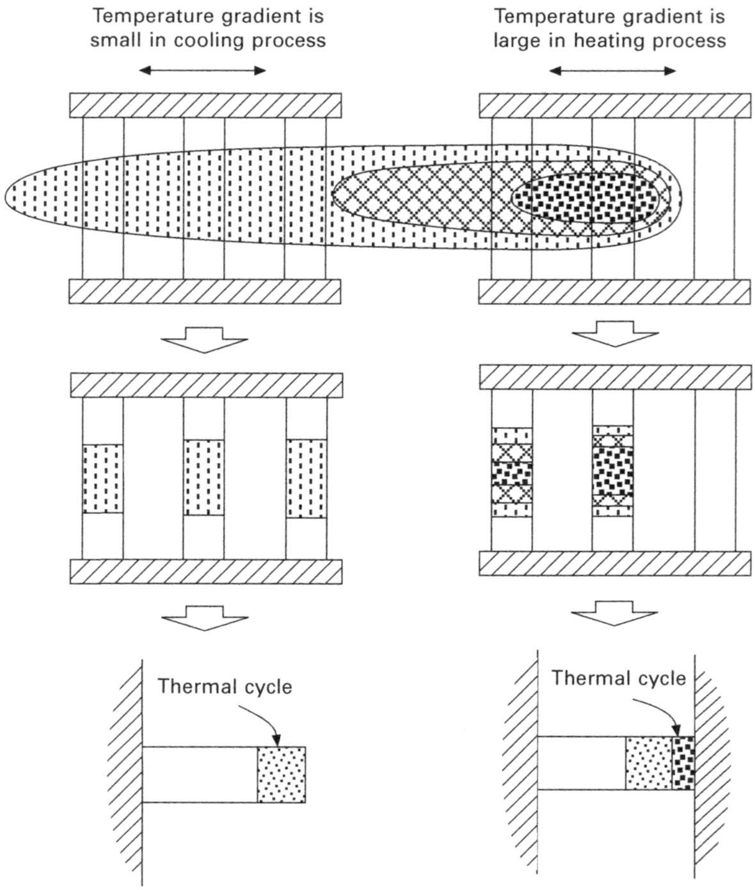

13.8 Temperature distribution and restraint in transverse direction.

process is small. Since the restraint in the heating process is large and the yield stress of the bars in the center and that just behind the heat source is small, the thermal expansion of these bars is fully transformed into the compressive plastic deformation. Thus, the shrinkage of these bars due to the plastic deformation δ_T^* is equal to the thermal expansion, i.e.:

$$\delta_T^* = \alpha Q/c\rho h \qquad [13.32]$$

where Q is heat input, ρ is density, c is specific heat, and h is thickness of plate. In the cooling process, the restraint becomes small due to the small temperature gradient in the longitudinal direction. Since the restraint and the stress are small, additional plastic deformation is not produced and the width of the plate becomes smaller by $\delta_T^* = \alpha Q/c\rho h$ after welding.

As is demonstrated through a simple three-bar model, the processes by which the stress and the deformation are formed are different between the welding and the transverse directions.

13.4 Influential factors on welding distortion and residual stress

The fundamental components of local deformations or the inherent deformations are the transverse shrinkage, the longitudinal shrinkage and the angular distortion shown in Fig. 13.1. The deformations, such as the longitudinal bending of beam, the buckling deformation and the twisting deformation shown in Fig. 13.9, are produced by the combination of the above fundamental components. Thus, if these inherent deformations are known, the distortion of the welded structure can be predicted by an elastic analysis (Murakawa *et al.*, 1998).

13.4.1 Transverse shrinkage and angular distortions

In case of arc welding, Satoh and Terasaki (1976) showed that the transverse shrinkage δ_T^* can be consistently related to the heat input parameter Q/h^2, where Q and h are welding heat input per unit length and thickness of plate, respectively. Figure 13.10 shows the relation between the transverse shrinkage and the heat input parameter Q/h^2 for mild steel under arc welding. The heat input Q can be estimated by the following formula:

$$Q = \eta IU/v \qquad [13.33]$$

where η is heat efficiency, I is welding current, U is welding voltage and v is welding speed.

The parameter Q/h^2 indicates the relative size of heat input compared to the plate thickness. When this parameter is small, only the part of the metal close to the surface along the weld bead is heated above the mechanical melting point (in the case of mild steel, it is roughly 700–800°C). When the parameter is greater than 20 J/mm^3, the temperature reaches the mechanical melting point throughout the thickness. In such cases, the measured points

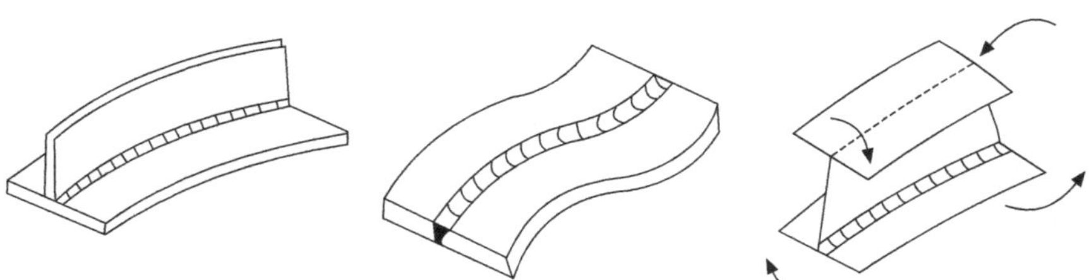

13.9 Longitudinal bending of beam, buckling deformation and twisting deformation.

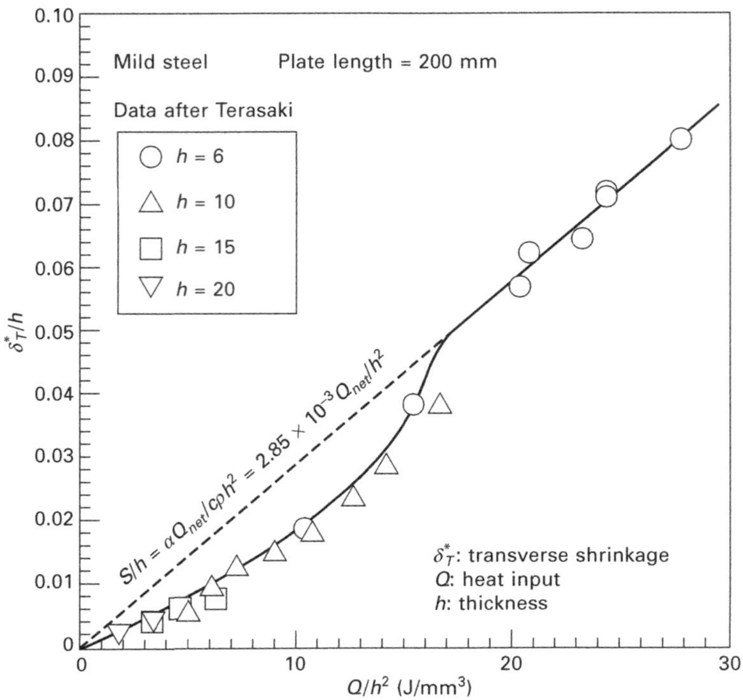

13.10 Influence of heat input on transverse shrinkage under bead on plate welding.

are plotted along a line and the linear relation holds between the transverse shrinkage δ_T^* and the parameter Q/h^2. This corresponds to the fact that the transverse shrinkage is roughly equal to the thermal expansion caused by the heat input. If mild steel is taken as an example:

$$\alpha \approx 1.2 \times 10^{-5}\ 1/°\text{C}$$

$$c \approx 4.6 \times 10^{2}\ \text{J/kg.°C}$$

$$\rho \approx 7.8 \times 10^{-6}\ \text{kg/mm}^3$$

By substituting the above material constants into Eq. [13.32], the following relation is obtained:

$$\delta_T^* / h = \alpha Q / c\rho h^2 = (3.3 \times 10^{-3})\ Q/h^2 \qquad [13.34]$$

The proportional constant 3.3×10^{-3} is close to the slope of the line in Fig. 13.10, which is 2.85×10^{-3}. If the heat input parameter is smaller than 20 J/mm^3, the shrinkage becomes smaller than the value corresponding to this line. This can be explained by the fact that the restraint from the region near the back surface of the plate, where the temperature is below the mechanical melting temperature, becomes large.

Similarly, Fig. 13.11 shows the relation between the angular distortion θ_T^* and the net heat input Q. The angular distortion is produced by the through-thickness temperature gradient. This is observed when the heat input is small and the plate is melted partially in the thickness direction. The angular distortion increases almost linearly with the heat input parameter up to $Q/h^2 = 10$ GJ/m^3. When the parameter is larger than this, the angular distortion decreases with the parameter because the through-thickness temperature gradient becomes small.

13.4.2 Longitudinal shrinkage

Unlike the transverse shrinkage and the angular distortion, which can be directly represented as deformations, the longitudinal shrinkage is represented as the tendon force F_L^*. As explained, the inherent deformation is reflected directly to the deformation when the restraint is small, while it is reflected mainly to the force or the stress when the restraint is large. Thus, it is quite natural to describe the longitudinal shrinkage in terms of the force.

According to White *et al.* (1980), the tendon force F_L^* is proportional to the net heat input Q, i.e.:

$$F_L^* = aQ \quad \text{(N)} \tag{13.35}$$

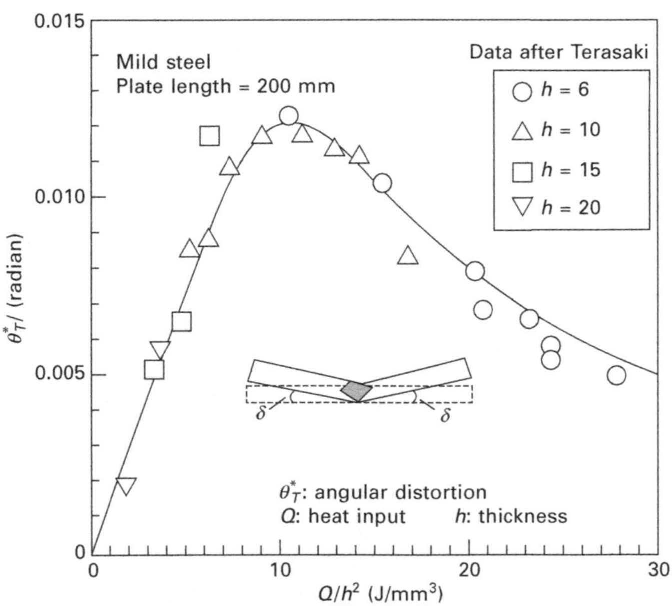

13.11 Influence of heat input on angular distortion under bead on plate welding.

White *et al.* proposed 200 as the proportional constant a. As discussed in the above, the transverse shrinkage, the angular distortion and the longitudinal shrinkage are basically proportional to the heat input Q.

The formula given by Eq. [13.35] can be derived from Eqs [13.24], [13.26] and [13.28]. When the butt welding of two plates with large enough width is considered, the restraint in the welding direction is very large so that the plane normal to the weld line can be assumed to be fully fixed. When the plate is thin, the highest temperature reached at the point y away from the center line of the weld $T_{max}(y)$ is approximated as:

$$T_{max}(y) = 0.242\frac{Q}{c\rho h}\frac{1}{y} \qquad [13.36]$$

Using Eqs [13.24], [13.26], [13.28] and [13.36], the distribution of the residual stress component in the welding direction can be drawn as in Fig. 13.12. Further its integration over the transverse cross section gives the tendon force F_L^*:

$$F_L^* = 0.335(E\alpha/c\rho)Q = aQ \qquad [13.37]$$

If the material is mild steel, the constant α becomes 233 and this value roughly agrees with the value proposed by White *et al.* Further dividing the tendon force by the stiffness of the cross section EBh, average strain in the longitudinal direction ε_L is obtained:

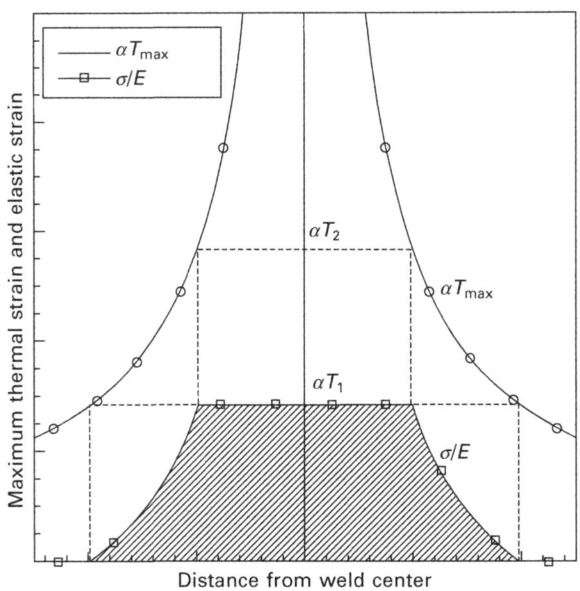

13.12 Distribution of longitudinal residual stress.

$$\varepsilon_L = F_T/EBh = 0.335(\alpha/c\rho Bh)Q \quad [13.38]$$

where B is the breadth of the plate. It is noticed from the above equation that the Young's modulus is involved in the tendon force but not in the longitudinal strain ε_L.

As shown in Fig. 13.12, the widths of the residual stress distribution B_1^* and B_2^* are determined by the temperature T_1 and T_2. From Eq. [13.36], T_1, T_2 and B_1^*, B_2^* are related through the following equations:

$$T_1^* = 0.242\frac{Q}{c\rho h}\frac{2}{B_1^*}, \qquad T_2^* = 0.242\frac{Q}{c\rho h}\frac{2}{B_2^*} \quad [13.39]$$

Noting that

$$T_1 = \sigma_Y/E\alpha \qquad \text{and} \qquad T_2 = 2\sigma_Y/E\alpha \quad [13.40]$$

the widths of residual stress distribution B_1^* and B_2^* are given by:

$$B_1^* = 0.242\frac{Q}{c\rho h}\frac{2E\alpha}{\sigma_Y} \qquad B_2^* = 0.242\frac{Q}{c\rho h}\frac{E\alpha}{\sigma_Y} = \frac{B_1^*}{2} \quad [13.41]$$

The above equation tells us that the width of the residual stress distribution is also proportional to the heat input Q. And they are proportional to the Young's modulus E and the thermal expansion ratio α, and they are inversely proportional to the yield stress σ_Y, specific heat c, density ρ and thickness h.

13.4.3 Buckling due to welding

Since the buckling is a nonlinear problem, it is difficult to explain it in a simple manner. Noting that the phenomena is linear up to the onset of buckling, it is relatively easy to discuss the critical heat input Q_{cr} at which the plate starts to buckle by the residual stress produced by welding, and it can be easily shown that Q_{cr} is proportional to $(c\rho/\alpha)(h^3/B)$. For simplicity, let's consider a square plate with stiffeners welded along four edges. The distribution of the welding residual stress in the plate is roughly approximated as shown in Fig. 13.13. The buckling is driven by the compressive residual stress σ_R distributing in the center part of the plate. When the equilibrium is assumed, σ_R must be in balance with the tensile stress distributing near the weld line or the tendon force F_L^*. Further, the tendon force can be related to the heat input Q through Eq. [13.35]. i.e.:

$$Bh\sigma_R = F_L^* = 0.335\frac{E\alpha}{c\rho}Q \quad [13.42]$$

On the other hand, buckling stress σ_{cr} is given by the following equation:

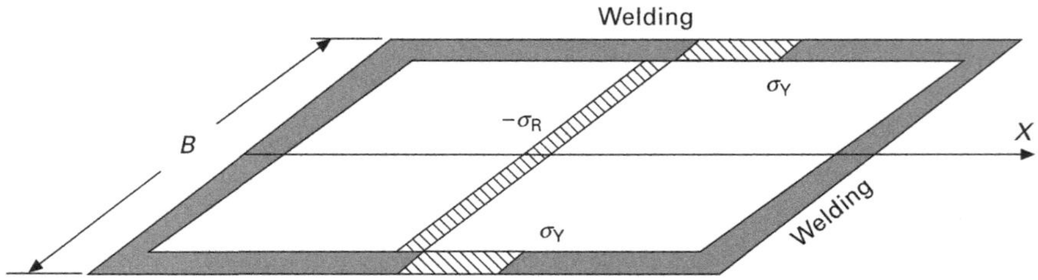

13.13 Distribution of residual stress in plate surrounded by welding lines.

Table 13.1 Factors influential to welding distortion and residual stress

Tendon force: F_L^*	$F_L^* = aQ \propto E\dfrac{\alpha}{c\rho}Q$
Longitudinal shrinkage: S_L	$S_L = \dfrac{F_L^* L}{EBh} \propto \dfrac{L}{Bh}\dfrac{\alpha}{c\rho}Q$
Transverse shrinkage: δ_T^*	$\delta_T^* = \dfrac{1}{h}\dfrac{\alpha}{c\rho}Q$
Transverse bending: θ_T^*	$\theta_T^* \propto \dfrac{1}{h^2}\dfrac{\alpha}{c\rho}Q$
Critical heat input: Q_{cr}	$Q_{cr} \propto k\dfrac{h^3}{B}\dfrac{c\rho}{\alpha}$
Width of residual stress: B_1^*	$B_1^* \propto \dfrac{E}{\sigma_Y}\dfrac{1}{h}\dfrac{\alpha}{c\rho}Q$

$$\sigma_{cr} = kE\left(\frac{h}{B}\right)^2 \qquad [13.43]$$

where k is a constant determined by the aspect ratio and the boundary condition of the plate. Noting that the plate buckles when the welding residual stress σ_R reaches the buckling stress σ_{cr}, the critical heat input Q_{cr} is shown to be:

$$Q_{cr} = 2.98k\frac{c\rho h^3}{\alpha B} \qquad [13.44]$$

The above equation tells us that Q_{cr} is proportional to $(c\rho/\alpha)(h^3/B)$.

As summarized in Table 13.1, the magnitudes of various types of welding deformations and the width of residual stress distribution can be related to the heat input Q and other parameters. All deformations and the width of residual stress distribution are linearly related to the heat input Q and the parameter $\alpha/c\rho$. It is also seen that the Young's modulus and the yield stress do not appear as primary factors except for the tendon force and the width of residual stress distribution.

13.5 Distortion and residual stress produced by laser welding

13.5.1 Welding distortion and its control

Laser welding is effective in reducing the distortion because of the small heat input resulting from deep and narrow weld seams compared with arc welding. For further improvement, influence of the welding power and welding speed on distortion and residual stress of laser beam welded shaft-hub joints was investigated by Buschenhenke *et al.* (2010), who reported that high brightness lasers generate deeper welds with higher aspect ratio but this leads to higher tensile stress levels and is not necessarily beneficial for deformation. The deformation can be reduced at the higher welding speed with large laser power as shown in Fig. 13.14. Also the effectiveness of the multi-beam laser shown in Fig. 13.15 has been investigated (Buschenhenke *et al.*, 2011). Figure 13.16 shows the distortions such as dishing slope and tilt measured and computed for single-beam and multi-beams. As seen from the figure, the tilt is significantly reduced by two beams on one side and

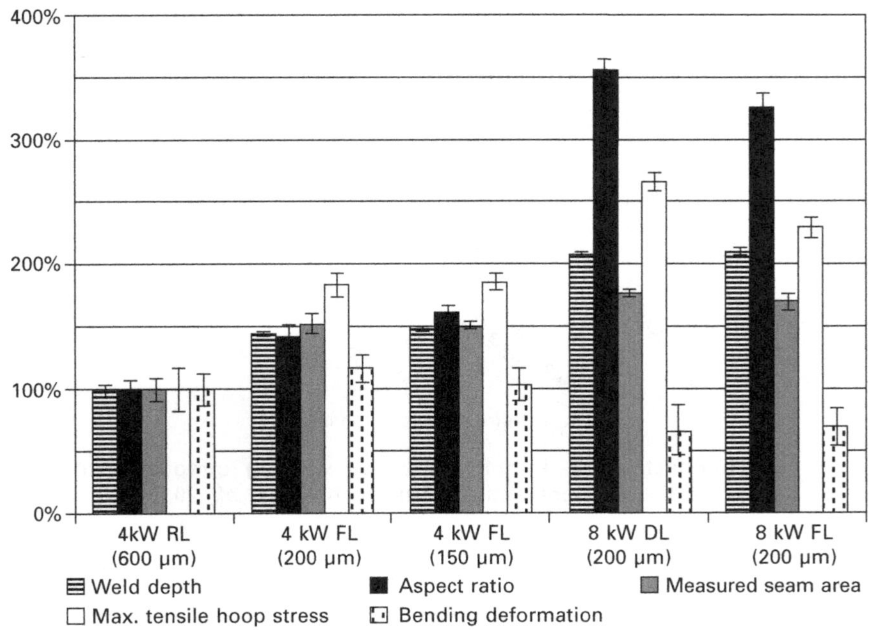

13.14 Influence of welding condition on residual stress and distortion (Buschenhenke *et al.*, 2010).

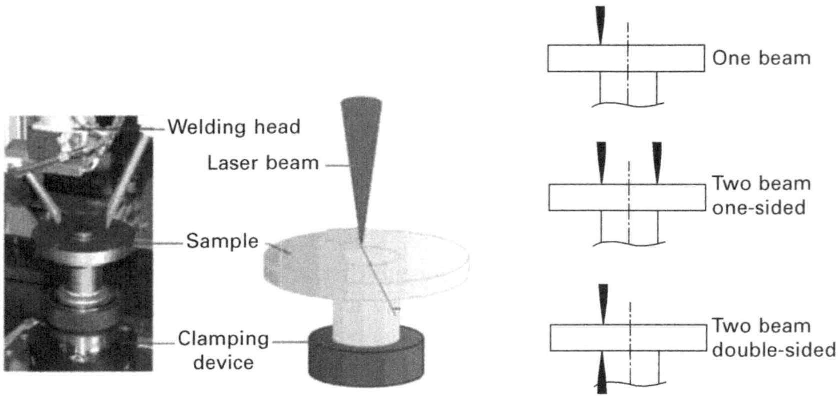

13.15 Welding set-up with one laser beam and the schematic overview of all beam positions (Buschenhenke *et al.*, 2011).

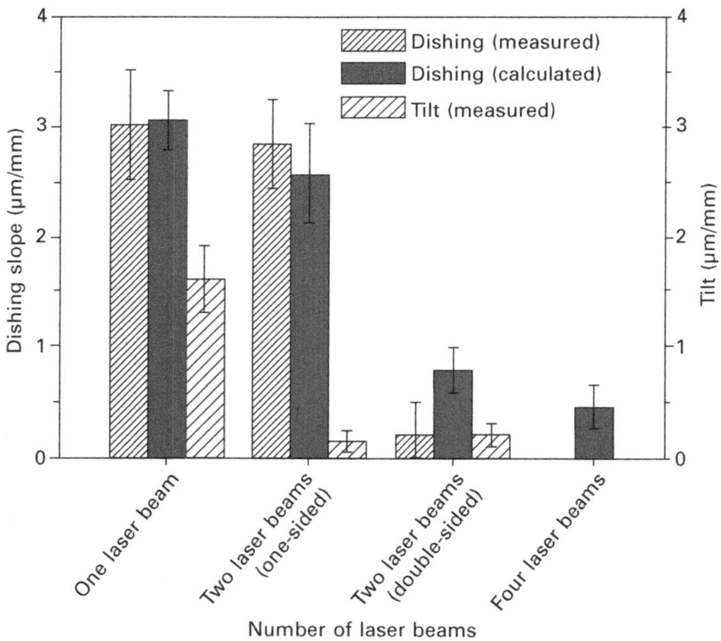

13.16 Effectiveness of multi-beam laser welding to reduce distortions such as dishing slope and tilt (Buschenhenke *et al.*, 2011).

the dishing can be reduced by two beams on the face and back sides. It is also shown in this report that dense tack welds are effective in controlling the tilt.

13.5.2 Residual stress

Zhang *et al.* (2010) investigated multi-pass narrow gap laser welding of austenitic stainless steels. Thick stainless steels are welded by three welding methods, namely LBW, GTAW and GTAW + SMAW. Figure 13.17 shows the distribution of residual stresses measured on the surface of the specimen. As seen from the figure, the maximal value of the residual stress perpendicular to the welding bead for the laser welded joint is lower than those of the conventional GTAW. As for the width of residual stress distribution, the width for the laser welding is smaller than GTAW in both residual stresses perpendicular and parallel to the welding. This is because the heat input for the laser welding is small and this agrees with Table 13.1, which shows that the width of residual stress distribution is proportional to the heat input.

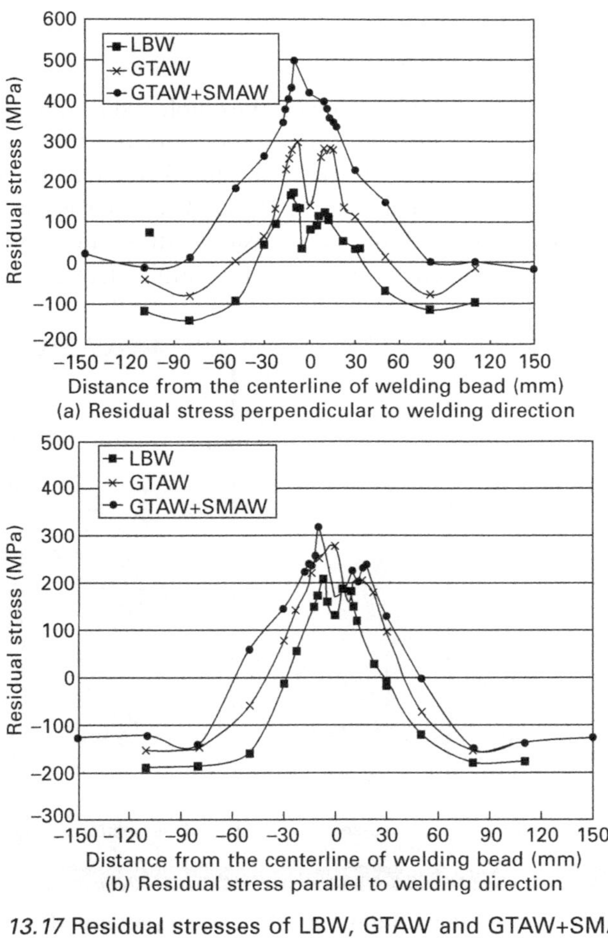

13.17 Residual stresses of LBW, GTAW and GTAW+SMAW (Zhang *et al.*, 2010).

In the case of laser powder deposition, the effect of the deposition pattern on the temperature and stress distribution has been studied by Foroozmehr and Kovacevic (2009). Figure 13.18 shows three deposition patterns tested and Plate XII (between pages 328 and 329) shows the influence of deposition pattern on the computed residual stress. It is reported that dividing the area under deposition into smaller areas results in less variation in temperature with a higher average temperature during the process. The stress distribution along the deposition direction and perpendicular to that direction are changed as a result of changing the deposition pattern. The results have a close agreement with the X-ray diffraction residual stress measurements. Similarly, Chae and Mazumder (2011) investigated how the residual stress in laser direct metal deposition is developed over time. Also the effects of the laser power on thermal behavior such as melt pool size and temperature field have been studied.

13.5.3 Prediction by simulation

There are two simulation methods which can be employed for the prediction of welding distortions and residual stresses. One is the thermal elastic plastic finite element analysis, in which the welding is treated as a transient nonlinear problem (Ueda and Yamakawa, 1971). The other is the inherent strain method, in which the distortion and residual stress are computed by elastic analysis using the inherent strain as initial strain (Ueda *et al.*, 1977;

13.18 Coupons made with different deposition patterns (Foroozmehr and Kavacevic, 2009).

Murakawa *et al.*, 1998). They have both advantages and disadvantages. The latter is advantageous in computational time and disadvantageous because the detail of the welding condition, such as the gap control and fitting, may not be fully considered in some cases. To predict the welding distortion and the residual stress of large-scale structures, methods combining these two are proposed as a practical approach.

A local–global finite element approach (Tsirkas *et al.*, 2003b) is one of the combined methods to predict welding distortions of large structures. Figure 13.19 shows the distortions of a member of a ship structure assembled in two different sequences by laser welding. The local approach takes into account thermal, metallurgical and mechanical aspects of the laser welding process. The simulation of the laser welding process was performed using a nonlinear heat transfer analysis, based on a keyhole formation model, and a coupled transient thermomechanical analysis. The global approach involves the transfer of residual plastic strains and the stiffness of the weld, obtained from the local model, to the global analysis, which provides the distortions

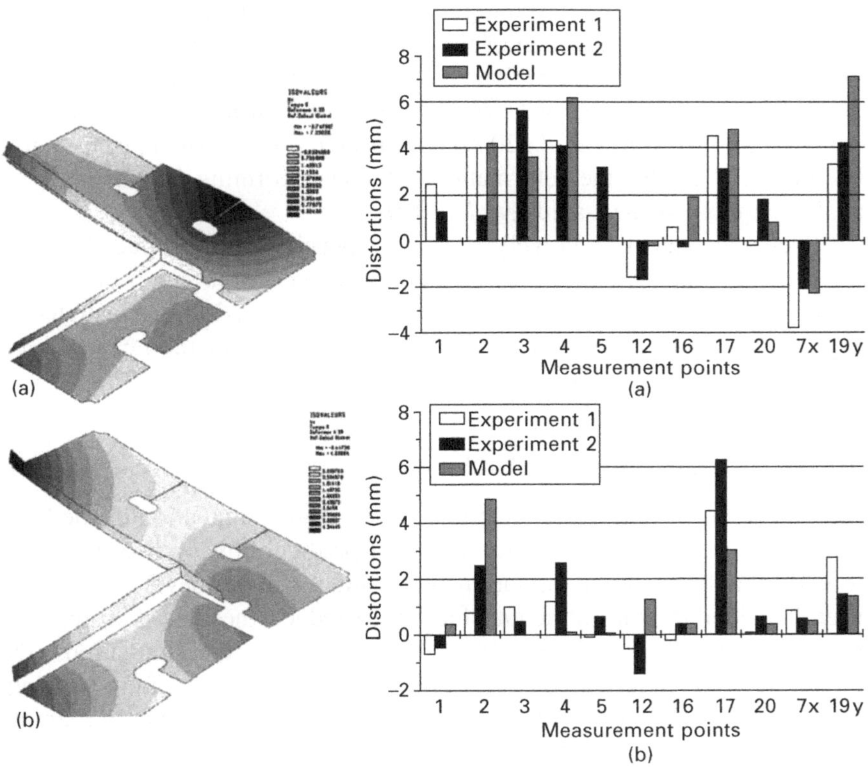

13.19 Predicted distortions of a ship structure member assembled in two different sequences by laser welding (Tsirkas *et al.*, 2003b).

in the member. The developed methodology was applied to the evaluation of global distortions due to laser welding of stiffeners on a plate.

The thermal elastic plastic finite element method was employed to predict distortions of various types of joints, such as butt joints (Tsirkas *et al.*, 2003a; Arai and Asano, 2009) and aluminum lap joints (Moraitis and Labeas, 2008).

13.5.4 Laser forming

Since laser is superior in controllability compared to other heat sources, such as gas frame and arc, it can be applied to forming three-dimensional doubly curved shapes from a flat plate. The mechanism of deformation during the forming process is the same as in the case of welding deformation. By selecting process parameters, such as the power, beam diameter and scanning speed, the laser can be applied to structures with a wide range of plate thicknesses from thick plate for ship structures (Romero *et al.*, 2011) to thin sheet metals (Jamil *et al.*, 2010). In the case of application to shipbuilding industries, Plate XIII (between pages 328 and 329) shows the computed deformation of plate after forming by CO_2 laser. Computer simulation has a great potential in the design of the forming procedure. Simulation of the process to form a bowl shape from a circular plate (Venkadeshwaran *et al.*, 2009) and the simulation of laser forming of the particle reinforced metal matrix composites (Liu *et al.*, 2009) have been reported. Attempts have also been made to predict the deformation in the laser forming process using an analytical method (Kyrsanidi *et al.*, 2000).

When the plate is relatively thick, the plastic strain produced by the thermal load can be employed. In the case of very thin plate, the plastic strain produced by micro shockwave (Bechtold *et al.*, 2009), illustrated in Fig. 13.20, can be employed and this type of forming method can be used for fine adjustment of geometry of thin metal parts.

13.5.5 Problems to be tackled

Judging from the amount of literature reporting the simulation of the laser welding process, it is clear that theoretical prediction is a powerful means to understand how the deformation and residual stresses develop. It is also useful to identify the influential factors to reduce the distortion and the residual stress. To obtain reliable conclusions from the simulations, the modeling of the heat source is very important as in the case of arc welding. In the case of the laser welding, development of an accurate and simple keyhole model is necessary.

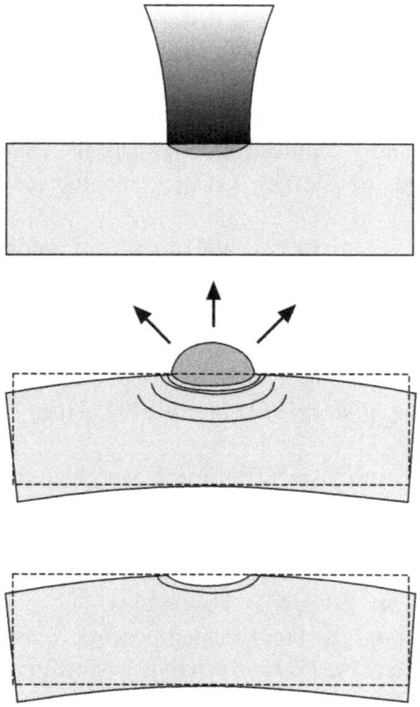

13.20 Mechanism of laser forming by micro shockwave (Bechtold *et al.*, 2009).

13.6 References

Arai T, Asano N (2009) 'Simulation of thin metal deformation by laser welding', *Proceedings of ICALEO'09*, 676–683.

Bechtold P, Roth S, Schmidt M (2009) 'Analysis and application of micro shockwave adjustment using ultrashort laser pulses', *Proceedings of ICALEO'09*, 1124–1133.

Buschenhenke F, Hofmann M, Seefeld T, Vollertsen F (2010) 'Distortion and residual stress in laser beam welded shaft-hub joints', *Proceedings of ICALEO'10*, 389–397.

Buschenhenke F, Seefeld T, Vollertsen F (2011) 'Distortion control during laser beam welding of shaft-hub joints within a process chain', *Proceedings of ICALEO'11*, 427–434.

Chae H M, Mazumder J (2011) 'Effects of melt pool size on transient and residual stress in DMD process', *Proceedings of ICALEO'11*, 267–274.

Foroozmehr F, Kovacevic R (2009) 'Effect of path planning on residual stress distribution in laser powder deposition process', *Proceedings of ICALEO'09*, 830–839.

Jamil M S C, Pinkerton A J, Li L, Sheikh M A (2010) 'The effect of beam geometry on diode laser forming of sheet metals', *Proceedings of ICALEO'10*, 495–502.

Kyrsanidi An K, Kermanidis Th B, Pantelakis Sp G (2000) 'An analytical model for the prediction of distortions caused by the laser forming process', *Journal of Materials Processing Technology*, 104, 94–102.

Liu F R, Chan K C, Tang C Y (2009) 'Investigations on interface debonding of the particle reinforced metal matrix composites in laser forming', *Proceedings of ICALEO'09*, Paper No. 503.

Moraitis G A, Labeas G N (2008) 'Residual stress and distortion calculation of laser beam welding for aluminum lap joints', *Journal of Materials Processing Technology*, 19(8), 260–269.

Murakawa H, Luo Y and Ueda Y (1998) 'Inherent strain as an interface between computational welding mechanics and its industrial application', *Mathematical Modelling of Weld Phenomena 4* (ed. H. Cerjak), London: Institute of Materials, Minerals and Mining, 597–619.

Romero P, Otero N, Pintos V, Arias R (2011) 'Experimental analysis of automated laser forming for shipbuilding industry', *Proceedings of ICALEO'11*, 1272–1279.

Satoh K, Terasaki T (1976) 'Effect of welding conditions on welding deformations in welded structural materials', *Journal of the Japan Welding Society*, 45(4), 302–308.

Tsirkas S A, Papanikos P, Kermanidis T (2003a) 'Numerical simulation of the laser welding process in butt-joint specimens', *Journal of Materials Processing Technology*, 134, 59–69.

Tsirkas S A, Papanikos P, Pericleous K, Strusevich N, Boitout F, Bergheau J M (2003b) 'Evaluation of distortions in laser welded shipbuilding parts using local–global finite element approach', *Science and Technology of Welding and Joining*, 8(2), 79–88.

Ueda Y, Yamakawa T (1971) 'Thermal stress analysis of metals with temperature dependent mechanical properties', *Trans. JWS*, 2(2), 90–100.

Ueda Y, Fukuda K, Nakacho K, Endo S (1977), 'Fundamental concept in measurement of residual stresses based on finite element method and reliability of estimated values', *Theoretical and Applied Mechanics*, 25, 539–554.

Venkadeshwaran K, Das S, Misra D (2009) '3D laser forming through circle line heating', *Proceedings of ICALEO'09*, 233–239.

White J D, Leggatt R H, Dwight J B (1980) 'Weld shrinkage prediction', *Welding and Meal Fabrication*, 11, 587–596.

Zhang X, Ashida E, Tarasawa S (2010) 'Properties of welded joint for narrow gap laser welding of austenitic stainless steels', *Proceedings of ICALEO'10*, 632–637.

Part III

Developments in emerging laser welding technologies

14
Applications of robotics in laser welding

M. DE GRAAF, Cargotec ACT/University of Twente, The Netherlands and R. AARTS, University of Twente, The Netherlands

DOI: 10.1533/9780857098771.3.401

Abstract: This chapter describes how seam-tracking sensors can be integrated in a robotic laser welding system for automatic teaching of the seam trajectory as well as for correcting small errors from a pre-defined seam trajectory. Calibration procedures are required to derive accurate transformations of laser and sensor tool frames relative to each other or relative to the robot flange. A tool calibration procedure is described that involves movements of the tool above a special calibration object. The accuracy of this procedure is influenced by errors in the robot movement. Therefore, another calibration procedure is developed, which is a direct measurement of the sensor tool frame relative to the laser tool frame. A trajectory-based control approach is presented for real-time seam tracking, where a seam-tracking sensor measures ahead of the laser focal point. In this control approach, sensor measurements are related to the robot position to build a geometric seam trajectory, which is followed by the laser focal point.

Key words: robotics, sensors, laser welding, hand-eye calibration, seam-tracking, trajectory-based control.

14.1 Introduction: key issues in robotic laser welding

In the manufacturing industry, many types of joints are used for laser welding. This chapter considers continuous seam welding of 3D seams, where the handling of the laser welding head is done with robotic manipulators. Six-axis robotic manipulators are relatively cheap and flexible as they can reach more complicated seams, opening a wider range of applications. A typical example of a 3D product where robotic laser welding may be applied can be found in Fig. 14.1.

Unfortunately, the absolute accuracy of robotic manipulators does not meet the accuracy demands of laser welding. Therefore, welding of 3D seams is not a trivial extension to seam welding of two-dimensional seams. Accurate manual programming of these welding trajectories is time-consuming. CAD/CAM or off-line programming software may be used to generate the welding trajectories, but it remains difficult to accurately match such a trajectory with the actual seam trajectory in the work cell. The important accuracy

14.1 Example of a complex 3D product that is very suitable for robotic laser welding.

requirement is the accuracy of the laser focal point relative to the product. This accuracy is influenced by the robotic manipulator and the product, e.g. product tolerances and clamping errors. For laser welding of steel sheets using a laser spot diameter of 0.45 mm and a focal length of 150 mm, it was found that a lateral tolerance of ±0.2 mm has to be satisfied in order to avoid weld quality degradation (Duley, 1999; Kang *et al.*, 2008), which has to be achieved at welding speeds up to 250 mm/s.

To cope with these issues, seam-tracking sensors that measure the seam trajectory close to the laser focal spot are required (Huissoon, 2002). The requirements for the positioning accuracy of the laser beam on the seam, the integration and the automation of the process will be fulfilled using seam-tracking sensors. Many different sensing techniques for measuring three-dimensional information can be distinguished, e.g. inductive measurements, range measurement using different patterns, surface orientation measurement using different surfaces, optical triangulation, etc. (Regaard *et al.*, 2009). Sensors based on optical triangulation with structured light are used most often as they provide accurate 3D information at a considerable distance from the product surface. Optical triangulation can be divided into two parts: beam scanning and pattern projecting. In beam scanning, a projector and detector are used that are simultaneously scanning the surface. Pattern projecting is done with different structures, e.g. one or multiple lines, circles, crosses or triangles. A two-dimensional camera observes the structures, and features are detected using image processing. The sensor uses a laser diode to project a single line of laser light on the surface at a certain angle. A camera observes the diffuse reflection of the laser diode on the surface at a different angle. Deviations on the surface will result in deviations on the camera image, which are detected. Several features are detected from the camera, e.g. the 3D position and an orientation angle.

A considerable amount of work preparation time consists of programming the robot for a new welding job. This can be decreased by using a seam-tracking sensor for automatic teaching of the seam-trajectory. Once a seam has been taught in this way, it can be welded in any number of products as long as no modifications of this trajectory occur, e.g. due to clamping errors, product tolerances, heat deviation errors, etc. In cases where modifications of this trajectory do occur, the sensor can be used to correct small errors from a pre-defined seam trajectory, using sensor measurements obtained some distance ahead of the laser focal point. Both the sensor and laser welding head are tools that are attached to the robot flange. The sensor measures relative to its own coordinate frame, but the corrections need to be applied in the laser coordinate frame. The transformations between these coordinate frames and the robot flange need to be accurately calibrated.

This chapter describes how such sensors can be integrated in a robotic laser welding system. A typical connection topology with the main components used in a sensor-guided robotic laser welding system is given in Section 14.2. Different coordinate frames are described in Section 14.3. From a scientific point of view, two major subjects are of particular interest: tool calibration (Section 14.4) and seam teaching/tracking (Section 14.5); for the latter a trajectory-based control architecture is presented (Section 14.6), which allows the construction of the robot trajectory during the robot motion. Finally, integration issues and future trends are discussed in Section 14.7.

14.2 Connection topology

Common components in a sensor-guided robotic laser welding system are a robot controller, seam-tracking sensor, laser source and a fibre to guide the laser light to a welding head, where it is focused on the product. Process gas protects the melt pool from oxidation and a crossjet protects the focusing lens in the welding head from process spatters. Figure 14.2 shows the components used in a typical sensor-guided robotic laser welding system.

All the components inside the dotted box require real-time execution. The robot controller controls not only the robot motion but also the other equipment during laser welding. The laser, crossjet and process gas have to be switched on and off synchronously with the robot motion. An industrial Devicenet bus system is used here, but in general many different bus systems exist that are equally well suited. The seam-tracking sensor is located within the real-time execution box, as its measurements must be synchronised with the measurements of the robot joint angles to get accurate results at the

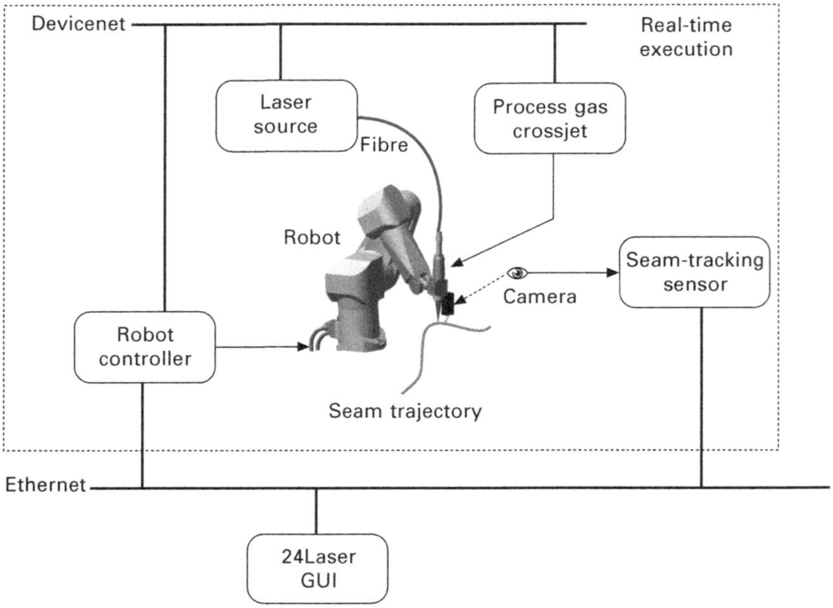

14.2 Typical connection topology in a sensor-guided robotic laser welding system.

high speeds that occur in a robotic laser welding system (De Graaf *et al.*, 2005).

The system operator can control the system using a (graphical) user interface (GUI) called '24Laser', which may be present on the teach pendent of the robot controller or on a separate computer. The GUI is used to prepare and start a laser welding job. Furthermore, it may contain various algorithms for seam teaching, real-time seam tracking and tool calibration.

14.3 Coordinate frames and transformations

An overview of the coordinate frames and transformations that can be distinguished in a sensor-guided robotic welding system can be found in Nayak and Ray (1993). Their overview is generalised, by adding a station frame M and a product frame H.

A transformation describes the location (position and orientation) of a frame with respect to a reference frame. Transformations are indicated by the symbol **T** with a leading superscript, that defines the reference frame they refer to. The leading subscript defines the frame they describe, e.g. transformation ${}^{A}_{B}\mathbf{T}$ describes frame B with respect to frame A.

In the literature, 4 × 4 homogeneous transformation matrices are often used to describe transformations (Craig, 1986; Khalil and Dombre, 2002).

They consist of a 3 × 3 rotation matrix and 3D position vector. Homogeneous transformation matrices have some useful properties, e.g. different transformations can be added using matrix multiplication. The following frames are defined in Fig. 14.3:

- **B:** Base or world frame. This frame is attached to the robot base and does not move with respect to the environment.
- **N:** Null frame. The null frame is located at the robot end-effector or the end of the robot flange. The null frame is described with respect to the base frame by coordinate transformation ${}^{B}_{N}\mathbf{T}$, which is a function of the joint positions of the robot arm (forward kinematics or direct geometric model).
- **L:** Laser tool frame. The laser tool frame is located at the focal point of the laser beam. The *z*-axis of this frame coincides with the laser beam axis. Because the laser beam is axi-symmetric, the direction of the *x*-axis is arbitrary, unless the symmetry is broken due to, e.g., the presence of peripheral equipment. Without loss of generality it will be chosen in the direction of the sensor tool frame. The transformation ${}^{N}_{L}\mathbf{T}$ describes the laser tool frame with respect to the null frame. This

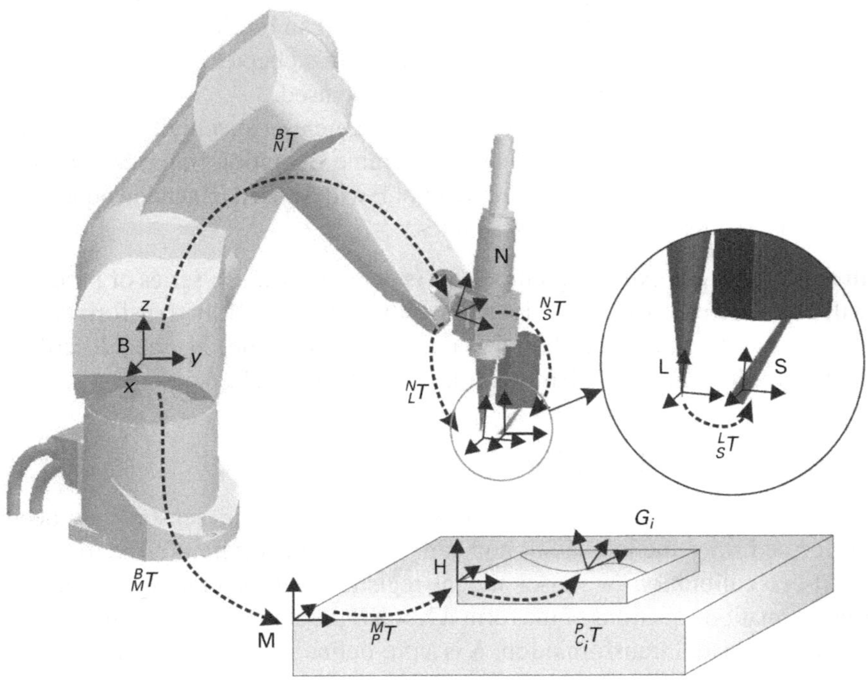

14.3 Frames and transformations for sensor-guided robotic laser welding.

is a fixed transformation determined by the geometry of the welding head.

- **S:** Sensor tool frame. The seam-tracking sensor is fixed to the welding head and therefore indirectly to the robot flange. The transformation ${}^{N}_{S}\mathbf{T}$ describes the sensor tool frame with respect to the null frame. Note that, because both transformations are fixed, this transformation can also be described with respect to the Laser tool frame instead of the Null frame by transformation ${}^{L}_{S}\mathbf{T}$.
- **M:** Station frame. The station frame is the base of the station or worktable a product is attached to. It is possible that the product is clamped on a manipulator which moves the product with respect to the base frame. In that case the transformation ${}^{B}_{M}\mathbf{T}$ describes the station frame with respect to the base frame and depends on the joint values of the manipulator.
- **H:** Product frame. This frame is located on a product. The transformation ${}^{M}_{H}\mathbf{T}$ describes the product frame with respect to the station frame. This frame is useful if a series of similar products are welded on different locations of a station.
- **G:** Seam frame. Every discrete point on a seam can be described with a different coordinate frame. The transformation ${}^{H}_{G}\mathbf{T}$ describes a seam frame with respect to the product frame.
- **T:** Robotic tool frame. A general robot movement is specified by the movement of a robotic tool frame T, which can be either the sensor tool S or laser tool L. Some equations used in this chapter account for either S or L. In that case, the symbol T will be used.
- **F:** Robotic frame. A general robot movement is specified with respect to a frame F, which can be the base frame B, station frame M, product frame H or seam frame G. The symbol F will be used in general equations that account for all of these frames.

In many cases, an external manipulator is not present and a series of products will only be welded at a fixed location in the work cell. Both transformations ${}^{B}_{M}\mathbf{T}$ and ${}^{M}_{H}\mathbf{T}$ can then be chosen as unity. A seam frame G is then described with respect to the robot base with transformation ${}^{B}_{G}\mathbf{T}$.

14.4 Tool calibration

Generic work on calibration of wrist-mounted robotic sensors has been performed since the late 1980s and early 1990s. It is often referred to as the hand-eye calibration or sensor mount registration problem. Homogeneous transformation equations of the form $\mathbf{AX} = \mathbf{XB}$ are solved, where $\mathbf{X}$ represents the unknown tool transformation, $\mathbf{A}$ is a pre-defined movement relative to the robot flange and $\mathbf{B}$ is the measured change in tool coordinates. It was first addressed by Chou and Kamel (1988), although several other approaches for obtaining a solution have been presented by Shiu and Ahmad (1989), Tsai and Lenz (1989), Zhuang and Shiu (1993), Park and Martin (1994), Lee and Ro (1996), Lu *et al.* (1996), Thorne (1999), Daniilidis (1999) and Huissoon (2002).

In most of the above-mentioned references, mathematical procedures are presented to solve the generic calibration problem of wrist-mounted sensors. In these publications, the procedures are verified using simulation data only. Daniilidis (1999) and Huissoon (2002) are the only ones who present experimental results. A major drawback in the presented simulations is that in all publications, except Lee and Ro (1996), only the influence of stochastic errors (measurement noise) is considered and not the influence of systematic errors originating from robot geometric nonlinearities, optical deviations, sensor reflections, etc. In De Graaf (2007), the influence of such systematic errors on the tool calibration procedures is analysed in more detail, which shows that the accuracy that can be achieved with these tool calibration procedures is limited to approximately 1 mm.

Huissoon (2002) is the only author to directly apply the tool calibration methods to robotic laser welding. However, he does not give results of the sensor tool calibration procedure he uses. No publications were found that apply the calibration procedures to a laser welding head without a sensor. A calibration procedure for a laser welding head is presented in Section 14.4.1, which uses the coaxial camera attached to the welding head and a calibration object.

In a sensor-guided robotic laser welding system, the sensor measures relative to its coordinate system (S), but the measurements are applied in the laser coordinate frame (L). According to Fig. 14.2 the transformation ${}^{L}_{G}\mathbf{T}$ that describes the transformation between laser frame L and seam frame G, representing the deviation of the laser focal point and the seam trajectory, can be computed as:

$$ {}^{L}_{G}\mathbf{T} = {}^{L}_{S}\mathbf{T} \cdot {}^{S}_{G}\mathbf{T}, \qquad [14.1] $$

where ${}^{L}_{S}\mathbf{T}$ is a fixed geometrical transformation from the laser frame to the sensor frame and ${}^{S}_{G}\mathbf{T}$ represents a measured location on the seam trajectory relative to the sensor frame **S**. To position the laser spot accurately on the seam, both of these transformations need to be accurately known (< 0.2 mm).

Robot movements can be defined relative to the flange frame **N** or relative to any tool frame **T**, which can be **S** or **L**. When robot movements are defined relative to the sensor or the laser tool frame, the transformations ${}^{N}_{L}\mathbf{T}$ (which represents the transformation from flange to laser frame) and ${}^{N}_{S}\mathbf{T}$ (which represents the transformation from flange to sensor frame) need to be known respectively. According to Fig. 14.2, one of the transformations ${}^{L}_{S}\mathbf{T}$, ${}^{N}_{L}\mathbf{T}$ and ${}^{N}_{S}\mathbf{T}$ can be computed if the other two are known using:

$$^{N}_{S}\mathbf{T} = {}^{N}_{L}\mathbf{T} \cdot {}^{L}_{S}\mathbf{T} \qquad [14.2]$$

The most important transformation that needs to be accurately known is the transformation ${}^{L}_{S}\mathbf{T}$, which can be computed using Eq. [14.2] from the result of a laser tool calibration procedure and a sensor tool calibration procedure. Because both ${}^{N}_{L}\mathbf{T}$ and ${}^{N}_{S}\mathbf{T}$ cannot be accurately determined due to geometric robot errors, a combined calibration procedure is presented in Section 14.4.2, where the transformation ${}^{L}_{S}\mathbf{T}$ is determined by means of a direct measurement that does not involve robot movements. Therefore, errors in the robot geometric model do not influence the measurement, which makes this method much more accurate.

14.4.1 Laser tool calibration

The laser tool calibration procedure uses a coaxial CCD camera, which is attached to the welding head. The setup with coordinate frames used are shown in Fig. 14.4.

For the calibration procedure, the following additional frames are defined:

- **C:** Camera frame. The surface normal of the camera chip coincides with the *z*-axis of frame **C** and with the optical axis of the laser beam.
- **R:** Reference frame. A calibration object with a mirror, a pinhole with a light emitting diode (LED) and a plate of black anodised aluminium is placed in the robot work space. The origin of frame **R** is located at the pinhole providing a fixed point in robot space that can be measured by the camera.
- **P:** Pilot laser frame. A low-power pilot laser will be applied, using the

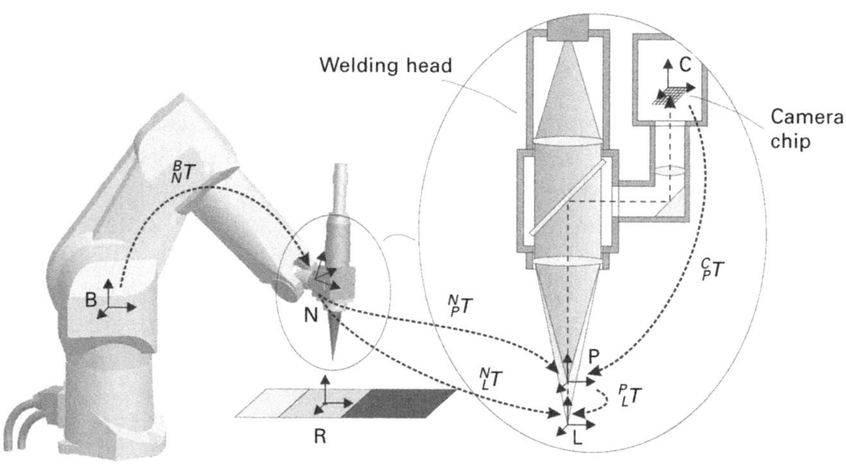

14.4 Frames for laser tool calibration.

same optical fibre as the high power Nd:YAG laser. The projection of the laser light spot from this pilot laser on the camera chip will be used for focal measurements.

In the calibration procedure, the pilot laser will be used. The transformation ${}^{N}_{L}\mathbf{T}$ can be split according to:

$$ {}^{N}_{L}\mathbf{T} = {}^{N}_{P}\mathbf{T} \cdot {}^{P}_{L}\mathbf{T} \qquad [14.3] $$

The pilot laser and Nd:YAG laser share the same optical axis. The rotation of these frames around the optical axis is assumed to be identical and ${}^{P}_{L}\mathbf{T}$ is a transformation that represents a pure translation with a distance d along the z-axis. The distance d only depends on the optical system of the welding head. If the optical system does not change, it needs to be determined once. In the calibration procedure the transformation ${}^{N}_{P}\mathbf{T}$ from flange frame to pilot frame will be determined and finally a procedure is outlined to determine distance d. Using Eq. [14.3] and known distance d, the transformation ${}^{N}_{L}\mathbf{T}$ from flange frame to laser frame is easily computed.

The camera may be rotated freely around the optical axis. The orientation between the laser frame or the pilot frame and the camera frame is therefore related as:

$$ {}^{L}_{C}\mathbf{R} = {}^{P}_{C}\mathbf{R} = \mathbf{rot}(z, \phi) \qquad [14.4] $$

where $\mathbf{rot}(z, \phi)$ is a rotation matrix that represents a pure rotation around the z-axis with an angle ϕ.

The camera is used to observe the pinhole in the calibration object or a reflection of the pilot laser beam in the mirror. In either case, the camera image shows a light spot in a dark background (Fig. 14.5). Using basic image processing (Van der Heijden, 1994), the centre of gravity of the light spot and the number of pixels of the light spot are determined. Details of the image processing algorithms used can be found in Van Tienhoven (2004). The position ${}^{C}\mathbf{P}$ in pixels of the centre of this light spot relative to the camera frame C is denoted by

$$ {}^{C}\mathbf{P} = [{}^{C}P_x \; {}^{C}P_y \; 0]^T. \qquad [14.5] $$

The camera measures two degrees-of-freedom. The orientation of the camera frame and the pilot frame are related through Eq. [14.4]. If frame $\mathbf{P}$ moves from location P_i to location P_{i+1}, the camera frame moves from location C_i to location C_{i+1}. The positions between the frame movements are then related as

$$ {}^{Pi}_{Pi+1}\mathbf{P} = \mathbf{rot}(z, \phi) \cdot {}^{Ci}_{Ci+1}\mathbf{P} \cdot 1/c \qquad [14.6] $$

where $1/c$ is a scaling factor to convert from camera pixels to millimetres. The camera pixels may not be square in which case different ratios $1/c_x$ and $1/c_y$ are used for the camera x-direction and y-direction.

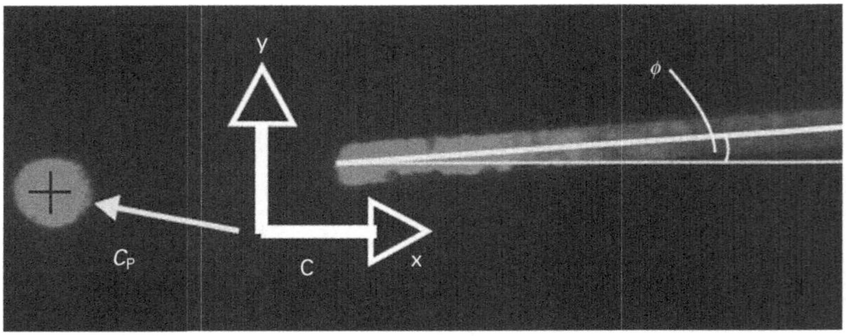

14.5 Typical camera measurement during the laser tool calibration procedure. Note that the angle of the line is irrelevant for the laser tool calibration, but used later in the combined calibration procedure of Section 14.4.2.

To make sure that the observed object stays within the range of the camera, movements are defined relative to a nominal pilot frame P_n, which is an estimation of the actual pilot laser frame P_a. The maximum allowable difference between the nominal frame and the actual frame depends on the field-of-view of the camera. In practice, this is large enough to start the calibration procedure with a user-defined nominal tool transformation ${}^{N}_{Pn}\mathbf{T}$ that is known from the dimensions of the welding head.

The generic **AX=XB** equation is written relative to the nominal pilot frame P_n, instead of the flange frame N, where **A** is a prescribed transformation (movement) in the nominal pilot frame and **B** is a measured transformation in the actual pilot frame. Transformation **B** cannot be completely measured because a 2D camera is used. By choosing the prescribed movement **A** in a suitable way, parts of the **AX=XB** equation can be solved in subsequent steps. During the calibration procedure, the translation and rotation components of **X** are solved separately using the following procedure:

- Determination of the pilot frame orientation: A number of robot movements are made that are defined as pure translations in the nominal pilot frame, while observing the pinhole in the camera image to find the rotation matrix ${}^{Pn}_{Pa}\mathbf{R}$. During this calibration step, the ratios c_x and c_y to convert from camera pixels to millimetres are computed as well as the camera rotation angle ϕ.
- Determination of the surface normal and focal position: The calibration object is placed arbitrarily in the robot workspace. In this step the optical axis of the welding head and the surface normal of the calibration object are aligned by projecting the pilot laser on the camera image using the mirror and moving the pilot frame along its optical axis. Furthermore, the focal position of the pilot laser on the camera will be determined.
- Determination of pilot frame position. A number of robot movements

are made that are defined as pure rotations in the nominal pilot frame, while observing the pinhole in the camera image to find the position vector ${}^{Pn}_{Pa}\mathbf{P}$.
- Determination of the difference *d* in focal height between the pilot frame as it is focused on the camera chip and the Nd:YAG frame as it is focused on the surface. This is done by making spot welds on a plate of anodised aluminium attached to the calibration object (in the same plane as the mirror) and determining the height corresponding to the smallest spot relative to the pilot focal position.

Details of the algorithms used and experimental results can be found in Van Tienhoven (2004) and De Graaf (2007).

14.4.2 Combined calibration

To be able to compute the sensor pose relative to a calibration object from a single sensor measurement, Huissoon (2002) proposes to add edges to such a calibration object. He describes a calibration object with three edges and a mathematical approach for computing the sensor pose relative to the calibration object. For this task, a calibration object with two edges is used as this is sufficient to determine the complete sensor pose. This calibration object offers improved accuracy and noise suppression over Huissoon's object. A more complex nonlinear matrix equation has to be solved however, which Huissoon avoids by using an additional edge feature.

The two upper edges of the calibration object give three lines in a sensor image (Fig. 14.6). From the lines in the sensor image, the intersections of the edges of the calibration object with the laser plane can be computed. This is done by fitting lines through the sensor data and computing their intersections P_1, P_2 and P_3 as indicated in Fig. 14.6(b). Note that P_1 is a virtual intersection that is constructed by extrapolating the outer line parts. These intersections are 3D vectors in the sensor frame, where two coordinates are measured in the CMOS image and the third coordinate is computed using the known diode angle of the sensor. Hence, there are a total number of nine parameters, where six of them are independent. Every sensor pose yields a unique combination of these parameters. The algorithms for computing the sensor pose from the intersections as well as experimental results can be found in De Graaf (2007).

In order to measure the transformation ${}^{L}_{S}\mathbf{T}$, the two calibration objects that are used in the sensor and laser tool calibration procedure are combined in a single calibration object (Fig. 14.7). From this figure, it can be derived that the transformation ${}^{L}_{S}\mathbf{T}$ is computed as:

$$ {}^{L}_{S}\mathbf{T} = {}^{L}_{R}\mathbf{T} \cdot {}^{R}_{E}\mathbf{T} \cdot {}^{E}_{S}\mathbf{T} \qquad [14.7] $$

(a)

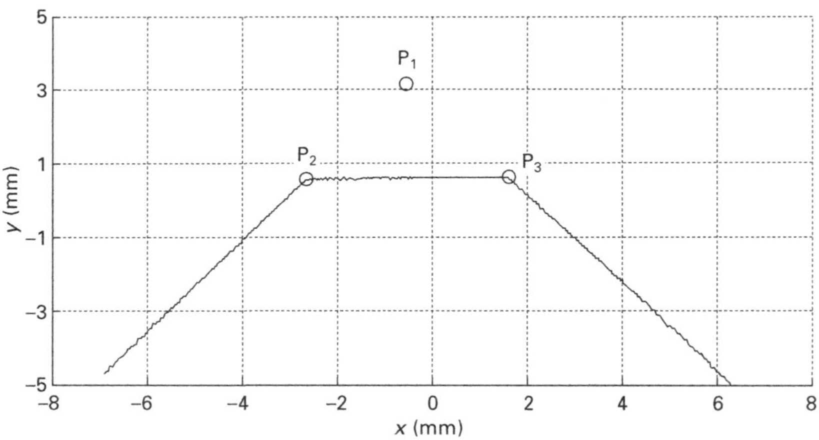

(b)

14.6 (a) Sensor positioned above the calibration object. (b) Typical sensor profile on the calibration object.

The sensor calibration object has its own frame E and is mounted in such a way that the transformation ${}^{R}{}_{E}\mathbf{T}$ is known from the geometry of the object. The transformation ${}^{E}{}_{S}\mathbf{T}$ can be measured using the measurement method described in De Graaf (2007). The only unknown transformation is ${}^{L}{}_{R}\mathbf{T}$, which will be derived next.

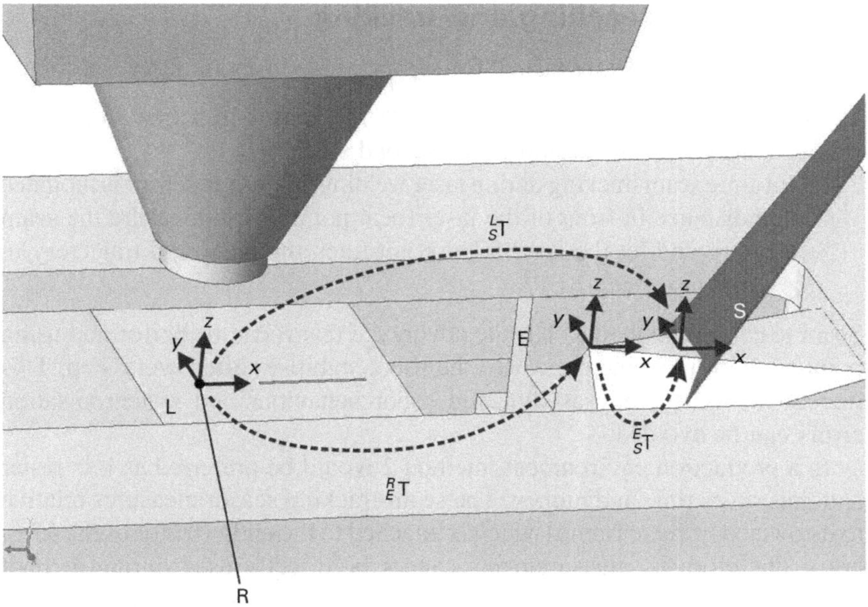

14.7 Laser and sensor positioned above the combined calibration object. Note that in this figure, frame L (attached to the welding head) overlaps frame R (attached to a pinhole in the calibration object), which is not generally the case.

After the laser tool calibration procedure has been performed, a number of important conditions are determined:

1. The welding head is positioned normal to the calibration object.
2. The origin of the laser frame L has been determined in the camera image, so the position of the light spot can be measured relative to frame L.
3. The focal position of the welding head is positioned at the surface of the calibration object.
4. The scaling factors c_x and c_y from camera pixels to millimetres are determined.

These conditions will be used to determine transformation $^{L}_{R}\mathbf{T}$. The optical axis of the welding head is normal to the calibration object (condition 1). Using condition 2, 3 and 4, it is known that transformation $^{L}_{R}\mathbf{T}$ can be determined from several features in the camera image, using the scaling factor c_x and c_y that were obtained during the laser tool calibration procedure.

The position of a spot and the orientation of a line in the camera image (Fig. 14.7) are needed to determine transformation $^{L}_{R}\mathbf{T}$. Image processing algorithms and experimental results are beyond the scope of this chapter and can be found in De Graaf (2007).

14.5 Seam teaching and tracking

Two strategies are distinguished for sensor-guided robotic laser welding:

1. Teaching of the seam trajectory with the sensor in a first step (seam teaching) and laser welding in a second step.
2. Real-time seam tracking during laser welding. When the sensor is mounted some distance in front of the laser focal point, it can measure the seam trajectory and let the laser focal point track the measured trajectory in the same movement.

Seam teaching (method 1) has the advantage that it can be performed using point-to-point movements, where the robot stabilises after every step. This increases the accuracy as dynamic robot behaviour and synchronisation errors can be avoided.

In a production environment, method 2 would be preferred as it is faster and thus saves time and money. The seam-tracking sensor measures relative to its own coordinate frame (which is attached to the end-effector of the robot arm). Therefore its measurements cannot be directly used during a robot movement. The sensor measurements will only be useful if the Cartesian location of the sensor frame in the robot workspace (computed from the robot joint angles) is known at the same time. This can be accomplished in two ways:

- Let the robot make a movement and wait until it stabilises. Because the robot is stabilised the location of the sensor frame in the robot workspace does not change in time. Then, a sensor measurement can be easily related to the corresponding robot position.
- When a sensor measurement is obtained during the robot motion, the time axes of the robot and the sensor need to be synchronised. If these are synchronised the robot joints can be interpolated to match the sensor measurement with the robot joints or vice versa.

Robot-sensor synchronisation can easily be achieved for a simple external sensor (like a LVDT), where the sensor operates continuously and is interfaced directly to the robot, because the sensor data is immediately available when the robot needs it. Synchronisation becomes more challenging when complex sensor systems, like camera-based sensors with their own real-time clock, are used (De Graaf *et al.*, 2005).

Seam teaching for arc welding has been regarded as being solved by many authors (e.g. Nayak and Ray, 1993). They focus on ways to use sensor information for real-time seam tracking and job planning. Most publications apply seam tracking to robotic arc welding, where the accuracy requirements (±1 mm) are less stringent than for laser welding. Two publications (Andersen, 2001; Fridenfalk and Bolmsjö, 2003) apply seam tracking to the application of robotic laser welding. Unfortunately, Andersen (2001) gives no technical details of his algorithms. Fridenfalk and Bolmsjö (2003) describe their algorithm but do not pay much attention to robot integration aspects (e.g., robot–sensor synchronisation). The inaccuracy of their system (±2 mm) is much larger than the 0.2 mm that is aimed at for laser welding of seams.

No publications were found that analyse the influence of various error sources (e.g., robot geometric errors) on seam teaching and seam tracking. The influence of these error sources on seam teaching and seam tracking cannot be underestimated, as shown in De Graaf (2007).

14.6 Trajectory-based control

To use a sensor for real-time seam tracking, a deep integration in the robot controller is needed. Unfortunately, most robot manufacturers do not allow the robot trajectory to be changed on-the-fly. Others do allow it, but no generic interfaces exist, and for each different robot manufacturer, different strategies have to be applied. In this section, a trajectory-based control strategy is proposed as a generic way of generating and extending the robot trajectory when the robot motion is already in progress.

The trajectory-based control approach can be used for a number of different operations:

- teaching a seam trajectory, with prior knowledge of its geometry, e.g. from CAD data
- teaching an unknown seam trajectory
- real-time tracking a seam trajectory, with prior knowledge of its geometry
- real-time tracking an unknown seam trajectory.

The prior knowledge of the seam trajectory is included in the nominal seam trajectory, which is an approximation of the actual seam trajectory that the laser focal point has to track during laser welding. The nominal seam trajectory can be obtained in several ways:

- by manual programming
- from a previous seam-teaching procedure
- from a CAD file or off-line programming software.

The seam locations of the nominal and actual seam trajectories are both stored in buffers that contain a number of seam locations. In practice, the nominal seam locations are expected to be within a few millimetres of the actual seam trajectory, which implies that if the sensor is positioned on the nominal seam trajectory, the actual seam trajectory is in the sensor's field of view.

Figure 14.8 shows a block diagram of the trajectory-based control strategy.

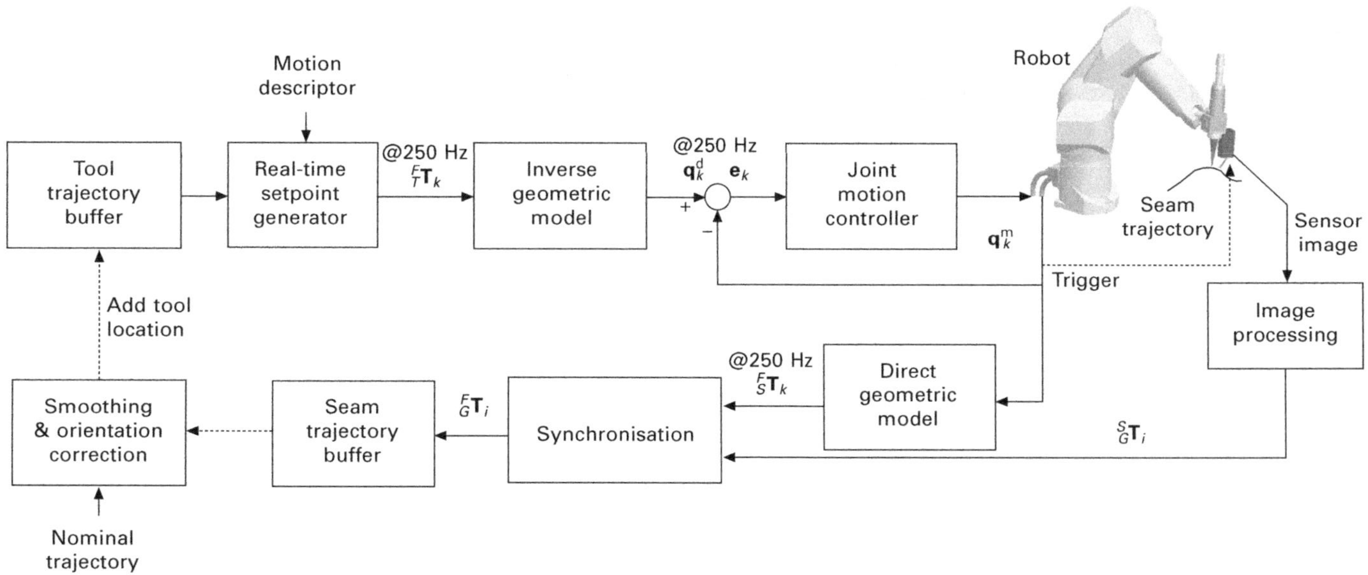

14.8 Trajectory-based control architecture.

The top part of this figure contains the robot trajectory generation, the bottom part contains the sensor integration part. The figure will be explained, starting from the top left.

The tool trajectory buffer contains a list of locations (positions and orientations) that the robot tool T (which can be S or L) has to pass through during the robot motion. These locations do not need to be equidistant. During the robot motion, new locations can be added to the tool trajectory buffer, thus extending the robot trajectory.

A real-time setpoint generator interpolates the locations in the tool trajectory buffer and computes location setpoints ${}^{F}{}_{T}\mathbf{T}_{k}$ for every *k*th fixed time interval (4 ms in our case). The movement should be smooth as defined by the maximum acceleration, velocity and deceleration specified in the motion descriptor. A real-time setpoint generator that uses cubic interpolation based on quaternions is used (De Graaf, 2007).

From a location setpoint ${}^{F}{}_{T}\mathbf{T}_{k}$, the known tool transformation ${}^{N}{}_{T}\mathbf{T}$ and the known frame transformation ${}^{B}{}_{F}\mathbf{T}$, transformation ${}^{B}{}_{N}\mathbf{T}_{k}$ is calculated. Next, a robot joint angle setpoint $q^{d}{}_{k}$ is calculated using the inverse geometric model (Khalil and Dombre, 2002).

The robot joint angle setpoints $q^{d}{}_{k}$ are the reference input for the joint motion controller, which is proprietary to Stäubli. It tracks the joint measurements $q^{m}{}_{k}$ in such a way that the tracking error $|q^{d}{}_{k}-q^{m}{}_{k}|$ on the specified path remains small.

The seam-tracking sensor computes the transformation ${}^{S}{}_{G}\mathbf{T}_{i}$ from a camera image using image-processing, where *i* is a measurement index. If properly synchronised, ${}^{S}{}_{G}\mathbf{T}_{i}$ can be related to the measured joint angles $q^{m}{}_{k}$ to compute a seam location ${}^{F}{}_{G}\mathbf{T}_{i}$. The result is stored in the seam trajectory buffer. After the robot joints and sensor image are synchronised, the actual time upon which the seam locations were measured is no longer relevant as the seam is defined by its geometric description only. Of course, the order in which the seam locations are obtained is of importance and it has to be assured that the number of points measured on the seam trajectory is sufficient in relation to the complexity of the seam. By moving the sensor tool along the seam trajectory and storing the seam locations obtained in the seam trajectory buffer, the complete geometry of the seam trajectory is obtained. For real-time seam tracking, the smoothing and orientation correction block is needed. In particular for this operation, the trajectory-based control approach yields some major advantages, compared to the common time-based control approaches:

- The measured locations in the seam trajectory buffer can be filtered in the position domain instead of in the time domain. This is more logical as seam trajectories are characterised by their radius of curvature. Curvature is meaningful in the position domain, not in the time or frequency domain.

- Both historical and future information is available for filtering as the sensor measures ahead of the laser focal point. Phase-coherent or central filtering can be easily applied.
- The sensor image processing is not part of the robot motion control loop. The control structure is therefore independent of a variable delay after processing of the sensor image, which may be caused by the image processing algorithm. Furthermore, stability issues are properly decoupled this way.
- It is easy to remove outliers. The sensor measurements may be incorrect (due to dust, spatter from the process, etc.) or completely missing (due to sensor–robot communication errors). As long as the robot tool frame has not reached the end of the tool trajectory buffer, the movement will continue.

The trajectory-based control structure only differs slightly for the mentioned seam teaching and seam tracking operations. In the case of teaching of a seam trajectory, with prior knowledge of its geometry, the smoothing and orientation correction block in Fig. 14.8 is not used, because the tool trajectory locations are known beforehand and the seam trajectory only needs to be recorded. For the other three procedures, the control loop has to be closed by on-line calculation and addition of locations to the tool trajectory buffer. A proper smoothing must be taken care of to prevent oscillatory motion behaviour, because the trajectory-based control strategy contains no other bandwidth limitations. In the case of teaching an unknown seam trajectory, the tool trajectory buffer needs to be filled with estimated seam locations, somewhere ahead of the current sensor location, which are extrapolated from the measured seam locations.

For real-time tracking of a seam trajectory, the sensor is used to obtain the location of the actual seam trajectory. The laser spot is kept on this seam trajectory. Obviously, the seam trajectory has to remain in the sensor's field of view, e.g. by rotating slightly around the laser tool. Real-time seam tracking without any prior knowledge of the geometry of the seam trajectory is considered to be impractical for safety reasons. We suggest that real-time seam tracking is only performed when a nominal seam trajectory is already present to prevent the welding head from hitting obstacles in the work cell due to erroneous sensor measurements, e.g. scratches. Only small deviations from a nominal trajectory are allowed. The nominal seam trajectory will be kept in the sensor's field of view to make sure its measurements can be used.

A detailed description of the trajectory-based control architecture is beyond the scope of this chapter. Experimental results and a description of the algorithms used can be found in De Graaf (2007).

14.7 Conclusion

To conclude this chapter, an overview is given of reasons that limit the current applicability of industrial robots for laser welding, which have been partly solved in this chapter:

- Serial link robots have a poor absolute accuracy at tip level, as the inaccuracies of each link accumulate towards the tip. The repeatability is usually an order higher. By making smart use of sensor information, the welding accuracy can reach the repeatability levels.
- The robot controller interface is not designed for the complex application of sensor-guided robotic laser welding. Many robot controllers allow operators or system integrators to program the robot controller with custom robot programming languages. These languages are usually fine for simple applications, like pick-and-place, but lack functionality for complex applications where modifications of the robot trajectory are needed on the fly. Furthermore, the programming languages differ between each robot manufacturer.
- Accurate synchronisation between the robot joint angles and external sensors is needed if tip-mounted sensors are used at high speeds. Most robot controllers have this information available at a low level, but few of them have a generic interface to synchronise with external sensors.
- Errors in the robot geometric model limit the accuracy of the tool calibration procedures, the seam teaching procedures and the real-time seam tracking procedures. It is expected that a considerable improvement can be achieved when more accurate geometric robot models are available, e.g. by calibrating the robot.

A number of promising techniques and trends are recognised to further improve the accuracy and usability of robotic laser welding:

- Development of integrated and compact welding heads (Iakovou, 2009; Falldorf Sensor, 2006): Welding heads with integrated sensors that measure close to or at the laser focal point give a great increase in welding accuracy. Besides this advantage, the welding head that is developed by Iakovou measures with a triangle around the laser focal point and does not require a specific orientation of the sensor relative to the seam trajectory. Therefore, it can be used to measure and weld specially shaped trajectories like rectangles, circles, spirals, etc.
- Iterative learning control (Hakvoort, 2009): ILC is a control strategy used to increase the tracking accuracy of the robot joint motion controller, by repeating the trajectory and learning from errors made in previous runs. For laser welding this has the advantage that the tracking accuracy of the laser spot is considerably increased (better than 0.05 mm), even during welding of complex trajectories, e.g. sharp corners. The accuracy

increase comes at the expense of a decrease in cycle time, as the welding trajectory needs to be taught at least once before it can be welded. Furthermore, a welding head with an integrated sensor that measures close to or at the laser focal point is needed.

- Improved laser beam manipulation (Hardeman, 2008): In this work, laser welding scanning heads are considered, where the laser spot can be moved in one or more directions by means of a moving mirror. These scanning heads contain integrated seam-tracking sensors and a motorised mirror. Using an integrated feedback controller, the laser spot is kept on the seam trajectory.
- Off-line programming for laser welding (Waiboer, 2007): Off-line programming software can be used to generate welding trajectories for complex 3D products. In the software, 3D models of the welding cell, robot, external manipulator, welding head, worktable, product and clamping materials can be used to create a collision-free welding trajectory. A disadvantage of off-line programming software is the deviation between the product position in the software and in the real work cell. Off-line programming software is very useful to generate complex welding trajectories, without spending a lot of time and money on the real welding equipment. Next, the work from this thesis can be applied using sensors to teach the actual seam with the accuracy that is needed for laser welding.
- Synchronous external axes support: By positioning the product on an external manipulator that moves synchronously with the robot arm, fast robot arm movements can be decreased considerably for complex products by using a combined motion of the robot and product on the manipulator, thus decreasing the dynamic tracking errors. Furthermore, the external manipulator can decrease gravity effects during welding by keeping the melt pool horizontal. The combination of external axes and sensors for seam teaching and seam tracking is considered a challenge.

14.8 References

Andersen, H.J., *Sensor Based Robotic Laser Welding* – Based on Feed Forward and Gain Scheduling Algorithms. PhD thesis, Aalborg University, 2001.

Chou, J.C.K. and Kamel, M., Quaternions approach to solve the kinematic equation of rotation AaAx = AxAb of a sensor mounted robotic manipulator. In *Proceedings of the 1988 IEEE International Conference on Robotics and Automation*, pp. 656–662, Philadelphia, 1988.

Craig, J.J., *Introduction to Robotics: Mechanics & Control*. Reading, MA: Addison-Wesley, 1986.

Daniilidis, K., Hand-eye calibration using dual quaternions. *The International Journal of Robotics Research*, 18(3): 286–298, 1999.

De Graaf, M.W., *Sensor-Guided Robotic Laser Welding*. PhD thesis, University of Twente, 2007.

De Graaf, M.W., Aarts, R.G. K.M., Meijer, J. and Jonker, J.B., Robot-sensor synchronization for real-time seam tracking in robotic laser welding. In *Proceedings of the Third International WLT-Conference on Lasers in Manufacturing*, Munich, 2005.

Duley, W., *Laser Welding*. New York: Wiley Interscience, 1999.

Falldorf Sensor GmbH, http://www.falldorfsensor.com, 2006.

Fridenfalk, M. and Bolmsjö, G., Design and validation of a universal 6D seam tracking system in robotic welding based on laser scanning. *Industrial Robot: An International Journal*, 30(5): 437–448, 2003.

Hakvoort, W.B.J., Iterative Learning Control for LTV systems with Applications to an Industrial Robot, PhD thesis, University of Twente, 2009.

Hardeman, T., Modelling and Identification of Industrial Robots including Drive and Joint Flexibilities, PhD thesis, University of Twente, 2008.

Huissoon, J.P., Robotic laser welding: seam sensor and laser focal frame registration. *Robotica*, 20: 261–268, 2002.

Iakovou, D. Sensor Development and Integration for Robotized Laser Welding. PhD thesis, University of Twente, 2009.

Kang, H.S., Suh, J. and Cho, T.D., Robot based laser welding technology. *Materials Science Forum*, 580–582, 565–568, 2008.

Khalil, W. and Dombre, E., *Modeling, Identification & Control of Robots*, Stanmore, Middlesex: Hermes Penton, 2002.

Lee, S. and Ro, S., A self-calibration model for hand-eye systems with motion estimation. *Mathematical and Computer Modelling*, 24(5/6), 49–77, 1996.

Lu, C.P., Mjolsness, E. and Hager, G.D., Online computation of exterior orientation with application to hand-eye calibration. *Mathematical and Computer Modelling*, 24(5/6): 121–143, 1996.

Nayak, N. and Ray, A., *Intelligent Seam Tracking for Robotic Welding*. Advances in Industrial Control. Berlin: Springer-Verlag, 1993.

Park, F.C. and Martin, B.J., Robot sensor calibration: Solving AX equals XB on the Euclidian group. In *IEEE Transactions on Robotics and Automation*, pp. 717–721, 1994.

Regaard, B., Kaierle, S. and Poprawe R., Seam-tracking for high precision laser welding applications – methods, restrictions and enhanced concepts, *Journal of Laser Applications*, 21(4), 2009.

Shiu, Y.C. and Ahmad, S., Calibration of wrist-mounted robotic sensors by solving homongeneous transform equations of the form AX=XB. In *IEEE Transactions on Robotics and Automation*, pp. 16–29, 1989.

Thorne, H.F., Tool center point calibration for spot welding guns. US Patent US5910719, 1999.

Tsai, R.Y. and Lenz, R.K., A new technique for fully autonomous and efficient 3D robotics hand/eye calibration. In *IEEE Transactions on Robotics and Automation*, pp. 345–357, 1989.

Van der Heijden, F., *Image Based Measurement Systems: Object Recognition and Parameter Estimation*. New York: John Wiley & Sons, 1994.

Van Tienhoven, J., Automatic Tool Centre Point Calibration for Robotised Laserwelding. Master's thesis, University of Twente, 2004.

Waiboer, R.R., Dynamic Modelling, Identification and Simulation of Industrial Robots – for Off-line Programming of Robotised Laser Welding. PhD thesis, University of Twente – Netherlands Institute for Metals Research, 2007.

Zhuang, H. and Shiu, Y.C., A noise-tolerant algorithm for robotic hand-eye calibration with or without orientation measurement. In *IEEE Transactions on Systems, Man and Cybernetics*, pp. 1168–1175, 1993.

15

Developments in beam scanning (remote) technologies and smart beam processing

A. WETZIG, Fraunhofer IWS Dresden, Germany

DOI: 10.1533/9780857098771.3.422

Abstract: This chapter describes the state of the art as well as the future trends of fast laser beam deflection and shaping for welding applications. The key principles of fast laser beam manipulation like galvanometer-driven and conventionally-driven mirrors are introduced. The beam movement over the workpiece as the first purpose and beam shaping as the second purpose are discussed in detail and explained by means of welding examples. Finally, an outlook for future trends and forthcoming requirements for the beam scanning equipment is given.

Key words: laser beam scanning, smart beam processing, fast laser beam deflection, beam shaping for welding applications, galvanometer scanning.

15.1 Introduction

A relative motion between the focused laser beam and the workpiece is needed in order to generate a welding seam. Usually, three axes are sufficient for a 2D and rotation-symmetric application like laser beam welding of a flat panel or drive shaft. Five axes are only needed in the case of three-dimensional welding. These axes can be designed completely as workpiece axes, split into workpiece and tool axes, or designed completely as tool axes. For conventional machines, the most dynamic set-up is achieved by a so-called split axes concept. However, maximum welding speed and acceleration, which are important for positioning motions, are limited by the mass inertia of the mechanical axis. A more elegant way to provide fast motion for welding and positioning is only to move the laser beam in lateral direction x and y by one or two deflection mirrors and to move the focusing optics in one direction where required. The focusing optics can be arranged in front of the bending mirrors or behind them (in the direction of beam propagation) (Kögel, 2011; Lütke, 2011).

The most common way to move the laser in the lateral direction is to utilize galvanometer-driven mirrors. It is the fastest method whereas the working field is limited to a size of typically 300 mm × 300 mm and the transmitted laser power is limited to 8 kW (Fig. 15.1). An alternative solution consists of using conventional direct rotary drives which has the advantage to be able to move bigger masses, e.g. larger mirrors that are providing a larger

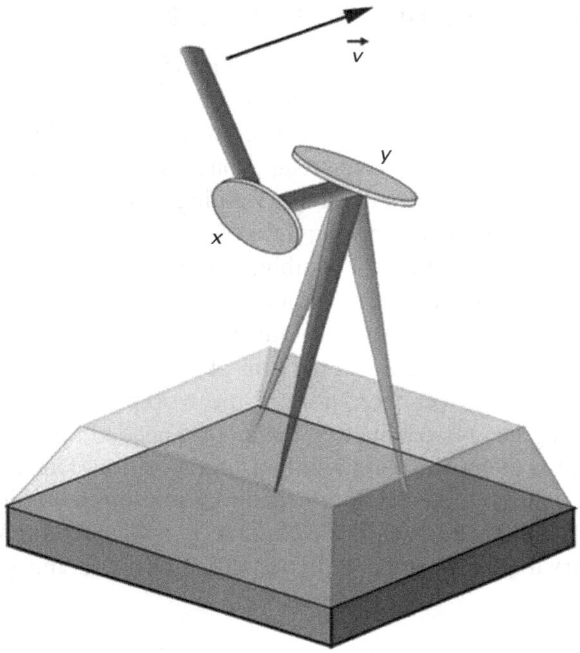

15.1 Principle of beam deflection by using scanning mirrors.

working envelope. There are two options available for these so-called large field scanners: usage of two mirrors (one for each direction) or usage of one gimbal mounted mirror (SEF, 2011).

In the case of beam shaping, where only small amplitudes are needed, an alternative way to move the laser beam is conducted by means of rotating mirrors or rotating prisms (Meier, 2006; Harris, 2001). In order to focus the laser beam, an objective is necessary which can be arranged either in front of the scanning unit or behind it.

CO_2 lasers are established as laser sources for remote welding processes since these lasers are available on the market with single mode beam quality (Bachmann, 2007). Nowadays, solid-state lasers, e.g. fiber lasers and disc lasers are representing an alternative laser source (Hammer and Brockmann, 2006). Since the beam parameter product of single mode CO_2 laser and multi-mode fiber or disc lasers are in the same range (2–4 mm mmrad) mirror sizes, focal length, and focal size of the laser beam are similar. The main difference concerns the coating of the mirrors as well as the material of the focusing optics. Another challenge is the required surface roughness of the mirrors which must be better than a tenth of the wavelength. That means in the case of a fiber or disc laser, a surface roughness of 0.1 μm is needed.

15.2 Beam movement over the workpiece

A typical example of a scanning head is shown in Fig. 15.2. The typical welding speed of less than 20 m/min can be achieved easily by this experimental set-up. The biggest advantage compared with conventional welding systems, besides the more cost-effective system set-up, consists of the fast positioning speed that can be provided between the single welding seams (Munzert, 2009). The disadvantage is the limited working field of about 300 mm × 300 mm. This drawback can be overcome by a combination with a robotic movement (Fig. 15.3). Figure 15.4 shows a typical example of welding of an automotive heat exchanger made of stainless steel by means of a CO_2 laser.

The scanning head is hooked up to the arm of the robot which provides a smooth path movement over the whole of the workpiece to be welded, whereas the scanning head is conducting the motion needed for the welding process as well as for the positioning within a superposition. Due to the fact that the high speed and the precision of scanning optics are combined with the flexibility of a robot, this welding technique is widespread among the automotive industry to join sheet metal. In particular it has become a standard

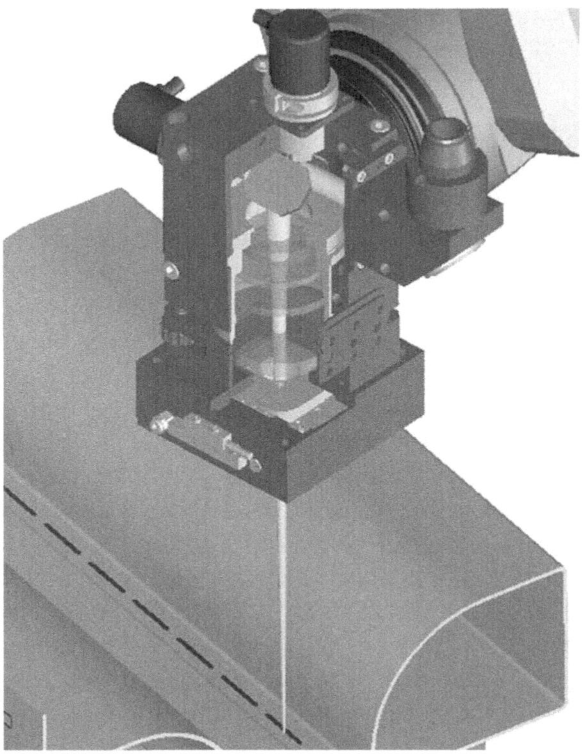

15.2 Small field scanner.

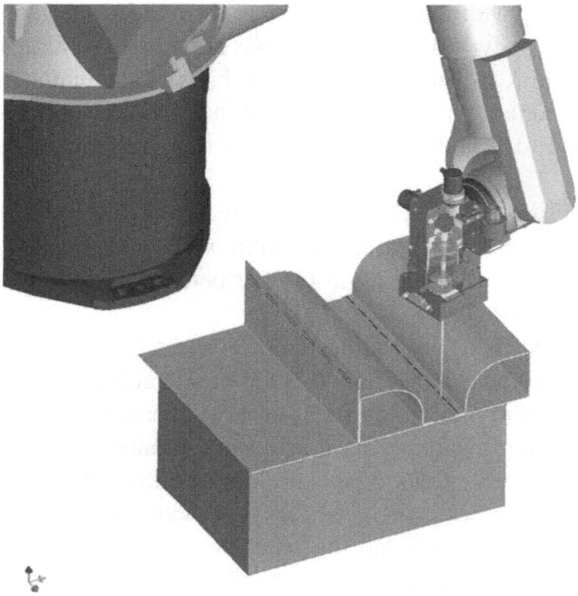

15.3 Superposition of scanning head with robot for welding of large-scale 3D parts.

15.4 Welding seam of automotive heat exchanger.

production technology within the body-in-white manufacturing (Hopf, 2010). Typical examples of body-in-white applications are remote welding of doors, inner side walls, and rear center sections. The biggest advantage of remote welding by using scanning technology in combination with robots is to raise productivity by up to 100% through reduction of non-productive positioning times.

Another option is to combine the scanning head with a Cartesian system, in particular when the welding seams are within one level (2D) and therefore an axes extension is needed, mainly in two, or better only in one, direction (Fig. 15.5) (Bartels, 2011).

Another alternative is a large-scale remote welding system (Fig. 15.6). The working field of about 2,000 mm × 2,000 mm is achieved by means of focal length up to 2,000 mm which requires a raw beam diameter of about 100 mm. Such a system is mainly used as a stand-alone system (Fig. 15.7) (SEF, 2011). Figure 15.8 shows typical welding seams, whereas the seam on a 2D-part in Fig. 15.8(a) was generated by means of superposition of scanning head with linear axes and the seam on a 3D part in Fig. 15.8(b) was generated by means of a large-scale remote welding system.

15.3 Beam shaping

A different goal of using fast beam scanning methods is to provide beam shaping. Beam shaping in the case of laser welding is necessary to provide

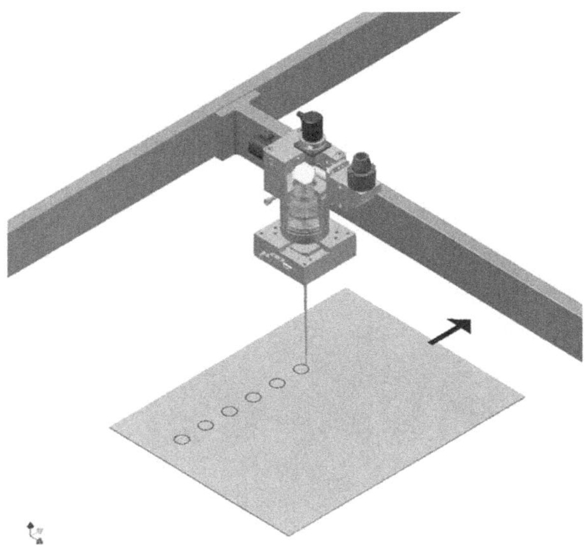

15.5 Superposition of scanning head with linear axes for welding of large-scale 2D parts.

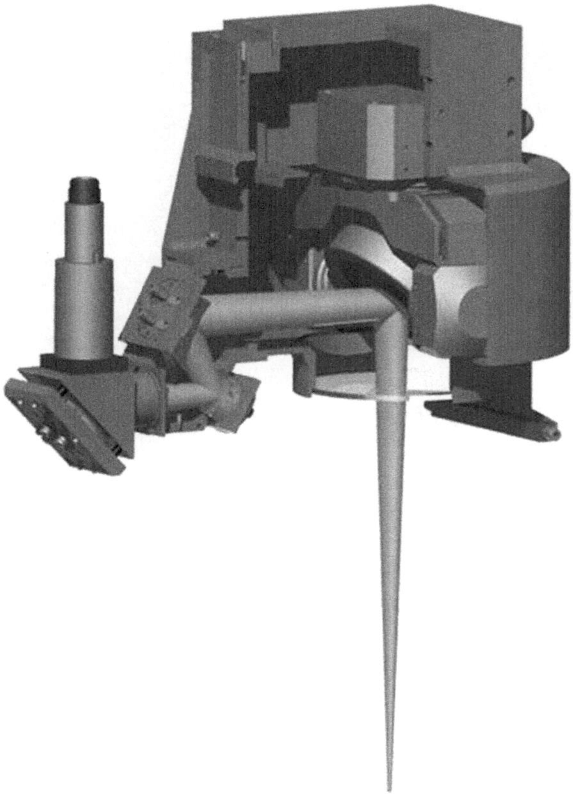

15.6 Principle of large-scale remote welding system.

15.7 Stand-alone system of large-scale remote welding station from Rofin (by courtesy of ROFIN-SINAR Laser GmbH, Hamburg).

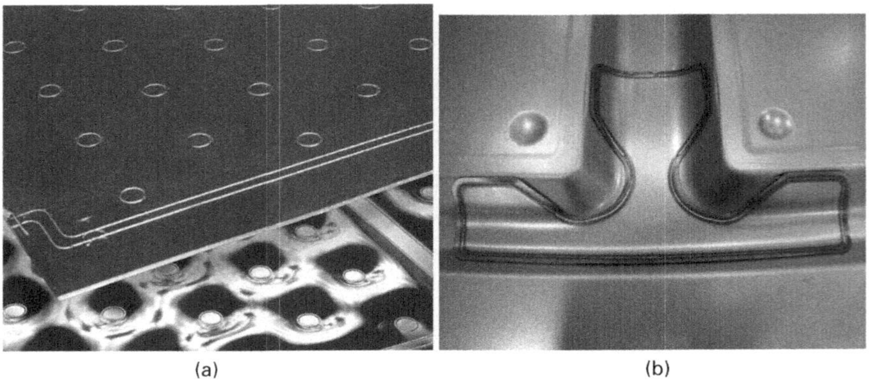

15.8 (a) Typical 2D-part welded; (b) typical 3D-part welded.

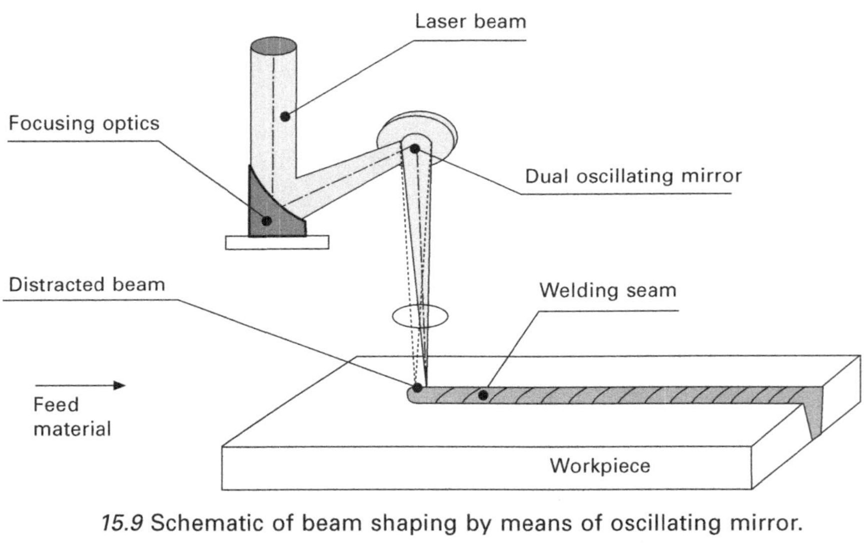

15.9 Schematic of beam shaping by means of oscillating mirror.

a fast keyhole movement which enables a variable shape of the welding seam by affecting the metallurgy, the energy input per unit length, and the solidification behavior. In order to achieve these effects, oscillations of the laser beam in the range of a maximum of 1 mm on the workpiece are needed. The method of choice is to use two galvanometer-driven mirrors which will allow a superposition with the feed motion (Fig. 15.9). Due to the fact that both mirrors can be operated independently, circular, ellipsoidal, and linear oscillations along and across the feed direction are possible (Fig. 15.10) (Mahrle and Beyer, 2007).

State-of-the-art scanning heads are approaching maximum frequencies of about 4 kHz (Krätzsch *et al.*, 2011), which is sufficient for most welding processes. The beam shaping can be utilized to improve the weldability

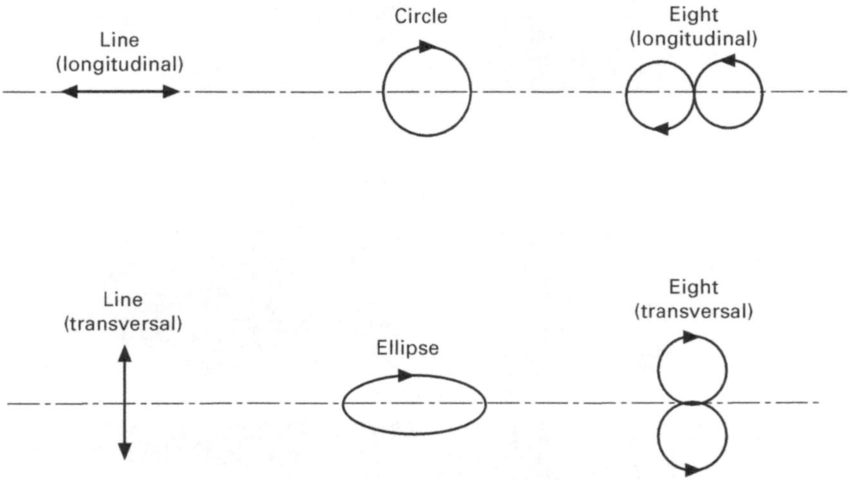

15.10 Various possibilities of oscillations.

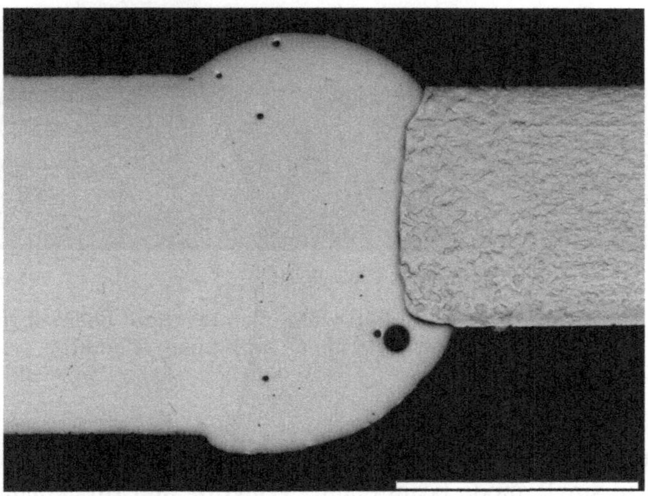

15.11 Cross section of welding seam of aluminum against copper achieved by beam oscillation.

of dissimilar metallic materials like copper and aluminum (Fig. 15.11), to enhance the joining capability for butt welding of thin material (Fig. 15.12) as well as to enable narrow-gap welding (Fig. 15.13).

15.4 Future trends

Remote laser welding is already an established manufacturing process. Momentum to further develop this welding technique will come from the

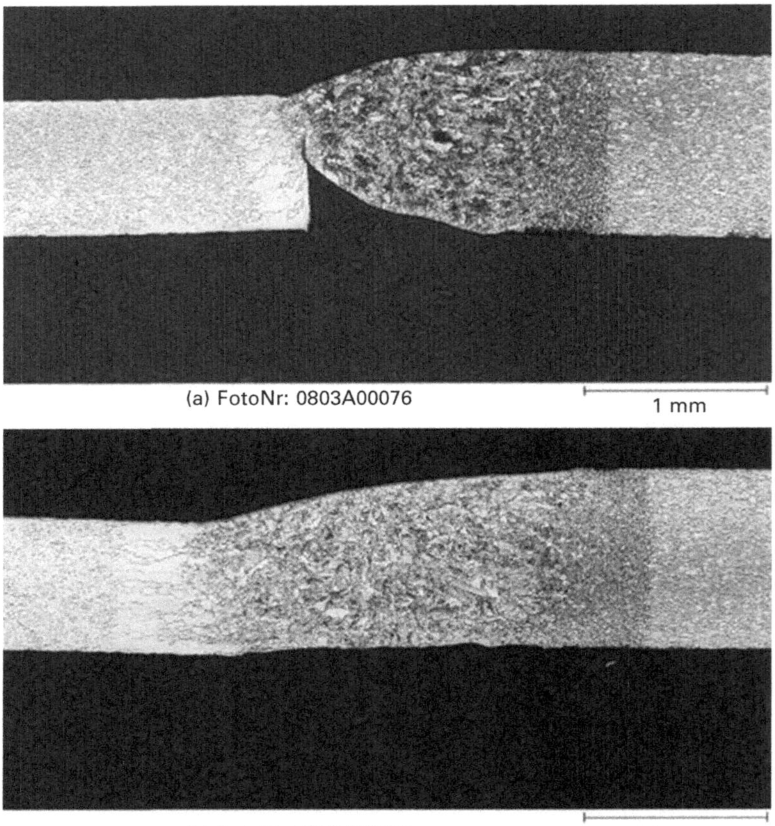

15.12 (a) Cross section of butt weld (steel) without beam scanning; (b) cross section of butt weld (steel) with beam scanning.

requirements of the joining process itself and from the availability of solid-state lasers in single mode beam quality (Havrilla and Brockmann, 2009, 2010; Gapontsev, 2009). The latter will lead to smaller laser spot sizes and therewith to narrower welding joints, if the same optical set-up is used. By maintaining the same focus size, the usage of fraction limited lasers will allow the size of the optics and deflections mirrors to be reduced, which will enhance the dynamics (e.g. positioning speed) of the whole system. Thus, diffraction limited solid-state lasers enable narrower welding joints, reduction of laser power, and faster positioning between separate joints. One of the future challenges that comes from the process is the requirement to join dissimilar materials like aluminum to copper. Therefore, scanning mirrors as shaping systems will be combined with deflection systems for beam movement over the workpiece. One option is to utilize only one deflection

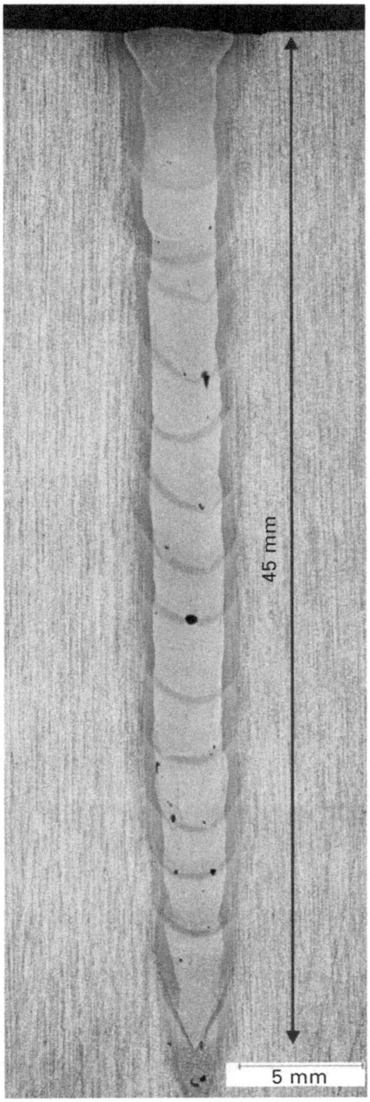

15.13 Narrow gap welding by means of beam scanning.

system, whereas the scanning mirrors provide the oscillation motion, which is necessary for beam shaping and the deflection of the laser beam within a larger angle range for guiding the beam on the workpiece. The other option would be to combine two scanning heads as illustrated in Fig. 15.14. Another trend will result from more dynamic scanning mirrors. This can be achieved by using lightweight carrier material for the mirrors like SiC or Be and by using resonance scanners.

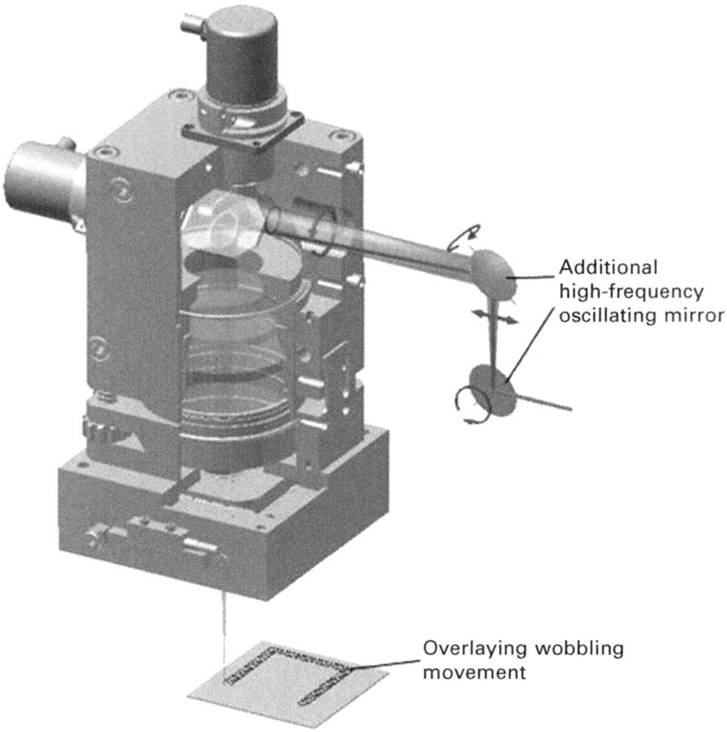

15.14 Combining of two different scanning heads to provide beam shaping while laser welding.

15.5 References

Bachmann F (2007), High Power Laser Sources for Industry and their Applications, *Proceedings of SPIE*, 6735, 67350T.

Bartels F (2011), *Beitrag zur Erhöhung der mittleren Bearbeitungsgeschwindigkeit beim Laserstrahlschneiden komplexer Bauteilgeometrien*, Fraunhofer Verlag, Stuttgart.

Gapontsev (2009), Recent Progress in Scaling High Power Fibre Lasers at IPG PHOTONICS, *22nd Annual Solid State and Diode Laser Technology Review*, Newton, MA, June 29–July 2.

Hammer T and Brockmann R (2006), Scannerschweißen mit dem Scheibenlaser – Auf dem Weg zur Serienanwendung, *LaserTechnikJournal*, 3(3), 36–38.

Harris J (2001), The Cutting of Thick Steel Plate using a Spinning Nd:YAG Laser Beam, *20th ICALEO*, pp. 221–228.

Havrilla D and Brockmann R (2009), Third Generation of Disk Lasers, *LaserTechnikJournal*, 6(3), 26–31.

Havrilla D and Brockmann R (2010), Latest Advances in High Power Disk Lasers, *2010. SPIE*. San Francisco, CA, C1–C11.

Hopf B (2010), Remote Laser Welding: Future Developments and Quality Assurance, *AKL'10 International Laser Technology Congress, Proceedings*, p. 111.

Kögel G (2011), High Power Scanning, *Journal EuroLaser*, 4, 28–31.

Krätzsch M, Standfuß J, Klotzbach A, Kaspar J, Brenner B and Beyer E (2011), Laser Beam Welding with High-Frequency Beam Oscillation: Welding of Dissimilar Materials with Brillant Fiber Laser, *Lasers in Manufacturing 2011, Proceedings*, pp. 142–149.

Lütke M (2011), *Entwicklung des Remote-Laserschneidens metallischer Werkstoffe*, TU Dresden, Dissertation.

Mahrle A and Beyer E (2007), Control of energy deposition during laser beam welding by oscillation techniques, *Lasers in manufacturing 2007, Proceedings*, pp. 97–104.

Meier O (2006), Hochfrequentes Strahlpendeln zur Erhöhung der Prozessstabilität beim Laserstrahlschweißen, Abschlussbericht AIF-Vorhaben 13660N.

Munzert U (2009), Bahnplanungsalgorithmen für das robotergesteuerte Remote-Laserstrahlschweißen, Technische Univerität München.

SEF (2011), *Die Laser Remote-Technik, Grundlagen-Lieferprogramm*, SEF Roboter GmbH, Scharnebeck.

16
Developments in twin-beam laser welding technology

Y. MIYASHITA, Nagaoka University of Technology, Japan

DOI: 10.1533/9780857098771.3.434

Abstract: In this chapter the phenomenon, mechanism and method of twin-beam laser irradiation are introduced. Welding, including dissimilar materials welding, surface modification and cutting of a brittle material, are shown as particular applications of twin-beam laser welding. The advantages and effects of the technique are described. An analytical study on the convection behavior occurring during twin-beam laser irradiation is also shown as an example of using a numerical approach to understand the mechanism of twin-beam laser welding.

Key words: twin-beam laser, welding, dissimilar material welding, surface modification, cutting.

16.1 Introduction

There are many advantages to laser welding, such as its small heat-affected zone, deep penetration, precision welding, high applicability for robotics technology, etc. There are, however, some limitations to the optimization of single-beam welding parameters for improving weldability, and weld quality, such as laser power, pulse condition and spot size, because they strongly depend on the kind of laser beam used. Twin-beam or dual-beam laser irradiation has been developed to overcome the limitations of single-beam irradiation welding. Twin-beam or dual-beam laser irradiation means that two laser beams are used at the same time in one process. Two effects are expected from the use of twin-beam irradiation: an increase in the number of heat sources compared to single-beam irradiation and the ability to realize different phenomena. The main result of applying a twin beam instead of a single one may simply be an increase in the power of the laser. The power of twin-beam irradiation is not, however, the simple sum of two beams. Sakamoto *et al.* [1] studied energy coupling in twin-beam laser irradiation. According to their calculation of the energy coupling rate based on the Gouffe–Dausinger model, the keyhole opening size, which is not simply the sum of the diameters of the two laser spots, was assumed to be a spot diameter when the distance between two beams was narrow. Furthermore, it is also expected that two heat sources can produce different temperature distributions and histories during the process, even if two of the same type of laser beam are used. In cases where two different types of laser beams are applied, complex and interesting phenomena can also be expected. This chapter introduces twin-beam laser welding techniques and describes methods of twin-beam laser irradiation and its phenomena and mechanism. Examples of the applications of twin-beam laser irradiation are also described to demonstrate its advantages.

16.2 Numerical study on molten metal flow behavior during twin-beam irradiation

The use of two heat sources as opposed to one, i.e. applying a twin beam instead of a single beam, has a significant effect on the flow behavior of molten metal. In this section, a comparison of the convection behavior in the molten pool during both single- and twin-beam laser irradiation is provided [2]. A magnesium alloy was the metal alloy used in the analysis, the physical properties of which are shown in Table 16.1.

Plate XIV (between pages 328 and 328) shows molten pool shapes in the cross section perpendicular to the laser scanning line, which were created using both single- and parallel twin-beam irradiation. For the purposes of the analysis, the region with a temperature above the melting temperature of a magnesium alloy was assumed to be the molten pool region. The total applied energy was the same for both the single and the twin-beam in this case, meaning that during the twin-beam irradiation, each beam used about half of its total applied energy. The maximum temperature at the laser irradiated point in the twin beam was, therefore, lower than that in the single beam. Plate XV (a) (between pages 328 and 329) shows the convection behavior in the cross section perpendicular to the laser scanning line in the case of single-beam irradiation. This was calculated according to the temperature distribution as shown in Plate XIV. Plate XV (a) shows the flow direction travelling from the inside to the outside around the laser irradiation point. The middle of the molten pool shows almost no flow. Flow behavior following irradiation using parallel twin beams is shown in Plate XV (b). In this case,

Table 16.1 Thermo-physical properties of the material used for the analysis

Heat conductivity, kW $m^{-1} K^{-1}$	155
Specific heat capacity, J $kg^{-1} K^{-1}$	1360
Liquid metal density, kg m^{-3}	1738
Expansion coefficient of volume, K^{-1}	1.523×10^{-4}
Surface tension gradient, kg s^{-2} K^{-1}	-1.49×10^{-4}
Liquid metal viscosity, m^2 s^{-1}	1.23×10^{-3}
Melting temperature, °C	650
Ambient temperature, °C	25
Coefficient of heat emission, $Wm^{-2}K^{-1}$	125

surface tension from the irradiation point of each beam directs the flow to the center between the two beams, and flow from the top to the bottom occurs at the center of the molten pool.

Plate XIV (between pages 328 and 329) shows molten pool shapes obtained in the cross section parallel to laser scanning line following both single- and tandem twin-beam irradiation. It is clear from the figure that the molten pool shapes created by the two types of irradiation are completely different. Plate XVII (between pages 328 and 329) shows flow behavior in the cross section parallel to laser scanning line by both single- and tandem twin-beam irradiation. In the case of single-beam irradiation, convection is not significantly observed in the middle part of the molten pool. In the case of the tandem twin-beam irradiation, however, flows are significantly observed at the end and middle sections of the molten pool. The two different convection flows induced by two-beam irradiation affect one another and cause the flows to become active in the middle section of the molten pool. Additionally, the total duration for melting of the material is longer in the tandem beam irradiation.

16.3 Apparatus and procedure of twin-beam laser technique

There are two methods of creating a twin-beam laser for welding: splitting one laser beam into two beams using a prism or a mirror, or using two laser oscillators. The former method has the advantages of a simple structure and low cost, whilst the latter allows the operator to control the two beams separately and precisely to create the phenomena caused by mixing different beams.

Figure 16.1 shows an example of a twin beam apparatus. When prisms are used to split a laser beam, the beam distance and intensity ratio can be controlled by altering the wedge angle and insert distance of the prism used [3]. Figure 16.2 shows a cross section of an aluminum alloy sheet following twin-beam laser irradiation. The distance between the two beams is changed by altering the wedge angle of the prism.

When two oscillators are used, two or more beams can be applied either in the same optical line or a different line. Figure 16.3 shows a three oscillator laser welding system [4]. Energy distribution obtained by applying three oscillators is shown in Fig. 16.4. In this case, two CWs and one pulse laser beam are applied through one single optical line. Figure 16.4 shows that three beams are not separated from one another and it seems like a one-beam irradiation. The actual waveform is almost following the command waveform as shown in Fig. 16.5.

Figures 16.6 and 16.7 show schematics of a twin-beam laser welding system and the distribution of beam intensity observed by using the system

16.1 Twin-beam laser head created using a prism to split an original beam.

Beam condition		Beam condition	
Single beam	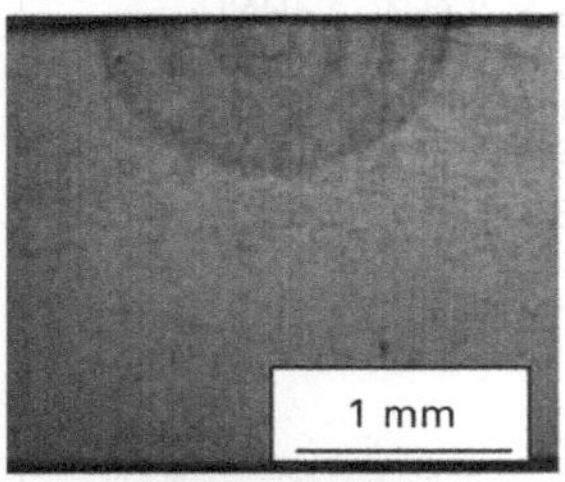	Wedge 3[deg]	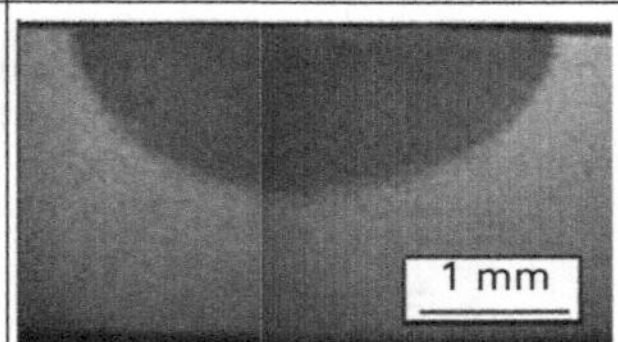
Wedge 2[deg]	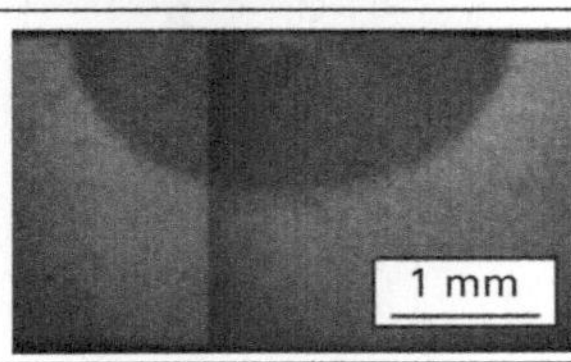	Wedge 4[deg]	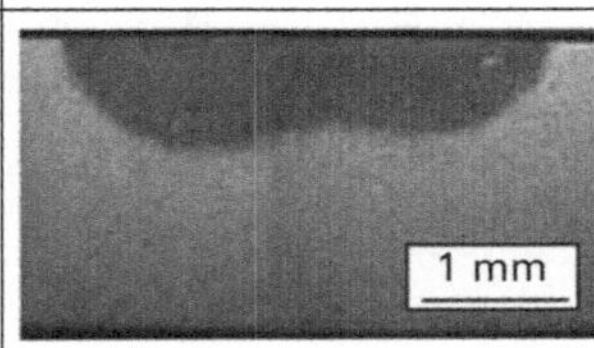

16.2 Cross-sectional observations applied to the twin-beam laser on A5052 with power of 1 kW.

[5]. In this case, the beam distance is controlled by the distance between the optic fibers.

16.4 Application of twin-laser beam

16.4.1 Welding

Jonghyan *et al.* [6] reported the melting characteristics of various metals following irradiation by combined lasers with different wavelengths. A

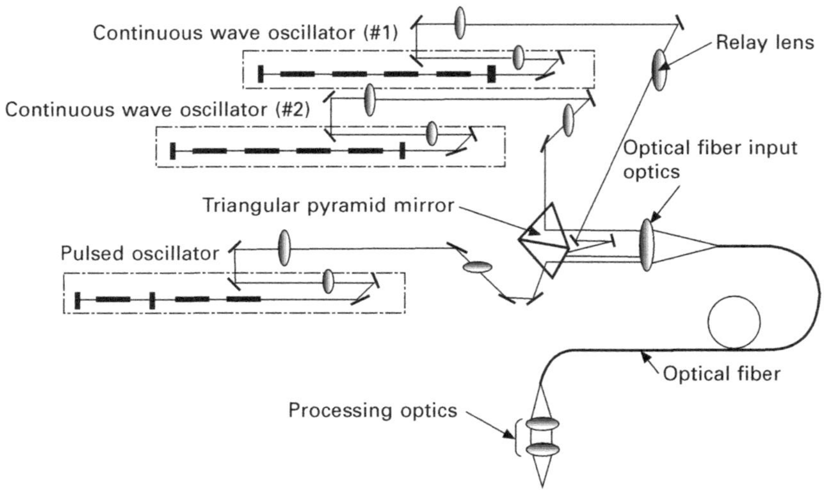

16.3 Schematic of the three-beam laser welding system.

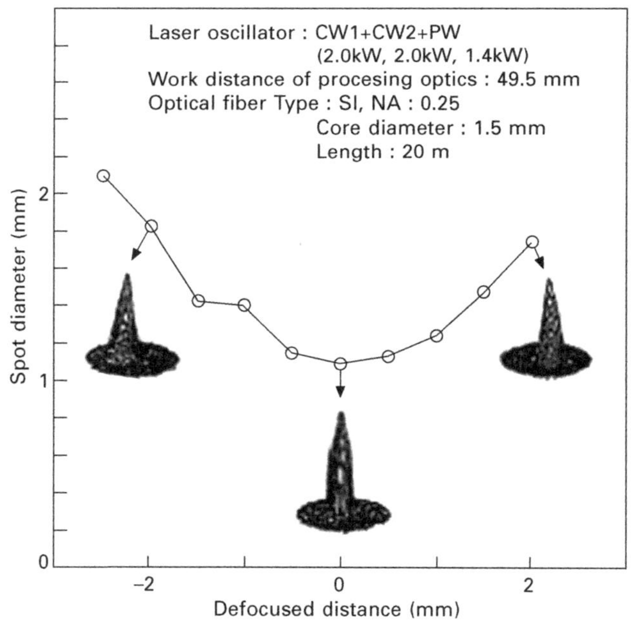

16.4 Energy distribution of the laser spot using the three-beam laser welding system.

Q-switched YAG laser and pulsed YAG laser were combined both through different optics lines and the same optics line as shown in Figs 16.8 (a) and (b), respectively. Figure 16.9 shows cross sections of a weld bead obtained both by single-beam irradiation and twin-beam irradiation. In this example,

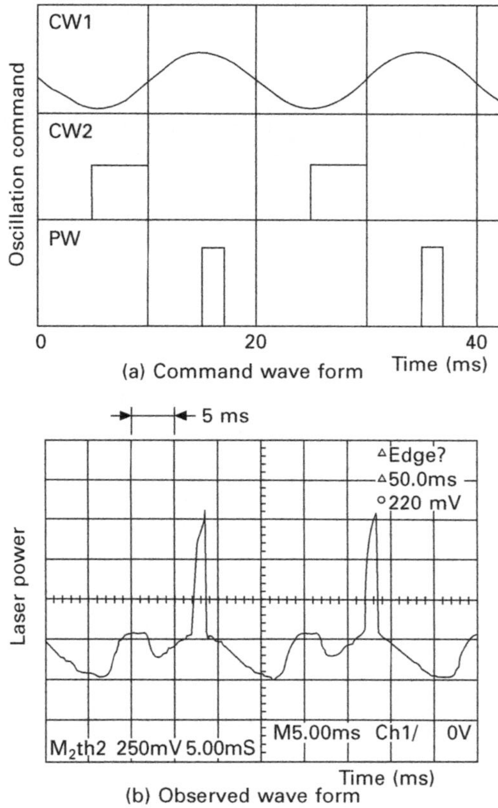

16.5 Laser power profile using the three-beam laser welding system: (a) command wave form, (b) observed wave form.

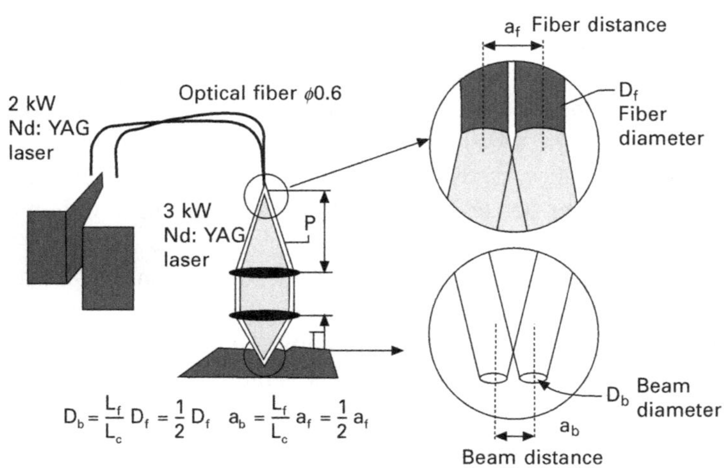

16.6 Twin-beam laser produced two laser oscillators.

16.7 Laser spot profile of twin beam produced by using two oscillators.

the Q-switch and pulse YAG laser were applied through different optical lines. The size and depth of the molten pool were increased after twin-beam irradiation. Figure 16.10 shows a cross section of weld bead obtained using twin-beam irradiation in different metals. Keyhole type penetration was observed and the molten area and depth of the molten pool increased following twin-beam laser irradiation (Fig. 16.10). The effect of twin-beam laser irradiation is schematically shown in Fig. 16.11. In cases where a highly reflective and thermally conductive material was used, two layers of molten pool were observed. Contrastingly, in cases where materials with a low thermal conductivity were used, deeper penetration was obtained. Jonghyan *et al.* [6] reported that the absorptivity of the laser increased when twin-beam laser irradiation was used by 21%, 29% and 25% for a pure copper, A5083 and SUS304, respectively.

Shibata *et al.* [7] studied the effect of twin spot beam configuration on the overlap joint welding of aluminum alloys. Two different Nd:YAG laser beams were separately applied as shown in Fig. 16.6. During this study the distance between the two beams was changed. Figure 16.7 shows the beam profile obtained at different beam distances. Tandem beam and parallel beam configurations were applied as shown in Fig. 16.12. Figure 16.13 shows a cross section of a weld bead. The penetration depth decreased with the increase in beam distance for both beam configurations. When the beam distance became too short, however, the welding process became unstable. The configuration of twin beams should be selected depending on material thickness, laser power and scanning speed in order to obtain higher quality welding.

The effect of the twin-beam laser irradiation on porosity formation was studied by Shibata *et al.* [5], who changed the welding configuration and applied the beam to an aluminum alloy. Two different Nd:YAG lasers were applied with different optical lines. Figure 16.14 shows the relationship between the level of porosity and the beam distance in both the tandem and the parallel beam configurations. The figure shows that the level of porosity decreased as the beam distance increased, regardless of the beam configuration. A large opening distance and a shallow keyhole depth might,

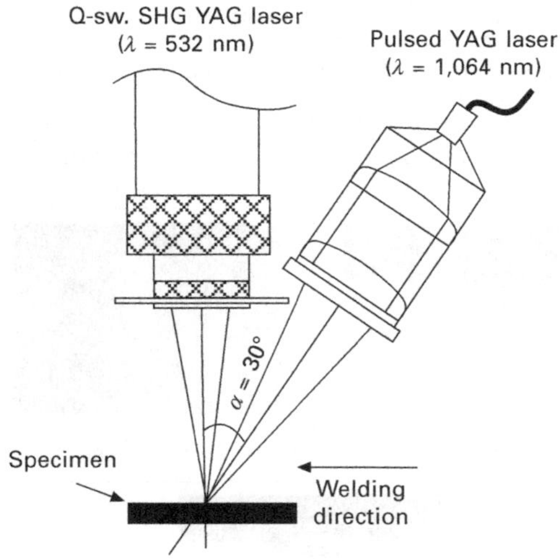

(a) Different optics line

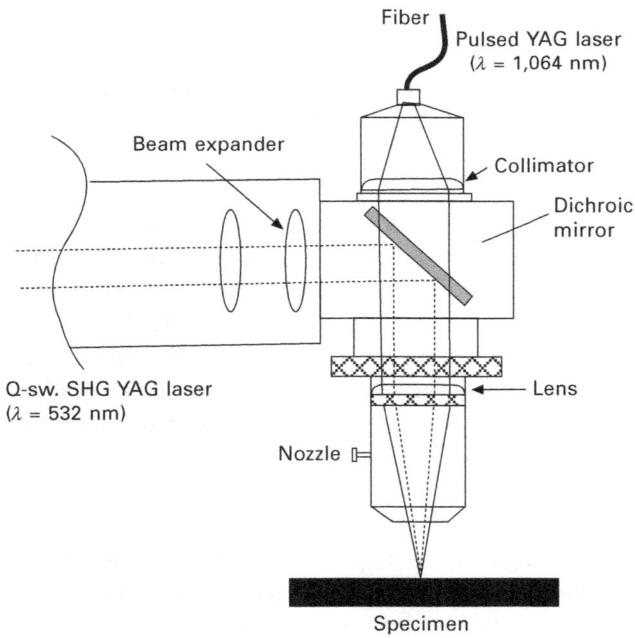

(b) Same optics line

16.8 Combination between Q-switched YAG laser and a pulsed YAG laser: (a) different optics line, (b) same optics line.

therefore, suppress the occurrence of porosity. Twin laser beam irradiation proved to be an effective method of controlling keyhole shape. Hayashi *et al.* [8] also studied the effect of twin-beam irradiation on porosity formation

A5083, Bead-on-plate welding, In air Pulsed 1 ω YAG laser (λ = 1,064 nm): E = 55 J/P, τ = 10 ms, f = 5 Hz, f_d = 0 mm (f = 120 mm) Q-sw. 2 ω YAG laser (λ = 532 nm): P_a = 49.5 W, f = 10 kHz, f_d = 0 mm (f = 100 mm)	
Pulsed 1ω YAG laser (λ = 1,064 nm)	Q-sw. 2ω YAG laser (λ = 532 nm)
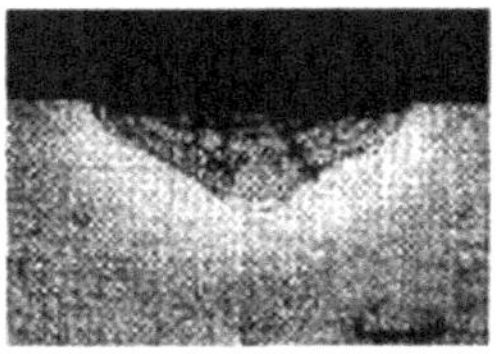	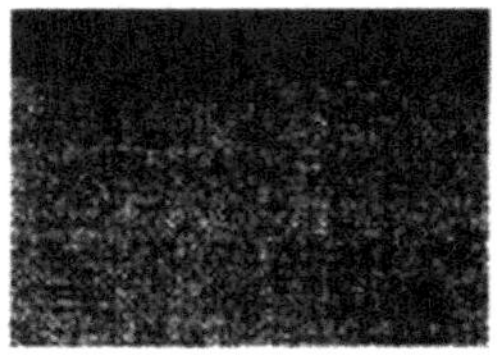
Pulsed 1ω YAG laser + Q-sw. 2ω YAG laser	
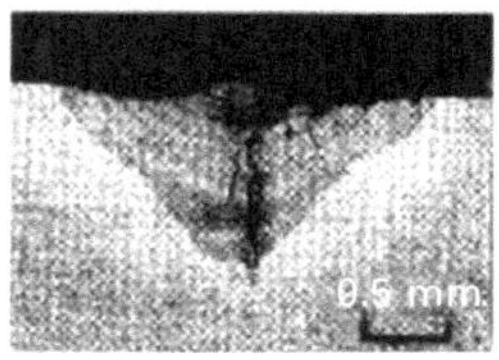	

16.9 Cross section of weld bead obtained using single-beam irradiation and twin-beam irradiation.

using stainless steel. A CO_2 laser beam was split into two beams using a mirror to obtain the tandem beam configuration. According to the X-ray transmission *in-situ* observation during the welding process, two types of keyhole were observed as shown in Fig. 16.15. Where the two beams were separately applied, as shown in Fig. 16.15(a), porosity formed at the bottom of the keyhole. Where one large keyhole formed, as shown in Fig. 16.15(b), however, the formation of porosity was suppressed due to the change in the flow behavior.

Another expected advantage of applying twin beams is their effect on gap tolerance as reported by Ohashi *et al.* [9]. In this study, two YAG lasers were used and applied through different optical lines and both tandem and parallel laser beam configurations were applied. In the case without a filler wire, the parallel beam configuration had a greater tolerance than the tandem beam configuration, as shown in Fig. 16.16. Figure 16.17 shows cross sections obtained by applying the twin beam to a filler wire. Welding was available until the gap dropped below 1.6 mm for both the tandem beam and the parallel beam configurations. Where the gap was much larger, however, the energy of the laser beam was spent in heating the wire and the heat input into the base material therefore decreased. One solution to the problem is shown in Fig. 16.18. In cases using the parallel beam configuration, the irradiation area can

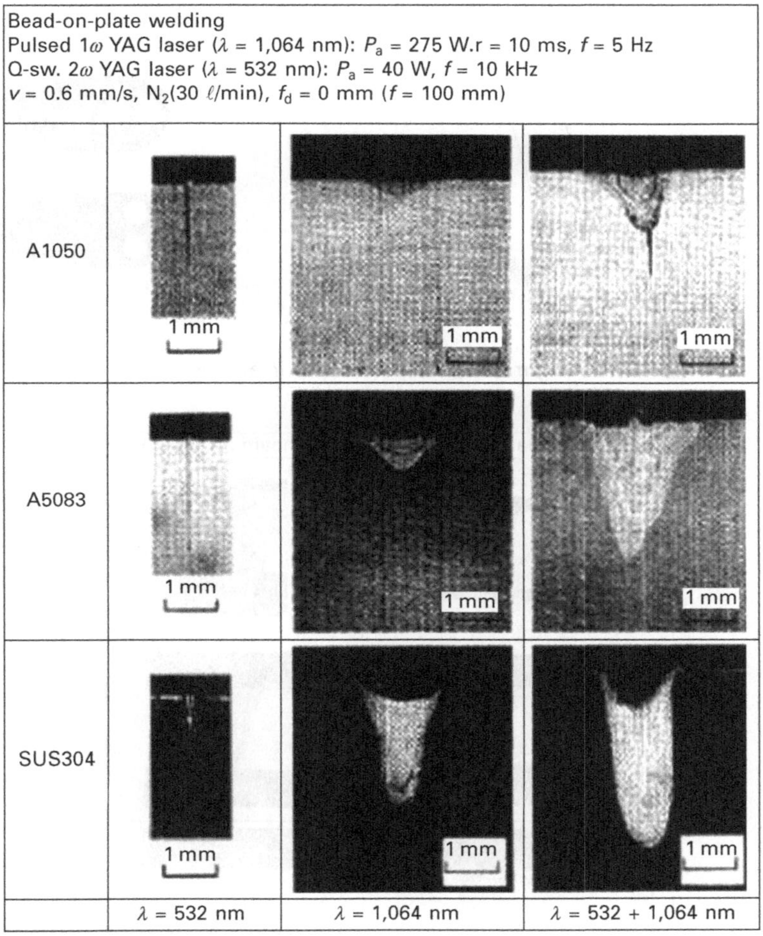

16.10 Cross sections of weld bead obtained by single-beam irradiation and twin-beam irradiation in different metallic materials.

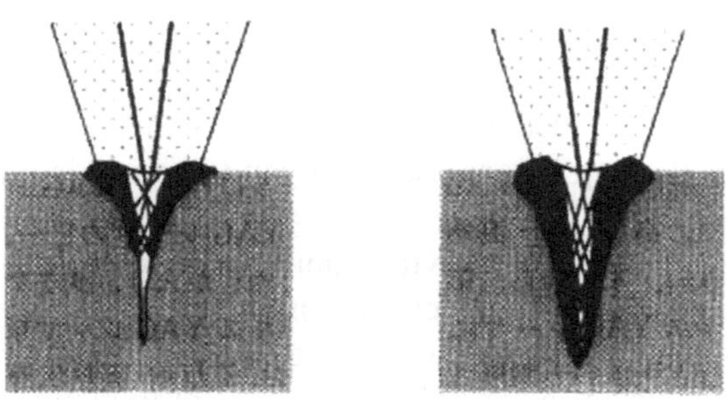

(a) (b)

16.11 Effect of twin-beam laser irradiation in (a) high reflective and high thermal conductive material, (b) low thermal conductive material.

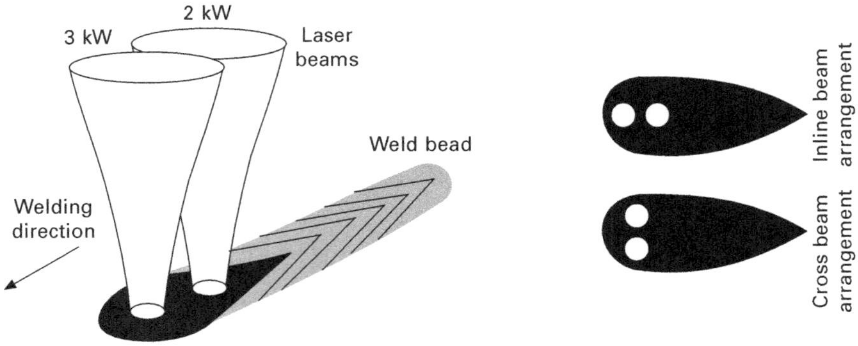

16.12 Configurations of twin-beam irradiation.

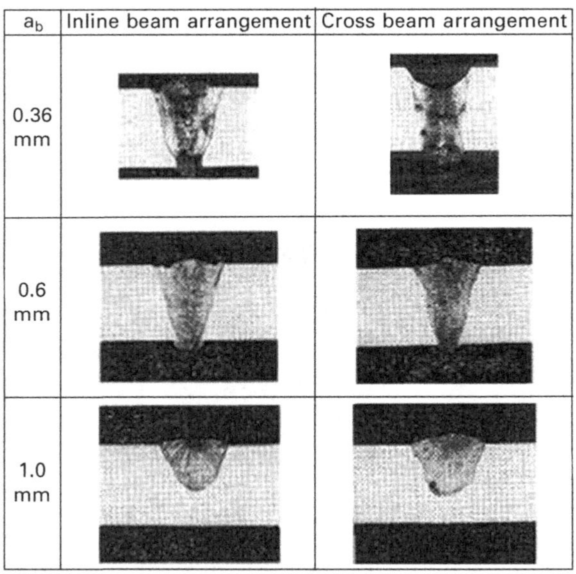

16.13 Effect of distance between laser beam spots on molten pool shape (Al-Mg alloy).

be controlled by changing the beam distance. By using this technique, heat input at the bottom of the gap is increased and gap tolerance is effectively improved.

Iqbal *et al.* [10], [11] applied the twin-beam technique for the welding of zinc coated steel. An Nd:YAG laser was used as an advancing beam and a CO_2 laser was used as the following one in the tandem twin-beam configuration. The front beam (Nd:YAG laser) cut a slot to remove the zinc coating and the back beam (CO_2 laser) welded a material as shown in Fig. 16.19. Figure 16.20 shows a schematic of the proposed welding method by the tandem twin-beam irradiation.

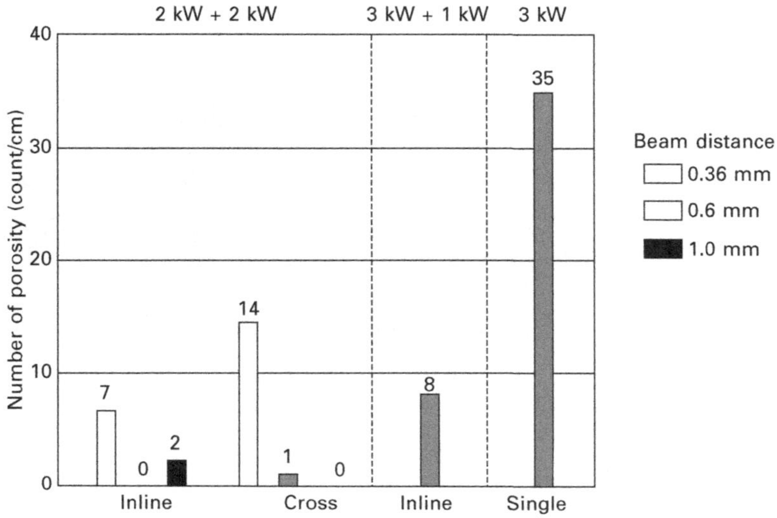

16.14 Effect of beam distance on formation of porosity in different beam configurations (aluminum alloy).

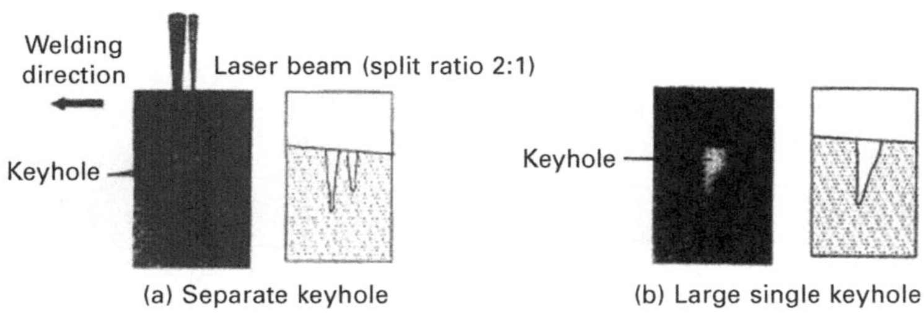

16.15 Schematic illustration of formation of porosity in type 304 stainless steel.

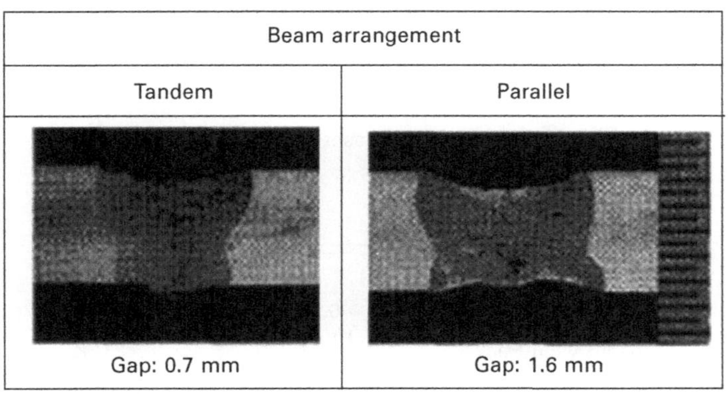

16.16 Effect of configuration of twin-beam laser on gap tolerance (type 304 stainless steel).

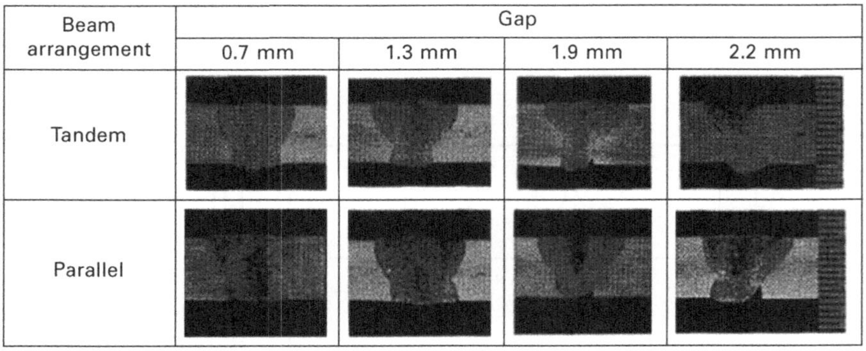

16.17 Effect of filler wire on gap tolerance in different configurations of twin-beam laser (base metal; type 304 stainless steel, filler wire; SUS308Y, dia = 12 mm).

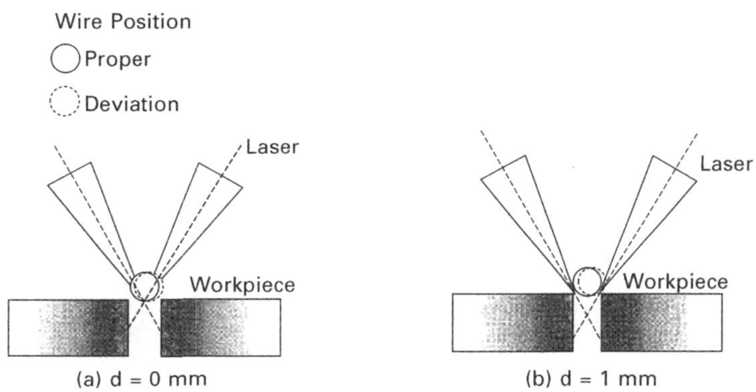

16.18 Increasing gap tolerance by changing the irradiation point of the laser beams.

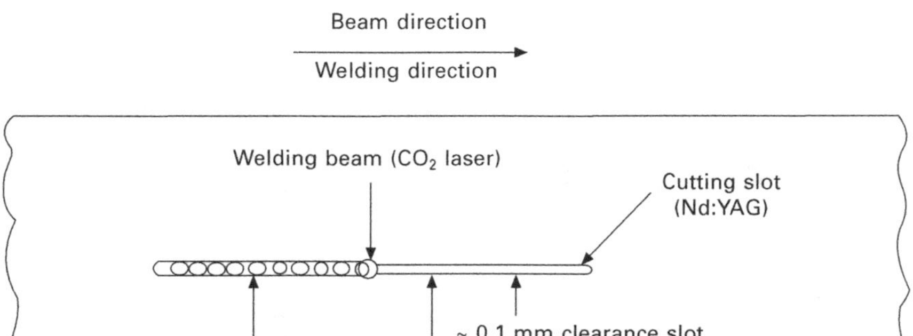

16.19 Laser lap welding by the tandem twin-beam laser in zinc-coated steel.

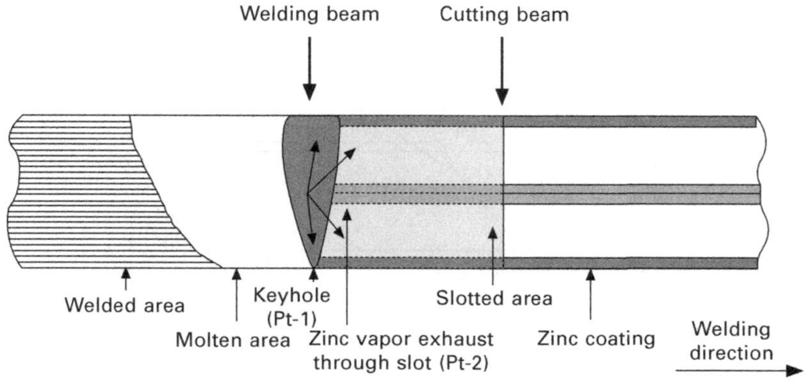

16.20 Schematic illustration of the proposed lap welding method by using the tandem twin-beam laser in zinc-coated steel.

16.4.2 Dissimilar materials welding

The application of the twin-beam laser technique for dissimilar materials welding is described below [3]. In the majority of cases involving the welding of dissimilar materials, intermetallic phases are formed, which reduce the strength of the joint. In cases where the materials have very different melting temperatures, the formation of the intermetallic phase could be controlled using a solid-state–liquid-state reaction at the interface [12]. When there is only a minor difference in the melting temperature of the materials, however, such as in magnesium alloy/aluminum alloy dissimilar metals welding, the solid-state–liquid-state reaction cannot be achieved and the intermetallic phase can be easily formed in the molten pool. An alternative method for controlling the formation of the intermetallic phases will therefore be required.

AZ31B magnesium alloy and A5052-O aluminum alloy plates of 1 mm in thickness were used in the present study. The Nd:YAG laser was used for welding with both single-beam and twin-beam with the parallel beam configuration that both beams were placed in perpendicular to the welding direction. Twin laser beams were obtained by splitting an original beam using a prism as shown in Fig. 16.1. In this study, the two wedge angles of the prism were 2° and 4°, which separated the beam distances into 0.87 and 1.60 mm, respectively. The welding configuration used is shown in Fig. 16.21. The laser was irradiated on the AZ31 plate, which was the upper plate of the lap joint configuration in the direct welding. The laser power used was 3 kW with CW.

Plate XVIII (between pages 328 and 329) shows cross sections of a weld pool for all welding configurations. The joint width and depth of the molten metal for the joints obtained are shown in Fig. 16.22. The weld bead width and depth of the molten metal slightly increased with a decrease in welding

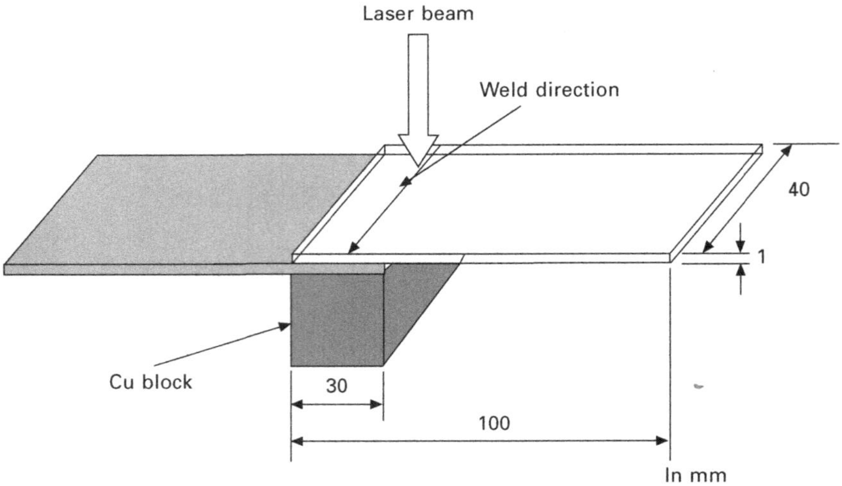

16.21 Welding and configuration in A5052/Z31.

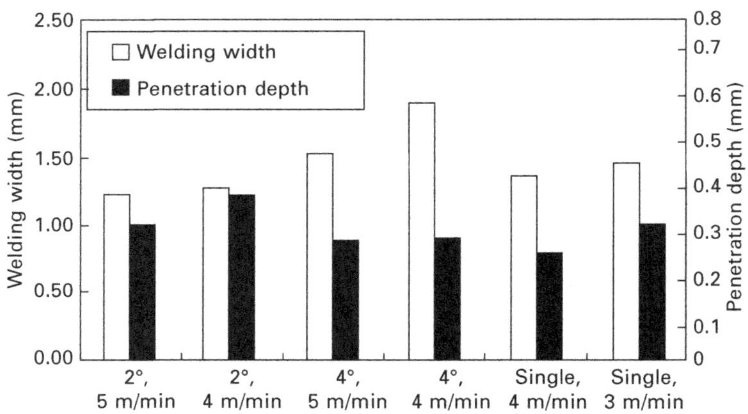

16.22 Effect of welding configuration on width and penetration depth of the weld pool in AZ31/A5052 dissimilar materials direct joint.

speed for single-beam welding and twin-beam welding with a 2° wedge prism. In the case of the twin beam with a 4° wedge prism, however, an increase in the depth of the molten metal was not significantly observed, but the weld width increased. Figure 16.23 shows the failure load and joining strength of the joints. It was found that an application of single-beam and twin-beam welding with a 2° wedge prism resulted in decreasing joining strength with a decrease in welding speed because deep penetration of the molten metal might cause the formation of thick intermetallic compounds. In the case of the twin-beam welding with a 4° wedge prism, however, the change in depth of the molten metal is very small resulting in an increase in

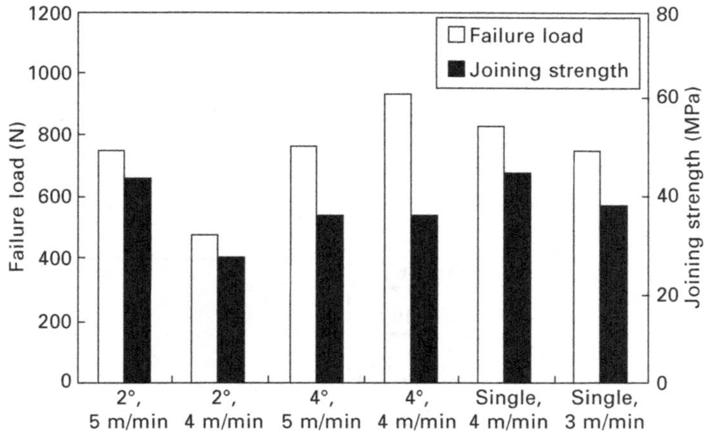

16.23 Effect of welding configuration on failure load and strength in AZ31/A5052 dissimilar materials direct joint.

the failure load and a negligible change in the joining strength even when the welding speed decreased.

The failure load of the joints could be improved by applying the twin beam to direct welding as just described. An intermetallic phase was still formed at the interface, however, which affected the strength of the joint. In order to control the composition of the material at the interface, a filler metal was applied. The filler wire used was an Al12Si eutectic alloy with a diameter of 1.2 mm. The wire was positioned at the edge of the upper plate. Both the single beam and twin beam were used. In the case of twin beam when welding with a filler metal, the tandem beam configuration with two beams placed parallel to the welding direction was used. A5052 was placed as an upper plate irradiated by the laser beam. The laser power used was 3 kW. For the twin-beam welding, a 4° wedge prism was used to split an original beam. The intensity ratio between the forward beam and the following beam was 1:2. A schematic illustration of the welding configuration for twin-beam laser welding with a filler wire is shown in Fig. 16.24.

According to observations made by a high-speed camera during the twin-beam laser welding process, it was confirmed that the forward beam was used only for pre-heating without melting the base metal. Plate XIX (between pages 328 and 329) shows cross-sectional observations of the joints welded with the filler wire for both single- and twin-beam welding. Full penetration always occurred during single-beam irradiation, whereas only partial penetration was observed during twin-beam irradiation. At the interface between the filler metal and the base metals, Mg_2Si and Al-Mg intermetallic phase was observed. Figure 16.25 shows the failure load of the joints welded with a filler wire. The failure load of the joints welded with the twin beam was higher than that of those welded with a single beam.

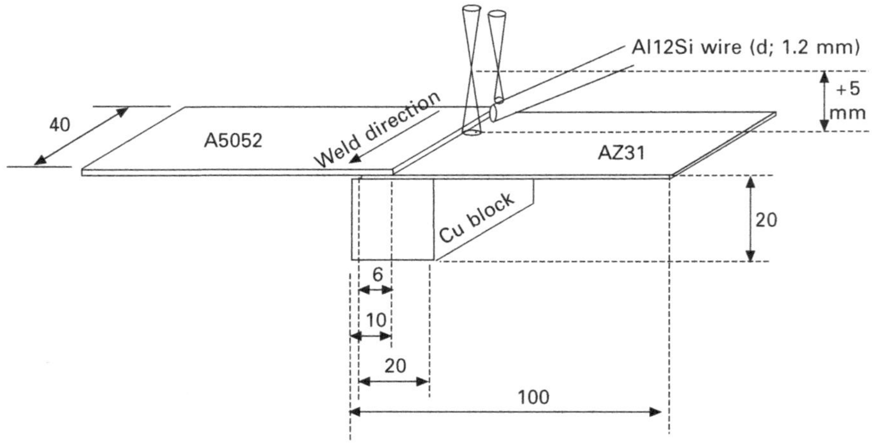

16.24 Welding configuration in A5052/AZ31 dissimilar materials welding with a filler wire.

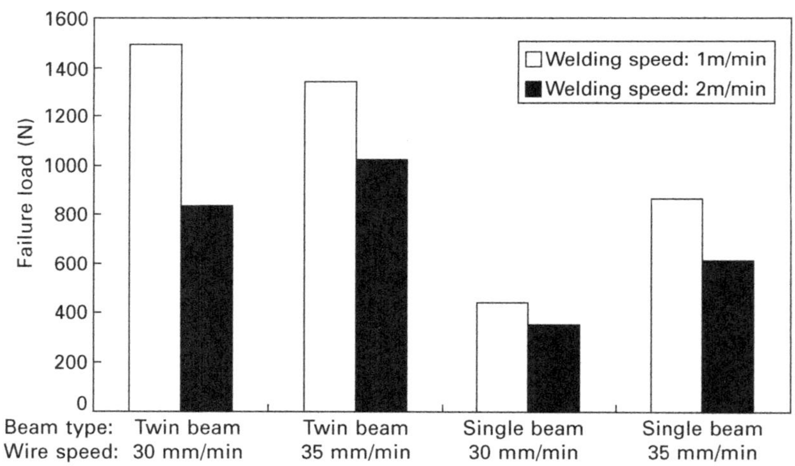

16.25 Failure load of the AZ31/A5052 dissimilar materials joints welded with a filler wire.

16.4.3 Surface modification

Laser cladding is a surface modification technique and has been used in applications such as preparing the valve sheet of an engine component in the motor industry. In the process, particles are mixed together with the molten metal at the surface. A composite or alloying material is then formed to improve the properties of the original material such as its ability to withstand corrosion, temperature and wear, etc. In this section, laser cladding using twin-beam irradiation on a magnesium alloy is discussed.

The material and particle used for the surface modification were an AZ31 magnesium alloy and Si. These can be expected to form Mg_2Si, which will increase the hardness and wear resistance of the alloy. An Nd:YAG laser was used and the beam was split by a prism with a wedge angle of 4°. Two configurations of twin beams were applied: (1) the parallel beam in which two beams were placed perpendicular to the laser scanning direction, and (2) the tandem beam in which two beams were placed parallel to the laser scanning direction. A Si powder layer with a thickness of 0.3 mm was placed on the surface of the base metal before irradiation of the laser beam. A schematic illustration of the laser cladding experiment is shown in Fig. 16.26.

The modified cladded surfaces following the single-beam and twin-beam irradiation are shown in Fig. 16.27. Two welding beads can clearly be observed in the case of the surface cladding using parallel beams. This separation of the beads becomes more apparent as the scanning speed increases. The bead width of the parallel beam surface was also small compared with that in the single beam case. Cross-sectional observations of the materials modified with single- and twin-beam irradiation are shown in Plate XX (between pages 328 and 329). In the case of the parallel twin-beam irradiation with lower laser scanning speed, Si powders were mixed well. The amount of non-melt Si powders increased with the laser scanning speed, however. The cross sections show that the depth of the molten pool created by the twin-beam irradiation was shallow compared with that created by single-beam irradiation. This was due to the low energy density of the laser beams in the twin-beam irradiation. If attention was paid to the one molten line, the Si compound was distributed along the molten pool, but the mixing was not well done inside the molten pool. In the middle section of the irradiation points of the two beams, granular-like Si compound was observed. It is speculated that the duration time for maintaining the molten pool decreases in the case of parallel twin-beam irradiation under the higher scanning speed conditions.

Observation of a weld bead created with the tandem-beam irradiation is shown in Fig. 16.27(c). The bead width was narrower with the parallel-beam irradiation when compared with the single-beam irradiation as previously mentioned. Bead width was also narrower in the case of the tandem-beam

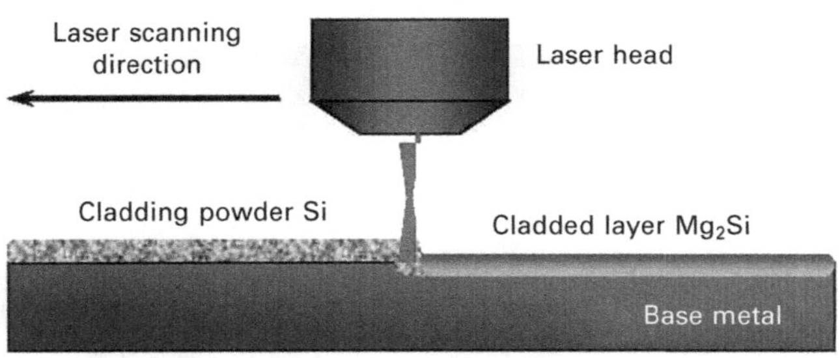

16.26 Schematic illustration of the laser cladding experiment.

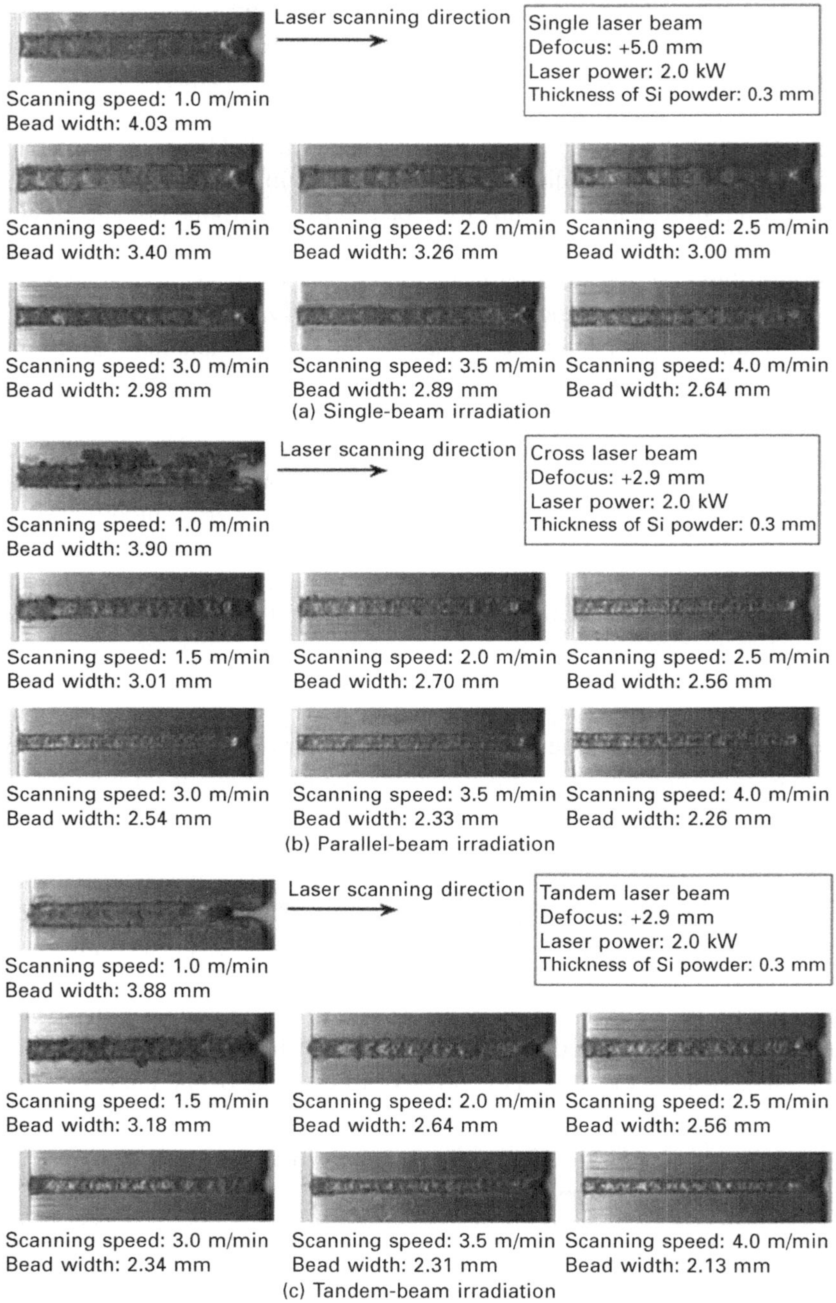

16.27 Appearance of cladding surface: (a) single-beam irradiation, (b) parallel-beam irradiation, (c) tandem-beam irradiation.

irradiation. The smooth surface of the material following irradiation with the tandem beams was obtained due to the long laser irradiation time. Plate XX (between pages 328 and 329) shows the cross section of the part subjected to the tandem laser beam cladding. The molten pool shape was completely different between the single-beam irradiation and the parallel twin-beam irradiation. In the cases of the single-beam irradiation and the parallel twin-beam irradiation, Si particles were found around the bottom of the molten pool and a lack of Si was also observed in the middle section. Contrastingly, in the case of the tandem-beam irradiation, Si particles were uniformly distributed in the molten pool. However, the increase in scanning speed meant that the molten metal and Si particles were not thoroughly mixed. Inside the molten pool, Mg_2Si was formed significantly. In the case of the tandem-beam irradiation, the following beam effectively mixed the material melted by the irradiation of the advancing beam and the total duration time of melting was much longer than that of either the single or parallel twin-beam irradiation. The tandem twin-beam irradiation is an effective method of inducing flow activity in the molten pool. Clusters of Si particles were not found around the bottom of the molten pool in laser clad material which had been irradiated using the tandem twin beam. Plate XXI (between pages 328 and 329) shows the cross section and Si distribution of the material after laser cladding using the tandem twin-beam irradiation. Around the interface between the base metal and the molten pool, the Si content continuously increased from the base metal to inside the molten pool. Plate XXII (between pages 328 and 329) shows the enlarged images of Plate XXI. A large number of Mg_2Si particles were formed. This was due to effective and thorough mixing of the molten metal induced by the tandem twin-beam irradiation.

16.4.4 Cutting

A diamond tool is conventionally used to cut brittle materials, since it provides a high cutting quality and productivity [13]. The unavoidable disadvantages of diamond tools, however, include the occurrence of pollution for the material and working environment, micro-cracking and a loss of the material cut due to finite thickness of the tool. Conventional laser cutting of metals usually causes the melting and vaporization of the material, which not only would not solve the conventional mechanical cutting process caused by diamond tools, but would add thermal distortion and residual stress in the cut material to the problem. The laser cutting method studied in this section avoids this melting process, however. A basic idea for the method is schematically shown in Fig. 16.28. The local thermal stress is generated ahead of the crack tip by laser irradiation without melting the material, which is a driving force for crack propagation. The method is a possible solution to the aforementioned brittle material cutting issues, which have been studied analytically as well

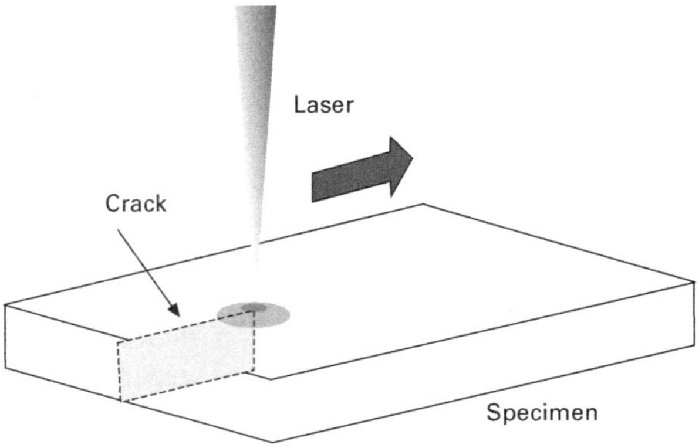

16.28 Schematic illustration of the laser cutting method.

as experimentally in previous works. Practical application of this new laser method is still limited, however, due to the difficulty of controlling crack propagation behavior. In this section, the applicability of twin-beam irradiation for laser cutting as a method of controlling crack propagation behavior is presented based on the results of FEM analysis.

A soda-lime glass with a thickness of 1.77 mm was used as an example of a brittle material. A starter notch was introduced using a diamond cutter at the starting point of laser irradiation. The laser equipment used was a Nd:YAG laser. The laser power and scanning speed were fixed at 500 W with continuous wave (CW) and 20 mm/s, respectively. The focusing point was adjusted to 3 mm up from the specimen surface to avoid the melting of the material. The laser beam began irradiating from the edge of specimen and continued to scan straight up to 10 mm and then changed the scanning direction of the laser beam to angles, ψ = 30, 60 and 90 degrees.

Figure 16.29 shows an example of a laser cutting line for $\psi = 90°$. The figure shows that the laser cutting line was swerved from the original laser scanning line. The crack propagation direction did not follow the laser scanning direction from 1.4 mm before the turning point of laser scanning direction. Similar results were obtained at other angles too. It is considered that the crack follows 1.4 mm behind the laser spot and the crack propagation direction is changed immediately when the laser scanning direction is changed.

The crack propagation behavior was estimated using FEM analysis under the same condition of laser irradiation as the experiment shown above. In the crack path estimation, a criterion for crack propagation is assumed as $K_{\theta max} > K_c$, where $K_{\theta max}$ is a stress intensity factor for the maximum tangential stress generated by the laser irradiation, and K_c is the fracture toughness of the material. The crack propagation direction is also assumed to be the direction of $K_{\theta max}$. The crack propagation path obtained using FEM analysis is also shown in Fig. 16.29. The crack propagation behavior estimated by the FEM

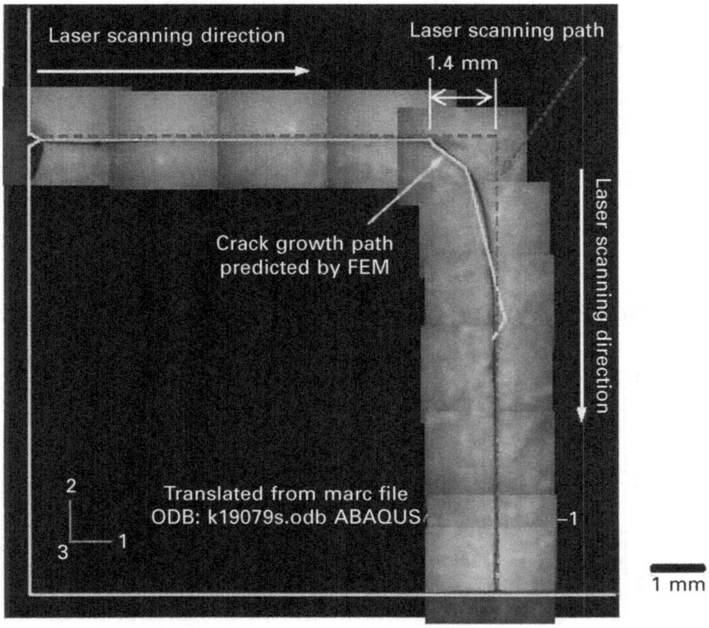

16.29 Comparison of crack path between experimental and analytical results for single-beam laser cutting at $\psi = 90°$.

analysis corresponds well with the experimental results. A stress intensity factor for the maximum tangential stress, $K_{\theta max}$, is therefore considered to be a dominant parameter for crack propagation. Crack growth direction can be controlled by controlling the maximum tangential stress, i.e. controlling the thermal stress distribution induced by the laser irradiation.

In order to control the crack growth direction during the laser cutting, a twin-beam laser technique was studied using FEM analysis. Laser cutting using the tandem twin beam is schematically shown in Fig. 16.30. The two beams are placed parallel to the laser scanning direction, which is called the tandem twin-beam configuration as shown in the figure, and the position of the crack tip is in between the front and the back beams. The back beam moves toward ϕ, which is equal to the expected angle for the change of crack path, when the crack tip reaches the expected position for changing the crack propagation direction. The expected angle for the change of crack path was 30° in this analysis. The distance between the front and the back beams was 1 mm with a laser spot radius of 0.35 mm. Figure 16.31 shows the change in stress intensity factors for Mode I and Mode II, i.e. K_I and K_{II}, due to the change in the intensity ratio between the front and the back beams. The mode ratio K_I/K_{II} is changed depending on the beam intensity ratio. This means that the direction of the maximum tangential stress can be controlled by the twin-beam irradiation. Figure 16.32 shows an example

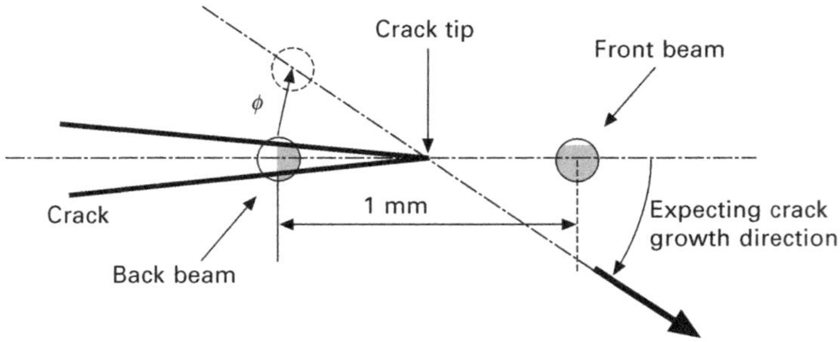

16.30 Tandem-beam irradiation in the laser cutting of a glass.

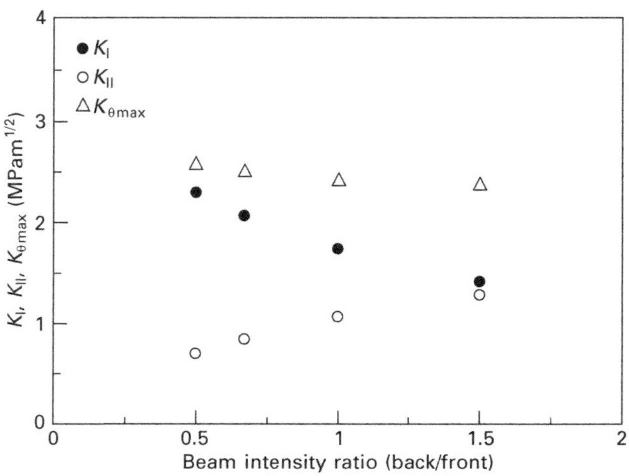

16.31 Effect of the beam intensity ratio on change in stress intensity factors, K_I, K_{II} and $K_{\theta max}$ at an expecting angle for crack growth of 30°.

of crack growth behavior estimated by FEM analysis for the tandem-beam irradiation with intensity ratio of 1:2 (front:back) at an expected angle for crack growth of 30°. As can be seen from the figure, the crack propagation direction followed the laser scanning direction. The twin-beam method could therefore control the crack propagation direction by changing the beam intensity ratio. Parallel twin-beam laser cutting, where two beams lined perpendicular to the laser scanning direction, was also studied [14], but the result indicated that it was not an effective method of controlling the crack path.

16.5 Conclusion

The application of twin laser beams allows for the possibility of overcoming limitations in laser processing with single-beam laser irradiation due to the

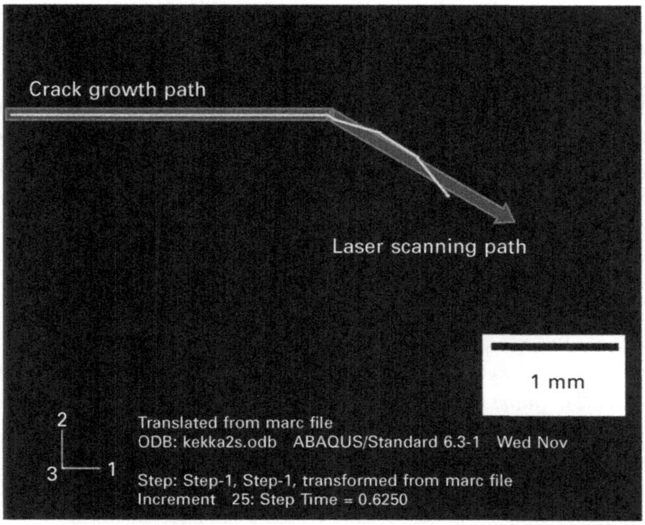

16.32 Crack path for the twin beam irradiation obtained by FEM analysis at an angle for crack growth of 30°.

increase in processing parameters, such as a combination of the wavelengths of two beams, beam distance and beam configuration, etc. A simple increase in laser power by using two beams is less important as an application than the twin beam, because the advances in the technology of laser welding apparatus with its cost reduction have been significant in recent years. Particular phenomena induced by the twin-beam irradiation, such as the increase in laser absorptivity, the active convection of molten metal, the increase in gap tolerance, and the control of thermal stress are all introduced in this chapter. Combining a different heat source with a laser beam is also being actively developed as a hybrid welding technique. Theoretical and numerical studies as well as experimental study are required to develop new applications for twin-beam laser welding and laser processing technology.

16.6 References

1. H. Sakamoto, T. Iwase and K. Shibata, Effect of Twin Spot Beam Arrangement on Energy Coupling During Welding – Study of Twin Spot Nd:YAG Laser Welding of Aluminum Alloys (Rep.3), *Quarterly Journal of The Japan Welding Society*, 21(4), (2003) 501–506.
2. J. Chen, Y. Mutoh, Y. Miyashita, D. Koide and R. Borrisutthekul, Computational Investigation on Molten Pool of Single and Twin Beam Laser Welding, *Journal of Solid Mechanics and Materials Engineering*, 1(5), (2007) 624–634.
3. Y. Miyashita, R. Borrisutthekul, J. Chen and Y. Mutoh, Application of Twin Laser Beam on AZ31/A5052 Dissimilar Metals Welding, *Key Engineering Materials*, 353–358 (2007) 1956–1959.

4. S. Fujinaga, H. Miura, T. Narikiyo, S. Katayama and A. Ohmori, Welding Characteristics with Three Nd:YAG Laser Beams Transmitted through Single Optical Fiber, *Quarterly Journal of The Japan Welding Society*, 16(4), (1998) 445–452.
5. K. Shibata, T. Iwase, H. Sakamoto, N. Seto, A. Matsunawa, B. Hohenberger, M. Mueller and F. Dausinger, Study on Keyhole Behavior in Twin Spot Beam Welding of Aluminum Alloys by Means of X-ray Observation – Study of Twin Spot Nd:YAG Laser Welding of Aluminum Alloys (Rep. 1), *Quarterly Journal of The Japan Welding Society*, 21(2), (2003) 204–212.
6. M. Jonghyan, M. Mizutani, S. Katayama and A. Matsunawa, Melting Characteristics of Various Metals with Combined Lasers of Different Wavelength, *Quarterly Journal of The Japan Welding Society*, 20(4), (2002) 449–457.
7. K. Shibata, T. Iwase and H. Sakamoto, Effect of Twin Spot Beam Configration on Overlap Joints Welding – Study of Twin Spot Nd:YAG Laser Welding of Aluminum Alloys (Rep.2), *Quarterly Journal of The Japan Welding Society*, 21(2), (2003) 213–218.
8. T. Hayashi, K. Matsubayashi, S. Katayama, N. Abe, A. Matsunawa and A. Ohmori, Reduction Mechanism of Porosity in Tandem Twin-Spot Laser Welding of Stainless Steel, *Quarterly Journal of The Japan Welding Society*, 20(2), (2002) 228–236.
9. R. Ohashi, S. Fujinaga, S. Katayama and A. Matsunawa, Extension of Gap Tolerance in Square Butt Joint Welding with Nd:YAG Laser, *Quarterly Journal of The Japan Welding Society*, 21(1), (2003) 25–32.
10. S. Iqbal, M. M. S. Gualini and F. Grassi, Laser Welding of Zinc-Coated Steel with Tandem Beams: Analysis And Comparison, *Journal of Materials Processing Technology*, 184, (2007) 12–18.
11. S. Iqbal, M. M. S. Gualini and A. ur Rehman, Dual Beam Method for Laser Welding of Galvanized Steel: Experimentation and Prospect, *Optics & Laser Technology*, 42, (2010) 93–98.
12. Y. Miyashita, I. Nakagawa, J.-Q. Xu, Y. Mutoh, M. Akahori and H. Okumura, Laser Welding of Dissimilar Metals Joint Aided by Unsteady Thermal Convection Boundary Element Method Analysis, *Quarterly Journal of the Japan Welding Society*, 23(1), (2005) 16–24.
13. Y. Miyashita, M. Mogi, H. Hasegawa, S. Sujatanond and Y. Mutoh, Study on a Controlling Method for Crack Nucleation and Propagation Behavior in Laser Cutting of Glass, *Journal of Solid Mechanics and Materials Engineering*, 2(12), (2008) 1555–1566.
14. Y. Miyashita, H. Hasegawa and Y. Mutoh, Laser Cutting of a Brittle Material by Controlling Crack Growth Behavior based on Fracture Mechanics (in Japanese), *Proceedings of 53rd Annual Meeting of Society of Materials Science, Japan* (2004) 93–94.

17
Developments in multi-pass laser welding technology with filler wire

X. ZHANG, Hitachi Research Laboratory, Hitachi Ltd, Japan

DOI: 10.1533/9780857098771.3.459

Abstract: This chapter presents the results of multi-pass laser welding of thick plates for type 316L stainless steels with filler wire. The lack of fusion can be prevented by optimizing the relationship between laser power intensity and the deposited metal volume. The solidification crack, which usually occurs in the weld metal of the build-up passes, can be suppressed by controlling the chemical composition of the filler wire. The maximal tensile residual stress and the range of the tensile stress of the laser welded joint are lower than those of the conventional gas tungsten arc welding (GTAW).

Key words: laser welding, filler wire, stainless steel, lack of fusion, solidification crack.

17.1 Introduction

Many thick-section structures are fabricated using welding in several industrial fields such as power plants, shipbuilding and pipelines. The welding processes for fabrication of these thick-section structures include traditional multi-pass arc welding, laser-arc hybrid welding, one-pass electron beam welding and laser welding. For arc welding, such as the application of narrow-gap arc welding, submerged arc welding and hot-wire tungsten inert gas welding, the production efficiency is limited by the low welding speed and too many weld passes. Furthermore, welding distortion is also a problem, and distortion correction is often needed after welding. In the past decade, electron beam welding[1–3] and laser welding have been paid more attention for welding of thick-section stainless steels.[4–9] They both are very high-efficiency and low-distortion process methods due to the features of ultra-deep penetration and very narrow weld bead width. For electron beam welding, the thickness to be penetrated for single-pass electron beam welding of stainless steels has been over 100 mm,[2,3] As for laser welding, it can produce weld joint with smaller welding distortion and higher welding speed than arc welding, and the process can be performed in atmosphere with appropriate shielding gas. Although the penetration depth is still lower than that of electron beam at present, the process flexibility is a great advantage by combination of a laser beam with additional materials like a filler wire or powder, or by combination with an arc, which can permit larger gap tolerance and adjust the composition of the weld metal. Especially for thick-section steel plates of 30–60 mm or greater, multi-pass laser welding with a filler wire is one of the appropriate methods.

This chapter presents the process principle, welding parameters and welding results for multi-pass laser welding with a filler wire. All the experimental results are for laser welding of SUS316L steels with thicknesses of 30–60 mm.[10–14] The exact results obtained for a welding operation depend on a wide variety of process parameters and vary with different laser welding systems. These parameters are not usually specified well enough to allow one to duplicate exactly the conditions for a particular process. Thus the tabulated and graphed data must be interpreted as indicating an approximate range of welding results that may be obtained. In order to obtain the best results, the welding parameters for a particular processing operation must be optimized experimentally.

17.2 Principle of multi-pass welding with filler wire

Multi-pass welding with a filler wire is one of the efficient methods for joining thick-section materials with a laser beam. It has been reported that 150 mm thick low carbon steels could be butt-jointed by narrow-gap laser welding with a filler wire.[15] Multi-pass laser welding can be performed from one side or both sides. Figure 17.1 shows the schematic diagram of the experimental setup for multi-pass laser welding from both sides. A groove with a thick root face and small groove angle is often prepared. In order to reduce the total weld passes, the thickness of root face can be prepared as thick as possible, for the laser beam welding has high penetrating ability. As the smaller the angle of groove is, the less is the total deposited metal, the angle of groove should be set as small as possible. The root pass is usually performed without filler wire, but it is also possible for filler wire to be added. The subsequent build-up passes are performed by adding a filler wire. One of the major aspects that concerns the laser beam properties is that the laser beam must be assured to irradiate the groove bottom, so it is necessary to use a laser beam with small beam parameter product and a focusing lens with long focal length.

As an example, Fig. 17.2 shows the groove geometries of the workpieces for one side (Fig. 17.2(a)) and both sides (Fig. 17.2(b)) welding when using an 8 kW disk laser to perform welding. The thickness of the plate is 30–50 mm for one-side welding, and 60 mm for both-sides welding. The thickness of root face R is from 8 mm to 10 mm for one-side welding, and 10 mm to 15 mm for both-sides welding. The groove angle α is 4–5° and the width of root face W is 2–4 mm. The groove depth is from about 20 mm to 50 mm for different thick plates.

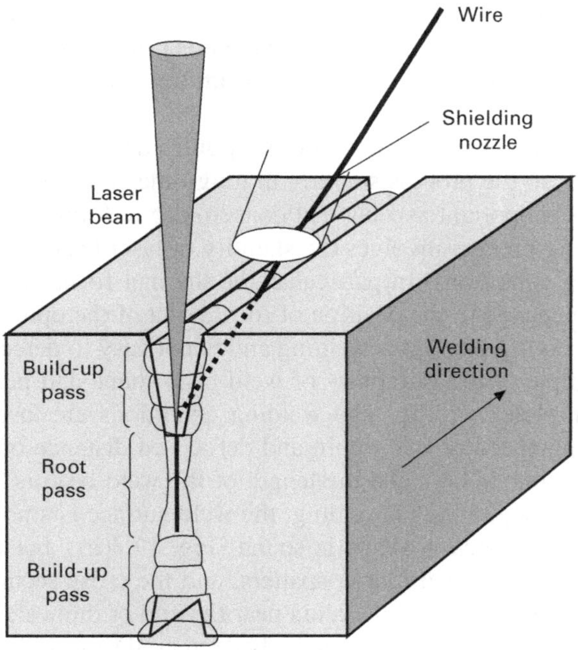

17.1 Schematic diagram of multi-pass laser welding with filler wire from both sides.

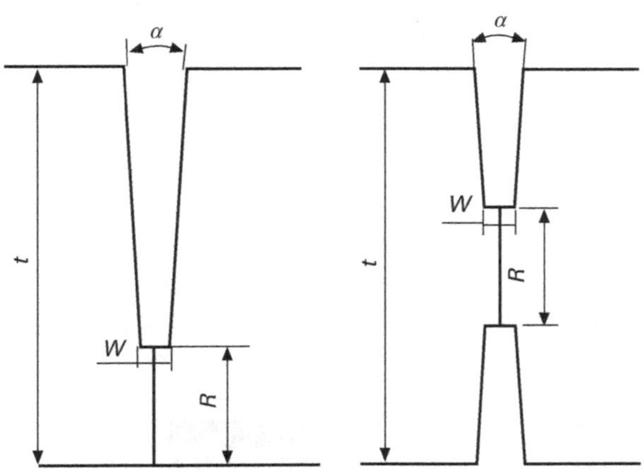

17.2 Groove geometry of workpiece for multi-pass laser welding (t = 30–60 mm).

17.3 Developments in technology

17.3.1 Welding systems

The welding system for multi-pass laser welding with a filler wire consists of at least a laser oscillator, an optical system for transferring a laser beam, a laser process head, a shielding gas nozzle, and a system for feeding filler wire into the groove. The other additional equipment includes a monitoring system to check the welding process, and a seam-tracking system to guide the welding position.

Multi-pass welding with a filler wire is usually utilized for large-scale thick-section structures, so the process stability in long-time continuous welding is one of the most important aspects that concern the welding quality. The stability of welding process involves the stability of laser beam quality and the other welding conditions. In particular, the thermal focus shift, which is induced by the change in the position of focal point of the optical lenses, often happens in high-power laser welding and is not easy to detect. Figure 17.3 shows a sample of the instability of weld bead shape and penetration depth in bead-on-plate welding. The welding conditions are laser power of 10 kW, welding speed of 0.3 m/min and defocused distance of –2 mm. The total welding time is 46 s and the length of the weld bead is 230 mm. It shows that at the beginning of welding, the weld surface is smooth (Fig. 17.3(a)) and the cross section shape is sound (Fig. 17.3(b)), but the weld surface becomes unstable, with large spatters, and the cross section shape also becomes narrow and underfill occurs near the end of the welding (Fig. 17.3(c)). The penetration depth is respectively decreased from 16.7 mm near the welding beginning to 14.7 mm near the welding end. Such instability in penetration depth and weld bead shape is considered to be related to the thermally induced focus shift of the focusing lens. Measures to prevent the thermal lens effect have to be taken for applications of high-power laser welding.

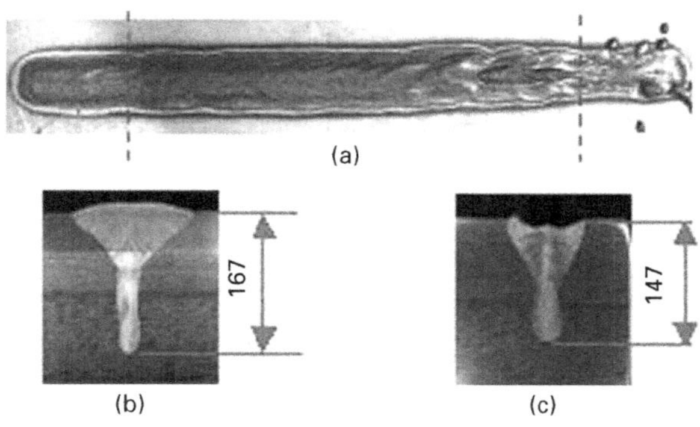

17.3 Instability of weld bead shape induced by thermal focus shift in laser welding[10] (laser power 10 kW, welding speed 0.3 m/min): (a) weld surface; (b) cross section of weld bead near weld start; (c) weld cross section near weld end.

17.3.2 Welding parameters

Root-pass welding

The effect of welding conditions on the weld geometry of the root pass has been investigated with penetration welding of 30 mm thick plates without filler wire. The thickness of the root face is 10 mm. The result is shown in Fig. 17.4. Instability or variation of penetration depth occurs under some conditions if no backing plate is used in welding. The appropriate welding conditions for good weld bead geometry are 6 kW laser power and 0.5 m/min welding speed. When the welding speed is increased to 0.75 m/min at 6 kW laser power, the weld bead becomes partial penetration. When the laser power is increased to 8 kW, the penetration bead becomes unstable at the welding speed of 0.8–1.2 m/min due to the excessive penetration, and partial penetration occurs at the welding speed of 1.5 m/min. It is predictable that the penetration bead is unstable when the welding speed is less than 0.8 m/min for 8 kW laser power. Thus the appropriate welding parameter range is too narrow and, moreover, many spatters are stuck on the groove surface and the backside of the workpieces in the experiment when the laser power is 8 kW. Therefore, the welding parameters of 6 kW laser power and 0.5 m/min welding speed are adapted for the root-pass welding.

Power (kW)	Speed (m/min)	Backside of weld	Cross section
6	0.5		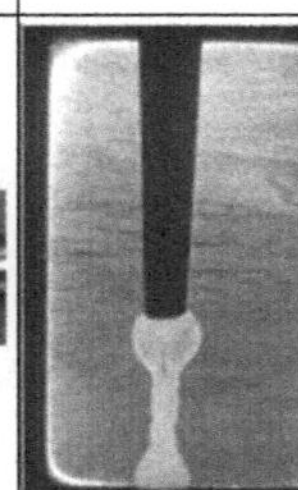
6	0.75	Partial penetration	—
8	0.8	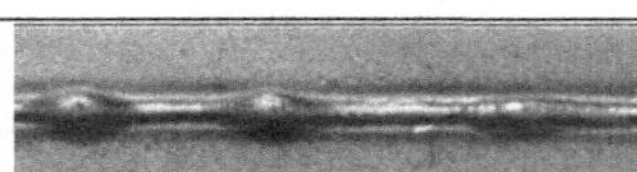	—
8	1.0		—
8	1.2	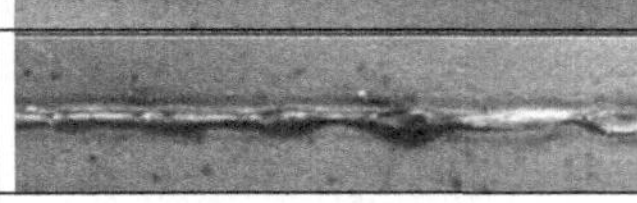	—
8	1.5	Partial penetration	—

17.4 Results of root pass welding.[13]

Build-up welding

One of the most important aspects that concerns the weld quality in build-up multi-pass welding is 'lack of fusion'. Build-up welding is performed under various welding conditions so as to study the relationship between welding conditions and weld bead geometry. The welding parameters are laser power 6 kW and 8 kW, welding speed 0.3 m/min and 0.4 m/min, and beam spot size at the welding point varying from 0.69 mm to 1.56 mm. The feeding speed of the filler wire is varied from 3.5 m/min to 8.0 m/min to get different build-up heights and deposition areas. The welding results show that lack of fusion occurs easily if the welding conditions are not appropriate, which is described in detail as follows.

The typical weld bead geometries for the build-up welding, and the relationship between the welding conditions and the welding results are shown in Fig. 17.5. Two types of lack of fusion, lack of fusion A and lack of fusion B, are observed in the weld joint. In Fig. 17.5, the horizontal axis is labeled laser power intensity, and the vertical axis is labeled deposited area, which is the result of the wire feeding speed multiplied by the cross area of wire. It shows that when the laser power intensity is low and the deposited area is high, lack of fusion A tends to occur. The feature of lack of fusion A is that the weld bead formed in the initial root-pass welding is not re-melted during the following build-up welding, and as a result, lack of fusion occurs between the root weld bead and the build-up bead. In this case, non-fully melted wire is also observed. This is because the volume of wire fed into the welding zone is excessive and the melting ability of the laser beam is insufficient; as a result, even the filler wire cannot be fully melted. As the laser power intensity increases, lack of fusion A can be suppressed due to increases in the melting ability of laser energy, and good weld bead can be obtained in an appropriate range. However, if the laser power intensity is increased to very high level, lack of fusion B will readily occur.

The feature of lack of fusion B is that the side wall of the groove could not be melted and, as a result, there is lack of fusion occurring between the side wall of the groove and the deposited metal. The reason for lack of fusion B can be explained as follows: the volume of filler wire fed in the welding zone is not enough, resulting in the weld bead formed in root welding being re-melted by a deep and very narrow keyhole. The width of the penetration part induced by the keyhole is narrower than that of the groove, thus the groove wall near the upper of the root bead (or the bottom of the build-up bead) is not melted, leading to lack of fusion. In this case, the non-fully melted wire observed in lack of fusion A is not observed. Therefore, the relationship between deposited area and laser power intensity should be considered so as to prevent the lack of fusion.

Figure 17.6 shows a sample of multi-pass laser welding of 50 mm thick stainless steel plates. The length of weld bead is 400 mm. The number of

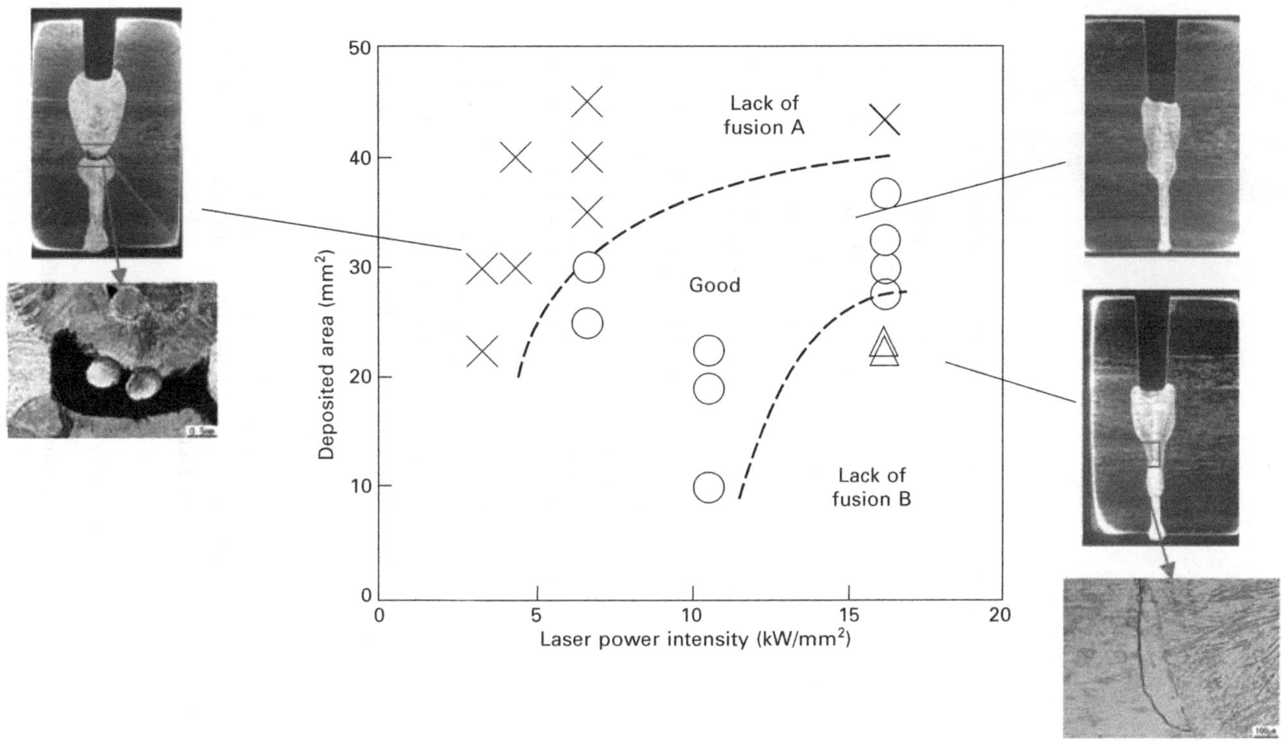

17.5 Effect of laser power intensity and deposited area on lack of fusion.

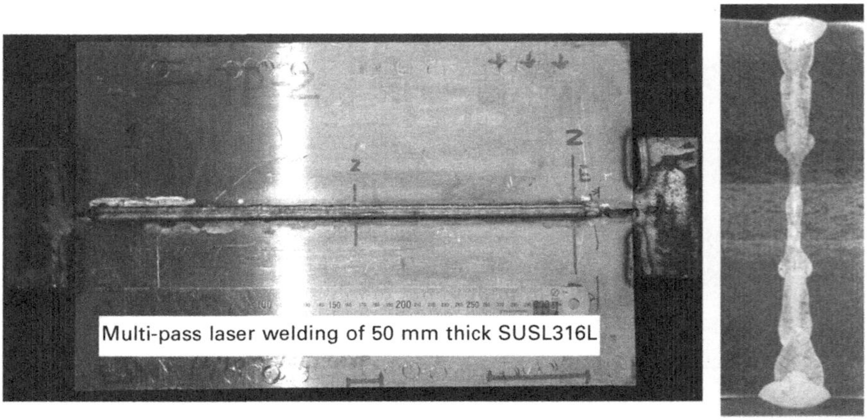

17.6 Sample of multi-pass laser welding of 50 mm thick plates.

Table 17.1 Chemical compositions of base metal and welding wires (mass%)

	C	Si	Mn	P	S	Ni	Cr	Mo	N	Fe	Ferrite content (%)
Base metal	0.011	0.48	0.87	0.023	0.005	12.37	17.48	2.24	–	Bal.	–
Wire A	0.016	0.39	1.78	0.017	0.001	13.35	19.30	2.32	0.055	Bal.	5.9
Wire B	0.009	0.42	1.83	0.011	0.003	12.42	18.95	2.15	0.0262	Bal.	9.2
Wire C	0.007	0.31	1.86	0.017	0.005	10.15	19.95	0.02	0.0406	Bal.	9.4
Wire D	0.014	0.5	1.5	0.006	0.0095	9.87	20.79	0.01	0.0022	Bal.	16.8

welding layers is 8, including two layers of root welding without filler wire, 4 layers of normal build-up welding, and 2 layers (4 passes) of reinforcement welding. No lack of fusion and porosity is found by optical microscope observation of the cross section of weld bead and X-ray inspection.

17.3.3 Example applications

This section presents welding results in terms of solidification cracking, microstructure and mechanical properties in applications of multi-pass welding for austenite stainless steel.

Welding crack and microstructure

Table 17.1 shows the compositions of the base metal and filler wires used in the welding experiment. Figure 17.7 shows the cross section of weld bead with crack and the microstructure of the weld metal near the crack. The crack usually occurs in the center of the build-up pass weld and is solidification

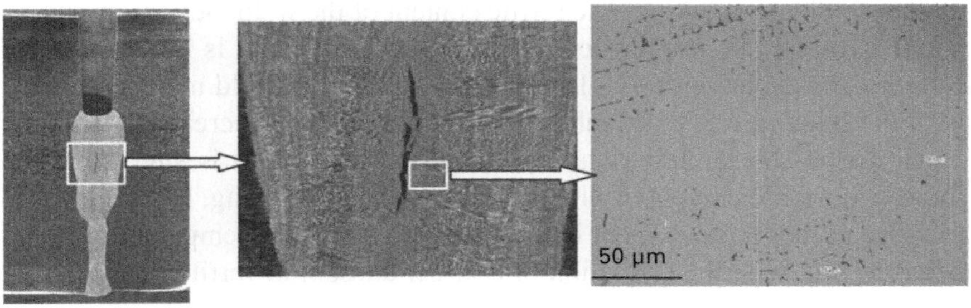

17.7 Bead shape, crack and microstructure of deposited weld.

Table 17.2 Effect of welding conditions and wire composition on welding crack

Welding conditions			Wire composition (ferrite %)			
Laser power (kW)	Welding speed (m/min)	Defocused distance (mm)	Wire A (5.9%)	Wire B (9.2%)	Wire C (9.4%)	Wire D (16.8%)
6	0.225	7	–	–	×	○
	0.3	7	×	–	×	○
		10	×	×	×	○
		15	×	–	×	○
	0.4	10	–	–	×	○

○: No crack ×: Crack –: no data

cracking according to SEM observation. For this weld, the filler wire is wire A (Table 17.1) and its theoretical ferrite content should be about 5%. In the experiment, nitrogen is used as the shielding gas in order to suppress keyhole type porosity; however, the nitrogen content of the weld metal is much higher than that of the filler wire and the base metal, resulting in low ferrite content in the weld metal. Figure 17.7 indicates that the ratio of the ferrite microstructure near the crack is less than 1%, much less than that of the other zone. An increase of 300–400 ppm nitrogen of the weld metal compared with that of the base metal is also apparent. Considering that the nitrogen shielding gas decreases the ferrite content, filler wire with higher ferrite content should be used.

Table 17.2 shows the effect of welding conditions and wire composition (mainly the ferrite content) on the welding crack. The ferrite contents of the filler wires A, B, C and D are 5.9%, 9.2%, 9.4% and 16.8%, respectively. Under any conditions of welding speed and defocused distance of a laser beam at a laser power of 6 kW, cracks are observed in all the weld metals produced with wire A, wire B and wire C. When filler wire D with the ferrite content of 16.8% is used, however, no crack is found in the weld metal for all the welding conditions. The ferrite content of the weld metal is measured and it is found that the ferrite content of the weld metal is in the range of 6–11%. The decrease of 5.8–10.8% ferrite content of weld metal compared with the filler wire is the result of the nitrogen content increase in the weld metal due to the shielding gas of nitrogen. Therefore, it is suggested that increasing ferrite content of filler wire can prevent cracking. In addition, the increase in the ferrite content in the weld metal at room temperature should be correlated with the formation of a larger amount of ferrite phase during solidification.

Microstructure and mechanical properties

Figure 17.8 shows the microstructure of the weld metal for multi-pass welded 50 mm thick plates. The ferrite contents in the weld metal differ with welding pass, but the ferrite content is in the range of 7–11% for all layers.

The results of the tensile test for the welded joint are shown in Table 17.3. It shows that the laser welded joints have enough tensile strength compared with that of the base metal, and also satisfy the specification of the base metal.

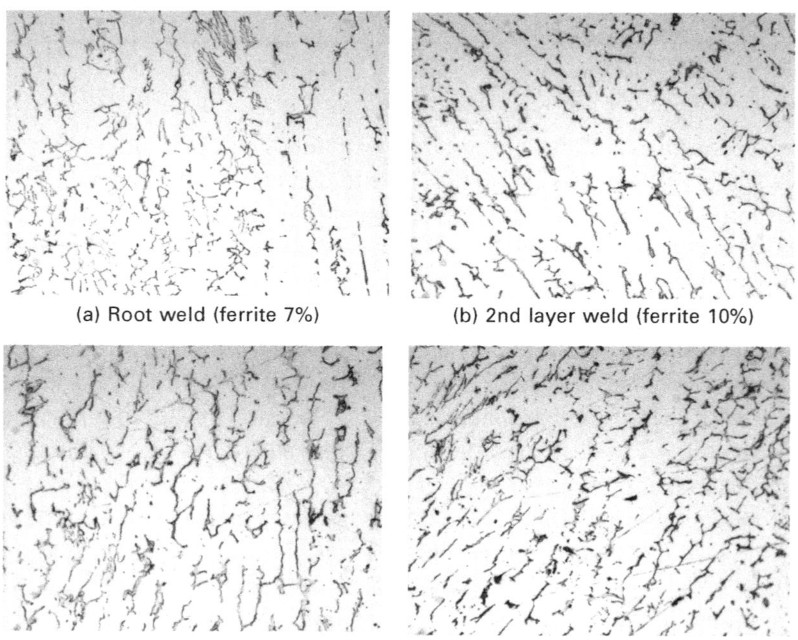

(a) Root weld (ferrite 7%) (b) 2nd layer weld (ferrite 10%) (c) 3rd layer weld (ferrite 10.5%) (d) Reinforcement weld (ferrite 11%)

17.8 Microstructure of weld metal for multi-pass laser welding of 50 mm thick SUS316L plates.[14]

Table 17.3 Results of tensile tests

	Welded joint	Base metal
Tensile strength (MPa)	577	551
0.2% yield strength (MPa)	294	236

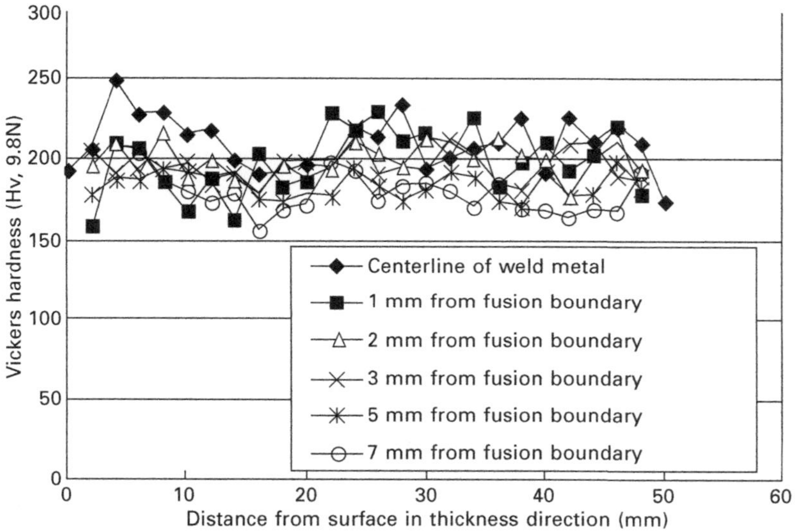

17.9 Vickers hardness distribution of laser welded joint.

Figure 17.9 shows the distribution of hardness of weld metal and the base metal near the weld bead. The hardness of the weld bead and the hardness at 3 mm from the fusion boundary is in the range from 180 Hv to 220 Hv, which is greater than the hardness of the base metal (150–160 Hv). As the distance from the fusion boundary increases, the hardness decreases. The value of hardness shows the same level as that of the conventional GTAW.

Residual stresses

The residual stresses on the surface of the laser welded joint with 50 mm thickness are measured. The distribution of residual stresses perpendicular to the welding direction and parallel to welding direction are shown in Fig. 17.10(a) and (b), respectively. To compare the residual stresses of laser beam welding (LBW) with GTAW and shielded metal arc welding (SMAW), the residual stresses of GTAW and GTAW + SMAW joints[16] are also shown in Fig. 17.10. It shows that the residual stresses of laser welded joints have the same tendency as that of GTAW and GTAW + SMAW joints. The maximum value of tensile stress is near the centerline of the welding bead. The tensile

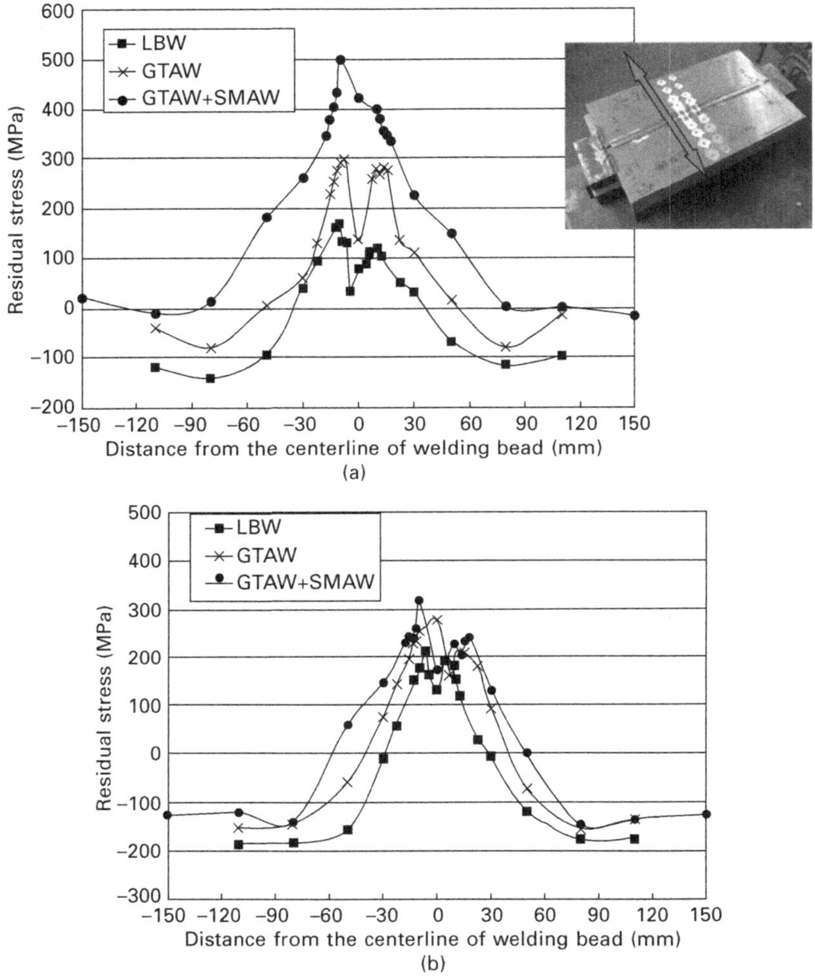

17.10 Residual stresses of LBW, GTAW and GTAW+SMAW.[14] (a) Residual stress perpendicular to welding direction (b) Residual stress parallel to welding direction.

stress decreases as the distance from the centerline of the welding bead increases, and the stress changes to be compressive at the areas far away from the welding bead. The maximum value of tensile residual stress and the tensile range of stresses are different in LBW, GTAW and SMAW joints. Compared with GTAW, the maximal tensile residual stress is lower and the tensile range (above 0 MPa of stress) is narrower for a laser welded joint. This is probably because of the different angles of the grooves for different welding methods, which results in different total heat inputs and the volume of the weld bead. The narrower welding groove, lower heat input and fewer weld passes for laser welding lead to lower residual stress.

17.4 Future trends: further improvement of welding efficiency

The number of welding passes need to be reduced further from the standpoint of process productivity; however, the thickness of the root pass is dependent on the penetration capability of the laser beam. For instance, the penetration is about 18 mm for welding stainless steel under the conditions of laser power 10 kW and 0.3 m/min welding speed.[10] One of the possible methods to increase the penetration is to use a side gas jet.[11–13,17] Figure 17.11 shows the schematic diagram of gas jet assisted laser welding. The main feature of this method is that the keyhole opening is deepened and widened by high pressure gas flow, so that the laser beam can enter the keyhole inside directly under no interaction with the surface of the molten pool at the keyhole opening. In this case, the keyhole depth can be increased and the penetration depth can be improved greatly, which has the same meaning as the use of higher power laser. In the next subsections, the difference between the gas jet assisted laser welding and the conventional laser welding is discussed, and then the results of the butt-joint welding of 40 mm thick plates and 50 mm thick plates are presented.[11–13]

17.4.1 Comparison between conventional and gas jet assisted laser welding

In order to understand the mechanisms of the penetration depth improvement induced by side gas jet, the keyhole behavior and the metal flow in gas jet assisted laser welding are investigated with high-speed cameras and X-ray

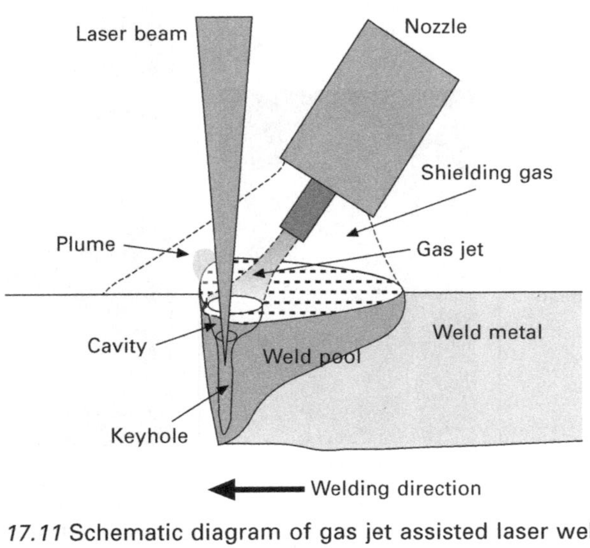

17.11 Schematic diagram of gas jet assisted laser welding.

transmission observation. Figure 17.12 shows a comparison between the conventional laser welding and gas jet assisted laser welding for bead-on-plate welding of 30 mm thick plates. The welding conditions are the same

	Conventional laser welding	Gas jet assisted laser welding
Weld pool	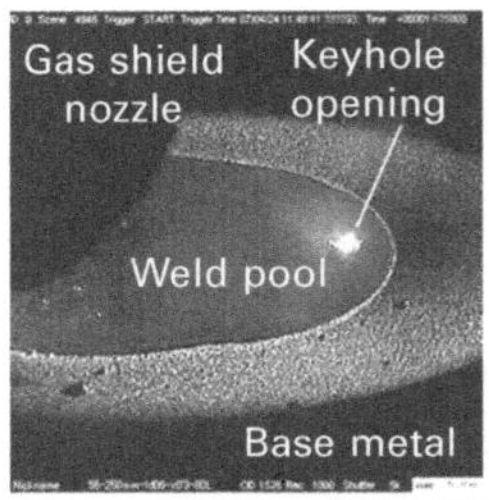	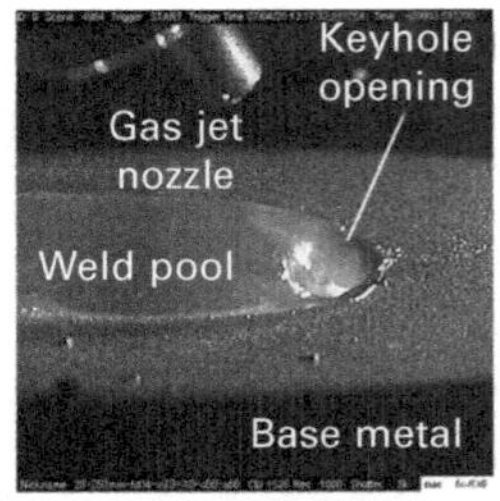
Plume	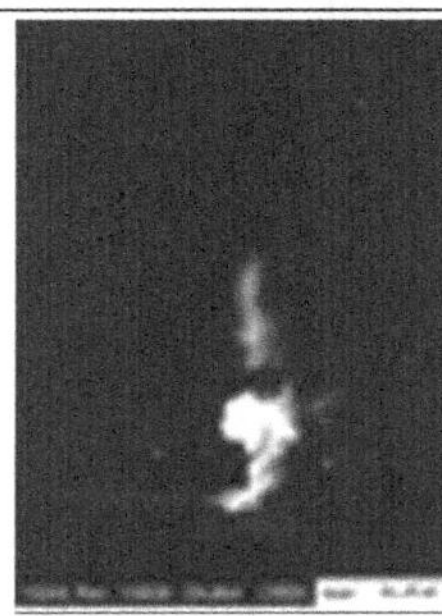	
Keyhole behavior		
Weld bead	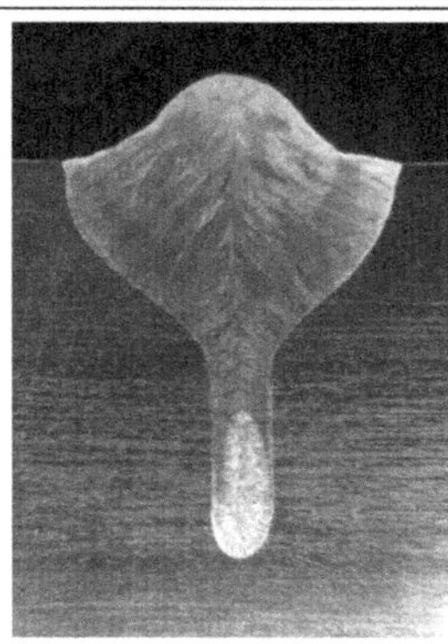 Penetration (18 mm)	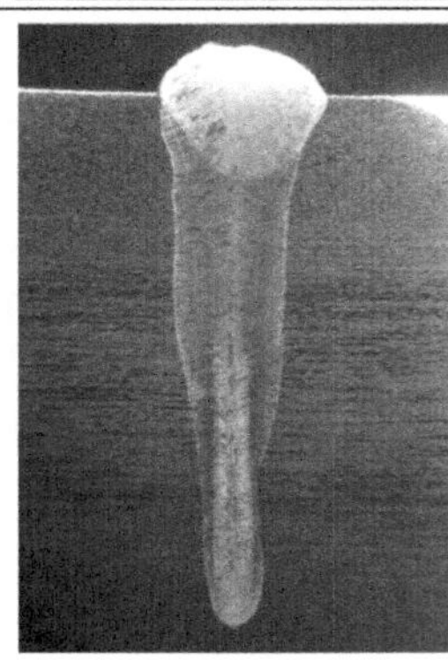 Penetration (25 mm)

17.12 Comparison of conventional and gas jet assisted laser welding.

10 kW laser power, 0.3 m/min welding speed and argon as the shielding gas for both processes. The different parameters between the two processes are the gas flow rate of the inner gas jet. The gas flow rate of the inner gas jet nozzle is set to 0 for the conventional laser welding, and 12 l/min for the gas jet assisted laser welding.

The welding phenomena and results for the conventional laser welding are shown in the left of Fig. 17.12. The photo of the weld pool, which is taken by high-speed camera, shows that for conventional laser welding without gas jet, a wide and flat surface of the molten pool is formed during welding (in the case of good shielding condition), but the keyhole opening is very small and melt metal with some extent of thickness exists between the front edges of the keyhole and the molten pool. The photo of the plume shows that a very strong and severely fluctuating plume is injected from the keyhole during welding. The keyhole behavior observed by the X-ray transmission apparatus shows that the keyhole presents a very narrow line shape, many bubbles form near the bottom of the keyhole, and some bubbles result in porosities during solidification. The photo of the cross section of weld bead shows that a wine-cup shape weld of about 18 mm in penetration and 17 mm in width is formed.

The welding phenomena and results for gas jet assisted laser welding are shown in the right of Fig. 17.12. For gas assisted laser welding, the weld pool becomes narrower and the keyhole opening is enlarged and deepened considerably, and the thickness of the melt metal between the keyhole front and the molten pool front also becomes thinner. The plume induced in the process is much weaker and more stable, and the height of the plume is also much lower than that of conventional laser welding without gas jet. From the X-ray *in-situ* observation, the keyhole shows different behavior, and the opening of the keyhole is significantly deepened and enlarged to form a cavity of about 3 mm in diameter and 3 mm in depth. In this case, the laser beam could be directly irradiated on the bottom of the keyhole opening under the conditions of no obstacles by removing the molten metal, resulting in a considerably deepened keyhole. As a result, the penetration depth is greatly increased to 25 mm at the same laser power and welding speed. Because of the formation of a deep keyhole opening, the defocused distance can be set to a greater minus value in order to get a deeper keyhole.

Another difference between the conventional laser welding and gas jet assisted laser welding is the weld appearance. The reinforcement height of the weld in conventional laser welding is less than 0.3 mm in the experiment. For gas jet assisted laser welding, however, the reinforcement height is about 3.0 mm even if no filler wire is used in the process. That is because the gas jet pushed the molten metal around the keyhole to flow backward and a weld bead with excessively high reinforcement is formed.

17.4.2 I-butt joint welding of 40 mm thick plates without filler wire

I-butt joint welding is undertaken on type 316L steel plates of 200 mm length, 100 mm width and 40 mm thickness, at the laser power of 10 kW. Two joints are welded: one is welded at 0.2 m/min welding speed, and the other at 0.3 m/min. Figure 17.13 shows the photos of the cross sections of weld beads. At the welding speed of either 0.2 m/min or 0.3 m/min, a regularly shaped bead without any lack of fusion and porosity is obtained in the welded joint. The width of the bead surface at 0.2 m/min welding speed is wider than that at 0.3 m/min. It shows that for butt-joint welding of 40 mm thick plates, a fully penetrated good joint can be obtained with a 10 kW laser in two passes, and penetrated from both sides without filler wire, by using the gas jet assisted laser welding method.

17.4.3 I-butt joint welding of 50 mm thick plates with filler wire

I-butt joint welding of 50 mm thick plates is tried by combining gas jet assisted welding and multi-pass welding with a filler wire. The geometry of

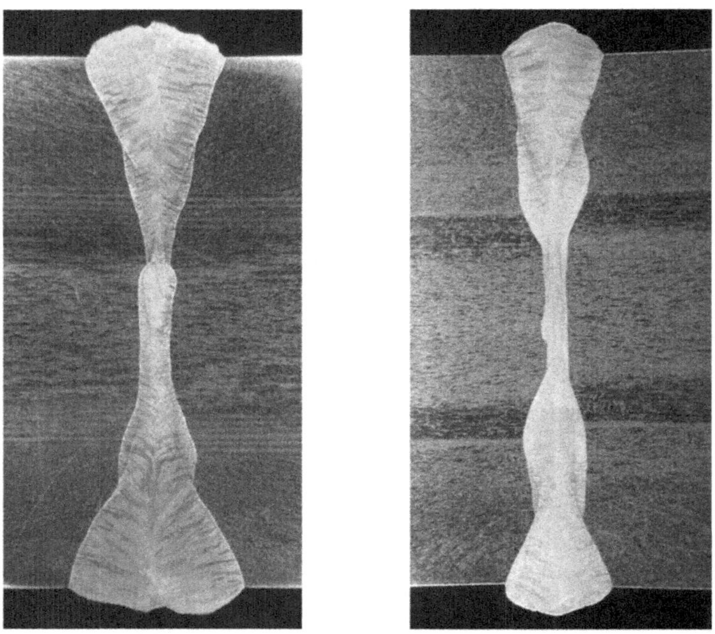

(a) (b)

17.13 Butt-joint laser welding of 40 mm thick plates with gas jet (two passes, without filler wire): (a) welding speed: 0.2 m/min; (b) welding speed: 0.3 m/min.

the test piece used in the experiment is shown in Fig. 17.14. The welding is performed from both sides. The welding conditions and the welding results are shown in Fig. 17.15. The number of passes is four, including two passes

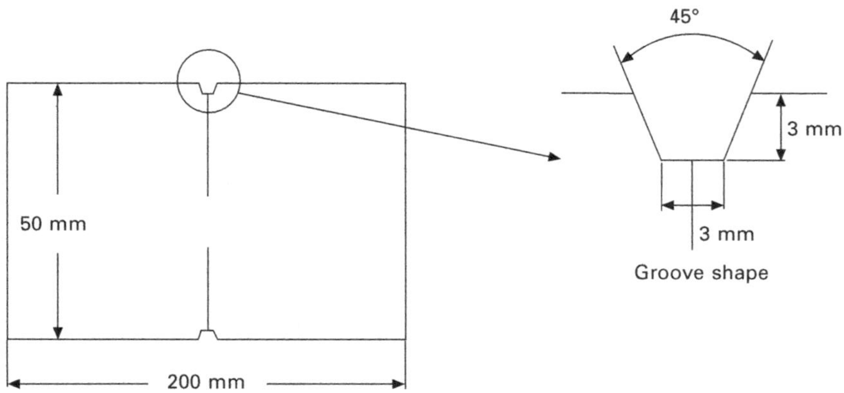

17.14 Groove geometry of 50 mm thick plates for gas jet assisted laser welding (butt joint).

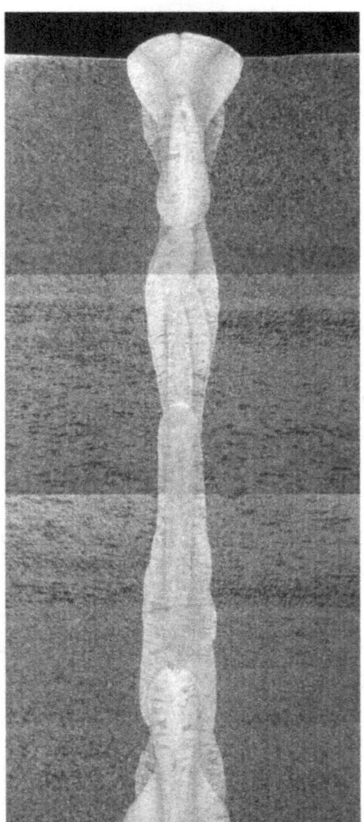

17.15 Butt-joint laser welding of 50 mm thick plates with gas jet (four passes, laser power: 10 kW, welding speed: 0.2 m/min).

of deep penetration laser welding with gas jet and two passes of repair welding with a filler wire. It shows a regularly shaped weld bead and no lack of fusion and porosity.

17.5 References

1. G. Takano, T. Shimoyama, Y. Uratani, S. Kuri, 'Study on weldability of electron beam welding for thick plate of austenitic stainless steel (Report 1): Selection of welding conditions and prevention of weld defects', *Transactions of The Japan Welding Society*, **14**(2), 10–19 (1983).
2. D.H. Kim, B.Y. Kim, S. Cho, K.J. Jung, I.S. Choi, M.H. Choi, I.H. Kim, G.Y. Lee, 'Welding technology development for the fabrication of ITER blanket shield block in Korea', *Fusion Science and Technology*, **56**(1), 43–47 (2009).
3. L. Jones, J.-P. Alfile, P. Aubert, C. Punshon, W. Dänner, V. Kujanpää, D. Maisonnier, M. Serre, G. Schreck, M. Wykes, 'Advanced cutting, welding and inspection methods for vacuum vessel assembly and maintenance', *Fusion Engineering and Design*, **51–52**, 985–991 (2002).
4. F. Vollertsen, F. Thomy, 'Welding with fiber laser from 200W to 17000W', *ICALEO2005*, 254–263 (2007).
5. K. Kinishita, M. Mizutani, Y. Kawahito, S. Katayama, 'Phenomena of welding with high-power fiber laser', *ICALEO2006*, paper 902 (2006).
6. Y. Arata, H. Maruo, I. Miyamoto, R. Nishio, 'High power CO_2 laser welding of thick plate', *Transactions of JWRI*, **15**(2), 27–34 (1986).
7. F. Vollertsen, S. Grunenwald, 'Defects and process tolerances in welding of thick plates', *ICALEO2008*, paper 1004 (2008).
8. T. Jokinen, M. Karhu, V. Kujanpää, 'Welding of thick austenitic stainless steel using Nd:yttrium-aluminum-garnet laser with filler wire and hybrid process', *Journal of Laser Applications*, **15**(4), 220–224 (2008).
9. F. Coste, L. Sabatier, O. Dubet, P. Aubert, 'Nd:Yag laser welding of 60 mm thick 316L parts using multiple passes', *ICALEO2001*, paper 514 (2001).
10. X. Zhang, E. Ashida, M. Mizutani, S. Katayama, Y. Anma, X. Luo, 'High-power fiber laser welding of thick steel at low welding speed', *ICALEO2007*, paper 1601 (2007).
11. X. Zhang, E. Ashida, S. Katayama, M. Mizutani, 'Deep penetration welding of thick section steels with 10 kW fiber laser', *Quarterly Journal of the Japan Welding Society*, **27**(2), 64–68 (2009).
12. X. Zhang, E. Ashida, M. Mizutani, S. Katayama, 'Development of ultra deep penetration welding with 10 kW fiber laser', *ICALEO2008*, paper 705 (2008).
13. X. Zhang, E. Ashida, S. Tarasawa, Y. Anma, M. Okada, S. Katayama, M. Mizutani, 'Welding of thick stainless steel plates up to 50 mm with high brightness lasers', *Journal of Laser Applications*, **23**(2), 022002 (2011).
14. X. Zhang, E. Ashida, S. Tarasawa, 'Properties of welded joint for narrow gap laser welding of austenitic stainless steels', *ICALEO2010*, paper 1805 (2010).
15. T. Tukamoto, H. Kawanaka, Y. Maeda, 'Laser narrow gap welding of thick carbon steels using high brightness laser with beam oscillation', *ICALEO2011*, paper 305 (2011).
16. Japan Nuclear Energy Safety Organization, Annual report on verification test of

SCC in stainless steel used in nuclear power plant, Report No. J-000-2006-00160 (2005) (in Japanese).

17. R. Fabbro, S. Slimani, I. Doudet, F. Coste, F. Briand, 'Experimental study of the dynamical coupling between the induced vapor plume and the melt pool for Nd-Yag CW laser welding', *Journal of Physics D: Applied Physics*, **39**, 394–400 (2006).

18
Developments in hybridisation and combined laser beam welding technologies

D. PETRING, Fraunhofer Institute for Laser Technology (ILT), Germany

DOI: 10.1533/9780857098771.3.478

Abstract: The laser welding process is the most flexible fusion welding process industrially available. The degree of flexibility is further enhanced and process limits are expanded if hybridisation or combination techniques are applied. After a general introduction of various related concepts, the enhancement of welding capabilities is illustrated in more detail by the state-of-the-art and latest advancements in laser-MAG (metal active gas) hybrid welding of up to 25 mm thick high-strength steel components. In addition, the manufacturing benefits of integrated cutting and welding with a multifunctional laser combi-head are described.

Key words: laser-arc hybrid welding, high-strength steels, aluminium alloys, laser combi-head, combined laser cutting and laser welding.

18.1 Introduction

Within the last three decades industrial lasers have become established as *the* tools for flexible manufacturing. Recent developments in fibre and disk lasers with multi-kW output power at beam qualities with single-digit mm*mrad values push the trend to laser assisted production even further. They open up new vistas to fibre-coupled laser systems, with multiple manufacturing cells sharing a single laser source. The increased electrical to optical conversion efficiency of the new diode-pumped sources and their better utilisation by laser networks leads to highly productive, flexible, and cost-efficient production solutions.

A further chance to enhance the capabilities of lasers is offered by their combination with other processes or tools. In fact, this idea is not new and already well implemented in various industrial applications. In order to systematise and illustrate the high variety of possibilities, Plate XXIII

This chapter is a revised and updated version of Chapter 10 'Enhancing laser welding capabilities by hybridisation or combination with other processes' by Dirk Petring, originally published in *Advances in laser materials processing* by J. Lawrence, J. Pou, D. K. Y Low and E. Toyserkani, Woodhead Publishing Limited, 2010, ISBN 978-1-84569-474-6.

(between pages 328 and 329) presents a map of combined laser processes and tools. While the abscissa accounts for the diversity of involved processes, the ordinate displays the diversity of involved tools. The combination is specified as hybridisation if the combined tools are different and the processes are acting simultaneously in a common processing zone. Obviously this specification and the scales of the coordinates in Plate XXIII are rather qualitative. However, the distribution of the depicted examples in the map demonstrates their relative position in the context of the selected systematisation.

Let us start at the origin, which is at the left bottom corner of the map in Plate XXIII. It represents the area with the lowest diversity of the contributing tools and processes. The picture shows two laterally displaced laser foci – normally produced by a beam-splitting element in the focusing head. It is a well known method to improve the capabilities of laser welding aluminium [1] or overlapped zinc-coated steel sheets [2]. The two laser spots may be identical or slightly different in power or power density. In the latter case, the more powerful spot is arranged as the leading part of the dual-spot tool, which is responsible for promoting the welding process at the keyhole front. The trailing spot supports the action of the first one by increasing the welding depth and, more importantly, it prevents the capillary back wall from approaching the front wall and thus from choking the capillary. The capillary is elongated in the welding direction and the top and bottom capillary apertures are increased. The proper shaping of the capillary stabilises the welding process and facilitates the escape of gas (such as insoluble hydrogen in aluminium) and vapour (such as vaporised zinc of coated steel sheets). Thereby, porosity and melt expulsions are significantly reduced. Another well-established application for the 'dual-spot laser welding' technique is using the two spots displaced transversally to the weld track, in order to increase the bridging capabilities (e.g. during welding of tailored blanks). Already this simple example at the origin of the map illustrates the benefits which can be achieved by changing from a standard system configuration to a combination of, in this case, even identical laser tools supporting the same process.

Consistently following this idea, we move only a little diagonally on the map, and arrive at the combination of different laser distributions, originating from the same or different lasers. The 'dual-laser hybrid processing' can be used, for example, by combining standard laser beams with diode laser beams, arranged in advance, subsequently or coaxially. Examples of this approach are the overlap welding of zinc-coated steels [3] and the burn-off stabilised thick section cutting of steel plates [4]. The assisting diode laser radiation provides additional heat beside the actually processing laser beam. The induced lateral temperature field serves, for example, for preheating, decoating, premelting or post-processing.

In the centre of the map, 'laser-arc hybrid welding' represents one of the most established laser hybrid processes. It will be discussed in more detail below, as well as 'cutting and welding with the combi-head', using a single tool for processes as different as cutting and welding, thus being well located in the bottom right corner of the map in Plate XXIII (between pages 328 and 329).

Exactly at the opposite corner of the map, the 'water jet guided laser cutting' is an impressive example of how a sophisticated combination of laser radiation with a completely different tool can expand the application range of lasers, both figuratively and literally speaking. A pressurised water jet with a diameter of 100 μm or less is used as a liquid laser waveguide into the cutting zone with a range of up to 100 mm. Additionally, the water jet supports material ejection out of the cutting kerf. The so-called micro-jet technique achieves an unrivalled focal position tolerance, effective cooling of the cutting zone and safe protection of the material surface from debris by the formation of a self-contained water film [5]. It is a laser hybrid process par excellence.

A special sequence of three different processes, the so-called 'eSIE. CONNECT laser cutting, riveting and laser welding' is positioned at the largest distance from the map's origin in Plate XXIII. Here, a rather large diversity of processes and tools is combined in a process chain for the manufacturing of modular railway vehicles. Laser cutting is used for preparing the parts including the drill holes for the riveting process, riveting itself is used for pre-assembling the sheet metal structures without the requirement of expensive clamping devices, and laser welding is used as the final joining technique to accomplish the strength and endurance which are essential for railway vehicles [6].

The above described spectrum of combinations, some of them also representing hybridisations, illustrates the wide variety already existing in the laser field and demonstrates the potential for more in the future. The following sections deal with the principles, state-of-the-art and recent progress in two of the most important representatives of hybridisation and combination techniques, namely laser-arc hybrid welding and combined laser cutting and laser welding.

18.2 Laser and arc hybrid welding

18.2.1 Principles and state-of-the-art

The expression 'laser-arc hybrid welding' represents welding processes which use – besides the laser beam – an arc as the second source, acting in the same melt pool as the laser beam. The arc provides additional heat and, as the case may be, also filler material. Laser-arc hybrid welding processes are meanwhile the most established hybrid welding variants and well recognised for their robustness, efficiency and flexibility. Especially the coupling of a deep-penetrating laser beam with the heat and molten metal supplying gas metal arc (GMA) is a proven hybrid technique. It significantly expands the original welding application range of lasers.

Laser-arc hybrid welding has been investigated since 1978, when Professor Steen and co-workers in the UK published their first paper about TIG (tungsten inert gas) augmented laser welding [7]. Replacing the tungsten electrode by a wire-fed electrode, heat of the arc as well as molten filler material of the wire are transferred to the welding zone by the so-called MIG/MAG (metal inert/active gas) process. Since the first industrial laser-MIG hybrid system was put into operation by Fraunhofer ILT in the year 2000 at a German oil tank manufacturer [8], a hybrid boom can be observed and various new installations followed. Laser hybrid welding systems have been installed in shipbuilding [9, 10], as well as in the automotive industry [11].

The laser-arc hybrid process is characterised by the simultaneous application of a focused laser beam and an arc, creating and moving a common melt pool along the weld pass (see Fig. 18.1). The combination offers an increased number of parameters compared to the single processes, thus allowing flexible control of the welding process adapted to the demands of material, construction and manufacturing conditions.

The main benefits of the hybrid technique compared to laser beam welding with or without filler wire are:

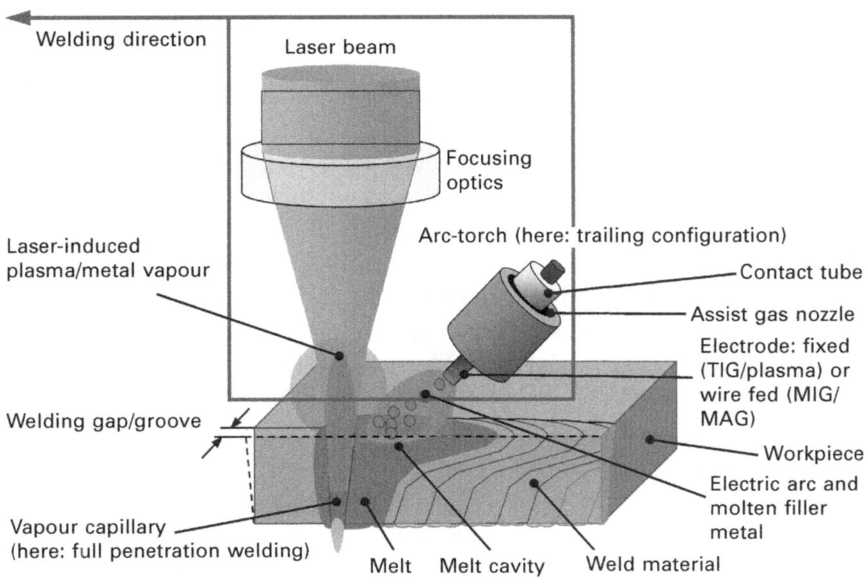

18.1 Principle of laser-arc hybrid welding [8].

- better gap bridging capability at lower laser beam power
- better levelling of edge offset
- lower demands on edge preparation and clamping
- improved adjustment of the thermal cycle.

There are also significant benefits compared to arc welding techniques:

- higher welding speed
- weldability of zero-gap and I-seam at lap joint
- single-pass full-penetration welding even at high welding depth
- lower heat input
- lower distortion
- at T-joints or in corners: smaller fillet, more clearance.

The result is a more flexible and robust welding process providing higher productivity and quality.

The arrangement of the arc relative to the laser beam axis (leading, trailing or coaxial type, inclination and distance between laser and arc) depends on the material to be welded and its surface properties. Also important are type of joint, edge preparation and welding position. The laser and arc power, the laser focusing parameters, and the laser wavelength, as well as the striven metal-transfer mode (MIG/MAG) and special boundary conditions (e.g. the accessibility for a seam-tracking sensor) are other factors with strong influence on the design of hybrid process and equipment.

Normally, the smallest possible arc inclination is desired. Angles in the range of 15–30° relative to the laser axis work with technically acceptable effort. Nd:YAG, disk and fibre laser radiation, due to a lower interaction with the arc's plasma, allows a closer approach to the arc than CO_2 laser radiation, as long as the laser induced vapour jet is not leading to detrimental interaction effects.

The currently most preferred laser-arc hybrid welding process is using MIG/MAG. The process can be controlled in such a way that the MIG/MAG part provides the appropriate amount of molten filler material to bridge the gap or fill the groove, while the laser is generating a vapour capillary within the molten pool to ensure the desired welding depth at high speed. The coupled process increases the welding speed above the sum of the single speeds, and produces an increased regularity of the weld bead. Improvements of metallurgical properties regarding hardness and toughness as well as diminished porosity due to promoted escaping of gas out of the enlarged melt pool are noted [8, 12].

The correct setting of gas parameters is an important factor in hybrid welding, where laser and arc-specific criteria have to be taken into account simultaneously. Using Nd:YAG, disk or fibre laser radiation, beam absorption within the plasma and resulting plasma shielding can be neglected due to the shorter wavelength compared to CO_2 lasers. Thus, with these lasers the selection of process gas can be determined according to the arc stability demands and bead shielding properties, for MIG/MAG also droplet detachment and spatter-free metal-transfer have to be considered. In this case argon will be the dominant portion of the gas used. Small addition of oxygen promotes the droplet detachment and reduces spatter. Mixtures of helium lead to higher arc voltage and the corresponding power increase results in wider seams, but also to a destabilising of the arc. Nevertheless, using CO_2 lasers, a helium mixture is necessary to avoid plasma shielding. Fortunately, the presence of the laser beam enables acceptable arc stability even with a significant helium portion [8].

To demonstrate the general capabilities of laser-MIG/MAG hybrid welding, typical macro sections of seams in aluminium and steel are shown in Fig. 18.2 [12]. The examples provide evidence of fast speed, wide gap bridging, smooth levelling of misalignment and high quality. A further two examples

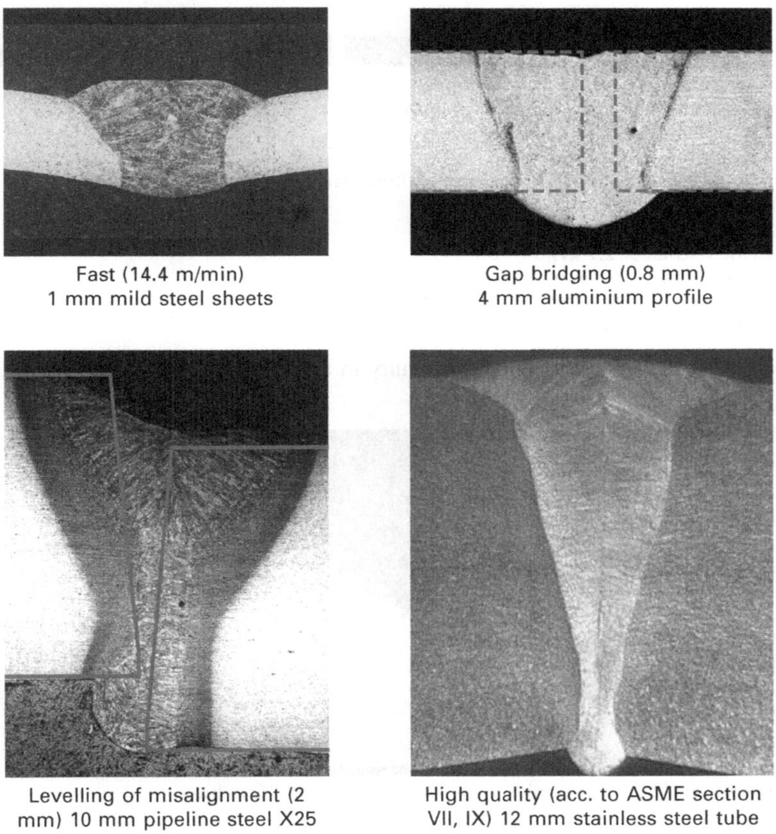

18.2 Cross sections and related parameters showing benefits of hybrid welding [12].

in aluminium illustrate important hybrid welding features in more detail. In Fig. 18.3 the gap variation at butt joint in 4 mm aluminium sheets could be bridged even without any backing melt support or root protection [8]. Fillet welds at lap joint (see Fig. 18.4) get an effective reinforcement by the contribution of the MIG process [12].

18.2.2 Physical model of the root formation

The appropriate amount of filler wire deposition depends on groove and gap volume as well as on welding speed and can easily be estimated and adjusted. The amount and the distribution of the energy input due to the laser beam power and the focal position influence the root formation and

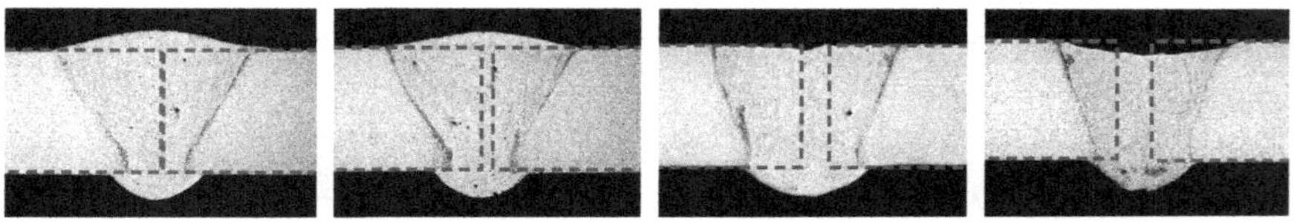

b_s = 0 mm, V_D = 8.3 m/min | b_s = 0.4 mm, V_D = 8.1 m/min | b_s = 0.8 mm, V_D = 7.8 m/min | b_s = 1.2 mm, V_D = 7.1 m/min

- Aluminium 6000 series alloy
- Sheet thickness 4 mm
- Square butt weld in flat position without backing bar or root protection
- Gap width b_s
- Welding speed 2.5 m/min
- Nd: YAG-laser 2.7 kW
- MIG impulse arc in trailing configuration
- Wire material S-AlSi12, diameter 1.2 mm, wire feed rate V_D
- Assist gas argon

18.3 Gap bridging capability in an aluminium alloy [8].

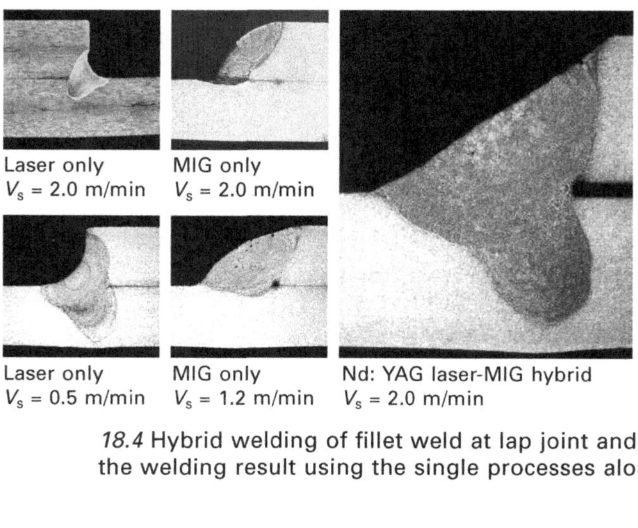

Laser only, V_s = 2.0 m/min | MIG only, V_s = 2.0 m/min
Laser only, V_s = 0.5 m/min | MIG only, V_s = 1.2 m/min | Nd: YAG laser-MIG hybrid, V_s = 2.0 m/min

- 6000 series alloy
- Thickness 4 mm
- Nd: YAG laser 2.7 kW
- MIG impulse arc (trailing)
- Wire material S-AlSi12
- Diameter 1.2 mm
- Assist gas argon

18.4 Hybrid welding of fillet weld at lap joint and comparison with the welding result using the single processes alone [12].

is handled experimentally. A remaining important question is: Which gap width can maximally be bridged and what are the determining factors for existing limits especially in thick-section welding?

As mass and energy balance have already been treated above, a momentum or pressure balance, respectively, shall give the answer (see Fig. 18.5). The pressure balance determines, for example, whether the root becomes concave with perhaps even incomplete penetration, whether it becomes convex and sound with regular appearance, or whether excessive penetration with drop through and sagging melt occurs. With a simple physical model the pressure balance at the root can be calculated [13]. The outcome will reveal suitable measures to improve gap bridging capabilities.

The model illustrated in Fig. 18.5 considers the gravitational force (gravitational constant g) due to the mass of the melt column (density ρ and height t) above the root with width w_m, the dynamic pressure of a 'downward' melt flow component with a velocity v_m and the supporting capillary forces due to the surface tension σ of the root melt. The arc pressure could also be included in the balance equation, but during thick-section welding of steels its contribution can be neglected compared to the other terms. The gap width w determines the minimum possible root width. On the other hand, the root width determines the minimum possible root radius $w_m/2$, which occurs at a contact angle $\alpha = 90°$ (see Fig. 18.5).

By setting gap and root width identical, the balance equation can be resolved for the maximum gap width w_{max} [13]:

$$w_{max} = 2\sigma\,[\rho\,(g\,t + v_m^2/2)]^{-1} \qquad [18.1]$$

To adjust the process, first of all an appropriate wire feed rate has to be set for properly filling the missing volume. It is also clear, that in order to fulfil

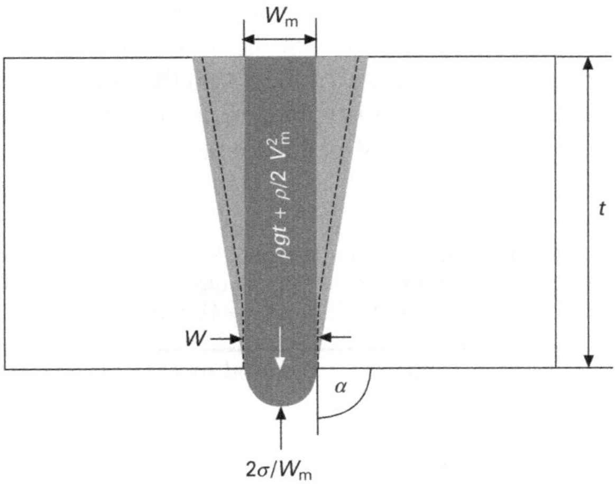

18.5 The pressure balance at the root [13].

the above condition of minimising the root width w_m to the gap width w, the laser beam power has to be adapted, namely minimised, accordingly:

- Adapted laser power reduces root width w_m to gap width w.

This is of course a somewhat idealised condition, but as the absolutely limiting case it perfectly leads to the maximum allowable gap.

Equation 18.1 presents three possibilities to maximise the allowable gap width w_{max}, which can be implemented by corresponding measures:

- increasing surface tension σ by root protection with inert gas (if two-sided access is accepted)
- reducing downward melt velocity v_m by ensuring a stable process with low melt dynamics, mainly achieved by a proper basic parameter configuration
- avoiding gravitational effects by using horizontal position PC.

An example is calculated in Table 18.1, where three different theoretical cases for laser-arc hybrid butt welding of structural steel plates at a thickness of 15 mm in flat position PA are compared. A corresponding experimental value for the maximum allowable gap size with properly adapted wire feed rate and laser power has also been determined. The first case in Table 18.1 implies a clean root, free of oxides and with correspondingly high surface tension. Furthermore, an ideal situation with no downward melt flow is assumed. In the second case, an oxidised steel melt surface reduces the surface tension at the root, and in the third case additionally a significant melt flow velocity typical for high dynamics is indicated. Following these steps, the maximum gap width which can be bridged is decreasing from 1.7 mm to 0.5 mm. An experimental value of 0.7 mm has been achieved without root protection, which means with an oxidised root. The process was adjusted quite stable, which should result in a moderate downward melt velocity component. This condition meets the situation between the second and third case and corresponds very well with the average of these two theoretical values. This rather simple model provides a surprisingly accurate description of the gap bridging capability and allows a well-defined course of action to optimise the process.

Table 18.1 Three different cases for the gap bridging capability w_{max} during welding steel in flat position PA, calculated according to Eq. [18.1] with ρ(steel) = 7,800 kg/m^3 and thickness t = 15 mm

	Clean root	Oxidised root	Oxidised root and melt dynamics
σ [N/m]	1.0	0.5	0.5
v_m [m/s]	0	0	0.5
w_{max} [mm]	1.7	0.9	0.5

18.2.3 Advanced technical equipment

With the standard approach of combining a discrete arc-torch in off-axis configuration with a laser focusing head, there are certain limitations on the possible position and orientation of the arc. In order to prevent the torch nozzle from interfering with the laser beam, it has to be positioned at sufficient distance and inclination. Another problem with this off-axis configuration approach as sketched in Fig. 18.1 is that it promotes entrainment of air into the weld by the Venturi effect.

To address these problems, a more sophisticated approach uses a welding head where the laser beam and arc are surrounded by a common water-cooled nozzle device with integrated contact tube for contacting and stable guiding of the wire electrode (see Fig. 18.6) [8]. This arrangement provides the closest laser and arc proximity at the steepest arc inclination. The process gas flows out of an annular channel coaxially to the laser beam. A diffusing aperture within the channel enables a homogeneously distributed stream of the assist gas onto the welding zone. Thus, a transverse suction of air by the Venturi effect is avoided and effective protection of the weld bead is ensured (see Fig. 18.6). Moreover, a minimal but sufficient leak gas flow in the upward direction avoids process gas contamination by air entrainment via the laser beam entrance.

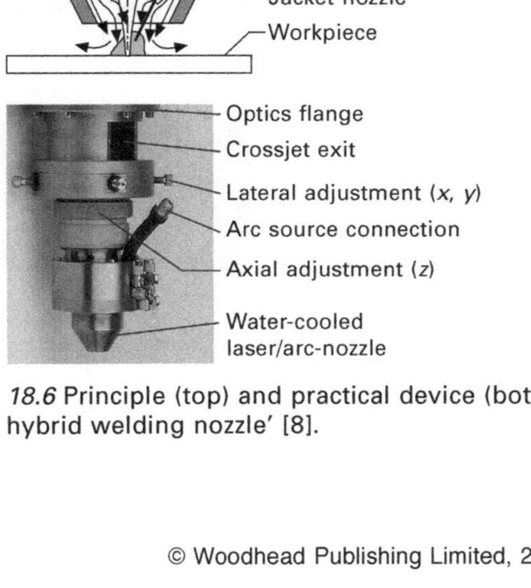

18.6 Principle (top) and practical device (bottom) of the 'integrated hybrid welding nozzle' [8].

18.2.4 Welding of heavy section high-strength steel plates

As with any other welding process, also the capabilities of laser-MAG hybrid welding are essentially determined by the appropriate selection of the system setup and the basic parameter configuration. If these boundary conditions are well chosen, hybrid welding proves to be a really stable, efficient and flexible technology. In the following, all welds are carried out with the above described integrated hybrid welding nozzle with an Ar-He-O_2 mixture, a 20 kW CO_2 laser and a programmable MAG welding source operated in the pulsed mode. In that case, the control parameters of laser, MAG and machine can be varied freely in a rather wide range in order to adjust the welding process to the required performance regarding welding depth, gap bridging, seam shape and metallurgical properties. As an example, the framed cross sections in Figs 18.7 and 18.8 indicate the parameters with the most efficient welding processes out of the respective variation, also to be recognised by the widest roots. On the other hand, in fact all cross sections in Figs 18.7

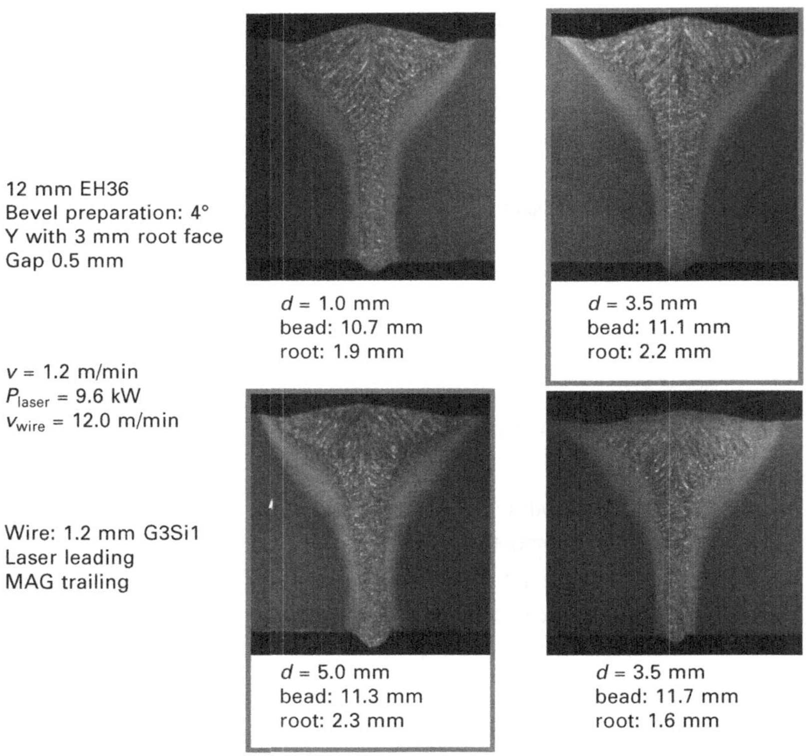

18.7 Moderate influence of distance d between laser and arc with an optimum for weld efficiency in the range of 3–5 mm [13].

and 18.8 show very acceptable properties, which demonstrate the robustness of the selected system setup and parameter configuration.

There are well-defined control functions for setting welding speed, laser power and wire feed rate adapted to material thickness, groove preparation, gap width and welding position (e.g. for 15 mm plate thickness in Figs 18.9 and 18.10). It is worth mentioning, that the allowable laser power ranges to be read from Fig. 18.9 are quite wide (more than 3 kW) and even overlapping in the case of welding 15 mm in flat position PA. A control function for the laser beam power depending on gap width could in this case be dispensable and the welding speed remains constant at 1.2 m/min. Only a control of the wire feed rate adapted to the gap width is recommendable as indicated in Fig. 18.9, if an excessive weld reinforcement or underfill shall be avoided at varying gap.

In Fig. 18.10 a phenomenon is demonstrated, which again confirms the robustness of the hybrid configuration used. With increasing gap width, the process has to be and can be continuously varied from laser-MAG hybrid welding to pure MAG welding. The transition to wide gaps up to 3 mm is achieved successfully by reducing the laser power, increasing the filler wire deposition and decreasing the welding speed in position PC. These are in fact the control functions to be considered also in practice when gap variations occur.

In the thickness range up to 25 mm, welds without any hot cracks and, if any, with only very few small pores, compliant with the highest assessment group B of EN ISO 13919-1 can be produced in a single pass [13]. Cross sections of optimised hybrid welds are presented in Fig. 18.11. In order to achieve crack-free welds, V- or Y-shaped groove preparation in the range

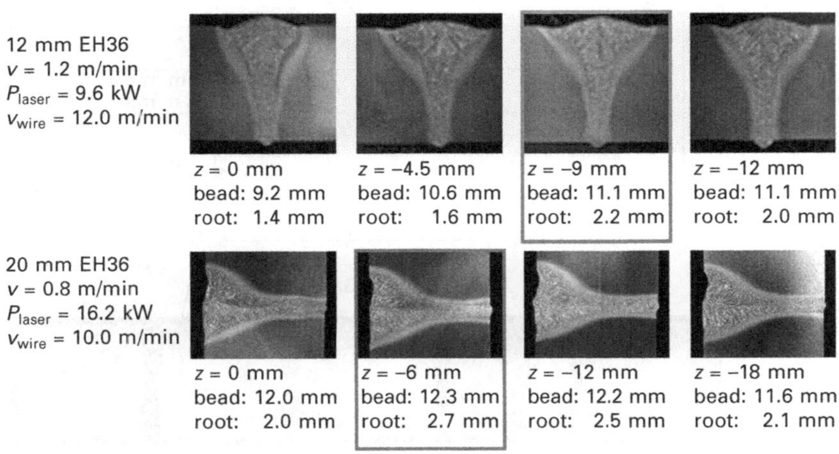

18.8 Moderate influence of focal position with an optimum for weld efficiency at focal positions within the plate [13].

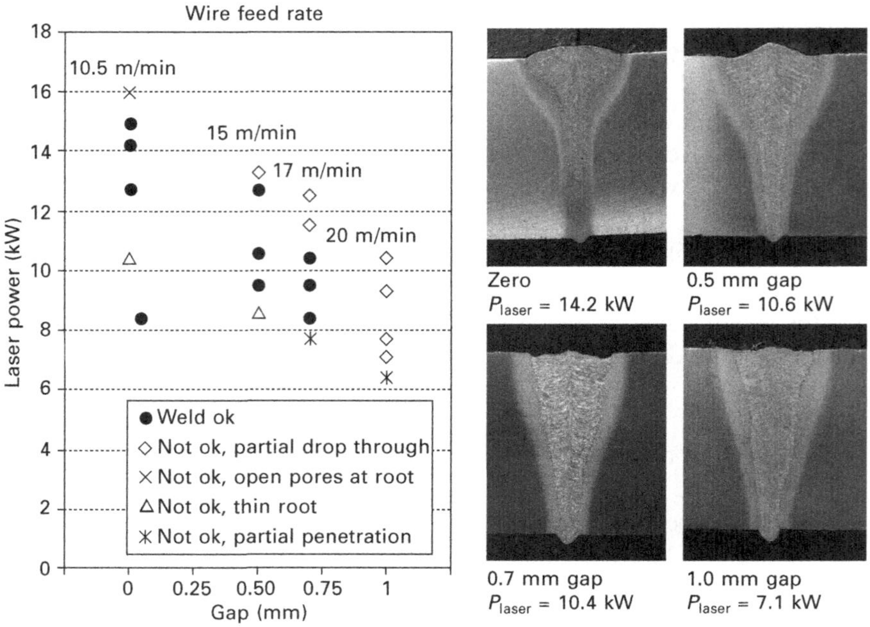

18.9 Process windows and gap bridging capability for laser-MAG hybrid welding of 15 mm structural steel in flat welding position (6° V, PA) with leading MAG at a constant welding speed of 1.2 m/min [13].

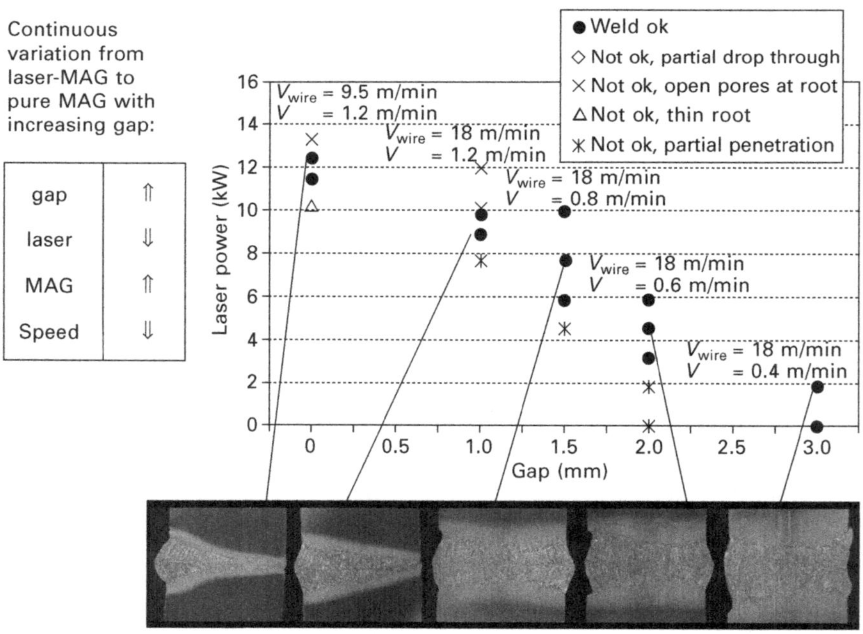

18.10 Expanded process windows and improved gap bridging capability for laser-MAG hybrid welding of 15 mm structural steel in horizontal position (6° V, PC) with leading MAG [13].

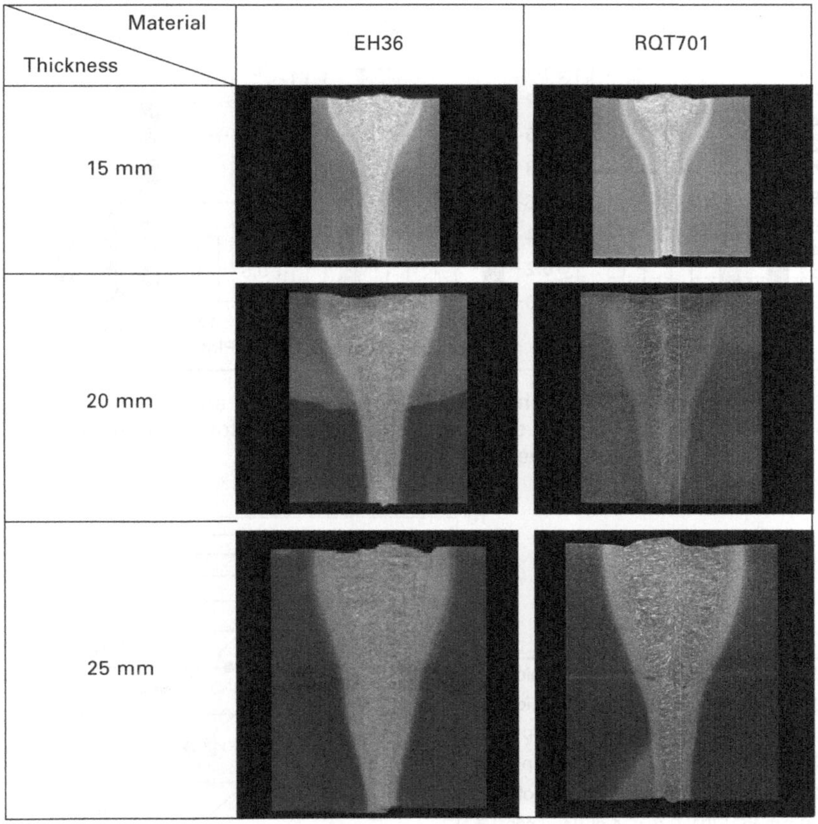

18.11 Cross sections of hybrid welded high-strength steel plates up to 25 mm thickness using a 20 kW CO_2 laser at adapted power level [13].

between 4 and 8° full angle and an appropriate welding speed together with the right energy input per unit length are the crucial points to be considered. These measures ensure an ascending columnar dendrite structure upwardly curved and partly converging against the centreline in the upper MAG-dominated cup-shaped part of the seam. In the lower laser-dominated 'trunk' of the seam, an equiaxial grain structure is preferred, which possibly changes in the root to small columnar dendrites downwardly curved and converging against the centre line or growing down to the root surface [13].

The laser beam power as well as the welding speed must be adapted to the plate thickness and the gap width. Figure 18.12 shows a general overview of the required beam power and the achieved process capabilities regarding weldable plate thickness, welding speed and gap width for butt joints in positions PA and PC, respectively. In the thickness range between 12 and 25 mm, excellent fatigue properties of high-strength structural steel butt

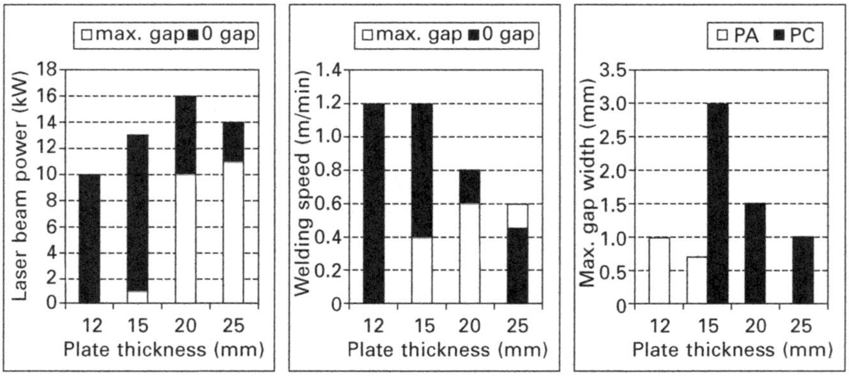

18.12 CO_2-laser-MAG hybrid welding parameters and capabilities versus thickness, gap size and welding position for single-pass butt joints in structural steel [13].

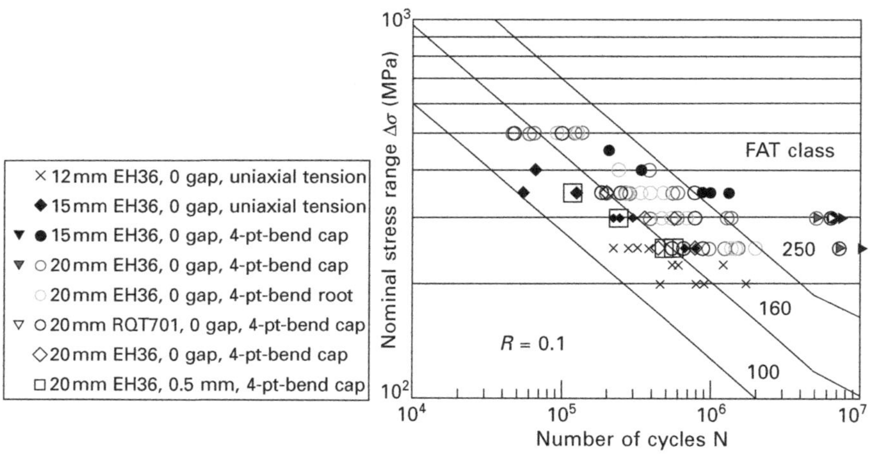

18.13 Fatigue results of heavy section laser hybrid welds [13].

joints have been determined by extensive uniaxial tension and four-point bend tests (Fig. 18.13) [13].

18.2.5 Applications

The advantages of laser-MAG hybrid welding are increasing the thickness range of laser welding significantly. Important benefits are high welding speed, low distortion, the ability to bridge gaps, and the capability of single-pass full-penetration welding. Applications with thick wall thicknesses can be found in pipeline construction, shipbuilding, load-bearing structures, offshore engineering, special constructions and in heavy vehicle construction such as earth-moving equipment.

18.3 Combining laser welding and laser cutting

18.3.1 Multi-functional processing

Demands for innovative manufacturing techniques are high. The main targets for these techniques are greater flexibility and higher productivity at lower costs. In many cases, this seems to be impossible; nevertheless, possibilities can be discovered if process chains are analysed and consequential methods to shorten or integrate them are found [14]. Sheet metal fabrication is one example of a complex process chain. Nearly all component assemblies being manufactured from sheet metal are produced by a sequence of cutting and joining techniques. Hitherto, these different processes were carried out on different machines. The sheet components are successively prepared by cutting, joined together and, for instance, completed by cut-outs or weld-ons. If in a single machine, both cutting and welding could be performed, by this integration and shortening of the process chain, the cost of production would be reduced substantially and products could be manufactured much more flexibly.

Even the particularly flexible laser systems are usually either dedicated to a single process or valuable time is spent on changing between applications such as cutting and welding. In practice, different processes are mainly carried out sequentially on different systems. Accordingly, machine expenditure is high. Within these classical process chains a high potential of flexibility and cost savings inherent to lasers remains unutilised hitherto in industrial production.

In order to overcome these limits a so-called 'combi-head' allows quick switching from one process to the other and vice versa in a temporally and spatially optimised, integrated process chain. Thus, a completely new degree of flexibility is achieved in production. New designs and a large number of variants can be produced much more cost effectively by combining cutting and welding capabilities in one tool. Process and tolerance chains are shorter due to minimised material and tool handling within a single machine and due to a combi-head with a process spanning, constant tool centre point (TCP) [15].

Laser manufacturing with a combi-head is a new, highly flexible solution based on old ideas [16–20]. Some years ago, these ideas were taken up again by several groups [21–23] who were motivated by the availability of laser sources with simultaneously high power and high beam quality and by increasing market demands for flexible production. In a multifunctional laser cell, the combi-head is able to perform 3D cutting and welding tasks in an arbitrary sequence without retooling. Part handling, positioning and clamping steps are omitted. This results in shorter and more flexible process chains and with it reduced production time and costs as well as improved manufacturing accuracy [24, 25]. Moreover, the combi-head opens up efficient solutions for innovative products from sheet metal with a wide range of variants.

18.3.2 The laser combi-head

In cutting heads, coaxially to the focused laser beam, an intense cutting gas flow has to provide for effective melt ejection out of the kerf. The laser beam is guided through a high pressure nozzle in such a way that the beam focus together with the gas jet meets the workpiece below the nozzle exit (see Fig. 18.14). A lens in the laser entrance aperture of the nozzle provides for a gas-tight sealing of the nozzle chamber and focuses the laser beam. In addition, the nozzle tip serves as a capacitive clearance sensor, ensuring a constant distance to the sheet surface of about 1 mm by a closed-loop control of the machine axes.

In welding heads, with solid-state and diode lasers also lens optics accomplish focusing, but with CO_2 lasers usually mirror optics are used. To create smooth shielding and process gas flow, as required for welding, mainly off-axis nozzles are applied some millimeters from the laser impingement point (see Fig. 18.14). In addition, a compressed air crossjet is arranged between the optics and the process gas nozzle that blows transversally to the laser axis in order to reliably protect the optics from smoke or spatters emitted from the weld zone. Accordingly, in contrast to the gas-tight optics of standard cutting heads, a welding head requires an open flow section below the optics, which is gas dynamically decoupled from the processing zone.

This conflict between the requirements of cutting and welding heads can be resolved with the autonomous nozzle [14] (see Plate XXIV between pages 328 and 329), which has best proved itself already in industrial applications with high-power mirror cutting heads up to 12 kW. Besides various job shops, it has also been installed at the Blohm & Voss shipyard, where it is used in a mirror cutting head for precise cutting of ship panels, which are later welded together with the same laser to produce large ship sections. In this

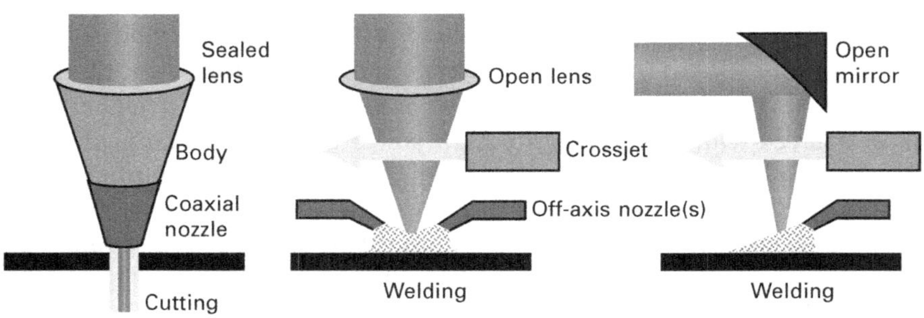

18.14 Difference of cutting heads and welding heads.

application, a head exchange was still accepted for switching from cutting to welding and vice versa.

In principle, the idea of the autonomous nozzle is rather simple and it seems predestined for solving the conflict described in Fig. 18.14. By shifting the laser entrance aperture in the vicinity of the nozzle exit, a patented flow design enforces a minimum process gas leakage and simultaneously prevents suction of air (see Plate XXIV). That way the nozzle can produce a coaxial process gas flow in a wide pressure range, independent from the focusing optics and without a sealing window. Being 'autonomous', it easily permits the integration of a crossjet between itself and the focusing optics. This is necessary because the smooth process gas jet used during welding cannot protect the optics from smoke and spatters. Consequentially, a crossjet with compressed air supply, an alternating check valve for cutting and welding gas, and an appropriate numerical control program to change process parameters suffice to also make a welding head out of the cutting head. The autonomous nozzle with its rotationally symmetrical, slim design and an integrated capacitive clearance sensor represents the key component of a universal processing head for three-dimensional cutting and welding. The combi-head, as sketched in Plate XXIV, can optionally be equipped with lens or mirror optics and is suited for solid-state lasers as well as for CO_2 lasers.

The main technical features of a laser combi-head as shown in Plate XXIV can be summarised as follows [14]:

- quick, software-controlled switching between cutting and welding in arbitrary sequences
- one machine, one setting, one tool for cutting (N_2 and O_2) and welding processes
- constant tool centre point (TCP)
- rotationally symmetrical, coaxial, and slim head design with the autonomous nozzle, integrated capacitive clearance sensor, and effective crossjet
- gas functions decoupled from the optical system
- no principle limits regarding beam power and gas pressure.

18.3.3 Impact of laser brightness

The manufacturing benefits of integrated cutting and welding with a multifunctional laser combi-head are evident. In principle, there are no differences between the capabilities and parameters of standard cutting and welding heads and the combi-head. Nevertheless, some details are worth mentioning in order to avoid needless confusion or scepticism. It is sometimes believed that adaptive optics or motorised nozzles for changing nozzle distance and focal position independently are obligatory during switching between cutting and welding. Of course these are possible options, but with the autonomous nozzle for many applications, it turned out to be appropriate to use an identical focal distance from the nozzle tip for both cutting and welding. That means focal position and nozzle distance relative to the workpiece surface are changed simultaneously, simply by lifting the complete head, when switching from cutting to welding. Fewer optical and electrical elements reduce the complexity of the combi-head to the required minimum and ensure maximum robustness. Of course the head distance as well as gas type and flow rate, laser beam power, and processing speed can be adapted automatically by the machine control being programmed accordingly.

What makes brightness relevant for combined cutting and welding is the enlarged operating window allowed for changing head distance. A 'slim' focal zone of a bright enough laser allows for cutting narrow kerfs with a small nozzle stand off and for excellent welding conditions with a larger stand off with the same nozzle, without changing the focal position relative to the nozzle exit [26]. A suitable beam quality and the correct layout of the collimation and focusing optics according to the demands of the combined processes are the crucial boundary conditions to be successful with the above-described concept.

18.3.4 Solutions and applications

Use of the combi-head allows precise cut edges to be produced and reliable welding of the resulting perfectly matched parts, without the need to change tools or employ a seam-tracking sensor [26]. Examples of the use of this technology include tailor-welded blanks (TWB) (Fig. 18.15) and thicker components from structural steel (Fig. 18.16).

Particularly in the production of nonlinear tailored blanks, lack of precision in the prefabrication of the blanks causes gaps to form between the edges being joined. This can lead to either increased costs during the prefabrication phase, greater demands being made on the accuracy of seam tracking and gap measurement during the welding process, or even the use of filler material. In this context, integrated cutting and welding with the combi-head opens up new possibilities.

Figure 18.15 illustrates how a test series of blanks is prepared with the help of an industrial combi-head from Laserfact GmbH. A 4 kW fibre laser from IPG is connected to the combi-head by a process fibre with a diameter of 100 μm. The combi-head is mounted on a six-axis industrial robot. The sheets are clamped on movable linear slides, used for positioning. First of all, the edges of the blanks out of two differently galvanised car body materials with thicknesses of 1.0 and 1.2 mm are laser cut at 8 m/min and 1.5 kW,

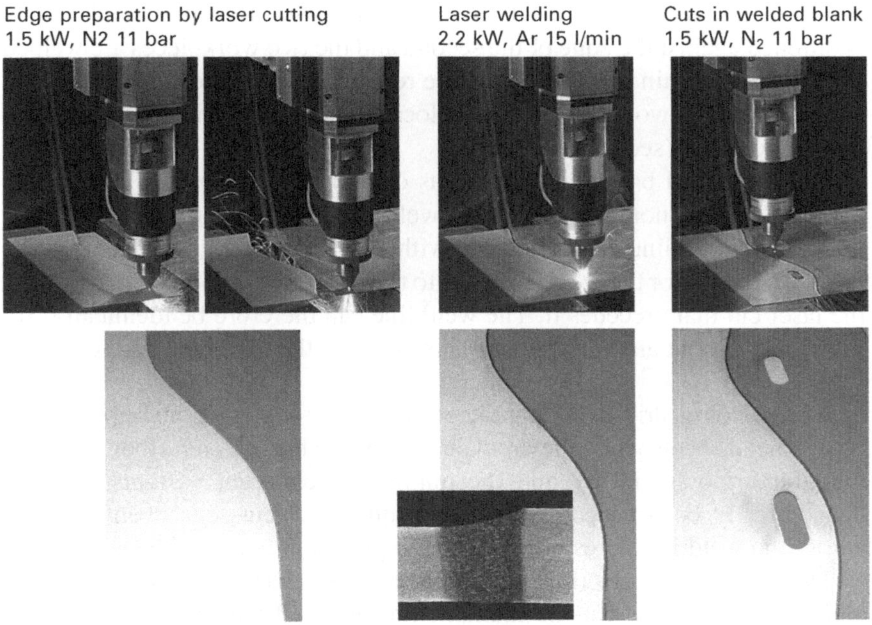

18.15 Integrated process chain for nonlinear tailored blank production (zinc-coated high-strength steel 1.0 mm/1.2 mm) using integrated laser cutting and welding with identical path concept at a speed of 8 m/min with moderate laser power; bottom pictures show processing results after the different process steps. (Photos: courtesy Fraunhofer ILT and Laserfact.)

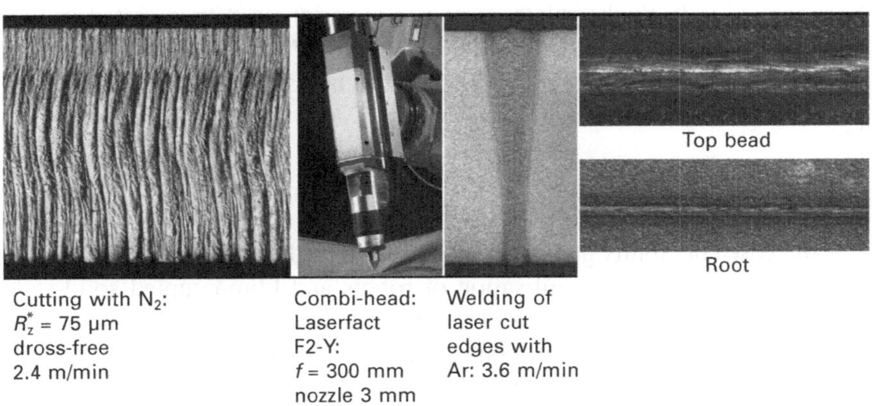

18.16 Edge preparation (oxide-free cutting with nitrogen) and welding of 8 mm structural steel plates at 8 kW with the combi-head Laserfact F2-Y, fibre-coupled to a disk laser TruDisk 8002 with a 200 µm fibre.

resulting in two perfectly matching edges. The edge of the first blank is then positioned against the edge of the second and the two workpieces are welded together at 8 m/min and 2.2 kW, while remaining in their original clamping device. Finally, two elongated holes located exactly to one another are cut out of the blank (see Fig. 18.15).

The integrated production flow was demonstrated to produce very high quality, even in nonlinear cuts and welds. It is particularly worth noting that the welding line can be located without a seam-tracking system, as the precise position for the weld is known to the machine coordinate system from the laser cut that precedes it. The weld line can therefore be identically (re) produced without any difficulties. This is called the 'identical path concept' [26].

After welding, drill-holes, contoured cut-outs or edge cuts can be performed on the metal sheets while they remain clamped. This option further enhances manufacturing flexibility and the number of different variants that can economically be produced. These subsequent cut-outs can even be made across the weld itself, resulting in a high precision component.

The integrated production of tailored blanks is a particularly cost-effective and flexible solution for the manufacture of small and mid-sized batches and can cope even with ultra high strength steels due to the integrated laser cutting process. Another industrial application of the combi-head is the processing of coils to produce 'endless' coil material in coating lines by laser trimming the ends of coils, laser welding them together and subsequent laser cutting of a smooth coil transition at the start and the end of the welded joint track. ThyssenKrupp Steel has been using a laser combi-head for this application in coil production lines in Germany and the United States since 2009. With enough laser power, the combined cutting and welding can also be managed with thicker plate or profile material, as demonstrated in Fig. 18.16. Achievable processing speeds (Fig. 18.17) and the wide spectrum of possible applications (Fig. 18.18) are the big drivers of multifunctional processing with lasers.

18.3.5 Future trends

The increased flexibility gained through combining the processes is particularly brought to bear with the utilisation of robots and fibre-coupled solid-state lasers. Combi-processing is used to manufacture 3D assemblies from 3D raw parts: deep-drawn sheets, blanks, profiles or tubes, etc. For example, the upper and lower shells of closed structures are fixed in a single clamping setup and can be welded together, sections can be cut out, and clips can be welded on in a rapid sequence of processes.

The impact of the features that support 3D performance (high beam quality, an optimised combi-head design and speed adapted laser power

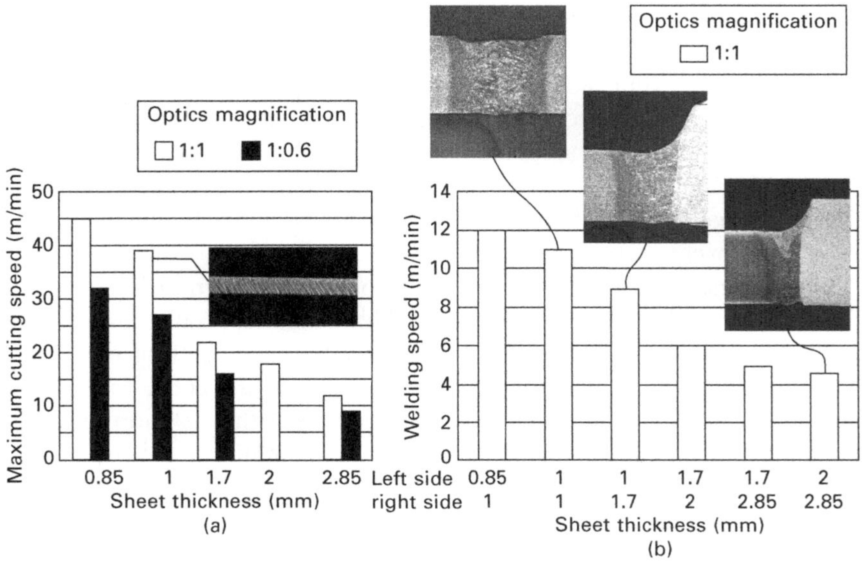

18.17 Cutting (a) and welding (b) speeds versus thickness during multifunctional processing of galvanised steel sheets at 4 kW with the combi-head Laserfact F2-Y, fibre-coupled to a fibre laser IPG YLR4000S with a 150 μm processing fibre.

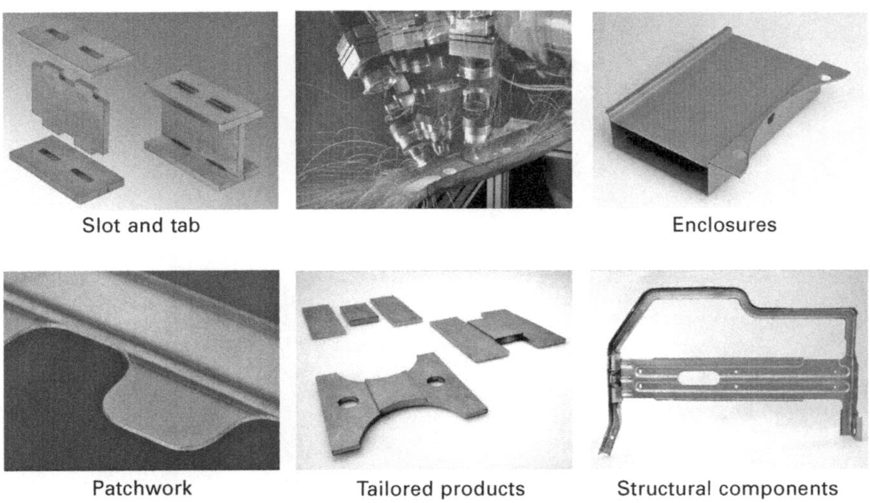

18.18 Various case studies of combined laser cutting and welding. (Photos: courtesy Fraunhofer ILT.)

modulation) is demonstrated in an automotive application example, namely combi-processing of a B-pillar [26]. Driven by the use of modern high-strength materials in cars, laser trimming and cutting of holes in hot-formed parts is well established, because those materials are difficult to cut mechanically.

With the availability of combi-processing, it is only a small step to integrate the cutting process with welding operations and thus shortening the process chain. At first, several holes are cut into the B-pillar. Secondly, the final dimensions of the outer contour are cut. Next, a reinforcing sheet is welded on the pillar and, finally, holes are cut through the reinforcing sheet and the B-pillar (Fig. 18.19). All operations are performed in one clamping, thus high positional accuracy between the outer contour and the holes, including those in the weld-on part, is guaranteed.

The processing was done on a gantry robot (RLP16 from Reis) with a Laserfact combi-head F2-X (Fig. 18.20). This head is specially designed for being optimally mountable to the axes-integrated beam path of the gantry robot – without a stressed fibre connection near to the end effector. The laser source was an IPG fibre laser YLR4000 SS with a 100 μm diameter process fibre. The large contours were cut at 15 m/min, the holes at 3–9 m/min, depending on their diameter and the material thickness (thickness of the B-pillar material at least 1.4 mm). During cutting of small radii and accordingly reduced speeds of the TCP, the pulsed mode was automatically activated by properly programmed control characteristics for the speed adapted laser power modulation. The welding speed was 3 m/min for the lap weld through the reinforcement plate (1.3 mm) and the pillar (1.4 mm). The maximum laser power was 2.5 kW. Depending on the details of the cut

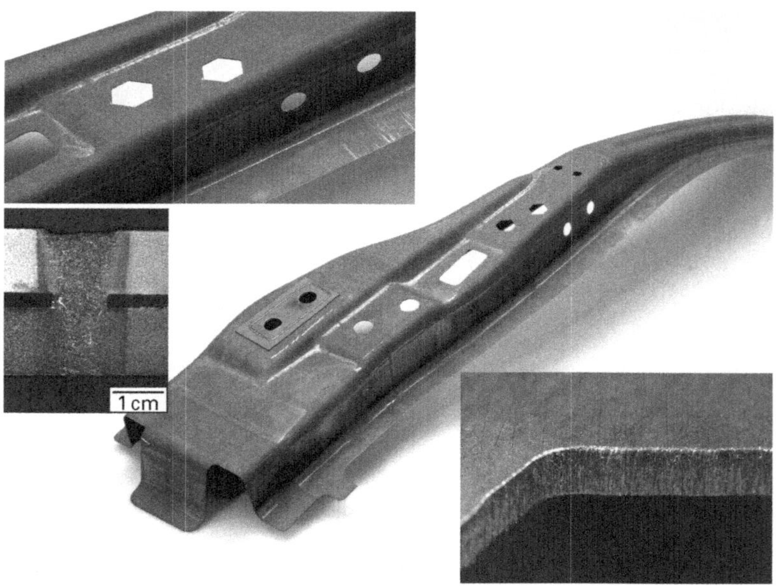

18.19 Combi-processed B-pillar with details of the welded reinforcing sheet and cut edges in small contour (hole, speed 6 m/min) and big contour (trim-cut, speed 15 m/min). (Photos: courtesy Fraunhofer ILT.)

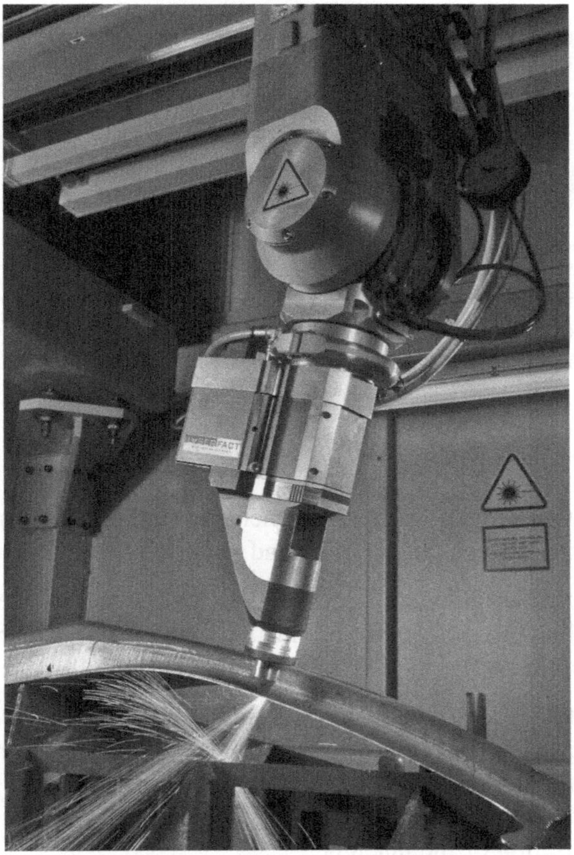

18.20 Combi-head at work. Cutting process on the B-pillar with a gantry robot. (Photo: courtesy Laserfact.)

contour, the overall processing time for cutting and welding a B-pillar as in Fig. 18.19 is in the range of 1 min.

Another demonstration of combi-processing capabilities is the possibility of overlap welding a circular seam just before cutting the hole precisely at the edge of the weld seam to produce a gap-free hole, preventing subsequent crevice corrosion (Fig. 18.21). The coordinates of the weld are known in the machine and with an identical path plus an offset, a precise position of the cut relative to the weld is possible thanks to the common TCP for cutting and welding. As both processes can be done right one after the other, there is no additional positioning.

The combi-head particularly shows advantages in the following three product groups:

- complex component assemblies requiring multiple quick changes between welding and cutting operations,

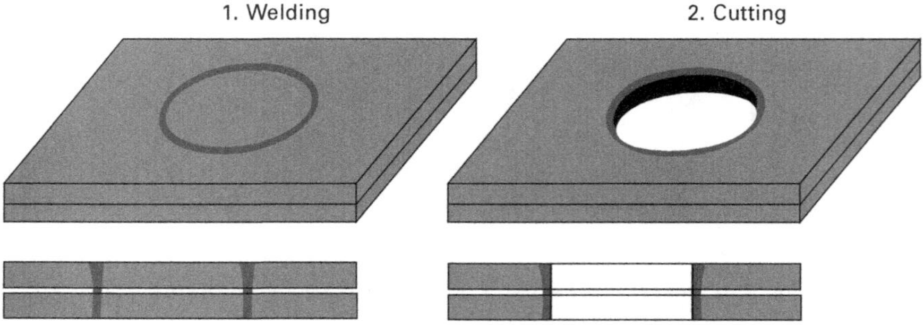

18.21 Producing a sealed overlap drill hole by laser cutting along a previously laser welded circular lap joint.

- products with a wide range of variants, i.e., customised, optional cut-outs and weld-ons,
- components with a functionality that benefits from the exact positioning and orientation between cut contours and weld joints.

Combination of laser welding and laser cutting with one tool in one system can lead to the following benefits for production when manufacturing sheet component assemblies:

- short, integrated process chains for high productivity,
- high flexibility and cost-efficient production of options and variants,
- savings in handling, positioning and clamping operations,
- savings in machine investment and floor space,
- increased utilisation of the machine,
- easy reconfiguration of the system for new products,
- short tolerance chain improving accuracy of parts and components.

Taking into account the combi-head's technical features and capabilities, it becomes obvious that investment, organisation and operation costs can be reduced by a factor of two or more in spite of equally improved productivity, time to market and flexibility. The innovative possibilities opened up by multifunctional processing are expanded as product designers and production people jointly take up this concept and combine their expertise and ideas.

18.4 References

1. Iwase, T., Sakamoto, H., Shibata, K., Hohenberger, B., and Dausinger, F. (2000) Dual focus technique for high-power Nd:YAG laser welding of aluminium alloys. *Proc. SPIE* 3888, 348–358.
2. Forrest, M.G., and Lu, F. (2007) Joining new auto body materials. *Industrial Laser Solutions*, Vol. 22, March.
3. Bonss, S., Seifert, M., Barthel, K., Brenner, B., and Beyer, E. (2001) New

developments in high power diode laser welding. *Proc. 20th International Congress on Applications of Lasers & Electro-Optics*, ICALEO 2001, Jacksonville, FL, October 15–18, 2001.
4. Petring, D., and Seme, B. (2007) Oxygen cutting with annular beam. In: *Diode Lasers.* (ed. Bachmann, F., Poprawe, R., Loosen, P.) Berlin: Springer, pp. 139–147.
5. Perrottet, D., Green, S., and Richerzhagen, B. (2006) Clean dicing of compound semiconductors using water-jet guided laser technology. *Advanced Seminconductor Manufacturing Conference, 2006, ASMC 2006.* The 17th Annual SEMI/IEEE Volume, pp. 233–236.
6. Petring, D., and Wolf, N. (2004) Flexible laser process and laser system technology for the accurate manufacturing in modular rail vehicle design. *ZEVrail Glasers Annalen* 128, 196–201.
7. Eboo, M., Steen, W. M., and Clark, J. (1978) Arc-augmented laser welding. *Proc. 4th Int. Conf. on Advances in Welding Processes*, 9–11 May, 257–265.
8. Petring, D. (2001) Hybrid laser welding. *Industrial Laser Solutions*, December, 12–16.
9. Lembeck, H. (2002) Laser-Hybrid-Schweißen im Schiffbau. *Proceedings Aachener Kolloquium für Lasertechnik 2002*, Aachen, Germany, September, 177–192.
10. Jasnau, U., Hoffmann, J., Seyffarth, P., Reipa, R., and Milbradt, G. (2003) Laser-MSG-Hybridschweißen im Schiffbau. *Proc. European Automotive Laser Application 2003*, Bad Nauheim, Germany, January.
11. Graf, T., and Staufer, H. (2003) Laser-hybrid-welding drives VW improvements. *Welding Journal*, January, 42–48.
12. Petring, D., and Fuhrmann, C. (2003) Hybrid laser welding: laser and arc in concert. *The Industrial Laser User*, Issue 33, December, 34–36.
13. Petring, D., Fuhrmann, C., Wolf, N., and Poprawe, R. (2007) Progress in laser-MAG hybrid welding of high strength steels up to 30 mm thickness, *Proceedings of 26th International Congress on Applications of Lasers & Electro-Optics ICALEO 2007*, Orlando, FL, 300–307.
14. Petring, D. (2005) One head does it all – new technology allows 3-D laser beam cutting and joining. *Welding Journal*, August, 49–51.
15. Petring, D. (2005) Laser cutting and welding with one tool. *The FABRICATOR*, October, 40–41.
16. JP 60 108 191 A, Japanese patent application: Working head of laser working device, publ. 1985.
17. GB 2 163 692 A, UK patent application: Laser apparatus, publ. 1986.
18. Felleisen, R., and Kessler, B. (1994) Higher productivity by combining CO_2-laser beam welding and cutting in one processing head. *Proc. of ECLAT '94, 5th European Conference on Laser Treatment of Materials*, Bremen, Germany, pp. 468–474.
19. Geiger, M., Neubauer, N., and Hoffmann, P. (1994) Intelligent processing head for CO_2 laser material processing. *Production Engineering*, 1(2), 93–98.
20. Kaplan, A.F., Zimmermann, J., and Schuöcker, D. (1997) Combined laser welding, cutting and scribing. *Proc. of LANE '97, Laser Assisted Net Shape Engineering 2*, Erlangen, Germany, pp. 757–766.
21. Appendino, D. (2003) Saldatura laser flessibile di Tailored Blanks. *Proc. of Expo Laser 2003*, Ancona, Italy.
22. NN: Cutting and welding with one processing optic. *TRUMPF EXPRESS*, July, p. 13.

23. Schneider, F., Wolf, N., and Petring, D. (2003) Cutting and welding with the 'Autonomous Nozzle'. Annual Report, Fraunhofer ILT, Aachen, Germany, p. 68.
24. Schneider, F., and Petring, D. (2004) Cutting and welding with a combined processing head without retooling. Annual Report, Fraunhofer ILT, Aachen, Germany, p. 64.
25. Schneider, F., Petring, D., and Poprawe, R. (2005) Integrated laser processing – cutting and welding with a combined processing head. *Proc. of LIM 2005, 3rd International WLT-Conference on Lasers in Manufacturing*, Munich, Germany, pp. 133–135.
26. Petring, D., and Schneider, F. (2009) One machine does it all for laser beam welding and cutting. *Welding Journal*, March, 38–41.

19
Developments in hybrid laser-arc welding technology

A. GUMENYUK and M. RETHMEIER, BAM Federal Institute for Materials Research and Testing, Germany

DOI: 10.1533/9780857098771.3.505

Abstract: Laser hybrid welding technology provides a number of advantages, which in the previous decade have promoted its implementation in different branches of industry such as automotive manufacturing, shipbuilding and power generation. Some examples of the applications of laser hybrid welding are reviewed to demonstrate the technological and economic potential of this technology. The main objective of this chapter is to show some of the benefits and limitations of the process and to give some examples of how the number of applications could rise, taking into account the advantages of the new laser systems, optical components, arc power sources and sensor systems available nowadays.

Key words: laser beam welding, laser hybrid welding, high power laser, shipbuilding.

19.1 Introduction

Global trends in manufacturing of welded components are characterised by a growing demand for reduced welding operation costs and welding cycle times on the one hand, and consistent, or even improved, quality and mechanical integrity of welded components on the other. In this situation, the substitution of a new welding technology for the existing traditional ones with a view to increasing productivity could be beneficial, if it guarantees the required quality levels. Laser hybrid welding technology provides a number of advantages, which over the last decade have seen it applied across a broad range of industrial branches such as automotive manufacturing, shipbuilding and power generation. The use of combined laser beam and arc welding in one process zone has been recognised since the early 1980s (Eboo *et al.*, 1978) but broad industrial application of this technology only began around the year 2000 (Möller and Koczera, 2003). The principle of laser hybrid welding based on the combination of laser and arc welding in one welding process zone is characterised not only by an increase in the overall process efficiency, resulting in an augmented melting capability compared with a single process, but also by an improvement in process stability, which allows it to realise a reliable welding process at significantly high welding speed.

The application of a variety of laser hybrid welding processes is wide and ranges from thin sheet metal welding to heavy section welding with plate thicknesses of up to 20 mm. Depending on the particular welding task, the choice of the appropriate laser or arc processes could be crucial to the successful application of the technology in terms of the welding process requirements (e.g. minimal or maximal required heat input, high melting efficiency of the welding process, required penetration depth, special filler materials, necessary dilution in weld metal, etc.). Recent reports state that hybrid processes with different laser sources like CO_2 and solid-state lasers as well as different arc processes (GMA, TIG, PAW, SAW) can be used in combination with lasers. All of these technologies provide certain benefits according to the above requirements. The purpose of the present contribution is not to give a full overview of all the possible applications of laser hybrid welding, but rather to focus on some of the advantages and limitations of the process. At the same time, the aim of the authors is to show using examples how the potential applications of this technology could be broadened and its limitations overcome by exploiting the advantages offered by recent developments of new laser systems, optical components, arc power sources and sensor systems.

19.2 Developments in technology

Recent developments in laser hybrid welding technology are closely connected with progress in equipment development, and first of all with the emergence of new laser sources. Until the beginning of the year 2000, the industrial application of continuous wave (CW) solid-state lasers was restricted by a power limit of about 6 kW. One reason for this was increasing energy dissipation in the active medium, which led to inhomogeneous heating, thermal lens formation and, as a result, to beam quality loss. New developments in solid-state laser design have helped to overcome this problem and have led to a number of innovative laser system concepts such as high power fibre lasers, disk lasers and high power direct diode lasers with high beam quality. This new era in the development and application of high power lasers began with the market launch of scalable high power fibre lasers by the company IPG around 2003. This boosted the utilisation of those systems in the high power segment, especially for welding applications. In contrast to the former generation of high power CO_2 lasers, the new solid-state laser systems have a number of advantages, making them very attractive for applications where heavy sections have to be manufactured (such as shipbuilding, power generation, mobile crane and railway wagon fabrication). The high available power ranges of solid-state lasers, along with their extremely high beam qualities, result in a very high power density in combination with a focal depth that enables their application for heavy plate welding, making them competitive with the electron beam. The typical BPP value of a multimode fibre laser operating in the power range of 15–30 kW corresponds to 11 mm*mrad, which allows the laser beam to be sharply focused at a very high power density along the beam axis. This parameter is far ahead of that of conventional solid-state laser systems (25 mm*mrad). Comparable beam quality is provided by disk lasers, even though their available power is currently restricted to 16 kW (Havrilla and Brockmann, 2009).

Another big advantage of the new generation laser systems, apart from their high brilliance, is their extremely high overall plug efficiency. This can reach a mark of 30% for modern fibre and disk lasers, whereas the hitherto available plug efficiency of CO_2 lasers was about 10% and of solid-state Nd:YAG lasers, between 3% and 12%. This extremely high energy efficiency in combination with compactness and robustness transforms the modern laser into a kind of mobile system for manufacture.

However, the high brilliance of the new laser sources imposes special requirements on the peripheries of optical systems like optical fibres, fibre connectors and optical components of the processing heads. Due to the thermal effects of the laser beams, the heating of the optical elements can lead to temporary instability of the optical characteristics, resulting in, for example, focus shift (Reithemeyer *et al.*, 2009) or in the worst case to damage of the parts of the optical system operating at high power. Manufacturers of optical components have made great progress in meeting these requirements in the last decade. There are currently several manufacturers who provide optical heads for laser material processing and particularly for welding applications designed for laser powers in the range of 10–30 kW. Apart from the traditionally used transmissive optical components (lenses) for shorter wavelengths, mirror based optics are now used more and more for high power applications (companies include Scannsonic, Trumpf, Kugler and Arnold). Special coatings provide a very high degree of reflectivity at 1 micrometer wavelength. Due to better cooling possibilities, these optics have the advantage of being more stable regarding the heating effects and not being influenced by the alteration of the shape or optical properties during operation, thus providing very stable beam characteristics.

Temporary stability of the focus position during the welding operation is one (but not the only) requirement for the design of the optics as a part of the welding head. Compactness, along with protectiveness against different kinds of emissions from the welding process in the form of evaporation products, spatters, etc., are two additional factors to be considered in determining the sometimes contradictory requirements for the design. Since laser hybrid welding implies a number of additional parameters to be adjusted, such as vertical and transverse distances between focal spot and wire tip of a GMA torch and between the spatial orientation of the GMA torch and the laser axis, the requirements for welding head design become even more complex.

One of the most important properties can be referred to as 'flexibility' which includes several criteria:

- the dimensions and weight of the welding head;
- the capacity to change and adjust the aforementioned geometrical parameters;
- in certain cases, especially for 3D welds, the existence of an integrated mechanical axis allowing precise and fast reorientation of the whole system and its alignment according to the weld trajectory.

An interesting solution to the last requirement has been provided by the company Cloos, who have designed a welding head with an additional rotational axis integrated in the optical head. This allows circular motion of the GMA torch around the optical axis of the focusing module (Paul, 2010).

One of the disadvantages of autogenous laser welding is its very high propensity to gaps. Successful application of laser welding requires in most cases that there be no gap. This can be achieved by precise preparation of the edges to be welded and careful application of the mechanical fixtures. During the assembly of heavy sections for joining, for example, fitting of the joint edges exactly together is often difficult, and poor fitting results in gaps occurring between the parts to be welded. Laser hybrid welding can improve the situation due to the application of filler material. Further advancements can be obtained by the application of special scanning modules that provide laser beam weaving transversely to the welding direction, with amplitudes up to 1 mm and frequencies up to 1 kHz. An example of such a system is the scanner module made by ILV. This system is based on the principle of a motor-driven mirror, which deflects the laser beam at a certain angle according to its initial optical axis. In contrast to the galvanometric systems, it uses a massive copper water-cooled mirror that is able to sustain extremely high laser beam powers up to 20 kW and, at the same time, allow a frequency of oscillation up to 1 kHz. Gook *et al.* (2010) have applied this system to improve ability to bridge the gap for single-run laser hybrid welding of 16 mm thick structural steel. It was found that when using 16 mm structural steel, gap sizes of above 0.3 mm could result in weld imperfections such as lack of side wall fusion, as shown on the left-hand side of Fig. 19.1. When the scanner optics module is applied at a frequency of 200 Hz and at an amplitude below 1 mm, the tolerable gap size can be effectively increased to up to 0.7 mm, as shown on the right-hand side of the figure. Another advantage of the application of the scanning system in this study was an improved formation of the root side of the weld, especially in the case of orbital welds produced in an overhead position.

Another important element of laser welding systems is an appropriate seam-tracking sensor, which becomes especially significant in the case of a fully automatic welding process. As the parts to be welded increase

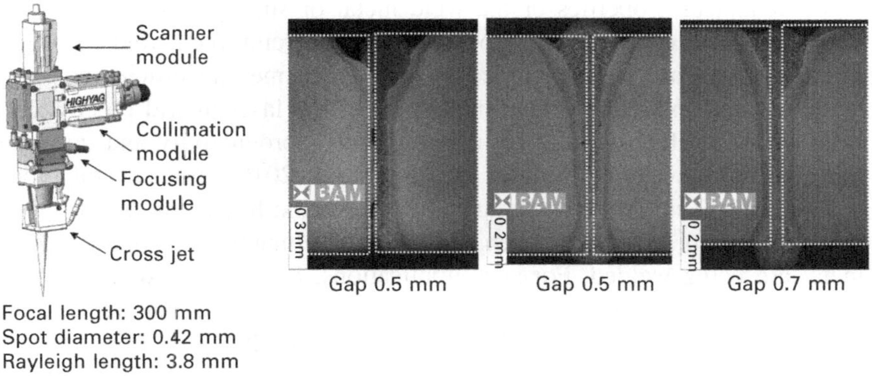

19.1 Optics with scanner module used in experiments: weld with 0.5 mm gap (left) without scanning compared to 0.5 mm gap (middle) and 0.7 mm gap (right) both with scan width of 0.7 mm at 200 Hz, 16 mm thick plates.

in thickness and the weld trajectory becomes more complex, the risk of misalignment and therefore lack of fusion due to improper laser head positioning increases. There are a large number of varied sensor solutions to this problem. Such systems can be either integrated into the optical system, such as CCD cameras, or they can be mounted externally, for example when using tactile sensors, laser triangulation-based sensors, etc. The latter group can be used not only for seam tracking but also for precise measurement of some of the quality-relevant parameters before and after welding, such as gap size, edge misalignment and weld profile. On-line systems that determine the misalignment during welding and allow correction of the welding head position have significant advantages in comparison to offline systems that program the motion of robots or other manipulators in the stage prior to welding. Stability and insensitivity to the intensive light emission from the welding zone are important requirements for online systems, especially in the case of laser hybrid welding processes. From this point of view, optical-based sensors are at a disadvantage compared to other types of sensor. For laser hybrid welding of T-joints, it is possible to use the arc as a seam-tracking sensor by measuring the arc voltage signal, as reported by Reisgen and Olschock (2008).

New developments in lasers and optics are not the only recent achievements in laser hybrid technology. Progress in the development and application of new equipment for arc welding has also had an impact on laser hybrid welding. Depending on the requirements of particular applications, and first of all on the metallurgical peculiarities of the materials to be welded, different modifications of the classical laser-GMA hybrid process can have certain advantages. In various applications it is necessary to increase the heat input of the laser hybrid process in order to enhance the metallurgical and mechanical properties of the weld metal or simply to increase metal deposition, i.e. to fill the rest groove for smooth reinforced welds. In such a situation, an arc welding process with a high melting efficiency (such as GMA-tandem welding) in combination with a laser hybrid process can be used as an effective tool to increase overall productivity and improve the quality of welding (Staufer and Egerland, 2010). For several other applications, it might be necessary to decrease the heat input during filler wire deposition, especially when crack-sensitive materials or thinner sheet metal are being welded. For such applications, Reisgen has demonstrated a successful combination of laser welding with low heat input arc welding processes such as cold arc or CMT (Reisgen *et al.*, 2011).

19.3 Examples of applications

The shipbuilding industry was a pioneer in the application of laser hybrid welding. Demand for new ships of lightweight design required the application of thinner steel sheets for their construction. This resulted in much higher levels of welding deformation and consequently in a high amount of rework in the pre-production stage. The main requirement for the introduction of a new welding technology was to significantly decrease welding distortions occurring during the manufacture of the stiffened panels, which require immense effort straightening procedures, as well as high costs (Roland and Lembeck, 2001). At the same time, the new welding technology needed to provide an appropriate quality at gap tolerances typical for the 20 m long sections. Laser hybrid welding provided the best prerequisites to fulfil these requirements, owing to its three times lower heat input compared to conventional arc welding and to its ability to bridge gaps of up to 1 mm in the thickness range relevant to shipbuilding. As a result of the substitution of laser hybrid welding for the conventional SAW in the production of stiffened steel panels at Meyer Werft in 2001, the productivity of this line was doubled and the amount of necessary straightening work reduced by 60%. Low levels of weld distortion allowed the manufacturers to increase the panel size from 10 m × 10 m to 20 m × 20 m. This resulted in the complete reorganisation of the production line, with four 12 kW CO_2 lasers executing all welding operations in stiffened panel manufacture. Today 20 m × 20 m large panels can be welded in 6 hours and a total length of 3 km of laser hybrid welds is produced per day. The total length of laser-hybrid welds in the construction of a 300 m long cruise ship is estimated to be about 400 km. The first generation of laser hybrid equipment used 12 kW CO_2 lasers in combination with a GMAW process. Recently the company purchased two 10 kW disk lasers, which will be integrated into the new production line for manufacturing parts of sections. The flexibility of the disk lasers due to the fibre delivery of the beam allows their use for more complex 3D welds, including welding in different positions.

Since 2008, a 10 kW fibre laser has been in use at the Fincantieri shipyard in Monfalcone, Italy. The 600 m long production line at Fincantieri was designed and realised by the company IMG (Seyffarth, 2010) for nearly deformation-free welding of 36 m long and 16 m wide panels with thicknesses in the range 4 mm to 20 mm. The production line also includes a highly precise milling machine for preparing the edges of the 16 m long steel sheets, directly before clamping them in the hydraulic fixture for welding. This provides a high level of precision with zero gap between the welded parts, minimising the risk of weld imperfections. For the first time an additional trailing tandem GMAW process was used in combination with laser hybrid welding, making a capping run. This combination enables the thickness of the welded sheets to be increased to up to 20 mm by preparing a Y-groove with a root face size adapted for 10 kW laser hybrid root run welding.

The compact design and high efficiency of high power fibre lasers provide the optimal conditions for their use in field pipeline construction. The first patented CO_2 laser system for offshore pipeline welding was tested in 1998. More recent solutions (Begg *et al.*, 2008; Harris and Norfolk, 2008; Neubert and Keitel, 2010) are based on the use of high power fibre or disk lasers in combination with one or several arc processes, which significantly contribute to the productivity and cost effectiveness of this welding technology. The idea of using lasers for pipeline welding is not a new one. Several research studies have been published since 2000 where the authors proposed CO_2 (Gainand *et al.*, 2000) or Nd:YAG (Fujinaga *et al.*, 2005) laser applications to establish this kind of welding technology for orbital pipeline welding. The weld thickness was mostly restricted to the value of 10 mm by producing either a single run or a root run for multi-run welding of line pipes of C-Mn steels. One reason for such a restriction was the power of the laser systems available at that time: either 12 kW CO_2 or 4.4 kW Nd:YAG. With the availability of high power solid-state laser systems like fibre and disk lasers, new possibilities have been opened up for establishing a single-run welding process for thicknesses up to 20 mm, as has recently been reported by Rethmeier *et al.* (2009).

The new laser tailored design of welded joints can result in significant savings on welding consumables. The higher welding speed of the laser beam welding process compared with conventional welding would result in a reduction of the production cycle time and in the long run in a reduction in the number of welding stations working in pipeline construction, while providing the same productivity as a conventional system. Direct comparison of three welding methods shows that the consumption of consumables can be reduced by a factor of five by the application of laser-hybrid welding to a 20 mm thick pipe compared with gas metal arc (GMA)-tandem welding, which is considered to be the most productive welding technology for pipelines, by a factor of ten compared with manual metal arc welding (MMAW), which is the most popular welding technology used today in pipeline construction. The difference is even more dramatic when studying the production time for one weld. For example, producing a complete weld will take about 1.5 min using laser hybrid welding, about 12 min using GMA-tandem welding and about 190 min using manual metal arc welding. At the same time, the energy consumption when using laser hybrid welding, despite the lower energy plug efficiency of modern solid-state lasers (approximately 30%), is comparable to that of GMA-tandem welding and an order of magnitude lower than that of manual metal arc welding. This is the reason why laser-hybrid welding can be considered a sustainable method of production, preserving resources and contributing to public health protection. On the other hand, in order to implement this technology in industry, it is necessary to meet the requirements of quality standards for the prevention of weld imperfections or at least for their reduction to an acceptable level, as well as for the achievement of high mechanical weld properties.

The possible applications of laser hybrid technology using high-power solid-state lasers for welding of transmission pipelines have been studied recently by several research teams. Different approaches have been suggested to suitably realise the orbital welding process for application in a field environment.

Taking advantage of the potential of modern lasers with output powers of 15 kW and more is an attractive option for welding of long distance transmission pipelines designed for high pressure operation and with a wall thickness of up to 42 mm. In this case, the economic impact of implementing the new welding technology becomes most remarkable, owing to the significant welding depths it is able to reach.

The first progress in this area was achieved by demonstration of the half orbital welding process, operating the fibre laser with an output power of 20 kW, for joining of pipe segments with a 36″ outer diameter and 16 mm wall thickness. Orbital welding equipment, as shown in Fig. 19.2, has been used by the BAM Federal Institute for Materials Research and Testing in Berlin to perform these experiments. It was shown that the laboratory-developed welding process can fit the requirements of field pipeline welding, although some modification of the equipment has to be carried out to suit field conditions. In general, the laser hybrid process meets the tolerances originating from misalignment and irregularity of the pipelines.

19.4 Quality issues

Successful application of laser-hybrid welding in the mass production of structural components requires, on the one hand, very high levels of integrity and safety of welded joints and, on the other, a high degree of confidence from customers and end product users.

19.2 Orbital welding equipment with laser arc hybrid weld head for large diameter pipes at BAM.

The acceptance of this technology by the potential user can be supported by development of new national and international standards and technical regulations based on long-term experience of the industrial users of this technology and on a high level of R&D expertise. The first international document dealing with quality standards of laser beam welding was developed under the European research project 'Laser Welding in Ship Construction' in the form of 'Guidelines for the Approval of CO_2-Laser Welding' and adopted by five European Classification Societies such as Germanischer Lloyd and Det Norske Veritas among others (Guidelines, 1996). Around the year 2000 a German national research project resulted in a number of EN standards specifying the quality requirements for laser welding applications. Recently, several classification societies have developed approval guidelines for laser hybrid welding based on numerous applications of this technology in different shipyards and on previous positive experience in the application of laser beam welding standards (DNV, 2006; ClassNK, 2009).

Since 2008, a working group of international experts in the field of laser hybrid welding (ISO/TC044/SC10/WG09) has been involved in the development of three new interrelated standards for laser hybrid welding:

- ISO DIS 12932 Welding – Laser-arc hybrid welding – Quality levels for imperfections.

- ISO DIS 15609-6: 2010 Specification and qualification of welding procedures for metallic materials – Welding procedure specification – Part 6: Laser beam arc hybrid welding.
- ISO DIS 15614-14 Specification and qualification of welding procedures for metallic materials – Welding procedure test – Part 14: Laser-arc hybrid welding of steels, nickel and nickel alloys.

These three standards are largely based on similar existing standards for arc welding and laser welding.

The problems of solidification flaw formation, which typically occurs in laser beam welding of heavy structural steel sections as a consequence of extremely directed solidification, do not seem to be an issue for laser hybrid welding. They are not evident when welding low carbon steels of up to 15–20 mm thickness, which require a certain degree of purity in terms of their chemical compositions. The critical values of carbon, sulphur and phosphor for laser welds have previously been determined (Guidelines, 1996). It is assumed that due to the higher heat input in the case of laser hybrid welding, the critical amounts of sulphur and phosphor are higher (Laitinen *et al.*, 2003), but direct experimental evidence has not yet been found. Another important factor to be considered as having an impact on solidification flaw formation is the mechanical restraint conditions affecting the stress and strain redistribution near the solidification zone. Unfavourable *local* restraint factors may originate from the design of the welded joint, the geometry of the welded parts or the external mechanical clamping. In Gebhardt *et al.* (2010), experimental evidence of such influence on the amount of solidification cracks was provided for the case of full penetration laser beam welding of 16 mm S690 steel. The experiments were conducted using specimens fixed in a specially designed rigid frame allowing gradual three-step variation of the restraint intensity by external fixture modification. The results of the restrained specimens were compared with unrestrained specimens freely shrinking upon cooling. The evaluation of the results showed a direct dependency between the number and the total length of solidification cracks appearing in the restrained specimens. In Fig. 19.3 the total crack length is represented as a function of the applied restraint intensity, i.e. the stiffness of the construction. As can be seen from the figure, the highest number of cracks during welding appears at the highest restraint intensity that could be realised in the experiments, whereas no cracks occurred when the unrestrained specimens were welded under the same conditions.

Another point of concern relates to the extremely short temperature cycles of laser hybrid structural steel welding. The typical cooling conditions in the heat-affected zone (HAZ) correspond to the heating rates of several thousands and to the cooling rates of several hundreds of Kelvin per second, which correlates with a typical characteristic cooling time between 800°C and 500°C ($t_{8/5}$) of less than one second. This results in most cases in brittle

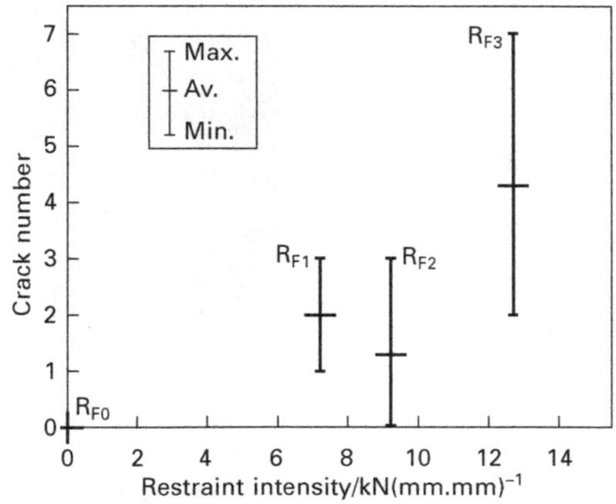

(a)

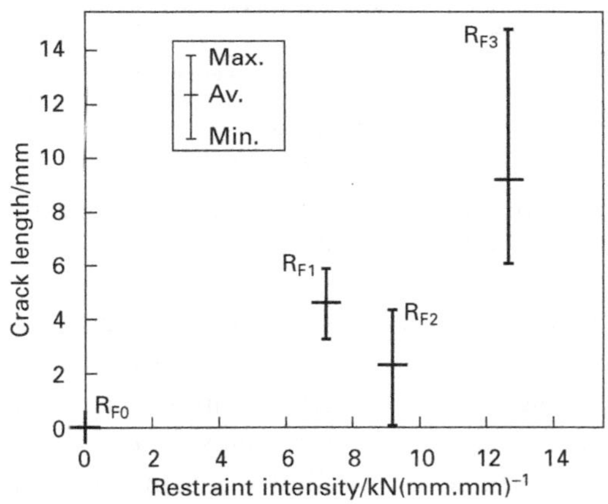

(b)

19.3 Increase of hot crack susceptibility for S690 with increasing restraint intensity.

martensitic weld metal and HAZ microstructures of steels, even for those which are typically considered to have good weldability properties. Even for steels with a reasonable amount of carbon at 0.18 wt%, this would result in hardness profile peaks of above 400 HV as shown in Fig. 19.4. This is the reason why pre-heating or post-weld heat treatment might be necessary to reduce the HAZ hardness values for some steel grades. As a consequence of the fully martensitic structure, reduced impact toughness values, especially at deep temperatures, may be observed in Charpy tests of laser hybrid welds. In contrast to the well-known properties of arc welded joints determined by critical Charpy test values of notched HAZ or fusion lines, the most critical zone for the notch position in laser hybrid welds is normally the weld centreline. An effective method of improving the weld metal toughness is to use a special filler wire containing grain refining

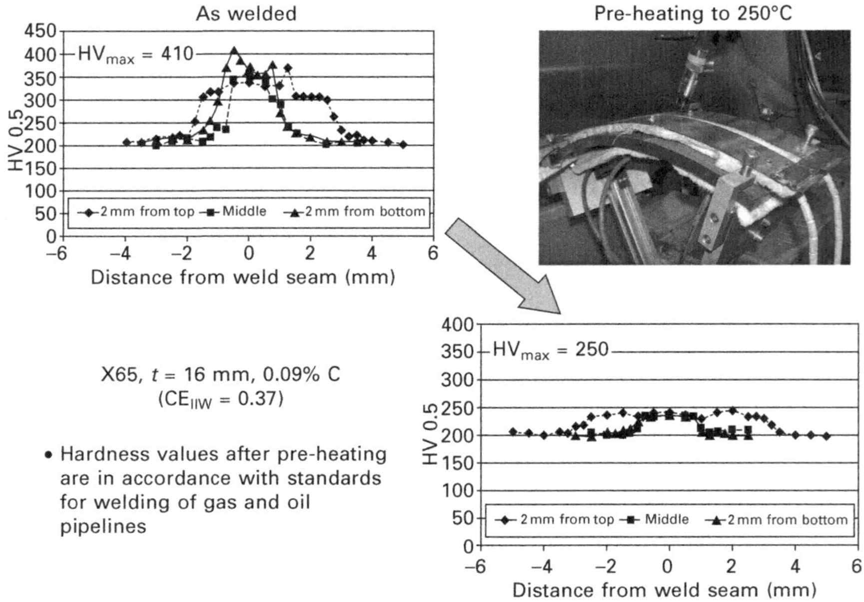

19.4 Hardness distribution in laser-hybrid welded X65 steel without (left) and with (right) pre-heating to 250°C.

elements such as Ti in combination with Ni. In several recent reports, the successful application of flux cored filler wire for laser hybrid welding was demonstrated. The achievable level of impact toughness was reported to be within or even exceeding the standard requirements. The thickness of the material can significantly determine the possibility of using the metallurgical potential for enhancing the mechanical weld metal properties, as it governs the through-thickness distribution of the elements supplied to the weld metal by filler material. It has been proven that the number of trace elements transported into the weld pool by the addition of filler wire during laser hybrid welding dramatically decreases over the depth of the weld in case of deep and narrow laser hybrid welds. The measured concentration of Ni (provided by the filler wire) at a depth of 10 mm changes by the factor of 3.8 compared with the superficial concentration (Gook *et al.*, 2012), as shown in Fig. 19.5. Other factors identified as having an effect on the distribution of alloying elements are, among others, the mutual orientation of the laser and GMA process (leading or trailing arc orientation) as well as the joint configuration (square butt joint, V-groove, Y-like joint, etc.).

19.5 Future trends

A major potential application of laser hybrid welding technology in shipbuilding is that of production re-engineering in the manufacture of complete ship

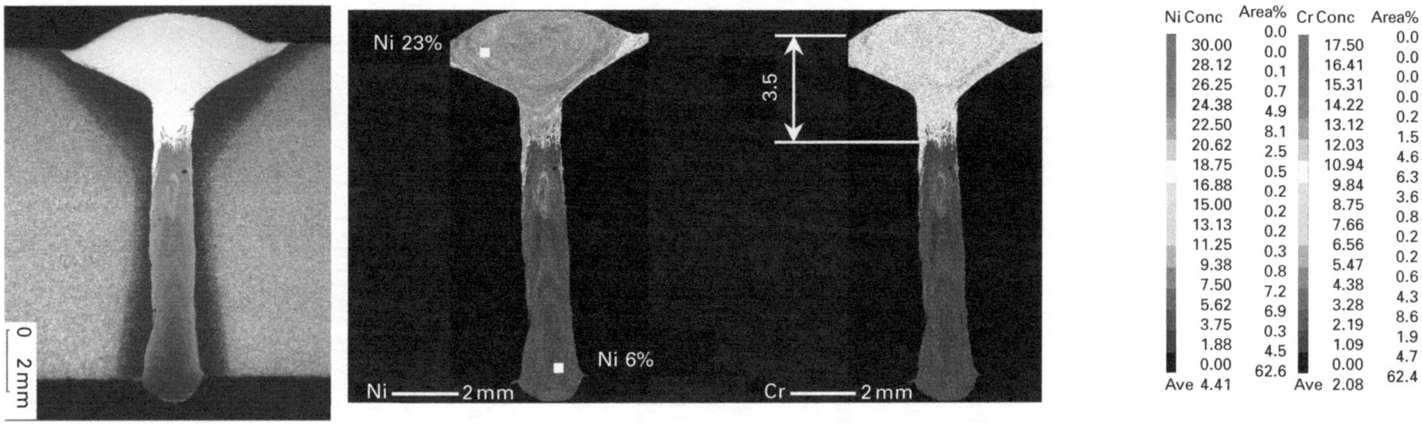

19.5 Element distribution (Ni, Cr) in laser-hybrid weld of S690 steel welded with Ni base filler wire (NiCr22Co12Mo9).

sections and blocks. The use of laser hybrid welding here would allow further reduction of welding deformations and a switch to near net shape fabrication. Another instance in which the use of laser hybrid welding could be advantageous would be fabrication of ship structures made of tailored blanks like those typical for the automotive industry. Using laser hybrid welding would allow the structural design of the sections to be fitted to the load requirements, resulting in high savings on material costs. On the other hand, the manufacture of hulls with higher plate thickness ranging at 20–25 mm (e.g., in container ships or LNG tankers) represents a further application field where the potential of laser hybrid welding technology could be deployed. A very important issue in this case would be the assessment of the weld for hot cracking resistance, as well as ensuring that the mechanical properties were according to the quality requirements.

Laser hybrid welding also has high economic potential in the oil and gas industry. Two sectors in which a breakthrough similar to that recently occurring in shipbuilding is possible in the near future are large diameter pipe fabrication for transmission pipelines and pipeline construction for on-shore and off-shore applications. Current research projects are focused on the possibility of applying laser hybrid welding to pipeline construction and similar applications in the field of oil and gas exploration. Most of these research programmes are either directly conducted by industrial companies or by an industrial consortium consisting of several companies. This technology could therefore be implemented in the near future. It is more likely that the first practical application will take place offshore, where the welding process would take place in less harsh environmental conditions and with fewer unpredictable factors, compared to those encountered in cross-country pipeline laying. On the other hand, the extremely high rental cost of pipe-laying barges, combined with the restricted number of welding stations available at those barges, means that the productivity of the welding process is a key factor in determining the overall cost of offshore pipeline construction. The economic impact of the application of laser hybrid arc welding would be remarkable if it were to become a competitor against the current high productivity welding methods used in pipe-laying barges. The latter enable welding of large diameter pipes with up to four double torch welding heads, operating at maximal welding speeds of up to 1.5 m/s (Serimax, 2012). The advantages of using the new welding technology could thus become especially significant through the realisation of single-run welding processes in 2G- or 5G-position.

The current applications of laser and laser hybrid welding in pipe mills (Outokumpu, 2007) are still restricted to stainless steel grades and pipes with relatively small wall thickness. Different research groups are working to implement laser hybrid welding technology for the manufacture of longitudinally welded large diameter pipes. In other words, progress can also be expected in this field, especially in connection with the implementation of high strength steel grades beyond X80. In the latter case, this technology not only provides an economical advantage compared with well-established conventional welding techniques, but has a real chance to compete with them in terms of the mechanical properties and reliability of the weld joints.

19.6 Sources of further information and advice

In the past, several reports have been published reflecting the state-of-the-art in technology and application of laser hybrid welding. Readers may find further information in the following works: (Dilthey *et al.*, 1999; Abe and Hayashi, 2002; Bagger and Olsen, 2005; Mahrle and Beyer, 2006; Olsen, 2009; Egerland *et al.*, 2012). A comprehensive report of physical mechanisms regulating the process of laser arc interaction during hybrid welding is given in the monograph (Krivtsun and Seyffarth 2002). A number of PhD theses covering certain aspects of laser hybrid welding have also been published recently, of which the following can be referred to as most important: (Neuenhahn, 1999; Wieschemann, 2001; Keller, 2003; Fuhrman, 2007; Olschok, 2008; Stelling, 2008; Lorenz, 2011; Kah, 2011).

19.6.1 Further reading

Abe N and Hayashi M (2002) Trends in laser arc combination welding methods. *Welding International*, 16, 94–98.

Bagger C and Olsen F (2005) Review of laser hybrid welding. *Journal of Laser Applications*, 17, 2–14.

Beyer E, Dilthey U, Imhoff R, Maier C, Neuenhahn J and Behler K (1994) New aspects in laser welding with increased efficiency. *Proc. of the Laser Materials Processing Conf. ICALEO'94*, Orlando, FL, 183–192.

Dilthey U, Lueder F and Wischemann A (1999) Technical and economical advantages by synergies in laser arc hybrid welding. *Welding in the World*, 43, 141–152.

Egerland S, Staufer H and Kammerhuber C (2012) Entwicklung und Anwendung des diodengepumpten Festkörperlaser-MSG Hybridschweißens zum Fügen von Wulstprofilen bei einseitigem Kehlnaht-Vollanschluss. *Schweiss- und Prüftechnik*, 4, 80–88.

Fuhrman C (2007) Laser-Lichtbogen-Hybridschweißen bis zu Blechdicken von 20 mm. PhD Thesis, RWTH Aachen.

Ishide T, Tsubota S and Watanabe M (2003) Latest MIG, TIG Arc-YAG laser hybrid welding systems for various welding products. *Welding International*, 17, 775–780.

Kah P (2011) Usability of Laser-Arc-Hybrid Welding Processes in Industrial Applications. PhD Thesis, Lappeenranta University of Technology.

Keller H. (2003) CO_2-Laserstrahl-MSG-Hybridschweißen von Baustählen im Blechdickenbereich von 12 bis 25 mm. PhD Thesis, RWTH-Aachen.

Krivtsun I and Seyffarth P (2002) *Laser-Arc Processes and Their Applications in Welding and Material Treatment.* Taylor and Francis, London.

Lorenz S (2011) Zur Qualifikation moderner Fülldraht- und Massivdrahtelektroden für das Laserstrahl-MSG-Hybridschweißen. PhD Thesis, University Magdeburg.

Mahrle A and Beyer E (2006) Hybrid laser beam welding – classification, characteristics and applications. *Journal of Laser Applications*, 18, 169–178.

Matsuda , Utsumi A, Katsumura M and Hamasaki M (1988) TIG or MIG arc augmented laser welding of thick mild steel plate. *Joining and Materials*, 1, 31–34.

Neuenhahn J-C (1999) Hybridschweißen als Kopplung von CO_2 Hochleistungslasern mit Lichtbogenschweißverfahren. PhD Thesis, RWTH-Aachen.

Olschok S (2008) Laserstrahl-Lichtbogen Hybridschweißen von Stahl im Dickblechbereich. PhD Thesis, RWTH Aachen.

Olsen F (2009) *Hybrid Laser Arc Welding.* Woodhead Publishing, Cambridge.

Stelling K (2008) Laserstrahl-Plasma-Hybridschweißen austenitischer Stähle. PhD Thesis, TU Berlin.

Wieschemann A (2001) Entwicklung des Hybrid- und Hydraschweißverfahren am Beispiel des Schiffbaus. PhD Thesis, RWTH Aachen.

19.7 References

Begg D, Benyon G, Hansen E, Deflco J and Light K (2008) Development of a hybrid laser arc welding system for pipeline construction. *Proc . 7th Int. Pipeline Conf.* Alberta, Canada, September–October.

ClassNK (2009) Guidelines on laser beam arc hybrid welding, Nippon Kaiji Koyokai.

DNV (2006) *Qualification and Approval of Hybrid Laser-Arc Welding in Shipbuilding*, Det Norske Veritas.

Eboo M, Steen W M and Clarke J (1978) Arc-augmented laser welding, Advances in Welding Processes, in *Proc. of the 4th Int. Conf.*, Harrogate, Paper 17, 257–265.

Fujinaga S, Ohashi R, Urakami T, Katayama S and Matsunawa A (2005) Development of an all-position YAG laser butt welding process with addition of filler wire. *Welding Journal*, 19, 441–446.

Gainand Y, Mas J P, Jansen J P, Coiffier J C, Dupont J C and Vauthier C (2000) Laser orbital welding applied to offshore pipeline construction. *Pipeline technology* Vol. II Elsevier Science, Amsterdam, 327–343.

Gebhardt M, Quiroz V, Gumenyuk A and Rethmeier M (2010) Restraint effects on stresses and strains in single-run high power laser beam welding of thick plates. In H Cerjak and N Enzinger (eds) *Mathematical Modelling of Weld Phenomena 9*, Verlag TU, Graz, 1011–1033.

Gook S, Gumenyuk A and Rethmeier M (2010) Orbital laser-hybrid welding of pipelines using a 20 kW fibre laser. *Proc. 5th Int. Conf. Laser in Manufacturing*, Munich, Germany, 65–70.

Gook S, Gumenyuk A and Rethmeier M (2012) Laser beam GMA hybrid welding on high strength pipeline steel for large diameter pipe production. Final Report FOSTA-Project P822, Düsseldorf.

Guidelines (1996) Guidelines for the approval of CO_2-laser welding. Laser welding in ship hull construction. Classification Societies.

Harris ID and Norfolk MI (2008) Hybrid laser/gas metal arc welding of high strength steel gas transmission pipelines. *Proc. 7th Int. Pipeline Conf.* Alberta, Canada, September–October.

Havrilla D and Brockmann R (2009) Third generation of disk lasers. *Laser Technik Journal*, 6, 26–31.

Laitinen R, Kujala P, Remes H and Nielsen S E (2003) CO_2-laser MAG weldability of laser cutting LASER RAEX steels, hull structural steel grade A and high strength formable steel OPTIM RAEX 700 MC. In Halmøy, E., *Proc. 9th Conf. on Laser Materials Processing in the Nordic Countries*. Trondheim, Norway. Trondheim: Norwegian University of Technology, 51–62.

Möller R and Koczera S (2003) Shipyard uses laser-GMA hybrid welding to achieve one-sided welding. *The Fabricator*, November.

Neubert J and Keitel S (2010) Influence of tolerances to weld formation and quality of laser-GMA-hybrid girth welded pipe joints. *Welding in the World*, 55, 50–57.

Outokumpu (2007) New investments provide increased capacity in process pipes. *Tubular News*, Issue 1, August, 7.

Paul C (2010) Laser beam GMA hybrid welding of large plate thicknesses at industrial application. *Proc. 8th Int. Conf on Beam Technology*, Halle (Saale), Germany, 95–99.

Reisgen U and Olschock S (2008) Using of a robot for laserGMA-hybrid welding for application in shipbuilding. *Proc. 9th Conf. Welding in Shipbuilding Civil Engineering*, Hamburg, 32–45.

Reisgen U, Olschock S and Mavany M (2011) Hybrid welding with controlled short arcs. *Proc. 30th Int. Cong. on Applications of Lasers & Electro–Optics ICALEO*, Orlando, FL, 587–591.

Reithemeyer D, Seefeld T, Vollertsen F and Bergmann J P (2009) Influences on the laser induced focus shift in high power fibre laser welding. *Laser in Manufacturing 2009, Proc. 5th Int. Conf.* Munich, Germany, June, pp. 293–298.

Rethmeier M, Gook S, Lammers M and Gumenyuk A (2009) Laser-hybrid welding of thick plates up to 32 mm using a 20 kW fibre laser. *Transactions of Japan Welding Research Institute*, 27, 74–79.

Roland F and Lembeck H (2001) Laser beam welding in shipbuilding. *Proc 7th Int. Aachen Welding Conference*, Aachen, 463–475.

Serimax (2012) Technical Datasheet SATURNAX 09. Available at: www.serimax.com (accessed 18 May 2012).

Seyffarth P (2010) Laser hybrid welding in shipbuilding using high performance fibre laser. *Proc. 8th Int. Conf on Beam Technology*, Halle (Saale), 84–89.

Staufer H and Egerland S (2010) Three arc laser hybrid gas shielded metal arc welding. *63rd IIW Annual Assembly*, Istanbul, July, IIW-Document IV-1028-10.

20
Developments in modelling and simulation of laser and hybrid laser welding

S-J. NA and W-I. CHO, KAIST, Korea

DOI: 10.1533/9780857098771.3.522

Abstract: Laser and hybrid welding processes involve very complex transport phenomena and thus rather limited modelling and simulation are available yet. With the advances in computing power and efficient computation techniques, however, it is expected that comprehensive modelling and simulation can be achieved and used to determine the optimal welding parameters resulting in minimum weld defects. This chapter presents the research trends and issues on modelling and simulation of laser and hybrid welding.

Key wards: laser welding, hybrid welding, arc welding, modelling, simulation.

20.1 Introduction: the role of modelling in laser welding

The CO_2 laser and the lamp-pumped Nd:YAG laser have been commonly used in industrial welding applications for many years. High power welding is realized mainly with a CO_2 laser but its 10.6 μm wavelength means it cannot be delivered via fibre-optic cable. The lamp-pumped Nd:YAG laser is applied in various fields because it enables the beam to be delivered by fibre; however, it is limited to relatively low power welding and has the disadvantages of low efficiency and low beam quality due to problems with elevated temperatures, thermal lensing and depolarization loss. Fibre lasers and disk lasers have recently been commercialized; by using a diode-pumping method as well as a fine fibre or a thin disk (as the gain medium), fibre and disk lasers can achieve a high level of efficiency and a high beam quality at high power levels. Accordingly, besides being used for microprocessing, fibre and disk lasers can now be used as an alternative to CO_2 lasers in multi-kW applications for a thick plate range. In this case, Yb:YAG is used as the gain medium and the beam can therefore be delivered by fibre.

One of the goals of welding research is to determine optimal conditions by analysing the effects of the welding conditions from the perspective of the process, metallurgy and mechanics. The focus of such analysis should be the final weldment after cooling and the molten pool behaviour during the welding process, as is done for welding of zinc-coated steels where a well-directed control of the process has reduced the chaotic behaviour [1]. Also empirical methods like the Taguchi method can be used to optimize the laser welding process [2], but a high number of experiments are needed to get a deeper insight into the process characteristics. A better understanding of the latter could be helpful for determining the direct effect of driving forces on the molten pool under various welding conditions. Thus, the molten pool behaviour is an important research topic. The related studies can be divided into experimental and numerical approaches. The experimental approach can be used to observe the molten pool flow and surface shape, though environmental limitations of the strong light and high temperature limit the availability of precise information. The X-ray transmission *in-situ* observation method of the Joining and Welding Research Institute is the most developed way of observing the keyhole shape inside a molten pool [3]. However, the application of this method is limited because the equipment is expensive and bulky. Therefore, a numerical approach is favourable for comprehending the molten pool flow.

Rosenthal used a simple heat conduction model to estimate the fusion zone shape [4]; and heat conduction simulations were conducted until the early 1980s. Numerical simulations of the molten pool flow began in the mid-1980s for arc welding and in the late 1980s for laser welding. In laser welding, the recoil pressure forms a unique keyhole, and estimating the keyhole shape is a major research topic. In the early stages of laser welding simulation, however, the focus was on the conduction mode of laser welding and the simulation was similar to that of arc welding [5]. Depending on the aim of the simulation, the usage of a Goldak heat source model can be sufficient in laser welding simulation, e.g. in determination of the residual stresses and distortion [6]. For some special tasks, even more simplified models than finite element modelling (FEM) can be helpful (see e.g. [7]). On the other hand, other defects, like pores, demand more detailed description of the keyhole. To estimate the laser keyhole shape, Kaplan used an energy balance equation with a simple multiple reflection model [8]. Multiple reflections were considered in a conical shape with an average angle of the keyhole wall; thus, only rough predictions could be made of the keyhole shape. Since then, various numerical techniques have been used to track the free surface so that the precise keyhole shape can be calculated. Ki *et al.* [9] used a level-set method [10] to calculate the free surface of the keyhole for the purpose of realizing the multiple reflections inside the calculated keyhole shape. Lee *et al.* [11] used a volume-of-fluid (VOF) method [12] to understand the mechanism of keyhole instability. Cho and Na [13] also used the VOF method in conjunction with a ray-tracing algorithm to calculate the keyhole shape and realize the real-time multiple reflections. The same algorithm was used to estimate the results of the drilling process vis-à-vis the polarization of the laser beam [14] as well as the results of laser-arc hybrid welding [15].

In addition, a laser model based on the optical geometry of a laser system and the theoretical laser- and material-dependent value which affects the reflectivity in the simplified Fresnel's reflection model was recently used in simulations of the laser-arc hybrid welding process [16] and laser welding process [17].

In welding simulations, it is important to formulate the reliable models based on actual welding phenomena. However, practical welding involves complex multi-physical phenomena, such as heat transfer, diffusion and electromagnetism, as well as solid, liquid, gas and plasma phases. Thus, researchers have attempted to simplify the physical phenomena of welding through various assumptions. This chapter features the numerical simulation of molten pool dynamics in laser and hybrid welding.

20.2 Key issues in modelling laser welding processes

20.2.1 Laser-matter interaction

If light moves between media of differing refractive indices, both reflection and refraction (or absorption) of the light can be described by the Fresnel reflection model. The Fresnel reflection model determines the reflectivity of a material surface according to the polarization of the incidence beam. It has already been defined in previous reports and textbooks that the plane of incidence is a plane determined by two vectors of the incidence ray and surface normal. p-Polarization refers to the case where the polarization direction of the beam is parallel to the plane of incidence. On the other hand, s-polarization means that the polarization direction is perpendicular to the incidence plane. If a beam ray is irradiated vertically on the material surface, however, one cannot define whether it is s-polarization or p-polarization. Determination of the absorption rate of the material surface starts with dividing the electric field of the incoming beam ray, $\vec{E}_i$, into two components of the directional unit vector, as shown in Fig. 20.1 and as given in the following equations [18, 19]:

$$E_{i\parallel} = \vec{E}_i \cdot \hat{e}_{i\parallel},\ R_{i\perp} = \vec{E}_i \cdot \hat{e}_{i\perp} \qquad [20.1]$$

$$\vec{E}_i = E_{i\parallel}\hat{e}_{i\parallel} + E_{i\perp}\hat{e}_{i\perp} \qquad [20.2]$$

$$\hat{e}_{i\perp} = -\frac{\hat{s}_i \times \hat{n}}{\sin\theta},\ \hat{e}_{i\parallel} = \hat{e}_{i\perp} \times \hat{s}_i \qquad [20.3]$$

One of the directional components $\hat{e}_{i\parallel}$ is parallel to the plane of incidence, which is determined by the incoming ray's unit vector $\hat{s}_i$ and the negative surface normal $\hat{n}$. This component represents the p-polarization case. The

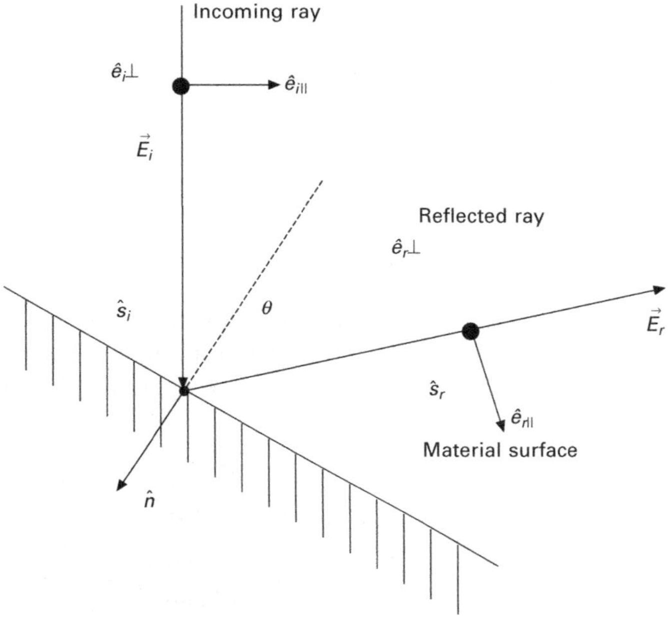

20.1 Schematic diagram of the reflection model.

other, denoted as $\hat{e}_{i\perp}$, is perpendicular to the plane of incidence, and results in s-polarization. Each direction is decided by the polar angle of incidence θ, as delineated in Eq. [20.3]. The reflected ray vector $\hat{s}_r$ lying on the plane can be derived by the following simple equation, and the electric field of the reflected beam ray also should be split into two directions according to a $\hat{e}_{r||}$ nd $\hat{e}_{r\perp}$ as described above:

$$\cos\theta = \hat{s}_i \cdot \hat{n} \qquad [20.4]$$

$$\hat{s}_r = -2\hat{n}\cos\theta + \hat{s}_i \qquad [20.5]$$

$$\hat{e}_{r\perp} = \hat{e}_{i\perp} \qquad [20.6]$$

$$\hat{e}_{r||} = -\hat{s}_r \times (\hat{e}_{i||} \times \hat{s}_i) \qquad [20.7]$$

$$\vec{E}_r = E_{r||}\hat{e}_{r||} + E_{r\perp}\hat{e}_{r\perp} \qquad [20.8]$$

$E_{r\perp}$ and $E_{r\perp}$ are calculated by the Fresnel's formula as follows:

$$E_{r\perp} = E_{i\perp}\frac{n_1\cos\theta - p + qi}{n_1\cos\theta + p - qi} = E_{i\perp}r_\perp \qquad [20.9]$$

$$E_{r||} = E_{i||}\frac{p - n_1\sin\theta\tan\theta + qi}{p + n_1\sin\theta\tan\theta - qi}r_\perp \qquad [20.10]$$

where p and q are expressed as:

$$p^2 = \frac{1}{2}[\{(n_2^2 - k_2^2 - n_1^2\sin^2\theta)^2 + 4n_2^2k_2^2\}^{1/2} + (n_2^2 - k_2^2 - n_1^2\sin^2\theta)] \quad [20.11]$$

$$q^2 = \frac{1}{2}[\{(n_2^2 - k_2^2 - n_1^2\sin^2\theta)^2 + 4n_2^2k_2^2\}^{1/2} - (n_2^2 - k_2^2 - n_1^2\sin^2\theta)] \quad [20.12]$$

In the above equations, n_1 is the refractive index of gas, n_2 and k_2 are the refractive and absorptive index of the material, respectively. The reflectivity c_r and the absorption rate of the material surface c_a are finally determined as follows:

$$c_r = \frac{E_{r\parallel}E_{r\parallel}^* + E_{r\perp}E_{r\perp}^*}{E_{i\parallel}E_{j\parallel}^* + E_{j\perp}E_{i\perp}^*} = \frac{1}{2}\left[\frac{(n_1\cos\theta - p)^2 + q^2}{(n_1\cos\theta + p)^2 + q^2} + \frac{(p - n_1\sin\theta\tan\theta)^2 + q^2}{(p + n_1\sin\theta\tan\theta)^2 + q^2}\frac{(n_1\cos\theta - p)^2 + q^2}{(n_1\cos\theta + p)^2 + q^2}\right] \quad [20.13]$$

$$c_a = 1 - c_r \quad [20.14]$$

where the asterisk superscript in the equation denotes a complex conjugate.

In Eq. [20.13], the reflectivity depends on the wavelength of the beam, the temperature of the surface, and the incident angle of the beam θ, because the complex refractive index of the material, $m = n_2 - ik_2$, varies with the wavelength of the beam and the temperature of the surface. The complex refractive index of the metal such as steel can be obtained by Drude theory. Because metal is generally a good conductor and free electrons move freely, the Lorentz model can have a special form: the Drude model with a vanishing spring constant leading to a resonance frequency of 0. Therefore, the complex refractive index can be obtained by solving the following equations:

$$n^2 - k^2 = 1 - \omega_p^2(\omega^2 + \tau^{-2}) \quad [20.15]$$

$$2nk = \frac{\omega_p^2}{\omega\tau(\omega^2 + \tau^{-2})} \quad [20.16]$$

where $\tau = \dfrac{m^*\sigma_0}{Ne^2} = \dfrac{1}{\gamma}$.

Here, ω_p is the plasma frequency (rad/s), ω is the angular frequency of the laser (rad/s), τ is the relaxation time of electrons (s), N is the number of free electrons per cubic centimetre, e is the electron electric charge (C), m* is the optical mass of the electron (kg), ε_0 is the permittivity of free space (F/m), σ_0 is the conductivity of material (Ω/m) and γ is the damping frequency (Hz). Based on the Fresnel reflection model and Drude theory, Fig. 20.2 shows the absorptivity according to the incident angle between the incoming ray and the surface normal for a CO_2 laser (10.6 μm) and a disk laser (1.03 μm).

The Fresnel reflection model of Eq. [20.13] can be simplified to Eq. [20.17] [20] by the Hagen–Rubens relation, $n_2 \cong k_2 \cong \sqrt{30\lambda\sigma_{dc}}$ (σ_{dc} is the *dc*-conductivity of the material), derived under the assumptions that the material is a good conductor and the wavelength of the beam is sufficiently long (infrared). In this equation, the reflectivity c_{rs} is a function of the incident angle of the beam and a laser- and material-dependent quantity, ε ($\varepsilon^2 \approx 1/n_2^2 \approx 1/k_2^2$) [21, 22]:

$$c_{rs} = \frac{1}{2}\left(\frac{1+(1-\varepsilon\cos\theta)^2}{1+(1+\varepsilon\cos\theta)^2} + \frac{\varepsilon^2 - 2\varepsilon\cos\theta + 2\cos^2\theta}{\varepsilon^2 + 2\varepsilon\cos\theta + 2\cos^2\theta}\right) \quad [20.17]$$

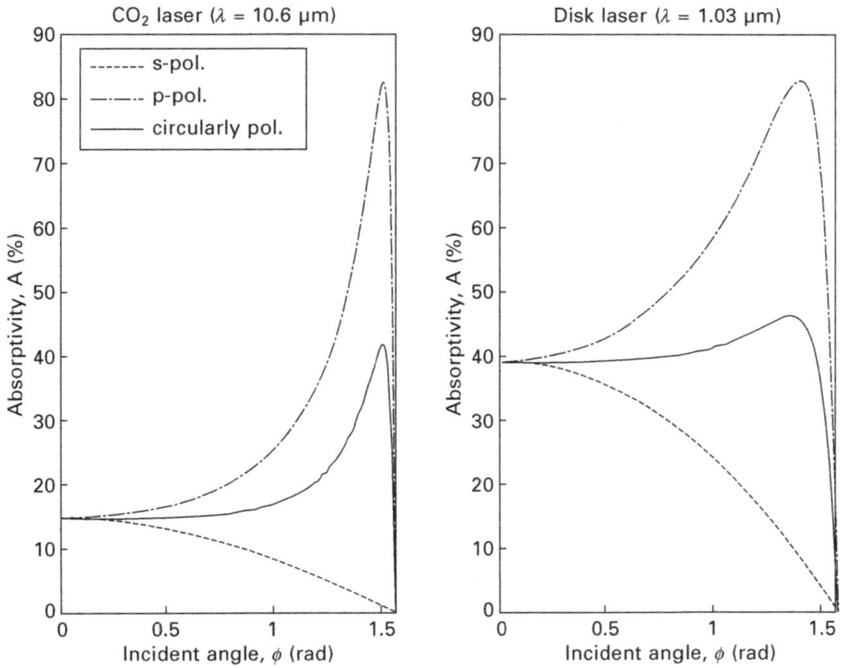

20.2 Absorptivity vs. incident angle based on the Fresnel reflection model and Drude theory.

where $\varepsilon^2 = \dfrac{2\varepsilon_2}{\varepsilon_1 + [\varepsilon_1^2 + (\sigma_{dc}/\omega\varepsilon_0)^2]^{1/2}}$.

Here, ε_1 and ε_2 denote the real parts of the dielectric constants of metal and plasma, respectively, and ε_0 denotes the permittivity of free space. Based on the simplified Fresnel reflection model and Hagen–Rubens relation, Fig. 20.3 shows the absorptivity according to the incident angle between the incoming ray and the surface normal for a CO_2 laser (10.6 μm) and a disk laser (1.03 μm).

20.2.2 Laser heat source model

In terms of energy transfer, the laser beam undergoes rather complex processes. However, in the simplest case, the absorbed laser beam power at the keyhole wall might be described by considering only the Fresnel absorption, using a Gaussian-like surface heat flux [8]. Figure 20.4 shows one important problem when the surface heat flux is used without consideration of an axisymmetric laser beam in nature. If bundle 1 and bundle 2 are symmetric, for example, the energies of bundle 1 and bundle 2 should be the same. However, because the projected area of the free surface in cell 1, A_1, is larger than

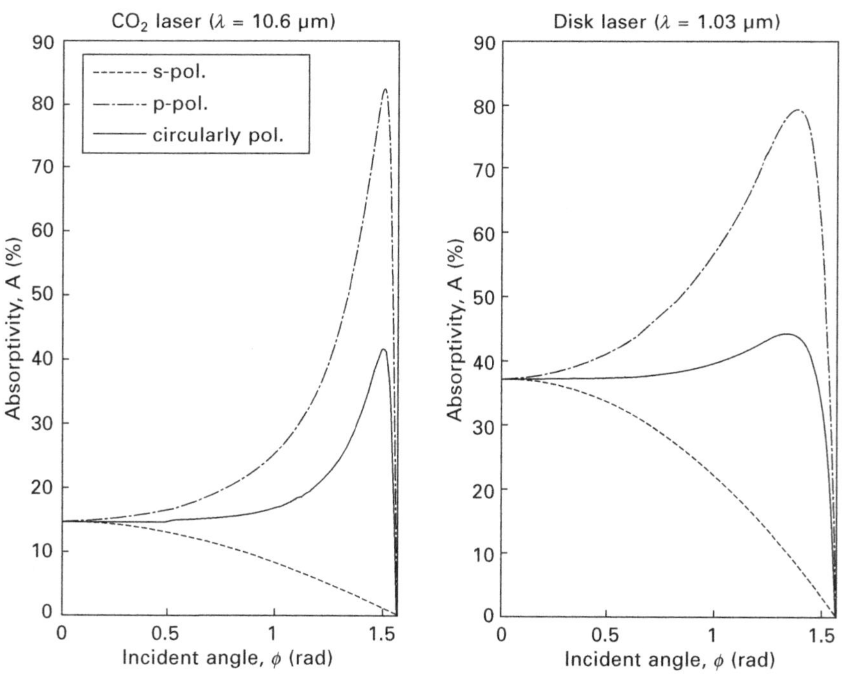

20.3 Absorptivity vs. incident angle based on the simplified Fresnel reflection model and Hagen–Rubens relation.

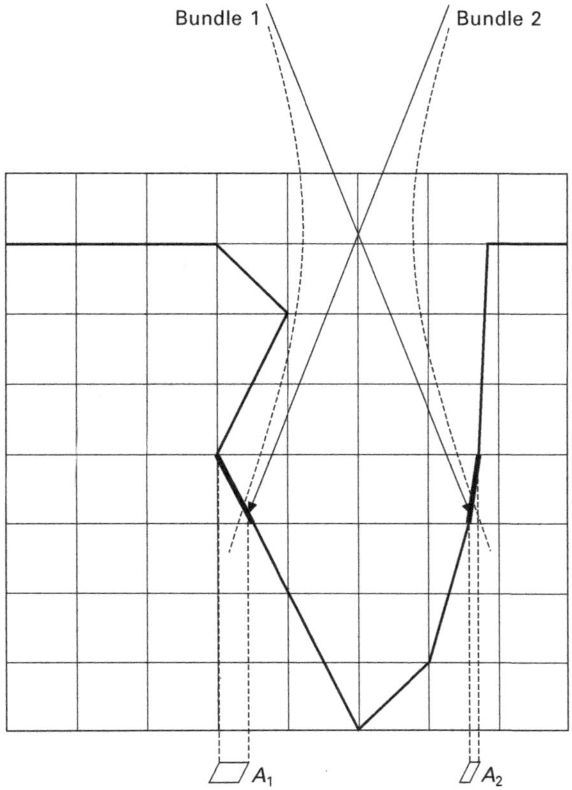

20.4 Schematic diagram of incoming two bundles of rays in the keyhole.

the projected area of the free surface in cell 2, A_2, the energy of bundle 1 is greater than that of bundle 2, a physically incorrect outcome. Therefore, it is necessary to define the bundles of rays whose energy distribution is Gaussian-like axisymmetric as well as whose number is finite but large enough to reproduce the real laser beam. To define the initial energies and directions of bundles of rays, the original mesh is divided at the focal plane into sub-meshes having smaller size, as schematically shown in Fig. 20.5. Then, the Gaussian-like energy distributions of the bundles at the focal plane can be predefined as given by:

$$e_{L0}(x, y, z_0) = \frac{2P_L}{\pi r_{f0}^2} \exp\left(-2\frac{x^2 + y^2}{r_{f0}^2}\right) \times A_{sm} \times \Delta t \qquad [20.18]$$

where $r_{f0} = \dfrac{2\lambda F M^2}{\pi}$.

Here, A_{sm} is the area of a sub-mesh and Δt is the time increment. After

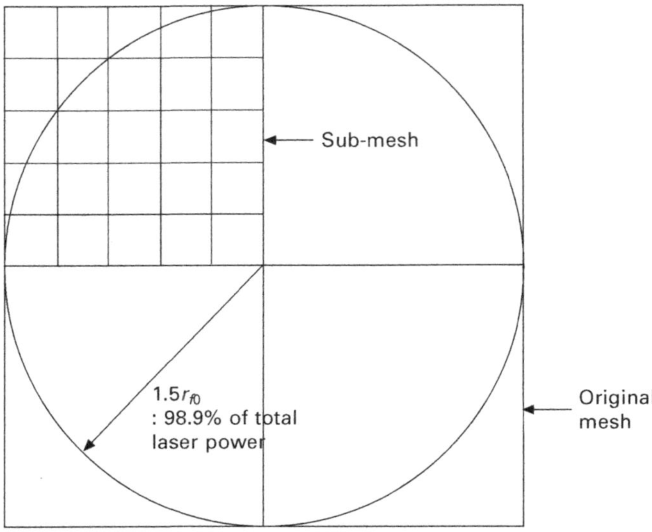

20.5 Schematic diagram of original mesh and sub-mesh.

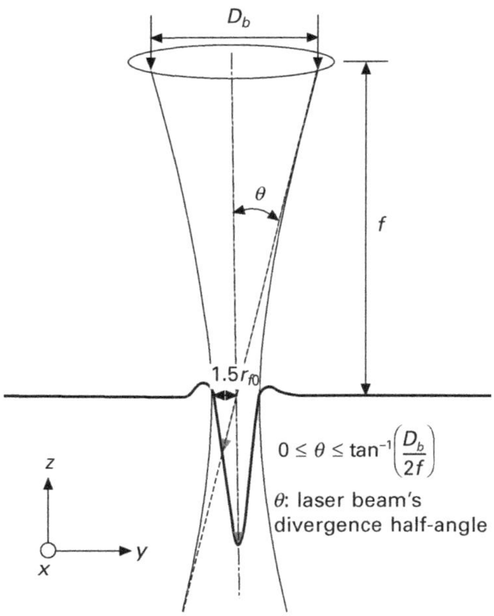

20.6 Schematic diagram of optical geometry of laser beam.

defining the initial energies of the bundles, the initial directions of the bundles can be defined according to the fact that bundles are limited by the optical geometry, which is related to the focal length f and the beam diameter on the optics D_b, as shown in Fig. 20.6. That is, if the divergence half-angle of the laser beam is limited by Eq. [20.19] and all bundles pass through the focal point, the initial directions of the bundles can be generated.

$$0 \le \theta \le \tan^{-1}\left(\frac{D_b}{2f}\right) \qquad [20.19]$$

Besides the focal plane, any plane perpendicular to the axis of the laser beam can be used to define the initial energy values of the rays. The previous work assumed that the beam was focused on an infinitesimal point that defines the initial directions of the rays, but this step is practically impossible in laser welding. In the practical laser processing, however, the calculation of the focal spot size and the corresponding laser beam ray directions are based on the diffraction parameter. Because the disk laser used in welding is not diffraction-limited, the beam radius at position z along the axis of the beam can be expressed as follows:

$$r_f(z) = r_{f0}\left[1 + \left(\frac{z - z_0}{z_r}\right)^2\right]^{1/2} \qquad [20.20]$$

where $z_r = \pm 2r_{f0}F$.

Equation [20.20] implies that the beam is diffracted in the manner shown in Fig. 20.7(a), where the direction vector $\vec{r}$ at any arbitrary point is calculated as shown in Fig. 20.7(b). In this calculation, the value of Δz is given as the value of the original mesh size.

20.2.3 Multiple-reflection model of keyhole

The portion of total energy transfer from laser beam to flat material surface is quite small compared to other welding methods. However, it increases dramatically after a keyhole is generated because of the multiple reflection

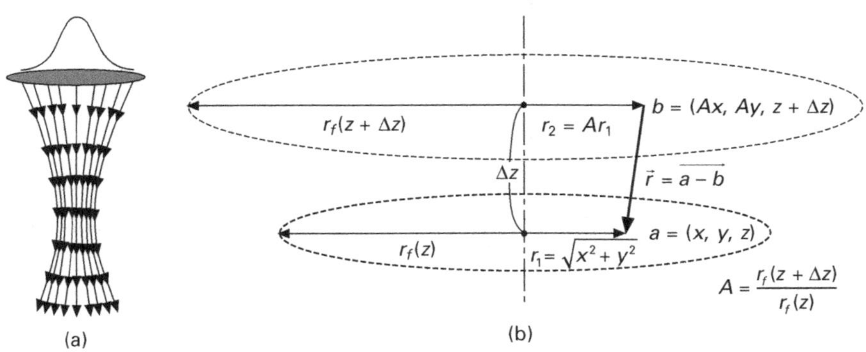

20.7 Schematics of (a) the diffracted laser beam and (b) the direction vector of a ray at an arbitrary point.

effect on the keyhole wall. Its basic concept is that a fixed portion of energy of the laser beam is absorbed at the irradiation point on the wall and the rest is reflected to the opposite wall and acts as a new energy beam. Most of the traveling rays in the keyhole are superposed at the keyhole bottom, which results in a deep penetration and an increase in the total energy absorption rate.

To implement the multiple reflections in analysis, a ray-tracing technique with the discrete grid cell system is proposed, as shown in Fig. 20.8. First of all, a ray vector is formed from the focal point to an arbitrary cell denoted as $\vec{I}$. Some portion of the energy contained in the ray is absorbed according to the Fresnel absorption model, which will be explained later, and the rest is delivered by the reflected ray $\vec{R}$ to the point P_r on the opposite side by the following simple vector equation:

$$\vec{R} = \vec{I} + 2(-\vec{I} \cdot \vec{N})\vec{N} \quad [20.21]$$

where $\vec{N}$ indicates the surface normal at the irradiated point P_i. In a discrete grid cell system, it is not possible to find an exact point consistent with a given $\vec{R}$ vector. Therefore, it is necessary to make a temporary vector

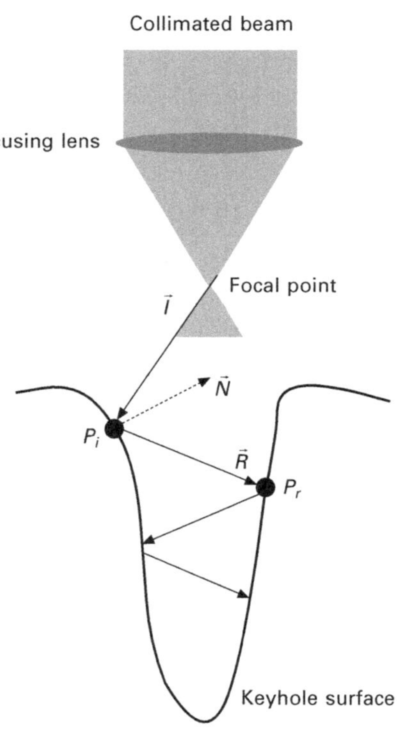

20.8 Schematic diagram of methodology of multiple reflection effect on the keyhole.

$\overrightarrow{P_r - P_i}$ in Fig. 20.8 and select a point which makes the vector most similar to $\vec{R}$. Its criterion is expressed in the following equation:

$$\frac{(\overrightarrow{P_r - P_i})\cdot\vec{R}}{|\overrightarrow{P_r - P_i}||\vec{R}|} \le \arctan\left(\frac{\Delta x}{|\overrightarrow{P_r - P_i}|}\right) \qquad [20.22]$$

In this equation, $\vec{P}_r$ indicates the point to be irradiated by the reflected ray and $\vec{P}_i$ the irradiated point by the incoming ray. Δx is the distance between grids along the x axis and Δy, Δz are set to be equal to Δx throughout the simulation. Proposed ray tracing is conducted for each ray incoming into the keyhole until there is no more point satisfying Eq. [20.22]. The whole of this process is expressed as a flowchart in Fig. 20.9. By using the suggested inverse tangent criterion, it is possible to trace the ray path even at the narrow bottom region in the keyhole as shown in Fig. 20.10 [13, 14].

In addition, the aforementioned algorithms for searching the free surfaces and the multiple-reflection model are modified as follows [16, 17]:

1. In the algorithm for searching the free surface cells, these are searched by comparing the fluid fraction F of 26 neighbouring cells with the critical fluid fraction F_{cr}. In the unmodified algorithm, however, many free surface cells are discarded at the narrow bottom region. In the modified algorithm, therefore, the searching algorithm for free surface

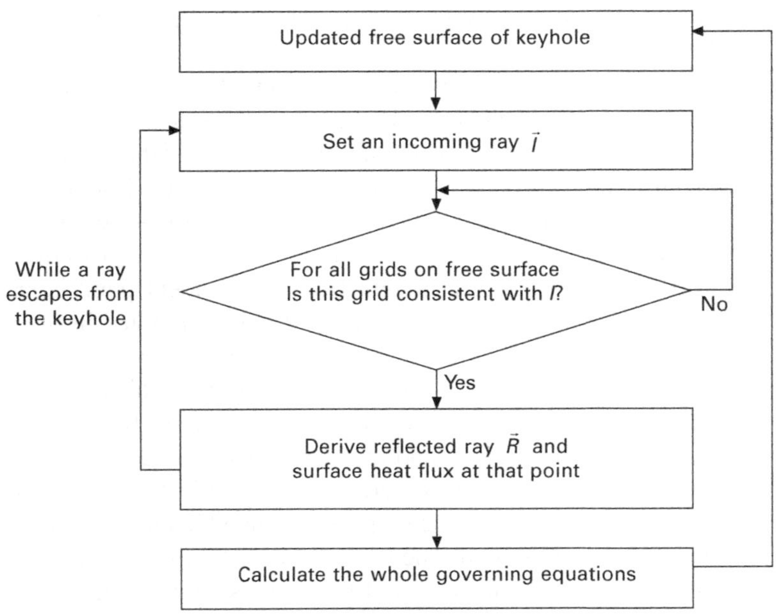

20.9 Flowchart of molten pool simulation process including multiple reflection iterations until a ray escapes from the keyhole.

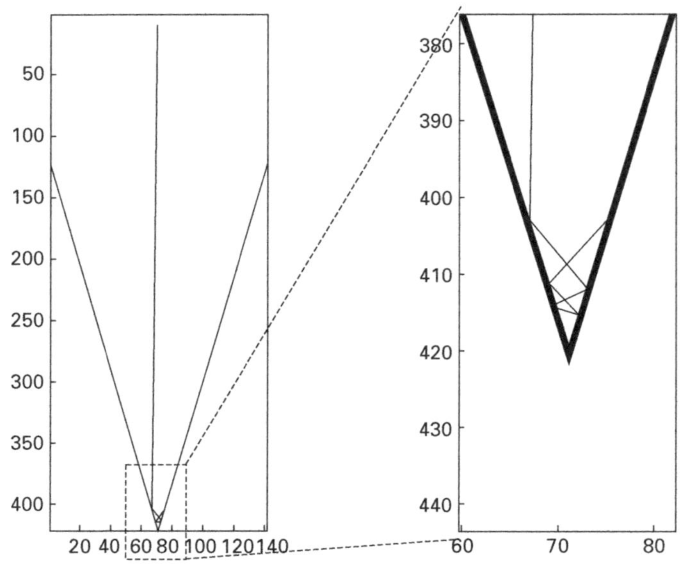

20.10 Example of ray-tracing technique proposed for an assumed keyhole. On the right is an enlarged picture of the keyhole bottom, which shows that the ray can reach to the bottom.

cells at the narrow keyhole region is improved by increasing the critical fluid fraction F_{cr}, from 10^{-6} to 0.5 and then by excluding the non-free surface cells indirectly by selecting the cell that has the shortest beam flight.

2. Figure 20.11 shows that the critical length, which is represented by the radius of the sphere, is decreased from Δl to $\Delta l \times \sqrt{3}/2$ to improve the search accuracy; that is, the second cell vanishes and the first cell, which is more reliable, is selected automatically. For the case where multiple cells are candidates, the selected cell can be changed by the scan direction in the solution domain. Thus, a new procedure wherein a cell having the shorter length from the incident point is selected by comparing the length of each cell and is incorporated in the algorithm to guarantee the consistency of multiple reflections.
3. As shown in Fig. 20.12, the large original cell makes it impossible to accurately select a cell where the ray is incident. In this example, the ray is captured at cell ① and therefore unable to reach cell ②, which is wrong if the original cell is used in multiple reflections. Greater accuracy can be attained in the search for the cell on which the reflection occurs if the original cell is subdivided and if the sub-cells with fluid are selected on the basis of a comparison of the lengths between the sub-cell centre and a position which is sufficiently far (in the surface normal direction) from the centre of the original cell. The selected sub-cells can be used to

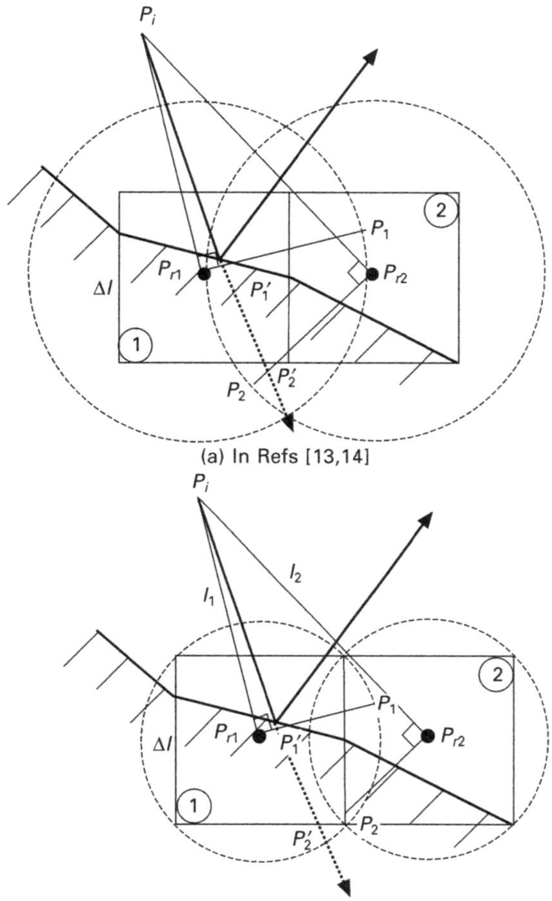

20.11 Schematic diagrams for the algorithm searching a free surface where reflection occurs.

test whether the ray is incident; cell ② can then be chosen for the next reflection cell in this example. As in other studies, the energy absorbed in a cell is redistributed by means of the weighting function to increase the computational efficiency.

20.2.4 Scattering model of keyhole

When an electromagnetic wave or a photon interacts with containing small particles, the radiative intensity may be changed by absorption and/or scattering. How much and in which direction a particle scatters an electromagnetic wave passing through its vicinity depends on (i) the shape of the particle, (ii) the material of the particle (i.e., the complex index of refraction, $m = n{-}ik$),

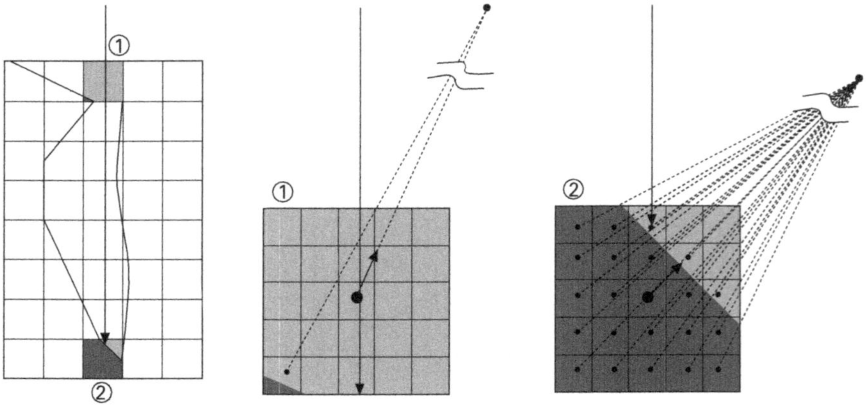

20.12 Search schematic for the cell on which the ray is incident.

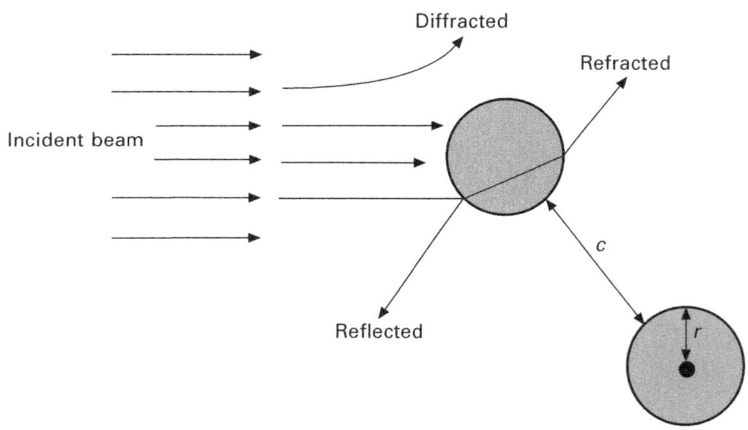

20.13 Interaction between electromagnetic waves and spherical particles [18].

(iii) its relative size, and (iv) the clearance between particles. In radiative analyses, the shape of particles is usually assumed to be spherical. These simplifying assumptions generally give excellent results.

In the following discussion, absorption and scattering by spherical particles, as shown in Fig. 20.13, are considered. An electromagnetic photon wave passing through the immediate vicinity of spherical particles will be absorbed or scattered. The scattering is due to three separate phenomena:

- diffraction (waves never come into contact with the particle, but their direction of propagation is altered by the presence of the particle),
- reflection by a particle (wave reflected from the surface of the sphere), and

- refraction in a particle (waves that penetrate into the sphere and, after partial absorption, re-emerge traveling in a different direction).

The vast majority of photons are scattered elastically. A tiny fraction undergo inelastic Raman scattering. If scattering by one particle is not affected by the presence of surrounding particles, it is considered as independent scattering, otherwise it is dependent scattering. Thus, the radiative properties of a cloud of spherical particles of radius r, interacting with an electromagnetic wave of wavelength λ, are governed by three independent non-dimensional parameters: complex index of refraction, size parameter, and clearance-to-wavelength ratio. If scattering is independent ($c/\lambda >> 1$), then only the first two parameters are needed. The nature of the interaction between electromagnetic waves and particles is determined by the relative size of the particles compared with the wavelength of the radiation. Let us define a size parameter

$$x = 2\pi r/\lambda^2 \qquad [20.23]$$

where r is the effective radius of the particle. It is distinguished among three different regimes:

1. $x << 1$, Rayleigh scattering, named after Lord Rayleigh, who studied the interaction of atmospheric air (whose molecules are, in fact, very small particles) with sunlight [23]. He observed that, for very small particles, scattering is proportional to $1/\lambda^4$. Thus, within the visible part of the spectrum, blue light is scattered the most (accounting for blue skies) and red the least (resulting in red sunsets), as shown in Fig. 20.14.

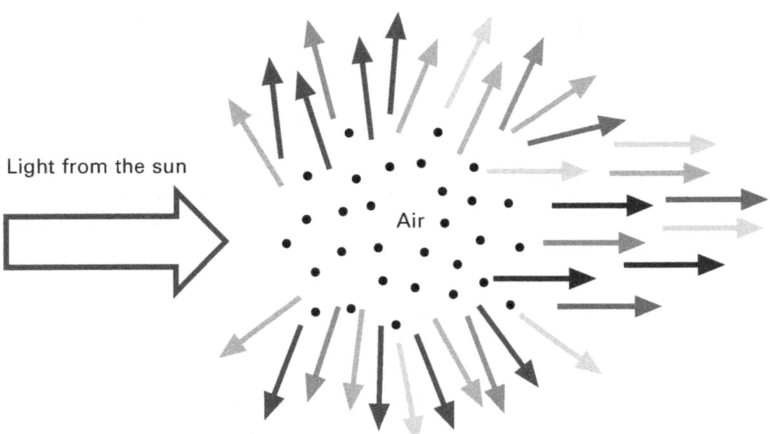

20.14 Scattering (why the sky is blue). Shorter-wavelength light is scattered out of the beam, leaving longer-wavelength light behind, so the sun appears yellow. In space, there's no scattering, so the sun is white, and the sky is black (Georgia Tech. Prof. Rick Trebino's lecture note).

2. $x \sim 1$, Mie scattering, named after Gustav Mie, who developed a comprehensive (and difficult) theory for the interaction between electromagnetic waves and particles [24].
3. $x >> 1$, normal surface, in this case the surface of the particle may be treated as a normal surface, and properties may be found through geometric optics. The high particle number density generated by laser causes the scattering of the incident light. And these particles lead to low degrees of transmission. Table 20.1 shows the classification of light-scattering phenomena.

20.2.5 Weld pool dynamics in laser welding

For analysis of the heat transfer and fluid flow in relation to the tracking free surface in the weld, solutions are found for the governing equations of mass conservation, momentum conservation (Navier–Stokes), energy conservation and the VOF. In the simulation, the molten metal is assumed to be an incompressible laminar flow with Newtonian viscosity.

With these assumptions, the governing equations can be expressed in Cartesian coordinates as follows:

- Mass conservation equation:

$$\nabla \cdot \mathbf{v} = 0 \qquad [20.24]$$

where $\mathbf{v}$ is the velocity vector.

- Navier–Stokes equation:

$$\frac{\partial \mathbf{v}}{\partial t} + \mathbf{v} \cdot \nabla \mathbf{v} = -\frac{1}{\rho} \nabla P + \nu \nabla^2 \mathbf{v} - K\mathbf{v} + \mathbf{G} \qquad [20.25]$$

Table 20.1 Classification of light-scattering phenomena

1. Frequency shift/Energy transfer
   (1) Elastic scattering: Rayleigh, Mie scattering (negligible energy transfer)
   (2) Inelastic scattering: Brillouin, Raman, inelastic X-ray and Compton scattering
2. Particle size
   (1) Rayleigh scattering: Size parameter << 1
   (2) Mie scattering: Size parameter ~ 1
3. Density
   (1) Independent scattering: Disregard 'interference'
   (2) Dependent scattering: Consider 'interference'
4. Interference
   (1) Coherent scattering: Constructive with in-phase, Destructive with out-of-phase
   (2) Incoherent scattering: Waves adding with random phase, Partially cancelling

where $K = C\dfrac{F_s^2}{(1-F_s)^3 + B}$ and $\nu = \nu_0\left(1.0 - \dfrac{F_s}{F_{s,cr}}\right)$.

In Eq. [20.25], ρ is the fluid density, P is the pressure, ν is the dynamic viscosity, K is the drag coefficient for a porous media model in the mushy zone, **G** is the body acceleration due to body force, C is a constant representing the mushy zone morphology, F_s is the solid fraction, B is the positive zero used to avoid the division by zero, ν_0 is the dynamic viscosity of the liquid phase and $F_{s,cr}$ is a user-defined constant denoting the critical solid fraction value at which all fluid in the mushy zone freezes. It is assumed that the drag coefficient is zero and the dynamic viscosity is constant in the molten metal (over liquidus temperature), but in the mushy zone those are dependent on the solid fraction as shown in Eq. [20.25]. The third term on the right-hand side of Eq. [20.25] is the source term induced by the frictional dissipation in the mushy zone. The Carman–Kozeny equation derived from the Darcy model is used to model the flow in the mushy zone, with the assumption that this zone can be treated as a porous medium [25].

- Energy conservation equation:

$$\frac{\partial h}{\partial t} + \mathbf{v}\cdot\nabla h = \frac{1}{\rho}\nabla\cdot(k\nabla T) \qquad [20.26]$$

where h is the enthalpy, k is the thermal conductivity, and T is the temperature. The following enthalpy-based continuum model is used in many cases to consider the solid–liquid phase transition:

$$h = \begin{cases} \rho_s C_s T & (T \le T_s) \\ h(T_s) + h_{sl}\dfrac{T - T_s}{T_l - T_s} & (T_s \le T \le T_l) \\ h(T_l) + \rho_l C_l (T - T_l) & (T_l \le T) \end{cases} \qquad [20.27]$$

where h is the enthalpy, ρ_s and ρ_l are the solid and liquid density, C_s and C_l are the specific heat at a constant volume of the solid and liquid phase, T_s and T_l are the solidus and liquidus temperature respectively, and h_{sl} is the latent heat of fusion.

- VOF equation:

$$\frac{\partial F}{\partial t} + \nabla\cdot(\mathbf{v}F) = 0 \qquad [20.28]$$

where F is the volume of fluid.

The keyhole formation process is a phenomenon of particular interest in laser welding. Its main driving force, called recoil pressure, involves the evaporation of material. The following model can be used in many cases to calculate the recoil pressure on the keyhole wall [26]:

$$P_R(T) \cong 0.54P_{sat}(T) = 0.54P_0 \exp\left(L_v \frac{T - T_b}{\overline{R}TT_b}\right) \qquad [20.29]$$

where P_{sat} is the saturated pressure, P_0 is the atmospheric pressure, L_v is the latent heat of vaporization, T_b is the boiling temperature, and $\overline{R}$ is the universal gas constant.

A surface tension model that Sahoo *et al.* [27] developed for a binary Fe-S system is used to model the Marangoni flow. Thus, the surface tension can be expressed as follows:

$$\gamma(T) = \gamma_m^o - A(T - T_m) - \overline{R}T\Gamma_s \ln(1 + k_1 a_i e^{-\Delta H^o/\overline{R}T}) \qquad [20.30]$$

where γ_m^o is the surface tension of pure metal at the melting point, A is the negative surface tension gradient for pure metal, T_m is the melting point, Γ_s is the surface excess at saturation, k_1 is the constant related to the entropy of segregation, a_i is the weight percent of sulphur, and ΔH^o is the standard heat of adsorption.

The buoyancy force (body force), which is modelled on the Boussinesq approximation, can be expressed as follows:

$$F_b = \rho g \beta(T - T_0) \qquad [20.31]$$

where g is the gravity, β is the thermal expansion rate, and T_0 is the reference temperature.

For the simulation of bubble formation through a partial keyhole collapse, the transient distributions of the metal vapour and plasma or the shielding gas mixture in the keyhole should be determined by solving the governing equations of a multiphase problem. For the sake of simplicity, however, a bubble is assumed to be filled with an ideal gas and the heat transfer and mass transfer between the gas and molten metal are ignored. The bubble can be maintained by using the energy conservation equation and state equation; specifically, the pressure is increased as the volume is reduced [16, 17].

In laser welding, metallic vapour is generated and ejected through the keyhole entrance. A spectroscopic measurement of a plume in 10 kW fibre laser welding confirms that the Inverse Bremsstrahlung effect is negligible; that the plume is weakly ionized; and that almost all the elements in the plume consist of neutral atoms of iron [28]. On account of the metallic vapour effect, the additional driving force and heat source are modelled on the assumptions that metallic vapour consists of iron vapour at a temperature of 6000 K [28] and that iron vapour is an ideal gas. The power absorbed during the welding process is calculated. The velocity of the metallic vapour at the keyhole entrance depends on the absorbed laser power density [29]. Inside the keyhole, the vapour velocity is assumed to increase linearly along the depth from zero velocity at the bottom of the keyhole to its calculated velocity at the entrance of the keyhole. The calculated velocity and average diameter of the keyhole suggest that the vapour flow is a laminar flow ($Re < 2000$); thus, the shear stress induced at the keyhole walls by the vapour can be expressed as follows [30]:

$$\tau_w = \frac{1}{8} f \rho V^2 \tag{20.32}$$

where $f = \dfrac{64}{Re}$ for the laminar flow.

With regard to the heat source of the metallic vapour plume, the calculation region is divided as shown in Fig. 20.15. In region 1 (inside the keyhole), only the convective heat transfer is considered. The convective heat transfer coefficient can be expressed as follows, provided that the keyhole has a circular tube shape with a laminar flow of metallic vapour, and provided that there is a fully developed condition and a uniform surface heat flux [31]:

$$h = \frac{48}{11}\left(\frac{k}{D}\right) \tag{20.33}$$

where D is the average diameter of the keyhole. In region 2 (above the keyhole), the convective heat transfer and the radiative heat transfer are both

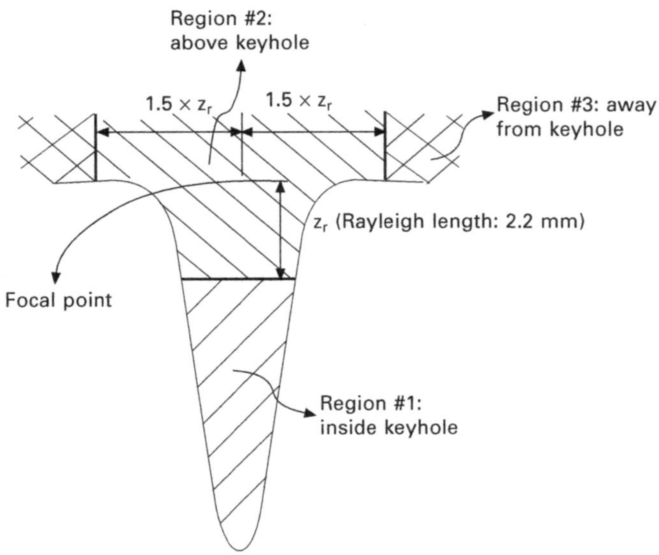

20.15 Schematic diagram of the regions for calculating the heat transfer by metallic vapour.

considered because the plume is observed to remain above the keyhole. Thus, an ambient temperature of 6,000 K is used to calculate the heat transfer in this region. This region is assumed to extend from the laser focal point to the bottom of the keyhole by a value of Rayleigh length. In region 3 (away from the keyhole), the convective heat transfer and the radiative heat transfer are calculated with respect to the ambient room temperature.

20.2.6 Laser–arc interaction

Laser arc hybrid welding, which simultaneously uses a high power laser and a conventional arc, has been developed to improve the welding speed and quality. After the development of hybrid welding techniques, a considerable number of studies have been carried out on system configurations, parameter optimization and the characteristics of weld beads. Although the hybrid welding plasma operates in a quite different way from the conventional one, little attention has been given to the new welding heat source. When the laser and arc plasma are used simultaneously, each heat source mutually assists and influences the other. The reason why the characteristics of hybrid plasma are improved is the generation of intensive metal vapour over the laser irradiated position. When the laser is irradiated on the metal surface, sufficient energy of light can cause metal vaporization and metal vapour can be further developed to laser induced plasma (LIP) plume by continuous supply of more laser energy. Because LIP is generated inside the arc plasma in laser arc hybrid welding, two kinds of plasma interact with and affect each other. In the coupling of laser and plasma, previous studies [32–34] have shown that the absorption of the laser beam is an important phenomenon; simulations have been used to understand this interaction effect. For the absorption mechanism of laser beams in the plasma, Inverse Bremsstrahlung, photoionization, and Mie absorption are considered in general [32, 33]. In the case of Nd:YAG and CO_2 lasers, Inverse Bremsstrahlung is dominant because of its relatively long wavelength [34], whereas it is known that for ultraviolet lasers the photoionization effect is dominant [33]. Additionally, the Mie absorption is dominant at low plasma temperature [35], when the effect of Inverse Bremsstrahlung and photoionization is low.

For CO_2 laser-TIG hybrid welding, Startsev *et al.* [36] conducted a numerical simulation for a co-axial axisymmetric hybrid configuration. They considered the absorption of laser beams in the Ar arc plasma. The results show a significant temperature rise in the core of the arc, but LIP is not considered. LIP has been considered in laser welding simulation [33] but it has not yet been considered with arc plasma in hybrid welding. Cho *et al.* [37] focused on an investigation of the interaction mechanism between Nd:YAG laser and gas tungsten arc/LIP mixture plasma using numerical analysis. In addition to this, the characteristics of hybrid plasma such as temperature distribution, current density distribution and flow pattern are explained. In numerical simulations, the other absorption mechanisms, except for inverse Bremsstrahlung, are neglected for Nd:YAG LIP and Ar arc plasma. First, the temperature distribution of the base metal due to laser irradiation is obtained by solving the corresponding heat conduction equations. By fixing this result as the boundary condition, the distribution of metal vapour induced by the laser can be obtained. The distribution of temperature in the base metal and metal vapour are set as the boundary condition and analysis domain, respectively, in order to conduct the analysis of the hybrid plasma. The analysis is completed after checking the absorption of laser light in the plasma by Inverse Bremsstrahlung. The simulation results of temperature distributions for various laser powers are shown in Fig. 20.16. In the case of a laser with low power, the amount of evaporation is relatively small and the effect of current concentration decreases. Consequently, the local increase

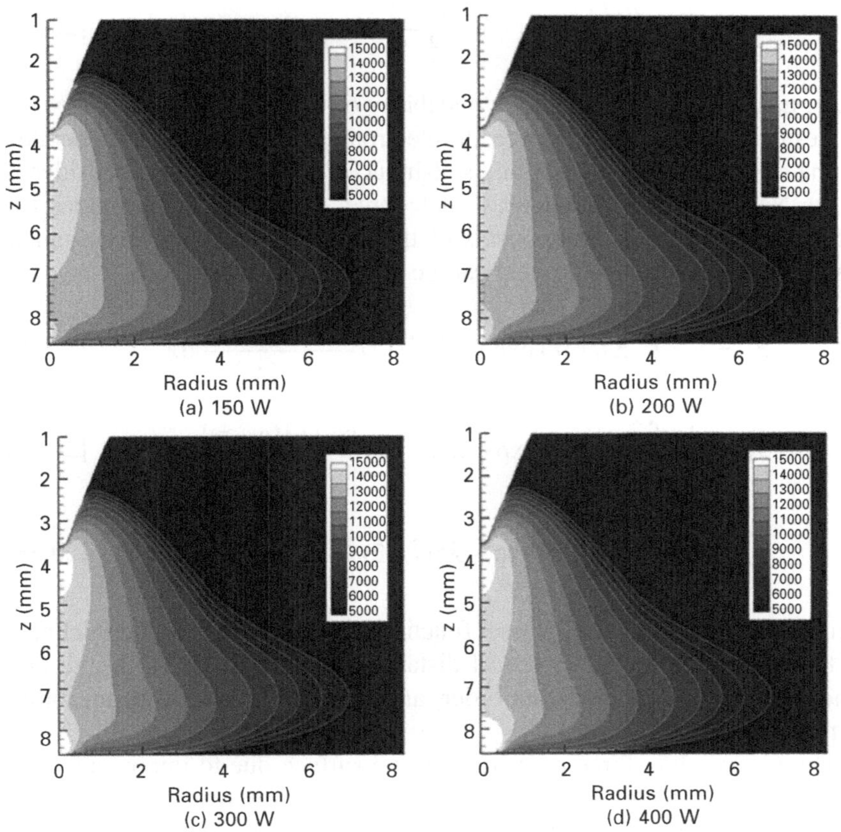

20.16 Temperature distribution of laser arc hybrid plasma for various laser powers (effective radius of the laser: 0.6 mm, arc current: 100 A, arc length: 5 mm).

in plasma temperature gradually disappears. Conversely, the local increase in plasma temperature in the vicinity of laser irradiation is extended and the absolute value of temperature also increases when the laser power is high.

20.2.7 Modelling of arc welding process

The heat source of the electric arc can be modelled as a surface heat flux with a Gaussian function:

$$\eta_A q_A(x, y) = \eta_A \frac{VI}{2\pi r_A^2} \exp\left(-\frac{x^2 + y^2}{2r_A^2}\right) \qquad [20.34]$$

Here, η_A is the arc efficiency and V and I are the electric voltage and current of the arc, respectively. The pressure boundary condition of the arc plasma can be modelled by a Gaussian-like function [38]:

$$P_A(x, y) = \frac{\mu_0 I^2}{4\pi^2 r_A^2} \exp\left(-\frac{x^2 + y^2}{2r_A^2}\right) \qquad [20.35]$$

Here, μ_0 denotes the magnetic permeability of free space.

Besides the gravitational force, the electromagnetic force that is generated by the electric current density and self-induced magnetic field is considered in the Navier–Stokes equation as body force. The analytical solutions for the vertical and radial components of the current density and the angular component of the magnetic field can be expressed as [39]:

$$J_z = \frac{I}{2\pi} \int_0^\infty \lambda J_0(\lambda r) \exp\left(-\lambda^2 r_A^2/12\right) \frac{\sinh[\lambda(c - z)]}{\sinh(\lambda c)} \, d\lambda \qquad [20.36]$$

$$J_r = \frac{I}{2\pi} \int_0^\infty \lambda J_1(\lambda r) \exp\left(-\lambda^2 r_A^2/12\right) \frac{\cosh[\lambda(c - z)]}{\sinh(\lambda c)} \, d\lambda \qquad [20.37]$$

$$B_\theta = \frac{\mu_m I}{2\pi} \int_0^\infty J_1(\lambda r) \exp\left(-\lambda^2 r_A^2/12\right) \frac{\sinh[\lambda(c - z)]}{\sinh(\lambda c)} \, d\lambda \qquad [20.38]$$

Here, J_0 is the first kind of Bessel function of zero order, c is the thickness of the workpiece, z is the vertical distance from the origin, J_1 is the first kind of Bessel function of first order, and μ_m is the magnetic permeability of the material.

The plasma drag force is exerted on the surface due to impingement of the arc plasma jet flow and it causes an outward flow from the centre. To model the plasma drag force, the following analytical solution of the wall shear stress produced by the normal impingement of the jet on the flat surface was used [40]:

$$\frac{\tau}{\rho_p u_0^2} Re_0^{1/2} \left(\frac{H}{D}\right)^2 = g_2\left(\frac{r}{H}\right) \qquad [20.39]$$

Here, τ is the shear stress, ρ_p is the plasma density, u_0 is the initial plasma velocity, Re_0 is the Reynolds number, H is the nozzle height, D is the nozzle diameter, r is the radius from the centre, and g_2 is the universal function. In the calculation, it is assumed that H and D are the same as the arc length and electrode diameter, respectively. Finally, models of the surface tension and the buoyancy force are shown in Eqs [20.30] and [20.31].

As mentioned above, in addition to the models of GMA welding, the additional conservation equation is used for calculating the distributions of alloying elements:

$$\frac{\partial \Phi}{\partial t} + \nabla \cdot (v\Phi) = \dot{\Phi}_s \qquad [20.40]$$

Here, $\dot{\Phi}_s$ is the scalar source term added by an alloying element in molten drops. By Eq. [20.40], a scalar quantity indicating the weight percentage of the alloying element in a cell can not only be conserved but also is affected by the molten pool flow.

The arc power is about 6.5 kW (231 A, 28 V). For adding the droplet in the simulation, the droplet is moved at an angle of 61° to the horizontal with measured values of velocity and generation rate, and its temperature is assumed to be 2,400K. In addition, the droplet diameter can be calculated using the measured droplet generation rate and wire feed rate of 8 m/min. For calculating the alloying element distributions, chromium and nickel are addressed and their concentrations are all zero in the case of a SS400 workpiece, while they are 20% and 10% in the case of droplets of Y308 solid wire. Plates XXV and XXVI (between pages 328 and 329) show the temperature and Cr distributions with the flow patterns in GMA welding. At the weld start, where only a limited amount of base metal is melted and mixed with the filler metal, the Cr concentration is relatively high and decreases gradually as the welding progresses, before it reaches a stationary distribution condition after some time.

20.2.8 Weld pool dynamics in hybrid laser welding

In the numerical simulation of the hybrid welding process, three-dimensional transient simulations are conducted by combining the arc welding model and the laser welding model without any interactive effect between the two. The temperature profile characteristics of the weld bead and molten pool flow and alloying element distribution in the molten pool are analysed. It is found that fusion zones and alloying element distributions predicted by numerical simulations are quite similar to the experimental ones [16]. It shows that the assumptions and models used in this study are appropriate. Despite those fair prediction results, it is important to note that some other effects have not been taken into consideration in the proposed algorithm, including:

- the Inverse Bremsstrahlung representing the laser-arc plasma interactions;
- the Rayleigh scattering representing the laser–metal vapour interactions; and
- the heat and mass exchanges between the molten metal and the bubble.

Accordingly, it is deemed that further study that takes into account the overlooked effects noted above is necessary.

20.3 Applications for improving the laser welding technique and the quality of laser welded components

Three-dimensional simulation of molten pool dynamics in multi-kW laser welding is conducted. Basically, in a three-dimensional transient simulation of laser welding, the mass conservation equation, the momentum conservation equation, referred to here as the Navier–Stokes equation, the energy conservation equation and the VOF equation were used as governing equations to execute the molten metal flow analysis. Molten metal is assumed to be an incompressible laminar fluid with Newtonian viscosity. The simulation considers the buoyancy force, the Marangoni force, the recoil pressure, the vapour-induced shear stress, and the vapour-induced heat transfer, and the bubble pressure model is used to simulate the pore generation with a keyhole collapse.

Structural carbon steel (namely A36 (ASTM) or SS400 (JIS)) with a thickness of 12 mm is used in both the simulation and experiment. The disk laser power is 6 kW, and the travel speed is 1 m/min. Plates XXVII (a) to (c) (between pages 328 and 329) show the three-dimensional cross-sectional views of the simulation results. The red part is the molten region which has the temperature over the solidus temperature. At the beginning of the welding process, the upper part of the molten region is extended slightly in opposite, welding direction. The region is gradually extended over time, but ends up with almost the same length. With regard to the shape of the keyhole, it reaches its widest point at the opening, becomes narrow at the middle, and then becomes wider again at the end. Accordingly, the keyhole may collapse at the middle. In the simulation results, however, only a few pores came up, and this behaviour is acceptable in terms of the industrial standard (ISO 13919-1). The keyhole is considered to be relatively stable because the laser energy is distributed evenly by multiple reflections rather than being focused at a specific region.

Plates XXVII (d) to (f) show the calculated temperature profiles and flow patterns from a cross-sectional side view. Clearly, as mentioned above, the molten pool is stretched at the upper part. Except for the upper part, however, the shape of the molten pool remains almost unchanged over time, probably due to the fact that the same flow pattern, namely a clockwise vortex, is maintained in the molten pool. This flow pattern helps transport the molten metal heated by the laser to the rear upper part of the molten pool. Because of this phenomenon, the rear upper part is extended over time in the opposite welding direction. The molten pool flow in this part oscillates back and forth, so that no specific flow pattern can be defined. The bottom of the keyhole is cooled relatively quickly because the strong flow pushes the molten metal in this region, which is bound by a large solid region. A narrow molten pool is consequently generated and acceptably small pores appear in this region.

Plate XXVIII (between pages 328 and 329) shows the temperature profiles and fluid flows from a cross-sectional front view. After a dimpled shape appears near the surface, as shown in Plate XXVIII(a), a keyhole is generated and a small amount of molten metal exists around the keyhole. As shown in Plate XXVIII(b), a downward flow was observed for about 0.1 s in a small region behind the keyhole, most likely due to the recoil pressure. At $t = 1.0$ s, the downward flow disappears and an upward flow appears. The upward flow is dominant for a long time until the solidification is completed. The solidification starts at the bottom and, as a result, the last solidification occurs in the upper molten region.

Plate XXIX (between pages 328 and 329) compares the simulation results of the weld bead cross section with the experimental results to verify the welding models and algorithms used in simulations. Plate XXIX(a) and (b) shows the simulation results of the fusion zones in the transverse and longitudinal cross sections after the solidification of the molten pool. In the experiment, the weld bead was polished and etched to show the fusion zone, and a transverse cross section is shown in Plate XXIX(c). The profile of the calculated fusion zone Plate XXIX(b) is in considerably good agreement with the experimental results Plate XXIX(c).

As mentioned above, in addition to the models of the GMAW and laser welding, the additional conservation equation is used for calculating the distributions of alloying elements. The separation between the arc and laser is 3 mm with the GMA leading the laser, and the laser and arc power is 6 kW and about 6.5 kW (231 A, 28 V), respectively. For adding the droplet in the simulation, the droplet is assumed to move at an angle of 61° to the horizontal plane with measured values of velocity and generation rate. Its temperature is assumed to be 2,400 K. In addition, the droplet diameter can be calculated using the measured droplet generation rate and wire feed rate of 8m/min. For calculating the alloying element distributions, chromium and nickel are addressed and their concentrations are all zero in the case of a SS400 workpiece, while they are 20% and 10% in the case of droplets of Y308 solid wire. The case of 1.8 s laser-GMA hybrid welding and a 1.2 s cooling process is simulated and analysed.

Plate XXX (between pages 328 and 329) shows the temperature and Cr distributions with the flow patterns in the longitudinal cross sections during hybrid welding. In the figures, maximum temperature and concentration values of Cr are set to 1,768 K and 8%, respectively. First, the molten pool with the flow patterns is analysed. At the keyhole front, where a high intensity laser beam is directly irradiated, the material is removed without a thin molten layer, a process that resembles the laser cutting and drilling process. The laser beam reflected at the keyhole front reaches the other position of the keyhole wall and is then absorbed. As shown in the figures, not only can the incident beam reach the keyhole wall easily, but also the reflected beam can be collected around a specific point, if the shape of the keyhole is not smooth but rather has a bulge. In particular, the region is rapidly expanded by the excessive recoil pressure, if the reflected laser beam is concentrated at the rear part of the keyhole surrounded by the molten metal. For this reason, complex flows with relatively high velocity are observed near the keyhole and can make the keyhole unstable. Meanwhile, it is observed that clockwise-rotating vortices form a quasi-steady state flow pattern in the wide molten region located away from the keyhole. It can be inferred that these vortices are driven by the strong flow at the keyhole. Because the vortices extending a relatively wide range of the molten pool not only have high Cr content but also rotate in the same direction, they help mix Cr in the molten pool and thus Cr is distributed uniformly in the wide molten region. However, the outer clockwise-rotating vortex with relatively high or low Cr content passes through around the solid–liquid interface, and thus Cr can be entrapped and make a pattern like the teeth of a comb. In this simulation, this type of alloying element distribution pattern can be observed but is not severe, when compared to the results of full penetration hybrid welding experiments, although the experiments were carried out with different welding parameters and filler metal [41]. For the region around the keyhole bottom, strong up-and-back flow arises and pushes out the flow having high content of alloying elements because of the fast cooling rate and small accompanying molten region. Therefore, low content of alloying elements exists in the keyhole bottom.

Plate XXXI (between pages 328 and 329) shows the simulation results of fusion zone and Cr and Ni distributions in the transverse and longitudinal cross sections after solidification of the molten pool. In these figures, maximum content values of Cr and Ni are set to 8 and 4%, respectively, for clear comparison with experiments. In the experiment, the weld was polished and etched to analyse the fusion zone. An electron prove micro analyzer (EPMA) was used to investigate the alloying elements of Cr and Ni in the weld, and the experimental results are shown in Plate XXXII (between pages 328 and 329). For the calculated fusion zone in Plate XXXI(a), the penetration depth and melting width are predicted considerably accurately, while the fusion zone is a little wider at the middle region in comparison with the experimental result of Plate XXXII(a). Given the observation of a fluctuating fusion zone in both Plate XXXI(b) and Plate XXXII (b), it appears that the laser induced keyhole is rather unstable, which can be explained by the unstable nature of the keyhole. From the simulation results for alloying element distributions presented in Plates XXXI (c) to (f), it is seen that the lower part of the molten region is very narrow and has a relatively low content of alloying elements. On the other hand, the upper part of the molten region is considerably wide and has a relatively high content of alloying elements. Between these two parts, a transitional region, where high and low contents of alloying elements coexist, is observed and wave-shaped patterns are extended from the bottom to the up-and-opposite direction of welding. Plate XXXI shows that strong up-and-back flow with low Cr content and outer clockwise-rotating vortex with high Cr content is intersected in this region. However, the pattern like the teeth of a comb is not remarkable because the alloying element is distributed relatively uniformly, when compared to the results of full penetration experiments [41]. These simulation results are in fairly good agreement with the experimental results in Plates XXXII(c)–(f). Shown as dashed lines in Plates XXXII(a), (c), and (e), the boundaries of the fusion zone and alloying elements are quite similar, indicating that the alloying elements can be reached only at the molten region under these welding conditions. Although the absolute content of Cr is twice that of Ni in the solid wire, the distribution characteristics of Cr in Plates XXXII(c) and (d) are quite similar with those of Ni in Plates XXXII (e) and (f). From the aforementioned results, it is thought that the distributions of the alloying element such as Cr and Ni are affected mainly by the molten metal flow and also that their diffusive effects are negligible in this case. By the simulation where the diffusive effect of each alloying element is not considered, identical patterns of Cr and Ni are calculated as shown in Plates XXXI(c)–(f).

20.4 Future trends

Using the aforementioned models for laser and arc welding, reasonable predictions can be made based on the comparison of fusion zone shape and alloying element distribution between the numerical calculation and the experimental observation. However, it should be pointed out that the developed models are far from the full solutions to the welding process, and developing a united model for the welding is very challenging. In particular, in the hybrid process the new phenomena such as the synergistic interaction of the laser and electric arc and the effect of the coupled parameters should be considered. The origins of those phenomena are not well understood and many important questions related to such modelling are still unanswered, which bring more difficulties for modelling of the hybrid welding. The main challenges for such modelling are as follows [42].

First, the synergistic interaction of the laser and electric arc is not well understood in terms of the following phenomena: the electric arc preheats the base metal that enhances the absorption of the laser energy by the target metal; the electric arc dilutes the laser induced plasma and hence reduces the ability of the plasma to absorb and reflect the laser energy; the laser beam stabilizes the electric arc; the laser induced metal vapour and plasma distort the electric arc structure; and the synergistic interaction of laser–arc changes the energy transfer and hence the welding process.

Second, the laser beam and the laser induced metal vapour significantly influence the features of metal transfer, but the underlying physics is still unclear. The transient arc plasma and droplet generation were ignored, and/or the droplets were assumed to impinge into the weld pool at a fixed size and frequency in the existing models, which fail to couple the generation of arc plasma, droplet formation, detachment and transfer during hybrid welding.

Third, the effects of shielding gas, including its composition, flow rate and injecting direction, are still unclear. The shielding gas has significant influence on the plasma formation, the arc stability and hence the weld quality in hybrid welding, which depends on the intrinsic properties of the shielding gas. However, the effects of the shielding gas in hybrid welding are contradictory in some functions for the individual welding process, which requires further research to reduce the plasma shielding effect for the laser and also enhance the electric arc stability.

Finally, a real united model of hybrid laser–GMA welding will need an excess computation time. Therefore, more advanced computer technology and a more time-efficient computation method are required.

20.5 References

1. Schmidt, M., Otto, A., Geiger, M., 2008. Analysis of YAG laser lap-welding of zinc coated steel sheets. *CIRP Ann – Manuf. Technol.*, 57, 213–216.
2. Sathiya, P., Abdul Jaleel, M. Y., Katherasan, D., 2010. Optimization of welding parameters for laser bead-on-plate welding using Taguchi method. *Prod. Eng. Res. Devel.*, 4, 465–476.
3. Seto, N., Katayama, S., Matsunawa, A., 2000. High-speed simultaneous observation of plasma and keyhole behavior during high power CO_2 laser welding: effect of shielding gas on porosity formation. *J. Laser Appl.*, 12, 245–250.
4. Rosenthal, D., 1941. Mathematical theory of heat distribution during welding and cutting. *Weld. J.*, 20, 220s–234s.

5. Zacharia, T., David, S. A., Vitek, J. M., Debroy, T., 1989. Weld pool development during GTA and laser beam welding of type 304 stainless steel, part I – theoretical analysis. *Weld. J.*, 68, 499s–509s.
6. Huan, J., Nie, P., Deng, D., Li, Z., 2011. Finite element analysis of laser fillet welding T-joint of thin mild steel plates. In: Vollertsen, F. (ed.) *Proceedings of the IWOTE'11: International Workshop on Thermal Forming and Welding Distortion*, Bremen, Germany, 255–261.
7. van der Aa, E. M., Hermans, M. J. M., Richardson, I. M., 2008. Simplified models for the investigation of longitudinal welding stress and strain development. In: Vollertsen, F., Sakkiettibutra, J. (eds) *Proceedings of the IWOTE'08: International Workshop on Thermal Forming and Welding Distortion*, Bremen, Germany, 171–180.
8. Kaplan, A., 1994. A model of deep penetration laser welding based on calculation of the keyhole profile. *J. Phys. D-Appl. Phys.*, 27, 1805–1814.
9. Ki, H., Mohanty, P. S., Mazumder, J., 2001. Modelling of high-density laser-material interaction using fast level set method. *J. Phys. D-Appl. Phys.*, 34, 364–372.
10. Osher, S., Sethian, J. A., 1988. Fronts propagating with curvature-dependent speed: algorithms based on Hamilton-Jacobi formulations. *J. Comput. Phys.*, 79, 12–49.
11. Lee, J. Y., Ko, S. H., Farson, D. F., Yoo, C. D., 2002. Mechanism of keyhole formation and stability in stationary laser welding. *J. Phys. D-Appl. Phys.*, 35, 1570–1576.
12. Hirt, C. W., Nichols, B. D., 1981. Volume of fluid method for the dynamics of free boundaries. *J. Comput. Phys.*, 39, 201–225.
13. Cho, J. H., Na, S. J., 2006. Implementation of real-time multiple reflection and Fresnel absorption of laser beam in keyhole. *J. Phys. D-Appl. Phys.*, 39, 5372–5378.
14. Cho, J. H., Na, S. J., 2007. Theoretical analysis of keyhole dynamics in laser drilling considering the polarization of laser. *J. Phys. D-Appl. Phys.*, 40, 7638–7647.
15. Cho, J. H., Na, S. J., 2009. Three-dimensional analysis of molten pool in GMA-laser hybrid welding. *Weld. J.*, 88, 35s–43s.
16. Cho, W. I., Na, S. J., Cho, M. H., Lee, J. S., 2010. Numerical study of alloying element distribution in CO_2 laser-GMA hybrid welding. *Comput. Mater. Sci.*, 49, 792–800.
17. Cho, W. I., Na, S. J., Thomy, C., Vollertsen, F., 2012. Numerical simulation of molten pool dynamics in high power disk laser welding. *J. Mater. Process. Technol.*, 212, 262–275.
18. Bang, S. Y., 1991. Effects of multiple reflections and beam polarization on laser grooving of hard ceramics. PhD thesis, Pennsylvania State University.
19. Modest, M. F., 1992. *Radiative Heat Transfer*, Mc-Graw-Hill, New York.
20. Stratton, J. A., 1941. *Electromagnetic Theory*, McGraw-Hill, New York.
21. Schulz, W., Simon, G., Urbassek, H. M., Decker, I., 1987. On laser fusion cutting of metals. *J. Phys. D: Appl. Phys.*, 20, 481–488.
22. Ducharme, R., Williams, K., Kapadia, P., Dowden, J., Steen, B., Glowacki, M., 1994. The laser welding of thin metal sheets: an integrated keyhole and weld pool model with supporting experiments. *J. Phys. D: Appl. Phys.*, 27, 1619–1627.
23. Rayleigh, L., 1881. On the electromagnetic theory of light. *Phil. Mag.*, 12, 81–101.
24. Mie, G. A., 1908. Beiträge zur Optik trüber Medien. speziell kolloidaler Metallösungen. *Annalen der Physik*, 25, 377–445.
25. Voller, V. R., Prakash, C., 1987. A fixed grid numerical modelling methodology for convection-diffusion mushy region phase-change problems. *Int. J. Heat Mass Transf.*, 30, 1709–1719.

26. Allmen, M., Blatter, A., 1995. *Laser-Beam Interactions with Materials*, 2nd edn. Springer, New York.
27. Sahoo, P., Debroy, T., McNallan, M. J., 1988. Surface tension of binary metal-surface active solute systems under conditions relevant to welding metallurgy. *Metall. Mater. Trans. B-Proc. Metall. Mater. Proc. Sci.*, 19, 483–491.
28. Katayama, S., Kawahito, Y., Mizutani, M., 2010. Elucidation of laser welding phenomena and factors affecting weld penetration and welding defects. *Physics Procedia*, 5, 9–17.
29. Amara, E. H., Bendib, A., 2002. Modelling of vapor flow in deep penetration laser welding. *J. Phys. D-Appl. Phys.*, 35, 272–280.
30. White, F. M., 1999. *Fluid Mechanics*, 4th edn. McGraw-Hill, New York.
31. Incropera, F. P., Dewitt, D. P., 1996. *Introduction to Heat Transfer*, 3rd edn. John Wiley & Sons, New York.
32. Schittenhelm, H., Callies, G., Straub, A., Berger, P., Hugel, H., 1998. Measurements of wavelength-dependent transmission in excimer laser-induced plasma plumes and their interpretation. *J. Phys. D: Appl. Phys.*, 31, 418–427.
33. Rozman, R., Grabec, I., Govekar, E., 2008. Influence of absorption mechanisms on laser-induced plasma plume. *Appl. Surf. Sci.*, 254, 3295–3305.
34. Yilbas, B. S., Yilbas, Z., Akcakoyun, N., 1996. Investigation into absorption of the incident laser beam during Nd:YAG laser processing of metals. *Opt. Laser Technol.*, 28, 503–511.
35. Callies, G., Schittenhelm, H., Berger, P., Hugel, H., 1998. Modeling of the expansion of laser-evaporated matter in argon, helium and nitrogen and the condensation of clusters. *Appl. Surf. Sci.*, 127–129, 134–141.
36. Startsev, V. N., Martynenko, D. P., Leonov, A. F., 2000. Investigation of characteristics of an arc column in laser arc welding using numerical simulation. *High Temp.*, 38, 20–25.
37. Cho, Y. T., Cho, W. I., Na, S. J., 2011. Numerical analysis of hybrid plasma generated by Nd:YAG laser and gas tungsten arc. *Opt. Laser Technol.*, 43, 711–720.
38. Cao, Z., Yang, Z., Chen, X. L., 2004. Three-dimensional simulation of transient GMA weld pool with free surface. *Weld. J.*, 83(6), 169s–176s.
39. Kou, S., Sun, D. K., 1985. Fluid flow and weld penetration in stationary arc welds. *Metall. Trans. A*, 16(A), 203–213.
40. Phares, D. J., Smedley, G. T., Flagan, R. C., 2000. The wall shear stress produced by the normal impingement of a jet on a flat surface. *J. Fluid Mech.*, 418, 351–375.
41. Zhao, L., Sugino, T., Arakane, G., Tsukamoto, S., 2009. Influence of welding parameters on distribution of wire feeding elements in CO_2 laser GMA hybrid welding. *Sci. Technol. Weld. Join.*, 14(5), 457–467.
42. Rao, Z. H., Liao, S. M., Tsai, H. L., 2011. Modelling of hybrid laser-GMA welding: review and challenges. *Sci. Technol. Weld. Join.*, 16(4), 300–305.

Part IV

Applications of laser welding

21
Applications of laser welding in the automotive industry

M. GRAUDENZ and M. BAUR, Audi AG, Germany

DOI: 10.1533/9780857098771.4.555

Abstract: This chapter introduces the key laser welding technologies and their applications in the car body production (hang-on parts included) at Audi AG. The successful implementations of laser welding technologies for steel and aluminium are presented at the instance of the laser remote steel door application of the Audi A3/A4/Q5 and the aluminium door and divided tailgate application of the Audi A6/A8.

Key words: laser, remote, aluminium, steel, welding, joining, car body, hang-on parts, seam quality, surface coating, laser source.

21.1 Introduction

Modern high brightness laser sources are the key to highly efficient production engineering in the automotive industry. The enhancements in beam quality and energy efficiency make laser applications more and more competitive with other established joining technologies (e.g. resistive spot welding) in cases where different joining technologies can be realized.

In most cases, laser welding technology is used when there are special demands from design and construction that no other joining technology can fulfil. The most important advantages of laser welding applications are the realization of very short flanges which means weight reduction and additional freedom for design, single-sided accessibility and contact-free joining in combination with a highly efficient productivity.

21.2 Production targets and challenges

21.2.1 Targets

Due to the demands of modern lightweight construction, design language, economic efficiency and environmental compatibility, the challenges in modern automotive production are enormous. The key to making the functionalities of a virtual car body concept real is the efficient use of new materials. The right material used in the right place – that is the philosophy of Audi. That can only be initiated by highly efficient and reliable existing joining technologies and by new joining technologies that have to be developed.

In the field of joining technologies, laser applications, especially laser welding applications, have some very important advantages compared to all other joining technologies and therefore are, and will become, increasingly important in the future.

21.2.2 Challenges

In modern automotive production there are different challenges during the development process from the first sketch to a real volume-produced vehicle with hundreds of cars or thousands of parts per day. The main challenge is to conflate design language and lightweight material design by means of different joining technologies and to integrate new applications and processes into running production.

In a modern car body concept like the body in white of the Audi A6/A7, a wide mixture of different material classes and mixtures are used, including aluminium panels, cast components and extruded profiles with steel panels. What is the reason for that?

On the one hand, there are increased safety requirements that mandate high strength materials in the crash-relevant areas. On the other hand, there are constantly rising environmental requirements such as the reduction of CO_2 emissions, leading to a reduction in fuel consumption or the electrification of the powertrain.

From our point of view, the most interesting approach to reach that aim is to reduce car weight. The body in white is one major part in total car weight and is the initialization point for inverting the weight spiral. Therefore it is very important to reduce weight to a minimum. All these requirements lead to car body constructions with reduced sheet thickness. As a logical conclusion, the materials used tend to be higher strength materials (e.g., high or ultra high strength steel) with specific properties. Additionally, multi-material design with combinations of different strength and lightweight materials (e.g., aluminium) are the consequence. The right material used in the right place – that also describes requirements for the future car body lightweight architecture at Audi. The joining of these different material mixtures in an economic and reliable fashion is now the great challenge.

Laser appropriate construction is the basis for exploiting synergies of laser applications and improving process productivity to the maximum. Minimal flange width for joining means an additional weight reduction in general and, in the case of the door application, an improved all-around view, which is a direct benefit to the customer. In the laser remote door application of the Audi A3/A4/Q5, flanges shorter than 6 mm could be realized. These requirements could not be realized by any other joining technology like standard resistant spot welding, where a minimal welding flange width of over 10 mm is necessary.

Another advantage of laser welding applications is the excellent process stability that is reflected in the high process reliability. Due to the low space requirements for a laser remote welding system with regard to the high output quantity, it is a preferred joining technology when there is only limited space available for a plant. Further advantages will be discussed in detail in Section 21.3.1. These are some of the most important unique characteristics of laser joining applications in the body shop.

With that in mind, laser welding applications enhance the philosophy of Audi 'Vorsprung durch Technik' ('Truth in Engineering') as a technological pioneer and one of the leading premium car manufacturers.

21.2.3 Economic efficiency

Besides the technical advantages, the economic efficiency is especially important for the success of the laser in the automotive industry. The most important figures for the use of laser applications are the investment and operating costs. Both are dependent on the reliability and flexibility of an application in combination with its space requirements. A modern body shop is a highly automated system, which consists of hundreds of robots with working tools, like the laser process optics or a spot welding gun, and every single component's reliability directly influences the overall system reliability.

In the body shop, the laser is often used for applications that cannot be realized with other joining technologies. When there are alternative joining technologies available, like in most cases, the laser must compete. An example of such a unique feature is the zero gap joint, between the side panel and the roof of a car, but normally this unique feature is not required.

For the development of laser invest costs, there has been a tremendous price reduction per kilowatt output power over the last 10 years of nearly 50 per cent. This could be realized through new technological approaches and solutions in laser concepts. One of the most important improvements was the development of much more efficient laser pumping processes, e.g. the replacement of lamp-pumped by diode-pumped systems. These improvements also enabled a massive reduction of the operating costs. The wall plug efficiency of today's most efficient laser sources could be improved by a factor of ten compared to a former lamp-pumped Nd:YAG laser. In concrete, the wall plug efficiency went from about 3 per cent to nearly 40 per cent and as a result of that the energy consumption could be reduced even more. As a direct consequence, the space requirements for laser cooling systems were reduced to a fraction. For a 4 kW laser system, the space requirements could be reduced by a factor of four, and this opens up the opportunity to place the laser source directly into the laser cabin and not onto the top of the laser cabin as often happened before.

The next important advantage of the new generation of lasers is the reduced maintenance expenses. In a lamp-pumped system, the lamps had to be changed every 1,200 working hours; in a diode-pumped systems, the diodes normally do not need to be changed over the complete cycle period of a car model.

All these improvements connected with new laser process monitoring systems and quality inspection systems result in the increased attractiveness and competitiveness of the laser in the automotive industry.

21.3 Laser applications in the body shop

With further development of modern laser sources the number of industrial laser applications is continuously growing. Classical laser welding, remote laser welding, cutting with scanner optics and laser brazing are well-known industrial applications. The current trends for laser applications also tend to specific functionalizing of materials and surfaces properties, like local hardening or softening of components, depending on the specified component requirements. Modern high brightness laser sources like disk or fibre lasers with best beam quality are available in the multi kilowatt power region and high efficiency laser concepts like direct diode lasers have reached wall plug efficiencies in the order of magnitude of 40 per cent. This means an improvement of a factor greater than ten compared to a conventional lamp-pumped solid-state Nd:YAG laser.

For every application one can find the perfect laser source for best performance and quality. Laser remote welding and cutting with a small spot diameter and a long focal length, which means a great distance from the workpiece, are predestined for disk and fibre lasers with a beam parameter product of 2–8 mm*mrad or better.

Laser remote welding with scanner optics is the laser joining application with the highest performance in the body shop. It is used very efficiently in steel door applications. A detailed overview of the very successful implementation of the laser remote door application in the Audi A3/A4/Q5 will be given in the next section.

In laser brazing applications, where only moderate beam quality is necessary, the direct diode laser with highest energy efficiency is the state of the art.

Typical fields of application for classical laser welding processes of steel or aluminium are structural components in the body in white or hang-on parts like doors or divided tailgates. Because of the high optical quality demands for aluminium hang-on parts like doors, the aluminium alloy group EN AW-6016-T4 is commonly used due to its better deep-drawing properties, but additional filler wire has to be used to reduce the increased hot shortness of that specific aluminium alloy (Ostermann, 2007).

21.3.1 Laser welding of steel

For laser welding, the light from the laser source is transferred and focused with optical components (fibre optics, laser optics, lenses) onto the workpiece. If energy density is high enough and impact time is long enough, the workpieces are melted locally and after solidification they are joined together.

In classical laser welding, the laser optics are guided with an industrial robot locally along the seam geometry. For the welding of steel, laser remote welding with scanner optics is also suitable, where the laser beam position can be varied between different positions in the working area nearly without delay. The laser remote process is not yet available for laser welding applications where an additional filler wire is required, because the filler wire cannot be tracked with the same speed as the laser beam. This is the case for some specific aluminium alloys, like the alloy group specified in DIN EN AW-6016-T4 (Ostermann, 2007), where additional filler wire is necessary.

Laser remote door application of Audi A4/Q5

For the implementation of laser remote welding of the A4/Q5 doors application, the following conditions had to be complied with in the project. The quality demands for door applications are generally very high because the joints are partly in direct view to the customer. For that reason, the joining technology must not be visible to the customer and single-sided accessibility of the joining technology is essential (see Fig. 21.1). Another direct customer benefit in door applications is the window opening that determines the maximum viewing angle. For safety and comfort aspects, the all-around view should be maximal. A maximum viewing angle can only be realized if the door

21.1 Side view of the Audi A4 with open door.

flanges are as short as possible. The joining technology that enables the shortest flange width is laser (remote) welding. The flange geometry of the Audi A4 door is shown in Fig. 21.2. In the Audi A4/Q5 door application, flanges shorter than 6 mm could be realized – at the moment no other joining technology can compare with that. For standard resistant spot welding, if flange geometry is optimal, a minimum flange width of more than 10 mm is required. By means of flange width reduction, a weight reduction of more than 100 g per door was achieved with laser remote technology. The implementation of laser remote welding was realized with a 4 kW TruDisk laser system with two-dimensional TRUMPF scanner optics (PFO-33).

For performance reasons the distance from the workpiece should be at least 500 mm. Short variations of the orientation of the scanner mirrors in the optics then lead to great variations of laser spot position on the workpiece and the classical non-welding times for transferring the optics from one seam position to the next are reduced drastically. The laser spot diameter on the workpiece was chosen to be 0.6 mm, which was realized by using a 0.2 mm diameter fibre-optic cable.

Another benefit of a long focal length is the effective reduction of contaminants on the optical components from the welding process itself. Steam and spatter from the welding process can affect the protection glass or, in the worst case, destroy expensive components like optical lenses. For a constant and stable laser welding process, different cross jet modules blow away the accruing metal vapour above the keyhole. If not removed, the metal vapour will absorb parts of the laser light, resulting in an alteration of the laser light intensity on the workpiece. That can affect the seam quality negatively.

All seams in the door application were performed as overlap joints with 0.7 mm on 0.6 mm zinc-coated metal sheets. Depending on the door variations (e.g. front or rear door), the number of joints varies from 45 to 49 laser seam

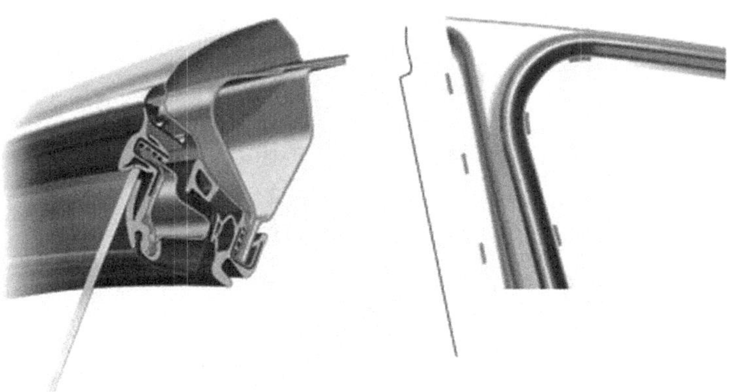

21.2 Door frame of an Audi A4 door (right) with section view (left).

steps per door. The length of each linear seam varies from 25 mm to 40 mm (see Fig. 21.3). Seams no. 1–34 are welded in flat position and the seams no. 35–45 are performed in a perpendicular orientation. To obtain the best seam quality, one has to ensure a constant and precise clamping over the whole flange geometry. A well-defined gap between the two metal sheets is one of the most important prerequisites in laser welding applications with zinc-coated metal sheets contrary to aluminium, where a zero gap is required. The steel sheets used in the Audi A4/Q5 door application are double-sided zinc-coated for corrosion requirements. The zinc coating leads to an additional challenge in the laser welding process. The melting point of steel is over 600°C higher than the vaporization temperature of the zinc coating. This leads to sudden vaporization of the coating during the welding process of steel. The zinc vapour that is generated between the metal sheets then leaks under high pressure through the molten metal. As a result of uncontrolled vapour expansion, the liquid metal will be ejected and spatters and pores will occur. The special preliminary treatments that are necessary to guarantee a controlled degassing for zinc-coated steel sheets are presented in Section 21.4.2.

The door production of the Audi A4/Q5 family was set up to a maximum quantity of 2,000 doors per day and welding station. That required four parallel welding stations, two for the front doors and two for the rear doors with a cycle time for producing a door of less than 33 seconds (arrangement, handling and joining included). The sole laser welding time per door is 19.5 seconds. This performance was only possible through the advantages of laser remote welding.

Moreover, laser remote welding in the door application is a very stable joining technology even without the use of laser process monitoring tools.

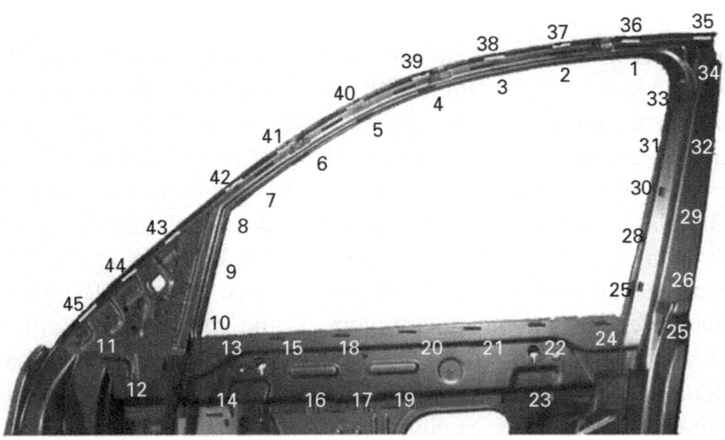

21.3 Seam positions on the Audi A4 door of two different welding orientations.

For that reason, the overall process reliability per laser cell, including laser source, cooler, scanner optics and robotics, reaches values beyond 98 per cent.

21.3.2 Laser welding of aluminium

As noted in Section 21.3 the laser welding of aluminium in the body shop of Audi differs from remote steel welding in terms of the additional filler wire that is needed. The filler wire is necessary to ensure the requested strength properties of the laser seam and to reduce hot shortness. The filler wire that is used for the 6000 aluminium compound has an alloy content of 12 per cent silicon (AlSi12). With the additional filler wire, the alloy content of the melt is raised above the critical value for crack susceptibility and a minimum volume fraction of 2 per cent should be obtained. The filler wire is also used as tactile guidance in the laser welding process. The wire is attached below the process optics so that the wire guides the process optics along the seam geometry. The additional material is fed into the laser spot where it melts and contributes to the welding process. To produce a better seam quality and to raise welding speed, a hot wire feed system is used, where the filler wire can be preheated separately.

The high optical reflectivity of aluminium in the 1 μm wavelength regime can lead to problems when the laser welding process is initiated. When the laser beam strikes nearly perpendicular onto a plain surface, an amount between 80 and 99%, depending on the surface finish of an aluminium sheet, is reflected. When the highly focused laser light is directly scattered back into the optical system, it can lead to complete destruction of laser components or transport/feeding fibre. For that reason, the wavelength-dependent absorption of laser light can be used to optimize the initiation of the laser welding process for aluminium. A higher absorption means that the energy from the laser source can be used more efficiently for melting the metal. As described in Fahrenwaldt (2009), the absorption curve of aluminium has a local maximum at the wavelength of 800 nm and the slope falls steeply at higher wavelengths. That leads to different laser welding efficiencies for lasers with higher operating wavelengths. The operating wavelengths of disk and fibre lasers are around 1,030 and 1,070 nm, which leads to an absorption of about 5%. For wavelength-coupled direct diode lasers, a wavelength mixture between 900 nm and 1,000 nm is used to generate multi-kilowatt output power. In that wavelength domain, the absorption for aluminium is a factor of two higher and the laser welding process can be initiated more efficiently. This is one reason why direct diode lasers can be favourable for aluminium laser welding applications.

Laser welding door application of Audi A6: demands of the project

Besides the general advantages of laser welding applications, as discussed in Section 21.3.1 for the steel doors of the Audi A4/Q5, there are additional potentials for the reduction of car weight. The use of lightweight materials for structural components in the body in white and for hang-on parts is one possibility. With a specific weight of 2.71 g/cm^3 (Hornbogen *et al.*, 2008), aluminium is roughly a factor of three lighter than steel.

For laser welding of aluminium, there are some specific requirements that should be discussed in detail. First of all the aluminium metal sheets used in the door application are not zinc-coated as in the case of steel doors, due to corrosion demands. The surface of a pure aluminium sheet reacts with atmospheric oxygen and an amorphous oxide layer accrues. That Al_2O_3 layer is very stable and resistive and naturally protects the aluminium surface from external influences. Under normal conditions, a very thin oxide layer starts to grow immediately. For the laser welding process, this oxide layer acts as a kind of thermal shield. To melt the aluminium underneath, the laser energy has to be transferred through the oxide layer into the aluminium volume. The melting temperature of the oxide layer lies at 2,054°C (Lide, 2010) and is much higher than the melting point of pure aluminium, which is 660.32°C (Lide, 2010). Due to the 'activation energy' that is caused by the sheet thickness independent oxide layer, the thickness of the aluminium sheet itself is limited to a minimum for a stable laser welding process. The minimum upper sheet thickness for the Audi A6 doors was chosen to be 1.2 mm. If the sheet thickness is below 1.2 mm, then the activation heat for breaking up the oxide layer cannot be deposited in the sheet volume without causing hot cracks or other defects.

Another aim was the reduction of the number of single parts for the door assembly, but there are restrictions from the produceability of the single parts. The deep drawing degree of the inner section part was too high to be produced as a single component. For that reason, the inner door section consists of two parts (1.2 mm and 1.5 mm) that are joined by laser welding. That welded sub-assembly is then joined with the outer doorframe element. The number of structural single parts in the Audi A6 door concept could thereby be reduced to three. With that new door concept, a total weight saving of over 10 kg, compared to the predecessor, could be achieved. The overall laser seam length in the Audi A6 door application reaches almost 10 m. The seam geometries used in this door application are fillet and overlap joints. For optimal welding, the surface of the aluminium sheets should be cleaned of oil and drying lubricant that are normally used in the manufacturing process of the single part. If not cleaned properly, surface contaminations will lead to an increased number of pores that will reduce the strength of the joint and pores can also be the origin of corrosion. One possibility to clean the aluminium surface is the so-called 'laser-clean' process which is presented in Section 21.4.2.

Laser welding door application of Audi A6: implementation

For the aluminium door application of the Audi A6, a direct diode laser (Laserline, LDF 4000-30) with a wall plug efficiency of nearly 40 per cent and a beam parameter product of 33 mm · mrad with a maximum output power of 4 kW was chosen. This beam quality is good enough for a classic laser welding application with a working distance of about 200 mm. For a higher optical seam quality, argon was used as a shielding gas in the laser welding process.

The high wall plug efficiency and the relative low investment costs in combination with the preferred wavelength window (900–1,000 nm) for aluminium welding make this laser source the ideal tool for that application. The laser optics in use is an adaptive laser processing optical system (ALO3) from Scansonic IPT GmbH. Several features are integrated in the ALO3 device, such as the functionality of beam-shaping optics with optional autofocus module and seam-tracking sensors. With these features, a weld-seam deviation in the millimetre range from the programmed robot path can be automatically compensated.

The challenge of the Audi A6 door application was to weld the three single door parts with 49 laser seams within a cycle time of 300 s per door. Due to high optical and functional quality demands of the welds, the welding speed was limited to 3 m/min. To make all 49 seam positions accessible for a single six-axis robot, a rotatable clamping device was used. A highly dynamic robotic path could be achieved for the welding process by avoiding the robotic movement through the single-axis singularities. The so-called axis singularities are fixed locations in the robotic frame of reference where two or more joints no longer independently control the position and orientation of the tool. In the American National Standard for Industrial Robots and Robot System – Safety Requirements (ANSI/RIA R15.06-1999) a singularity is defined as 'a condition caused by the collinear alignment of two or more robot axes resulting in unpredictable robot motion and velocities.'

Laser welding divided tailgate application of Audi A8

The demands of this project were similar to that of the laser welding door application. The tailgate of the Audi A8 could also not be produced as a single part due to high deep drawing degree in the region of the badge hollow. As a consequence, a divided tailgate was transcribed with the guideline that the joint, in that case a zero gap laser joint, must not be visible to the customer as a junction of two single parts and additionally no post-treatment (e.g., a finish process), was acceptable.

Therefore the laser welding process with its very high optical seam quality was chosen. To raise the optical seam quality to the maximum, an additional shielding gas (He) was used. The basic conditions for the divided tailgate application of the Audi A8 were similar to that of the door application of the Audi A6. The cycle time was also 300 s and the welding speed 3 m/min. The same tactile guided process optics (ALO3) in combination with a 4 kW Laserline direct diode laser (LDF 4000-40) has been used. The joining length for the divided tailgate laser welding application was roughly 1 m.

21.4 Quality issues

The quality issues for joining technologies can roughly be distinguished into two different groups: the group of inner and the group of outer findings of a joint. Quality criteria like the optical impression of the seam, seam geometry, edge notches, spatters, outer cracks and surface pores among the surface are included in the outer findings. They can normally be identified non-destructively by visual/optical inspection, which can be done manually or via automated process control. Optical camera systems can generate a three-dimensional profile of the seam surface that can be analysed in a further analysis step. The inner findings are characterized through the strength properties of the joint. Typical volume imperfections are inner pores and cracks or a reduced binding length.

21.4.1 Stability

The binding or weld length is the minimal joint length where two metal sheets are fused together in a way that an intact metallic compound is formed. Inner pores reduce the effective binding length of a laser joint and the maximum force that a laser joint can transmit is reduced. The minimal binding length for a correctly performed laser joint depends on the thickness of the metal sheets used and the joint geometry. As a coarse estimate, a minimal binding length of more than 70% of the thinner sheet thickness should be achieved. Microcracks directly in the weld or hot cracks in the heat-affected zone lead to crack propagation under stress that can cause complete failure of the joint.

Seam geometries used in laser door applications are mostly overlap and fillet joints. Laser processing parameters like laser power, focusing conditions or welding speed are well known and can be controlled during the welding process. External process parameters (e.g., surface coatings or contaminants like oils and drying lubricants), the dimensional stability of the single assembly parts and the gap situation between the metal sheets in the clamped situation are not normally known exactly during the welding process and they also cannot be controlled that easily, but they greatly influence the quality of the joint. For laser welds used in crash-relevant areas, preliminary treatment before the laser welding process is therefore essential. A detailed overview of pre-treatment methods for zinc-coated metal and aluminium sheets is given in the next section.

21.4.2 Preliminary treatments

Surface cleaning of aluminium metal sheets

The surface conditions of an aluminium metal sheet play a major role in the quality of the laser welding process. The stability of the laser process and the formation of defects, like pores, are massively influenced by the surface conditions of an aluminium metal sheet. To guarantee a constant and stable surface quality for the laser welding process, the clean surface conditions have to be 'recovered' in a separate cleaning process. That should be done as close as possible to the laser welding process to guarantee the best surface conditions for joining.

The cleaning of aluminium can be implemented with different pre-treatments. The most efficient treatment is a chemical washing process, where the aluminium parts are handled through different immersion baths. In the first treatment step, surface contaminants like oils and lubricants are washed away, in a second step the oxide layer is chemically etched, and finally the aluminium surface is passivated. The chemical cleaning of aluminium parts is very space and cost-intensive, because the cleaning chemicals have to be renewed at regular intervals. Additionally, the cleaned parts have to be transferred to and from the chemical cleaning facility, which requires a high logistical effort.

An alternative method is the so-called laser-clean process. The laser-clean process is industrially available (commercially distributed by Clean-Lasersysteme GmbH) and established. It is implemented in several Audi car projects, e.g. the pre-cleaning of the zero gap joint of the Audi TT and the divided tailgate of the Audi Q5. The physical principle that is used for this cleaning process arises from the different optical properties of the aluminium surface and the oxide or dirt layer for a laser radiation of 1 μm wavelength. The oxide layer absorbs that wavelength particularly well and blasts away from the surface. Existing dirt on top of the oxide layer will also be ejected and can then be removed easily. Blasting away the oxide layer is realized through very short but intense laser pulses.

The aluminium surface is highly reflective to the laser light used and therefore the cleaning process does not remove the aluminium surface underneath the oxide layer. The result is a bare metallic and oxide-free surface. The cleaning speed is limited by the laser track width and the ablation rate. In fact, a cleaning speed of 65 mm/s can be achieved. For best seam quality, all relevant flange surfaces that are affected in the joining process have to be cleaned by that method and, at worst, that can be up to four in an overlap joint situation with two metal sheets. Due to the use of highly intense laser radiation, generally increased security requirements, e.g. laser safety cabins, are required.

Pre-treatment of zinc-coated metal sheets

As stated in Section 21.3.1, a special pre-treatment is necessary for laser welding applications with zinc-coated metal sheets. To consider the situation for an overlap joint geometry with two double-sided zinc-coated metal sheets, two different sections have to be differentiated: the outer surfaces of the overlap joint on the one hand, and the inner surfaces, where the two metal sheets are in direct contact, on the other. When the joint is performed, the steel is melted and the zinc coating is vaporized suddenly due to the high welding temperatures required. For the outer surfaces the zinc coating can freely expand when vaporization temperature is reached. That only barely affects the quality of the welded seam.

The situation is completely different for the inner surfaces of the joint. For a zero-gap situation, the two zinc layers will be vaporized under high local pressure when the laser beam strikes the intermediate zone through the keyhole. As a result of the zero-gap situation, the zinc vapour can only leak through the molten metal, because the lower sheet is still in the solid phase. The consequence is an uncontrolled vaporization of the zinc layer with the side effect of spatters, pores and holes that negatively affect the quality of the joint (see Fig. 21.4).

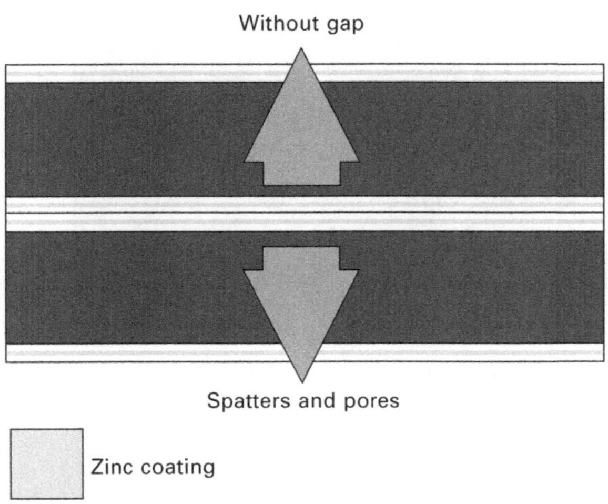

21.4 Uncontrolled zinc expansion without degassing gap.

A technological solution is obtained when a controlled degassing situation can be guaranteed where the zinc vapour must not pass through the molten metal. How can such a situation be generated? With a defined gap between the two metal sheets, the zinc vapour that is generated through vaporization of the inner layers can leak out to the sides and the quality of the joint is not affected (Fig. 21.5). The easiest solution is a 'spacer' between the two metal sheets. With this spacer in between, the metal sheets can be clamped with a constant gap. These spacers are normally performed in the form of small, so-called humps. These humps can be produced directly in the manufacturing process of the single part by stamping them into the surface of the sheet metal or they are produced in a separate process afterwards. For the door application of the Audi A4/Q5, the latter option was chosen as it has some advantages over a mechanical process. The best gap distance for a controlled degassing in the laser welding process is only approximately 0.15 mm. Therefore, sub-millimetre fluctuations in the humping height affect the quality of the welding process considerably. Influencing factors, such as a continuing tool wear during the single part production, would lead to a continuing deterioration of the degassing gap in the mechanical process, which should be avoided.

An alternative generating process for the humps is laser pulsing. For this purpose, the laser optics are guided over the welding flange of the underlying part before clamping the two parts together for the welding process. With short but highly intensive laser pulses, the humps, are 'shot' into the surface of the component. The right choice of pulse energy and pulse duration melts

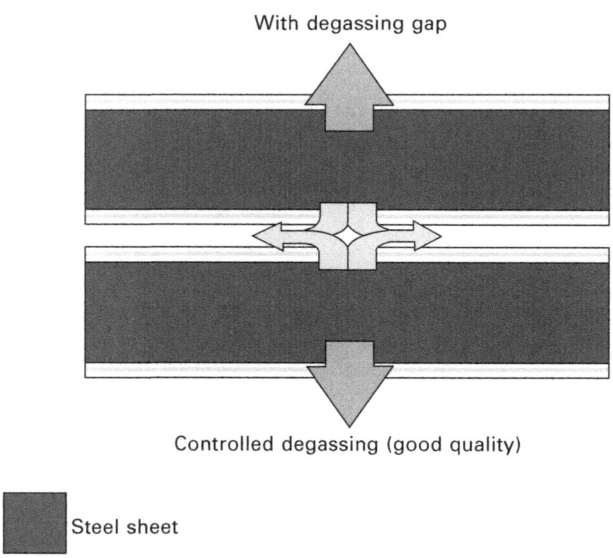

21.5 Controlled zinc expansion with degassing gap.

the surface locally as the melt is swapped aside. Solidification of the melt leads to a crater in the centre of the laser focus and to a 'hump' at the edges. These humps generate the necessary degassing gap for the following welding process between the zinc-coated sheet metals. Through the high quality and precision of the laser tool, the heat quantity and the pulse duration can be controlled very exactly. This results in a very reproducible and stable humping process and, as a direct consequence, in the high stability of the laser welding process.

21.5 Future trends

The most important objective targets in automotive car body manufacture are stable and highly efficient production processes with highest reliability and quality output. To ensure that these targets can be reached requires a lot of effort and expense from the first car body concepts to the final automobile.

For a new car body concept, all potential metal sheet combinations for construction have to be validated before the first prototypes can be produced. The starting point of the validation of new materials or metal sheet combinations for a laser welding application is normally a joining test in the laboratory. After successful welding tests, individual prototypes of structural components are produced. If all pre-tests have been successfully finished, more validation with a larger number of parts, produced with prototype tools and finally with the tools for volume productions follows.

Additional to the evaluation of the materials used and the joining process itself, there are a lot more relevant technical parameters that can only be estimated in the real volume production process. The variation of material quality, temperature effects, local contaminations, abrasions, fitting accuracy of the component parts, vibrations and additional effects that only appear after the production of a large number of parts cannot be safeguarded without the production of original parts in the serial plant. This time-intensive and costly safeguarding process is necessary to guarantee an optimal start of production (SOP). The basis for an efficient validation process is a detailed knowledge of the laser process and all the relevant explanatory variables. This is essential for stable processes and a high quality.

21.5.1 Car body construction

Enhanced environmental responsibility and the fact that fossil fuels are limited induce alternative mobility concepts like electro mobility and CO_2 reduction in terms of lowering fuel consumption. New concepts for energy storage have to be developed and integrated into the car structure. Additional components induce additional car weights that have to be compensated if the driving dynamics are to be maintained. For that reason, future car body concepts will be driven by lightweight design. This trend will also affect more and more hang-on parts like doors and tailgates. Lightweight metals will increasingly substitute classical steel hang-on parts in the future.

Weight reduction in the car body can be reached by functional use of multi-material design. Therefore the Audi philosophy for car body design is the use of materials with specific properties exactly where they are needed and where they generate maximum benefit for car body functionality.

This leads to the challenge to join different materials with different physical properties. At the moment, laser welding is only possible for a few different material combinations. For example, laser joining of aluminium with steel is principally possible, but problems like contact corrosion in the presence of an electrolyte and the huge difference in the coefficients of thermal expansion lead to physical problems of the joint during later thermal treatment of aluminium parts (temperatures up to 180°C are possible). Also laser remote welding, which is the most efficient laser welding technology, is not yet available for lightweight materials like aluminium, where an additional filler wire is essential.

On the one hand, the developments of new laser welding processes have the potential to overcome those limitations and to transfer the steel remote welding performance to aluminium. The active control of welding dynamics, for example keyhole formation/stability and temperature control of the melt, is a promising approach. On the other hand, the development of new materials (e.g. sandwich structures) with better welding characteristics and optimized physical properties has continuously to be pursued.

21.5.2 Laser process monitoring and control

Laser joining technologies are very successful applications, if prerequisites of the technical process are considered correctly. Every process device in the laser application has technical limitations in accuracy. For every single car body component there are tolerances due to the production process that lead to variations in the assembly parts. The surface quality of the metal sheet varies and surface impurities can occur. These inaccuracies affect process stability and the quality of the joining process. In the worst case, these inaccuracies produce imperfections that make the component or the assembly part useless.

In order to be able to secure uniform quality demands at worldwide production sites, objective assessment criteria are necessary. These criteria have to be interpreted fully automated and non-destructive during the production process. This enables an automated control of construction parts of different quality for post-stored process steps, e.g. special rework procedures. Due to the complexity of quality criteria, the manual evaluation of component quality is strongly restricted.

The simplest form of process monitoring is a quality rating after the laser welding process where the corresponding quality criteria are checked. Optical camera systems for three-dimensional (3D) seam visualization are state-of-the-art for laser welding applications. A measuring principle for the visual process analysis is laser triangulation, for example. For this purpose, the laser seam geometry is optically analysed and the three-dimensional information of the seam surface is generated and compared with defined quality criteria like seam length or seam height, for example. General quality criteria like seaming location, welding sink, edge notches or surface pores and cracks can only be detected within certain limits. For volume information optical systems are normally not usable. Volume defects like volume pores and inner microcracks or the binding length that is representative of the maximum force level cannot be detected that way.

For these quality criteria, non-destructive analysis for industrial applications is not yet available and alternatives have to be developed. Methods like computer tomography, ultrasonic or thermo-graphic procedures are promising approaches.

The success of a monitoring method is also determined by local conditions like the seam variation (e.g. fillet or overlap joints), and of the component contour. The quality of a process monitoring system is decided through the reliable detection of actual imperfections in combination with minimal pseudo error rate. Pseudo errors are events where the process monitoring system reports an imperfection that is not, however, present according to the pre-defined assessment criteria.

Therefore an intensive qualification phase is necessary for each process monitoring system when maximum reliability should be reached. These quality limits are manual prerequisites for margins of error that are verified in a regular qualification process.

In the next degree of complexity of process monitoring systems, inline process detection of characteristic process variables (e.g., the temperature or the length of the melting bath), is required. Certain process variables are then correlated with characteristic quality features. A typical example is the correlation between thermal conductance and the reduction of effective weld width when two metal sheets are not joined correctly. With defects in the welding seam, the effective weld width is reduced and the thermal conductance through that cross section drops. As a result, the heat that the laser beam induces into the upper metal sheet cannot drain off freely into the lower sheet and the temperature of the upper sheet rises due to heat accumulation. That can be measured inline during the welding process and thus deliver additional information for the process quality.

However, the real goals are process-monitoring systems that contribute actively to error prevention. For this purpose, it is no longer sufficient to only detect a defect, but process variables have to be identified that are good quality regulating variables at the timescale of the welding process.

The regulation of laser processes is very difficult due to their high process complexity and the high process dynamics. Nevertheless, there are first approaches such as the full penetration hole extraction in laser welding processes for overlap joints (Nicolosi *et al.*, 2009). The future challenges in this field lie in the development of new, at this time unknown, error prevention strategies for laser welding applications on one hand, and the qualification of existing approaches for industrial practicability on the other.

21.5.3 Simulation of laser processes

To reduce development times and costs, the need for virtual validation concepts that allow a detailed variation of technical parameters to locate the stable process window for joining technologies is essential. For laser joining technologies, process simulations are not yet available because of the complexity of the laser welding process. For that reason, the further development of laser process simulations is essential to completely understand the laser welding process in detail. With a deeper process understanding, the quality of laser applications can be improved if the causes of seam imperfections like pores, cracks and spatters can be identified and avoided by specific process optimization.

Even for a quite simple laser weld model, like an overlap joint with two plain metal sheets, process simulation has to factor a lot of physical properties and dynamic effects: wavelength-dependent absorption of laser light for different incident angles, polarization, the temperature-dependent properties of the material in different phases (solid, liquid and gas phase/plasma) and pressures which determine the dynamics in the melt and in the keyhole/vapour plume to take only a few into consideration. Many of those physical properties are only available for concrete temperatures under idealized preparation and laboratory conditions, but for the relevant temperature range of the real laser welding process, they are not universally known. The unknown values can only be estimated or calculated if suitable physical models exist. Nevertheless, there are approaches of process simulation models (Otto and Schmidt, 2010) that can compete with real laser welding processes under special circumstances but computational effort is enormous.

In the next step, laser process simulation models have to be developed further to steadily predict real laser welding processes more exactly. When laser welding process simulation can make reliable predictions for real laser applications, the question whether new multi-material car body designs can be realized with laser welding technology can then be estimated before the first hardware prototypes of structural elements exist. That will reduce the development period, raise the quality of laser welding applications and massively reduce costs and capacities for expensive safeguard experiments. At this point, virtual technologies like laser process simulation will become an essential tool for all future laser application in the body shop and beyond.

21.5.4 Environmental compatibility

Environmental compatibility is one of the main targets for Audi as a premium automotive manufacturer. The current CO_2 debate and new regulations will lead to further developments of new technologies that aim to reduce energy consumption. One approach is the improvement of existing combustion engine concepts, but in future the focus will shift increasingly to alternative concepts such as electric engines or fuel cells. In addition to the improvement of existing engine concepts, the overall weight of the car has to be reduced to fulfil the environmental demands. Moreover, an integrated approach with total life cycle energy demands per car is important. Taking this approach one step further, the necessary energy for producing a car becomes as important as the fuel consumption. For that reason Audi board member for production, Dr Frank Dreves, has stated that a CO_2-neutral production plant is one of the major targets for Audi and 'Ingolstadt should be the first CO_2 neutral automotive plant world wide'.

Mr Arne Lakeit, Head of Manufacturing Engineering, explains how to reach that challenging target: 'All sectors of the production plant, from the primary energy production over the complete production process to intelligent concepts for waste-heat recovery have to be taken into account. In the primary energy production process, Audi uses the heat of a waste incinerating plant and renewable energy concepts like photovoltaic cells on the roof of buildings and wind turbines for the production plant in Ingolstadt. Furthermore every single energy consumer in the production process is analysed and energy consumption will be optimised.'

With the developments in laser source efficiency and the technological advantages of laser applications in terms of weight saving, laser technology has become a very important part on the way to CO_2-neutral production. The nearly ten times higher wall plug efficiency when replacing a lamp-pumped Nd:YAG laser with a direct diode laser in combination with the huge number of laser sources in use in the Audi body shop, lead to massive energy and CO_2 savings per year. Besides the direct energy saving as a result of the higher wall plug efficiency, the reduced waste heat enables smaller and more efficient laser cooling systems, which conserves cooling water. For Mr Lakeit, energy saving is an important contribution for an environmentally correct production but he is establishing a more holistic approach: 'The reduction and recycling of waste, air pollution and water consumption are equally important to realize a worldwide environmental correct production for Audi'.

21.6 References

Fahrenwaldt, H.J. (2009) *Praxiswissen Schweißtechnik*, Wiesbaden: Vieweg+Teubner.

Hornbogen, E., Eggeler, G. and Werner, E. (2008) *Werkstoffe*, Berlin-Heidelberg: Springer.

Lide, D. R. (ed.) (2010) *CRC Handbook of Chemistry and Physics*, 90th Edition (Internet Version 2010), Boca Raton, FL: CRC Press/Taylor & Francis.

Nicolosi, L., Tetzlaff, R., Abt, F., Hofler, H., Blug, A. and Carl, D. (2009) New CNN based algorithms for the full penetration hole extraction in laser welding processes, *IEEE International Symposium on Circuits and Systems ISCAS*, Taipei, Taiwan.

Ostermann, F. (2007) *Anwendungstechnologie Aluminium*, Berlin-Heidelberg: Springer.

Otto, A. and Schmidt, M. (2010) Towards a Universal Numerical Simulation Model for Laser Material Processing, *Physics Procedia*, 5, 35–46.

22
Applications of laser welding in the railway industry

H. WANG, Changchun Railway Vehicles Co. Ltd, China

DOI: 10.1533/9780857098771.4.575

Abstract: With so many advantages, such as lightweight, antirust, low cost of maintenance and simple manufacturing process, the production of high-level stainless steel railway vehicles has become the development strategy of European, American and other developed nations. High-quality surface state and structure strength can be achieved by the partial penetration lap laser welding of the side panel structure. Laser welding will be the new technology for the manufacture of railway vehicles. The laser welding process, influence of laser welding parameters on the surface and strength, the heat source model of lap laser welding, methods of quality control, and future trends in laser welding are discussed in this chapter.

Key words: laser lap welding, process parameters, quality control, heat source model.

22.1 Introduction: the role of laser welding in railway engineering

Stainless steel railway vehicles have many advantages such as lightweight body structure, high anti-corrosion, safety features, good environmental performance, etc. SUS301L is one of the most widely applied materials in the stainless steel vehicle structure on unpainted outside panels [1]. Because of its low thermal conductivity and large linear expansion coefficient, stainless steel is usually assembled with resistance spot welding (RSW). But due to the punched power and thermal input of resistance of the spot welding electrode, there will be a visible indentation 1 cm in diameter left on the surface of the stainless steel vehicle body, influencing the appearance quality and airtightness. It is necessary and urgent to find a new method to resolve these issues.

Recently, high power density welding processes have been increasingly used in industrial manufacturing, because of their enormous advantages, such as minor shape distortions, reduction of size of heat-affected zone (HAZ), and faster welding speed associated with high penetration [2]. Some studies [3,4] have investigated the possibility of applying laser technology to replace resistance spot welding for stiffener assembly on rail vehicle side panels. Japanese Kawasaki Heavy Industries have studied the method of lap laser welding of SUS304 stainless steel with no visible indentation on the outside surface, which was reported in Ref. [3]. The author also pointed out the problems discussed above and summarized the resistance spot welding joint as 'point' connections, and the laser welding joint as 'line' connections. Some experiments [5,6] indicated that the welding speed should match the laser power, namely high welding speed matching high laser power, to get a joint with no visible indentation on the surface. Various studies have indicated that the welding speed could be up to 5 m/min, which was too high for ordinary lasers [3,4].

In addition, laser welding technology will be applied to Siemens magnetic levitation train in Germany, DMU IC4 level high-speed train in Denmark, ETR500 high-speed commuter trains in Italy, and so on.

22.2 Laser welding technology for stainless steel railway vehicles

The disk laser is a brilliant combination of a solid-state laser and a diode laser. The diode laser provides a high degree of pumping efficiency and the solid-state laser, as a disk laser, provides the high beam quality [7]. The optimized combination of high laser power plus excellent beam quality is kept constant across the entire power range. YAG laser beam welding with a continuous wave is a high energy density and low heat input process. The result is a small HAZ, which cools very rapidly with very little distortion, and a high depth-to-width ratio for the fusion zone [8].

SUS301L is a kind of austenitic stainless steel with reliable structural strength. Because of its light weight and environment protection, SUS301L is widely applied in the stainless steel vehicle structure as the outside panel with no painting [9]. Due to deformation caused by the welding heat, the stainless steel is assembled with resistance spot welding with small thermal importation. But because of the punched power and thermal input of resistance of spot welding electrodes, there will be a 1 cm in diameter left indentation on the outside surface of the stainless steel vehicle body (Fig. 22.1(a)). This influences the appearance of public transport vehicles, and should not be ignored.

To laser weld SUS301L stainless steel to get an overlap joint with no welded traces on the outside surface (shown in Fig. 22.1(b)) and higher tensile strength than that of resistance spot welding is a new area of research and will be applied in railway vehicle manufacturing.

Ordinarily, laser beam welding involves many variables: laser power, welding speed, defocusing distance and type of shielding gas, any of which may have an important effect on heat flows and fluid flows in the weld pool. In turn, this will affect penetration depth, shape and microstructure of the weld fusion zone. Both the shape and microstructure of the fusion zone will

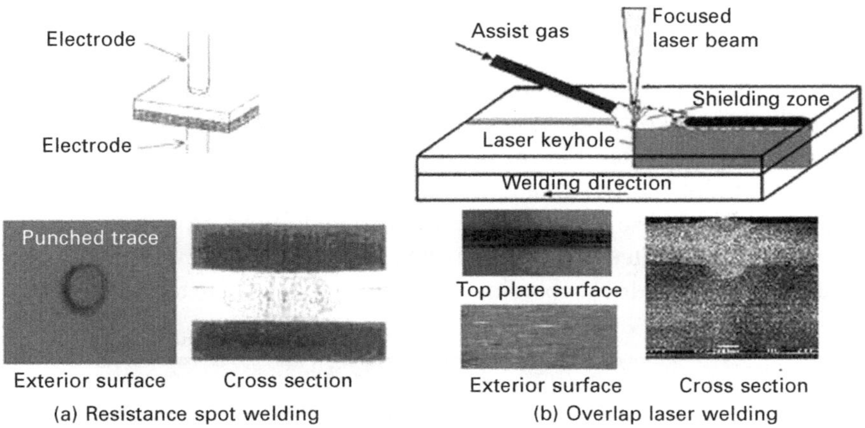

22.1 Appearance of cross section of (a) resistance spot welding and (b) laser welding joints.

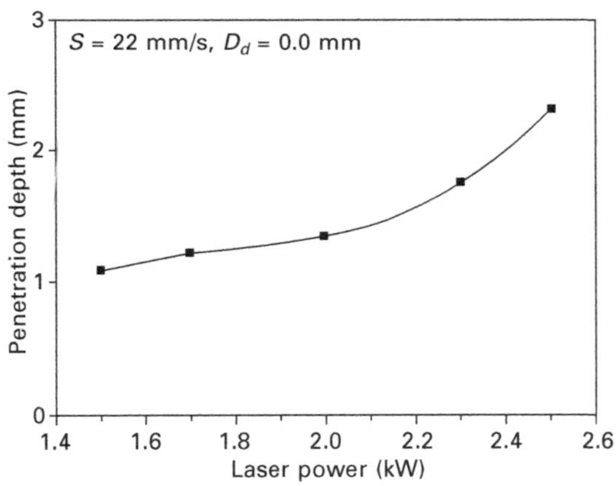

22.2 The influence of laser power on penetration depth of overlap joint.

have a considerable influence on the properties of the joints and the quality of the exterior surface.

22.2.1 Influence of laser welding parameters

Influence of laser power

The influence of laser power was investigated, keeping both welding speed and defocusing distance constant at 22 mm/s and zero, respectively. The penetration depth increased sharply with increasing laser power from 1.5 to 2.5 kW as shown in Fig. 22.2.

Figure 22.3 shows an example of a cross section of overlap weld joint using laser power of 2.0 kW. The microstructure of the cross section is shown with an inverted parabola shape, as the intensity of the laser beam has a Gaussian-like distribution (Fig. 22.4). The fusion zone is symmetrical about the axis of the laser beam. No welding cracks or pores (porosity) were found in any of the welds, and this may be partly due to the good crack resistance of the base metal and the welding conditions provided.

Mazumder and Steen [10] investigated a typical keyhole welding process model and the cylindrical coordinate system (r, θ, z) used in the model which

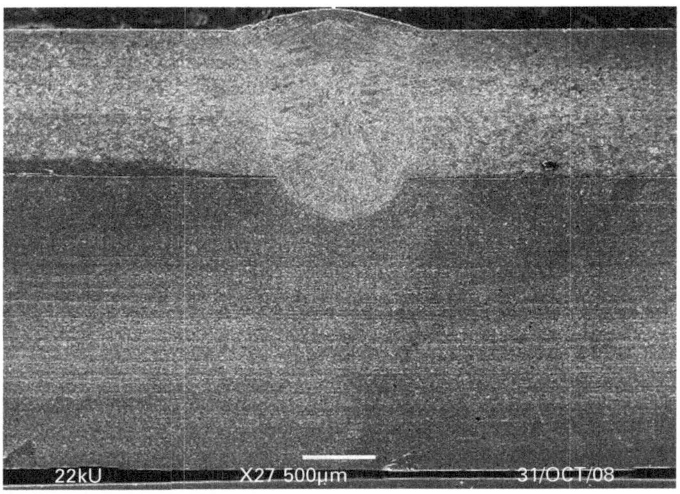

22.3 The microstructure of the cross section of overlap joint with the parameters P = 2.0 kW, v = 22 mm/s, D_d = 0 mm.

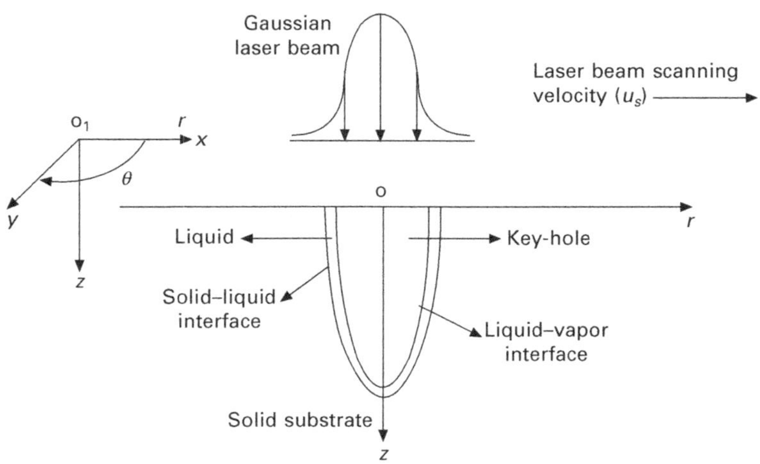

22.4 Gaussian-like distribution of the intensity of laser beam.

is shown in Fig. 22.4. According to this model, the results also indicate that the development of the weld pool is essentially symmetrical about the axis of the laser beam. Laser power has less of an influence on both weld profile and HAZ width in comparison with its effect on penetration depth. This is in agreement with other researchers' work, where they pointed out that changing laser power between 3 and 5 kW [11] did not result in any significant change in the size or shape of the weld.

When the laser power is lower than 2 kW and the penetration depth is below 1.4 mm, there are no welded traces on the exterior, but on the contrary distinct welded traces appear on the surface.

Influence of welding speed

The influence of welding speed was investigated at the optimum laser power (2 kW) and zero defocusing distance. Figure 22.5 shows the relationship between welding speed and fusion zone area. The fusion zone area is defined as half of the product of depth and width of the fusion zone. The fusion zone area decreased sharply from 2.26 to 1.59 mm^2 with the increase in welding speed from 16 to 28 mm/s.

The dependence of the fusion zone area on welding speed was confirmed at a different laser power (2.3 kW). A lower welding speed resulted in a considerable increase in the fusion zone area and consequently an increase in shear tensile strength of the overlap joint leading to unacceptable surface quality as shown in Fig. 22.6. When the weld speed is higher than 22 mm/s, there is no welded trace on the exterior surface. The shear tensile strength of the overlap joint at a speed of 25 mm/s is lower than that of the joint at a speed of 22 mm/s.

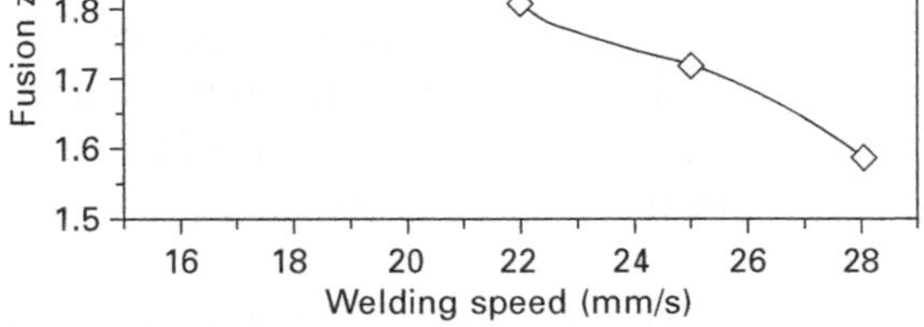

22.5 The influence of welding speed on area of fusion zone.

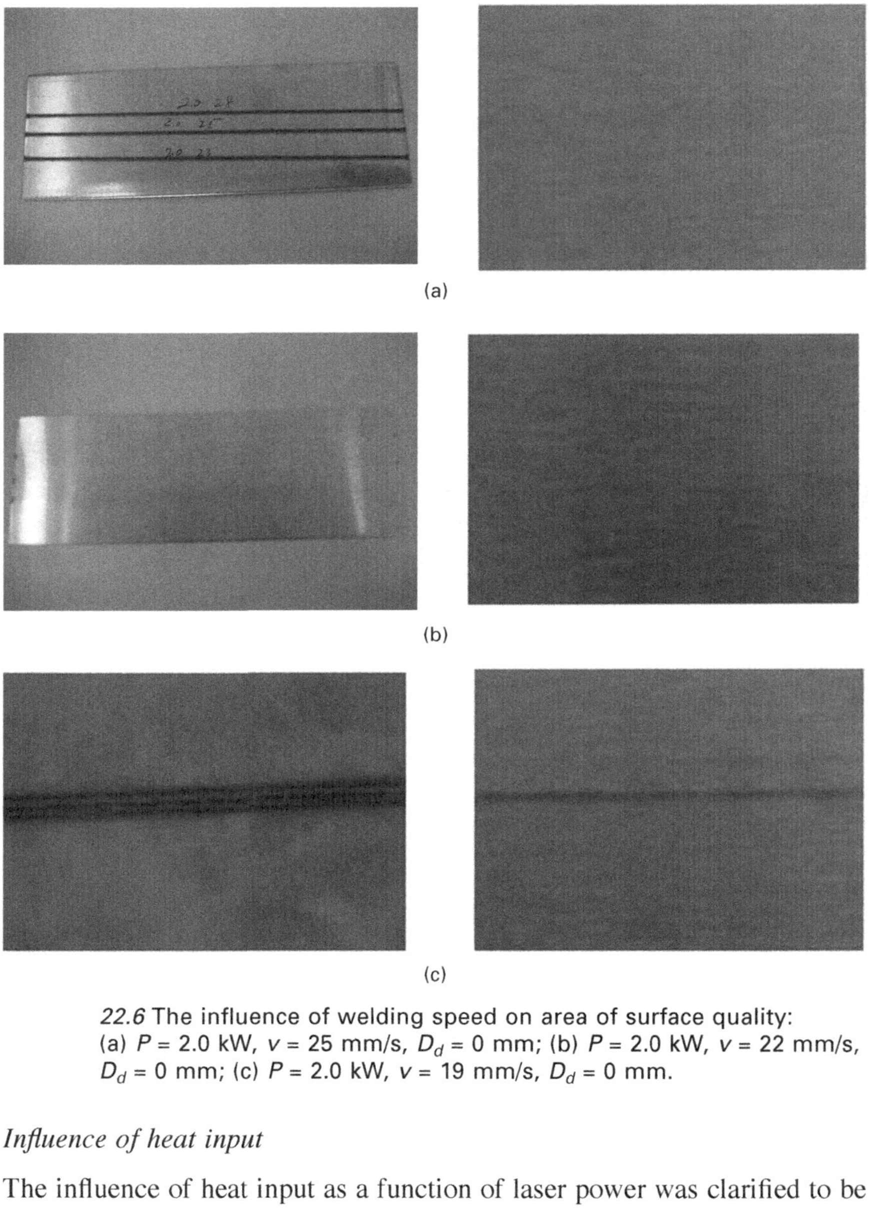

22.6 The influence of welding speed on area of surface quality: (a) $P = 2.0$ kW, $v = 25$ mm/s, $D_d = 0$ mm; (b) $P = 2.0$ kW, $v = 22$ mm/s, $D_d = 0$ mm; (c) $P = 2.0$ kW, $v = 19$ mm/s, $D_d = 0$ mm.

Influence of heat input

The influence of heat input as a function of laser power was clarified to be $H_I = P/S$. The above results have shown that the laser power and welding speed should be optimized in order to minimize heat input and to obtain a satisfactory weld with reliable quality. This reflects one of the most notable features of laser welding compared with other welding processes, namely low heat input.

Turning to the macrographs shown in Figs 22.3 and 22.7, incomplete penetration and no welded trace on the exterior surface with higher shear

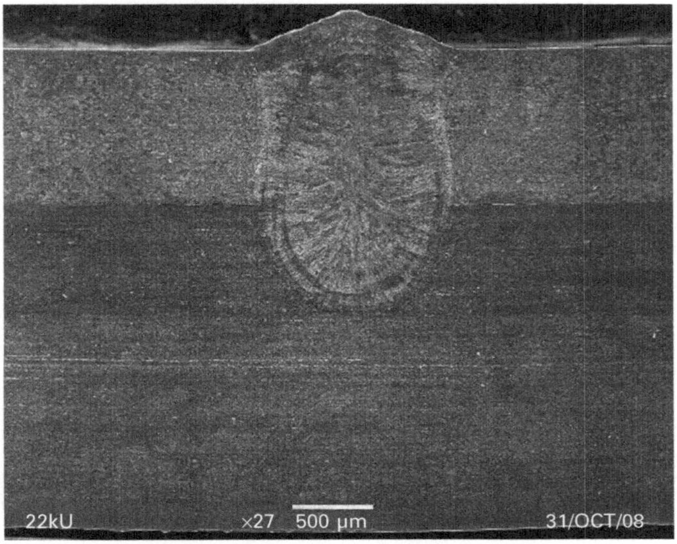

22.7 The microstructure of the cross section of overlap joint with the parameters P = 2.3 kW, v = 25 mm/s, D_d = 0 mm.

tensile strength than that of RSW fusion zone profile could be obtained using either P = 2 kW, S = 22 mm/s (Fig. 22.7) or P = 2.3 kW, S = 25 mm/s (Fig. 22.11). However, P = 2 kW, S = 22 mm/s resulted in a smaller welded trace on the exterior surface. In laser welding, if the laser power is too high, the spatter will increase to pollute the focus lens, increasing the risk of lens explosion [12]. When achieving the same result, the lower laser power should be selected.

Effect of defocusing distance

Defocusing distance is the distance between the specimen surface and the optical focal point. In order to study its influence on both penetration depth and surface quality, a lap joint was made by changing the defocusing distance between –2 and 2 mm. Laser power (2 kW) and welding speed (22 mm/s) were selected to obtain incomplete penetration.

The relationship of the defocusing distance and the penetration of the overlap joint is shown in Fig. 22.8. The penetration depth is considerably decreased by changing the defocused distance from –2 mm to 2 mm. In laser welding, when the focus is below the upper surface, the laser power density of the interior is higher than the surface, and the energy transfers deeper into the material, making it possible to weld thick materials with deep penetration. For thin materials, a positive focal position is usually used [13].

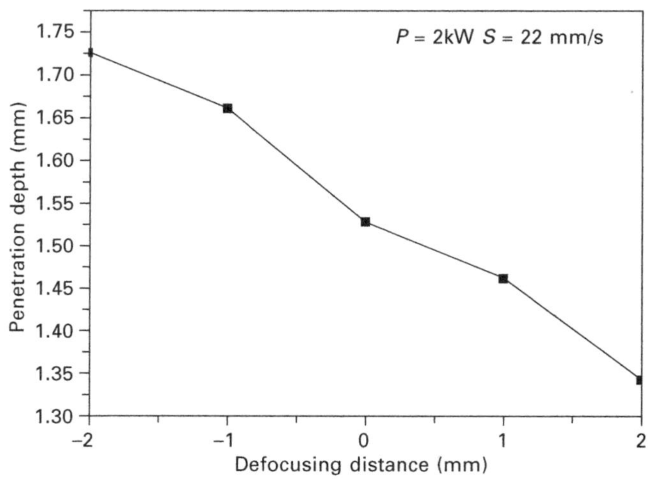

22.8 The influence of defocusing distance on penetration depth of overlap joint.

22.2.2 Features of laser welding joints in stainless steel railway vehicles

Microstructure of laser beam welds

In laser welding, the material is molten for a very short period and solidifies quickly, resulting in the difference between the microstructure of the laser welding and that of conventional welding methods. Molten metal first crystallized in the solid–liquid interface then grows up rapidly to the interior of the molten zone. The crystallized form in the fusion line extends to the center of the weld seam and finally combines there, and almost no HAZ can be seen as demonstrated in Fig. 22.9(a). No cracking or porosity was observed.

A clear dividing line marking the boundary of the fusion zone can be seen in the microstructure of the transition from the base material to the fusion zone (Fig. 22.9(b)). With the great degree of subcooling existing in the solid–liquid interface, the crystal of the fusion line grows upwards into columnar dentrites with rapid crystallization speed and distributed in parallel perpendicular to the fusion line. In the weld seam center, the gradient of temperature is smaller than that of its surroundings and the unfused suspended particle becomes the solidification surface of non-spontaneous nucleation, growing up freely from the cooling conditions into equiaxed dendrites (Fig. 22.9(c)). The equiaxed grains of the center were fine and shaped during the re-crystallization of the welding heat circulation.

Microstructures of type SUS301L steel weld metals made using two different welding speeds, 19 and 22 mm/s, with the same laser power, 2

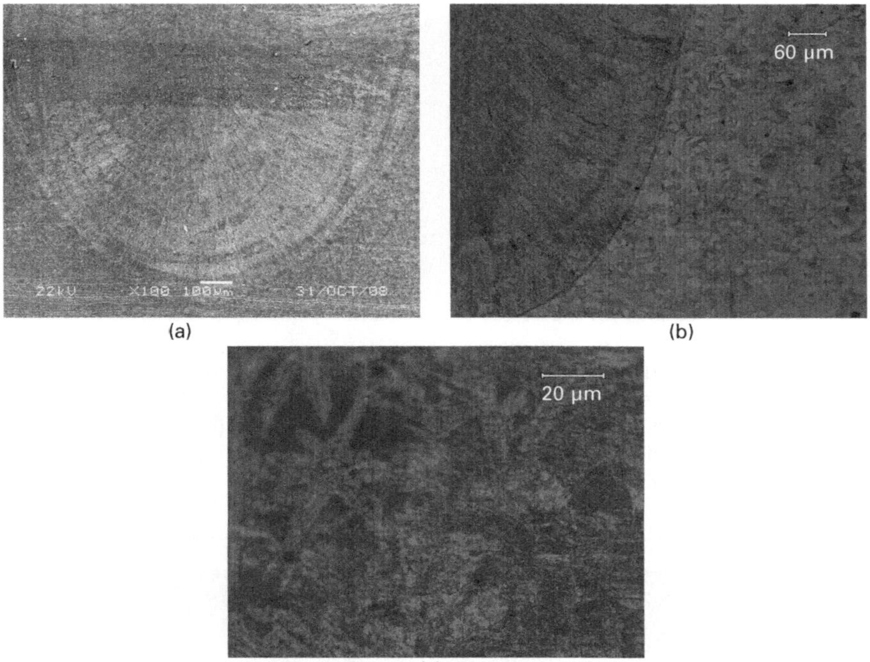

22.9 Microstructure of the overlap laser welding joint: (a) The microstructure of laser welding with parameters $P = 2.0$ kW, $v = 22$ mm/s, $D_d = 0$ mm; (b) the junction of weld seam and basement material with parameters $P = 2.0$ kW, $v = 22$ mm/s, $D_d = 0$ mm; (c) the equiaxes dendrites in the seam center with the parameters $P = 2.0$ kW, $v = 22$ mm/s, $D_d = 0$ mm.

kW, are shown in Fig. 22.10(a) and (b), respectively. The noticeable feature is the highly directional nature of the microstructure around the axis of the laser beam. This is due to solidification of the weld metal at high cooling rate compared with that of conventional GTA welding [14].

It can also be noticed that the higher the welding speed, the finer the dendritic structure (Fig. 22.10(b)). This is attributed to an increase in both solidification and cooling rates due to low heat input resulting from high welding speed. Concerning the effect of laser power, the higher the laser power, the coarser is the dendritic structure due to decreasing cooling rate. However, the effect of laser power was relatively smaller than that of welding speed.

Mechanical properties

Shear tensile test results of all type 301L laser lap welds with incomplete penetration showed that failure took place in the weld metal. Vickers hardness measurements with a 50 g loading force were carried out on the laser welding

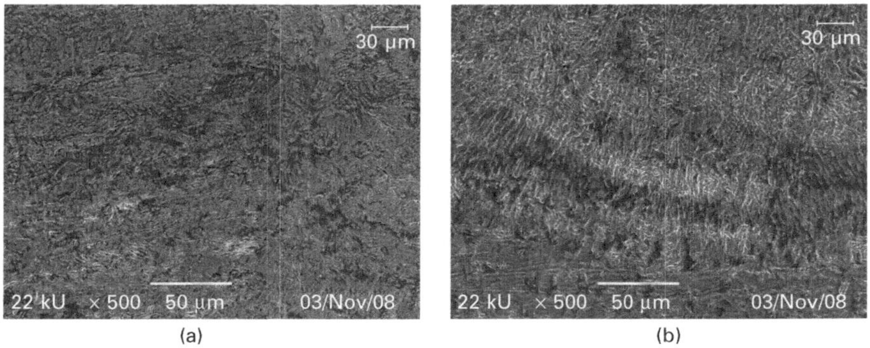

22.10 The dendritic structure of overlap joint with different welding speed: (a) the dendritic structure of laser welding with the parameters P = 2.0 kW, v = 19 mm/s, D_d = 0 mm; (b) the dendritic structure of laser welding with the parameters P = 2.0 kW, v = 22 mm/s, D_d = 0 mm.

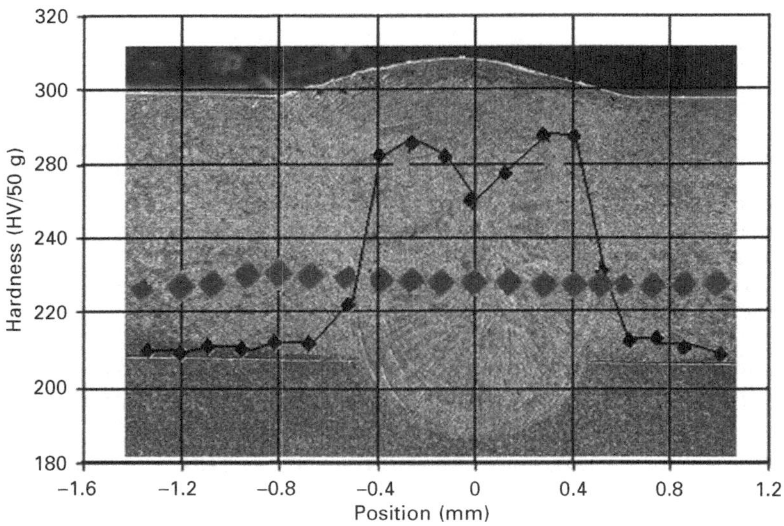

22.11 Cross section of an overlap laser welding joint with diagram of hardness P = 1.2 kW, V = 1.2 m min^{-1} Δf = 0 mm.

seams produced with different parameters. Figure 22.11 shows a cross section of a laser lap welded joint with the diagram of hardness. As expected, the hardness of the weld seam is higher than that of the base material. But the hardness of the weld seam center is lower than that of the part between the center and the fusion line.

The cooling rate of the weld seam center is more moderate than that of its surrounding part, and equiaxed dendrites appear in the center in the form of polygonal ferrite (PF) (as shown in Fig. 22.12(a)). In the zone between the fusion line and the weld seam center, the cooling rate is rapid and columnar

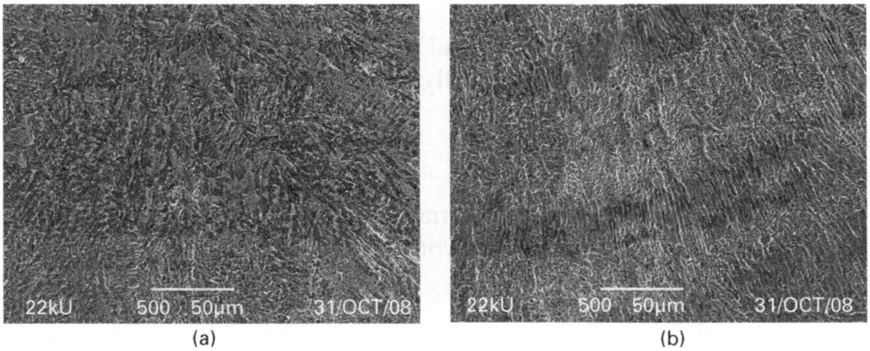

22.12 The phase organization of the weld: (a) the equiaxed dendrites in the weld seam center; (b) the columnate dentrites surrounding the center.

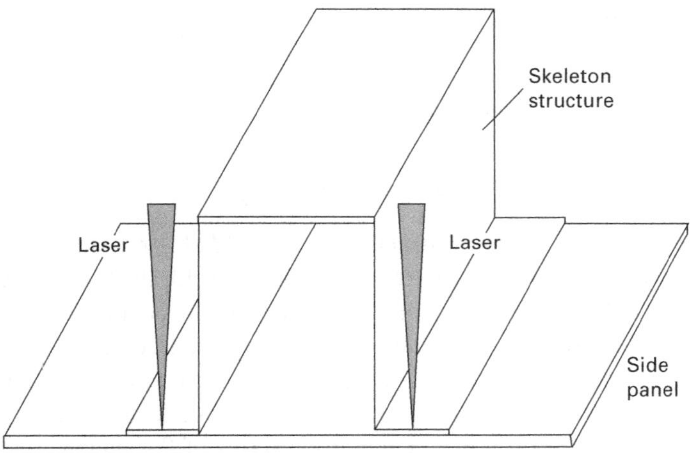

22.13 Physical model.

dentrites appear there in the form of acicular ferrite (AF) (as shown in Fig. 22.12(b)). Between the two adjacent acicular ferrite (AF), there is martensite (M-A constituent as the main element). The cementite or martensite and the acicular ferrite (AF) have higher hardness than polygonal ferrite (PF), explaining the lower hardness in the weld seam center.

22.3 Heat source model of lap laser welding of stainless steel vehicles

22.3.1 Model and mesh

The physical model of a stainless steel vehicle side panel is shown in Fig. 22.13. According to the features of the side structure, the side part is symmetrical. So half of the physical model is taken as the finite element model in the simulation, as shown in Fig. 22.14. Since the laser energy density is high, the mesh near the weld is smaller, and in order to reduce the computational cost, the mesh size increases gradually relatively far away from the seam.

22.3.2 Boundary conditions

The initial temperature of the material is 20°C. Taking into account the laser heating, the welding surface convection and radiation heat transfer [15] are shown by:

$$-k_e \frac{\partial T}{\partial z} + h(T_s - T_{ext}) + \sigma\varepsilon(T^4 - T_{ext}^4) = q \qquad [22.1]$$

where T_s is surface temperature, T_{ext} is the ambient temperature, h is the convective heat transfer, σ is the Stefan-Boltzmann constant, and ε is the thermal radiation coefficient [16].

SUS301L is much used in stainless steel car body material. Thermal conductivity and the specific heat coefficient of SUS301L are shown in Fig. 22.15 and other parameters are listed in Table 22.1.

22.3.3 Selecting a heat source model

Figure 22.16 shows the microstructure of the weld of partial lap laser welding of stainless steel. According to the shape of the weld, the heat source model is selected. The combination of three-dimensional positive cone + three-dimensional inverted cone + half-ellipsoid heat source model is shown in Fig. 22.17. The upper and central parts of the heat source model are two three-dimensional cones. The lower part is ellipsoid heat source. The three-dimensional cone heat model is actually a series of flat plate Gaussian heat sources superimposed along the thickness. After calculation, the energy distribution rate of the three heat source models is 0.25:0.3:0.45.

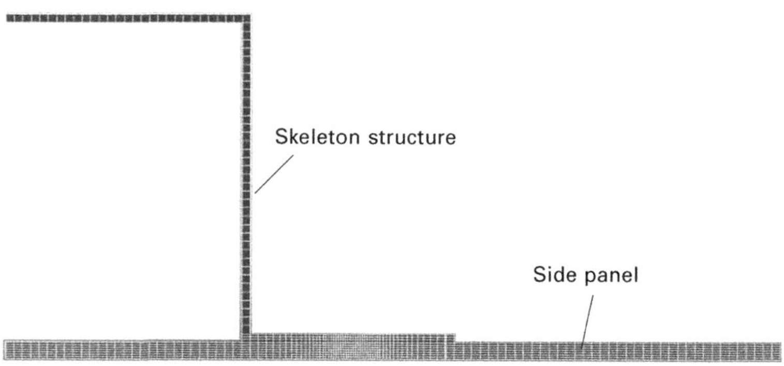

22.14 Simulated diagram.

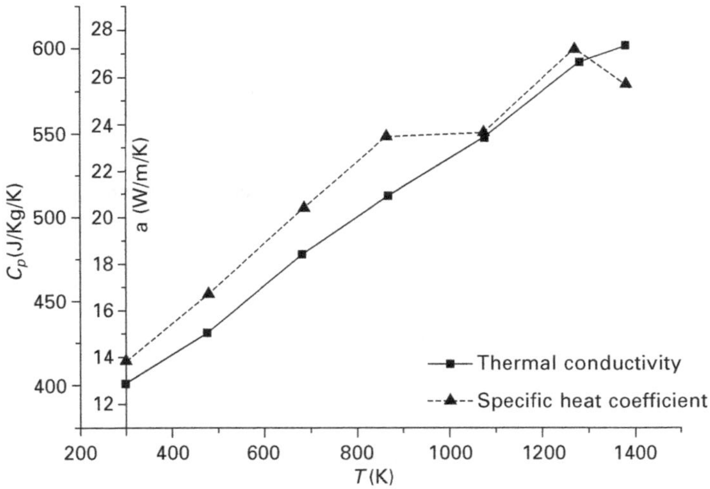

22.15 Thermal conductivity and specific heat coefficient.

Table 22.1 Relative physical properties of SUS301L material

Boiling point T_g (°C)	Liquefaction temperature T_l (°C)	Latent heat L_f ($J{\cdot}Kg^{-1}$)	Density p (kg/m^3)	Poisson's ratio μ
3073	1447	2.6×10^5	7800	0.3

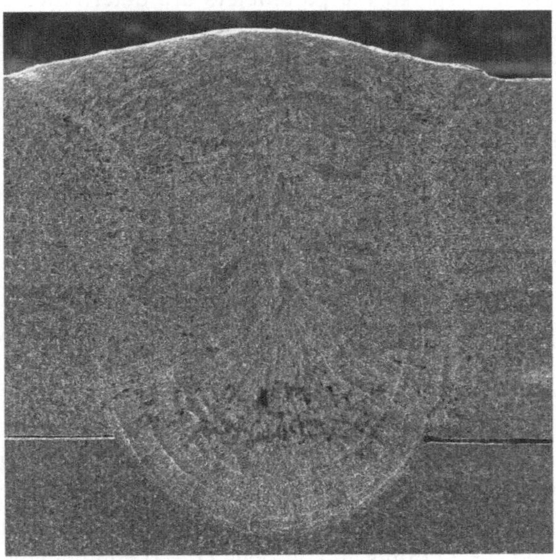

22.16 Cross section of laser gap weld.

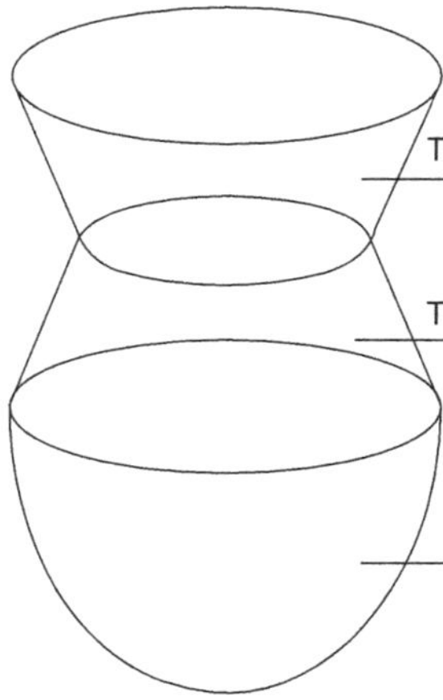

22.17 Combination of heat source models.

The three-dimensional cone heat equations are shown below [17]:

$$r_0(z) = r_e - (r_e - r_i)\frac{z_e - z}{z_e - z_i}$$

$$q(r, z) = \frac{9Qe^3}{\pi H(e^3 - 1)(r_e^2 + r_e r_i + r_i^2)} \exp\left(-\frac{3r^2}{r_0^2}\right) \quad [22.2]$$

In these equations, z_e, z_i are z coordinate positions where the heat source is working on the surface of the workpiece, and r_e, r_i are the radii of the heat source distribution [18]. All the parameters are determined based on experience and the test results.

The ellipsoid heat source formula is given by:

$$q(x, y, z) = \frac{6\sqrt{3}Q}{a_h b_h c_h \pi\sqrt{\pi}} \exp\left(-\frac{3x^2}{b_h^2} - \frac{3y^2}{a_h^2} - \frac{3z^2}{c_h^2}\right) \quad [22.3]$$

where, a_h, b_h and c_h, respectively, are the half axle lengths in the y, x and z directions, respectively.

22.3.4 Weld shapes under different parameters

Figure 22.18 shows the comparison of molten pools between simulation and experiment. It can be seen in the figure that the simulated pool shape is similar to a vase shape, and the trend of the fusion line is in good agreement with the experimental results.

Figure 22.19 shows the simulated pool shapes for different laser powers with the welding speed $S = 22$ mm/s and defocus $D_d = 0$ mm unchanged. The laser powers are 1.3, 1.5, 2.0, 2.5 and 2.7 kW. According to the experiment

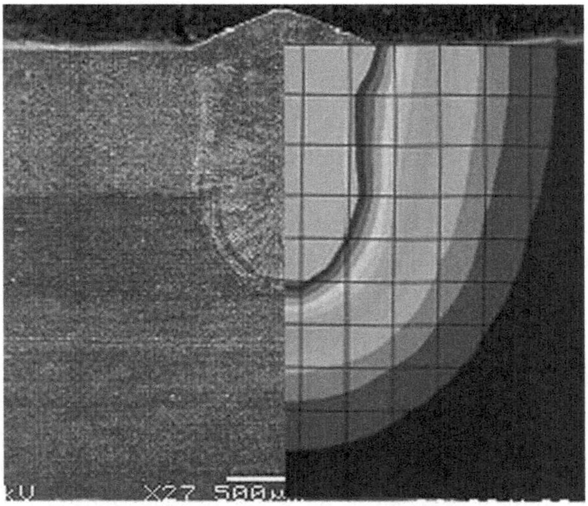

22.18 Comparison of molten pool between simulation and experiment.

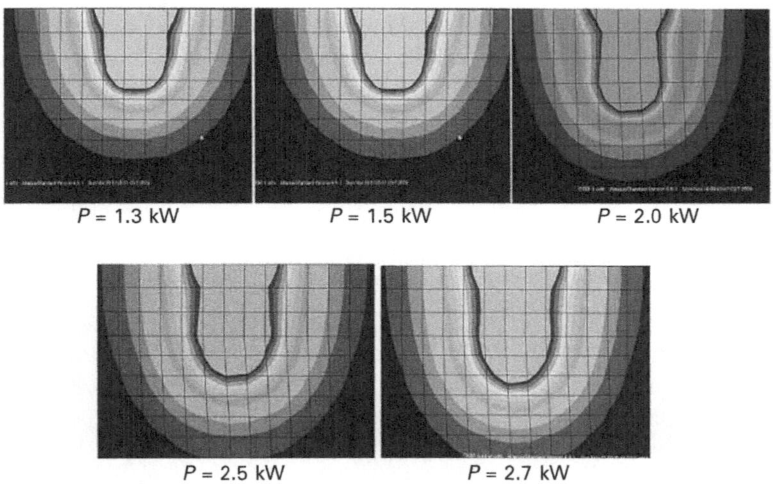

22.19 Simulated pool shapes for different laser powers.

results in Ref. [19], the simulated vase-shaped pool is consistent with the experimental results. As the laser power increases, so too does the weld penetration. By comparing the simulation and experimental results, when the laser power is higher than 2.5 kW, the weld penetration is more than 1.5 mm, and there are welded traces on the outer surface. When the laser power is less than 1.5 kW, the joint strength is too low to meet production requirements. Therefore, the range of power options for stainless steel vehicles is 1.5–2.5 kW.

Figure 22.20 shows the simulated pool shapes for different welding speeds with the laser power P = 2.0 kW and defocus D_d = 0 mm unchanged. The welding speeds are 15, 17, 22, 27 and 29 mm/s. As the welding speed is increased, the weld penetration decreases. By comparing the simulation and experimental results, when the welding speed is less than 17 mm/s, the weld marks on the outer surface are obvious. When the welding speed is higher than 27 mm/s, the joint strength is too low to meet production requirements. Therefore, the range of welding speed options for stainless steel vehicles is 17–27 mm/s.

22.4 Quality control of laser welding joints in stainless steel vehicles

As an alternative to spot welding for the vehicle body, laser partial penetration welding of two stainless steel sheets is used. In laser partial-penetration lap welding, the laser beam is focused on the surface of the sample and the other parameters are controlled to guarantee partial-penetration welds without welded traces on the exterior surface. Figure 22.21 shows the joint arrangement for laser welding. By changing the welding parameters, such as laser power and welding speed, different widths and penetrations should be achieved. In order to understand the laser weld defects, the metal structure, convex surface and pores of the laser partial-penetration lap weld are observed, and these defects on the fatigue strength should be explored. The extent of laser weld penetration has an important impact on the fatigue strength. Considering the impact, non-destructive inspection methods are necessary to evaluate the extent of penetration.

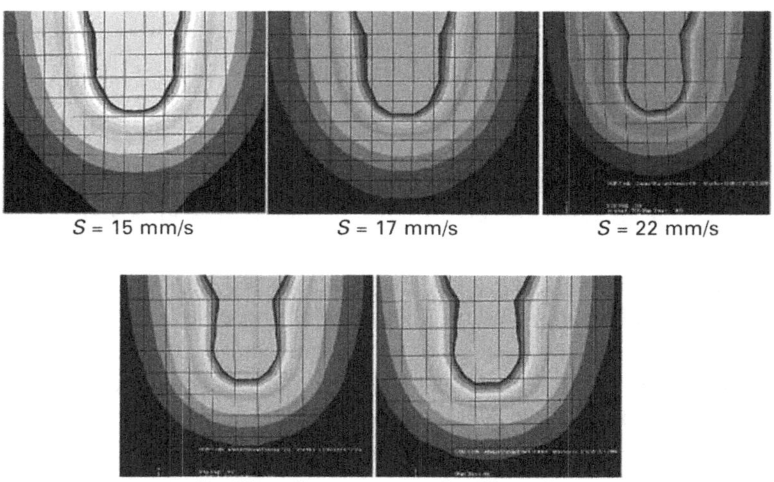

22.20 Simulated pool shapes for different welding speeds.

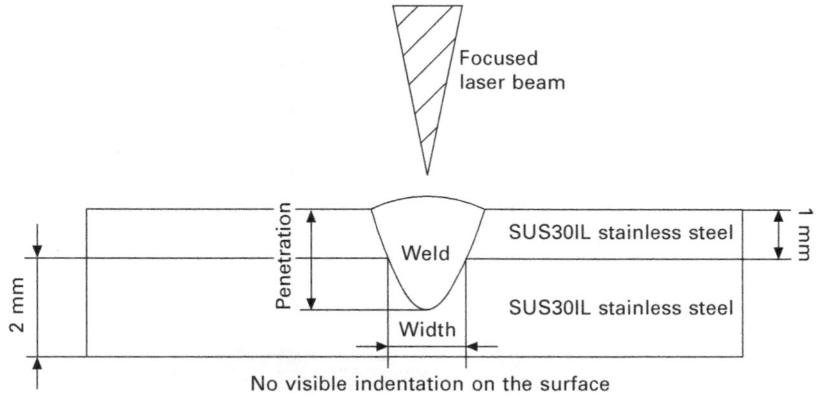

22.21 Schematic diagram of joint arrangement of laser welding.

22.4.1 Quality requests

The side wall structure of the railway vehicle is produced using a single-sided spot welding, which is an ideal way to ensure a good flatness of the outer surface of the side wall. Compared to arc welding, using resistance spot welding can reduce the body's welding deformation. But due to the large number of weld spots, local deformation is induced. According to the JISZ3140-1989 [20] spot welding test, the strength of the laser welding joint should be bigger than that of the resistance spot welding joint. The amount of deformation should be less than 2 mm/m, and there should be no trace of the weld discoloration.

Figures 22.21 and 22.22 show the mode of laser welding of railway vehicles and the structure of the side panel, respectively.

22.4.2 Ultrasonic inspection

In order to confirm the state of penetration of the laser weld, a separate and wedge-shaped high-frequency probe with oscillator front-end should be used. The ultrasonic flaw detection was carried out inside the laser weld. The results showed that a high-frequency wave detection probe could easily identify and confirm the weld, and the scanning was over a large area. Currently, the quality test of the laser welding of railway vehicles relies only on joint-destructive testing, and there is no quality assessment of laser weld non-destructive testing to evaluate weld quality control standards and methods. So it is urgent to explore the use of ultrasonic probes to detect joint penetration and the state of penetration of the laser welding joint and it is necessary to increase development of non-destructive testing devices.

Figure 22.23 shows the working principle of the ultrasonic probe.

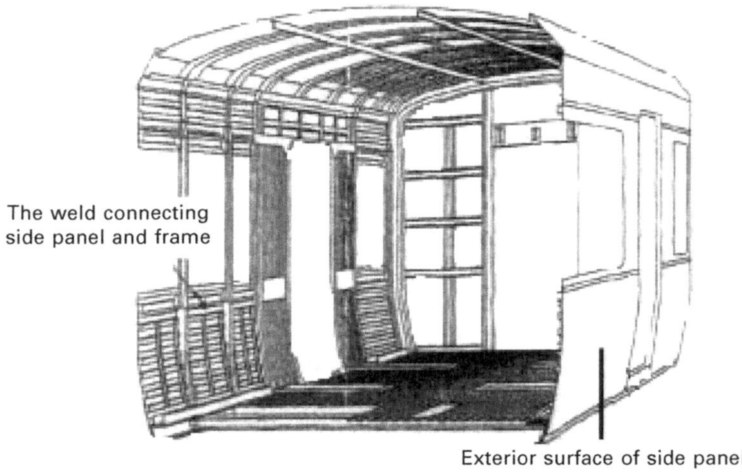

22.22 Lap laser welding of stainless steel vehicle structure.

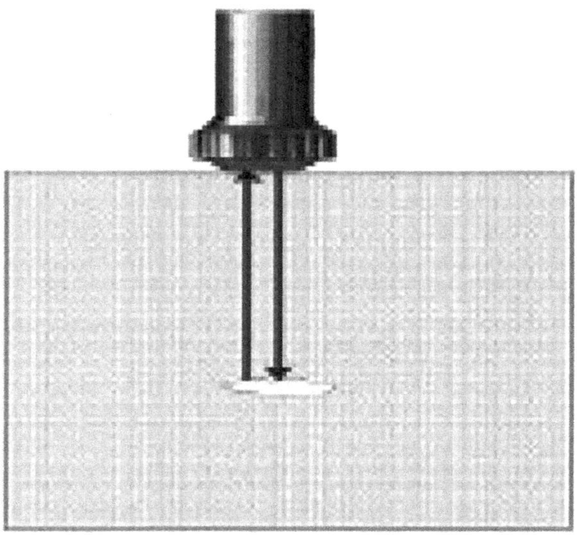

22.23 Ultrasonic probe.

22.5 Future trends

Use of laser welding of austenitic stainless steel, aluminum alloy and low alloy high strength steel for the manufacture of railway vehicles is growing. Increasing the application of laser welding technology to railway vehicles will play an important role in increasing the competitive environment in the future marketplace, and this technology has significant commercial value.

The application of laser welding technology to railway vehicles has the following four prospects:

1. Laser welding technology has been applied to high-speed train bodies in Japan. In addition, laser welding technology will be applied on Siemens magnetic levitation train in Germany, DMU IC4 level high-speed train in Denmark, ETR500 high-speed commuter trains in Italy, and so on.
2. The manufacturing standard of high-speed stainless steel bodies has seen a substantial increase, and there is a possibility for it to replace aluminum. It is necessary to develop the manufacturing technology.
3. The trends in advanced body structure design are for the overall load-bearing structures to have no frame, a car shell roof and floor, and both sides made of laminated sheet, which sandwich a light structure. This kind of structure can be achieved by laser welding technology, and also reflects the structure of the advanced design concept.
4. When the overall body is produced by laser welding technology replacing resistance spot welding, the fatigue strength can be increased by 15% and the vehicle's rigidity and strength are greatly improved, leading to a decrease in weight.

22.6 Sources of further information and advice

Laser welding technology of stainless steel vehicle car body

For the high-quality requirements of unpainted stainless steel body surface, although resistance spot welding can significantly improve the deformation of the surface state, there is still solid joint deformation at the surface. Major domestic and foreign companies that manufacture stainless steel rail passenger cars in recent years have begun to use laser welding to solve these problems. High-quality surface state and structural strength can be achieved by the partial penetration lap laser welding of the side panel structure.

Laser-MAG hybrid welding method for side body applications

The beam side body is currently welded by MAG welding methods, mostly using one-sided welding and surface rendering. MAG welding methods present the disadvantage that it is not easy to ensure adequate penetration. With a large welding current, welding speed is very low, resulting in thick deposited metal. Figure 22.24 shows that Laser-MAG hybrid welding can achieve a deeper penetration.

Laser welding of aluminum sheets

Currently, one of the main welding problems has been welding distortion of aluminum sheets. Especially in thin-walled aluminum Maglev car projects, welding of long seams is needed for 1 mm and 2 mm thick aluminum alloy

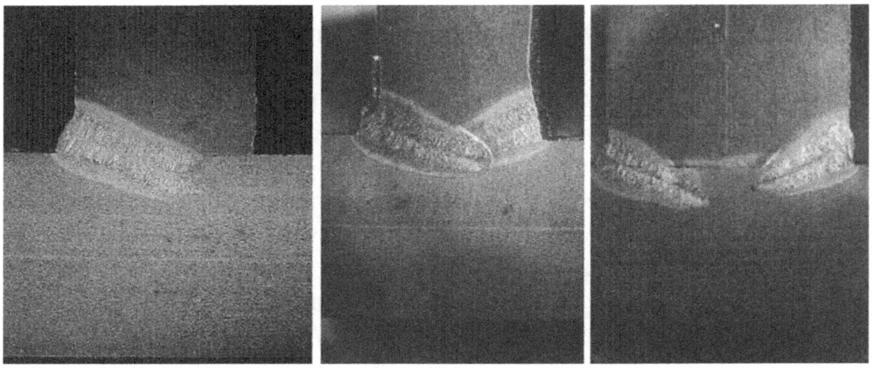

22.24 Laser-MAG hybrid welding.

sheets. The traditional method is MIG welding and the welding quality cannot meet the requirements of welding deformation. Laser welding and hybrid welding should be developed as new welding technologies to counter such problems in the future.

22.7 References

1. Xue K, The selection of material of city railway car body, *Reaching of Railway Communication*, Shanghai, China, 2003.
2. Wang J, Development and hope of laser welding technology, *Laser Technology*, Shanghai, China, 2001.
3. Yamadori, Kawasaki Heavy Industries Technical Report, 2006.
4. Dilthey U and Wieschemann A, Prospects by combining and coupling laser beam and arc welding processes, *Welding in the World*, 44(3): 37–46, 2000.
5. Samanta S and Kumar M *et al.*, Influence of welding speed on microstructure and oxidation behaviour of laser welded austenitic stainless steels, *Iron and Steel Institute of Japan*, 46(1): 100–105, 2006.
6. El-Batahgy A M, Effect of laser welding parameters on fusion zone shape and solidification structure of austenitic stainless steels, *J Mater Lett*, 32: 155–163, 1997.
7. Maren G, In-laminate laser soldering – a gentle method to assemble and interconnect silicon solar cells to modules, *21st European Photovoltaic Solar Energy Conference*, Dresden, Germany, 2006.
8. Wang H, Laser welding of 8 mm thick stainless steel plates. *Chinese Journal of Lasers*, 5: 463–466, 2003.
9. Arata Y, Challenge of laser advanced material processing, *Proceedings Conference on Laser Advanced Material Processing (LAMP'87)*, Osaka, May 1987.
10. Mazumder J and Steen W M, Heat transfer model for CW laser material processing, *J Appl Phys*, 51(2): 941–947, 1980.

11. Solana P and Ocana J, Mathematical model for penetration laser welding as a free boundary problem, *J Phys D: Appl Phys*, 30(9): 1300–1313, 1997.
12. Postacioglu N, Kapadia P and Dowden J, Theoretical model of thermocapillary flows in laser welding, *J Phys D: Appl Phys*, 24(1): 15–20, 1991.
13. Upadhyaya B N, A highly efficient 5 kW peak power Nd:YAG laser with time-shared fiber optic beam delivery, *Optics & Laser Technology*, 40: 337–342, 2008.
14. Choo RTC, Szekely J and David S A, On the calculation of the free surface temperature of gas-tungsten-arc weld pools from first principles: Part II modeling the weld pool and comparison with experiments, *Metallurgical Transaction B*, 23(6): 371–384, 1992.
15. Huzeng R and Zhou J, Application of ABAQUS simulation of laser welding temperature field, *Laser Technology*, 31(3): 326–329, 2007.
16. Tailor G A, Hushes M and Pericleous K, The application of three dimension finite vohme method to the modeling of welding phenomena, *Modeling of Casting, Welding and Advanced Solidification Process IX*, (ed. Sahm R), Aachen: Shaker, 852–859, 2002.
17. Hu Q, Finite element analysis of keyhole plasma arc welding temperature field, Jinan: Shandong University, 2007.
18. Charg W S and Na S J, A study on the prediction of the laser weld shape with vary heat source equations and the thermal distortion of a small structure in micro-joining, *Journal of Materials Processing Technology*, 120(1): 208–214, 2002.
19. Wang H and Wang C, The investigation of partial penetration lap laser welding applied on stainless steel railway vehicles. *Advanced Materials Research*, Vols. 97–101, 3832–3835, 2010.
20. JISZ3140-1989 spot welding test, Japan.

23
Applications of laser welding in the shipbuilding industry

J. KLÆSTRUP KRISTENSEN, The Technical University of Denmark (DTU), Denmark

DOI: 10.1533/9780857098771.4.596

Abstract: Laser welding, especially hybrid MAG-laser welding, shows great potential in the welding of structural steels in the shipbuilding industry. Advantageous properties include low distortion, high welding speed and easy automation. The shipbuilding industry has been one of the first to introduce high power laser and laser-hybrid welding for structural applications, with reducing distortion being the major motivation. However, many challenges had to be dealt with before laser welding processes could be introduced on an industrial scale. Today, laser-based welding is used quite extensively in European yards. This chapter reviews the advantages and challenges associated with laser welding, the mechanical properties that can be achieved, and its development and application in the shipbuilding industry.

Key words: laser welding, laser hybrid welding, mechanical properties, Classification Society guidelines, shipbuilding, ship yard applications.

23.1 Introduction

Laser and laser-hybrid welding offer many advantages over traditional welding processes due to their higher power density. The advantages include high-speed seam welding, low distortion, single pass welding of thick workpieces, easy automation, and positive effects on the working environment. For these reasons, the use of laser-based welding in structural applications has increased dramatically in heavier sections. In the fully synergetic hybrid processes, plasma formed at the interaction point of the laser on the workpiece surface controls the arc-root by reducing the cathode voltage drop, so that the interaction point becomes the same for the laser and the arc. When welding thick parts and/or in the presence of a gap, a hybrid laser welding process using the MAG process is most advantageous because, integrated in the process, it offers easy addition of filler material. In heavy-section

This chapter is a revised and updated version of Chapter 8 'Shipbuilding applications of hybrid laser-arc welding' by J. K. Kristensen originally published in *Hybrid laser-arc welding*, edited by Flemming Olsen, Woodhead Publishing Ltd, 2009, ISBN 978-1-84569-370-1.

laser-arc hybrid welding, penetration is to a first approximation determined by the laser alone, but a high welding speed may be maintained even in the case of a relatively large gap. Increased ability to bridge a gap as well as a significant increase in speed may thus be taken advantage of by using the hybrid process rather than traditional high power pure laser welding and, furthermore, the heat input per unit length (and therefore also the distortion) is typically not increased significantly.

Laser and especially laser-hybrid welding are now making an important impact on the medium- and heavy-section welding industry. The shipbuilding industry has been one of the first to introduce high-power laser and laser-hybrid welding, but this technology is also relevant to other areas, for example, pipe and boiler manufacturing. Many concrete initiatives have been taken during the last two decades, and several European shipyards have now introduced the process. The major motivation for this is reduced distortion; it is estimated that between 20 and 30% of the man-hours used in shipbuilding presently is due to reworking caused by welding distortions. Figure 23.1 illustrates this very convincingly.

23.2 The approval of laser-based welding in shipbuilding

In addition to physical defects produced during welding, a welded structure may develop flaws or completely fail in service owing to:

- limited strength or ductility,
- limited toughness (e.g. impact or quasi-static),
- environmentally assisted cracking (hydrogen cracking and/or stress corrosion cracking),
- general corrosion,
- fatigue.

23.1 Typical arc welded deck panels for a passenger ship; note the distortions.

For traditional arc welding, a number of mechanical tests have been developed for assuring that the ductility and toughness is sufficient and that serious cracks do not develop during the lifetime of a product. Evaluating a number of the following mechanical properties is usually recommended as part of welding procedure testing:

- hardness,
- strength and ductility,
- impact toughness, sometimes supplemented by crock tip opening displacement (CTOD) values,
- weld defects, such as geometrical defects, cracks or pores.

Compared to traditional arc welding techniques, laser-based welding offers many advantages, owing to the low distortion and high speed characterising the deep penetration laser welding process. However, the properties of the weldment pose a major challenge. The introduction of first pure laser welding and later hybrid laser-arc welding necessitated a number of general problems related to the properties of welded structures being solved. Developments during the previous three to four decades in materials and their response to welding had to be reconsidered in the light of this new technology.

The ductility of the weld could no longer be measured by the standard tensile testing techniques due to the narrow weld and potentially hard weld metal. Impact and toughness testing showed fracture path deviation, making evaluation of the tests difficult. Solidification cracking also turned out to be a challenge when welding thicker sections and the techniques for performing non-destructive evaluation of welds also had to be reconsidered. Related to fatigue, concern was raised about the potential sharpness of the geometrical stress concentration imperfections in, for example, the top bead and the weld toe regions, and similarly the fatigue crack growth rate in the potentially martensitic weld metal was questioned, as was the general and stress corrosion properties of the weldments.

23.2.1 Self-quenching

Comparing the heat input per unit length for typical arc and laser welds (for the same penetration or throat thickness, respectively) shows that the heat input in laser welds is typically almost an order of magnitude lower than in arc welds. It must therefore be taken into consideration that structural steels may harden significantly in both the weld metal (WM) and the heat-affected zone (HAZ) as a consequence of the fast thermal cycle inherently connected to high-power, high-speed laser welding. This immediately gave rise to concerns related to hardness, ductility, toughness, fatigue and corrosion. Three approaches to controlling this challenge may be identified:

- **Scenario 1:** Limiting the weld speed so that the microstructure and consequently also the mechanical properties become acceptable for the relevant steels.
- **Scenario 2:** Preheating in order to decrease the cooling time and thus the hardness. It is primarily the cooling rate in the temperature interval 800–500°C that influences the hardness in steels.
- **Scenario 3:** Limiting the steel composition so that the mechanical properties become acceptable for all welding speeds, i.e. also for a fully martensitic microstructure.

All three approaches are used industrially. Important consequences of scenario 1 are limitations to productivity as well as a need to use steel composition dependent welding parameters. Scenario 2 is of course difficult to use on large structures. The third scenario takes advantage of the fact that the maximum hardness of a fully martensitic structure to a first approximation is determined solely by the carbon content. This talks in favour of scenario 3, but it must be taken into account that a consequence of this scenario may be a limitation in carbon content to, e.g., approx. 0.12% in order to limit the maximum hardness to roughly 380 HV, and this limitation in composition of course sets further limits on the manufacturing of the steels.

23.2.2 The effects of mismatching

Compared to arc welds, where HAZ properties often control the performance of a welded joint, in laser welds the weld metal dominates the properties. Furthermore, the possibility of influencing the composition of the weld metal by the filler material is very limited, because the zero gap situation must normally be accepted. In general, a high degree of overmatching in strength will occur and an overmatch by a factor of two is not unusual. The high level of overmatching combined with the aforementioned narrowness of the weld makes laser-based welds of structural steels an extreme example of a mismatching joint of the 'sandwich-type'.

The effect of such heterogeneities has been much investigated and discussed. It is, however, generally found that weld metal notched wide-plate tests under transverse loading conditions show a beneficial effect of weld metal strength overmatching, by promoting failure in the lower strength, but higher toughness base plate. In contrast, undermatched welds require a high straining capacity in the soft weld metal due to the deformation being localised here. It may also be noted that crack propagation preferably takes place in the softer zones, i.e. the fracture path for a crack initiated in the weld metal of an overmatched weld deviates into the softer and tougher base metal. These conclusions are supported by experimental and numerical investigations on laser welds using smaller specimens as well as wide plates [1]–[4].

The impact toughness of the weld metal was at first very difficult to measure by the standard Charpy V-notch test technique due to fracture path deviation into the softer base metal. These test results were therefore originally disregarded but, based on the understanding described above, the traditional Charpy test is now considered valid even in the presence of fracture path deviation.

23.2.3 Solidification flaws

Solidification flaws constitute a type of weld imperfection which is occasionally found in laser welds in structural steels. The imperfection is a solidification defect but, unlike usual solidification cracks or hot cracks, solidification flaws are generally small, isolated imperfections occurring with a certain regularity along the weld. The flaws are normally embedded and seen only in the weld centreline. The causes of the flaws are manifold and their interactions complex, but controlling this phenomenon is a key challenge in thick plate structural steel laser welding. If total freedom from the flaws is to be achieved then the steel composition and also the welding parameters have to be controlled.

One approach has been to correlate the steel composition with the volume of the weld parameter tolerance box (weldability lobe). In this way, the steel is tested by a range of weld parameters, and the size of the weld parameter box obtained is taken as a measure of the weldability with respect to solidification flaw formation of the actual steel. For pure laser welding, this approach has been adopted since 2004 in many Classification Society rules for shipbuilding (e.g., [5]). The approach was later extended to use in laser hybrid welding [6].

23.2.4 Mechanical properties

Shipbuilding steels are typically ferritic steels strengthened by C, Mn and various microalloying elements such as Ti, V or Nb. Different production routes are used depending on the demands on the steels and the techniques available at given steelworks. Carbon is typically in the range 0.05–0.18% (weight), Mn in the range 0.5–1.5% and S & P in the range 0.005–0.03%. The guaranteed yield strength of structural steels for shipbuilding typically falls in the range 235–460 MPa, but higher grades are also used.

The heat input per unit length for typical laser-based (pure or hybrid) welds is typically almost an order of magnitude lower than for a similar arc weld. It must therefore be taken into consideration that structural steels may become hard in both the weld metal and the HAZ as a consequence of the fast thermal cycle. The resulting hardness for a given heat input depends very much on the composition of the steel, which determines hardenability, as well as on the carbon content alone, as this controls the maximum possible hardness in a fully martensitic structure. Figure 23.2 [7] thus shows three examples of hardness traverses in three different 15 mm steels, which were

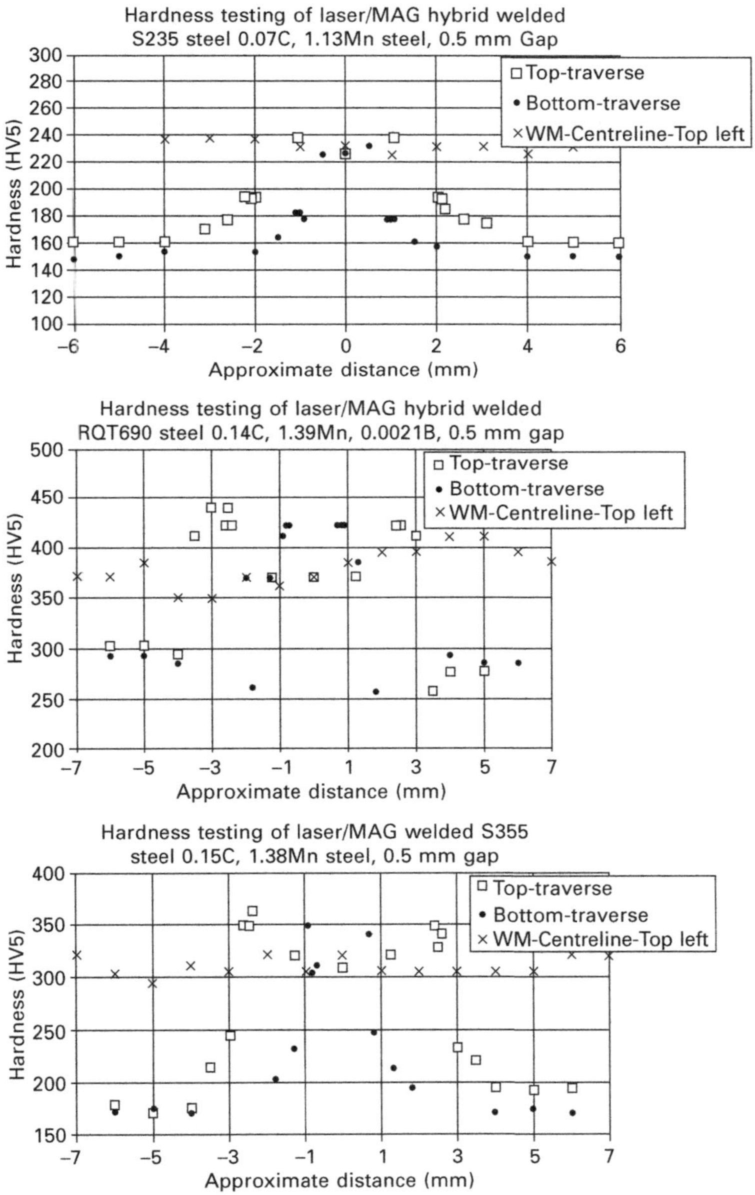

23.2 Hardness profiles for welds in S235 (0.07C, 1.13Mn) (top), S355 (0.15C, 1.38Mn) (bottom) and RQT 690 (0.14C, 1.39Mn, 0.0021B) (middle) steels. The welding data are given in the text. Source: Ref. [7].

welded with the hybrid MAG-laser process using an EN 440-G3Si1 (SG2) solid wire, 0.5 mm gap, a total power of 18.5 kW and a welding speed of 1.2 m/min.

Measuring the impact toughness of the weld metal was initially very challenging to using the standard Charpy V-notch test due to fracture path deviation into the softer base metal. These test results were disregarded at first, but based on a large number of wide plate and other large-scale tests as well as numerical simulation, the traditional Charpy test is now considered valid even in the presence of fracture path deviation. As an example, Fig. 23.3 [7] shows a full Charpy V-notch transition curve for a weld in the 12 mm S235 structural steel welded with an initial gap of 0.5 mm, a total power of 18.5 kW and a welding speed of 1.2 m/min. The transition temperature is below –40°C and, due to the very tough combined structure, the level of the high temperature shoulder is significantly higher than the level of the base metal alone.

There are inherent differences, such as weld profile, mismatch level, residual stress, and heterogeneity of the weld metal that may affect the fatigue behaviour of hybrid laser welds, perhaps for the better compared to conventional fusion welds. Fatigue behaviour has been extensively investigated using four-point bending and uniaxial loading, and Fig. 23.4 [7] shows as an example the results of a four-point bending test in laser hybrid welded 20 mm RQT690 steel. In general, fatigue tests show similar results, i.e. that the fatigue behaviour of laser-based welds in structural steels shows properties in excess of the FAT100 line.

23.2.5 Unified guidelines

The challenge of getting approval for the new welding approach that pure laser and laser hybrid welding represented from the ship classification

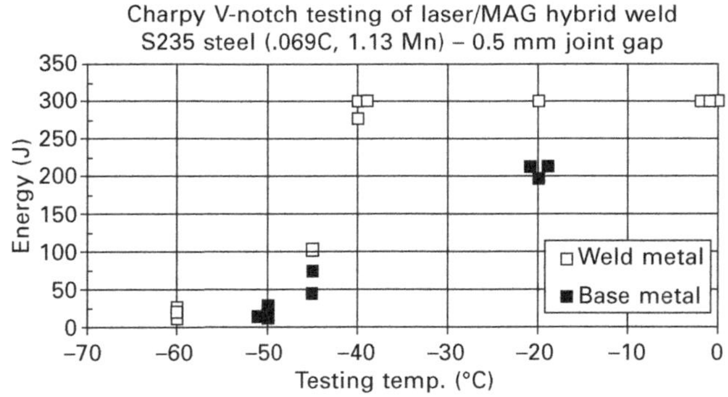

23.3 Charpy V-notch transition curve for the 0.5 mm gap weld made in 12 mm S235 structural steel. The maximum energy of the testing equipment is 300 J. Source: Ref [7].

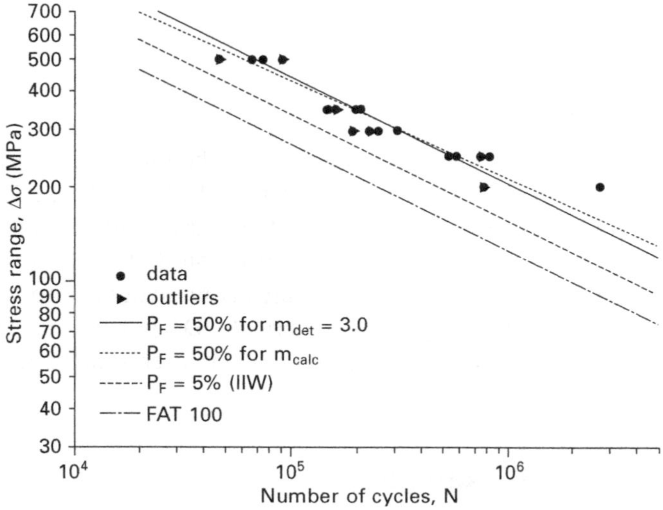

23.4 Four-point bending test of a 20 mm RQT690 steel. Source: Ref. [7].

societies was tremendous and could not be overcome by one shipyard or one country alone. A large number of tests have been performed in various European collaboration projects over more than two decades, which involved several yards and classification societies as well as many institutes and universities.

The work performed involved:

- process development (including filler metal addition),
- sensor-based seam tracking and real-time adaptive control,
- mechanical properties:
  - hardness and strength,
  - ductility,
  - toughness (impact and quasi-static),
  - fatigue,
- corrosion properties:
  - stress corrosion,
  - general corrosion,
- weld imperfections and defects:
  - surface defects,
- hydrogen introduced cracking (cold cracking),
  - solidification cracking (hot cracking),
  - porosity,
- health and safety aspects:
  - working environment,
  - external environment,

- NDT (non-destructive testing):
  - on-line and off-line,
- wide plate and large-scale testing,
- repair welding.

The European ship classification societies issued their first unified guidelines for the approval of pure CO_2 laser welding in 1996 [8]. These guidelines only considered pure CO_2 laser welding and, because solidification cracking was still a problem in thicker sections, 100% non-destructive testing was typically demanded, which made it economically unfavourable for shipyards to apply laser welding to a larger scale. Normal NDT practice has only been allowed in shipyards since 2004, as have other laser sources and hybrid laser/MAG welding [5].

23.3 Industrial examples

The shipbuilding industry was a pioneer in being first to adopt pure laser and later laser hybrid welding in the manufacture of large steel structures. In particular, cruise and passenger ships are well suited to this application, because the steelwork is dominated by the large amount of deck structures needed, which typically are made in 5–10 mm plate thickness. Figure 23.5 shows a large (the world's biggest – in 2008) cruise ship that was built at Aker Yard in Turku, Finland, now STX Finland Cruise Oy.

23.3.1 Meyer Werft, Germany

By 1995 Meyer Werft had included the first simple sandwich panels in ship structures, though these were of minor structural importance. From 1998, these panels were also used for passenger decks. The panels were of the type shown in Fig. 23.6.

Meyer Werft invested in a new and technically ambitious panel manufacturing line in 2001, which is equipped with a total of four high-power CO_2 laser (12 kW) hybrid welding machines. Since 20 m long plates were not available for the entire range of required plate thicknesses (4.5–30 mm),

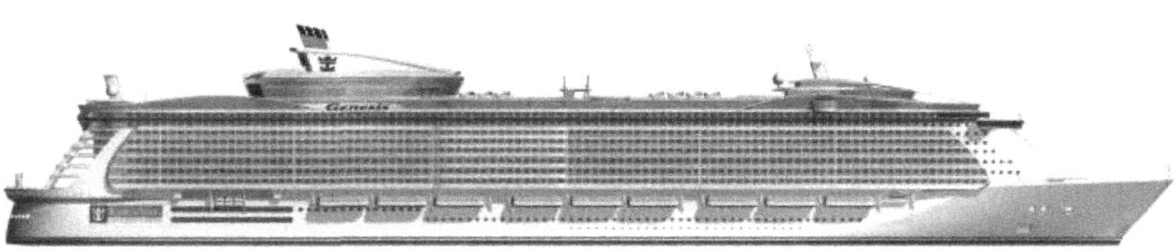

23.5 Large cruise ship built at Aker Yard in Turku, Finland, now STX Finland Cruise Oy. (Courtesy: Kari Laiho, STX Finland Cruise Oy, Turku.)

23.6 The first laser welded sandwich panels produced at Meyer Werft, Germany. (Courtesy: Guido Pethan, Meyer Werft.)

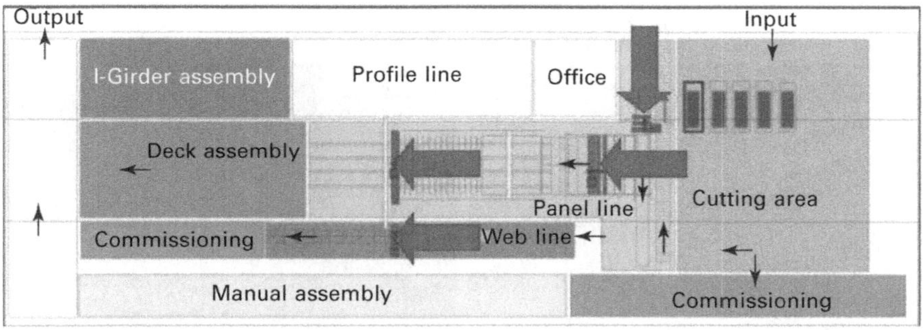

23.7 Layout of new pre-manufacturing workshop at Meyer Werft, arrows indicate laser hybrid welding positions. (Courtesy: Guido Pethan, Meyer Werft.)

and structural optimisation called for a high flexibility in plate thickness and materials used within one section ('tailored decks'), a butt welding station was needed to produce plate 'strips' of up to 20 m long and 4 m wide [9]. These 20 m long 'strips' delivered from the first butt welding station are then joined and strengthened with stiffeners in order to form deck panels, which can be up to 20 m × 20 m in size. In order to reduce man-hours, fully automated operation including material handling, edge preparation by grinding/milling and welding is used along the entire panel line [9]. The resulting principal layout of the steel pre-manufacturing workshop is shown in Fig. 23.7, and Fig. 23.8 shows the installation as well as a stiffener welding head and clamping tools.

In 2010 a further system was installed that can join deck sections with a width of up to 30 m and a length of 25 m (maximum 750 m^2). Interestingly, disc laser-based hybrid welding is used for the butt welds. The installation area in the workshop covers 2,300 m^2 and the longitudinal traverses, which the laser welding system moves upon, have a length of 43.5 m and a weight of 73 tons [10]. Laser hybrid welds in a large cruise ship now account for approximately 50% of the total weld length, which may be in the order of 400 km, and it is estimated that a total of 1,300 km laser weld is performed

23.8 Photos of the installation as well as of a stiffener welding head and clamping tools. (Courtesy: Guido Pethan, Meyer Werft.)

yearly at the yard [11]. The majority of the plates welded are 'thin' plates, i.e. in the order of 5–8 mm [9].

23.3.2 Fincantieri, Italy

The Fincantieri yard has also had a laser installation since around 1996. The original installation was for butt welding of plates for deck structures and uses a 17 kW laser source in combination with an advanced clamping system. This installation has, however, been used mainly for development work. In 2008 the yard invested in a new highly automated panel line installation of 4–20 mm thick panels, somewhat similar to the above at Meyer Werft, in which butt welding is performed using a 10 kW fibre laser for hybrid welding. This installation may produce panels up to 16 × 36 m and, due to the small distortion, the time and the cost of following ordinary filled welding of stiffeners are reduced as is the cost of the later fitting processes. The stiffener mounting and welding gantry only require one operator. The automated mounting process allows welding without initial tack welding [12].

23.3.3 Blohm + Voss, Germany

Blohm + Voss GmbH typically builds frigates, corvettes and mega-yachts, and the yard has used laser technology for more than ten years for manufacturing complete deck parts (including frames, and longitudinal and transversal stiffeners). These welds are typically T-joints in the plate thickness range $t = 3$–12 mm, which are welded with two 12 kW CO_2 lasers simultaneously without additional wire, with nitrogen laser cutting for the edge preparation [13]. Figure 23.9 shows a T-joint welding head and cross section.

23.3.4 STX Finland Cruise Oy, Turku, Finland

These yards mainly build cruise ships and ferries. STX has implemented laser hybrid welding into production since December 2006. The laser technology is installed on the existing panel line, which reduced the total costs of the investment. At this stage, laser hybrid is used for welding butt welds in panel fabrication. The laser beam is delivered by a 6 kW fibre laser [14]. Figure 23.10 shows an overview of the installation as well as various elements of it.

23.3.5 Odense Steel Shipyard, Denmark

Unlike the other yards mentioned, Odense Steel Shipyard builds mainly very large commercial ships, particularly container ships such as the one shown in Fig. 23.11, which in 2008 was the largest container ship in the world. At Odense Steel Shipyard, a combined laser welding and cutting installation was installed in 1997. Interestingly, the same 12 kW CO_2 laser source is used for both purposes. The first production of ship elements with laser welding took place in January 1998 followed by various instances of laser welding in production. In August 2000 the first trials using CO_2 laser/MAG hybrid welding was performed, followed by the production of parts for different blocks with laser hybrid welding [15]. Figure 23.12 shows the installation together with a welding head. Sadly the yard is in the process of being closed down and the last ship was delivered in January 2012.

23.4 Future trends

Laser welding in shipbuilding has so far mainly been a European speciality, but the technique will most likely also be taken up in other areas of the world

23.9 A T-joint welding head and cross section. (Courtesy: Jens Keil, Blohm + Voss.)

where shipbuilding is a major activity, such as Japan, Korea and China. A recent paper reviewing basic investigations was presented by authors from Nagasaki Shipyard of Mitsubishi Heavy Industries [16]. Many joint European projects have been carried out over the years and recently they have been including laser sources that allow fibre delivery of the laser light (e.g. [17]). Fibre and disc laser sources seem to have great potential for automated or robotised welding, and much effort is being devoted to solving safety challenges that arise when welding in open areas such as docks.

23.10 An overview of the installation as well as the laser head (bottom, left) and the laser source and MAG power supply (bottom, right). Due to the safety requirements caused by the laser light wavelength, the installation is encapsulated. (Courtesy: Kari Laiho, STX Finland Cruise Oy, Turku.)

23.11 11,000 TEU container vessel at Odense Steel Shipyard. With 11,000 TEU containers including 1,000 FEU reefers, the ship was in 2008 the world's largest container carrier and reefer vessel. (Courtesy: Odense Steel Shipyard.)

23.12 The laser installation at Odense Steel Shipyard (left) and the T-joint hybrid welding head (right). (Courtesy: Mads Elvang, Odense Steel Shipyard.)

23.5 Conclusion

Laser welding shows great potential in the welding of structural steels in general and in shipbuilding in particular. Its attractive properties include high welding speed, low distortion and easy automation. In comparison to pure laser welding, hybrid MAG-laser welding has the advantage of an increased ability to bridge a gap as well as a significant increase in speed when welding a wide gap.

As a result of intensive work during the last two decades, tests have been modified to suit the demands of laser welds. Furthermore, relevant standards and Classification Society guidelines have been made, allowing the use of laser hybrid welding in ship production.

Wide-plate and other large-scale tests using a pre-placed artificial weld defect have been performed. The findings of the large-scale tests are in accordance with the results of the small-scale tests, and it may be concluded that laser welds behave at least as well as arc welds. In all cases, the fracture deviated into the base metal immediately after initiation. In the narrow zone of a laser weld, overmatching seems to be beneficial in preventing flaws from developing in transverse loading conditions. The impact toughness properties also generally show acceptable results as does the ductility.

With regard to hardness and strength, pure and hybrid laser/MAG welding are very similar, owing to their similar weld geometries and degree of overmatching. The resulting hardness may, due to the fast thermal cycle inherently connected to the processes, become quite high. Besides the heat input caused by welding, the hardenability depends very much on steel composition and carbon content. Fatigue behaviour has also been extensively investigated, and tests typically show results in excess of the FAT100 line.

The weld imperfections most likely to occur are pores and solidification flaws, of which the latter are more critical owing to their flat and sharp character. An understanding of the influence of steel composition and weld parameters on the occurrence of solidification flaws has been achieved, and it seems possible to avoid solidification flaws entirely in laser-based welding of structural steels, by combining control of the steel composition and welding parameters.

Today, laser-based welding is used quite extensively in European yards, and fibre and disc laser sources have great potential for robotised welding as well as welding in the dock area.

23.6 References

1. Koçak M, Kim Y-J, Cam G, Dos Santos J, Riekehr S, Totster F, Instran A, Cardianl N, Webster S, Klæstrup Kristensen J and Borggreen K, Recommendations on Tensile and Fracture Toughness Testing Procedures for Power Beam Welds, *Proc. of European Symposium on Assessment of Power Beam Welds*, Geesthacht, Germany (February 1999).
2. Kim Y-C, Koçak M and Schwalbe K-H, Defect Assessment Procedures for Power Beam Welds, *Proc. of European Symposium on Assessment of Power Beam Welds*, Geesthacht, Germany (February 1999).
3. Koçak M, Riekehr S, dos Santos J, Cardinal N, Webster S, Klæstrup Kristensen J, Borggreen K, Klein R and Fischer R, Analysis of Fracture Behavior of Laser Beam Welded Wide Plates, *Proc. of European Symposium on Assessment of Power Beam Welds*, Geesthacht, Germany (February 1999).
4. Senogles D J, Harrison P L and Cardinal N, Analysis of Wide Plate Behavior of Steel Laser Welds, *Proc. of European Symposium on Assessment of Power Beam Welds*, Geesthacht, Germany (February 1999).
5. *Classification Guidelines for the Approval of Autogenous Laser Welding and Hybrid Laser Welding*, May 2004.
6. Klaestrup Kristensen J, Webster S and Petring D, Hybrid laser welding of thick section steels – The HYBLAS project, *Proc. of 12th NOLAMP Conference*, Copenhagen, Denmark (August 2009).
7. Klaestrup Kristensen J, State of the art in shipbuilding applications of hybrid laser-arc welding, *Proc. of 12th NOLAMP Conference*, Copenhagen, Denmark (August 2009).
8. *Laser Welding in Ship Construction – Classification Society Unified Guidelines for the Approval of CO_2 Laser Welding*, 1996.
9. Pethan G, Meyer Werft, personal communication (January 2008).
10. The website of the company Graebener Maschinenrechnik: http://www.graebener-maschinentechnik.de/Shipbuilding.gmt-schiffbau+M52087573ab0.0.html (accessed January 2012).
11. The website of the journal 'BänderBleche Rohrer', Henrich Publikationen: http://www.bbr.de/index.cfm?pid=1441&pk=92414 (accessed January 2012).
12. Seyffarth P. and Gaede R, Fully automated panel line with highest accuracy minimizes reworking and assemblying. *Proc. of 3rd European Conf. on Production Technologies in Shipbuilding*, p. 127, Stralsund, Germany (October 2008).
13. Keil J, Blohm + Voss, personal communication (January 2008).

14. Laiho K, STX Finland Cruise Oy – Turku, personal communication (January 2008).
15. Elvang M, Odense steel Shipyard, personal communication (2008).
16. Goda H and Koga H, *Application of laser-arc hybrid welding to shipbuilding*, IIW Document IV-1017-10, International Institute of Welding, 2010.
17. The official website of the European project: Breakthrough in European Ship and Shipbuilding Technologies (BEEST): http://www.besst.it/ (accessed January 2012).

Index

CPSIA information can be obtained at www.ICGtesting.com
Printed in the USA
LVOW012246210613

3585LVUK00001B/1/P

9 780857 092649